TUBULAR STEEL STRUCTURES —

Theory and Design

SECOND EDITION

by

M.S. TROITSKY, D.Sc.

Professor of Engineering
Concordia University
Montreal

To my wife Tania

Second Edition, 1990

Library of Congress Number — 81-80993

publisher's preface

As authors of major works in highly specialized areas of technology know, finding a publisher for the manuscript may be a difficult task. When the subject matter is of narrow interest and highly circumscribed, it is outside the province of the commercial publisher, who must realize a sales volume more than enough to cover his costs. The more comprehensive and scholarly the work becomes — and, thus, the more useful to the technology — the more difficult is the problem of finding a publisher. When publication requires hundreds of pages and a great mass of art work, it is usually beyond the scope of the journals of pertinent technical societies. In such cases, a patron may be required if the scholarly work is to be put in print, hard-covered, and made available to the public.

Fortunately, there are patrons in most areas of science, engineering, and technology — foundations, associations, endowments, and even individuals who serve to fill the gap in publication possibility. The James F. Lincoln Arc Welding Foundation is such a patron for comprehensive studies pertaining to arc-welding technology and arc-welded steel design. As a part of its broad function in providing educational services that enhance the progress of arc welding, the Foundation subsidizes the preparation, publication, and distribution of scholarly works that benefit designers, industrial managements, and others concerned with the technology and engineering of arc-welded steel. Dr. Troitsky's monumental work on *"Tubular Steel Structures, Theory and Design"* fits admirably into this function of the James F. Lincoln Arc Welding Foundation.

The growing importance of tubular steel in the design of structures is well known to engineers. The properties of the circular cross section and the achievements of strength with light weight that can be realized with it are intriguing to the designer. But even though tubular steel is being used extensively in structures, the technology of its use is widely distributed through the literature. Dr. Troitsky in this book brings together the scattered information on tubular design and analysis in a well organized manner that should make it a valuable text and reference source.

Dr. Troitsky is not unknown to beneficiaries of the James F. Lincoln Arc Welding Foundation's publishing ventures. He is also author of the work *"Orthotropic Bridges, Theory and Design"*, published under the aegis of the Foundation in 1967. This book, as the present one, also was considered as having too limited outlet possibilities for a commercial publisher — although the first printing of 5,000 copies sold out within a year. Such widespread distribution of volumes of highly specialized interests is facilitated by the Foundation's subsidized distribution policy, which supplements its patronage of the preparation and printing costs. In effect, the Foundation offers the books it publishes at but a fraction of costs — just a nominal sum to guarantee that the recipient has a real need and interest in the work.

Other well-known design books published by the Foundation include *"Design of Weldments"* and *"Design of Welded Structures"*, both by Omer W. Blodgett. In addition, the Foundation has sponsored many other works of more widespread interest, such as *"Metals and How to Weld Them"*, *"Design Ideas for Weldments"*, and *"Modern Welded Structures"*.

The James F. Lincoln Arc Welding Foundation was formed in 1936 by the Lincoln Electric Company to provide educational services in the burgeoning art of arc welding and arc-welded steel design. Although expending most of its efforts on publications, the Foundation has also sponsored various Award Programs to encourage architects, engineers, and designers to use arc-welded steel more efficiently and esthetically in weldment and structural designs.

We feel certain that *"Tubular Steel Structures, Theory and Design"* will be a worthy addition to any engineering library.

Richard S. Sabo,
Executive Director, The James F. Lincoln
Arc Welding Foundation

preface

In this second edition the contents have been thoroughly revised and every effort was made to correct errors and misprints detected in the first edition. Although much of the data of the first edition has been retained, the following changes and additions were made in the contents of the second edition.

In Chapter 7 "Bins and Bunkers" sections retained to the Ring Girders and Roofs for Circular Bins, were substituted by the new text for Circular Girders Supporting Bins and Roofs for Circular Bins. Also in Chapter 12 "Tubular Conveyor Galleries" new sections concerning the design of "Ring Girders at Supports, Method by Krupka" were added.

resume

M.S. Troitsky, D.Sc. is Professor in the Department of Civil Engineering at Concordia University, Montreal. He has taught at Ecole Polytechnique and McGill University, Montreal. He has varied industrial experience in a full-time and consulting capacity, and was responsible for the design and supervision of the construction on several major bridges. He is the author of the books: *Orthotropic Bridges; Cable-Stayed Bridges; Stiffened Plates: Bending, Stability and Vibrations; Design Guidelines for Rectangular Steel Bins; Tubular Steel Structures;* and *Prestressed Steel Structures;* as well as numerous papers on structural and bridge engineering. He is a member of a number of technical societies.

acknowledgements

In preparing a book in the field of engineering, an author must depend upon the publishing works of many engineers, scientists, mathematicians and engineering firms. The author and publisher of this book gratefully acknowledges the help and cooperation received from many sources.

Special acknowledgement is herewith made to the Canadian Steel Industries Construction Council and Imperial Oil Limited, who have financially supported the preparation of the research reports by the author on some topics treated in this book.

The author is also indebted to many organizations and individuals for their help, assistance and permission to reproduce the designs, tables and graphs.

The author also wishes to thank Mr. G. Korioth for the preparation of the drawings, and to Miss M. Stredder for typing the manuscript.

Special acknowledgement is herewith given to the Canadian Institute of Steel Construction for whom the author has prepared a number of research reports and which work inspired the author to prepare this book.

contents

CHAPTER 4 — THERMAL STRESSES IN TUBULAR STRUCTURES

CHAPTER 5 — SELF-SUPPORTING STACKS

CHAPTER 6 — MULTILEVEL GUYED STACKS

CHAPTER 7 — BINS AND BUNKERS

CHAPTER 8 — HORIZONTAL STORAGE TANKS

CHAPTER 9 — ABOVE GROUND PIPELINES

CHAPTER 10 — UNDERGROUND PIPELINES

CHAPTER 11 — UNDERWATER PIPELINES

11.13.5 Design of Ring Stiffeners 11-22

CHAPTER 12 — TUBULAR CONVEYOR GALLERIES

APPENDIX A

CHAPTER 1

introduction

1.1 General Concept

Tubular structures consist of stiffened and unstiffened shells, or cylinders constructed from steel sheets with large diameter-to-thickness ration D/t. Such structures are used for the storage of granular solids, liquids and gases, transportation of liquids and gases, as supporting galleries for conveyors and for discharge of cool or heated gases.

Tubular steel structures in most cases are space type structures such as: stacks, bins and bunkers, liquid storage tanks, pipelines and conveyor galleries.

Cylinders with large diameter-to-thickness ratios most often are called shells. Typically, shells are stiffened. When a single descriptive term is required, the "cylinder" or "tubular member" is utilized.

1.2 Advantages of Tubular Structures

Any advantages of tubular structures may be summarized as follows:

1.2.1 Optimum Utilization of Material

Structurally, the shell is very efficient, since it is a continuous medium and a high proportion of the material can be used to capacity. The cylindrical shell provides the most efficient cross-sectional shape under axial compression having equal lateral restraint in all directions normal to the cylinder axis. The material is conveniently distributed along the perimeter and the radius of gyration is constant for all axis passing through the centroid of cross-section. To insure the stability of compressed section in all directions the circular cylinder is the best by a comparison of all other sections and the most economical.

The particular advantages of some tubular structures consists in their capacity to perform technological functions and simultaneously to serve as a carrying structure, as for instance, in the case of pipelines.

Generally, it is possible to design tubular structures having a minimum weight of metal, simplicity in fabrication and erection, and durability in operation.

The second factor permitting the reduction of the weight of tubular structures is substantially smaller aerodynamic and hydrodynamic resistances under wind and water actions, respectively. For instance, for relatively high tubular structures where the stress often depends by 80% to 90% of wind pressure, the amount of metals in a structure is possible to be reduced more than twice by comparison with conventionally designed structure.

It should be mentioned also that the material of tubular structures by comparison with the equal cross-sectional areas has a minimal perimeter of contact under atmospheric influences. Therefore, tubular structures require relatively smaller quantities of paint or other protective measures. Also, experience indicates that the effect of corrosion at the exposed area of tubular structures is from 16% to 30% smaller than at conventional structures.

1.2.2 Fabrication

Although the advantages of the tubular structures have been known for a long time, it was the introduction of welding which made its extensive use possible. Welding has facilitated the design and construction of a great variety of tubular structures due to simplification in the fabrication of joints. Old types of joint connections of tubular members by bolts and rivets were expensive and did not provide simple and safe joints.

Fusion welding is the most widely used method of fabrication for the construction of tubular structures. This method of construction is virtually unlimited with regard to size and is extensively used for the fabrication and erection of large-size tubular structures in the field. Often such structures are fabricated by the method of sub-assembly. In this process, sections of the unit are shop welded and then assembled in the field. Members of tubular structures having a size sufficiently small to permit transportation by trucks, rail, or barge are usually completely shop welded because of the lower cost and greater control of the welding procedure in the shop.

There are two types of fusion welding that are extensively used for the fabrication of tubular structures. These are:

1. The gas welding process, in which a combustible mixture of acetylene and oxygen supply the necessary heat for fusion, and
2. The electric-arc welding process, in which the heat of fusion is supplied by an electric current.

Arc welding is the preferred process because of the reduction of heat in the material being welded, the reduction of oxidation and better control of the deposited weld metal. A wide range of arc-welding equipment is available, from the small portable welding units to the large automatic welding machines. Small arc-welding machines are widely used in welding shops that fabricate small equipment whereas automatic machines are better suited for the welding of heavy sections involving the deposition of a large quantity of weld metal.

One of the most recent and successful developments in the field of arc welding is the submerged-arc welding process. The process involves submerging of the arc beneath a blanket of granulated mineral flux. In addition to completely protecting the weld metal from the atmosphere, this process makes the weld metal virtually free of hydrogen.

The success of fabrication by welding is dependent upon the control of the welding variables such as experience and training of the welder, the use of proper materials, and welding procedures. A number of codes and standards have been established for control of such welding variables.

The American Welding Society (AWS) established the basic standards for qualifying operators and procedures. These standards of qualification form the basis for most of the standards in the various codes. For practical purposes, therefore, the rules for qualifying welders and welding procedures are essentially the same in the various codes and standards. Regardless of whether or not the welded tubular structure is intended to meet one of the codes or standards, it is advisable that the welding conform to one of the minimum standards.

A variety of types of welded joints are used in the fabrication of vessels. The selection of the type of joint depends upon the service, the thickness of the metal, fabrication procedures, and code requirements.

1.3 Structural Behavior of Tubular Structures

1.3.1 Application of Thin-Walled Shells Theory to Tubular Structures

Tubular structures which have found a wide application in modern industry consists generally of the cylindrical, conical and spherical shells of their combinations. For the practical analysis and design of tubular structures, we may apply the theory of thin-walled shells, however, with certain limitations. This is because the formulas for thin-walled shells were developed under the theoretical assumption that the ratio t/R is very small. However, in practice, in application to tubular structures, this ratio t/R may have greater values and the results obtained always have an approximate character. Therefore, it is important to find the maximum value of the ratio t/R when the deviation obtained by the theory of thin shells is practically permissible, or to determine the limiting value of ratio t/R at which the tubular structure may be considered as "thin-walled."

We approach this problem under certain consideration which may permit us to obtain a practical solution of the problem.

Let us consider a closed cylinder under internal pressure, having internal and external radii r_1 and R respectively, and $t = R - r_1$ is the thickness of the wall. The maximum and minimum values of the circumferential stresses applying the theory of thick wall cylinders are

$$\sigma_{max} = p\frac{R^2+r_1^2}{R^2-r^2} \quad ; \quad \sigma_{min} = 2p\frac{r^2}{R^2-r^2} \tag{1.1}$$

But from the theory of thin-walled cylinders, if r is the middle radius of the wall, we have

$$\sigma = p\frac{r}{t} \tag{1.2}$$

Because $R = r + \frac{t}{2}$ and $r = r - \frac{t}{2}$, after substituting these values into (1.1), we obtain

$$\begin{aligned} \sigma_{max} &= \frac{pr}{t} + \frac{pt}{4r} \\ \sigma_{min} &= \frac{pr}{t} + \frac{pt}{4r} - p = \sigma_{max} - p \end{aligned} \tag{1.3}$$

The deviation of the approximate value of σ from the exact will be given in per cent by the expressions

$$\begin{aligned} \alpha_1 &= 100\frac{\sigma_{max}-\sigma}{\sigma_{max}} = 25\frac{(t/r)^2}{1 + 4(t/r)^2} \\ \alpha_2 &= 100\frac{t/r - 1/4(t/r)^2}{1 - t/r + 1/4(t/r)^2} \end{aligned} \tag{1.4}$$

Assuming certain values of t in function or r, we may find corresponding values of α_1 and α_2 shown in Table 1.1.

TABLE 1.1 — Values of Coefficients α_1 and α_2.

t / α	0.1r	0.15r	0.2r	0.4r
α_1	0.25	0.56	0.99	3.8
α_2	7.8	17	23.5	56

This Table indicates that even at t = 0.2r the deviation from σ_{max} does not reach 1%. Regarding the deviation of 23.5% from σ_{min}, this deviation may be considered as the factor of greater safety. Therefore, for the design analysis of cylindrical shells, used in practice, formulas developed for thin-walled shells, may be applied up to the values of t = 0.2r.

For the analysis of cylindrical shells also of great interest is the determination of edge forces. These forces basically depend on the deformations at the shell's edges under external forces. The change of the $r \leqslant \rho \leqslant R$ at any point of the wall of closed cylinder according to the theory of thick-walled shells is given by the expression

$$\Delta = \frac{1-2\nu}{E} \cdot \frac{pr_1^2}{R^2-r_1^2}\rho + \frac{1+\nu}{E} \cdot \frac{pr_1^2R^2}{R^2-r_1^2} \cdot \frac{1}{\rho} \tag{1.5}$$

After substituting into this expression the values

$$r_1 = r - t/2; \quad R = r + t/2 \quad \text{and} \quad \rho = r \tag{1.6}$$

we obtain the change of the middle radius of the wall

$$\Delta_m = \frac{p}{2tE}\left[(1-2\nu)\left(r-\frac{t}{2}\right)^2 + (1+\nu)\left(r-\frac{t^2}{4r}\right)^2\right] \tag{1.7}$$

At small values of ratio t/r we have from (1.7)

$$\Delta'_m = \frac{2-\nu}{2tE}\,pr^2 \tag{1.8}$$

which represents the formula for thin-walled cylinders.

The deviation Δ'_m from Δ_m in per cent will be

$$\beta = 100\,\frac{\Delta_m - \Delta'_m}{\Delta_m} \tag{1.9}$$

Using values t = 0.1r, 0.2r and r = 0.3, we obtain from (1.9)

$$\beta = 2.4\% \text{ to } 5.6\% \tag{1.10}$$

Therefore, in this case the deviation is also small and practically acceptable.

Similar analysis for spherical type shells indicates an even better degree of approximation than for the cylindrical shells.

Regarding the conical shells where a more exact theory does not exist, we conclude for the general considerations that the above obtained results for a cylinder and sphere may also be used in the case of the cone.

Therefore, in the design of tubular structures, we may analyze them as "thin-walled shells" at least for the values of t = 0.1r, where r is the minimum radius of curvature of the middle surface and for the approximate designs, this limit may be increased even to the value of t = 0.2r.

1.3.2 Buckling Stability of Tubular Structures

Experience indicates that thin-walled shells under external pressure may deform, buckle and collapse irrespective of the fact that acting pressure produces relatively low stresses in the walls of the shell.

About the worst thing that can happen to a tubular structure either during construction or operation is for it to develop a buckle. The buckle may be defined as a localized failure in the form of a wrinkle or indentation caused by over-stress or instability of the wall of tubular structure on the compression side of the structure subjected to bending. Such local buckling is the result of a structural instability of the wall of the structure when subjected to compressive stress. Tubular structures can experience this type of buckling if the bending stresses exceed critical values.

The only thing worse than local buckling would be the collapse of the tubular structure. It is very important to understand the collapse mechanism, since there are indications that under certain conditions, the mode of failure may be progressive. A collapse initiated by a critical condition at one point in the tubular structure may progress or run along the structure causing collapse at locations where the stress would not otherwise have been critical.

Collapse is defined as a general failure usually in the form of a flattening of the shell cross-section over a considerable length as the result of the action of external pressure on the shell.

Both resistance to collapse and resistance to local buckling are functions of the diameter to thickness ratio D/t of the shell. The lower this ratio, the greater the resistance to failure in either case.

Collapse and local buckling become extremely important during the design of large diameter tubular structures, since the D/t ratio is apt to be quite high for economic reasons.

The pressure producing such deformation is called "the critical pressure". The value of this pressure depends exclusively on the geometrical shape and dimensions of the shell and from the physical properties of the material of the wall. Critical pressure may be defined as the pressure under which the shell will be deformed and does not return to the initial shell shape after removal of pressure. The shell, however, will take the initial shape if the pressure is less than the critical.

The value of safety factors or the ratio p_{cr}/p usually is taken in the limits of 3 to 5. The smaller values may be used for the vertical type of tubular structures. For horizontally placed tubular structures in which the initial ovality of cylindrical shell under its own weight and weight of the load which it carries, have a tendency with time, to change, greater factors of safety should be used.

It should be noted that in technical literature pertaining to the shell-type structures, many solutions are available for aerospace type structures and often these are not applicable to those structures used by civil engineers in the field of civil engineering.

Also, the computer solutions that have been proposed to-date, are often of little value in the analysis of the stability of shell-like civil engineering structures.

In order for a solution to be of much value to civil engineers, it must at least satisfy the following conditions: (1) deflections, (2) fabrication and erection tolerances, and (3) edge conditions. In addition, such items as restricted wave-lengths of buckles, joint details, local buckling and member buckling, should be considered.

Due to the importance of the buckling stability in the design of tubular structures, this problem will be treated in all details in the following chapter.

1.3.3 Junction of Shells of Different Geometry

One of the most common and important problems in tubular structures is the stress analysis of the shell at the junction of two shells of different geometry. The analysis can make use of membrane solutions and edge bending solutions.

When internal pressure is applied to the head-cylinder structures, Figure 1.1, membrane stresses are set up in both the heads and the cylinders.

These are well-known stresses and they can be computed by available formulas, and their distribution over the wall sections of the structures is assumed to be uniform.

However, differences between the natural flexibility of the heads, and that of the cylinder, cause these two parts to attempt to deform radially and angularly at different rates under the effect of internal pressure.

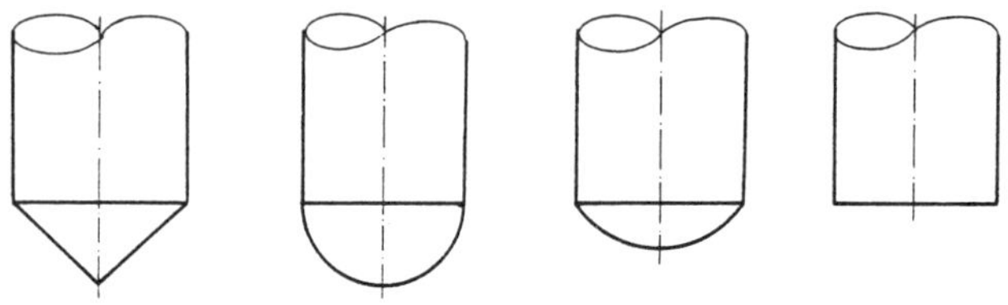

FIGURE 1.1 — Types of head-cylinder structures.

This is prevented at their junctions by the rigid attachment of the heads and cylinders, one to the other. Such a disturbance of the natural deformations, arising from internal pressure, produces shear and flexural stresses which obviously are most severe at or near the head-cylinder junctures. These stresses do not vary circumferentially because of the axial symmetry of the structures and decay ultimately, if not at once, both in the cylinders and in the heads as the axial distance from this point of stress disturbance increases. These juncture stresses exist in addition to the membrane stresses.

More particularly, the juncture stresses consist of axial shear stresses, axial flexural stresses, and circumferential membrane stresses. All vary in an axial direction only, as previously stated, due to the axial symmetry of the structure under consideration.

This same property of symmetry obviously excludes the existence of circumferential shear stresses, so that circumferentially only uniform membrane stresses and uniformly varying flexural stresses are obtained. These two types of stresses are additive, and the magnitudes of their algebraic sums, taken at any axial point on the structures, at both the inner and outer wall surfaces related to that point, represent the greatest and least combined circumferential stresses at the point.

It should be noted that when the greater stress is at the outer wall surface, then the lesser stress occurs at the inner wall surface, and, vice versa. Both stresses are principal stresses, due to the absence of any circumferential shear stresses to combine with them, and as axial distances from the head-cylinder juncture increase, the greater of the two ultimately decreases, both in the head and in the cylinder, down to the circumferential membrane stresses of those parts as a limit.

Likewise, the greater and lesser axial combined stresses are also derived as the algebraic sums of the axial membrane and axial flexural stresses, in the same manner as the greater and lesser circumferential stresses just described. They too, are principal stresses.

From the foregoing, it is seen that the important stresses, at any axial point on the middle surface in the head-cylinder structure, are the circumferential stresses on the inner and outer wall surfaces related to the point; the axial stresses on the inner and outer wall surfaces related to the point and the axial shear stress at the point.

A detailed stress analysis of the tubular structures at the junction will be given in Chapter 3.

CHAPTER 2

local and overall buckling of cylindrical shells

2.1 Introduction

The problem of determining the external pressure at which a thin-walled cylinder of large diameter will collapse confronts the designers of steel stacks, bins, tanks, pipelines, conveyor galleries, and similar structures.

In the design of the above structures the collapsing pressure and the evaluation of the effect of stiffeners upon the strength of tubular structure are frequently encountered. After determining the cylinder plate thickness in order to satisfy tensile stress requirements, the stability of the shell should be checked for compressive stresses against buckling. This analysis is more complex because the general and local buckling of thin-walled cylindrical shell under different loading conditions should be investigated.

In a linear shell theory, displacements are proportional to loads. The essence of shell buckling, however, is a disproportionate increase in displacement resulting from a small increase in load. It becomes obvious that a nonlinear shell theory is required. Thus, shell buckling is fundamentally a sub-topic of nonlinear shell theory.

The purpose of this Chapter is to discuss stress analysis of the local and overall buckling of thin-walled large diameter tubular steel structures. Although tubular structures are susceptible to buckling, most structural standards or codes do not give complete design information on the buckling analysis of such structures. Presumably, the column buckling formulas given in these standards may be applied to tubular structures, because the standards do not restrict the formulas to any particular shape. However, the tubular structures especially those having relatively large ration D/t, as experience shows, cannot be safely designed by formulas given in standards. Furthermore, no limitations based on local buckling of tubular structures under compression, bending, or combined loadings are given in these standards.

If the purpose of engineering analysis is to predict the behavior of structures under a variety of loading conditions, then experience tells us that the buckling is often the prime consideration in the design of tubular structures.

Considering the stability analysis of tubular structures it is necessary to clarify the situation regarding such terminology as buckling and collapse.

Since the terms "buckling" and "collapse" are often used interchangeably, we define a buckle as a localized failure in the form of a wrinkle or indentation caused by overstress or instability of the pipe wall on the compressive side of a pipe subjected to bending.

Collapse, on the other hand, is defined as a general failure usually in the form of a flattening of the pipe cross-section over a considerable length as the result of the action of external pressure on the pipe.

2.2 Overall and Local Buckling

In the design of tubular structures, after determining the shell plate thickness in order to satisfy tensile stress requirements, the stability of the shell should be checked for compressive stresses against buckling.

A thin-walled cylindrical shell subjected to compression in the direction of its longitudinal axis may fail either by the instability of the shell as a whole, involving bending of the axis, or by the local instability of the wall of the shell which may not at all involve lateral distortion of the axis. The former type of failure is that investigated by Euler, when the strength depends on the ratio of length to the radius of gyration of the shell. The latter type of failure has been called by various authors: secondary flexure, crinkling, wrinkling, or local buckling, and is often the governing consideration in the design of thin-walled cylinders.

The stability against local buckling depends on the ratio of thickness to the radius of the shell wall (t/R). Wrinkling is local in nature and depends upon the combined compressive stresses at the point under consideration. Failure of this type is due to the formation of characteristic wrinkles or bulges, circular or lobed in shape, Figure 2.1.

In studying thin-walled tubular structures, two considerations are of importance. First, local buckling should be prevented at stresses below yield strength; secondly, a more severe restriction is that the tendency to buckle locally should not reduce the general buckling load of a whole structure.

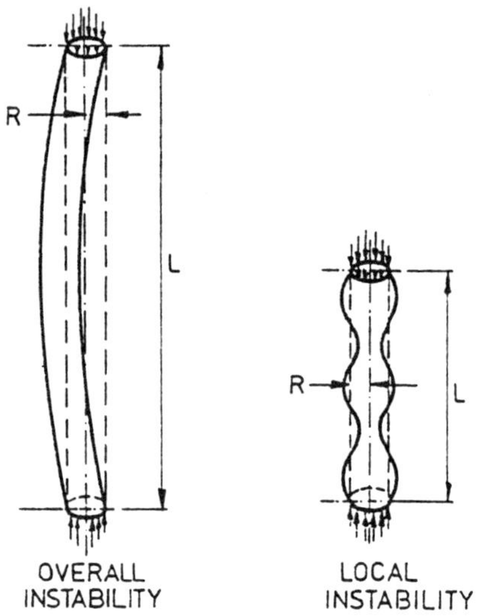

FIGURE 2.1 — Overall and local buckling.

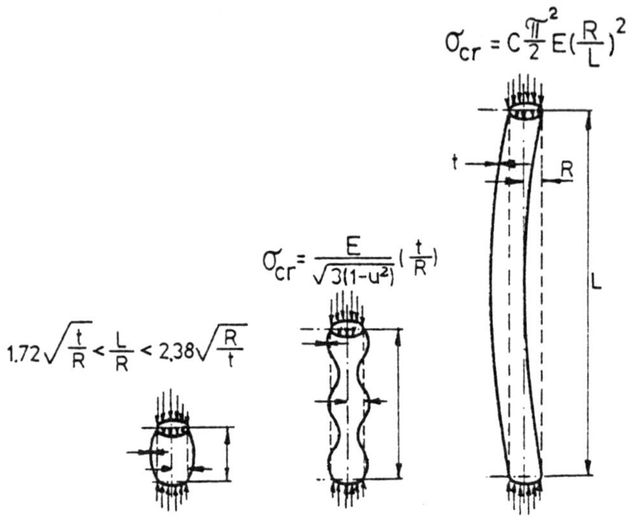

FIGURE 2.2 — Cylindrical shell subjected to axial compression.

2.3 A Paradox in the Buckling Analysis of Cylindrical Shell Under Axial Compression

To investigate the phenomena of the local and overall buckling of tubular structures, the classical approach was to investigate the fundamental case, namely — the buckling of cylindrical shell under axial compression. Solutions to this problem were obtained more than sixty years ago, first by Lorenz [2.1] in 1908, then by Timoshenko [2.2] in 1910 and by others. Lorenz used Euler's method and substantially simplified the problem, assuming that upon the loss of stability all generatrix of the cylinder were bent in equal manner, and that the axially-symmetric shape at the loss of stability had taken place, Figure 2.2.

He arrived at the classical expressions of buckling where the critical stress is equal

$$\sigma_{cr} = \frac{Et}{R\sqrt{3(1-\mu^2)}} = 0.6E(\frac{t}{R}) \qquad (2.1)$$

and the "classical buckling load" is

$$P_{c\ell} = \frac{Et}{\sqrt{3(1-\mu^2)}}(\frac{t^2}{R}) = 0.6E(\frac{t^2}{R}) \qquad (2.2)$$

where

R = is the radius of the shell, that is, the inside radius plus one-half of the wall thickness

t = thickness of the shell wall

E = modulus of elasticity of material

μ = 0.3 Poisson's ratio

Southwell [2.3], Dean [2.4] and Prescott [2.5] give for a lobed form of buckling the following formula

$$\sigma_{cr} = \frac{E}{\sqrt{3(1-\mu^2)}}(\frac{t}{R})\frac{n^2-1}{n^2+1} \qquad (2.3)$$

in which

n = the number of lobes in the wrinkle.

In these derivations, it is assumed that the elastic limit of the material is not exceeded. In general, although different methods of approach were used, the same results were obtained, namely, for a uniform circular bulge or wrinkle, as shown in Figure 2.1. In any case, where the number of lobes is greater than 3, Eq. (2.3) gives substantially the same critical stress as Eq. (2.1).

It may be noted that Eqs. (2.1) and (2.2) for local buckling do not involve the length of the shell. That is, the critical local buckling stress is independent of the length of the shell. Nevertheless, in the case of long slender shells, the total load-carrying capacity is affected by the ratio of the length to the radius of gyration. For, if there is a tendency to buckle, the stress will no longer be uniform over a section, and failure will occur when the maximum stress on the section becomes equal to the critical buckling stress. When we pass from the local loss of stability to the overall, the ratio L/R lies within the following limits [2.6], Figure 2.3.

$$1.72\sqrt{t/R} < L/R < 2.38\sqrt{R/t} \qquad (2.4)$$

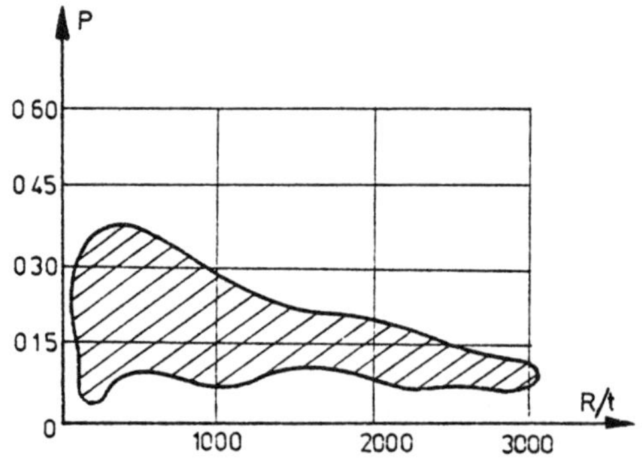

FIGURE 2.3 — Slenderness Ratios (L/R) for local buckling.

For a long slender cylindrical shell, failure does not occur by wrinkling, but by the type of buckling investigated by Euler. The unit stress at which buckling is likely to occur in this case is given by the formula

$$\sigma_{cr} = C \frac{\pi^2 E}{(L/r)^2} \qquad (2.5)$$

where

σ_{cr} = unit stress at failure

r = radius of gyration of the cross-section of the cylinder

C = a constant, depending on the end conditions

For a very thin shell $r^2 = R^2/2$, and Eq. (1.5) becomes

$$\sigma_{cr} = C \frac{\pi^2}{2} E (\frac{R}{L})^2 \qquad (2.6)$$

When the experiments were made to check the validity of Eq. (2.1), certain behavioral patterns were observed which were completely at variance with the theory.

There is a serious disagreement between the results of classical and experimental stress for the buckling of isotropic cylindrical shells under axial compression.

Similar discrepancies can be observed for other loading conditions. These experiments indicated critical stress levels in the order of ⅓ of those given by the classical linear theory.

This paradox puzzled the investigators for more than 30 years!

The modern phase of the investigation of the buckling of thin-walled cylindrical shells subjected to axial compression began in 1940, with two papers published by Von Karman [2.7], [2.8] and his collaborators. They showed that the significant difference between the buckling stresses predicted by the theory and those observed in experiment could be attributed to the fundamentally nonlinear nature of the buckling process.

They obtained a lower value which is three times smaller than that given in the classical theory, or

$$\sigma_{cr} = 0.195E(\frac{t}{R}) \qquad (2.7)$$

Von Karman and Tsien also suggested that imperfections which were inevitable in the fabrication, such as initial irregularities in shape in the test cylinders, might cause a round-off of the sharp peak between the linear and nonlinear branches of the load-displacement curve, and, thus, result in a lower maximum point.

TABLE 2.1 — Critical Buckling Stress σ_{cr}.

Author	Year	Buckling Stress	References
Michielson	1948	$\sigma_{cr} = 0.194E\left(\frac{t}{R}\right)$	[2.9]
Kirste	1954	$\sigma_{cr} = 0.187E\left(\frac{t}{R}\right)$	[2.10]
Kempner	1954	$\sigma_{cr} = 0.182E\left(\frac{t}{R}\right)$	[2.11]
Pogorelov	1967	$\sigma_{cr} = 0.160E\left(\frac{t}{R}\right)$	[2.12]

Several investigators extended the Von Karman-Tsien analysis and found the lowest values of the buckling stress as shown in Table 2.1 [2.9, 2.10, 2.11, 2.12].

2.4 Imperfections of the Shell Shape and Edge Effect

Theoretical analyses [2.12; 2.13; 2.14; 2.16] of the effect of imperfections on the buckling behavior of cylinders have clearly demonstrated that relatively small imperfection amplitudes can drastically reduce the critical load of the shell.

Despite the substantial theory available, few experimental data [2.17; 2.18; 2.19] exist describing the effects of specific imperfections in shape in reducing the static buckling load. Consequently, it was of particular interest to determine the buckling load reduction caused by an initial axissymmetric imperfection in shape defined by a simple trigonometric function. This problem was investigated by Koiter.

Koiter [2.20; 2.21] developed a rigorous theory of the maximum load and showed that thin-walled circular cylindrical shells are very sensitive to small deviations in the initial, unstressed state from the exact circular cylindrical shape.

The results of the Koiter analysis indicate that an initial imperfection amplitude equal to the shell thickness is sufficient to reduce the buckling load to only 20 percent of the corresponding value for the perfect cylinder.

Investigations by Stein [2.22], Ohira [2.23] and Hoff [2.24] showed lower buckling load due to the edge effect. In the classical analysis the influence of the edge restraint on the prebuckling deformation was neglected. However, in reality, the diameter of the restrained cylinder tends to increase under axial compression due to Poisson's effect. This increase is prevented at the ends of the cylinder by its boundary restraints.

Hence, the generators of the cylinder are distorted prior to buckling and axial forces in the cylinder at the ends are eccentric relative to portions of the shell wall near their mid-length. When this eccentricity is considered, the theoretical prebuckling equilibrium becomes nonlinear.

2.5 Practical Application of Experimental Data

A disagreement between the theory and experiments considering the buckling of thin cylindrical shells under axial compression has lasted a long time, causing much disappointment, and inducing a number of new theories. And yet, when the final clarification arrived, the answer was simple and clear: one should have made better experiments and more extensive calculations!

Small imperfections of the test specimen, nonuniformity in loading, and small uncertainties in the control of boundary conditions have had large effect. To compare theory with experiment, a very careful analysis and experimental control would have to be made.

When we turn to experimental data in considering the magnitude of critical stress, this data is somewhat contra-

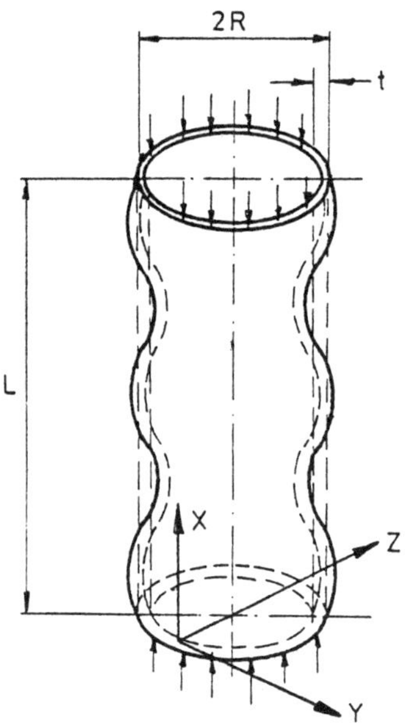

FIGURE 2.4 — Region of experimental data for critical compression stress.

dictory, since it strongly depends on initial imperfections in the form of the shell and in conditions of loading.

In Figure 2.4, a region of experimental values is shown of the buckling coefficient p in equation

$$\sigma_{cr} = pE\left(\frac{t}{R}\right) \tag{2.8}$$

based on experiments performed by different investigators [2.25].

A significant part of the experiments leads to the values of p lying above 0.18. However, certain values lie below this magnitude and in separate cases turn out to be equal to 0.06 — 0.15. Figure 2.4 indicates an evident tendency of p to drop during the growth of ratio R/t. It is necessary to recall that with an increase of R/t, the probability of the appearance of initial imperfections should increase. This, undoubtedly, should lead to a lowering of the average magnitude of real critical stresses.

2.6 Allowable Design Stresses

In view of discrepancy between theoretical and experimental results, it seems advisable to rely largely on test results for developing adequate design provisions to safeguard against local buckling.

Donnell [2.26] developed the following formula for the ultimate buckling stress of circular cylinders in compression

$$\sigma_c = E\,\frac{0.6\frac{t}{R} - 10^{-7}\frac{R}{t}}{1 + 0.004\,\frac{E}{F_y}} \tag{2.9}$$

where F_y is the yield stress of the material.

The formula is designed to give the average strengths to be expected, and if it is desired to know the minimum strength likely to be encountered under any circumstances, some factor must be used with it.

A systematic evaluation of test evidence obtained by a number of investigators was analyzed by Plantema [2.29]. The permissible compressive stress is given by the following formula:

$$\sigma_{cr} = \frac{662}{D/t} + 0.399\,F_y \tag{2.10}$$

The ratio D/t is valid for

$$\frac{3{,}300}{F_y} < \frac{D}{t} < \frac{13{,}000}{F_y} \tag{2.11}$$

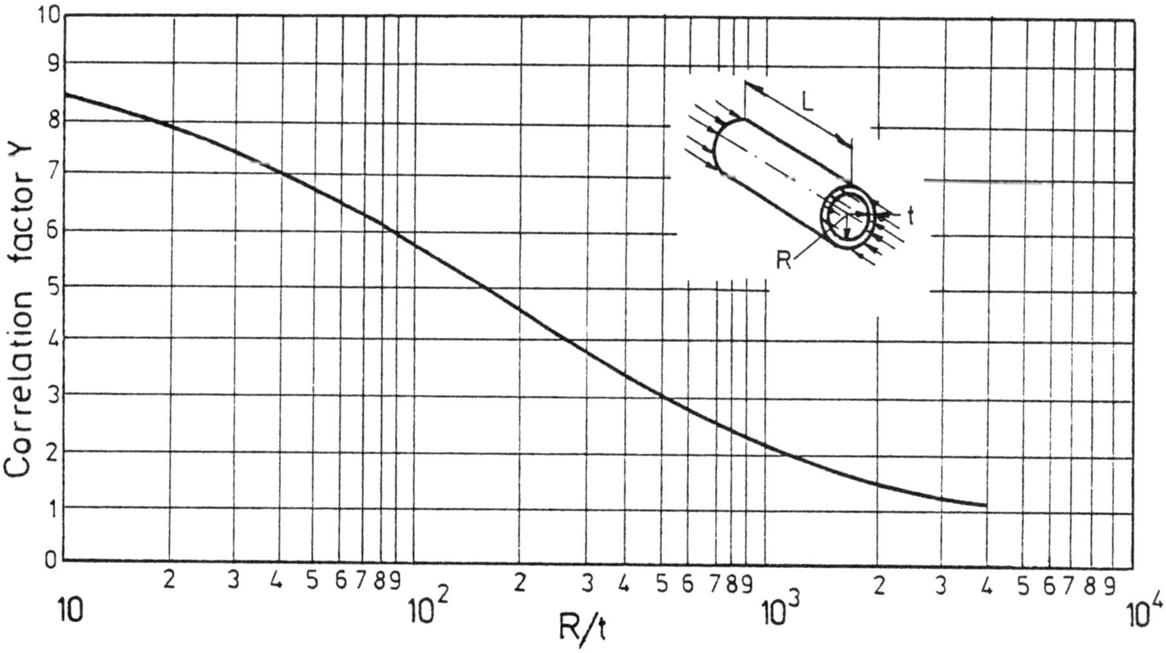

FIGURE 2.5 — Correlation factors for unstiffened circular cylinders subjected to axial compression.

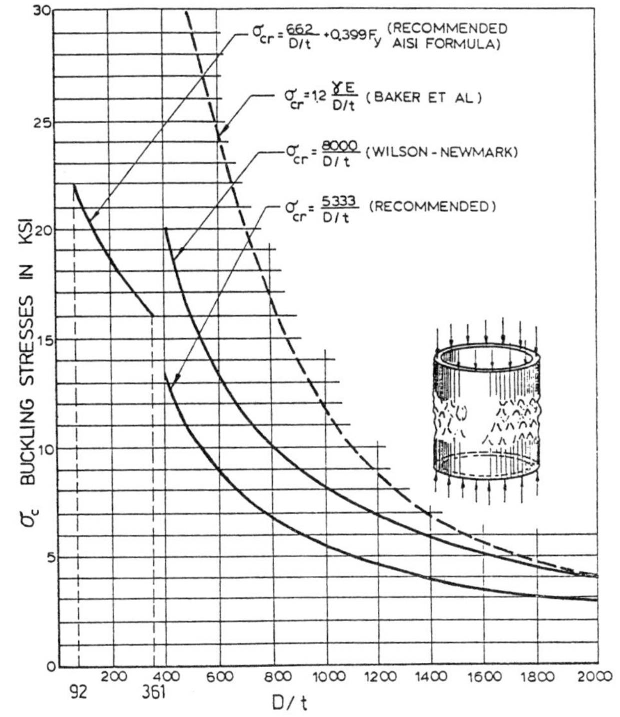

FIGURE 2.6 — The recommended allowable buckling stresses.

At yield point for mild steel Fy = 36 ksi, the limits are

$$92 < \frac{D}{t} < 361 \qquad (2.12)$$

Formula (2.10) is recommended by the American Iron and Steel Institute [2.28].

Baker et al [2.29] proposed the following formula for the determination of the local buckling stress for cylindrical shells of moderate length.

$$\sigma_{cr} = 0.6\gamma E\left(\frac{t}{R}\right) \qquad (2.13)$$

where the values of the correlation factor γ in the function of the ratio (R/t), are shown in Figure 2.5. The correlation γ is introduced to account for the difference between theoretical and experimental results.

Wilson and Newmark [2.30] carried tests using tubular steel compression members having a large D/t ratio. In the elastic range, the magnitude of the critical buckling stress is expressed as follows

$$\sigma_{cr} = \frac{8{,}000}{D/t}\text{, ksi} \qquad (2.14)$$

Assuming the factor of safety = 1.5, the allowable critical stress for local buckling is

$$\sigma_{cr} \quad \frac{5{,}333}{D/t}\text{, ksi} \qquad (2.15)$$

The recommended allowable stresses for local buckling in the function of D/t are shown in Figure 2.6 [2.31].

Long cylindrical shells must be checked for overall buckling as an Euler column, by the formula (2.6).

Buckling analysis of the cylindrical shell indicates that there has been little agreement between theoretical and experimental results for critical loads of shell structure, since apparently, infinitesimal deviations in boundary conditions and in the shape of the shell yield drastic reduction in critical loads. It is believed that accurate formulation of a problem in terms of nonlinear theory and exact solution of the equations would result in a close agreement between theoretical and experimental results. At present, for actual practice, however, this procedure is prohibitively difficult. Nonlinear theory serves to broaden our knowledge of shell buckling analysis and to clarify the meaning and limitations of linear stability theory, but at present, it is not a design tool for direct determination of the buckling load.

REFERENCES — CHAPTER 2

[2.1] Lorenz, R., "Achsymmetrische Verzerrungen in dunwandigen Hohlzylinder", Zeitschrift des Vereins Deutscher Ingenierus, Vol. 52, 1908, p. 1706. (in German).

[2.2] Timoshenko, S., "Einige Stabilitatsprobleme der Elastizitatstheorie", Zeitschrift fur Math. und Physik, Vol. 58, 1910, pp. 337-385. (in German).

[2.3] Southwell, R.V., "On the General Theory of Elastic Stability", Philos. Trans. Royal Soc., London, Series A, Vol. 213, 1913, pp. 187-244.

[2.4] Dean, W.R., Proc. Royal Society, London, Series A, Vol. 107, 1925, p. 734.

[2.5] Prescott, J., "Applied Elasticity", Longmans, Green and Co., London, 1924, pp. 530-564.

[2.6] McGuire, W., "Steel Structures", Prentice-Hall, Inc., Englewood Cliffs, N.J., 1968,p. 418.

[2.7] Von Karman, Th., Dunn, L.G., and Tsien, H.S., "The Influence of Curvature on the Buckling Characteristics of Structures", Journ. Aeron. Sci., Vol. 7, No. 7, 1940, p. 276.

[2.8] Von Karman, Th., and Tsien, H.S., "The Buckling of Thin Cylindrical Shells Under Axial Compression", Journ. Aeron. Sci., Vol. 8, No. 8, 1941, p. 303.

[2.9] Michielson, H.F., "The Behaviour of Thin Cylindrical Shells After Buckling", Journ. Aero. Sci., Vol. 15, No. 12, 1948.

[2.10] Kirste, L., "Abwickelbare Verformung dunnwandiger Kreiszylinder (Evoluable Formation of a Thin-Walled Ring Cylinder)", Oester. Ing.-Archiv. Vol. 8, No. 2-3, 1954. (in German).

[2.11] Kempner, J., "Postbuckling Behavior of Axially Compressed Circular Cylindrical Shells", Journ. Aero. Sci., Vol. 21, 1954, pp. 329-334.

[2.12] Pogorelov, A.V., "Geometric Methods in the Nonlinear Theory of Elastic Shells", Moscow, 1967, p. 232. (in Russian).

[2.13] Hutchinson, J., "Axial Buckling of Pressurized Imperfect Cylindrical Shells", AIAA Journal, Vol. 3, No. 8, Aug. 1965, pp. 1461-1466.

[2.14] Almroth, B.O., "Influence of Imperfections and Edge Restraint on the Buckling of Axially Compressed Cylinders", CR-432, NASA, April, 1966.

[2.15] Budiansky, B., and Hutchinson, J.W., "A Survey of Some Buckling Problems", AIAA Journal, Vol. 4, No. 9, Sept. 1966, pp. 1505-1510.

[2.16] Dym, C.L., and Hoff, N.J., "Perturbation Solutions for the Buckling Problems of Axially Compressed Thin Cylindrical Shells of Infinite or Finite Length", Journ. Appl. Mech., Dec. 1968.

[2.17] Babcock, C.D., and Sechler, E.E., "The Effect of Initial Imperfections on the Buckling Stress of Cylindrical Shells", TN D-2005, NASA, July, 1963.

[2.18] Babcock. C.D., "The Influence of a Local Imperfection on the Buckling Load of a Cylindrical Shell Under Axial Compression", SM-68-4, Feb. 1968, Graduate Aeron. Lab., Calif. Instit. of Technology.

[2.19] Arbocz, J., and Babcock. C.D., "Experimental Investigation of the Effect of General Imperfections on the Buckling of Cylindrical Shells, SM 68-7, Feb. 1968, Grad. Aeron. Lab., Calif. Instit. of Technology.

[2.20] Koiter, W.T., "On the Elastic Stability of Elastic Equilibrium", Thesis, Polytechnic Institute Delft, Translation by the National Aeronautics and Space Administration, Washington, D.C., March, 1967.

[2.21] Koiter, W.T., "Elastic Stability and Postbuckling Behaviour", Proc. of the Sympos. on Nonlinear Problems. Ed., Langer, R.E., Univ. of Wisconsin Press, 1963, p. 257.

[2.22] Stein, M., "Notes on the Influence of Prebuckling Deformations and Stresses on the Buckling of Perfect Cylinders", Lecture on the Static and Dynamic Stability of Structures, UCLA, Sept. 25-Oct. 4, 1963.

[2.23] Ohira, H., "Local Buckling Theory of Axially Compressed Cylinders", Proc. Eleventh Japan. Natl. Congr. for Appl. Mech., Tokyo, 1961.

[2.24] Hoff, N.J., and Tsai, C.S., "Buckling of Circular Cylindrical Shells in Axial Compression", N65-14484, Stanford University, Dept. of Aeronautics and Astronautics, SUDAER No. 204, 1964.

[2.25] Vol'mir, A.S., "Stability of Elastic Systems", National Technical Information Service, U.S. Department of Commerce, 1965 (Translation from Russian), pp. 590-591.

[2.26] Donnell, L.H., "A New Theory for the Buckling of Thin Cylinders Under Axial Compression and Bending", Trans. ASME, Vol. 56, 1934, p. 795.

[2.27] Plantema, F.J., "Collapsing Stresses of Circular Cylinders and Round Tubes", Report S. 280, Nat. Luchtvaartlaboratorium, Amsterdam, Netherlands, 1946.

[2.28] American Iron and Steel Institute, "Specification for the Design of Cold-Formed Steel Structural Members", Edition 1968, p. 25.

[2.29] Baker, E.H. et al., "Structural Analysis of Shells", McGraw-Hill Book Co., New York, 1972.

[2.30] Wilson, W.M., and Newmark, N.M., "The Strength of Thin Cylindrical Shells as Columns", Bulletin No. 255, University of Illinois, February, 1933.

[2.31] Troitsky, M.S., "On the Local and Overall Stability of Thin-Walled Large Diameter Tubular Structures", Proceedings Canadian Structural Engineering Conference, Montreal, February 22-24, 1976, pp. 1-32.

CHAPTER 3

edge effect at tubular structures

3.1 Physical Concept

Stresses and deformations in thin-walled shells, determined by applying the membrane theory are correct only in zones located at certain distances from the changes of such geometrical parameters as shape, dimensions and stiffness and also from the places of sharp changes in acting forces.

For sections having changes in geometry, apart from forces, stresses and deformations which are determined in applying the membrane theory, additional forces, stresses and deformation which also originate are called — the edge effect.

Due to the elastic resistance of the adjoining parts, the edge effect does not spread too far, but rather acts upon relatively narrow zones. The edge effect is spread by relatively fast diminishing waves, the general character of which is shown in Figure 3.1, where at the axis x are ordinates of the wave curve and along the axis y- are plotted lengths of the generatrix of the shell.

The physical causes of origination of the edge effect are:

a. An absence of free deformation of the shell, under membrane stresses in a circumferential direction.

b. Sudden changes or eccentricity of the generatrix, which lead under axially-symmetrical loading to the origin of local forces, distributed along the circumference of shell as projections of meridional forces on a plane, normal to the axis, or on moments due to its eccentricity.

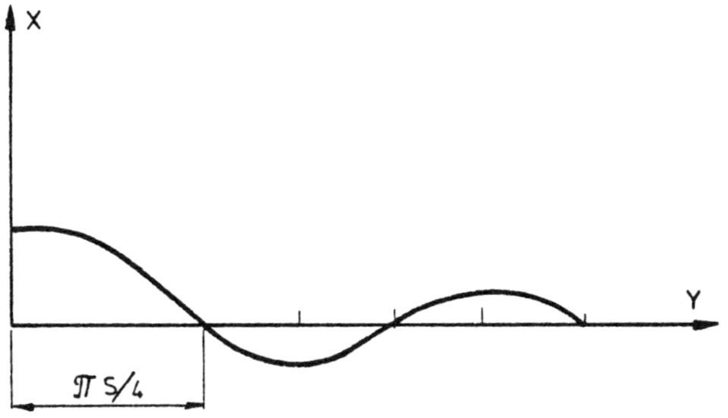

FIGURE 3.1 — Diagram of diminshing of edge effect. Axis Ox — location of origin of edge effect.

An example, shown in Figure 3.2 illustrates the original causes of edge effect.

At the elevation a-a under free deformation and loading P_r an increase of radii may be expressed as follows:

For a cylinder

$$\Delta r = \frac{r}{E}(\sigma_2 - \nu\sigma_1) = \frac{r^2 p}{Et_{cyl}}\left(1 - \frac{\nu}{2}\right) \qquad (3.1)$$

For a cone

$$\Delta R = \frac{R}{E}(\sigma_2 - \nu\sigma_1) = \frac{R^2 p}{Et_{cone}}\left(1 - \frac{\nu}{2}\right) \qquad (3.2)$$

where

σ_1 = longitudinal stress

σ_2 = circumferential stress

t_{cyl}, t_{con} = thicknesses of the walls of a cylinder and cone, respectively

ν = Poisson's ratio

E = modulus of elasticity of shell materials

In general cases

$$\Delta r \neq \Delta R \sin\beta \qquad (3.3)$$

and at the free elastic deformations results in a relative displacement at section a-a. However, the interconnection between the cylindrical and conical shells prevents free deformation at these shells at a level a-a which results in the origin of local bending at this level. Apart from this, the edge effect originates due to a break of the generatrix and the existence of the local circumferential forces in the plan a-a, which are projections of the meridional forces of conical shell. The presence of these forces cause different deformations and stresses to those of the parts of a shell located relatively at a distance from section a-a. This results in the local bending of shells at their interconnections. Therefore, in general, as shown in Figure 3.2, the edge effect is due to both causes.

In particular, at the ratio of a shell thickness

$$\frac{t_{cyl}}{t_{cone}} = \sin\beta \qquad (3.4)$$

and under internal or external uniform pressure, we have

$$\Delta r = \Delta R \sin\alpha \qquad (3.5)$$

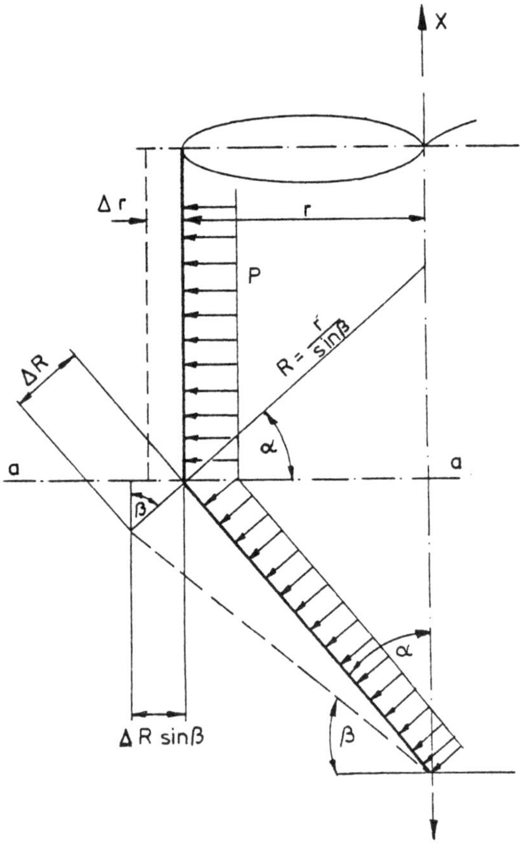

FIGURE 3.2 — Intersection of cylindrical and conical shells.

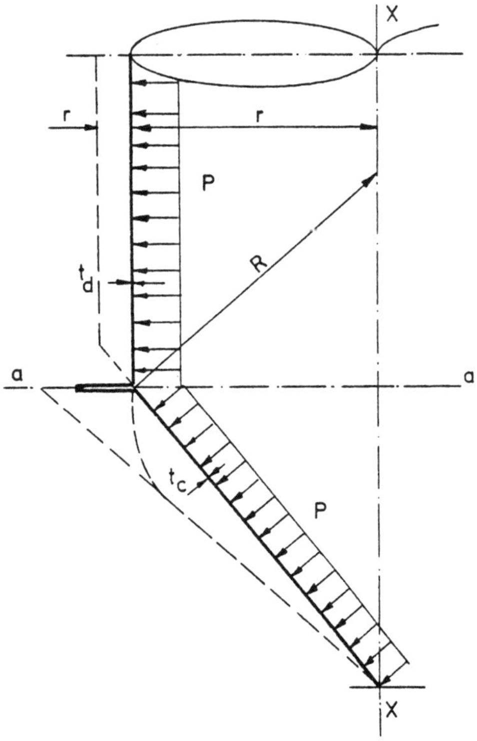

FIGURE 3.3 — Intersection of the cylinder, cone and stiffening ring.

which may be easily obtained from formulas (3.1) and (3.2).

In this case, the edge effect originates from the second cause.

However, it is possible to choose such a ratio of wall thicknesses of both shells at their intersection, to cancel the edge effect.

Also, by using the stiffening ring at the intersection, it is possible to prevent free deformations of shells and cancel the edge effect, Figure 3.3.

The increase of radius of stiffening ring is expressed by the value

$$\Delta r_{st} = (pc + H_{cyl} + H_{con})\frac{r_{st}^2}{EA} \qquad (3.6)$$

where

H_{cyl}, H_{cone} = thrusts of the cylinder and cone, respectively

A = cross-sectional area of the ring

The values of thrusts H_{cyl} and H_{cone} may be obtained after solving the problem of edge effect.

Figures 3.4 (a) and 3.4 (b) show examples of the appearance of an edge effect due only to the first cause of its origin.

In this case there is no break of the generatrix, but the deformations are restricted in case (a) due to the presence of the stiffening ring and in case (b) due to the junction of different thicknesses of the walls of both shells, which results in unequal values of the free deformations under the same loading p.

The example shown in Figure 3.5 is characterized by the presence of the second cause of the origin of the edge effect.

The brake producing the eccentricity of the generatrices cause the appearance along the ring at level a-a of the local moments due to the eccentricity originating at the edge bending of the shell.

3.2 Methods for the Solution of Edge Effect Problem

At intersections of the shells, where the edge effects originate, apart from the meridional and circumferential forces, determined by the membrane theory, the bending of shells also originate, accompanied by the bending moment, shear force and additional meridional and circumferential forces, caused by the bending of the shells.

These additional forces producing deformations, cannot be determined by using membrane theory only. For their determination, it is necessary to apply the moment theory and consider the displacements of a statically-indeterminate system consisting of connected shells with or without stiffening rings. It is possible then to solve the problem of edge effects by applying known methods of structural mechanics, namely a method of forces or a method of deformations.

3.2.1 Solution of the Problem by the Force Method

By the solution of the problem by the force method, we now apply conditions of continuity of deformations (relative displacement, relative angle of rotation and break of

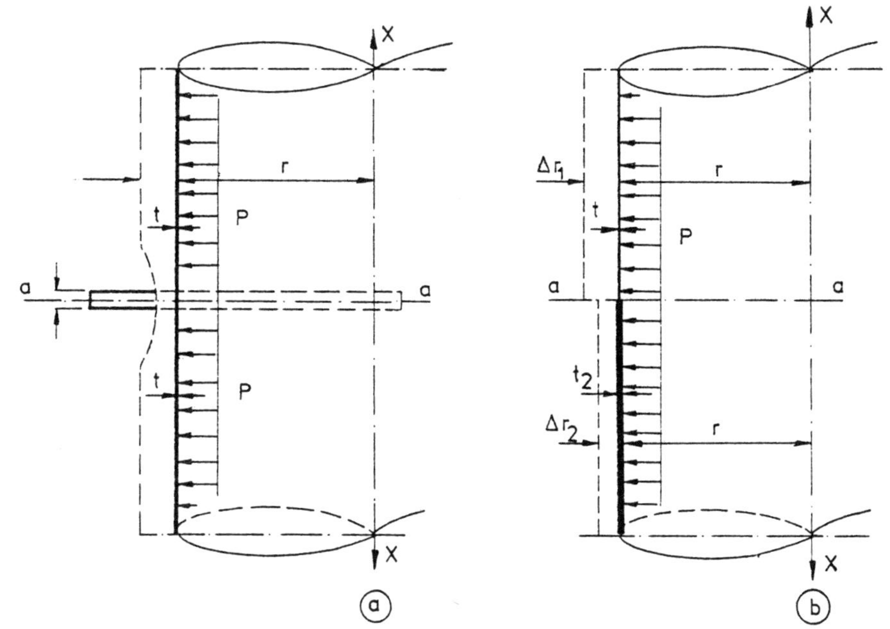

FIGURE 3.4 — Schemes of intersections of shells: (a) Cylindrical shell having stiffening ring; (b) Junction of two cylindrical shells having different thicknesses of the walls.

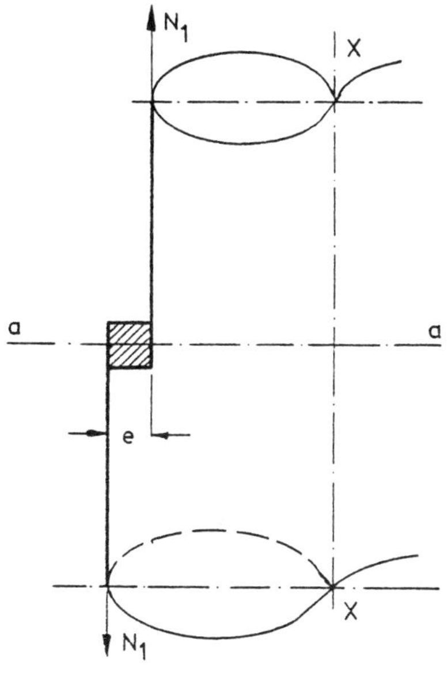

FIGURE 3.5 — Scheme of the intersection of shells in the case of eccentric function of two cylindrical shells.

the sections are equal to zero). The shell at the place of origin of the edge effect is cut by the plane, normal to the axis of the shell, and therefore, is transformed into the basic statically-determinate system.

To the separated parts of the shell, we shall apply given loadings and forces, which have been determined by the membrane theory and also any unknown forces, namely moments in a meridional direction and forces in the plane normal to the axis of the shell. Then, we construct the usual equations of the force method. In particular, having two unknowns-moment X_1 and force X_2 — the equations are

$$\left.\begin{aligned} X_1\delta_{11} + X_2\delta_{12} + \Delta_{1P} &= 0 \\ X_1\delta_{11} + X_2\delta_{22} + \Delta_{2P} &= 0 \end{aligned}\right\} \tag{3.7}$$

The values of the unit displacements δ_{ik} and displacements due to the loadings Δ_{iP} are given in Tables 3.1 to 3.12, and Table 3.13. After obtaining the solutions and finding the unknown factors, we then determine any stresses and deformations by the methods described below.

3.2.2 Solution of the Problem by the Method of Deformations

By using the method of deformations we apply the condition of equalization to zero or to the given value of external reactive forces acting on connected shells and stiffening rings at the original location of the edge effect, Figure 3.6.

At the location of the edge effect an assumed fixity under twist and displacement is given. The basic equations of deformations are

$$\left.\begin{aligned} y_1 a_{11} + y_2 a_{12} + A_{1P} &= 0 \\ y_1 a_{21} + y_2 a_{22} + A_{2P} &= 0 \end{aligned}\right\} \tag{3.8}$$

where the unknowns are an angle of twist of the joint y_1 and displacement of the joint y_2 along the radius in the plane normal to the axis of the shell.

The coefficients of equations a_{in} designate a summary of the external reactive forces which are necessary to

TABLE 3.1 — Values of Displacements δ and Δ in Basic Systems for the Analysis of Shells by the Force Method.

<table>
<tr><th>Basic System Circular Cylinder</th><th>Unit Displacements
δ_{in} (M = 1; H = 1; N = 1)</th></tr>
<tr><td>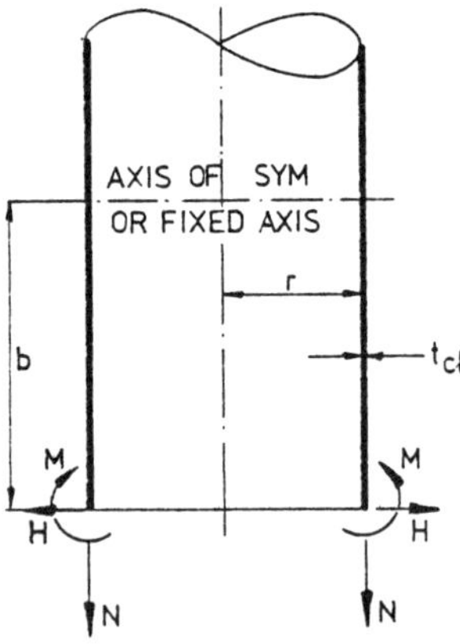</td><td>

$$\delta^{cy\ell}_{MM} = \frac{S_{cy\ell}K_1}{EJ^{cy\ell}} ; \quad \delta^{c\ell}_{HM} = \delta^{c\ell}_{MH} = \frac{S^2_{c\ell}K_2}{2EJ^{c\ell}_0}$$

$$\delta^{c\ell}_{HH} \frac{S^3_{c\ell}K_3}{2EJ^{c\ell}} ; \quad \delta^{c\ell}_{NN} = \frac{b}{Et_{c\ell}}$$

$$\delta^{c\ell}_{MN} = \delta^{c\ell}_{NM} = \delta^{c\ell}_{NH} = \delta^{c\ell}_{HN} = 0$$

$$S_{c\ell} = \frac{\sqrt{rt_{c\ell}}}{\sqrt[4]{3(1-\nu^2)}}$$

at $\nu = 0.3$; $S_{c\ell} = 0.78\sqrt{rt_{c\ell}}$

$$J^{c\ell}_0 = \frac{t^3_{c\ell}}{12(1-\nu^3)}$$

at $\nu = 0.3$; $J^{c\ell}_0 = \frac{t^3_{c\ell}}{10.92}$

K_1; K_2; K_3 — See Table 3.13; K is determined in the function of

$$\lambda = \frac{\ell_1}{S_{c\ell}}$$

where ℓ_1 = distance between the location of origin of the edge effect.

At $\lambda \geqslant 3$, K = 1

ν = Poisson's ratio

</td></tr>
</table>

TABLE 3.2 — Values of Displacements δ and Δ in Basic Systems for the Analysis of Shells by the Force Method.

Displacements due to the loadings Δ_{iP}		
Uniform Pressure P in the Closed Shell	**Uniform Pressure P in the Open Shell**	**Triangular Loading P in the Open Shell**
Basic System — Circular Cylinder		
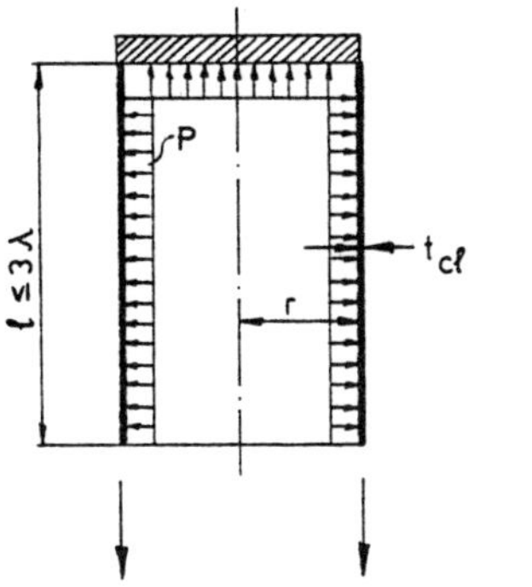	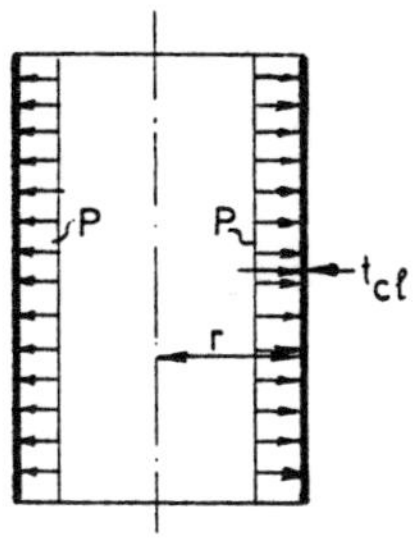	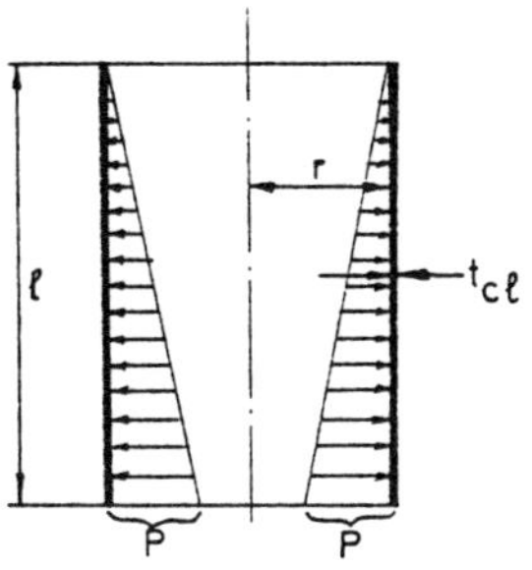
$N_1^{c\ell} = \frac{pr}{2}$; $N_1^{c\ell} = \frac{pr}{2}$ $\Delta_{MP}^{c\ell} = 0$ $\Delta_{HP}^{c\ell} = \frac{pr^2}{Et_{c\ell}}(1 - \frac{\nu}{2})$ $\Delta_{NP}^{\ell} = \frac{Pbr}{2ET_{c\ell}}(1 - 2\nu)$	$\Delta_{MP}^{c\ell} = 0$ $\Delta_{HP}^{c\ell} = \frac{pr^2}{t_{c\ell}E}$ $\Delta_{NP}^{c\ell} = 0$	**FOR UPPER PART** $\Delta_{MP}^{c\ell} = -\frac{pr^2}{t_{c\ell}E\ell}$ $\Delta_{HP}^{c\ell} = 0$; $\Delta_{NP}^{c\ell} = 0$ **FOR BOTTOM PART** $\Delta_{MP}^{c\ell} = \frac{pr}{t_{c\ell}E\ell}$ $\Delta_{HP}^{c\ell} = \frac{pr^2}{t_{c\ell}E}$ $\Delta_{NP}^{c\ell} = 0$

TABLE 3.3 — Values of Displacements δ and Δ in Basic Systems for the Analysis of Shells by the Force Method.

Displacements Due to the Loadings Δ_{iP}		
Vertical Uniformly Distributed on Surface Loading g	**Loading N_0, Applied Uniformly Along the Generatrix**	**Uniform Heating in t Degrees**
Basic System — Circular Cylinder		
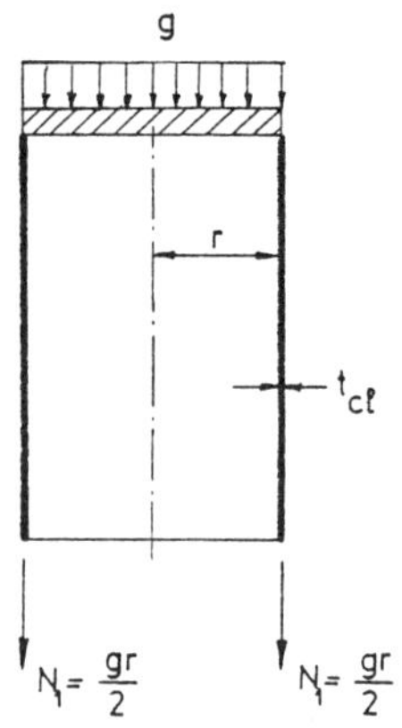 $\Delta^{c\ell}_{M_g} = 0$ $\Delta^{c\ell}_{H_g} = \frac{\nu g r^2}{2Et_{c\ell}}$ $\Delta^{c\ell}_{N_g} = -\frac{grb}{2t_{c\ell}E}$	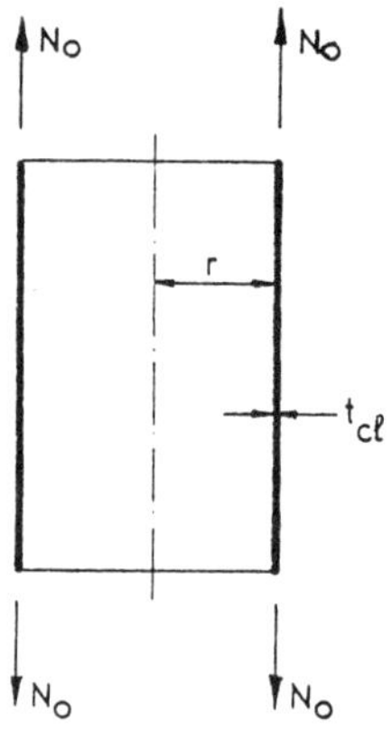 $\Delta^{c\ell}_{MN_0} = 0$ $\Delta^{c\ell}_{HN_0} = -\int \frac{dN_0 r}{Et_{c\ell}}$ $\Delta^{c\ell}_{NH_0} = \frac{N_0 b}{t_{c\ell}E}$	$\Delta^{c\ell}_{Mt} = 0$ $\Delta^{c\ell}_{Ht} = t\alpha r$ $\Delta^{c\ell}_{Nt} = t\alpha b$ where α = the coefficient of thermal expansion. For steel $\alpha = 0.000012$

TABLE 3.4 — Values of Displacements δ and Δ in a Basic System for an Analysis of Shells Using the Force Method.

<table>
<tr><th colspan="4">Basic System</th></tr>
<tr><th colspan="4">Unit Displacements — δ_{in} (M = 1; H = 1; N = 1)</th></tr>
<tr>
<td rowspan="2">Circular Cone</td>
<td>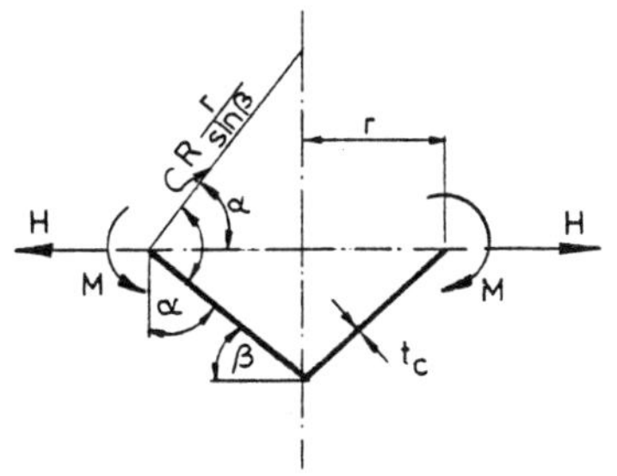
</td>
<td rowspan="2">Truncated Circular Cone</td>
<td>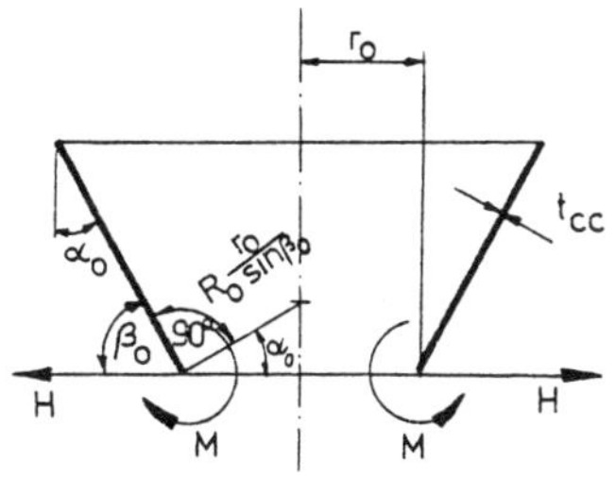
</td>
</tr>
<tr>
<td>$\delta^{c}_{MM} = \frac{S_c}{EJ^c_0 \nu_1}$; $\delta_{HM} = \delta^{c}_{MH} = \frac{S_c^2 \sin\beta}{2EJ^c_0 \nu_1}$

$\delta^{k}_{HH} = \frac{S_k^2 \sin^2\beta}{2EJ^c_0} \cdot \frac{\nu_2}{\nu_1}$

where

$S_k = 0.78\sqrt{Rt_c}$ (at ν=0.3); $J^c_0 = \frac{t_c^3}{12(1-\nu^2)}$

$\nu_1 = 1 + \frac{S_c \operatorname{ctg}\beta(4-\nu)}{4R} = 1 + \frac{S_c \cos\beta(4-\nu)}{4r}$

$\nu_2 = 1 - \frac{S_c \operatorname{ctg}\beta}{4R} = 1 - \frac{S_c \cos\beta}{4R}$</td>
<td>$\delta^{cc}_{MM} = \frac{S_{cc}}{EJ^{cc}_0 \nu_3}$; $\delta^{cc}_{MH} = \delta^{cc}_{HM} = \frac{S_{cc}^2 \sin\beta_0}{2J^{cc}_0 \nu_3}$

$\delta^{cc}_{HH} = \frac{S_{cc} \sin^2\beta_0}{2EJ^{cc}_0} \cdot \frac{\nu_4}{\nu_3}$

where

$S_{cc} = 0.78\sqrt{R_0 t_{cc}}$ (at ν=0.3); $J^{cc}_0 = \frac{t_{cc}^3}{12(1-\nu^2)}$

$\nu_3 = 1 - \frac{S_{cc} \operatorname{ctg}\beta_0 (1-\nu)}{4R_0} = 1 - \frac{S_{cc} \cos\beta_0 (1-\nu_0)}{4r_0}$

$\nu_4 = 1 - \frac{S_{cc} \operatorname{ctg}\beta_0}{4R_0} = 1 - \frac{S_{cc} \cos\beta_0}{4r_0}$</td>
</tr>
</table>

TABLE 3.5 — Values of Displacements δ and Δ in a Basic System for the Analysis of Shells by the Force Method.

<table>
<tr><th rowspan="2"></th><th colspan="2">Displacements Due to the Loadings Δ_{iP}</th></tr>
<tr><th>Uniform Pressure P in a Closed Shell</th><th>Uniform Pressure P in an Open Shell</th></tr>
<tr><td>Circular Cone</td><td>
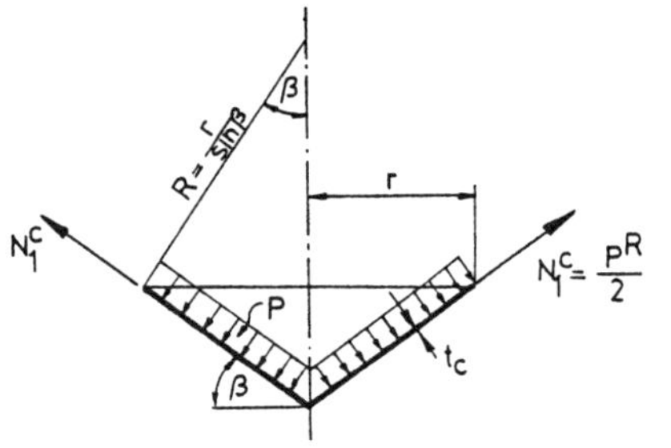

$\Delta^{c}_{MP} = \frac{3PRctg\beta}{2t_cE}$

$\Delta^{c}_{HP} = \frac{PR^2(1-\frac{\nu}{2})\sin\beta}{t_cE}$
</td><td>
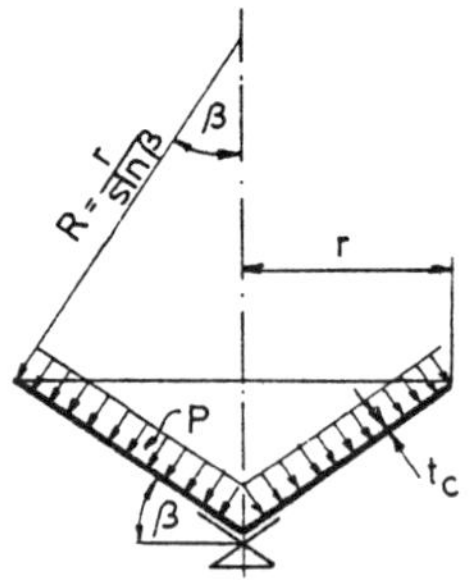

$\Delta^{c}_{MP} = \frac{3pRctg\beta}{2(1-\frac{\nu}{2})t_cE}$

$\Delta^{c}_{HP} \; \frac{pR^2\sin\beta}{t_cE}$
</td></tr>
<tr><td>Truncated Circular Cone</td><td>
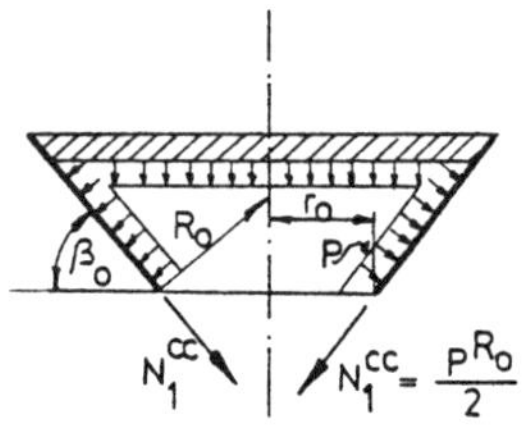

$\Delta^{cc}_{MP} = \frac{3pR_0ctg\beta_0}{2t_{cc}E}$

$\Delta^{cc}_{HP} = \frac{pR_0^2(1-\frac{\nu}{2})\sin\beta_0}{Et_{cc}}$
</td><td>
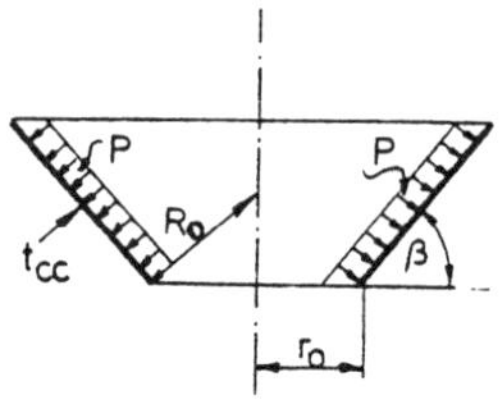

$\Delta^{cc}_{MP} = \frac{3pR_0ctg\beta_0}{2(1-\frac{\nu}{2})t_{cc}E}$

$\Delta^{cc}_{HP} = \frac{pR_0^2\sin\beta_0}{Et_{cc}}$
</td></tr>
</table>

TABLE 3.6 — Values of Displacements δ and Δ in a Basic System for the Analysis of Shells by the Force Method.

Displacements Due to the Loading Δ_{iP}				
	Vertical Uniformly Distributed on Surface Loading g	Vertical Uniformly Distributed on Surface Loading q		Uniform Heating in t Degrees
Circular Cone	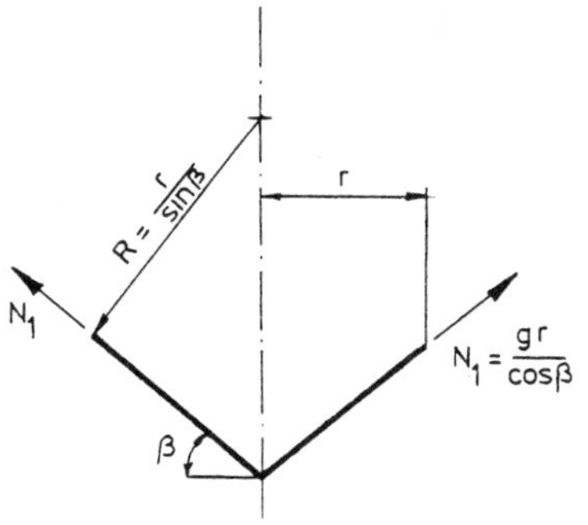	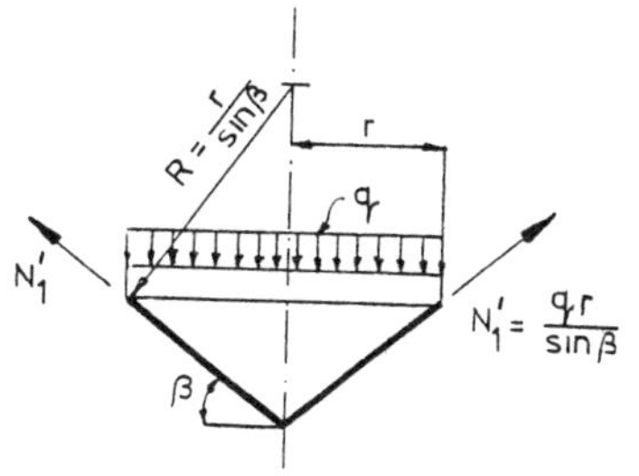	Circular Cone	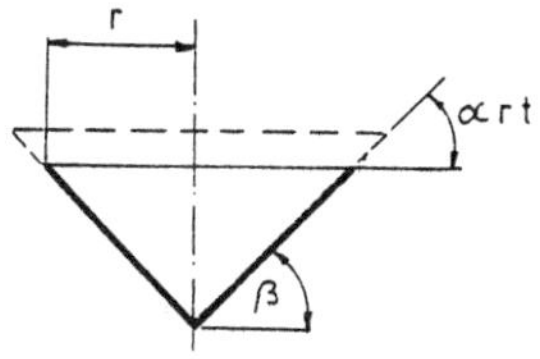 $\Delta^{c}_{Mt} = \theta$ $\Delta^{c}_{Ht} = \alpha r t = \frac{\alpha r t}{\cos\beta}$
Circular Cone	$\Delta^{c}_{MP} = \frac{gr}{t_c \sin\beta E}[(1 + \frac{\nu}{2}) \times \cos 2\beta + \frac{1-\nu}{2}]$ $\Delta^{c}_{HP} = \frac{gR^2}{2Et_c}[\sin 2\beta - \nu\tan\beta]$	$\Delta^{c}_{MP} = \frac{qR}{Et_c \tan\beta}[(1 + \frac{\nu}{2}) \times \cos 2\beta + \frac{1-\nu}{2}]$ $\Delta^{c}_{HP} = \frac{qR^2}{2Et_c}[\cos 2\beta + (1-\nu)]\sin\beta$	Truncated Circular Cone	$\Delta^{cc}_{Mt} = 0$ $\Delta^{cc}_{Ht} = \alpha r_0 t$

TABLE 3.7 — Values of Displacements δ and Δ in Basic System for the Analysis of Shells by the Force Method.

Unit Displacements δ_{in} (M = 1; H = 1; N = 1)

Basic System — Sphere

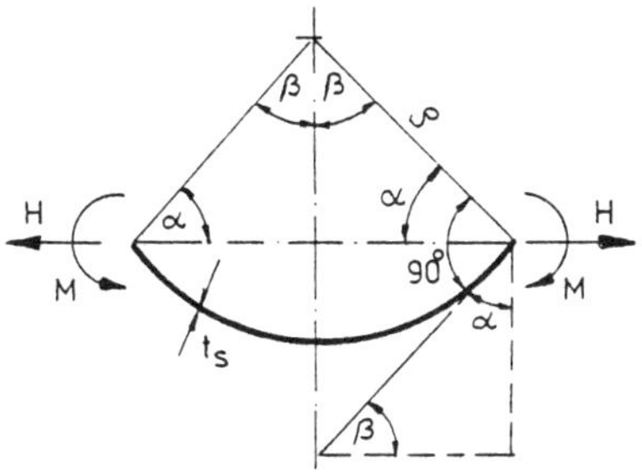

$$\delta^{s}_{MM} = \frac{S_s}{EJ^{s}_{0}W_1} = \frac{4\delta^2}{Et_s S_s^3 W_1} \; ; \; \delta^{s}_{MH} = \delta^{s}_{HM} = \frac{S_s^2 \sin\beta}{2EJ^{s}_{0}W_1} = \frac{2\rho^2 \sin\beta}{Et_s S_s^2 W_1}$$

$$\delta^{s}_{HH} = \frac{S_s^3}{2EJ^{s}_{0}}\left(\frac{\sin\beta}{2} \cdot \frac{W_2}{W_1} - \nu\frac{\cos\beta}{2}\right)\sin\beta = \frac{2\rho^2}{Et_s S_s}\left(\frac{\sin\beta}{2} \cdot \frac{W_2}{W_1} - \nu\frac{\cos\beta}{2}\right)\sin\beta$$

$$S_s = 0.78\sqrt{\rho t_s} \; ; \; S_s = \frac{\sqrt{\rho t_s}}{\sqrt[4]{3(1-\nu^2)}} \; ; \; J^{s}_{0} = \frac{t_s^3}{12(1-\nu^2)}$$

(at $\nu = 0.3$)

$$W_1 = 1 - \frac{S_s}{2\rho}(1-2\nu)\cot\beta \; ; \; W_2 = 2 - \frac{S_s}{\rho}(1-2\nu)\cot\beta$$

TABLE 3.8 — Values of Displacements δ and Δ in a Basic System for the Analysis of Shells by the Force System.

	Displacements Due to the Loading Δ_{ip}				
	Uniform Pressure P in Closed Shells	**Uniform Pressure P in Open Shells**	**Uniform Pressure P in Closed Shells**	**Uniform Pressure P in Open Shells**	**Triangular Pressure P in an Open Shell**
Sphere	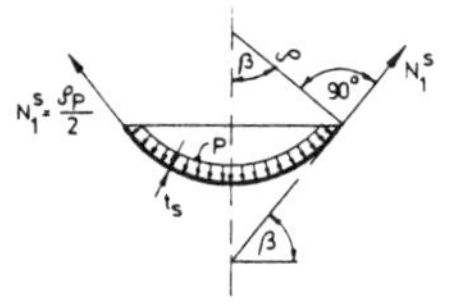	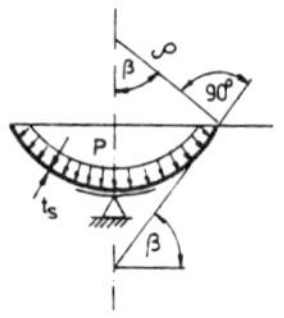	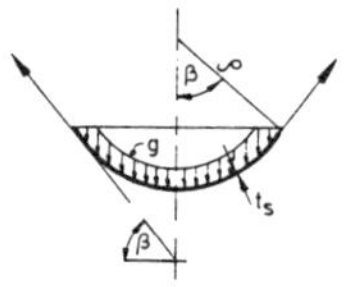	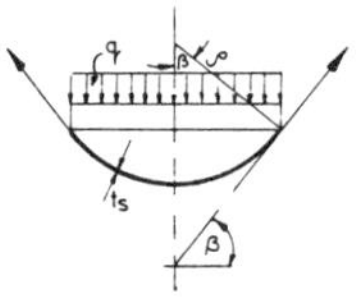	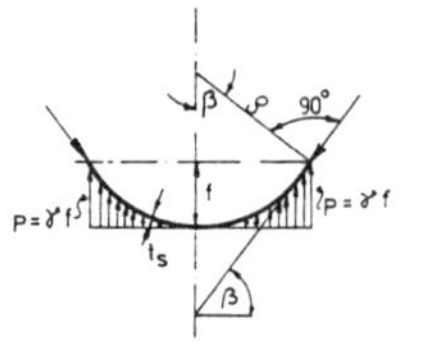 $\Delta^{s}_{MP} = -\frac{\gamma\rho^{2}\sin\beta}{Et_{s}}$ $\Delta^{s}_{HP} = \frac{\delta\rho^{3}\sin\beta}{Et_{s}}[(1-\cos\beta)\ x$ $x\ (\frac{1+\nu}{3}\cot^{2}\beta+1) - \frac{1+\nu}{6}$
	$\Delta^{s}_{MP} = 0$ $\Delta^{s}_{HP} = \frac{P\rho^{2}\sin\beta(1-\nu)}{2Et_{s}}$	$\Delta^{s}_{MP} = 0$ $\Delta^{s}_{HP} = \frac{P\rho^{2}\sin\beta}{2Et_{s}}$	$\Delta^{s}_{MP} = \frac{(2+\nu)\rho g\ \sin\beta}{Et_{s}}$ $\Delta^{s}_{HP} = +\frac{g\rho^{2}\sin\beta}{Et_{s}}\ x$ $x(\cos\beta - \frac{1-\nu}{1+\cos\beta})$	$\Delta^{s}_{MP} = \frac{q\rho\sin\beta\cos\beta(1+\nu)}{Et_{s}}$ $\Delta^{s}_{HP} = \frac{q\rho^{2}\delta\sin\beta(\cos 2-\nu)}{2Et_{s}}$	**Uniform Heating in t Degrees** $\Delta^{s}_{Mt} = 0$ $\Delta^{s}_{Ht} = \alpha t\rho\sin\beta$

TABLE 3.9 — Values of Displacements δ and Δ in Basic System for the Analysis of Shells by the Force Method.

Basic System — Plane Ring Plate

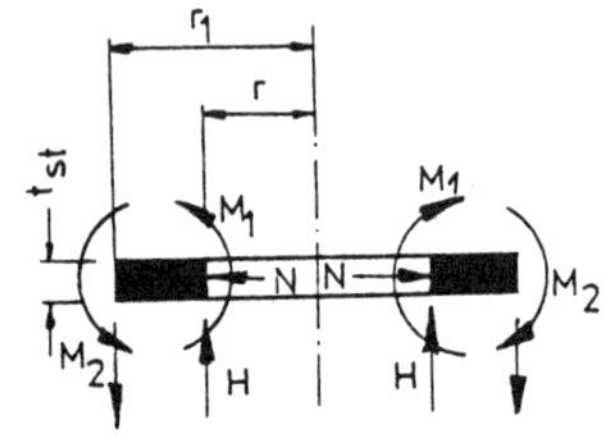

$H = \frac{r_1}{r_2}$ $\qquad$ $H = \frac{r_1}{r_2}$

Designations:

Index "st" — Stiffness

$$J^{st}_{0} = \frac{t^{3}_{st}}{12(1-\nu^{2})}$$

Unit Displacements — δ_{in} (M = 1; H = 1; N = 1)

$$\delta^{st}_{M_1M_1} = \frac{r_1^{3}r_2^{2}}{EJ^{st}(r_2^{2}-r_1^{2})}\left[\frac{1}{r_1^{2}(1-\nu)} + \frac{1}{r_2^{2}(1+\nu)}\right]$$

$$\delta^{st}_{M_1M_2} = \delta^{st}_{M_2M_1} = \frac{r_1r_2^{2}}{EJ^{st}(r_2^{2}-r_1^{2})}\left[\frac{1}{1+\nu} + \frac{1}{1+\nu}\right]$$

$$\delta^{st}_{M_1H} = \delta^{st}_{HM_1} = \frac{r_1^{2}}{EJ^{st}_{0}}\left[\frac{1}{2(1+\nu)} - \frac{r_2^{2}}{(1-\nu)(r_2^{2}-r_1^{2})}\ \ell n\ \frac{r_1}{r_2}\right]$$

Unit Displacements

δ_{in} (M = 1; H = 1; N = 1)

$$\delta^{st}_{M_1N} = \delta^{st}_{NM_1} = \delta^{st}_{NM} = \delta^{st}_{NM_2} = \delta^{st}_{HN} = \delta^{st}_{NH} = 0$$

$$\delta^{st}_{M_2M_2} = \frac{r_2}{EJ^{st}_{0}(r_2^{2}-r_1^{2})}\left[\frac{r_2^{2}}{(1\ \ \nu)} + \frac{r_1^{2}}{(1-\nu)}\right]$$

$$\delta^{st}_{M_2H} = \delta^{st}_{HM_2} = \frac{r_2^{2}}{EJ^{st}_{0}}\left[\frac{1}{2(1+\nu)} - \frac{r_1^{2}}{(1-\nu)(r_2^{2}-r_1^{2})}\ \ell n\frac{r_1}{r_2}\right]$$

$$\delta^{st}_{HH} = \frac{r_1}{4EJ^{st}_{0}}\left[\frac{(3+\nu)(r_2^{2}-r_1^{2})}{2(1+\nu)} + r_1^{2}\ell n\frac{r_1}{r_2}\right];$$

$$\delta^{st}_{NN} = \frac{(r_1+r_2)^{2}}{4EA}$$

where $\quad A = (r_2-r_1)t_{st}$

TABLE 3.10 — Values of Displacements δ and Δ in a Basic System for the Analysis of Shells by the Force Method.

	Displacements Due to the Loading Δ_{ip}	
	Uniform Pressure P in an Open Shell	**Uniform Heating in t Degrees**
Plane Ring Plate	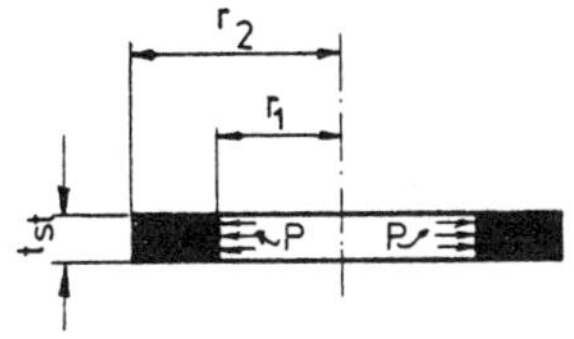 $\Delta^{st}_{M_1P} = \Delta^{st}_{M_2P} = \Delta^{st}_{HP} = 0$ $\Delta^{st}_{NP} = \frac{Pt_{st}(r_1+r_2)^2}{4EA}$ where $A = (r_2-r_1)t_{st}$	$\Delta^{st}_{Mt} = \Delta^{st}_{Ht} = 0$ $\Delta^{st}_{Nt} = \alpha t r_1$

TABLE 3.11 — Values of Displacements δ and Δ in a Basic System for the Analysis of Shells by the Force Method.

Plane Circular Plate	Unit Displacements δ_{in} (M = 1; H = 1; N = 1)
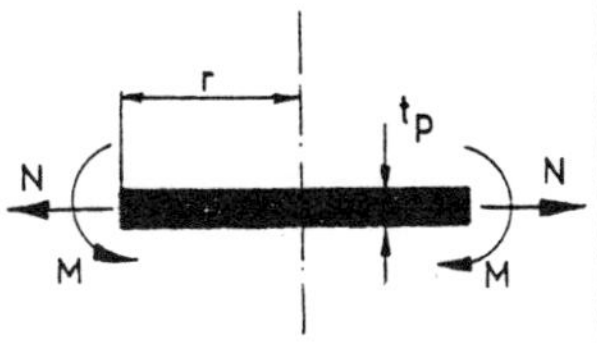 	$\delta^{p}_{MM} = \frac{12r(1-\nu)}{Et_p^3}$ $\delta^{p}_{MN} = \delta^{p}_{NM} = 0$ $\delta^{p}_{NN} = \frac{r(1-\nu)}{Et_p}$

TABLE 3.12 — Values of Displacements δ and Δ in a Basic System for the Analysis of Shells by the Force Method.

	Displacements Due to the Loading Δ_{ip}	
	Uniform Pressure p in a Closed Shell	**Uniform Heating for t Degrees**
Plane Circular Plate	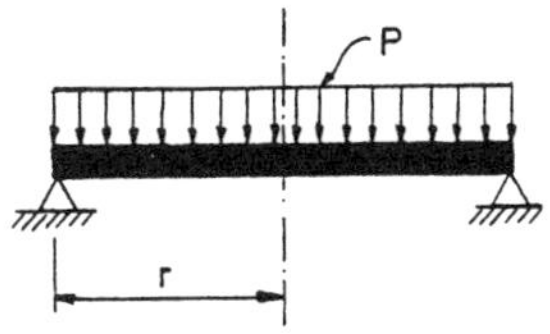 $\Delta^{p}_{MP} = -\frac{3p(1-\nu)r^3}{2Et_p^3}$ $\Delta^{p}_{NP} = 0$	$\Delta^{p}_{Mt} = 0$ $\Delta^{p}_{Nt} = \alpha t r$

TABLE 3.13 — Values K_1, K_2, K_3, for Formulas in Table 3.1.

λ	K_1	Difference	K_2	Difference	K_3	Difference
0,50	24,18661	5,95160	12,02662	2,07805	4,00520	0,36315
0,55	18,23501	4,12393	9,94857	1,57786	3,63937	0,30190
0.60	14,11108	2,94570	8,37071	1,22584	3,33747	0,25530
0,65	11,16538	2,15940	7,14487	0,97125	3,08217	0,21855
0,70	9,00598	1,61671	6,17362	0,78152	2,86362	0,18894
0,75	7,38927	1,23318	5,39210	0,63750	2,67468	0,16492
0,80	6,15609	0,95610	4,75460	0,52537	2,50976	0,14519
0,85	5,19999	0,75133	4,22923	0,44100	2,36457	0,12853
0,90	4,44866	0,59787	3,78823	0,37012	2,23604	0,11454
0,95	3,85079	0,48081	3,41811	0,31396	2,12150	0,10259
1,00	3,36998	0,39030	3,10415	0,26836	2,01891	0,09227
1,05	2,97968	0,31949	2,83579	0,23077	1,92664	0,08336
1,10	2,66019	0,26339	2,60502	0,19944	1,84328	0,07552
1,15	2,39680	0,21856	2,40558	0,17320	1,76776	0,06864
1,20	2,17824	0,18233	2,23238	0,15099	1,69912	0,06253
1,25	1,99591	0,15286	2,08139	0,13209	1,63659	0,05708
1,30	1,84305	0,12868	1,94930	0,11590	1,57951	0,05220
1,35	1,71437	0,10871	1,83340	0,10194	1,52731	0,04781
1,40	1,60566	0,09209	1,73146	0,08983	1,47950	0,04381
1,45	1,51357	0,08821	1,64163	0,07930	1,43569	0,04021
1,50	1,34536	0,06654	1,56233	0,07008	1,39548	0,03690
1,55	1,36882	0,05669	1,49225	0,06197	1,35858	0,03389
1,60	1,312213	0,04834	1,43028	0,05484	1,32469	0,03110
1,65	1,26379	0,04123	1,37544	0,04852	1,29359	0,02855
1,70	1,22256	0,03516	1,32692	0,04291	1,26504	0,02619
1,75	1,18740	0,02991	1,28401	0,03794	1,23885	0,02401
1,80	1,15743	0,02553	1,24607	0,03355	1,21484	0,02199
1,85	1,13190	0,02170	1,21257	0,02955	1,19285	0,02012
1,90	1,11020	0,01891	1,18302	0,02654	1,17273	0,01890
1,95	1,09129	0,01510	1,15648	0,02234	1,15383	0,01624
2,00	1,07619	0,01316	1,13414	0,02004	1,13759	0,01527
2,05	1,06303	0,01107	1,11410	0,01752	1,12232	0,01387
2,10	1,05196	0,00927	1,09658	0,01527	1,10845	0,01257
2,15	1,04269	0,00773	1,08131	0,01327	1,09587	0,01138
2,20	1,03496	0,00641	1,06805	0,01148	1,08449	0,01026
2,25	1,02855	0,00528	1,05658	0,00988	1,07423	0,00924
2,30	1,02327	0,00433	1,04669	0,00846	1.06499	0,00829
2,35	1,01894	0,00351	1,03823	0,00722	1,05670	0,00741
2,40	1,01543	0,00283	1,03101	0,00611	1,04929	0,00661
2,45	1,01260	0,00225	1,02490	0,00514	1,04268	0,00587
2,50	1,01035	0,00177	1,01976	0,00429	1,03681	0,00519
2,55	1,00858	0,00137	1,01547	0,00354	1,03162	0,00459
2,60	1,00721	0,00105	1,01193	0,00290	1,02703	0,00401
2,65	1,00616	0,00079	1,00903	0,00235	1,02302	0,00351
2,70	1,00537	0,00057	1,00668	0,00187	1,01951	0,00305
2,75	1,00480		1,00481		1,01648	

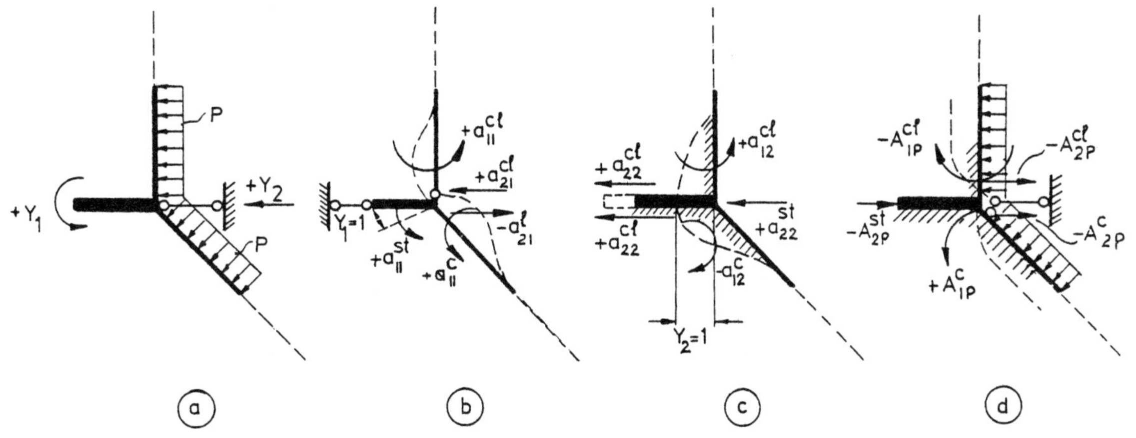

FIGURE 3.6 — Design scheme for the method of deformation for the intersection joint of cylinder, cone and ring: (a) Basic system; (b) Reaction due to the unit twist; (c) Reaction due to the unit displacement; (d) Reaction in the basic system under loading.

apply to adjoined shells and stiffening rings to give them the displacement or twist equal to the unity. The first index designates the direction of the reaction and the second — the cause. Coefficients A_{ip} designate a summary of the external reactive forces of the fixed ends of adjoining elements under external loading. The coefficients a_{in} and A_{ip} are called force coefficients. Their signs are considered as positive when the directions of the reactive forces are coinciding with the given directions of the displacements. The obtaining of positive signs of the displacements resulting after the solution of Equations (3.8) indicate that the directions were correctly accepted. The resulting negative sign of the displacement indicates a reverse direction by comparison to that assumed.

To illustrate, let us consider the analysis of the joint for the cone, cylinder and stiffening ring, Figure 3.4, under uniform internal pressure p

$$a_{11} = a_{11}^{cl} + a_{11}^{c} + a_{11}^{st} \; ; \; a_{12} = a_{12}^{cl} - a_{12}^{c}$$

$$a_{21} = a_{21}^{cl} + a_{21}^{c} \; ; \; a_{22} = a_{22}^{cl} + a_{22}^{c} + a_{22}^{st} \qquad (3.9)$$

$$A_{1p} = - A_{1p}^{cl} + A_{1p}^{c} \; ; \; A_{2p} = - A_{2p}^{cl} - A_{2p}^{c} - A_{2p}^{st}$$

The unit force coefficients are determined from the relation

$$a_{ik} = \frac{1}{\delta_{ik}} \qquad (3.10)$$

and, therefore, may be determined from Table 3.1 for the determination of the unit displacements δ_{ik} by the force method. The force coefficients under loading A_{ip} which are the support reactions under the loading at the shell and stiffening ring, belonging to the system under consideration are determined by the force method.

After the solution of Equations (3.8) and the finding of the unknowns y_1 and y_2 we determine the unknown forces M_1 and H as the summary of the forces in the basic system, for instance, as in the case shown in the example in Figure 3.4

$$\begin{aligned}
M_1^{cl} &= a_{11}^{cl} y_1 + a_{12}^{cl} y_2 + A_{1p}^{cl} \\
M_1^{c} &= a_{11}^{c} y_1 + a_{12}^{c} y_2 + A_{1p}^{c} \\
M_1^{st} &= a_{11}^{st} y_1 + a_{12}^{st} y_2 + A_{1p}^{st} \\
H^{cl} &= a_{21}^{cl} y_1 + a_{22}^{cl} y_2 + A_{2p}^{cl} \\
H^{c} &= a_{21}^{c} y_1 + a_{22}^{c} y_2 + A_{2p}^{c} \\
H^{st} &= a_{21}^{st} y_1 + a_{22}^{st} y_2 + A_{2p}^{st}
\end{aligned} \qquad (3.11)$$

In this case

$$A_{1p}^{st} = 0$$

3.3 Influence of Edge Effect on the Carrying Capacity of the Shell

When the shells are designed under the assumption of an unlimited elasticity of the material, very often we may obtain significant local stresses much greater, not only of the membrane stresses, but greater than the yield stresses.

In the case when the residual or substantial deformations cannot be tolerated, the thicknesses of the shells in the area of the edge effect of a shell shall be increased or

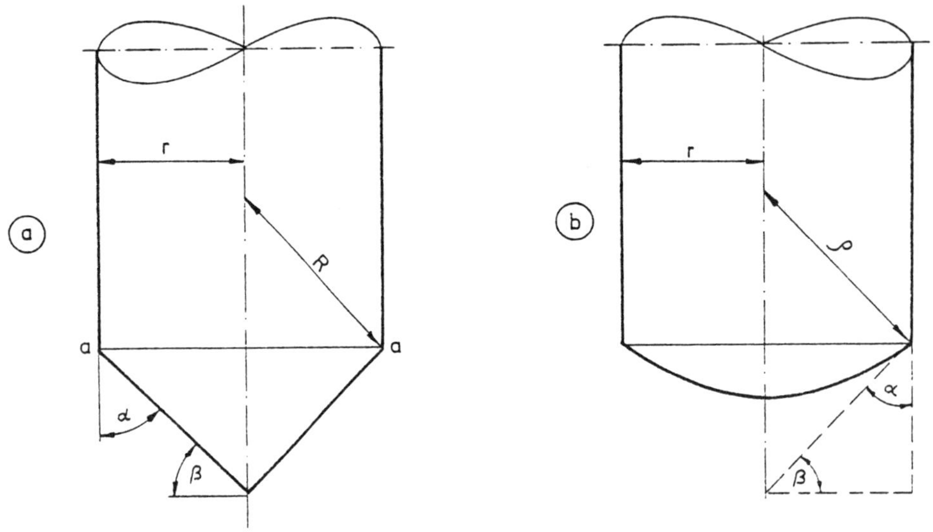

FIGURE 3.7 — Schemes of the intersections of the shells: (a) Intersection of cylinder and cone; (b) Intersection of cylinder and sphere.

the configurations of the shells should be chosen to achieve a smooth transfer from one form of a shell to the other.

Smoothness in the change of the shape of shells and the proper choice of the stiffening rings are the most effective structural measures for the reduction of the influence of an edge effect.

The stability of shells under compression in zona of an edge effect should be considered separately during their design. Stability conditions may require the use of stiffening rings in areas of an edge effect.

3.4 Local Stresses and Deformations Under Internal or External Pressure Along the Intersection of Closed Cylindrical Shells

We analyze this problem in the elastic stage for the cases shown in Figure 3.7.

For the considered cases, the simplest solution is the force method. All solutions are given for the internal pressure. For the transfer to external pressure, it is necessary to reverse the signs of the solutions.

3.4.1 The Following Designations are Introduced

cl = cylinder

c = cone

s = sphere

r, R and δ = are radii of cylinder, cone and sphere, respectively

p = internal pressure in lbs/in² or atm

t = thickness of the wall of the shell

σ_1 = membrane stress acting in a meridional direction

σ_2 = membrane stress acting in the circumferential direction

$\overline{\sigma}_1$ and $\overline{\sigma}_2$ = additional uniform edge stresses due to the forces $\overline{N}_1$ and $\overline{N}_2$

σ_1 and σ_2 = additional bending edge stresses due to moments M_1 and M_2

$\sigma_{(1)}$ = $\sigma_1 + \sigma_1 + \sigma_1$ = the total local membrane and due to edge effect stress in the meridional direction

$\sigma_{(2)}$ = $\sigma_2 + \overline{\sigma}_2 + \sigma_2$ = total local membrane and due to edge effect stress in the circumferential direction

N_1 = membrane force in the meridional direction per unit length of the circumferential section of the shell

N_2 = membrane force in the circumferential direction per unit length of the meridional section of the shell

$\overline{N}_1$ and N_2 = additional meridional and circumferential forces due to the origin of the circumferential sections of the shell of the edge forces M_1 and H

M_1 = bending moment in the meridional direction per unit length of the circumferential section of the shell

M_2 = bending moment in the circumferential direction per unit length of meridional section of the shell

Unknown edge forces per unit length of the circumferential section of the shell at the location of origin of the edge effect, which enter into the basic equations during the solution of the problem of the edge effect by the force method in which

X_1 = bending moment in the meridional direction

X_2 = force in the direction of radius in the plane normal to the axis of the shell

Q = transverse force of a shell at the location or origin of the edge effect per unit length of the circumferential section normal to the generatrix

H_0 = the projection of the membrane force H on the plane of the circumferential section of the shell

$H=H_2+H_0$ = total force in the plane normal to the axis of the shell per unit length of the circumferential section of the shell

δ_{11},δ_{12} = δ_{21},δ_{22}, etc = are displacements due to the unit unknowns in the basic system which are entering in the basic equation of the force method

$\Delta_1 p, \Delta_2 p$, etc = displacements under loadings in the basic system entering into the basic equations of the force method

Auxiliary values indicating the lengths of diminishing waves of the local stresses of the cylindrical closed shells.

a. *Cylindrical shell*

$$S_{cl} = \frac{\sqrt{rt_{cl}}}{\sqrt[4]{3(1-\nu)^2}}$$

For steel Poisson's ratio ν = 0.3, and from the expression (3.12), we obtain

$$S_{cl} = 0.78\sqrt{rt_{cl}} \tag{3.13}$$

By designating $\xi = \frac{r}{t_{cl}}$, we have

$$S_{cl} = 0.78\, t_{cl}\sqrt{\xi} \tag{3.14}$$

b. *Cone*

$$S_c = \frac{\sqrt{Rt_c}}{\sqrt[4]{3(1-\nu^2)}} \tag{3.15}$$

For ν = 0.3, we obtain

$$S_c = 0.78\sqrt{Rt_c} \tag{3.16}$$

c. *Sphere*

$$S_s = \frac{\sqrt{\rho t_s}}{\sqrt[4]{3(1-\nu^2)}} \tag{3.17}$$

For ν = 0.3, we obtain

$$S_s = 0.78\sqrt{\rho t_s} \tag{3.18}$$

$$J_0 = \frac{t^3}{12(1-\nu^2)}$$ = moment of inertia per unit length of the shell (3.19)

At ν = 0.3

$$J_0 = \frac{t^3}{10.92}$$

$$D = EJ_0 = \frac{Et^3}{12(1-\nu^2)}$$ = cylindrical stiffness of the shell (3.20)

At ν = 0.3

$$D = \frac{Et^3}{10.92}$$

3.4.2 Rule of Signs

a. *Stresses*

(+) tensile stresses and (–) the compressive stresses.

b. *Forces*

The transverse forces and loadings in the direction of the radius from the axis of the shell and the production of its tension are considered to be positive (+). The reverse direction of the transverse forces and the loadings is considered as negative (–). The bending moment in the circumferential sections of the shell directed outside and the producing tension of the internal fibres and compression of the external fibres are considered as positive (+). The reverse direction of the bending moments is considered as negative (–).

c. *Displacements*

The displacements in the basic system δ_{in} and Δ_{ip} applying force method are considered as positive (+) if during the displacement we achieve positive work, or when the directions of the displacements for both indexes coincide. The displacement is considered to be negative (–), when the directions of the displacements for the two indexes do not coincide.

If, as the result of a solution of basic equations of the force method, the sign of the unknown force X_1 is positive (+), this indicates that the actual direction of X_1 coincides with that assumed in the design scheme.

After solving the basic equations of the force method and establishing the actual directions and values of the forces X_1, they should have signs according to the rule given above for the forces corresponding to their actual directions.

The forces X_1, together with these finally established signs, should be used in all the following chapters.

3.5 Determination of the Stresses Under Internal Pressure in the Thin-Walled Shells According to the Membrane Theory

The membrane theory is applied in cases where the bending stresses in the shell are so small that they may be neglected in comparison to the uniform tensile or compressive stresses, acting in a tangential direction to the middle surface.

The stresses in the closed shell, having one axis of symmetry and circumferential cross-sections in the planes, normal to the axis, Figure 3.8 (a), will be determined from the equations of the equilibrium.

Let us cut from the shell element at arbitrary level a-a by two sections normal to the axis a-a and by two sections along the generatrix. To find the conditions of equilibrium of this element under the influence of the forces

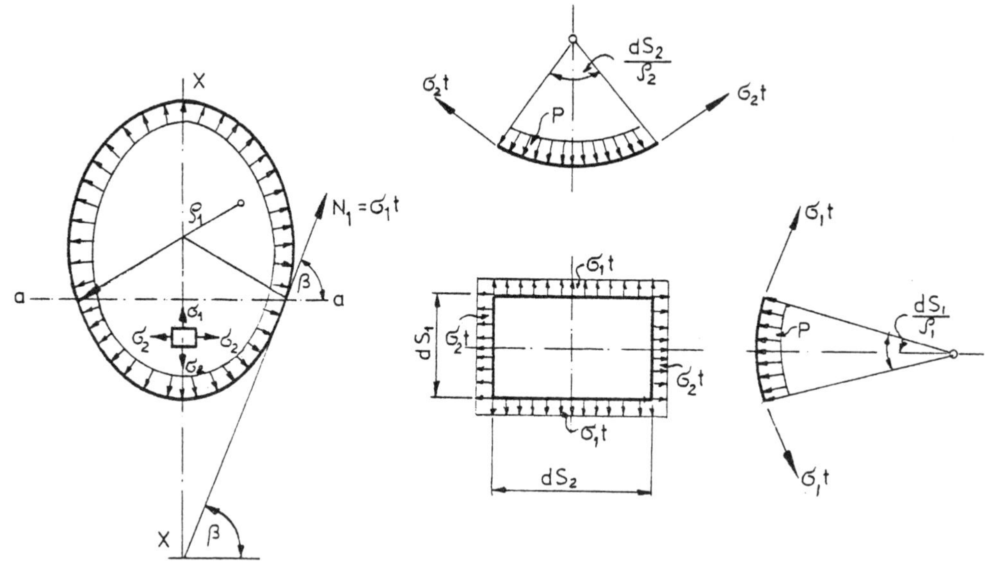

FIGURE 3.8 — Design scheme for the determination of the membrane forces and stresses at axis-symmetrical shell of arbitrary shape: (a) Geometrical characteristics; (b) Element of the shell and forces.

applied along the edges and given loading, it is necessary to find components of all the forces along the axis, normal to the surface of the shell, Figure 3.8 (b).

The equation of equilibrium is

$$p ds_1\, ds_2 = \sigma_1 t ds_2 \frac{ds_1}{\rho_1} + \sigma_2 t ds_1 \frac{ds_2}{\rho_2} \qquad (3.21)$$

From this follows

$$\frac{p}{t} = \frac{\sigma_1}{\rho_1} + \frac{\sigma_2}{\rho_2} \qquad (3.22)$$

The Equation (3.22) has two unknowns σ_1 and σ_2. To obtain the second equations of equilibrium we make a section at level a-a and consider the bottom part as a free body. Considering the components of all the forces along the axis x-x, we find the following equation of equilibrium

$$p\pi\rho_2^2\sin^2\beta = 2\pi\rho_2\sin^2\beta t\sigma_1 \qquad (3.23)$$

From this we obtain

$$\sigma_1 = \frac{p\rho_2}{2t} \qquad (3.24)$$

This represents the stress acting in the meridional direction.

After substituting expression (3.24) into Equation (3.23), we obtain

$$\sigma_2 = \frac{p\rho_2}{2t}\left(2 - \frac{\rho_2}{\rho_1}\right) \qquad (3.25)$$

This is the stress acting in the circumferential direction. The values of the membrane stresses for the shells of different shapes, we obtain as particular cases from the equations (3.24) and (3.25).

a. *Spherical shell*

In this case

$$\rho_1 = \rho_2 = \rho$$

and by designating the thickness of the sphere as t_S, we obtain from the equations (3.24) and (3.25)

$$\sigma_1^S = \sigma_2^S = \sigma^S = \frac{p\rho}{2t_S} \qquad (3.26)$$

b. *Cylindrical shell*, (Figure 3.9)

$$\rho_1 = \infty; \quad \rho = r$$

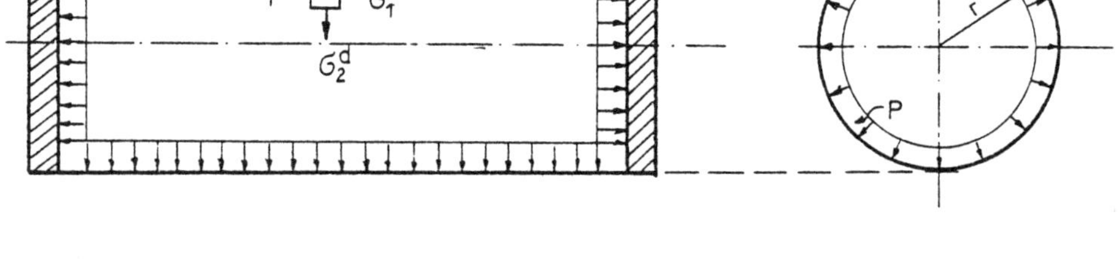

FIGURE 3.9 — Scheme of the closed cylindrical shell under internal pressure.

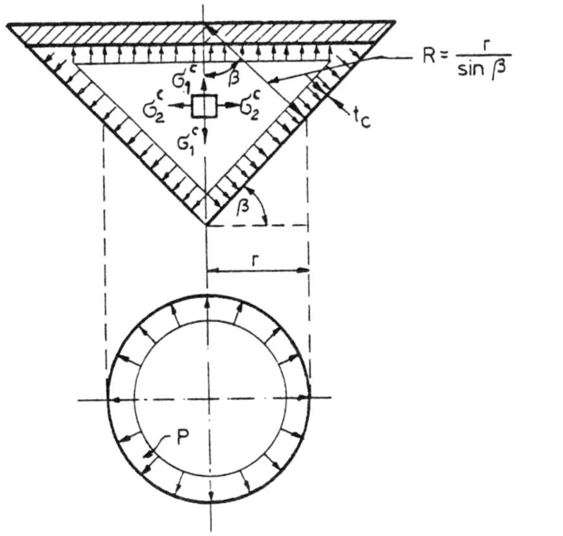

FIGURE 3.10 — Scheme of the closed conical shell under internal pressure.

From Equation (3.24) we obtain

$$\sigma_1^{c\ell} = \frac{pr}{2t_{c\ell}} \tag{3.27}$$

and from Equation (3.25) follows

$$\sigma_2^{c\ell} = \frac{pr}{t_{c\ell}}$$

c. *Conical shell*, (Figure 3.10)

$$\rho_1 = \infty; \quad \rho_2 = R$$

From the Equation (3.24), we obtain

$$\sigma_1^c = \frac{PR}{2t_c} \tag{3.28}$$

and from the Equation (3.25) we have

$$\sigma_2^c = \frac{PR}{t_c} \tag{3.29}$$

3.6 Edge Stresses and Deformations Along the Intersection of Closed Cylindrical and Conical Shells Under Internal Pressure

3.6.1 Design Loading

The given system is shown in Figure 3.11 (a) on the following page. $t_{cy\ell}$ and t_{cone} are thicknesses of the shells in the zona of the edge effect action. After cutting the shell on line a-a, we analyze a primary system by using the superposition equations.

The interaction of the parts which are cut, we substitute by the redundants X_1 and X_2 and also by the forces $N_1^{cy\ell}$, N_1^{cone} and H_0 which are found by the membrane theory.

In addition, on each primary part of the cylinder and cone are the acting given load — internal pressure P.

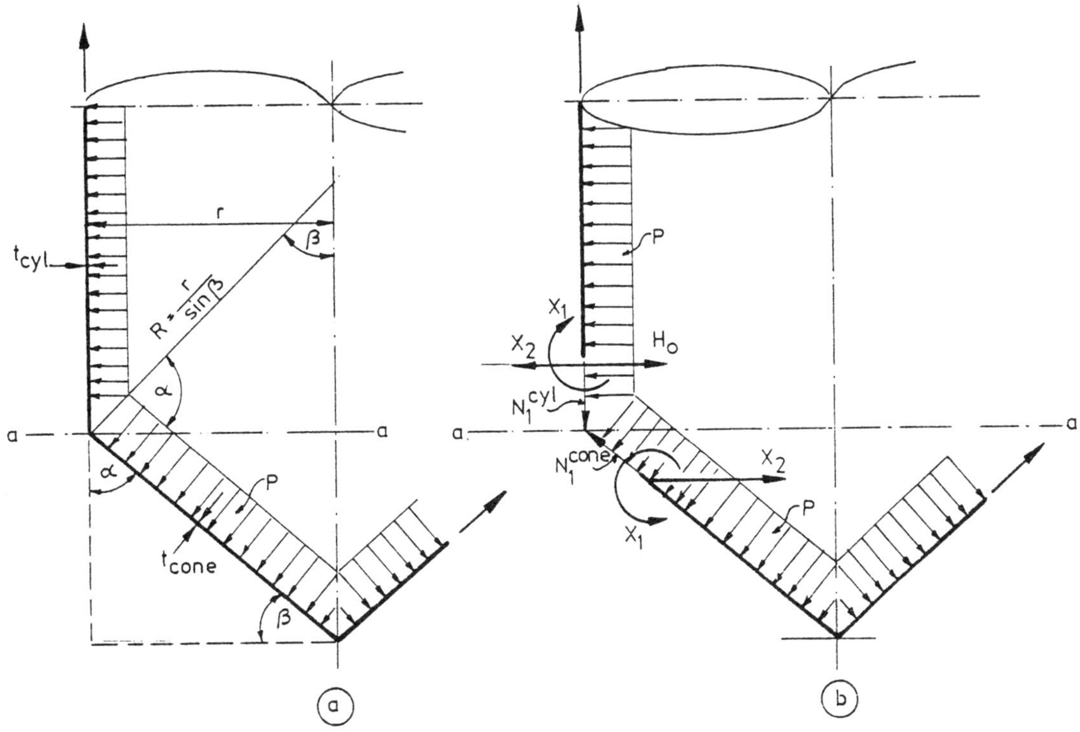

FIGURE 3.11 — Design scheme for intersection of cylinder and cone: (a) Given system; (b) Primary system.

3.6.2 Forces According to the Membrane Theory

The meridional force in the cylinder is

$$N_1^{cyl} = \frac{p\pi r^2}{2\pi r} = \frac{pr}{2} \quad (3.30)$$

The meridional force in the cone

$$N_1^{cone} = \frac{N_1^{cyl}}{\sin\beta} = \frac{pR}{2} \quad (3.31)$$

The horizontal projection of this force is

$$H_0 = N_1^{cone}\cos\beta = \frac{N_1^{cyl}\cos\beta}{\sin\beta} = N_1^{cyl}\,\mathrm{ctg}\beta \quad (3.32)$$

3.6.3 Determination of Redundants X_1 and X_2, and Values of M_1 and H

The superposition equations are

$$X_1(\delta_{11}^{cyl} + \delta_{11}^{cone}) + X_2(\delta_{12}^{cyl} - \delta_{12}^{cone}) + \Delta_{1P}^{cyl} + \Delta_{1P}^{cone} - \Delta_{1H_0} = 0$$

$$X_1(\delta_{21}^{cyl} - \delta_{21}^{cone}) + X_2(\delta_{22}^{cyl} + \delta_{22}^{cone}) + \Delta_{2P}^{cyl} - \Delta_{2P}^{cone} - \Delta_{2H} = 0 \quad (3.33)$$

The values of the unit displacements and the displacements under the loadings for their positive values are shown in Tables 1 to 12. We should substitute the displacements δ and Δ in Equations (3.33) with the signs indicated in these equations.

After solving Equations (3.33) and obtaining the values and directions of the unknown, it is necessary to accept the final signs, using them according to the signs rule for the forces.

In particular, for the basic system shown in Figure 3.11 (b) and after a solution of the basic equation when we obtain the positive values X_1 and X_2 (this indicates that their actual directions coincide with those which were assumed in the basic system) we obtain the final signs of the unknown forces: for the cylinder + X_1^{cl} and + X_2^{cl}; for the cone + X_1^{cone} and − X_2^{cone}.

Forces for the unit length of the circumferential cross-section at the location of the origin of the edge effect are as follows.

In the general case

$$M_1^{cl} = M_1^{c} = M_1 = X_1$$

$$H^{cl} = X_2^{cl} - H_0$$

$$H^{c} = -X_2^{c} \quad (3.34)$$

$$Q^{cl} = X_2^{cl} - H_0 \;;\quad Q^{c} = -X_2^{c}\sin\beta$$

In the particular case considering signs according to the above example and rule of the signs

$$M_1^{cl} = M_1^{c} = M_1 = X_1$$

$$H^{cl} = X_2^{cl} - H_0$$

$$H^{c} = -X_2^{c} \quad (3.35)$$

$$Q^{cl} = X_2^{cl} - H_0$$

$$Q^{c} = -X_2^{c}\sin\beta$$

3.6.4 Total Local Stresses of Cylindrical Shell (in the given system) at Level a-a, Figure 3.9 (a)

The meridional edge stresses of the cylindrical shell are

$$\sigma_{(1)}^{cl} = \frac{N_1^{cl}}{t_{cl}} \pm \frac{6M_1}{t_{cl}^2} \quad (3.36)$$

where N_1^{cl} and M_1 are given by the formulas (3.30) and (3.34) respectively.

The forces necessary to substitute in the expressions of stresses (3.36) and others given below, considering the signs after formula (3.35). Before term, deforming bending stress, the upper sign relates to the internal part of the wall, and bottom sign — to the external wall

$$\sigma_1^{cl} = \frac{N_1^{cl}}{t_{cl}} \quad (3.37)$$

is the membrane meridional stress in the cylinder.

The circumferential edge stresses of the cylindrical shell

$$\sigma_{(2)}^{cl} = \frac{N^{cl}}{t_{cl}} + \frac{N^{-cl}}{t_{cl}} \pm \frac{6M^{cl}}{t_{cl}^2} \quad (3.38)$$

where

$$N_2^{cl} = pr \quad (3.39)$$

The additional local circumferential force due to action of M_1 and H^{cl}, is

$$N_2^{-cl} = \frac{2r}{S_{cl}}\left(\frac{M_1}{S_{cl}} + H^{cl}\right) \quad (3.40)$$

where M_1 and H^{cl} are used after formulas (3.34) and

$$S_{cl} = 0.78\sqrt{rt_{cl}} \quad (3.41)$$

value of

$$M_2^{cl} = \nu M_1 \quad (3.42)$$

The tangential edge stresses of the cylindrical shell are

$$\tau_{c\ell} = \frac{Q_{c\ell}}{t_{c\ell}} \qquad (3.43)$$

3.6.5 Summary of the Edge Stresses of the Conical Shell at Level a-a, Figure 3, 9 (a)

Meridional edge stresses of the conical shell

$$\sigma^{c}_{(1)} = \frac{N_1^c}{t_c} + \frac{\bar{N}_1^{c}}{t_c} \pm \frac{6M_1}{t_c^2} \qquad (3.44)$$

The values of the terms in the formula (3.44) are

$$N_1^c = \frac{pR}{2} \qquad (3.45)$$

According to formula

$$\bar{N}_1^c = H^c \cos\beta \qquad (3.46)$$

and

$$\bar{N}_1^c = -X_2^c \cos\beta \qquad (3.47)$$

3.6.6 The Circumferential Edge Stresses of the Conical Shell

$$\sigma^{c}_{(2)} = \frac{N_2^c}{t_c} + \frac{\bar{N}_2^{c}}{t_c} \pm \frac{6M_2^c}{t_c^2} \qquad (3.48)$$

where

$$N_2^c = pR = \frac{pr}{\sin\beta} \qquad (3.49)$$

The additional local circumferential force due to M_1 and H_1^c has the value

$$\bar{N}_2^c = \frac{tc}{r}\left(-\delta^c_{12} M_1 + \frac{\delta^c_{22}}{\nu_2} H^c\right) \qquad (3.50)$$

where

δ^c_{12} and δ^c_{22} equal the unit displacements from the force equations (3.7) and (3.33) which are determined from the Tables 3.1 to 3.12.

M_1 and H^c we find from Equations (3.34) and (3.35)

$$\nu_2 = 1 - \frac{S_c \mathrm{ctg}\beta}{4R} = 1 - \frac{S_c \cos\beta}{4r} \qquad (3.51)$$

The additional local circumferential moment M_2^c due to the forces M_1 and H^c, is expressed by the formula

$$M_2^c = \nu M_1 + \frac{(1+\nu^2)\cos\beta t_c H^c}{\nu_1\sqrt{12(1-\nu^2)}} = \nu M_1 + \frac{0.34\cos\beta t_c H^c}{\nu_1} \qquad (3.52)$$

where

$$\nu_1 = 1 + \frac{S_c \mathrm{ctg}\beta(4-\nu)}{4R} = 1 + \frac{S_c \cos\beta(4-\nu)}{4r} \qquad (3.53)$$

$$S_c = 0.78\sqrt{Rt_c}$$

3.6.7 Tangential Edge Stress of the Concial Shell

$$\tau_c = \frac{Q_c}{t_c} \qquad (3.54)$$

3.6.8 Displacements in the Plane a-a of the Points at the Intersection Line of the Middle Surfaces of the Shells, Directed Along the Radius r From the Axis of the Shell, Figure 3.11 (b)

$$\bar{\Delta}r = \delta^{c\ell}_{22} H^{c\ell} + \delta^{c\ell}_{12} M_1 + \Delta^{c\ell}_{2P} = \delta^c_{22}(H^c + N_1^c \cos\beta) - \delta^c_{12} M_1 = \Delta^c_{2P} \qquad (3.55)$$

The values δ and Δ should be substituted into formula (3.55), considering its signs in the force equations (3.33) and forces M, N and H — considering their final signs for the given case.

3.6.9 Twisting of the Section at Level a-a Considering Clockwise Direction as the Positive, Figure 3.11 (b)

$$\bar{\psi} = \delta^{c\ell}_{11} M_1 + \delta^{c\ell}_{12} H^{c\ell} + \Delta^{c\ell}_{1P} = -\delta^c_{11} M_1 + \delta^c_{12} H^c - \Delta^c_{1P} \qquad (3.56)$$

3.7 Edge Effect at the Intersection of Closed Cylindrical and Spherical Shells Under Internal Pressure, Figure 3.12

3.7.1 Design Scheme and Determination of the Forces

This problem may be solved by the force method used before.

The forces in the basic system in section a-a are determined by the membrane theory

$$N_1^{c\ell} = \frac{pr}{2} \qquad (3.57)$$

$$N_1^s = \frac{p\rho}{2} \qquad (3.58)$$

$$H_0 = N_1^s \cos\beta = N_1^{c\ell} \mathrm{ctg}\beta \qquad (3.59)$$

Force equations (3.7) for the basic system, Figure 3.10 (b) considering signs are

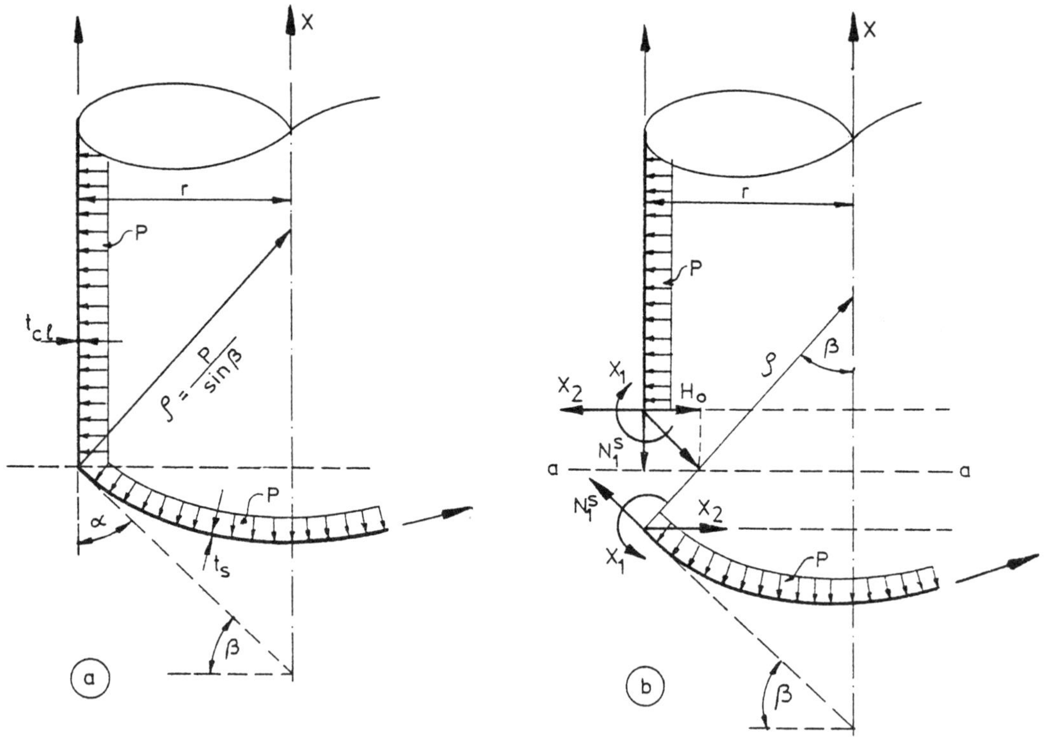

FIGURE 3.12 — Design of junction of cylinder and sphere: (a) Given system; (b) Basic system.

$$X_1(\delta^{c\ell}_{11}+\delta^{s}_{11}) + X_2(\delta^{c\ell}_{12}-\delta^{s}_{12}) + \Delta^{c\ell}_{1P} + \Delta^{s}_{1P} - \Delta_{1H_0} = 0$$

$$X_1(\delta^{c}_{21}-\delta^{s}_{21}) + X_2(\delta^{c}_{22}+\delta^{s}_{22}) + \Delta^{c\ell}_{2P} + \Delta_{2P} - \Delta_{2H_0} = 0$$

(3.60)

In the particular case after obtaining the positive signs X_1 and X_2 after the solution of the force equations (3.36) the final signs of the unknown forces will be

$$+X^{c\ell}_1; \quad +X^{s}_1; \quad +X^{c\ell}_2; \quad -X^{s}_2$$

The forces per unit length of the circumferential section a-a at the location of origin of the edge effect are

$$M^{c\ell}_1 = M^{s}_1 = M_1 = X_1$$

$$H_{c\ell} = X_2 + H_0$$

$$H_s = X_2 \qquad (3.61)$$

$$Q_{c\ell} = H_{c\ell}$$

$$Q_s = H_s \sin\beta$$

In the particular case when directions of the X_1 and X_2 coincide with those accepted in Figure 3.12 (b)

$$M^{c\ell}_1 = M^{s}_1 = M_1 = X_1$$

$$H_{c\ell} = X^{c\ell}_2 - H_0$$

$$H_s = -X^{s}_2 \qquad (3.62)$$

$$Q_{c\ell} = X^{c\ell}_2 - H_0$$

$$Q_s = X^{s}_2 \sin\beta$$

3.7.2 Summary of the Local Stresses of the Cylindrical Shell in a Given System at Level a-a, Figure 3.12 (a)

These stresses are determined as the summary of stresses due to the forces in the basic system, Figure 3.10 (b) at level a-a.

3.7.2.1 Meridional Edge Stresses of the Spherical Shell

$$\sigma^{s}_{(1)} = \frac{N^{s}_{1}}{t_s} + \frac{\bar{N}^{s}_{1}}{t_s} \pm \frac{6M_1}{t_s} \qquad (3.63)$$

where

N^s_1 is used after formula (3.58) and M_1 after formula (3.60).

The additional meridional force

$$\overline{N}^{s}_{1} = H_{s}\cos\beta \qquad (3.64)$$

and in the particular case under consideration at signs after formulas (3.62)

$$\overline{N}^{s}_{i} = -X^{s}_{2}\cos\beta \qquad (3.65)$$

3.7.2.2 The Circumferential Edge Stresses of the Spherical Shell

$$\sigma^{s}_{(2)} = \frac{N^{s}_{2}}{t_{s}} + \frac{\overline{N}^{s}_{2}}{t_{s}} \pm \frac{6M^{s}_{2}}{t^{2}_{s}} \qquad (3.66)$$

where

$$N^{s}_{2} = \frac{P\rho}{2} \qquad (3.67)$$

The additional local circumferential force due to the M_1 and H_S is

$$\overline{N}^{s}_{2} = \frac{2\rho\sin\beta}{S_{s}}\left(\frac{M_{1}}{S_{s}} + H_{s}\right) \qquad (3.68)$$

where

$$S_{s} = 0.78\sqrt{\rho t_{s}}$$

The additional local circumferential moment M^s_2 due to the M_1 and H_S is defined by the formula

$$M^{s}_{2} = \nu M_{1} - (1-\nu^{2})\frac{S_{s}\cos\beta}{\rho\sin^{2}\beta}\left(\frac{S_{s}H_{s}}{2} - M_{1}\right) \qquad (3.69)$$

The tangential edge stress of the spherical shell is

$$\tau_{s} = \frac{Q}{t_{s}} \qquad (3.70)$$

3.7.3 The Increase of the Radius r at Level a-a, Equal to ΔR, and turn of the Section at the Same Level Clockwise, $\overline{\psi}$, Figure 3.12

$$\overline{\Delta r} = \delta^{c\ell}_{22} + \delta^{c\ell}_{21}M_{1} + \Delta^{c\ell}_{2P} = \delta^{s}_{22}(H_{z} + N^{s}_{1}\cos\beta) - \delta_{21}M_{1} - \Delta^{s}_{2P} \qquad (3.71)$$

$$\overline{\psi} = \sigma^{c\ell}_{11} + \delta^{c\ell}_{12}H_{c\ell} + \Delta^{c\ell}_{1} = -\delta^{s}_{11}M_{1} + \delta^{s}_{12}H_{s} - \Delta^{s}_{1P} \qquad (3.72)$$

3.8 Edge Effect of Closed Cylindrical Shells Having Stiffening Ring at Their Junction

Let us consider shells under internal pressure, having stiffening rings as shown in Figure 3.13.

In the following we consider the solution for the case shown in Figure 3.13 (a), using the force method.

The system is four times statically indeterminate. The forces $N^{c\ell}_1$ and N^c_1 we determine applying the membrane theory, formulas (3.30) and (3.31).

The force equations (3.7) for this case are

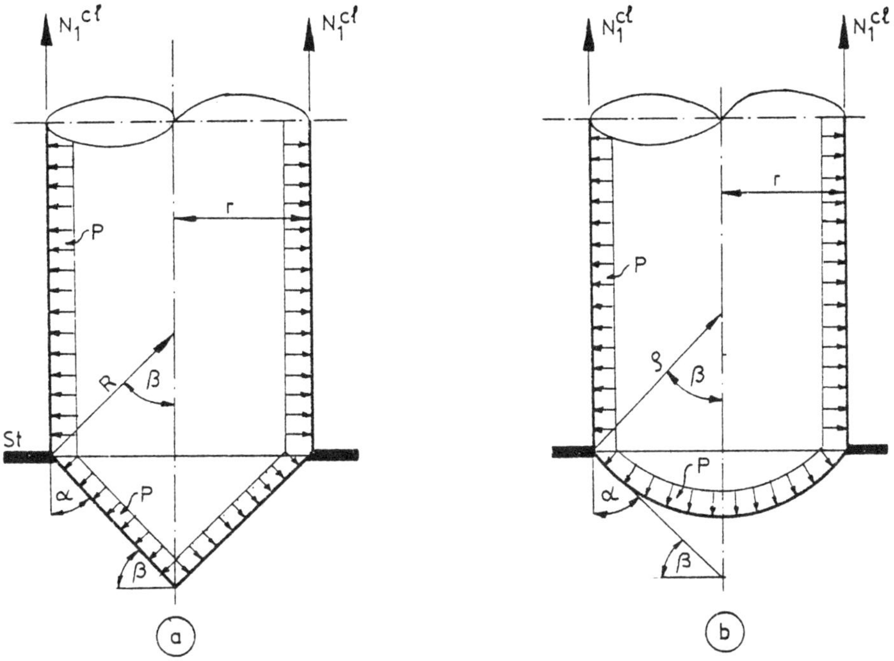

FIGURE 3.13 — Schemes of shells having stiffening rings: (a) Junction of cylinder, cone and ring; (b) Junction of cylinder, sphere and ring.

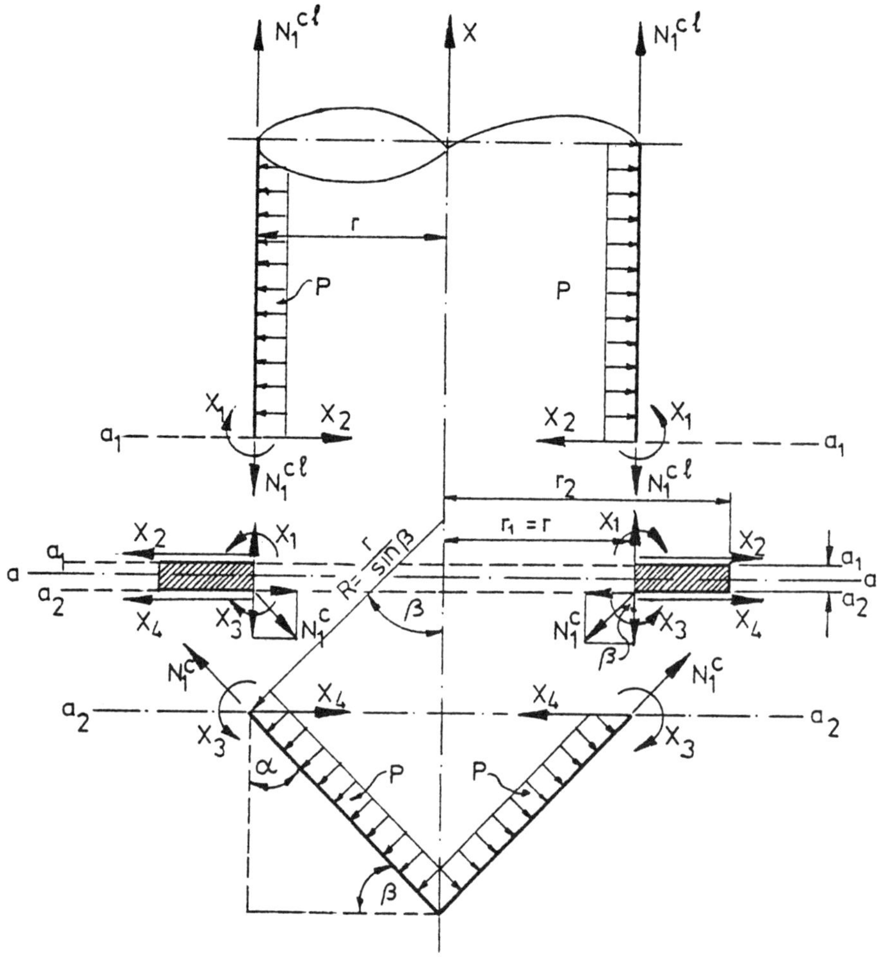

FIGURE 3.14 — Design scheme (basic scheme) for junction of the cylinder, cone and ring.

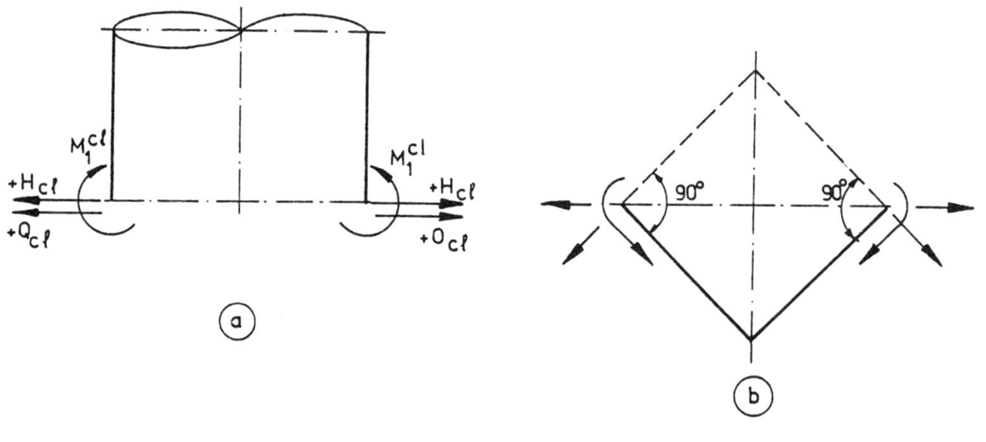

FIGURE 3.15 — The positive directions of the forces at cross-sections of the shells: (a) Cylindrical; (b) Conical.

$$X_1\delta_{11} + X_2\delta_{12} + X_3\delta_{13} + X_4\delta_{14} + \Delta_{1P} = 0$$

$$X_1\delta_{21} + X_2\delta_{22} + X_3\delta_{23} + X_4\delta_{24} + \Delta_{2P} = 0$$

$$X_1\delta_{31} + X_2\delta_{32} + X_3\delta_{33} + X_4\delta_{34} + \Delta_{3P} = 0$$

$$X_1\delta_{41} + X_2\delta_{42} + X_3\delta_{43} + X_4\delta_{44} + \Delta_{4P} = 0 \qquad (3.73)$$

The force equations (3.7) on the previous page were written considering the basic system and taking into account the signs for the displacements and directions of the forces and also loads in the basic system, are shown in Figure 3.14.

$$X_1(\delta^{c\ell}_{11}+\delta^{st}_{11}) + X_2(-\delta^{c\ell}_{12}+\delta^{st}\frac{t_{st}}{2}) + X_3(-\delta^{st}_{12}) + X_4(-\delta^{st}\frac{t_{st}}{2}) + \Delta^{st}_{IN_1c\ell} + \delta^{st}_{11}N^c_1\cos\beta\frac{t_{st}}{2} - \Delta^{st}_{1(N^c_1\sin\beta)} = 0$$

$$X_1(-\delta^{c\ell}_{21}+\delta^{st}_{11}\frac{t_{st}}{2}) + X_2[\delta^{c\ell}_{22}+\delta^{st}_{22}+\delta^{st}_{11}(\frac{t_{st}}{2})^2] + X_3(-\delta^{st}_{33}\frac{t_{st}}{2}) + X_4(\delta^{st}_{24}-\delta^{st}_{13}\frac{t_{st}}{2}) - \Delta^{c\ell}_{2P} + \Delta^{st}_{2P} + \Delta^{st}_{IN_1c\ell}\frac{t_{st}}{2} - \Delta^{st}_{2(N^c_1\cos\beta)} + \delta^{st}_{11}N^c_1\cos\beta(\frac{t_{st}}{2})^2 - \frac{t_{st}}{2}\Delta^{st}_{1(N^c_1\sin\beta)} = 0$$

$$X_1(-\delta^{st}_{31}) + X_2(-\delta^{st}_{33}\frac{t_{st}}{2}) + X_3(\delta^{st}_{33}\ \delta^{c\ell}_{33}) + X_4(\delta^{st}_{33}\frac{t_{st}}{2}-\delta^c_4) - \Delta^{st}_{3N^c_1} - \delta^{st}_{33}N^c_1\cos\beta\frac{t_{st}}{2} + \Delta^{st}_{3(N^c_1\sin\beta)} + \Delta^c_{3P} = 0$$

$$X_1(-\delta^{st}_{11}\frac{t_{st}}{2}) + X_2(\delta^{st}_{42}-\delta^{st}_{31}\frac{t_{st}}{2}) + X_3(\delta^{st}_{33}\frac{t_{st}}{2}-\delta^c_{43}) + X_4[\delta^{st}_{44}\ \delta^{st}_{33}(\frac{t_{st}}{2})^2 + \delta^c_{44}] + \Delta^{st}_{4P} - \Delta^{st}_{3Nc\ell}\frac{t_{st}}{2} - \Delta^{st}_{4(N^c_1\cos\beta)} - \delta^{st}_{31}N^c_1\cos\beta(\frac{t_{st}}{2})^2 + \Delta^{st}_{3(N^c_1\sin\beta)}\frac{t_{st}}{2} - \Delta^c_{4P} = 0 \qquad (3.74)$$

Using designations from Tables 3.1 to 3.12, the displacements from the equations (3.74) are connected by the following equalities:

$$\delta^{c\ell}_{11} = \delta^{c\ell}_{MM};\ \delta^{c\ell}_{12} = \delta^{c\ell}_{21} = \delta^{c\ell}_{HM} = \delta^{c\ell}_{MH};\ \delta^{c\ell}_{22} = \delta^{c\ell}_{HH}$$

$$\delta^{st}_{11} = \delta^{st}_{13} = \delta^{st}_{33} = \delta^{st}_{M_1M_1}$$

$$\delta^{st}_{22} = \delta^{st}_{44} = \delta^{st}_{24} = \delta^{st}_{NN};\ \delta^c_{33} = \delta^c_{MM}$$

$$\delta^c_{43} = \delta^c_{34} = \delta^c_{HM};\ \delta^c_{44} = \delta^c_{HH};$$

$$\Delta^{st}_{IN_1c\ell} = \Delta^{st}_{3N_1c\ell} = \delta^{st}_{HM_1N_1c\ell}$$

$$\Delta^{st}_{1(N^c_1\sin\beta)} = \delta^{st}_{HM_1}N^c_1\sin\beta;\ \Delta^{c\ell}_{2P} = \Delta^{c\ell}_{HP};$$

$$\Delta^{st}_{2P} = \Delta^{st}_{4P} = \Delta^{st}_{NP}$$

$$\Delta^{st}_{2(N^c_1\cos\beta)} = \Delta^{st}_{4(N^c_1\cos\beta)} = \delta^{st}_{NN}N^c_1\cos\beta$$

$$\Delta^{st}_{3(N_1\sin\beta)} = \delta^{st}_{HM_1}N^c_1\sin\beta;\ \Delta^c_{3P} = \Delta^c_{MP};\ \ \Delta^c_{4P} = \Delta^c_{HP} \qquad (3.75)$$

If we disregard the eccentricity $\frac{t_{ts}}{2}$ of the application of the forces X_2, X_4 and $N^c_1\cos\beta$, considering these forces as applied at the level a-a, or at the middle plane of the stiffening ring "st", and the loadings P, acting on the ring along the small depth t_{st}, then the equations (3.74) may be simplified substantially as follows:

$$X_1(\delta^{c\ell}_{11}+\delta^{st}_{11}) + X_2(-\delta^{c\ell}_{12}) + X_3(-\delta^{st}_{13}) + X_4 x0 + \Delta^{st}_{IN_1c\ell} - \Delta^{st}_{(N^c_1\sin\beta)} = 0$$

$$X_1(-\delta^{c\ell}_{21}) + X_2(\delta^{c\ell}\delta^{st}_{22}) + X_3 x0 + X_4\delta^{st}_{24} - \Delta^{c\ell}_{2P} - \Delta^{st}_{2(N_1\cos\beta)} = 0$$

$$X_1(-\delta^{st}_{31}) + X_2.0 + X_3(\delta^{st}_{33}+\delta^c_{33}) + X_4(-\delta^c_{34}) - \Delta^{st}_{3N_1c\ell} + \Delta^{st}_{(N^c_1\sin\beta)} + \Delta^c_{3P} = 0$$

$$X_1 x 0 + X_2 \delta^{st}_{42} + X_3(-\delta^{c}_{43}) + X_4(\delta^{st}_{44}\ \delta^{c}_{44}) -$$

$$- \delta^{st}_{4(N^c_1 \cos\beta)} - \Delta^{c}_{4P} = 0 \tag{3.76}$$

Using Tables 3.1 to 3.12 and Equation (3.75) we may determine the unknown factors X_1, X_2, X_3 and X_4, and after the forces M_1, H and Q, acting on the separated shells of the basic system, Figure 3.14, in Sections a_1-a_1 and a_2-a_2, apart from the membrane forces $N^{c\ell}_1$ and N^{c}_1. The positive directions of the forces are shown in Figure 3.15.

The forces M_1, H and Q per unit length of the circumferential sections a_1-a_1 and a_2-a_2, at the locations of the original edge effect, will be determined by the following expressions.

In general cases,

$$M^{c\ell}_1 = X_1; \quad M^{c}_1 = X_3$$

$$H_{c\ell} = X_2; \quad H_c = X_4 \tag{3.77}$$

$$Q_{c\ell} = H_{c\ell}; \quad Q_c = H_c \sin\beta$$

In particular cases when the directions of X_1, X_2, X_3 and X_4 are coinciding with the direction used in Figure 3.14,

$$M^{c\ell}_1 = X_1; \quad M^{c}_1 = X_4$$

$$H_{c\ell} = X_2; \quad H_c = X_4 \tag{3.78}$$

$$Q_{c\ell} = -X_2; \quad Q_c = -X_4 \sin\beta$$

When the forces N, M, H and Q are known, it is possible to determine the local stresses in the shells using the formulas (3.36) to (3.43) and (3.44) to (3.54).

The forces acting per unit length of the stiffening ring in the basic system at the positive directions, shown in Figure 3.16, have the following values.

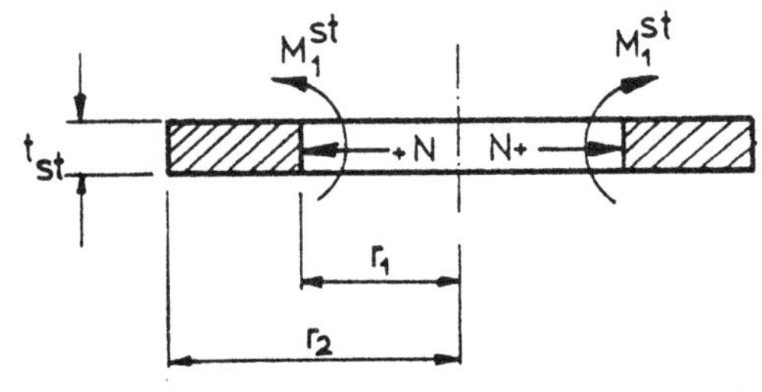

FIGURE 3.16 — The positive directions of the forces in the stiffening rings.

In the general case

$$N = X_2 + X_4 + pt_{st} + N^{c}_1 \cos\beta$$

$$M^{st}_1 = X_1 + X_3 + (X_2 + X_4 + N^{c}_1 \cos\beta)\frac{t_{st}}{2} \tag{3.79}$$

In the particular case with the coincidence of directions of X_1, X_2, X_3, X_4 and N with the directions shown in Figure 3.12

$$N = X_2 + X_4 + pt_{st} - N^{c}_1 \cos\beta$$

$$M^{st}_1 = X_1 - X_3 + (X_2 - X_4 + N^{c}_1 \cos\beta)\frac{t_{st}}{2} \tag{3.80}$$

It is necessary to remember that

$$N^{c\ell}_1 = N^{c}_1 \sin\beta$$

and, therefore, the force acting on the stiffening ring along the axis of the shell (from the plane of the ring) is equal to zero.

The displacement in the direction of the radius r from the axis of the shell, Figure 3.14 is

$$\overline{\Delta}r = \delta^{c\ell}_{22} H_{c\ell} - \delta^{c\ell}_{12} M^{c\ell}_1 - \Delta^{c\ell}_{2P} =$$

$$= \delta^{c}_{44}(H_c\ N^{c}_1 \cos\beta) - \delta^{c}_{43} M^{c}_1 - \Delta^{c}_{4P} \tag{3.81}$$

The turn of the section at level a-a (positive turn) is accepted clockwise for the left joint or unclockwise for the right joint.

$$\overline{\psi} = \delta^{c\ell}_{11} M^{c\ell}_1 = \delta^{c\ell}_{12} H_{c\ell} + \Delta^{c\ell}_{1P} = \delta^{st}_{11} M^{st}_1 =$$

$$= - \delta^{c}_{34} M^{c}_1 - \delta^{c}_{34} H_c - \Delta^{c}_{3P} \tag{3.82}$$

3.9 Edge Effect at the Stiffening Rings of Cylindrical Shells Under Internal Pressure p

This case is shown in Figure 3.17.

The increase of shell's radius in zona located at some distance from the original edge effect is

$$\Delta r = \frac{pr^2}{Et_c}\left(1 - \frac{\nu}{2}\right) \tag{3.83}$$

The increase of the radius of the stiffening ring

$$\Delta r_{st} = (pC + 2H)\frac{r^2}{EA} \tag{3.84}$$

where A is a cross-sectional area of the stiffening ring.

Force equations for the case under consideration are

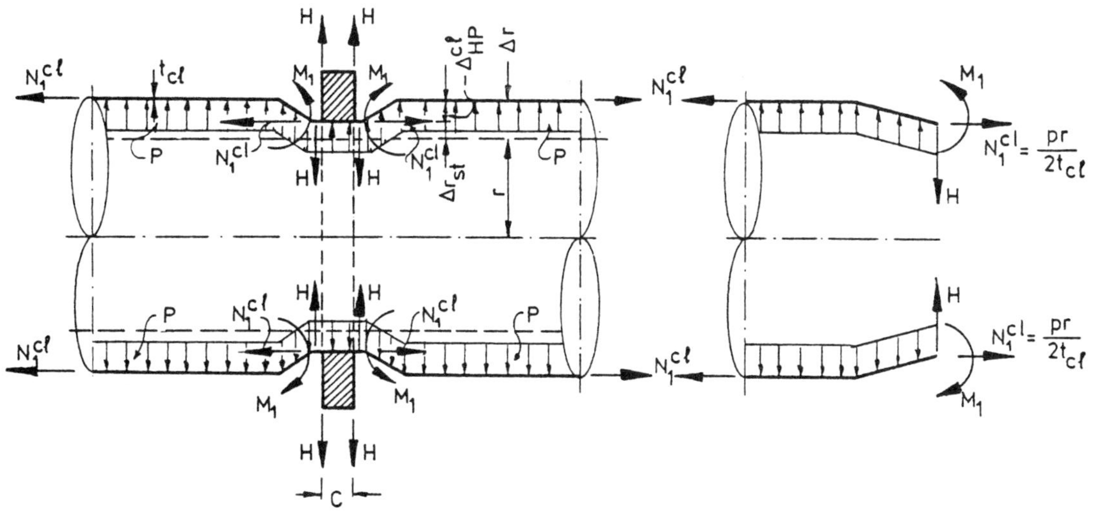

FIGURE 3.17 — Design scheme of cylindrical shell having stiffened ring, under internal pressure p.

$$M_1\delta_{MM}^{c\ell} + H\delta_{MH}^{c\ell} + \Delta_{Mp}^{c\ell} = 0 \quad (3.85)$$

$$M_1\delta_{HM}^{c\ell} + H\delta_{HH}^{c\ell} + \Delta_{HP}^{c\ell} = 0 \quad (3.86)$$

According to Tables 3.1 to 3.12, $\Delta_{MP}^{c\ell} = 0$, therefore

$$M_1\delta_{MM}^{c} = -H\delta_{MH}^{c} \quad (3.87)$$

After substituting into (3.87) the values of δ from Tables 3.1 to 3.12, we obtain

$$\frac{M_1 S_{c\ell}}{E J_0^{c\ell}} = \frac{HS_{c\ell}^2}{2EJ_0^{c\ell}} \quad (3.88)$$

from this, it follows that

$$M_1 = \frac{HS_{c\ell}}{2} \quad (3.89)$$

$$\Delta_{HP}^{c\ell} = -(\Delta r - \Delta r_{st}) \quad (3.90)$$

After substituting into Equation (3.86) the value (3.89), the values from Tables 3.1 to 3.12, and the value (3.90), considering also Equations (3.1) and (3.84), we obtain

$$\frac{HS_{c\ell}^3}{4EJ^{c\ell}} + (pC+2H)\frac{r^2}{EA} - \frac{pr^2}{ET_{c\ell}}\left(1-\frac{\nu}{2}\right) = 0 \quad (3.91)$$

and, considering Equation (3.12), we find

$$\frac{r^2H}{t_{c\ell}S_{c\ell}E} + (pC+2H)\frac{r^2}{EA} - \frac{pr^2}{ET_{c\ell}}\left(1-\frac{\nu}{2}\right) = 0 \quad (3.92)$$

From Equations (3.91) and (3.92) we obtain

$$H = \frac{pr^2J_0^{c\ell}[A(1-\frac{\nu}{2}) - Ct_{c\ell}]}{t_{c\ell}(0.25S_{c\ell}^3A + 2r^2J^{c\ell})} = \frac{pS_{c\ell}[A(1-\frac{\nu}{2}) - Ct_{c\ell}]}{A + 2t_{c\ell}S_{c\ell}} \quad (3.93)$$

At infinitely rigid rings, we may obtain from Equation (3.92)

$$H = PS_{c\ell}\left(1-\frac{\nu}{2}\right) \quad (3.94)$$

The stresses we may find from the formulas (3.36) to (3.43) using

$$H = H_{c\ell} = X_2 \quad \text{and} \quad H_0 = 0 \quad (3.95)$$

3.10 Edge Effect Due to the Eccentric Junction of the Shells

The values of the bending moments originating in the shell at an eccentric junction of separate elements, may be determined by the joint solution of the equations

$$m_\ell + m_r = m \quad (3.96)$$

$$\frac{m_\ell}{m_r} = A$$

where

- m = summary of the moments acting in both parts of the shell per unit length which is equal

$$m = N_e \quad (3.97)$$

where

- m_ℓ = moment per unit length in the left part of the shell
- m_r = moment per unit length in the right part of the shell
- A = value which depends on the configuration of those parts of the shells at the junction and used according to Table 3.2, assuming that the twisting angles of both parts of the shell are equal

TABLE 3.14 — Eccentric Junctions of Shells.

	Case of Junction	Angles of Twist of the Shells — Left	Angles of Twist of the Shells — Right	$A = \frac{m_\ell}{m_r}$
Cylindrical Shells	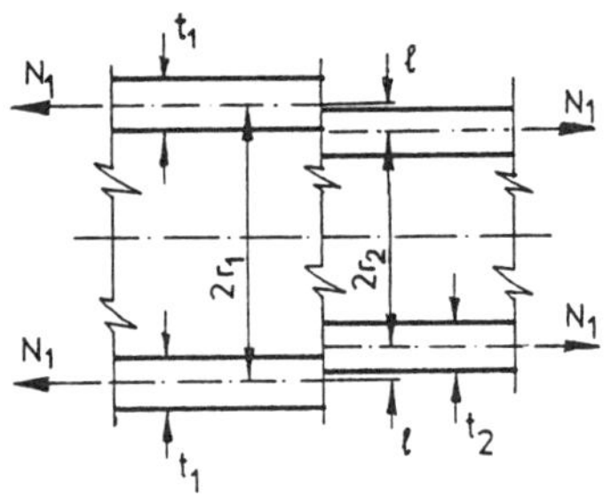	$\frac{4r_1^2 m_\ell}{t_1 S_{1c\ell}^3 E}$ $S_{c\ell} = \frac{\sqrt{rt}}{\sqrt[4]{3(1-\nu^2)}}$ at $\nu = 0.3 \quad S_{c\ell} = 0.78\sqrt{rt}$	$\frac{4r\ m_r}{t_2 S_{2c\ell} E}$	$\left[\frac{t_1}{t_2}\right]^{5/2}$
Spherical Shells	$\alpha < \frac{\pi}{2}$ 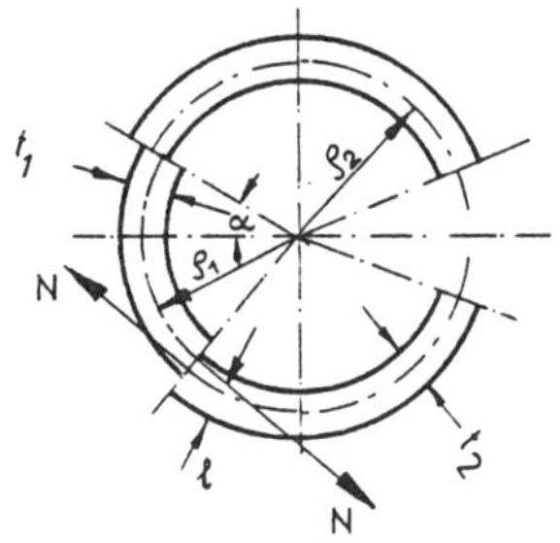	$\frac{4\rho_1^2 m_\ell}{t_1 S_{1s}^3 E W_1}$ $S_s = \frac{\sqrt{\rho t}}{\sqrt[4]{3(1-\nu^2)}}$ at $\nu = 0.3 \quad S_s = 0.78\sqrt{\rho t}$ $W_1 = 1 - 0.5\frac{S_{1s}}{\rho_1}(1-2\nu)\,\mathrm{ctg}\alpha$ $W_2 = 1 + 0.5\frac{S_{1s}}{\rho_2}(1-2\nu)\,\mathrm{ctg}\alpha$	$\frac{4\rho_2^2 m_r}{t_2 S_{2s}^2 E W_2}$	$\left[\frac{t_1}{t_2}\right]^{5/2}\frac{W_1}{W_2}$

REFERENCES — CHAPTER 3

[3.1] Timoshenko, S., Strength of Materials Part II, Advanced Theory and Problems, 2nd ed., D. Von Nostrand Company, Inc., New York, 1953, pp. 164-168.

[3.2] Roark, R.J., Formulas for Stress and Strain, 3rd ed., McGraw-Hill Book Company, Inc., New York, 1954, pp. 264-268.

[3.3] Biezeno, C.B. and Grammel, R., Engineering Dynamics, Vol. II, Blackie & Son Ltd., London, 1956, pp. 366-368.

[3.4] Timoshenko, S. and Woinowsky-Krieger, S., Theory of Plates and Shells, 2nd ed., McGraw-Hill Book Company, Inc., New York, 1959, pp. 482-485.

[3.5] Flugge, W., Stresses in Shells, 2nd ed., Springer-Verlag, New York, 1973, pp. 346-351.

[3.6] Brownell, L.E. and Young, E.H., Process Equipment Design—Vessel Design, John Wiley & Sons, Inc. 1968, pp. 113-119, 126-128.

[3.7] Lessig, E.N., Lileev, A.F. and Sokolov, A.G., "Steel Sheet Structures", State Edition of Literature on Constructions and Architecture, Moskow, 1956, (in Russian), pp. 372-412.

[3.8] Coates, W.M., "State of Stress in Full Heads of Pressure Vessels", Trans. Am. Soc. Mech. Engrs., Vol. 52, Paper No. APM-52-12, 1930, pp. 117-131.

[3.9] Watts, G.W., and Burrows, W.R., "The Basic Elastic Theory of Vessel Heads under Internal Pressure", J. Appl. Mechanics, Vol. 71, March 1949, pp. 55-73.

[3.10] Gill, S.S., Editor, The Stress Analysis of Pressure Vessels and Pressure Vessel Components, Pergamon Press, Oxford, 1970, pp. 36-50.

[3.11] **Watts, G.W. and Lang, H.A., "Stresses in a Pressure Vessel with a Conical Head", Transactions AM. Soc. Mech. Engrs., Vol. 74, 1952, p. 315.**

thermal stresses in tubular structures

4.1 Basic Assumptions and Designations

The change of steel temperature is assumed in limits of the constant modulus of elasticity and stresses in the limits of Hooke's Law.

The diagrams of change in temperature are shown in Figure 4.1, where

t_0 = initial temperature

t_{max}, t_{min} = are maximum and minimum temperatures, respectively

The signs are shown according to the positive change of temperature (heating).

In the case of reverse changes of temperature it is necessary to change the signs of the corresponding ordinates. The average temperature of the shell or ring is

$$t_i = \frac{t_{max} + t_{min}}{2} \tag{4.1}$$

The values of edge ordinates ± t of the diagram of temperature change from t_i (when t_i = t_0), are as follows

$$t = \frac{t_{max} - t_{min}}{2} \tag{4.2}$$

With a change of temperature the deformation may be expressed as

$$\Delta\ell = \alpha\ell\Delta t \tag{4.3}$$

where

α = coefficient of thermal expansion. For steel α 1.2 × 10^{-5}

ℓ = dimension of the member

Δt = general increase of temperature

The relative deformation is

$$\varepsilon = \frac{\Delta\ell}{\ell} = \alpha\Delta t \tag{4.4}$$

At free change of temperature deformations thermal stresses do not originate in the constructions. However, when the deformations are restricted, then the stresses originate, and according to Hooke's law

$$\sigma = \varepsilon E = \pm\alpha\Delta TE \tag{4.5}$$

At the two-dimensional state of stress of the shell, we have the known relations

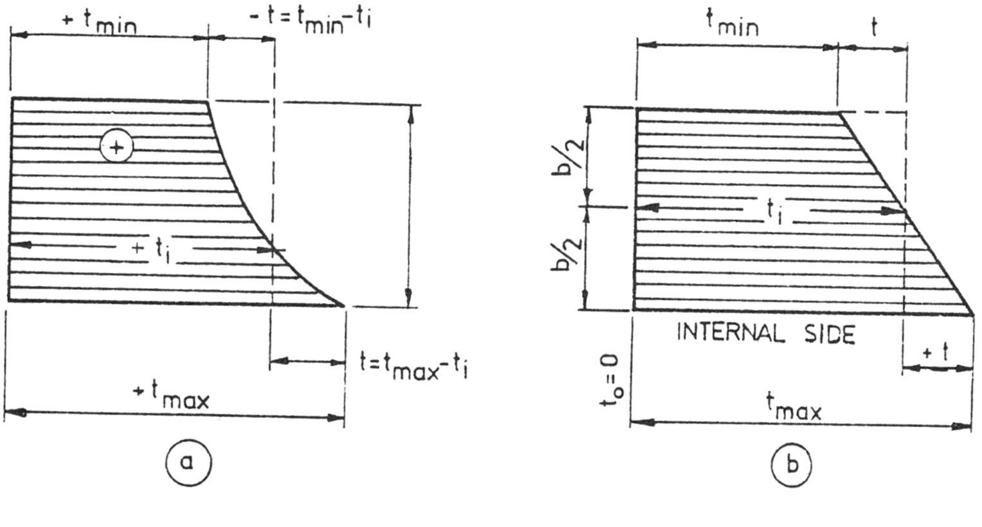

FIGURE 4.1 — Diagram of the change of temperature: (a) Arbitrary change; (b) Linear change.

$$\left.\begin{aligned}\sigma_1 &= \frac{E}{(1-\nu^2)}(\varepsilon_1+\gamma\varepsilon_2)\\ \sigma_2 &= \frac{E}{(1-\nu_1)}(\varepsilon_2+\gamma\varepsilon_1)\end{aligned}\right\} \quad (4.6)$$

and substituting into (4.6) the values from Equation (4.4) $\pm\epsilon_1 = \pm\epsilon_2 = \epsilon$, we obtain

$$\sigma_1 = \sigma_2 = \pm\frac{E\alpha\Delta t}{(1-\nu)} \quad (4.7)$$

The linear relations between the changes of temperature, deformations and stresses permit the superposition of the diagrams of temperature changes.

4.2 Thermal Stresses and Deformations in the Walls of Cylindrical Shells

Let us now consider the influence on the shell by the linearly changing temperature, Figure 4.1 (b).

The diagram of changeable temperature we resolve as follows: the t_1-diagram representing uniform distribution as defined by the expression (4.1) and the diagram indicating changeable t along the thickness of the wall, as defined by the expression (4.2).

4.2.1 Stresses and Deformations Under Uniform Heating — Diagram "t"

Let us consider the sections located far from the ends.

a. *The case when one end of the shell is free*

In this case the deformations are free and the stresses are

$$\sigma_1 = \sigma_2 = 0$$

The values of the displacements and deformations we may determine using formulas (4.3) and (4.4) after substituting into Δt the value t_i and instead of the corresponding dimension of the shell (radius, length of the circumference or the length of the generatrix.)

b. *The case when the external connections do not permit the increase of the lengths of the generatrix, but the radius may freely increase*

The value of the restrained deformation along the generatrix is

$$\varepsilon_1 = \alpha t_i \quad (4.8)$$

The value of the restrained deformation along the rine

$$\varepsilon_2 = 0 \quad (4.9)$$

After substituting these values ϵ_1 and ϵ_2 into the expression (4.6), we obtain

$$\sigma_1 = -\frac{E\alpha t_i}{(1-\nu^2)} \quad (4.10)$$

$$\overline{\sigma}_2 = -\frac{\nu E\alpha t_i}{(1-\nu^2)}$$

The elongation of the generatrix is equal to zero, and elongation of the radius is

$$\Delta r = \alpha r t_i \quad (4.11)$$

c. *The case when the external connections do not permit the elongation of the radii and in the direction of the generatrix there are no connections*

As in the previous case, $\Delta_r = 0$, and the elongation of the generatrix is

$$\Delta L = \alpha L t_i \quad (4.12)$$

Restrain of the deformations is

$$\varepsilon_1 = 0 \quad (4.13)$$

$$\varepsilon_2 = \alpha t_i$$

The stresses are

$$\sigma_1 = -\frac{\nu E\alpha t_i}{(1-\nu^2)}$$

$$\sigma_2 = -\frac{E\alpha T_i}{(1-\nu^2)} \quad (4.14)$$

d. *Case of the total restrain of the expansion at heating*

$$\varepsilon_1 = \varepsilon_2 = -\alpha t_i$$

$$\sigma_1 = \sigma_2 = -\frac{E\alpha t_i}{(1-\nu^2)} \quad (4.15)$$

This is according to the expressions (4.6) and (4.7).

4.2.2 Stresses and Deformations Under Uniform Heating-Diagram "t_i"

Let us consider edges of the shells.

a. *Case when the edge of the shell is free*

The stresses and deformations will be the same as in the case considered above, namely as at the sections remote from the edge at the similar edge conditions.

4.2.3 Stresses and Deformations Under Nonuniform Heating — Diagram "±t"

Let us consider section remote from the edges of the shell, Figure 4.1.

In this case there will be no deformations, because the shape of the circular shells and resistance of the adjoining sections do not permit their origin. The diagrams of stresses σ_1 and σ_2 will be the same and similar to the diagram t. The values of the stresses of the edge fibers according to Equation (4.7) are

$$\sigma_1 = \sigma_2 = \pm\frac{E\alpha t}{(1-\nu)} \quad (4.16)$$

During the linear change of the temperature, Figure 4.1 (b) the state of stress is equivalent to the case of the parallelepided having the dimensions 1 × 1 × t, cut from the shell along two generatrix and two radial sections of the bending moments

$$M_1 = M_2 = \pm \frac{\sigma_1 t_w^2}{6} = \pm \frac{\sigma_2 t_w^2}{6} = \pm \frac{E\sigma t t_w^2}{6(1-\nu)} = \pm \frac{2\alpha t D(1+\nu)}{t_w} \quad (4.17)$$

These moments M_1 and M_2 are acting in two mutually directions on the circumferential and meridional sections.

4.2.4 Stresses at Free Edge of Circular Shells Under Non-Uniform Heating-Diagram "±t"

For the transfer from the case considered above to the case under consideration, Figure 4.1 (b) and to satisfy conditions at the free edge

$$M_1 = 0$$

and

$$\sigma_1 = 0$$

it is necessary to impose on the bending moment, given by the expression (4.17), the equal moment but having the opposite sign.

If the shell is heated from the inside, Figure 4.2, then it is necessary to impose on the bending moment — M_1, the edge moment + M_1. Then at the edge will be

$$\Sigma M_1 = -M_1 + M_1 = 0 \quad (4.18)$$

The imposing of the moment + M_1 will initiate at the edge in the ring's direction. The appearance of the additional moment is equal to

$$\nu M_1 = \nu M_2$$

because $M_1 = M_2$, and also displacement of the edge outside and appearance of the additional ring force + $\overline{N}_2$ due to the displacement.

a. *Cylindrical shell, Figure 4.2 (a)*
According to Table 1, the displacement is

$$M_1\delta_{HM}^{c\ell} = \frac{S_{c\ell}^2 M_1}{2EJ_0^{c\ell}} = \frac{S_{c\ell}^2 M_1}{2D} \quad (4.19)$$

According to Hooke's law the displacement is

$$M_1\delta_{HM}^{c\ell} = \frac{\overline{\sigma}_2 r}{E} = \frac{\overline{N}_2 r}{tE} \quad (4.20)$$

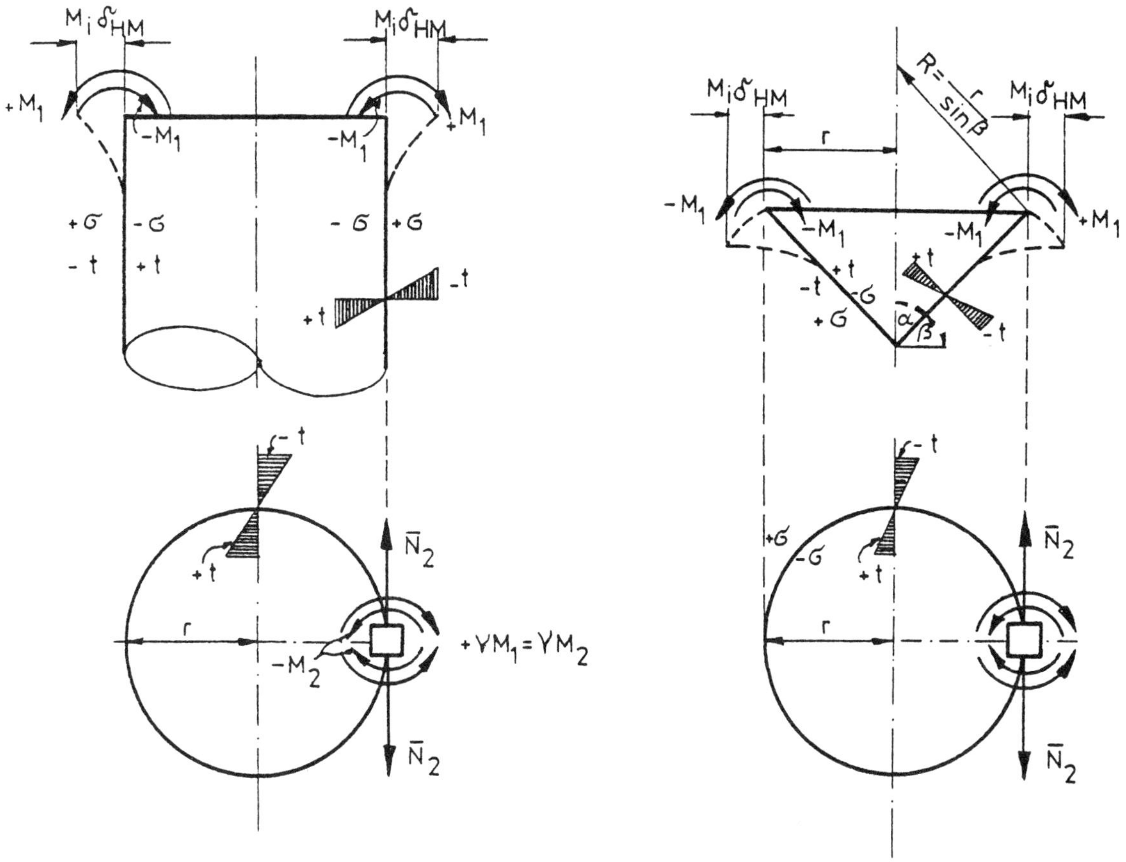

FIGURE 4-2 — Designing schemes of the shells under the influence of variable temperature ±t, which changes linearly through the thickness of the wall.

Be equalizing (4.19) and (4.20), we obtain

$$\overline{N}_2 = \frac{S_{c\ell}^2 M_1 E t_w}{2Dr} = \frac{2rM_1}{S_{c\ell}^2} = \frac{4r\alpha tD(1+\nu)}{S_{c\ell}^2 t_w} = \frac{E t_w t\alpha\sqrt{1-\nu^2}}{\sqrt{3}\,(1-\nu)} \quad (4.21)$$

The total ring moment at the edge is

$$\varepsilon M_2 = -M_2 + \nu M_2 = -M_2(1-\nu) = -\frac{2\alpha tD(1-\nu^2)}{t_w^2} = \frac{E t_w^2 \alpha t}{6} \quad (4.22)$$

The total ring stress at the free edge is

$$\sigma_2 = \overline{\sigma}_2 \pm \sigma_2 = \frac{\overline{N}_2}{t_w} \pm \frac{6\Sigma M_2}{t_w^2} = \frac{Et\alpha\sqrt{1-\nu^2}}{(1-\nu)\sqrt{3}} \pm E\alpha t \quad (4.23)$$

and the meridional stress is

$$\sigma_1 = 0 \quad (4.24)$$

In Equation (4.23) the sign plus is related to the external fibres and sign minus — to the internal. At the reversed thermal action signs should be changed.

At $\nu = 0.3$ the maximum ring stress at the edge according to formula (4.23) is approximately 25% greater than the stresses determined after formula (4.16) for the sections located far from the edges.

b. *Conical shells*

On the basis of similar consideration as that under (a), it is possible to find the total stresses at the free edge of the conical shell under heating, using diagram ±t, Figure 4.1 (b).

For the conical shell, Figure 4.2 (b)

$$\sigma_2 = \overline{\sigma}_2 \pm \sigma_2 = \frac{\overline{N}_2}{t_w} \pm \frac{6\Sigma M_2}{t_w^2} = \frac{Et\alpha\sqrt{1-\nu^2}}{(1-\nu)\nu_1\sqrt{3}} \pm E\alpha t \quad (4.25)$$

$$\sigma_1 = 0 \quad (4.26)$$

Stresses at cylindrical shells having hinged support under nonuniform linear heating, according to diagram ±t, Figure 4.1 (b).

With reference to the above case it is possible to conclude that $\Sigma M_1 = 0$ and $\sigma_1 = 0$.

The displacement $M_1 \delta HM = 0$ and from the Equation (4.20), it follows that $\overline{N}_2 = 0$.

The bending ring moment

$$\Sigma M_2 = \frac{E t_w^2 \alpha t}{6} \quad (4.27)$$

at change of temperature according to the scheme in Figure 4.2.

The ring stress at the edge

$$\sigma_2 = \pm E\alpha t \quad (4.28)$$

The support reaction at the edge H will be determined from the equation

$$M_1 \delta_{HM} + H\delta_{HH} = 0 \quad (4.29)$$

Reactions H are directed towards the centre at the flow of temperature according to Figure 4.2, and from the centre at the reverse of the flow.

The stresses at the points of zona of the edge effect, which are not at the edge itself, we may find by the superposition of stresses after the formula (4.16), stresses due to the forces H and M_2 applied at the edge. The signs of H and M are necessary to be used accordingly in correspondence with the actual direction of the thermal flow.

REFERENCES — CHAPTER 4

[4.1] Timoshenko, S.P. and Goodier, J.N., Theory of Elasticity, 3rd ed., McGraw-Hill Book Company, New York, 1970, pp. 433-439.

[4.2] Lessig, E.N., Lileev, A.F. and Sokolov, A.G., Steel Sheet Structures, State Edition of Literature on Constructions and Architecture, Moscow, 1956, (in Russian), pp. 417-419.

[4.3] Kent, C.H., "Thermal Stresses in Thin-Walled Cylinders", Transactions of the Am. Soc. of Mech. Engrs., Vol. 75, 1953, pp. 167-180.

self-supporting stacks

5.1 Types of Stacks

There are two basic types of steel stacks: self-supporting stacks which together with the foundation should remain stable under all working conditions without additional support, and guyed stacks in which all external applied loads are not totally carried by the stack shell and for which guys or stays are provided to ensure stability [5.1].

Self-supporting steel stacks are usually made with the upper part cylindrical and with the lower part flared or belled. The height of the flare usually varies from ⅛ to ¼ in height. Where the height of the flare is ¼ the height of the stack, the diameter of the base with a conical flare will be ⅓ greater than the diameter of the upper part of the stack. The ratio of the diameter of the base of the flare to the diameter of the stack in well designed stacks varies from 4/3 to 5/4 [5.2].

Figure 5.1 shows some of the more common types presently being employed.

Steel stacks are either lined or unlined. Independent, self-supporting brick lining should have enclosing steel bands at regular intervals in its height and, in particular, immediately below each change in lining thickness, where the weight of the lining above is applied eccentrically to the section below the offset. Brick lining may be supported at intervals by shelf angles in the steel shell. Linings of high-alumina cement and expanded shale aggregate have also been used extensively. This is pneumatically applied directly to the inside of the shell over mesh reinforcement attached by welded studs to the steel shell. Independent, continuous steel plate linings, using either carbon or corrosion-resistant steel, insulated on the outside of the lining may prove quite effective.

5.2 Design Loads

The stack shell should be designed to resist stresses resulting from the following loads and forces or their combinations when they exist [5.3, 5.4, 5.5, 5.6, 5.7, 5.8].

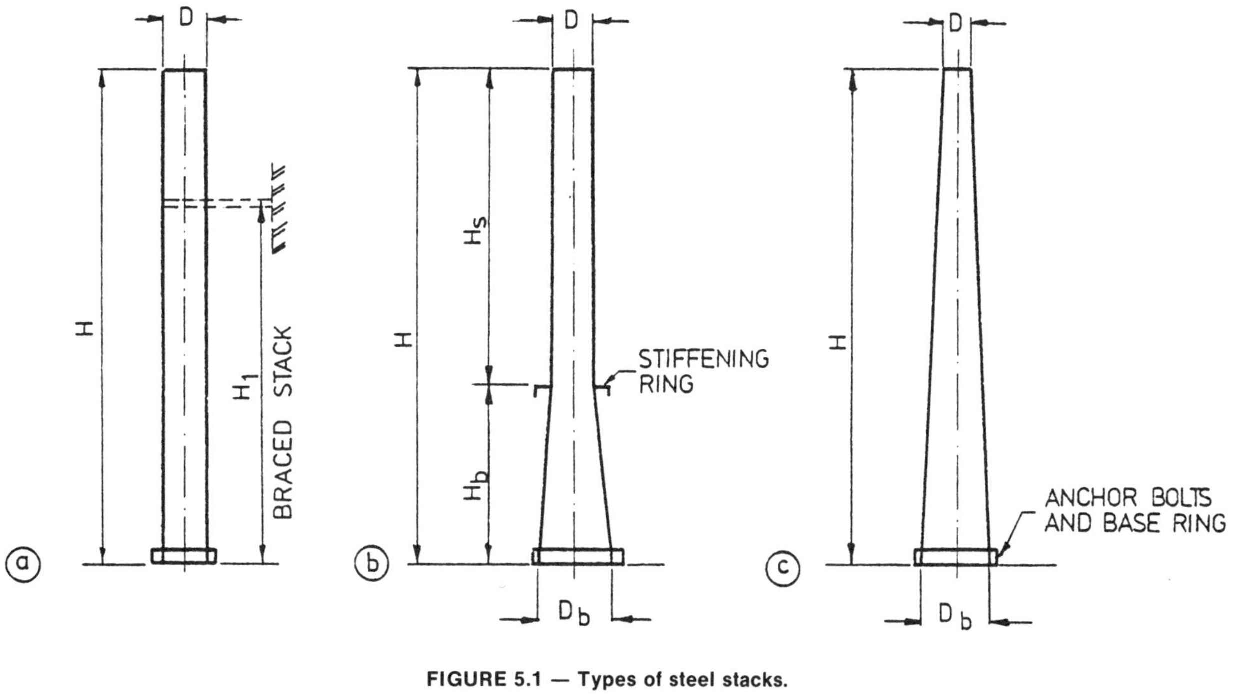

FIGURE 5.1 — Types of steel stacks.

a. Dead load
b. Live load
c. Icing
d. Wind or earthquake forces whichever are greater
e. Temperature effects, both vertically and circumferentially

In addition to these main loads and forces, secondary effects should also be considered. These are stresses resulting from support and erection procedures.

5.2.1 Dead Load

Dead loads are computed on the basis of the unit weights of the materials. Dead load consists of the particular weight of the shell, linings, ladders with safety rings and any permanent equipment. The structure should be divided into units or sections for weight tabulations. Divisions can be made at convenient changes in taper, cross-section, or plate thickness, at lining changes, and so as to avoid any one segment being too long.

5.2.2 Live Load

According to the German Specification DIN 4133 [5.5] live load, including snow load should be 50 psf, considering platforms and stairs. In addition, any part of the stack, platform and stairs may be loaded with the vertical concentrated load of 650 lbs at locations producing critical effect. All parts of the structure which have projections, such as canopies, which are at an angle of 30 degrees or less with the horizontal, with the exception of the ladder steps or rings, should be designed using a vertical concentrated load of 220 lbs, and a distributed loading of 6 psf, acting simultaneously. For the design of horizontal railings, the horizontal uniformly distributed load of 30 lbs/ft should be applied.

It is suggested that for American conditions, the applied snow load, s, be $s = C_s g$, where C_s is equal to 0.8 and g is the ground snow load as specified in American National Building Code [5.9]. S should be used to determine the applied snow load on catwalks, roofs, canopies, spiral access stairs, sampling platforms and whatever other structures are attached to the stack and are capable of transferring loads to the stack structure.

5.2.3 Icing

In locations where the strongest winds and icing may occur simultaneously, structural memebers, cables and ropes must be calculated assuming an ice covering based on climatic and local experience. The total force, F_n, for the guyed cable as specified in Supplement No. 4 of NBC [5.8] is given by the following equation

$$F_n = C_n \cdot q \cdot C_g \cdot C_e \cdot A \qquad (5.1)$$

where

C_n = force coefficient, depending on the Reynolds' number R_e which is expressed by $d\sqrt{q}$, where d is the diameter of cable in feet and q is the velocity wind pressure in psf. Figure 5.2 provides values of C_n for poles, rods and cables.

q = CV_W^2 in psf is the velocity wind pressure

where

C = 0.0027 if V_W in miles per hour

C_g = the gust effect factor

C_e = the exposure factor for the appropriate height of the surface or part of the surface

A = de = the projected area of the cable in elevation

5.2.4 Wind Forces

The stack should be designed for all loads and/or deflections caused by wind on the stack calculated in accordance with ACI 307 "Specification for the Design and Construction of Reinforced Concrete Chimneys" [5.6], the National Building Code of Canada [5.8], or other specified wind loads.

The force of the wind on a stack of circular cross-section shall be taken as the resultant horizontal pressure on its projected area considered as bands of uniform loading applied to the stack within height zones. This force should not be less than the design pressure specified by the Figure 5.3 [5.8].

Total wind force acting on the stack is

$$F = C_n \cdot q \cdot C_g \cdot C_e A \qquad (5.2)$$

where

C_n = force coefficient $d\sqrt{q} > 2.5$, used according to Figure 5.2

q = the reference velocity pressure

C_g = the gust effect factor

C_c = the exposure factor

A = dh = the projected area of the stack in elevation

d = diameter of the stack, ft.

h = height of the stack, ft.

5.2.5 Stack Vibration

Stack shells are prone to oscillations due to wind action [5.10, 5.11, 5.12, 5.13, 5.14, 5.15, 5.16, 5.17, 5.18, 5.19, 5.20, 5.21, 5.22, 5.23, 5.24, 5.25, 5.26, 5.27, 5.28, 5.29, 5.30].

In order to guard against these it is necessary to calculate the natural frequency of the shell and to compare it with the frequency of wind vortex shedding, inducted either by the stack itself, or by adjacent structures, at steady wind speeds of the order of 10 to 40 mph.

Vibration is manifested in rhythmic swaying to large amplitudes or transverse oscillations and in rhythmic ovalling of the chimney, both of which occur at right angles to the direction of the wind. This phenomenon is attributed to resonance of the chimney with respect to eddies, called von Karman vortices, formed in the airstream at regular intervals on alternate sides of the chimney. If the natural frequency of the stack as a cantilever approximates the frequency of the eddy formation, resonance may lead to major vibrations despite the small magnitude of the forces imposed by the vortex trail.

L/d > 100 C_n : FORCE COEFFICENTS

TOTAL FORCE $F_n = C_n \cdot q \cdot C_g \cdot C_e \cdot A$

$A = d \cdot l$

	$d\sqrt{q}$ < 2.5	$d\sqrt{q}$ > 2.5
SMOOTH WIRES, RODS, PIPES	1.2	0.5
MOD. SMOOTH WIRES, RODS	1.2	0.5
FINE WIRE CABLES	1.2	0.9
THICK WIRE CABLES	1.3	1.1

FIGURE 5.2 — Force coefficient C_n for poles, rods and wires.

TOTAL FORCE $F = C_n \cdot q \cdot C_g \cdot C_e \cdot A$ WHERE $A = h \cdot d$

C_n, FORCE COEFFICIENT FOR $d\sqrt{q} > 2.5$

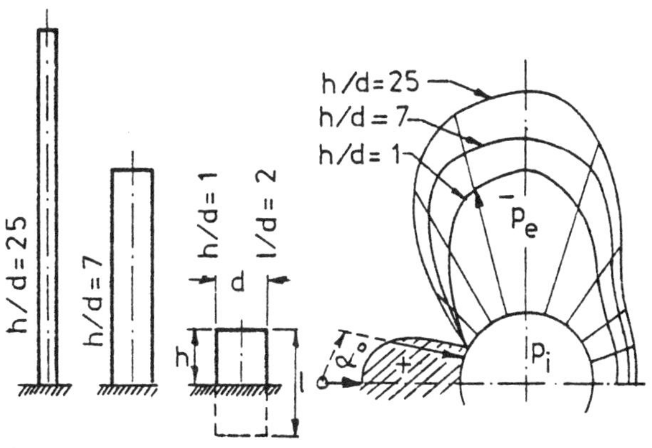

SLENDERNESS h/d = →	25	7	1
CROSS SECT. AND ROUGHNESS	C_n	C_n	C_n
MODERATELY SMOOTH (inetal, timber, concrete)	0.7	0.6	0.5
ROUGH SURFACE (rounded ribs h = 2% d)	0.9	0.8	0.7
VERY ROUGH SURFACE (sharp ribs h = 8% d)	1.2	1.0	0.8
SMOOTH AND ROUGH surface sharp edges	1.4	1.2	1.0

C_{pe}: EXTERNAL PRESS COEFF. FOR $d\sqrt{q} > 25$ AND MODERATELY SMOOTH SURFACE

h/d	l/d	α	0°	15°	30°	45°	60°	75°	90°	105°	120°	135°	150°	165°	180°
25	5	C_{pe}	+1.0	+0.8	+0.1	-0.9	-1.9	-2.5	-2.6	-1.9	-0.9	-0.7	-0.6	-0.6	-0.6
7	14	C_{pe}	+1.0	+0.8	+0.1	-0.8	-1.7	-2.2	-2.2	-1.7	-0.8	-0.6	-0.5	-0.5	-0.5
1	2	C	+1.0	+0.8	+0.1	-0.7	-1.2	-1.6	-1.7	-1.2	-0.7	-0.5	-0.4	-0.4	-0.4

$\Delta p = p_i - p_e$

$p_i = C_{pi} \cdot q \cdot C_g \cdot C_e$ STACK FULLY OPERATING $C_{pi} = +0.1$

$p_e = C_{pe} \cdot q \cdot C_g \cdot C_e$ STACK THROTTLED $C_{pi} = -0.8$

FIGURE 5.3 — Wind Coefficients.

5.2.5.1 Transverse Oscillations

British Standards of Steel Chimneys provides the following data [5.3]:

So long as the self weight of the stack is small compared to the critical buckling load (i.e. the critical buckling load for the stack considered as a vertical cantilever, without regard to the local buckling of the shell wall) the natural frequency f for a cantilever stack of constant section is given by:

$$f_n = \frac{C}{2\pi H}\sqrt{\frac{EIg}{W}} \tag{5.3}$$

where

- f = natural frequency of stack, cycles/sec.
- H = height of stack, ft.
- E = modulus of elasticity of shell material at operating temperature lbs/ft²
- I = moment of inertia of stack section, ft⁴
- W = weight per unit length of stack, lb/ft
- g = gravitational constant, 32 ft/sec²

C is a constant depending on the mode of vibration. For the first mode (the natural frequency),

$$C = 3.515$$

In calculation for lined stacks the section properties of an equivalent steel section may be calculated using a suitable value for the modular ratio, e.g.

$$\frac{E_{steel}}{E_{lining}} = 20$$

The natural frequency is

$$f_n = \sqrt{\frac{g \sum_i^n W_i x_i}{\sum_i^n W_i x_i^2}} \tag{5.4}$$

where

- W = the dead load weight of the section applied as a lateral load, and
- x_i = the deflection at its centroid, which will be produced if the stack were considered to be a vertical cantilever.

The natural frequency (1st mode) for a stack of varying cross-sections can be calculated by dividing the stack into a number of sections, Figure 5.4.

According to the British Standard [5.3], "The maximum deflection of steel chimney shall be limited to 1/200 of their unsupported length. This deflection shall be calculated on the design thickness alone without including the corrosion allowance."

An approximate formula is given by Dickey [5.31] for unlined steel stacks of straight or flared shape at normal temperature

$$f_n = 2520\ D(1+14.48t)/H_e^2 \tag{5.5}$$

where the equivalent height is

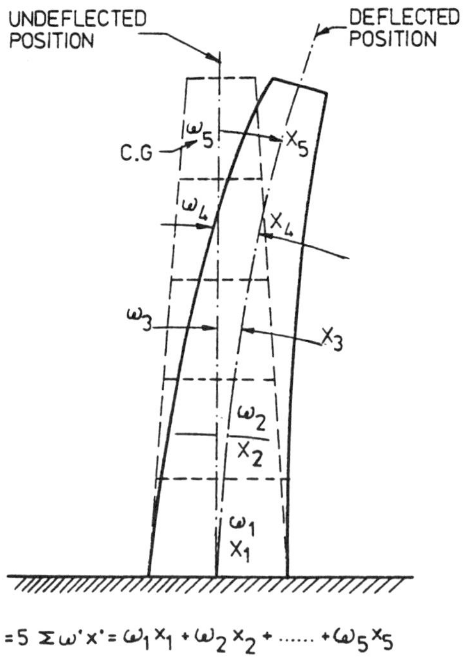

FIGURE 5.4 — Weights of sections and their deflections.

$$H_e = H_s + \frac{H_b}{3} + \frac{D}{2},\ \text{ft} \tag{5.6}$$

- D = diameter, ft
- t = thickness of shell wall at bottom of straight section, ft.

The natural frequency of a tapered stack, Figures 5.1 (b) and 5.1 (c) can be determined by the formula given by Dockstader [5.32]

$$F_n = \frac{3.52\ D_e}{48\ H_e^2}\sqrt{\frac{Eg}{2W_s}} \tag{5.7}$$

where

- F_n = natural frequency, cps
- g = gravity acceleration = 386 in/sec²
- E = modulus of elasticity, psi
- w_s = unit weight of steel shell = 0.283 lb/in³
- H_e = the equivalent height, ft
- D_e = the equivalent diameter, ft

The natural frequency is determined by using a straight cylinder of equivalent stiffness. The diameter D_e and height H_e of the equivalent cylinder are given by the following equations.

From Figure 5.1 (b)

$$H_e = H_s + H_b\left(\frac{2D_e}{D_e+D_b}\right)^2 \quad \text{where} \quad D_e = D \tag{5.8}$$

and from Figure 5.1 (c)

$$H_e = H\sqrt{\frac{2D_e}{D_e+D_b}} \quad \text{where} \quad D_e = \frac{D+D_b}{2} \tag{5.9}$$

5.2.5.2 *Effect of Vortex Shedding*

Slender exposed structural elements such as stacks should be designed to resist the effects of vortex shedding. A structure may be considered slender in this context if the ratio of height to diameter exceeds 5.

The critical wind speed at which vortex shedding may occur can be calculated from the following formula

$$V = \frac{fD}{S} \tag{5.10}$$

where

- f = natural frequency of stack, cycles/sec.
- D = stack diameter, ft.
- S = 0.20 Strouhal number
- V = critical wind speed, ft/sec.

The resonant wind velocity can be determined by

$$V_{res} = \frac{DF_n}{0.22} \cdot \frac{3,600}{5,280} = 3.1\ DF_n \tag{5.11}$$

Where a steel stack is lined and the weight of the lining is carried by the shell, an added mass must be set in vibration by the vortex trail. Since there is no major increase in the moment of inertia of the cantilever, the lined stack will have a lower natural frequency than an unlined stack of the same dimensions and will be relevant at lower wind velocities. The approximate natural frequency and resonant wind velocity of a stack with a uniform weight of lining throughout its height can be computed by Eqs. (5.10) and (5.11) provided the value of w_S is properly determined. For this purpose, w_S may be taken as the weight of the shell plus lining, divided by the thickness of the plate. The weight and thickness should be taken at about 1/4 the actual height above the base.

5.2.5.3 *Ovalling Oscillations*

In addition to swaying oscillations of stacks, vortices also tend to produce diametric oscillations. This phenomenon, referred to as "ovalling" or "breathing", has been observed with the largest amplitude at the top of the stack [5.33, 5.34, 5.35, 5.36, 5.37].

This type of motion has generally been seen on stacks before lining was in place and consists of a rhythmic change in the circular cross-section of the stack where the diameter alternately increases and decreases by several inches. This phenomenon occurs separately or in combination with a bending oscillation. The ovalling has occurred at twice the wind velocity indicated by the Strouhal number and may have involved "subharmonic excitation" in which the exciting force acted at twice the frequency of the oscillation. It is evident that two cycles of ovalling occurred during one vortex cycle.

According to the British Standard "Steel Chimneys" [5.3], the frequency of ovalling of a circular stack of constant section is given by

$$f_o = 170\ t/R^2 \tag{5.12}$$

where

- f_o = frequency of ovalling (cycles/sec)
- t = thickness of shell (ft)
- R = radius of stack (ft)

The critical wind speed at which ovalling may occur can be calculated from the following formula:

$$V = f_o \frac{D}{2S} \tag{5.13}$$

If the critical wind speed V is less than or equal to the steady wind speed, then it may be assumed that the shell will be prone to oscillations.

A simple practical cure against ovalling is to use stiffening rings near the top of the stack.

5.2.6 Earthquake Forces

Stacks in earthquake areas shall be designed and constructed to withstand as a minimum the lateral seismic forces, assuming that the forces can act in any direction [5.38, 5.39, 5.40, 5.41, 5.42, 5.43, 5.44, 5.45]. **The project specification should state applicable earthquake coefficient z [5.46].**

With the availability of the high-speed digital computer, stacks can now be analyzed for accelerograms of actual earthquake records. Although a rational analysis based on model response calculations is recommended for final design, certain procedures that are adaptable to hand computations are useful in preliminary design.

The most thoroughly studied and generally applied earthquake design criteria in the United States are those developed by the Structural Engineers Association of California in 1967 [5.47]. The SEAOC Code is of broad scope and has included criteria for "structures other than buildings". Stacks fall into this category. The Code as applied to stacks provides the basic criteria that define forces, shears, and moments.

5.2.6.1 *Base Shear*

The base shear or total lateral force, Figure 5.5, shall be computed according to the National Standard Building Code [5.46] by the following equation

$$V = SUCW \quad \text{or} \quad V = ZUCW_1 \tag{5.14}$$

where

- V = total shear at the base
- Z = zone coefficient for earthquake magnitude from the map of the United States showing the zones of approximately equal seismic probability in the latest edition of the Uniform Building Code. This coefficient shall not be less than 0.3 for Zone 1, 0.5 for Zone 2, and 1.0 for Zone 3.
- U = use factor and varies from 1.3 to 2.0.
- C = numerical coefficient for base shear which is equivalent to KC in the Uniform Building Code

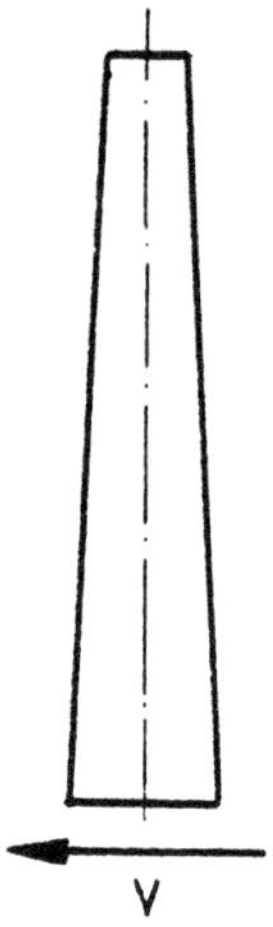

FIGURE 5.5 — Base shear.

W = total weight of stack without lining, lb.

W_l = total weight of stack including corbel supporting lining, lb.

The value of C shall be computed by Equation (5.15)

$$C = 0.1/\sqrt[3]{T} \tag{5.15}$$

The value of the fundamental period of vibration for a cantilevered structure of uniform cross-section may be expressed by the following formula

$$T = 1.79\sqrt{\frac{WH^4}{EIg}} \tag{5.16}$$

For steel stack this can be transformed as

$$T = 765 \times 10^{-6}\left(\frac{H}{D}\right)^2\sqrt{\frac{12wD}{t}} \tag{5.17}$$

where

T = fundamental period in seconds

H = height, ft

D = diameter, ft

w = weight per unit height, lbs/ft

t = shell thickness, inches

E = modulus of elasticity, psf

I = moment of inertia of the cross-section, ft^4

g = acceleration of gravity = 32.2 ft/sec^2

If the base of the steel stack is flared, the period of such a stack can be calculated by

$$T = 2\pi\sqrt{\frac{0.80\delta}{g}} \tag{5.18}$$

where

δ = the calculated deflection in feet at the top of the stack due to 100% of its weight applied as a lateral load

If the stack is lined, the weight per foot must include the weight of the lining, and the added stiffness due to the lining must be determined as an equivalent thickness of steel. The modified thickness of steel then should be used in the appropriate formula given above. This has a good correlation with measured periods of lined stacks. Gunite lining, for example, can reduce the period of steel stack in the order of 15%.

Mitchell [5.48] developed a form of solution which is especially useful in calculating fundamental periods of cylindrical, tapered-cylindrical, and step-tapered-cylindrical structures.

$$T = \left[\frac{H}{100}\right]^2\sqrt{\frac{\Sigma w\Delta\alpha + (1/H)\Sigma P\beta}{\Sigma\overline{E}D^3t\Delta\gamma}} \tag{5.19}$$

where

T = fundamental period in seconds

H = height in feet

w = weight per foot of height over a uniform, or assumed uniform section of the structure in pounds per foot

p = concentrated loads that may be attached to the structure at any levels, which add mass but do not contribute to the stiffness

$\overline{E}$ = modulus of elasticity in psi

D = diameter of each section in feet

t = thickness of each section of shell in inches

$\Delta\alpha$ = difference in the α values between the top and bottom of each section

β = constant at the section at which concentrated loads P are attached

$\Delta\gamma$ = difference in the γ values between the top and bottom of each section

Σ = summation of the product of the quantities shown over the height of the structure

Values of the α, β and γ in the function of the ratio h_x/H are given in Table 5.1.

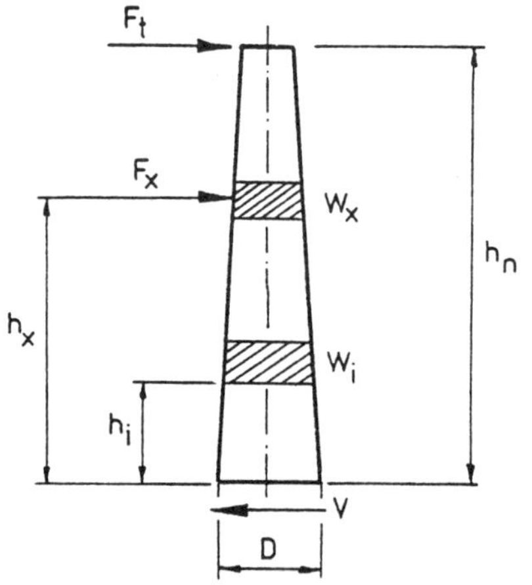

FIGURE 5.6 — Distribution of base shear.

TABLE 5.1 — Values of α, β, and γ in the function of h_x/H.

$\frac{h_x}{H}$	α	β	γ	$\frac{h_x}{H}$	α	β	γ
1.00	2.103	8.347	1.000000	0.50	0.1094	0.9863	0.95573
0.99	2.021	8.121	1.000000	0.49	0.0998	0.9210	0.95143
0.98	1.941	7.898	1.000000	0.48	0.0909	0.8584	0.94683
0.97	1.863	7.678	1.000000	0.47	0.0826	0.7987	0.94189
0.96	1.787	7.461	1.000000	0.46	0.0749	0.7418	0.93661
0.95	1.714	7.248	0.999999	0.45	0.0578	0.6876	0.93097
0.94	1.642	7.037	0.999998	0.44	0.0612	0.6361	0.92495
0.93	1.573	6.830	0.999997	0.43	0.0551	0.5372	0.91854
0.92	1.506	6.626	0.999994	0.42	0.0494	0.5409	0.91173
0.91	1.440	6.425	0.999989	0.41	0.0442	0.4971	0.90443
0.90	1.377	6.227	0.999982	0.40	0.0395	0.4557	0.89679
0.89	1.316	6.032	0.999971	0.39	0.0351	0.4167	0.88864
0.88	1.256	5.840	0.999956	0.38	0.0311	0.3801	0.88001
0.87	1.199	5.652	0.999934	0.37	0.0275	0.3456	0.87033
0.86	1.143	5.467	0.999905	0.36	0.0242	0.3134	0.86123
0.85	1.090	5.285	0.999867	0.35	0.0212	0.2833	0.85105
0.84	1.038	5.106	0.999317	0.34	0.0185	0.2552	0.34032
0.83	0.938	4.930	0.999754	0.33	0.0161	0.2291	0.82901
0.82	0.939	4.758	0.999674	0.32	0.0140	0.2050	0.81710
0.81	0.892	4.589	0.999576	0.31	0.0120	0.1826	0.80459
0.80	0.847	4.424	0.999455	0.30	0.010293	0.16200	0.7914
0.79	0.804	4.261	0.999309	0.29	0.008769	0.14308	0.7776
0.78	0.762	4.102	0.999133	0.28	0.007426	0.12576	0.7632
0.77	0.722	3.946	0.998923	0.27	0.006249	0.10997	0.7480
0.76	0.683	3.794	0.998676	0.26	0.005222	0.09564	0.7321
0.75	0.646	3.645	0.998385	0.25	0.004332	0.08267	0.7155
0.74	0.610	3.499	0.998047	0.24	0.003564	0.07101	0.6981
0.73	0.576	3.356	0.997656	0.23	0.002907	0.06056	0.6800
0.72	0.543	3.217	0.997205	0.22	0.002349	0.05126	0.6610
0.71	0.512	3.081	0.996689	0.21	0.001878	0.04303	0.6413
0.70	0.481	2.949	0.996101	0.20	0.001485	0.03579	0.6207
0.69	0.453	2.820	0.995434	0.19	0.001159	0.02948	0.5902
0.68	0.425	2.694	0.904681	0.18	0.000893	0.02400	0.5769
0.67	0.399	2.571	0.993834	0.17	0.000677	0.01931	0.5536
0.66	0.374	2.452	0.992885	0.16	0.000504	0.01531	0.5295
0.65	0.3497	2.3365	0.99183	0.15	0.000368	0.01196	0.5044
0.64	0.3269	2.2240	0.99065	0.14	0.000263	0.00917	0.4783
0.63	0.3052	2.1148	0.98934	0.13	0.000183	0.00689	0.4512
0.62	0.2846	2.0089	0.98739	0.12	0.000124	0.00506	0.4231
0.61	0.2650	1.9062	0.98630	0.11	0.000081	0.00361	0.3940
0.60	0.2464	1.8068	0.98455	0.10	0.000051	0.00249	0.3639
0.59	0.2288	1.7107	0.98262	0.09	0.000030	0.00165	0.3327
0.58	0.2122	1.6177	0.98052	0.08	0.000017	0.00104	0.3003
0.57	0.1965	1.5279	0.97823	0.07	0.000009	0.00062	0.2669
0.56	0.1816	1.4413	0.97573	0.06	0.000004	0.00034	0.2323
0.55	0.1676	1.3579	0.97301	0.05	0.000002	0.00016	0.1965
0.54	1.1545	1.2775	0.97007	0.04	0.000001	0.00007	0.1597
0.53	0.1421	1.2002	0.96683	0.03	0.000000	0.00002	0.1216
0.52	0.1305	1.1259	0.96344	0.02	0.000000	0.00000	0.0823
0.51	0.1196	1.0547	0.95973	0.01	0.000000	0.00000	0.0418

5.2.6.2 Distribution of Lateral Forces

The following Figure 5.6 indicates how the design base shear is distributed up through the structure as equivalent lateral static forces.

First, if the height-to-width ratio of the structure is greater than 3, then part of the base shear, up to 0.15 V maximum, is arbitrarily assigned to the top of the structure. The purpose of this is to increase the shear in the upper parts of the structure to account for the increasing participation of the higher modes of vibration in the response of taller, more flexible structures. It is also recognized that the fundamental mode deflection curve departs from the assumed straight line as the structure deflection is due more to bending action rather than shear deflection. This tends to cause greater shears in the top of the structure than would result from the so-called "triangular distribution."

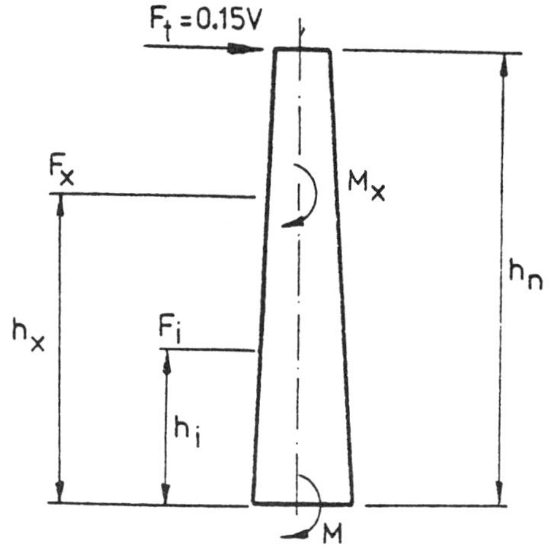

FIGURE 5.7 — Overturning moments.

$$F_t = 0.004\, V\left(\frac{h_n}{D}\right)^2 \qquad (5.20)$$

$$F_t = 0 \text{ for } \frac{h_n}{D} \leq 3 \qquad (5.21)$$

$$F_t = 9{,}k5V \text{ for } \frac{h_n}{D} \geq 6.12 \qquad (5.22)$$

$$F_x = (V - F_t)\frac{w_x h_x}{\sum_{i=1}^{n} w_i h_i} \qquad (5.23)$$

$$V = F_t + \sum_{i=1}^{n} F_i \qquad (5.24)$$

$$V_x = F_t + \sum_{i=1}^{n} F_i \qquad (5.25)$$

Most stack structures have height-to-width ratios greater than 6.6; hence, generally, the distribution of the base shear will be 0.15 V as F_t at the top, with the remaining 0.85 V distributed in accordance with the triangular distribution formula (5.23). In applying this formula, the structure arbitrarily can be divided into convenient increments, depending somewhat upon the weight distribution but usually not more than 10 sections. The transverse design shear at any horizontal plane in the structure is the cumulative sum of the lateral forces above that plane, resulting from the distribution of the base shear by the criteria given by formulas (5.20) and (5.23). Formula (5.24) states this for any plane x above the base, giving V_x. Formula (5.25) provides the corresponding value at the base V, which of course, should check the base shear that previously has been distributed and now is re-accumulated.

5.2.6.3 Overturning Moments

Having determined the base shear and the distribution of that base shear, the remaining basic design criteria relate to the overturning moment on the structure as a whole and at any horizontal plane x, Figure 5.7, **[5.49]**,

$$M_x = J_x\left[0.15V(h_n - h_x) + \sum_{i=x}^{n} F_i (h_i - h_x)\right] \qquad (5.26)$$

in which the value of J_x is determined from

$$J_x = J + (1 - J)\left(\frac{h_x}{h_n}\right)^3 \qquad (5.27)$$

where

$$J = \frac{0.6}{\sqrt[3]{T}} \qquad (5.28)$$

and J shall be not less than 0.45 nor greater than 1.0.

5.2.6.4 Supports

If it is expected that conditions at supports may change, this should be considered at the design stage [5.5].

5.2.6.5 Erection Conditions

Loads and forces which may appear under the erection conditions, for instance, temporary supports or jacking, should be considered at design stage [5.5].

5.2.7 Temperature Influence

The influence of the temperature variations on the design of steel chimneys is proposed following German Standards DIN 4133 [5.5]:

a. *Permissible temperature*
   The influence of the temperature should be considered in the analysis of shell if the temperature is higher than + 50° C.

b. *Thermal forces*
   Provision shall be made for stresses resulting in temperature. The rise of + 35° C and fall of – 35° C in temperature shall be figures from an assumed temperature at the time of erection.

   In the case of changes of temperature higher than + 50° C or lower than – 30° C, the difference of both temperatures should be considered in the design.

5.3 Stress Analysis for Self-Supporting Stacks

The self-supporting stack may be treated as a vertical cantilever beam subjected to loads caused by the horizontal wind pressure and the weight of the stack. It may be analyzed by beam theory for the purpose of determining the resulting longitudinal stresses. The only error introduced in the analysis of stack shell by the beam theory, as opposed to shell analysis, is the assumption that plane surfaces remain plane, and the stresses are proportional to the distances from the neutral axis. However, the error introduced by this assumption is small and conservative.

The maximum unit stress is usually compression and occurs due to the wind load. An allowance for possible earthquake stresses must be made for the stacks to be constructed in certain localities.

In designing a horizontal joint, the means of fastening must be able to transmit the maximum compressive stresses occurring in any part of the joint without exceeding the allowable unit stresses in bearing and shear. The gross thickness of the plate must be able to safely take the compressive stress, while the effective thickness of the plate must be able to safely take the maximum tensile stress in any part of the joint.

The stresses in the stack shell are [5.50, 5.51]:

a. *The Compressive Stress* resulting from dead loads including the weight of the shell itself, the weight of the lining and attached equipment. Also, stresses caused by an eccentricity resulting from irregular load distribution.
b. *Stresses resulting from bending moments* caused by wind or seismic loads acting on the stack.
c. *Shearing stresses* due to wind or earthquake.
d. *Thermal* stresses.
e. Stresses resulting from *fabrication procedures* such as cold forming and welding.
f. Stresses resulting from *erection.*
g. Allowance should be included for higher resulting stresses due to *corrosion.*

5.3.1 Dead Loads

Compressive stresses caused by dead loads may be considered in two parts for convenience:

a. *Stress induced by shell and lining*
Weight of steel in chimney having a constant shell thickness in pounds

$$W_s = \frac{\pi}{12} D_i t H \rho_s, \quad \text{lbs} \tag{5.29}$$

The stress due to the dead weight of shell

$$f_s = \frac{W_s}{\pi D_i (t-c)} \quad \text{psi} \tag{5.30}$$

where

W_s = weight of shell, lbs
D_i = inside diameter of shell, ft
t = thickness of shell, ins
H = height of the stack, ft
ρ_s = density of shell material, 490 lb/ft³ for steel
c = corrosion allowance

Weight of lining, in pounds

$$W_\ell = \frac{\pi}{12} D_\ell t_\ell H \rho_\ell, \quad \text{lbs} \tag{5.31}$$

The stress due to the dead weight of lining

$$f_\ell = \frac{W_\ell}{\pi D_i (t-c)} \quad \text{psi} \tag{5.32}$$

where

W_ℓ = weight of lining, lbs
D_ℓ = mean diameter of lining, ft
t_ℓ = lining thickness, ins.
H_ℓ = height of the lining, ft.
ρ_ℓ = density of lining material, lb/f⁶
t = thickness of shell, ins.

b. *Stress induced by eccentric loadings*
In stacks, externally attached equipment such as: platforms, ladders, pipes, etc., act eccentrically and may be reduced to vertical loads acting along the shell axis and moments acting eccentrically with respect to the shell axis. If the total vertical equipment load is ΣG, then the combined moment is $M_e = \Sigma Ge$ and the compressive and tensile stresses due to the eccentric loads are

$$f_e = \frac{\Sigma G}{\pi D_i (t-c)} \pm \frac{4M_e}{\pi D^2 (t-c)}, \quad \text{psi} \tag{5.33}$$

where

f_e = unit stress in extreme fiber or the shell, psi
e = eccentricity, the distance from the shell axis to the resulting vertical load ΣG

5.3.2 Wind or Earthquake Loads

a. *Stresses induced by wind loads*
In general, a self-supported steel stack may be considered as a vertical cantilever beam subjected to the horizontal wind pressure.

According to formula (5.2), the total wind force acting on the stack is $F = C_n \cdot q \cdot C_q \cdot C_e \cdot A$. Therefore, the maximum bending moment is

$$M_w = F.\frac{H}{2} \tag{5.34}$$

and the stress due to wind is

$$f_w = \pm \frac{2FH}{\pi D^2 (t-c)}, \quad \text{psi} \tag{5.35}$$

b. *Stresses induced by earthquake loads*
According to formula (5.26) the total bending moment under earthquake force acting on the stack is

$$M = J_x[0.15V(h_n - h_x) + \sum_{i=x}^{n} F_i(h_i - h_x)] \tag{5.36}$$

and the stress due to earthquake is

$$f_E = \pm \frac{4J_x}{\pi D^2 (t-c)} [0.15V(h_n - h_x) + \sum_{i=x}^{n} F_i(h_i - h_x)] \tag{5.37}$$

5.4 Design of Self-Supporting Stacks

5.4.1 Dead Load Stress

According to formulas (5.30), (5.32) and (5.33)

$$f_{D.L} = f_s + f_\ell + f_e =$$

$$= \frac{W_s + W_\ell + \Sigma G}{\pi D_i (t-c)} \pm \frac{4M_e}{\pi D^2 (t-c)}, \quad \text{psi} \qquad (5.38)$$

5.4.2 Dead Loads and Wind Stresses

According to formulas (5.30), (5.32), (5.33) and (5.35)

$$f_{D.L+W} = f_s + f_\ell + f_e + f_w$$

$$= \frac{W_s + W_\ell + \Sigma G}{\pi D_i (t-c)} \pm \frac{4M_e}{\pi D^2 (t-c)} \pm$$

$$\pm \frac{2FH}{\pi D^2 (t-c)} \qquad (5.39)$$

5.4.3 Dead Loads and Earthquake Force Stresses

According to formulas (5.30), (5.32), (5.33) and (5.37)

$$f_{D.L+E} = f_s + f_\ell + f_e + f_E =$$

$$= \frac{W_s + W_\ell + \Sigma G}{\pi D_i (t-c)} \pm \frac{4M_e}{\pi D^2 (t-c)} \qquad (5.40)$$

$$\pm \frac{4J_x}{\pi D^2 (t-c)} \left[0.15V(h_n - h_x) + \sum_{i=x}^{n} F_i (h_i - h_x) \right]$$

5.4.4 The Allowable Stresses

The allowable buckling stresses for mild steel are used according to Equation (2.10). At the condition

$$92 < d/t < 361$$

the allowable buckling stress for ASTM A-36 is

$$f_{allow} = \frac{662}{D/t} + 0.399\, F_y, \quad \text{ksi}$$

And when $D/t > 361$, then according to the formula (2.15)

$$f_{allow} = \frac{5333}{D/t}, \quad \text{ksi}$$

According to the German specification DIN for design of steel stacks [5.5], the allowable buckling stress can be estimated by using the following formula

$$\sigma_c = KE\left(\frac{t}{R}\right) \qquad (5.41)$$

where the coefficient K is

$$K = \frac{1}{2 + 5\frac{e}{t}} \qquad (5.42)$$

and "e" is the eccentricity of the cross-section of the stack due to fabrication imperfections

$$e = \frac{1}{25.4}\left(10 + \frac{R}{100}\right) \geq t, \text{ in} \qquad (5.43)$$

5.4.5 Checking of the Wind Induced Oscillations

The resonant wind velocity for the stack configuration of Figure 5.1 (a) and 5.1 (b) is determined according to the formula (5.11)

$$V_{res} = 3.1DF_n$$

where the first mode of vibration is after formula (5.7)

$$F_n = \frac{3.52D_e}{48\pi H_e^2} \sqrt{\frac{Eg}{2W_s}}$$

For the stack configuration of Figure 5.1 (c) D should be taken as the actual diameter at 1/8 the height below the top. The critical wind speed at which ovalling may occur is calculated following the formula (5.13)

$$V_{cr} = f_o \frac{D}{2S}$$

If the critical wind speed V is less than or equal to the steady wind speed, than it may be assumed that the shell will be prone to oscillations.

5.5 Base Design

To analyze the stresses between the base of the stack and its foundation, it is important to consider the degree of fixity of the base, which depends on the connection details. ϵ is a factor which allows for this degree of fixity and is graphically explained in Figure 5.8.

5.5.1 Design of Anchor Bolts

In designing anchor bolts for a self-supporting steel stack, the weight of the lining is not considered as the steel work is usually built first and the masonry lining added afterwards. Sometimes a considerable portion of the lining is removed and renewed during the life of the stack. This means that the anchor bolts must be large enough to keep the stack from overturning before the lining is removed and renewed during the life of the stack.

The load acting on the stack is transferred either as a compressive or tensile load to the concrete footing through the anchor bolts. The bending moment M and the weight of the stack W results in a loading condition in the concrete footing similar to that as shown in Figure 5.9.

In the calculation it is assumed that the bolt ring is the center of the bearing plate. The moment and weight of the

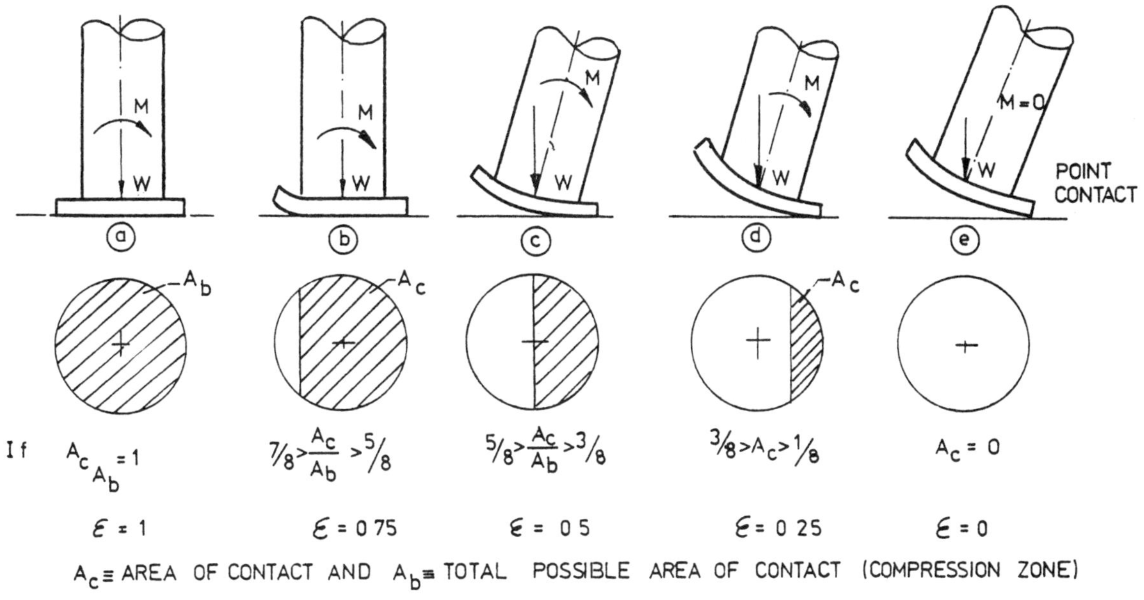

FIGURE 5.8 — ϵ as a function of base area contact of stack.

stack result in a tensile load on the left-side anchor bolts and a compression load on the right-side [5.52].

Calling κD the distance between the neutral axis and the mean circumference on compression side, as shown in Figure 5.10, we have by similar triangles

$$\kappa = \frac{nf_c}{nf_c + f_s} = \frac{1}{1 + \frac{f_s}{nf_c}} \tag{5.44}$$

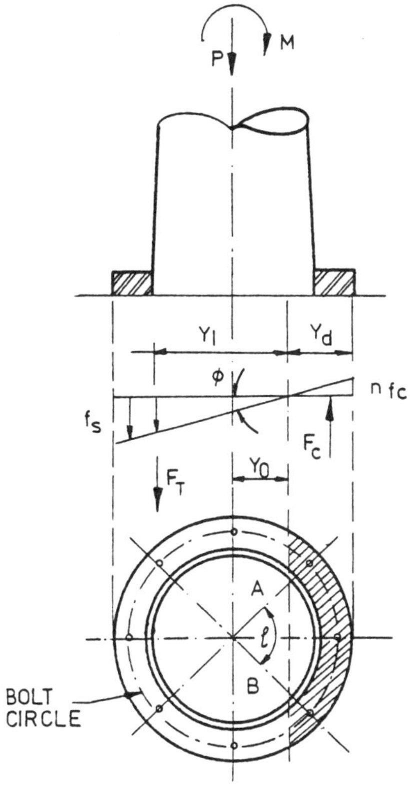

FIGURE 5.9 — Stress distribution at stack base.

where

$n = \frac{E_s}{E_c}$ = the ratio of moduli or elasticities of steel to the concrete.

From Figure 5.10, the location of the neutral axis n-n may be defined in terms of angle α

$$\cos\alpha = \frac{D/2 - \kappa D}{D/2} = 1 - 2\kappa \tag{5.45}$$

The total force, T, on the tension side of the section is

$$T = 2 \int_0^{(\pi-\alpha)} t_s r f_s \frac{(\cos\theta + \cos\alpha)}{(1+\cos\alpha)} d\theta$$

$$= f_s r t_s \frac{2}{(1+\cos\alpha)} [\sin\alpha + (\pi-\alpha)\cos\alpha] \tag{5.46}$$

Since any given position of the neutral axis determines α, this equation may take the form

$$T = C_T f_s r t_s \tag{5.47}$$

in which C_T is a constant for a given position of the neutral axis and shown in Table 5.2.

The moment of the total tensile force, T, about the neutral axis is

$$M_T = 2 \int_0^{(\pi-\alpha)} t_s r f_s \frac{r(\cos\theta + \cos\alpha)^2}{(1+\cos\alpha)} d\theta$$

$$= t_s r^2 f_s \frac{2}{(1 \ \cos\alpha)} [(\pi-\alpha)\cos^2\alpha + + \frac{3}{2}\sin\alpha\cos\alpha + \frac{1}{2}(\pi-\alpha)] \tag{5.48}$$

Dividing M_T by T, we have as the distance of the center of tension from the neutral axis

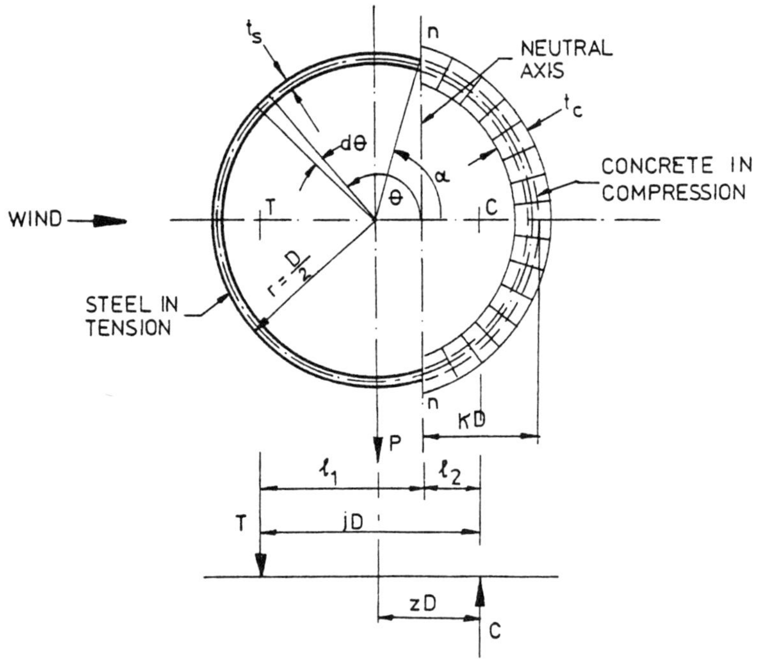

FIGURE 5.10 — Load on anchor bolts and bearing plate.

$$\ell_1 = \frac{[(\pi-\alpha)\cos^2\alpha + \frac{3}{2}\sin\alpha\cos\alpha + \frac{1}{2}(\pi-\alpha)]}{[\sin\alpha + (\pi-\alpha)\cos\alpha]} r \tag{5.49}$$

Similarly, we determine the total compression force

$$C = C_P f_c r\ (t_c + nt_s) \tag{5.50}$$

in which Cp is a constant for a given position of the neutral axis shown in Table 5.2.

The moment of the total compressive force C about the neutral axis is

$$M_c = (t_c + nt_s) f_c r^2 \frac{2}{(1-\cos\alpha)} [\alpha\cos^2\alpha - \frac{3}{2}\sin\alpha\cos\alpha + \frac{1}{2}\alpha] \tag{5.51}$$

TABLE 5.2 — Values of k, C_P, C_T, z and j.

	Function of k			
k	C_P	C_T	z	j
0.050	0.600	3.003	0.490	0.760
0.100	0.852	2.887	0.480	0.766
0.150	1.049	2.772	0.469	0.771
0.200	1.218	2.661	0.459	0.776
0.250	1.370	2.551	0.443	0.779
0.300	1.510	2.412	0.433	0.781
0.350	1.640	2.333	0.427	0.783
0.400	1.765	2.224	0.416	0.784
0.450	1.884	2.113	0.404	0.785
0.500	2.000	2.000	0.393	0.786
0.550	2.113	1.884	0.381	0.785
0.600	2.224	1.765	0.369	0.784

and the distance of the center of compression from the neutral axis is

$$\ell_2 = \frac{M_c}{C} = \frac{[\alpha\cos^2\alpha - \frac{3}{2}\sin\alpha\cos\alpha + \frac{1}{2}\alpha)]}{(\sin\alpha - \alpha\cos\alpha)} r \tag{5.52}$$

The system of forces, as shown in Figure 5.10, must be in equilibrium. Hence, taking moment ab out the force P, we may write

$$TjD = M - PzD$$

But

$$T = C_T f_s r t_s$$

Therefore

$$C_T f_s r t_s jD = M - PzD$$

Whence

$$rt_s = \frac{M - PzD}{C_T f_s jD}$$

The total area of steel required is

$$A_s = 2\pi r t_s$$

Therefore

$$A_s = \frac{2\pi(M-PzD)}{C_T f_s jD} \tag{5.53}$$

From Table 5.1, it may be seen that the constant j changes but slightly for a considerable variation in the position of the neutral axis. Taking $\frac{2\pi}{j}$ = 8 for all cases. Equation (5.53) may be

$$A_s = \frac{8(M-PzD)}{C_T f_s jD} \quad (5.54)$$

Applying now the condition that the summation of all vertical forces must be zero, we have

$$C - T = P$$

Substituting the values of C and T as previously found, we

$$C_P f_c r\,(t_c + nt_s) - C_T f_s r t_s = P$$

Solving for t_c, we obtain

$$t_c = \frac{P + (C_T f_s - C_P f_c n) r t_s}{C_P f_c r} \quad (5.55)$$

The number of anchor bolts for a self-supporting steel stack should never be less than 8 and should preferably be 10 or 12 or more depending on the size of the stack.

Generally, for given values of M and P, and assumed number and cross-sections of anchor bolts it is required to determine the maximum stresses in the anchor bolts f_s and the concrete under compression f_c. The problem is solved by a method of successive trials, since the position of the neutral axis is not known. The procedure is as follows:

1. Assume a position of the neutral axis, select the constants accordingly, substitute into Equations (5.54) and (5.55) and solve them for f_s and f_c.
2. Check the position of the neutral axis as fixed by these values of f_s and f_c is the same as the position assumed at the start. If the two positions agree, then f_s and f_c as found are the actual stresses.
3. If not, a new position of the neutral axis must be assumed, new constants selected, and new values of f_s and f_c computed. Thus a series of trials must be made until the location of the neutral axis as assumed is consistent with the computed values of f_c and f_s.

5.5.2 Base Plate

5.5.2.1 Base Plates Without Gussets

A base plate without gussets may be assumed to be a uniformly loaded cantilever beam with f_c the uniform load. The maximum bending moment for such a beam occurs at the junction of the stack shell and the base plate for unit circumferential length b = 1 in and is equal to

$$M_{max} = f_c b \ell \left(\frac{\ell}{2}\right) = \frac{f_c \ell^2}{2} \quad (5.56)$$

where ℓ is the base plate minus the other radius of the stack shell.

The maximum stress in an elemental strip of unit width is

$$f_{max} = \frac{6M_{max}}{bt_1^2} = \frac{3f_c \ell^2}{t_1^2} \quad (5.57)$$

where t_1 is the base-plate thickness in inches. Letting $f_{max} = f_{all}$ and solving for t_1 gives

$$t_1 = \ell\sqrt{3f_c/f_{all}} \quad (5.58)$$

5.5.2.2 Base Plates With Gussets

If gussets are used to stiffen the base plates, the loading conditions on the section of the plate between two gussets may be considered to act similarly to that of a rectangular uniformly loaded plate with two opposite edges simply supported by the gussets, the third edge joined to the shell, and the fourth and outer edge free. For this particular case Timoshenko and Woinowsky-Kreiger **[5.53]** have tabulated the deflections and bending moments as shown in Table 5.3.

To determine the base-plate thickness from the bending moment the following formula should be used

$$t_1 = \sqrt{\frac{6M_{max}}{f_{all}}} \quad (5.59)$$

Note that in Table 5.3, for the case where L/b = 0 (no gussets or gusset spacing b = ∞) the bending moment is reduced to Equation (5.56), and the thickness of the base plate is determined by Equation (5.58). Also when L/b is equal to or less than 3/2, the maximum bending moment occurs at the junction with the shell because of cantilever action. If L/b is greater than 3/2, the maximum bending moment occurs at the middle of the free edge.

5.5.2.3 Practical Considerations in Designing Base Plates

Rolled-angle base plates may be used for stacks if the calculated thickness of the base plates is 1/2 in. or less. The steel angle is rolled to fit as shown in Figure 5.11 **[5.54]**.

If the required base-plate thickness is 1/2 in. to 3/4 in., a design using a single-ring base plate may be used as shown in Figure 5.12.

TABLE 5.3 — Maximum Bending Moments in a Base Plate with Gussets.

l/b	$\left(M_x \; \begin{matrix} x = b/2 \\ y = \ell \end{matrix}\right)$	$\left(M_y \; \begin{matrix} x = b/2 \\ y = 0 \end{matrix}\right)$
0	0	$-0.500 f_c \ell$
1/3	$0.0078 f_c b^2$	$-0.428 f_c \ell$
1/2	$0.0293 f_c b^2$	$-0.319 f_c \ell$
2/3	$0.0558 f_c b^2$	$-0.227 f_c \ell$
1	$0.0972 f_c b^2$	$-0.119 f_c \ell$
3/2	$0.123 f_c b^2$	$-0.124 f_c \ell$
2	$0.131 f_c b^2$	$-0.125 f_c \ell$
3	$0.133 f_c b^2$	$-0.125 f_c \ell$
∞	$0.133 f_c b^2$	$-0.125 f_c \ell$

b = gusset spacing (x direction) inches.
δ = ℓ = bearing-plate outside radius minus skirt outside radius (y direction) inches.

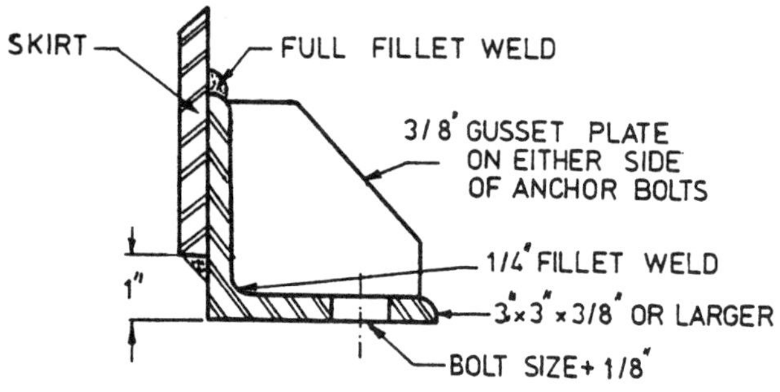

FIGURE 5.11 — Rolled-angle base plate.

If the required base-plate thickness is ¾ in. or greater a bolting "chair" may be used as shown in Figures 5.13 and 5.14.

Although the number and size of bolts required should be checked for each individual design, some typical values of maximum numbers of chairs can be obtained from Table 5.4 for a given stack base diameter.

When checking the base-plate thickness for a centered chair, Figure 5.13, the plate inside the stiffeners is considered to act as a concentrated loaded beam with fixed ends. The concentrated load, P, produced by the bolt is equal to the maximum bolt stress multiplied by the bolt root thread area.

The maximum moment in the base plate occurs on the line of symmetry centered inside the chair and is given by

$$M_{max} = \frac{Pb}{8} \quad (5.60)$$

The anchor bolt hole reduces the effective width of the plate. Taking this into consideration, the base-plate thickness, t_2, is

$$t_2 = \sqrt{\frac{6M_{max}}{(b_1 - bhd)\, f_{all}}} \quad (5.61)$$

where b_1 is the width of the base plate, bhd is the bolt-hole diameter and f_{all} is the allowable stress in psi.

If the number of bolts required is greater than that given in Table 5.4, an external bolting chair may be used, as shown in Figure 5.14.

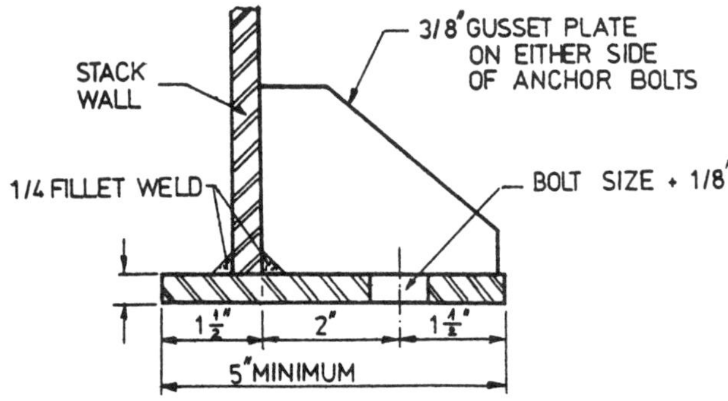

FIGURE 5.12 — Single ring base plate.

TABLE 5.4 — Number of Chairs for Various Size Stack Diameter.

Stack Diameter, ft	No. of Chairs
3	4
4	8
5	8
6	12
7	16
8	16
9	20
10	24

With reference to Figure 5.14, plan view, with y in the radial direction and z in the circumferential direction, the maximum bending moments M_Z and M_X are given by

$$M_z = \frac{P}{4\pi}\left[(1+\mu)\,\ell n\left(\frac{2\ell \sin\frac{\pi a}{\ell}}{\pi e}\right) + 1\right] - \left[\frac{\gamma_1 P}{4\pi}\right] \quad (5.62)$$

$$M_x = \frac{P}{4\pi}\left[(1+\mu)\,\ell n\left(\frac{2\ell \sin\frac{\pi a}{\ell}}{\pi e}\right) + 1\right] - \left[(1-\mu-\gamma_2)\frac{P}{4\pi}\right] \quad (5.63)$$

where

μ = Poisson's ratio (0.30 for steel)

ℓn = natural logarithm

α = radial distance from outside of skirt to bolt circle, inches

ℓ = radial distance from outside of skirt to outer edge of compression plate, inches

b = gusset spacing, inches

e = radius, of action of concentrated load, inches or one-half distance across flats of bolting nut, inches

γ_1, γ_2 = constants from Table 5.5

when b – unity, Mx = Mz, and when b > 1, Mz > Mx and therefore Mz controls.

For the case in which a = $\ell/2$ and Mz is controlling, Equation (5.62) reduced to

$$M_z = \frac{P}{4\pi}\left[(1+\mu)\;\ell n\frac{2\ell}{\pi e} + (1-\gamma_1)\right] \quad (5.64)$$

The maximum stress in the compression ring of unit width is

$$f_{max} = \frac{6M_y}{t_3^{\,2}} \quad (5.65)$$

TABLE 5.5 — Constants for Moment Calculations.

b/ℓ	1.0	1.2	1.4	1.6	1.8	2.0	α
γ_1	0.565	0.350	0.211	0.125	0.073	0.042	0
γ_2	0.135	0.115	0.085	0.057	0.037	0.023	0

Note: For a, b, ℓ less than 1.0 invert b,ℓ and rotate axes 90^2.

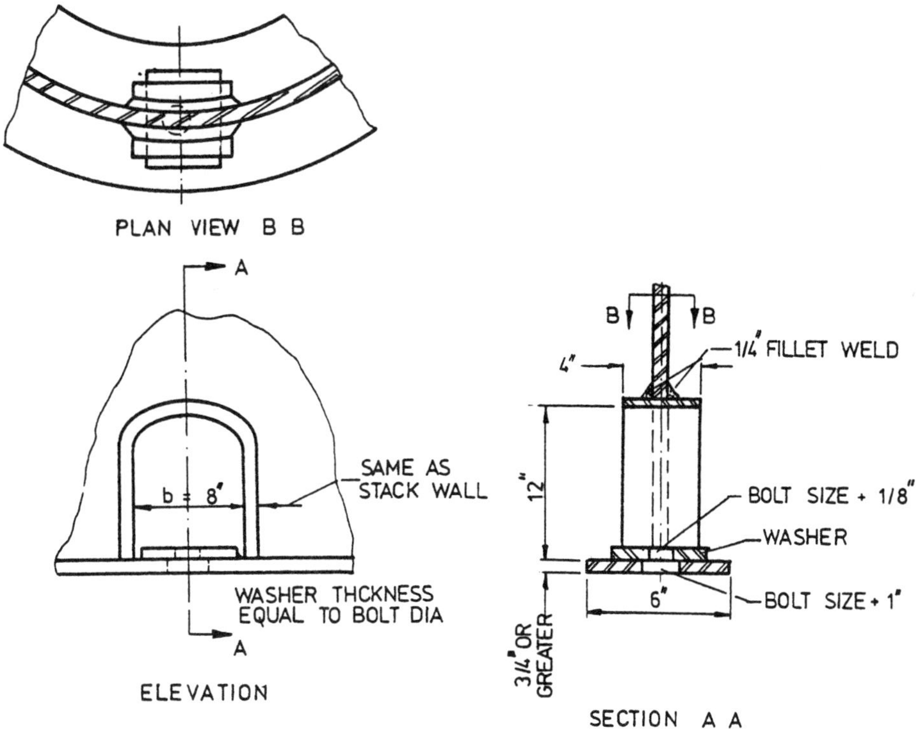

FIGURE 5.13 — Center anchor-bolt chair.

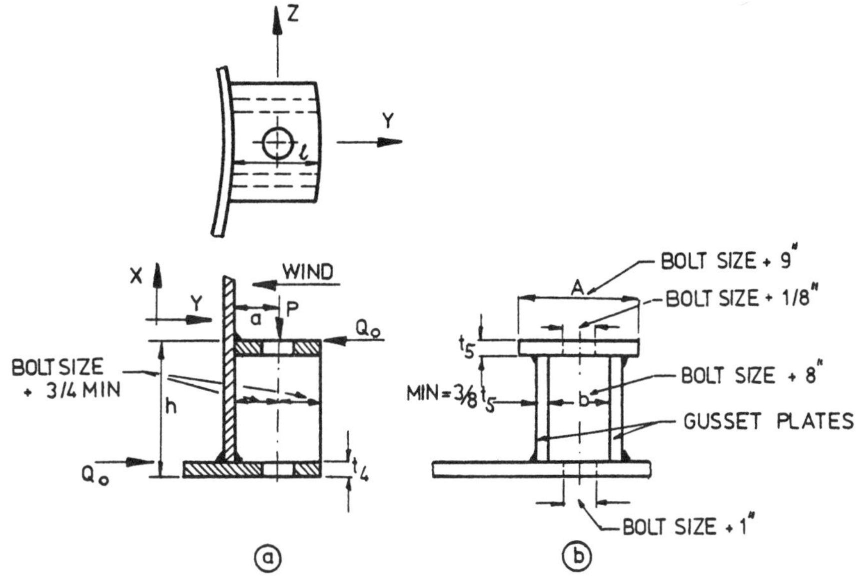

FIGURE 5.14 — External bolting chair.

where t_3 is the thickness of the compression ring.

5.5.2.4 Design of Gusset Plates for Compression Rings

If the gussets are spaced evenly as shown in Figure 5.15, they may be considered to behave as vertical columns [5.56].

The moment of inertia of the gusset about the axis having the radius of gyration is given by

$$I = \frac{\ell t_4^3}{12} = ar^2 = \ell t_4 r^2 \tag{5.66}$$

or

$$r^2 = \frac{t_4^2}{12} \tag{5.67}$$

where

a =area of cross-section, in

r = radius of gyration, in

t_4 = gusset-plate thickness, in

ℓ = width of gusset, in

and

$$f = \frac{P}{a} = \frac{18000}{1 + (h^2/18000r^2)} \quad \text{if } 60 < h/r < 200 \qquad (5.68)$$

where

h = height of gusset, in

From Equations (5.66), (5.67) and (5.68) we may obtain

$$18,000\,\ell t_4^3 - (\text{bolt load})t_4^2 - \frac{h^2(\text{bolt load})}{1,500} = 0 \qquad (5.69)$$

If h is small the third term in the equation may be disregarded, therefore simplifying Equation (5.69) to

$$t_4 = \frac{\text{bolt load}}{18,000\,\ell} \qquad (5.70)$$

When an external bolting chair is used the thickness of the stack shell, t, at the base should be checked. To determine the thickness Equation (5.71) can be used

$$t = 1.76\left(\frac{P_a}{mh\,f_{all}}\right)^{2/3} r^{1/3} \qquad (5.71)$$

where

r = radius of the stack at the point under consideration, inches

P = maximum bolt load, pounds

a = radial distance from outside of stack shell to the anchor bolt circle, inches

h = gusset height, inches

m = 2A (see Figure 5.14) or bolt spacing

5.6 Approximate Breech Opening Analysis

From a functional point of view, steel stacks have to resist the effects of wind and of their own weight. In the case of steel stacks the effects of the wind predominate by a comparison of the influence of their own weight.

With respect to their structural behavior there are two types of steel stacks: the self-supporting type and guyed stacks.

The self-supporting type acts as a column resisting its own weight, acting vertically, and the overturning effect of the wind, the base only being fixed. In other words, it is a cantilever subjected to a relatively large overturning moment and a relatively small vertical load.

Guyed stacks resist the same forces, but it is assisted by lateral supports in the form of guyed cables equally spaced around the stack and anchored at the other end to concrete blocks, some distance from the stack footings. The effect of those guy cables is to produce a somewhat smaller wind bending moment in the stack.

Near the base of the stack a relatively wide *breech opening* or *flue inlet* is cut. To facilitate the inspection and lining of stacks and to enable soot, etc., to be removed, it is usual to fit an *access door* near the base. As in the case of the breech opening, any opening of this nature should be adequately reinforced.

The breech opening weakens the wall of the stack considerably and some forms of compensation are required to take the loading and provide safety against local buckling.

In the following discussion the breech opening analysis and reinforcement design are considered for self-supporting stacks. However, it should be noted that such an analysis and design may be applied to guyed stacks. This is because guyed stacks are affected by such similar loadings as self-supporting stacks. Only the vertical load may be increased on account of the vertical reactions of guy cables.

In the following are given two methods for the analysis and design of breech openings: one approximate and another more rigorous method.

An approximate method may be useful for the preliminary analysis and design and a more rigorous method is recommended for the final design [5.55].

5.6.1 Conception of the Method

Experiments show that the applied load causing the general collapse of the thin-walled cylindrical shell axially-loaded is slightly greater than load initiating local buckling. Prior to local buckling, a shell is quite sensitive to slight disturbances. When the applied load is just below the local buckling load any lateral disturbance would cause local buckling to occur in the hole region cut in the shell.

Once the local buckling had occurred, the shell did not seem as sensitive to slight disturbances because these disturbances seldom led to its general collapse. Buckling loads continued to decrease as the hole dimensions increased.

During the experimental tests the loads were applied differently by the testing machines used in shell tests and

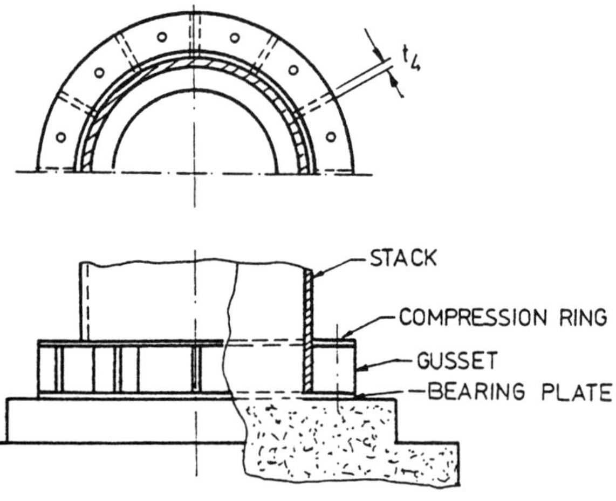

FIGURE 5.15 — Stack shell with external bolting chairs.

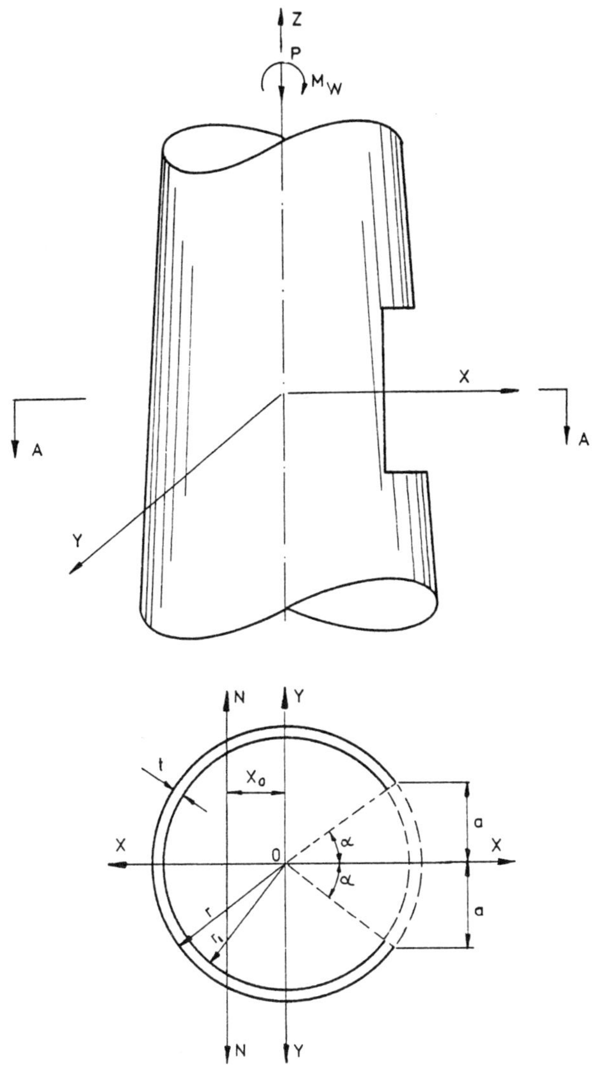

FIGURE 5.16 — Shell and hole geometry.

the measured shell buckling loads were reduced to approximate equivalent applied stresses in order to provide a common basis for comparison. This was done by replacing the applied load with a *statically-equivalent membrane stress system acting on the plane* which may be called *the applied stress plane* and which is perpendicular to the cylinder axis, Figure 5.16.

Therefore, the statically-equivalent applied membrane stress system must equilibrate the applied axial load plus a static bending moment due to wind action and eccentricity of axial load with respect to the neutral axis of cross-section.

To simplify the form of this applied membrane stress system the following assumptions were made:

a. The membrane stress applied at $y = \pm a$, Figure 5.16, on the hole edge can be used to represent the applied stress corresponding to local buckling of the shell, and

b. The hole is far enough from the ends of the shell, so that the assumed applied stresses are not influenced by the end support conditions.

5.6.2 Stresses at the Edge of Breech Opening

Reduced cross-section of the stack shell is under influence of axial load P and wind bending moment M_W.

Based on the geometry of the applied stress plane shown in Figure 5.16, the resulting stresses in the wall of the stack are given by

$$f = -\frac{P}{A} \pm \frac{(M_W + Px_c)x}{I} \qquad (5.72)$$

where

A = reduced cross-sectional area of the stack.

I = moment of inertia of the reduced cross-sectional area of the stack

If the resulting stresses remain negative (compression only), then there will be no anomaly, and the usual laws of

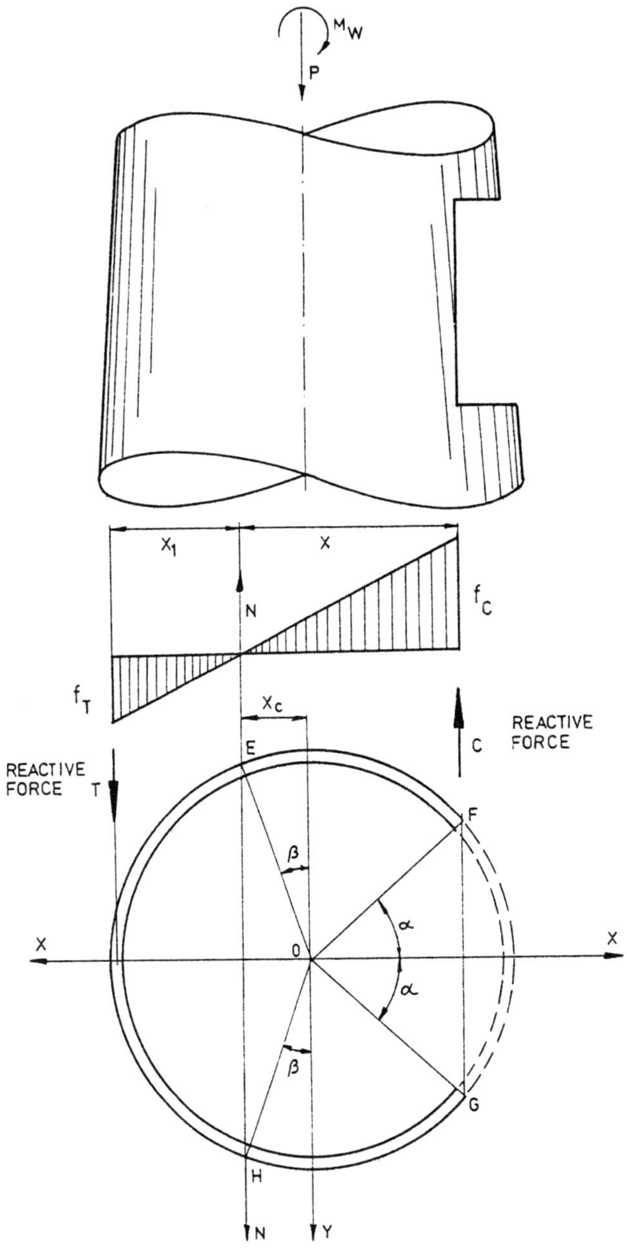

FIGURE 5.17 — Cross-section of the stack. Stress distribution diagram.

statics will be valid. If, however, the stress distribution diagram includes a tensile portion, this should be considered in further analysis, assuming that the position of neutral axis is known.

With reference to Figure 5.17, the following notations are introduced.

Sc,ST = statical moments of wall under compression and tension, respectively, referred to the neutral axis.

Ic,IT = moments of inertia of wall under compression and tension, respectively, referred to the neutral axis.

C, T = total compressive and tensile forces, respectively.

Mc,MT = moments capable of being resisted by the compressive and tensile sides, respectively.

ϕ = fT = the slope of the stress line and

$$f_c = x \frac{f_T}{x_1} \qquad (5.73)$$

Then, the compressive force is

$$C = \int f_c dA = \left(\frac{f_T}{x_1}\right) \int x\, da = \phi S_c \qquad (5.74)$$

By analogy, the value of tensile force is

$$T = \phi S_T \qquad (5.75)$$

Bending moment capable of being resisted by the compressive area is

$$M_c = \int x f_c dA = \left(\frac{f_T}{x_1}\right) \int x^2 dA = \phi I_c \qquad (5.76)$$

By analogy, the value of the bending moment capable of being resisted by the tensile are is

$$M_T = \phi I_T \qquad (5.77)$$

From the condition of the equilibrium of vertical forces, we have

$$P + T - C = 0$$

or

$$P = C - T = N \qquad (5.78)$$

Total bending moment is

$$M = M_c + M_T = M_w + Px_o \qquad (5.79)$$

After substituting into formula (5.78) values (5.74) and (5.75), we obtain

$$N = C - T = \phi\ (S_c = S_T) \qquad (5.80)$$

and formula (5.79), using (5.76) and (5.77) may be represented as

$$M = M_w + Px_o = \phi\ (I_c + I_T) \qquad (5.81)$$

From (5.81), the value of ϕ is

$$\phi = \frac{M_w + Px_o}{I_c + I_T} \qquad (5.82)$$

The value of N determined from the formula (5.80) should be equal to P, or

$$N = \phi\ (S_c - S_T) = P \qquad (5.83)$$

The average compressive stress in the wall may be obtained from the formula

$$f_{av} = \frac{C}{Lt} = \frac{\phi S_c}{Lt} \qquad (5.84)$$

where L is the total length of the compressed wall, Figure 5.17

$$L = \text{arc EF} + \text{arc GH}$$

or

$$L = 2 \times \frac{\pi R(90 - \alpha + \beta)}{180} \qquad (5.85)$$

Using midwall radius

$$R = \frac{r + r_1}{2} \qquad (5.86)$$

we obtain

$$L = \frac{\pi(r + r_1)(90 - \alpha + \beta)}{180} \qquad (5.87)$$

Therefore, the average compressive stress is

$$f_{av} = \frac{\phi S_c}{t}\ \frac{180}{\pi(r + r_1)(90 - \alpha + \beta)} \qquad (5.88)$$

The maximum value of the compressive stress occurring at the edge of the opening can be approximately determined by considering 1 inch wide strip at the edge, Figure 2.18.

From the diagram of stresses we may find the following ratios

$$\frac{f_1}{x_1} = \frac{f_3}{x} \ ;\ f_1 = x_1 \frac{f_3}{x} = \phi x_1$$

$$\frac{f_2}{x_2} = \frac{f_3}{x} \ ;\ f_2 = x_2 \frac{f_3}{x} = \phi x_2 \qquad (5.89)$$

The maximum stress at the middle of the wall is

$$f_{max} = \frac{f_1 + f_2}{2} = \frac{\phi}{2}\ (x_1 + x_2) \qquad (5.90)$$

By substituting

$$x_1 = x_6 + r \cos\alpha \ ;\ x_2 = x_6 + r_1 \cos\alpha$$

$$f_{max} = \frac{\phi}{2}\left[2x_c + (r + r_1)\cos\alpha\right] \qquad (5.91)$$

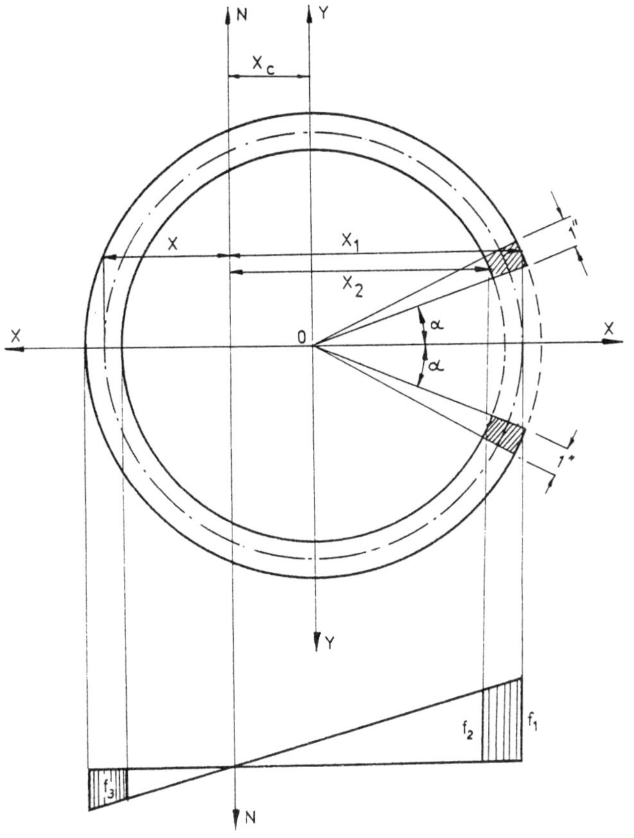

FIGURE 5.18 — Diagram of stresses in the wall of stack.

5.6.3 Geometrical Properties of the Reduced Cross-Section of the Stack

5.6.3.1 Statical Moment of the Compressed Area Figure 5.19

Ordinate of centroid of the sector AOB is

$$y_c = y_1 + y_2 = \frac{2r \sin\gamma_1 \ \sin\gamma_2}{3\gamma_1} + r \sin\beta \tag{5.92}$$

Area of the sector AOB

$$A = r^2\gamma_1 \tag{5.93}$$

Statical moment of the sector AOB is

$$S_1 = Ay_c = r^2\gamma_1 \left(\frac{2r \sin\gamma_1 \ \sin\gamma_2}{3\gamma_1} + r \sin\beta\right) \tag{5.94}$$

and for sector A_1OB_1

$$S_2 = r_1^2\gamma_1 \left(\frac{2r_1 \sin\gamma_1 \ \sin\gamma_2}{3\gamma_1} + r \sin\beta\right) \tag{5.95}$$

Total statical moment is

$$S_c = 2(S_1 - S_2) = \frac{4}{3}(r^3 - r_1^3) \sin\gamma_1 \sin\gamma_2 + 2r(r^2 - r_1^2)\gamma_1 \sin\beta \tag{5.96}$$

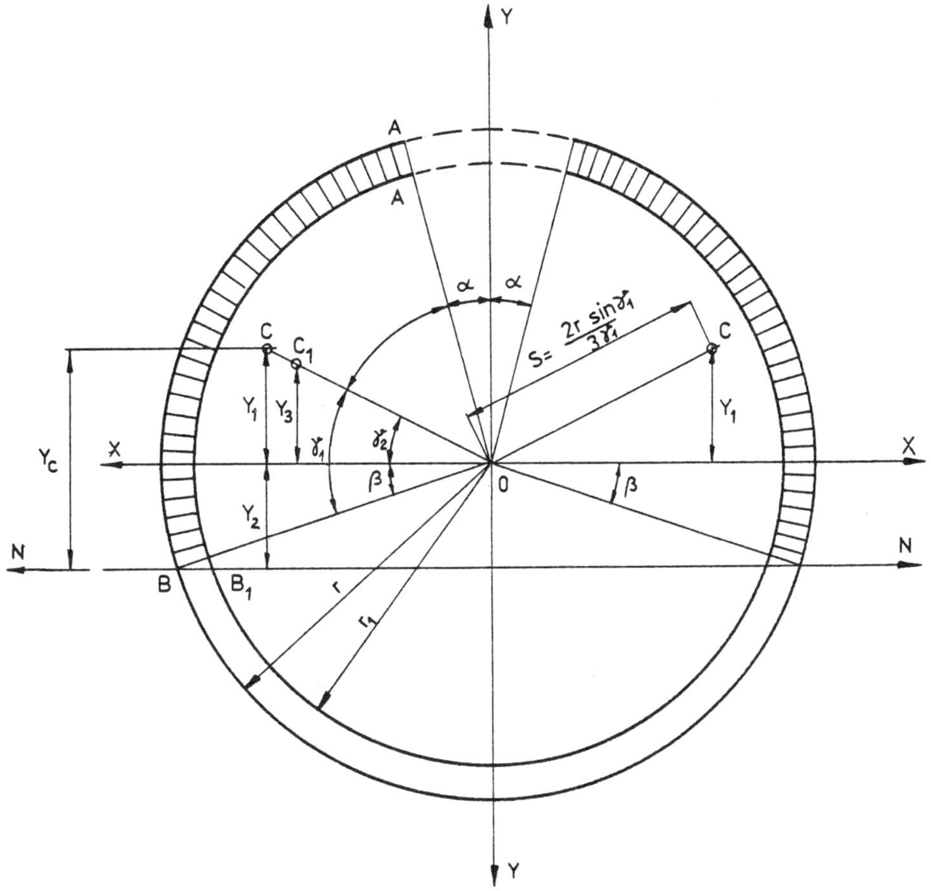

FIGURE 5.19 — Geometric data for the determination of statical moment of compressed area.

where

$$\gamma_1 = \frac{90 - \alpha + \beta}{2} \quad (5.97)$$

$$\gamma_2 = \frac{90 - \alpha - \beta}{2} \quad (5.98)$$

5.6.3.2 Statical Moment of Tensile Area: Figure 5.20

Ordinate of centroid of segment DEF

$$y_c = \frac{4\ r \cos^3\beta}{3\ [2(90-\beta) - \sin 2\beta]} \quad (5.99)$$

Area of the segment DEF is

$$A = \frac{r^2}{2}\left[2(90 - \beta) - \sin 2\beta\right] \quad (5.100)$$

Statical moment of the segment DEF with respect to the neutral axis N-N is

$$S_1 = A\ (y_c - r \sin \beta) =$$

$$= \frac{r^2}{2}\left[2\ (90-\beta) - \sin 2\beta\right] \times$$

$$\times \left\{\frac{4r \cos^3\beta}{3\ [2(90-\beta) - \sin 2\beta)]} - r \sin \beta\right\} =$$

$$S_1 = \frac{r^2}{2}\left\{\frac{4}{3}\ r \cos^3\beta - r \sin\beta\left[2(90-\beta) - \sin 2\beta\right]\right\} \quad (5.101)$$

By analogy, statical moment of the segment $D_1E_1F_1$ with respect to the neutral axis N-N is

$$S_2 = \frac{r_1^{\ 2}}{2}\left\{\frac{4}{3}\ r_1 \cos^3\beta - r \sin\beta\left[2(90-\beta) - \sin 2\beta\right]\right\} \quad (5.102)$$

Total statical moment of the tensile area with respect to neutral axis N-N is

$$S_T = S_1 - S_2 = \frac{2}{3}\ (r^3 - r_1^3) \cos^3\beta -$$

$$- \frac{r}{2}\ (r^2 - r_1^2) \sin \beta \times \left[2(90 - \beta) - \sin 2\beta\right] \quad (5.103)$$

5.6.3.3 Moment of Inertia of the Compressed Area; Figure 5.21

The moment of inertia of the compressed area with respect to the neutral axis N-N is

$$I_N = I_x + A\ (y_c^2 - y_1^2) \quad (5.104)$$

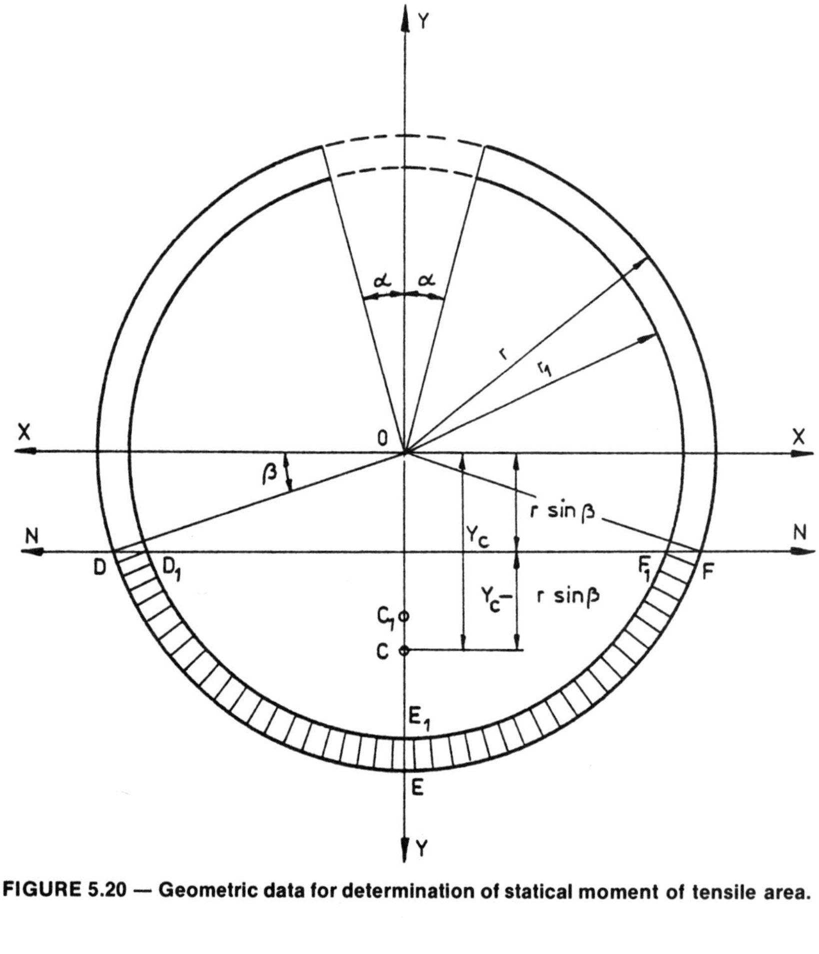

FIGURE 5.20 — Geometric data for determination of statical moment of tensile area.

a. *Moments of inertia of two sections AOB with respect to neutral axis x-x are*

$$I_{1x} = 2\int_{0}^{r}\int_{\alpha}^{90+\beta} (R d\phi dR)(R\cos\phi)^2 =$$

$$= 2\int_{0}^{r} R^3 dR \int_{\alpha}^{90+\beta} \cos^2\phi d\phi = \tag{5.105}$$

$$= \frac{r^4}{4}\left[2\gamma_1 - \sin(\alpha+\beta)\cos(\alpha-\beta)\right]$$

where

$$\gamma_1 = \frac{90 - \alpha + \beta}{2} \tag{5.106}$$

By analogy, the moments of inertia of two sectors A_1OB_1 with respect to neutral axis x-x are

$$I_{2x} = \frac{r_1^4}{4}\left[2\gamma_1 - \sin(\alpha+\beta)\cos(\alpha-\beta)\right] \tag{5.107}$$

Therefore, the moments of inertia of two rings AA_1BB_1 are

$$I_x = I_{1x} - I_{2x} =$$

$$= \frac{(r^4 - r_1^4)}{4}\left[2\gamma_1 - \sin(\alpha+\beta)\cos(\alpha-\beta)\right] \tag{5.108}$$

b. *Determination of the expression* $A_1(y_c^2 - y_1^2)$
Where A_1 is the area of the sector AOB.

$$A_1(y_c^2 - y_1^2) = r_1^2\gamma_1(y_c + y_1)(y_c - y_1) =$$

$$= r^2\gamma_1\left(\frac{4r\sin\gamma_1\sin\gamma_2}{3\gamma_1} + r\sin\beta\right) r\sin\beta =$$

$$= r^4\gamma_1\sin\beta\left(\frac{4\sin\gamma_1\sin\gamma_2}{3\gamma_1} + \sin\beta\right) \tag{5.109}$$

c. *Determination of the expression* $A_2(y_{cl}^2 - y_3^2)$
Where A_2 is the area of the sector A_1OB_1.

$$A_2(y_{cl}^2 - y_3^2) = r_1^2\gamma_1(y_{cl} + y_3)(y_{cl} - y_3) =$$

$$= r_1^2\gamma_1\left(\frac{4r_1\sin\gamma_1\sin\gamma_2}{3\gamma_1} + r_1\sin\beta\right) r\sin\beta =$$

$$= r r_1^3\gamma_1\sin\beta\left(\frac{4\sin\gamma_1\sin\gamma_2}{3\gamma_1} + \sin\beta\right) \tag{5.110}$$

The resulting value of the expressions (5.109) and (5.110) is

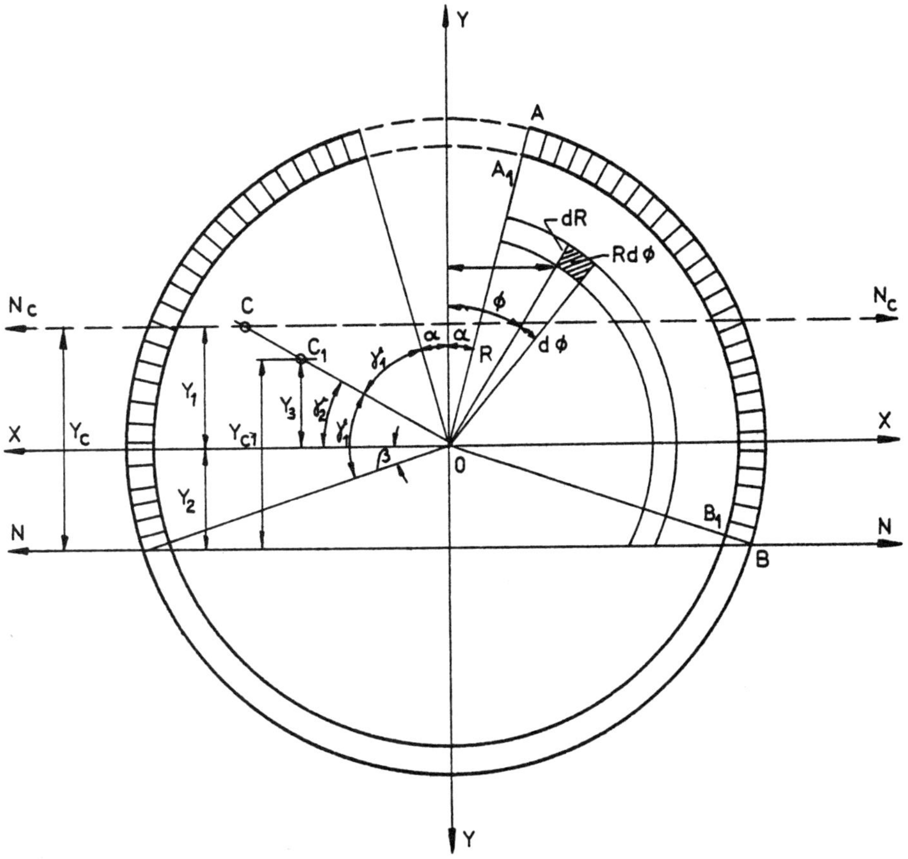

FIGURE 5.21 — Geometrical data for determination of moment of inertia of compressive area.

$$A_1(y_c^2 - y_1^2) - A_2(y_{c1}^2 - y_3^2) =$$

$$= r(r^3 - r_1^3)\gamma_1 \sin\beta \; x$$

$$x \left(\frac{4 \sin\gamma_1 \sin\gamma_2}{3\gamma_1} + \sin\beta\right) \tag{5.111}$$

Therefore, total moment of inertia of compressed area with respect to the axis N-N is

$$I_c = \frac{(r^4 - r_1^4)}{4}\left[2\gamma_1 - \sin(\alpha+\beta)\cos(\alpha-\beta)\right] +$$

$$+ 2r(r^3 - r_1^3)\gamma_1 \sin\beta \left(\frac{4 \sin\gamma_1 \sin\gamma_2}{3\gamma_1} + \sin\beta\right) \tag{5.112}$$

5.6.3.4 *Moment of Inertia of the Tensile Area, Figure 5.22*

Moment of inertia or ring EE_1GG_1 is with respect to the axis N-N is

$$I_N = I_{x_c} + Ay_N^2 \tag{5.113}$$

Ordinate of centroid C of segment EFG is

$$y_c = \frac{4r \sin^3\phi}{3\,(2\phi - \sin 2\phi)} = \frac{4r \sin^3 (90-\beta)}{3[2(90-\beta) - \sin 2\beta]} \tag{5.114}$$

Moment of inertia of segment EFG with respect to centroid axis $x_c - x_c$ is

$$I_{x_c} = \frac{r^4}{8}(2\phi - \sin 2\phi)\left(1 + \frac{4 \sin^3\phi \cos\phi}{2\phi - \sin 2\phi}\right) \tag{5.115}$$

Determination of the expression $A_1 y_{N_1}^2$, where A_1 is the area of the segment EFG

$$A_1 y_H^2 = \frac{r^2(2\phi - \sin 2\phi)}{2}(y_c - r\cos\phi)^2 =$$

$$= \frac{r^2(2\phi - \sin 2\phi)}{2}\left[\frac{4r \sin^3\phi}{3(2\phi - \sin 2\phi)} - \right.$$

$$\left. - r\cos\phi\right]^2 \tag{5.116}$$

From (5.115) and (5.116), we obtain the expression for moment of inertia of segment EFG

$$I_1 = \frac{r^4}{8}\left[(2\phi - \sin 2\phi) + 4\sin^3\phi \cos\phi\right] +$$

$$+ \frac{r^4}{2}(2\phi - \sin 2\phi)\left[\frac{4 \sin^3\phi}{3(2\phi - \sin 2\phi)} - \cos\phi\right]^2 \tag{5.117}$$

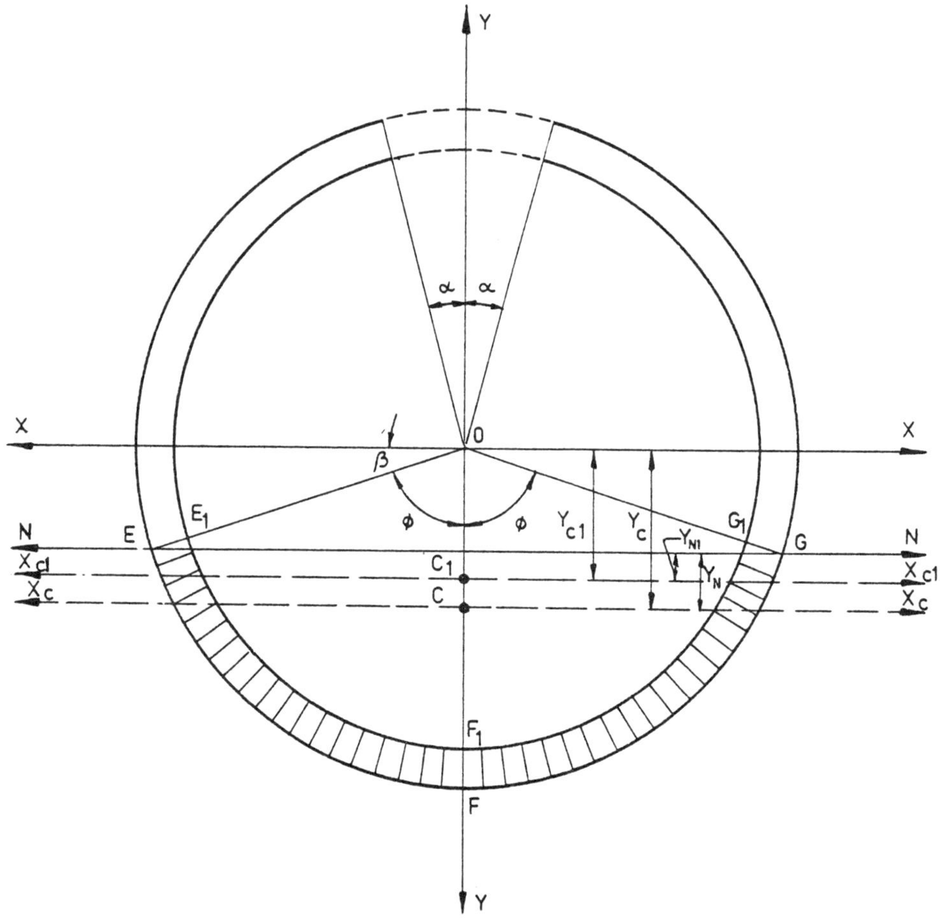

FIGURE 5.22 — Geometrical data for determination of moment of inertia of tensile area.

Moment of inertia of segment $E_1F_1G_1$ with respect to centroidal axis $x_{c_1} - x_{c_1}$

$$I_{c_1} = \frac{r_1^4}{8} (2\phi - \sin 2\phi)\left(1 + \frac{4 \sin^3\phi \cos\phi}{2\phi - \sin 2\phi}\right) \tag{5.118}$$

Determination of the expression $A_2y_{N_1}^2$ where A_2 is the area of the segment EF_1G_1.

$$A_2y_{N_1}^2 = \frac{r_1^2(2\phi - \sin 2\phi)}{2} (y_{c_1} - r\cos\phi)^2 = \frac{r_1^2(2\phi - \sin 2\phi)}{2}\left[\frac{4r_1 \sin^3\phi}{3(2\phi - \sin 2\phi)} - r\cos\phi\right]^2 \tag{5.119}$$

From (5.118) and (5.119), we obtain the expression for moment of inertia of segment $E_1F_1G_1$

$$I_2 = \frac{r_1^4}{8}\left[(2\phi - \sin 2\phi) + 4\sin^3\phi \cos\phi\right] + r_1^2 \frac{(2\phi - \sin 2\phi)}{2}\left[\frac{4r_1 \sin^3\phi}{2(2\phi - \sin 2\phi)} - r\cos\phi\right]^2 \tag{5.120}$$

Therefore, total moment of inertia of tensile area with respect to the neutral axis N-N is obtained by substructing from formula (5.117) the value (5.120), or

$$I_T = I_1 - I_2 = \frac{r^4}{8}\left[(2\phi - \sin 2\phi) + 4\sin^3\phi\cos\phi\right] + \frac{r^4}{8}(2\phi - \sin 2\phi)\left[\frac{4\sin^3\phi}{3(2\phi - \sin 2\phi)} - \cos\phi\right]^2 - \frac{r_1^4}{8}\left[(2\phi - \sin 2\phi) + 4\sin^3\phi\cos\phi\right] - \frac{r_1^2(2\phi - \sin 2\phi)}{2}\left[\frac{4r_1\sin^3\phi}{3(2\phi - \sin 2\phi)} - r\cos\phi\right]^2 = \frac{(r^4 - r_1^4)}{8}\left[(2\phi - \sin 2\phi) + 4\sin^3\phi\cos\phi\right] + \frac{8}{9}(r^4 - r_1^4)\frac{\sin^6\phi}{(2\phi - \sin 2\phi)} - \frac{4}{3}r(r^3 - r_1^3)\sin^3\phi\cos\phi + \frac{r^2(r^2 - r_1^2)}{2}\cos^2\phi(2\phi - \sin 2\phi)$$

$$I_T = (r^4 - r_1^4)\left\{\frac{1}{8}\left[(2\phi - \sin 2\phi) + 4\sin^3\phi\cos\phi\right] + \frac{8}{9}\frac{\sin^6\phi}{(2\phi - \sin^2\phi)}\right\} - \frac{4}{3}r(r^3 - r_1^3)\sin^3\phi\cos\phi + \frac{r^2(r^2 - r_1^2)}{2}\cos^2\phi(2\phi - \sin^2\phi) \tag{5.121}$$

After substituting $\phi = 90 - \beta$, we obtain

$$I_T = (r^4 - r_1^4)\left\{\frac{1}{8}\left[2(90-\beta) - \sin 2\beta + 4\cos^3\beta\sin\beta\right] + \frac{8}{9}\frac{\cos^6\beta}{2(90-\beta) - \sin 2\beta}\right\} - \frac{4}{3}r(r^3 - r_1^3)\cos^3\beta\sin\beta + \frac{r^2(r^2 - r_1^2)}{2}\left[2(90-\beta) - \sin 2\beta\right]\sin^2\beta \tag{5.122}$$

5.6.4 Stresses at Breech Opening. Experimental Tests

In this chapter the effect of rectangular cutouts on the buckling of circular cylinder is discussed based on experimental investigations.

Because of the nonsymmetric nature of a cylinder with a cutout, analytical solutions for the buckling loads of such cylinders are difficult to obtain. This is especially true for large cutouts where nonlinearities become more pronounced. Consequently, studies of the effect of cutouts on the buckling of cylindrical shells have generally been limited to a few analytical solutions and to experimental tests.

Brogan and Almroth [5.56, 5.57] carried out a theoretical and experimental investigation of the effect of rectangular cutouts on the buckling loads of cylinders. The analysis was based on a two-dimensional finite-difference scheme, and this numerical solution entails treatment of a large system of nonlinear algebraic equations.

5.6.4.1 The Effect of Square and Rectangular Cutouts

The effects of square and rectangular cutouts on the buckling of cylinders loaded by central axial compression force are summarized by the results which are bounded by the curves shown in Figure 5.23.

Test cylinders had shell radius to thickness of the wall ratios $r/t = 400$. The dashed curve in Figure 5.23 repre-

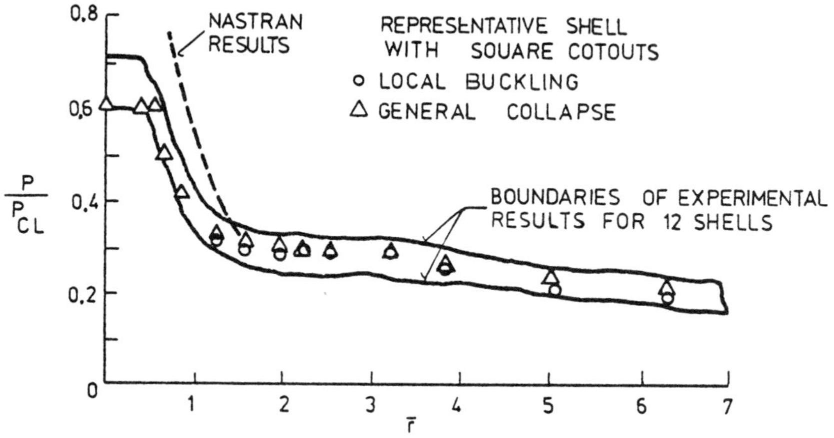

FIGURE 5-23 — The effect of square and rectangular cutouts on the buckling of circular cylinders loaded by central axial compression.

sents the results of a finite-element analysis. Most of the cylinders tested consisted of specimens with increasingly larger concentric square or rectangular cutouts in their sides. A limited number of cylinders had both square and rectangular cutouts with a common center.

The available prebuckling analyses are similar to that they all provide solutions that are dependent on a nondimensional geometric parameter

$$\bar{r} = \frac{r}{(Rt)^{\frac{1}{2}}} \qquad (5.123)$$

where

r = a characteristic hole dimension

R = the shell radius

t = the wall thickness

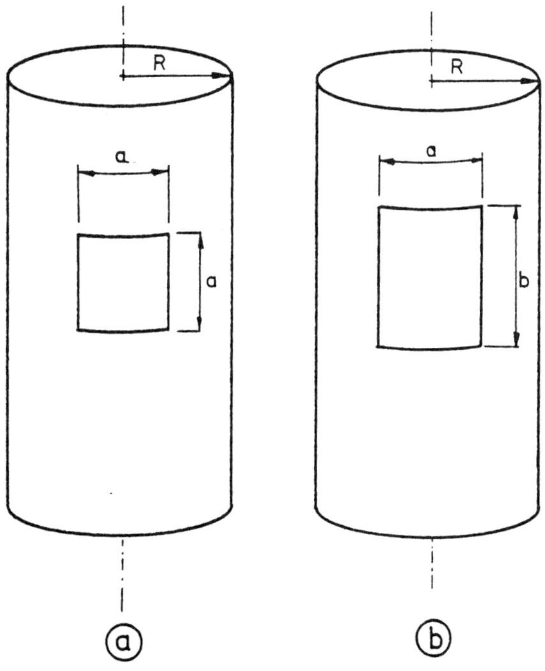

FIGURE 5.24 — Square and rectangular cuts in cylinders.

The characteristic hole dimension in the circulation parameter $\bar{r}$ is taken with reference to Figures 5.24 a and 5.24 b as follows:

The characteristic hole dimension as shown in Figure 5.24 a equal to one-half of the side length for the squares, or

$$\text{For } r = \frac{a}{2} \quad ; \quad \bar{r} = \frac{a}{2(Rt)^{\frac{1}{2}}} \qquad (5.124)$$

and for rectangular hole shown in Figure 5.24 b

$$\text{For } r = \frac{a+b}{4} \quad ; \quad \bar{r} = \frac{a+b}{4(Rt)^{\frac{1}{2}}} \qquad (5.125)$$

Even for such diversely different configurations as longitudinal and circumferential rectangles with aspect ratios equal to 2.0, all the experimental results for the cylinders with square and rectangular cutouts fill within the relatively narrow scatterband of Figure 3.1 when plotted with respect to r.

For values of $\bar{r}$ less than approximately 1.2, the buckling behavior of cylinders with square and rectangular cutouts was much the same as the behavior of cylinders with circular holes, Figure 5.25.

For values of $\bar{r}$ greater than about 1.2 the buckling loads continued to decrease with increasing $\bar{r}$, but at a much smaller rate of decline than that for the smaller values of $\bar{r}$.

For rectangular and square cutouts there was no transitional range as there was for circular cutouts, and the general collapse of the cylinders was always preceded by a stable local buckling made with an inward postbuckling deformation pattern.

For $\bar{r}$ between approximately 1.2 and 1.6, the stable local buckling mode always occurred in a symmetrical pattern approximately in the form of an ellipse with its semimajor axis tangent to the shell circumference. For $\bar{r}$ larger than about 1.6, the symmetrical stable local buckling mode was usually preceded by a nonsymmetrical mode aligned with one of the cutout diagonals which either snapped into its postbuckling form or just began growing inward as a larger local deformation.

For square and longitudinal rectangular cutouts with $\bar{r}$ greater than approximately 2.5, noticeable symmetric outward prebuckling deformations of the longitudinal

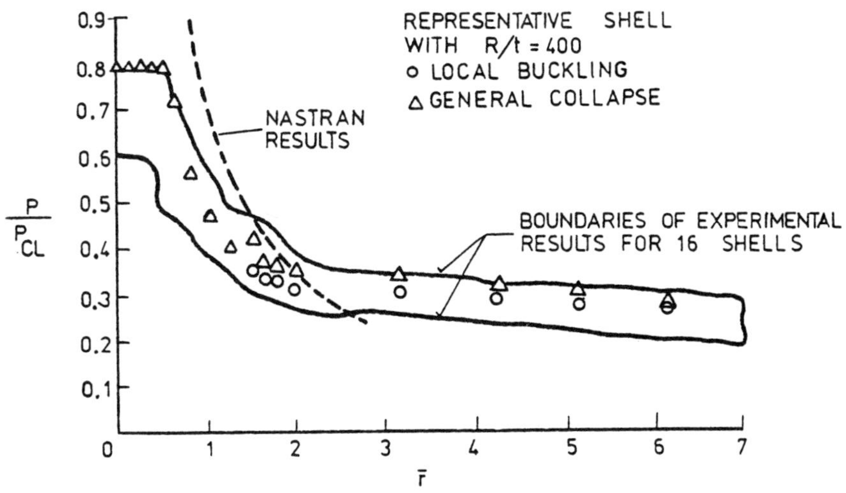

FIGURE 5.25 — The effect of circular cutouts on the buckling of circular cylinders loaded by central axial compression.

cutout sides preceded the above nonsymmetrical deformation.

For circumferential rectangles with $\bar{r}$ greater than about 3.2, noticeable symmetric outward prebuckling deformations of the circumferential cutout sides preceded the nonsymmetrical pattern. In all cases, the difference in the loads for the first noticeable local buckling and general collapse was small.

For $\bar{r}$ less than approximately 4.5, there was no detectable difference (beyond normal experimental scatter) between the general collapse loads for longitudinally and circumferentially oriented rectangles for the limited number of cylinders tested with each cutout configuration.

For $\bar{r}$ greater than 4.5, the general collapse loads of the circumferentially oriented rectangles tended to occupy the lower portion of the scatter bond in Figure 5.25. This phenomenon appears to be a result of the large prebuckling deformations and stable local buckling mode for the cylinder with circumferentially oriented rectangles which cause intensive stress redistribution to occur farther around the cylinder circumference.

Generally, for a given value of $\bar{r}$, the general collapse loads for cylinders with square and rectangular cutouts

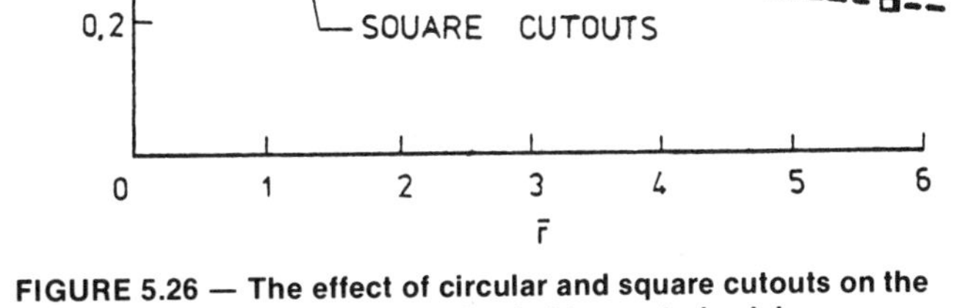

FIGURE 5.26 — The effect of circular and square cutouts on the buckling of a circular cylinder loaded by central axial compression.

were found to be slightly lower than those for cylinders with circular cutouts. As an example, the difference in the general collapse loads for circular and square cutouts in the same cylinder are shown in Figure 5.26.

5.7 Design of Stiffeners at Breech Opening

5.7.1 Design of Vertical Stiffeners

In order to avoid eccentric loading and a change in the direction of the stress, it is necessary to keep the neutral axis throughout the compensated cross-section on the centre line of the whole stack. It should be noted that it is not easy to obviate some of the eccentric loading on the compensating members. Such eccentric loading is offset in part by the fact that the line of action of the loading in the stack plates above and below the breech opening is often as far from the centre line of the stack, as the neutral axis of its compensating member.

The statical moment of the removed plate, where 2α is the angle in radians subtended at the stack center by the arc removed, Figure 5.27, is

$$S_x = \int_{-\alpha}^{+\alpha} (tR d\alpha)\, R\cos\alpha = 2tR^2\sin\alpha \qquad (5.126)$$

Using two vertical stiffeners, we may determine the required cross-sectional area of each stiffener by equalizing statical moments of the cross-sectional areas of the stiffeners $2A_S$ and the removed part of the stack, or

$$2A_s d = 2tR^2\sin\alpha \qquad (5.127)$$

from which

$$A_s = \frac{tR^2\sin\alpha}{d} \qquad (5.128)$$

where $\underline{d}$ is the distance between the centroid of cross-section of stiffener and axis x-x of the stack.

5.7.1.1 The Load Imposed on Each Stiffener by Steel Shell and Lining

$$G = \frac{\pi \alpha H}{360} (D_s t_s \rho_s + D_\ell t_\ell \rho_\ell) \quad (5.129)$$

where

G = total load on single stiffener

D_s, D_ℓ = inside diameter of the stack and lining, respectively

t_s, t_ℓ = the thickness of the wall and lining, respectively

ρ_s, ρ_ℓ = specific weights of the steel and lining materials, respectively

H = height of the stack above the breech opening

The axial compressive stress in each stiffener, under the weight of the wall and lining above the breech opening is

$$f_G = \frac{G}{A_s} = \frac{\pi d H}{t R^2 \sin\alpha}\left(\frac{\alpha}{360}\right)\left(D_s t_s \rho_s + D_\ell t_\ell \rho_\ell\right) \quad (5.130)$$

5.7.1.2 The Load Imposed on Each Stiffener Due to Wind Bending Moment

Compressive stress in the stack's wall due to the wind load moment is

$$\sigma_w = \frac{4M_w}{\pi D_m^2 t} \quad (5.131)$$

where

M_w = bending moment due to wind load

D_m = mean diameter of the stack at cross-section under consideration

Cross-sectional area of the wall of the stack is

$$\pi D_m t \quad (5.132)$$

Therefore, the approximate load imposed on each stiffener is

$$P_w = \frac{4M_w}{\pi D_m^2 t} \times \pi D_m t \times \frac{\alpha}{360} = \frac{M_w \alpha}{90\ D_m} \quad (5.133)$$

and the axial compressive stress in each stiffener under the wind pressure on the stack is

$$f_w = \frac{P_w}{A_s} = \frac{M_w \alpha}{90\ D_m} \times \frac{4d}{t D_m^2 \sin\alpha} = \frac{M_w d\alpha}{22.5\ t D_m^3 \sin\alpha} \quad (5.134)$$

5.7.1.3 The Eccentric Bending Moment Due to Eccentricity of Stiffener

By designating the distance between wall and centroid of the cross-section of the stiffener "e", the eccentric bending moment due to the loads $(G + P_w)$ is

$$M_e = (G + P_w)e \quad (5.135)$$

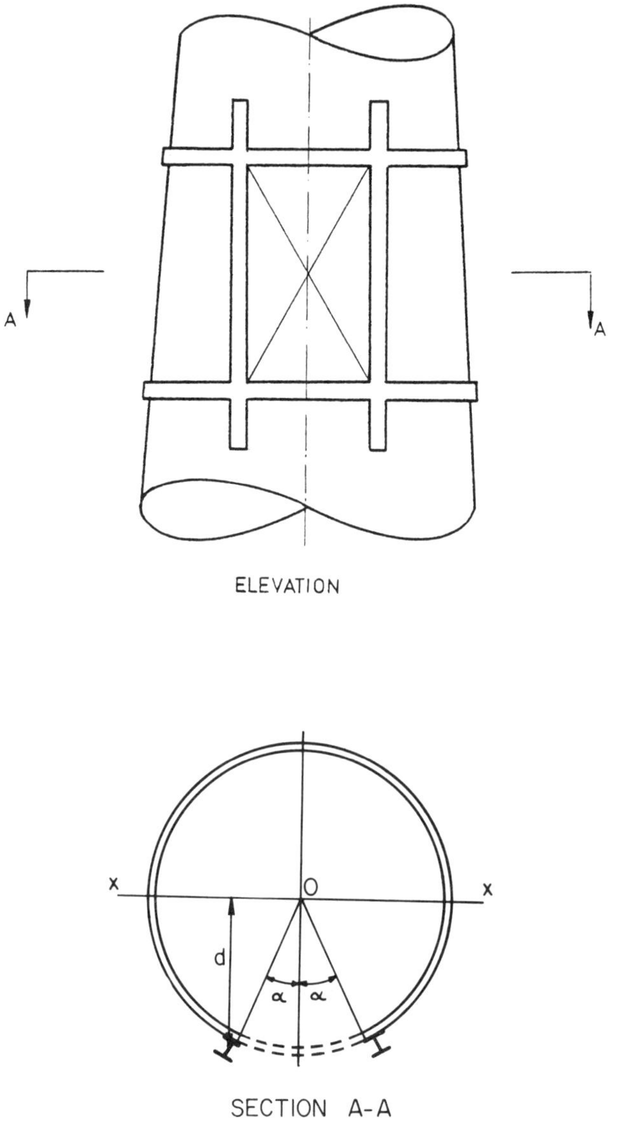

FIGURE 5.27 — Breech opening geometry.

and corresponding axial stress is

$$f_e = \frac{(G + P_w)e}{S_{stiff}} \quad (5.136)$$

where S_{stiff} is section modulus of the stiffener cross-section.

Total axial stress is

$$f_{tot} = f_G + f_w + f_e \quad (5.137)$$

5.7.1.4 The Effect of the Stiffened Breech Opening on the Strength of the Stack

To evaluate this effect it is necessary to determine the moment of inertia of the removed area of the arc with respect to the stack center line, as follows, Figure 5.28.

The moment of inertia of the area of the arc removed about a stack center line is

$$I_{rem} = t\int_{-\alpha}^{\alpha} R^3\cos^2\alpha d\alpha = R^3\left(\alpha + \frac{\sin 2\alpha}{2}\right)t \qquad (5.138)$$

The approximate moment of inertia of the stiffener cross-section with respect to the stack center line is

$$I_{stiff} = A_s d^2 \qquad (5.139)$$

where

A_S = cross-sectional area of the stiffener

d = distance between the center of gravity of the cross-section of stiffener and axis x-x.

Assuming

$$2I_{stiff} > I_{rem} \qquad (5.140)$$

then

$$2I_{stiff} - I_{rem} = 2A_s d^2 - I_{rem} = \Delta I \qquad (5.141)$$

where Δ I is an increase in the moment inertia of removed part of the cross-section of the stack.

The moment of inertia of the original cross-sectional area of the stack is I, therefore the compensated value is

$$I + \Delta I \qquad (5.142)$$

By designating the distance between the extreme edge of the stiffener and center of the stack as d_1, Figure 5.28, the section modulus of the cross-section of the stack is

$$S_{stiff} = \frac{I + \Delta I}{d_1} \qquad (5.143)$$

Assuming that

$$S_{orig} > S_{stiff} \qquad (5.144)$$

Therefore, the reduction in section modulus of stiffened cross-section of the stack in % is

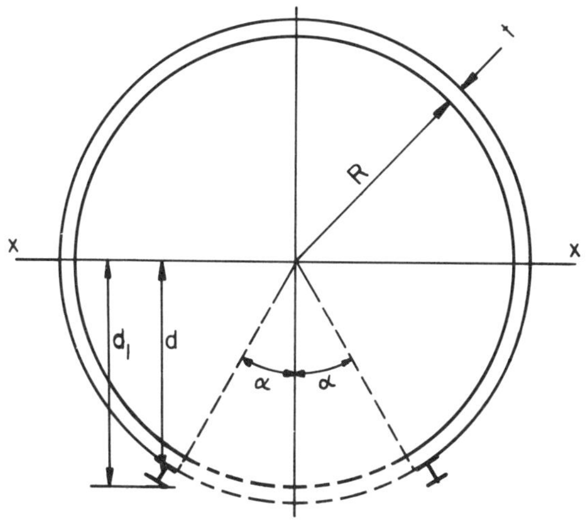

FIGURE 5.28 — Breech opening and stiffeners geometry.

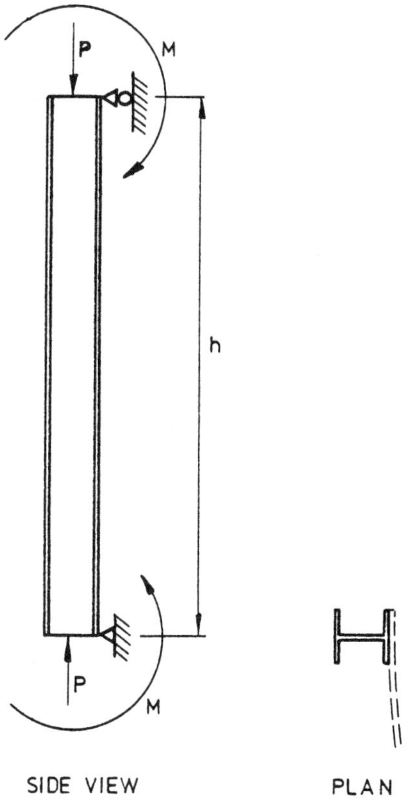

FIGURE 5.29 — Vertical stiffener as beam-column.

$$\frac{S_{stiff}}{S_{orig}} \times 100\% \qquad (5.145)$$

It should be remembered, however, that the stress in the extreme edge of the stiffener is

$$f_{stiff} = \frac{\Sigma M}{S_{stiff}} \qquad (5.146)$$

whereas the stress in the stack plates themselves is actually lessened in the ratio

$$\frac{I}{I + \Delta I} \qquad (5.147)$$

or, the original moment of inertia divided by the new moment of inertia. Thus, the original section modulus is not seriously weakened, and the plate stress is reduced.

5.7.1.5 Buckling Stability of Vertical Stiffeners

Vertical stiffeners under action of axial load and bending moments due to the eccentricity of axial loads at both ends, may be considered as beam-columns. To achieve the greater factor of safety, the stiffeners will be considered as isolated beam-columns, neglecting the plating of the wall to which the stiffeners are connected. Both ends of the stiffeners are assumed as hinges, Figure 5.29.

The maximum stress at mid length of the column is given by the secant formula [5.58]

$$\sigma_{max} = \frac{P}{A} + \frac{M_o c}{I} \sec\frac{h}{2}\sqrt{\frac{P}{EI}} \qquad (5.148)$$

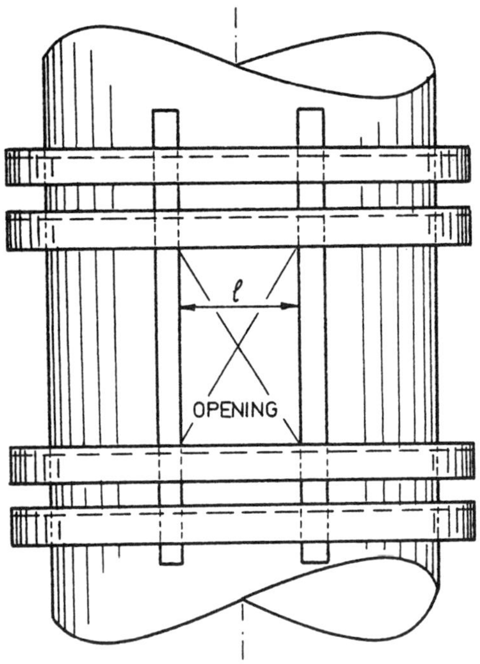

FIGURE 5.30 — Ring girders.

5.7.2 Ring Stiffeners

In addition to the compensation members which are placed as vertical stiffeners, horizontal reinforcement should be provided by using ring girders above and below the breech opening. These ring stiffeners should be designed as beams fixed to the wall of the stack to carry the unsupported parts of the stack above and below breech opening, Figure 5.30.

The span in bending is the width between the vertical stiffeners, but the girder should encircle the stack to preserve circularity at the opening.

To form each ring girder, steel plates should be placed to act as top and bottom flanges. The shell of the stack can serve as a web.

Around the stack, each ring girder must be capable of carrying a uniformly distributed load lb/in of arc due to axial vertical load

$$w_1 = \frac{G}{2\pi R} \qquad (5.149)$$

and due to the wind bending moment, when the stress is

$$f = \frac{M_w}{\pi R^2 t} \qquad (5.150)$$

Therefore, the corresponding vertical load is

$$ft = w_2 = \frac{M_w}{\pi R^2} \qquad (5.151)$$

The total distributed load in lb/in

$$w = w_1 + w_2 = \frac{G}{2\pi R} + \frac{M_w}{\pi R^2} \qquad (5.152)$$

The bending moment in the girder is

$$M = \frac{w\ell^2}{12} = \frac{\ell^2}{12\pi R}\left(\frac{G}{2} + \frac{M_w}{R}\right) \qquad (5.153)$$

5.8 Rigorous Breech Opening Analysis

Ever since the advent of plate structures, the structural arrangements in the vicinity of openings in them have been a matter of considerable concern.

When analyzing and designing such structures, various factors have to be taken into consideration. One of these is the "stress concentration" around openings, recesses and cutouts in plane and three dimensional parts of structures.

The initial theoretical work on this subject was performed by Inglis [5.59] who approximate the effect of a rectangular opening with rounded corners by a pair of ellipses intersecting obliquely. Savin [5.60] studied both the square and the rectangle openings with rounded corners. Greenspan [5.61] working independently and by another method, produced a solution for the square with rounded corners. None of these analytical approaches is applicable to a wide variety of rectangular openings. The first step to rectify this situation was taken by Brock [5.62] who presented a solution for the entire family of rectangles with rounded corners.

An outstanding theoretical study of stresses around rectangular openings in cylindrical shells appears by Muskhelishvili [5.63]. He makes use of the complex-variable method in conjunction with conformal mapping technique in investigating such a problem. His solution is based on the assumptions of plane elasticity: homogeneous, isotropic material within the elastic limit, uniform stress across the thickness of the shell with no stress normal to it, an opening "small" relative to the shell and "small" displacements.

An accurate analysis of the stresses around openings in stacks until recently has been beyond the state of the art in shell analysis. The large number of parameters involved makes in impossible to produce design charts for designing reinforcement around openings by use of a purely empirical approach and a theoretical analysis has been prohibited by the lengthy and complicated mathematics involved. Consequently, design of opening reinforcement has been based on rules of thumb which are generally quite conservative due to the uncertainty involved. However, recent advances in the Finite Element Method of Analysis coupled with improvements in computer technology and numerical analysis methods have brought the state of the art to a level where it now appears feasible to establish design procedures with a more solid foundation [5.64, 5.65, 5.66, 5.67, 5.68].

5.8.1 The Finite Element Program

The invention of digital computers has significantly improved the output of engineering profession. Manual methods were the only means of performing engineering calculations up to the end of the Second World War. They were not practical because of the enormous calculation work involved in solving a large number of simlutaneous equations encountered in analyzing any reasonably large structure. To overcome this difficulty, engineers resorted to comparatively easier but approximate methods, e.g. relaxation and successive approximation methods. These methods, though still in use at present, have been virtually replaced by computerized stiffness solution methods —

incorporating the finite element technique and matrix methods for structural analysis [5.69].

Computers seem to be ideally suited for modern structural analysis problems because of their versatility and tremendous speed resulting in substantial savings in time and man-hours required for a particular project. In addition, computer programs using finite elements have the following distinct advantages:

1. Their ability to use elements of various types, sizes and shapes and to model a structure of arbitrary geometry.
2. Their ability to accommodate arbitrary support conditions and arbitrary loading, including thermal loading.
3. Their ability to model composite structures involving different structural components such as stiffening members and a shell and combination of plates, bars and solids, etc.
4. The finite element structure closely resembles the actual structure instead of being quite a different abstraction that is hard to visualize.

Invariably, the finite element method of analysis entails a few disadvantages as well, as shown below:

1. A specific numerical result is obtained for a specific problem. A general close form solution, which would permit one to examine system response to changes in various parameters, is not produced.
2. Experience and judgement are needed in order to construct a good finite element model.
3. A large computer and a reliable computer program are essential.
4. Input and output data may be large and tedious to prepare and interpret.

5.8.2 Method of Finite Element Analysis

Finite element analysis has come about through the use of digital computers solving with matrix algebra methods the many equations of compatibility and equilibrium created by classical solution techniques such as slope deflection. Generally, most finite element analysis programs utilize displacement methods of solution, resolving deflections first, followed by a stressing routine, which solves for member forces (stresses) from the deflections — i.e., a stiffness solution method [5.70, 5.71].

What must be appreciated, however, is the fact that all finite element models are at best approximations of the actual structures they may represent. Firstly, the applicability of finite element analysis must be determined with respect to a problem in hand. If it is possible to model an actual structure under consideration, it then becomes necessary to carefully choose a proper element and grid pattern such that the errors, that incur when the actual structural continuum is replaced by the finite element model, may be minimized. Furthermore, though must be used to establish the appropriate constraints that must be applied about the model so that the model and actual structure tend to behave similarly.

Over the past years, as finite element technology grew, its popularity also increased. Subsequently, at the moment, good evidence exists verifying the viability of the finite element analysis method as quite reliable when applied properly.

Generally speaking the steel stack analyzed in this report is well ordered and no a complex problem to solve. It is anticipated that the results of the analysis presented in Appendix B, are very reflective of the forces and deflections that would exist if the modelled structure and the applied loads were in actual existence.

5.8.3 ANSYS Computer Program

ANSYS is a large scale general purpose computer program for the solution of several classes of engineering analysis problems. Its analyzing capabilities include static and dynamic; plastic, creep and swelling; small and large deflections; steady state and transient heat transfer and steady state fluid flow.

The matrix displacement method of analysis based upon finite element idealization is employed throughout the program. The library of finite elements available numbers more than forty for static dynamic analyses and ten for heat transfer analyses. This variety of elements gives the ANSYS program the capability of analyzing frame structures (two dimensional frames, grids and three dimensional frames), piping systems, two dimensional plane and axisymmetric solids, flat plates, three dimensional solids, axisymmetric and three dimensional shells and nonlinear problems including interfaces and cables.

Loading on the structure may be forces, displacements, pressures, temperatures or response spectra. Loading may be arbitrary time functions for linear and nonlinear dynamic functions for linear and nonlinear dynamic analyses. Loadings for heat transfer analyses include internal heat generation, convection and radiation boundaries, and specified temperatures or heat flows [5.72].

The ANSYS program uses the wave front (or "frontal") direct solution method for the system of simultaneous linear equations developed by the matrix displacement method, and gives results of high accuracy in a minimum of computer time. The program has the capability of solving large structures. There is no limit on the number of elements used in a problem. The number of nodes can be in excess of 2500 for three dimensional problems, and 5000 for two dimensional problems. There is no "band width" limitation in the problem definition, however, there is a "wave front" restriction. The "wave front" restriction depends on the amount of core storage available for a given problem. Up to 576 degrees of freedom on the wave front limitation tends to be restrictive only for analysis of arbitrary three dimensional solids or in the use of ANSYS on a small computer.

ANSYS has the capability of generating substructures (or super-elements). These substructures may be stored in a library file for use in other analyses. Substructuring portions of a model can result in considerable computer time savings for nonlinear analyses.

Geometry plotting is available for all elements in the ANSYS library, including isometric, perspective and section views of three dimensional structures. Plotting subroutines are also available for the plotting of stresses and displacements from two and three dimensional solid or

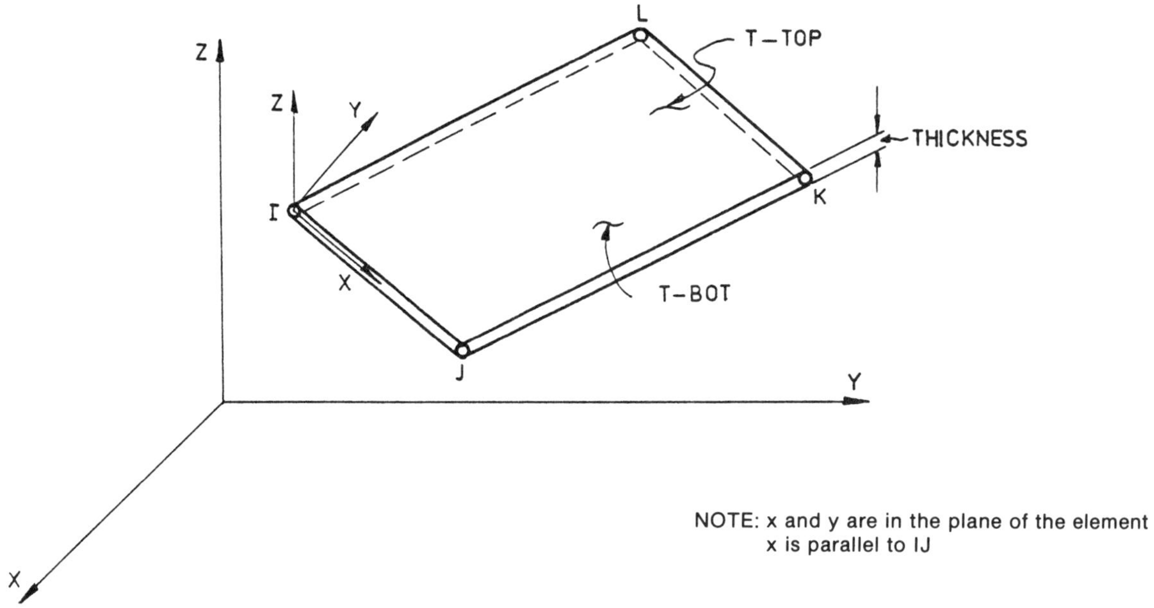

FIGURE 5.31 — Flat quadrilateral shell element.

shell analyses, node shapes from dynamic analyses, distorted geometries from static analyses, transient forces and displacements vs. time curves from transient dynamic analyses, and stress strain plots from plastic and creep analyses.

Post processing routines are available for algebraic modification, differentiation, and integration of calculated results. Root mean square operations may be performed on seismic modal results. Results from various loading nodes may be combined for harmonically loaded axisymmetric structures.

Options for multiple coordinate system in cartesian, cylindrical, or spherical coordinates are available, as well as multiple region generation capabilities to minimize the input data for repeating regions.

Sophisticated geometry generation capabilities are included for two dimensional plane and axisymmetric structures and for intersecting three dimensional shell structures [5.73, 5.74].

The quadrilateral shell element has bending and membrane capabilities. Both in-plane and normal loads are permitted. The element has six degrees of freedom at each node: translation in the nodal x, y, and z directions and rotations about the nodal x, y, and z axes, Figures 5.31 and 5.32.

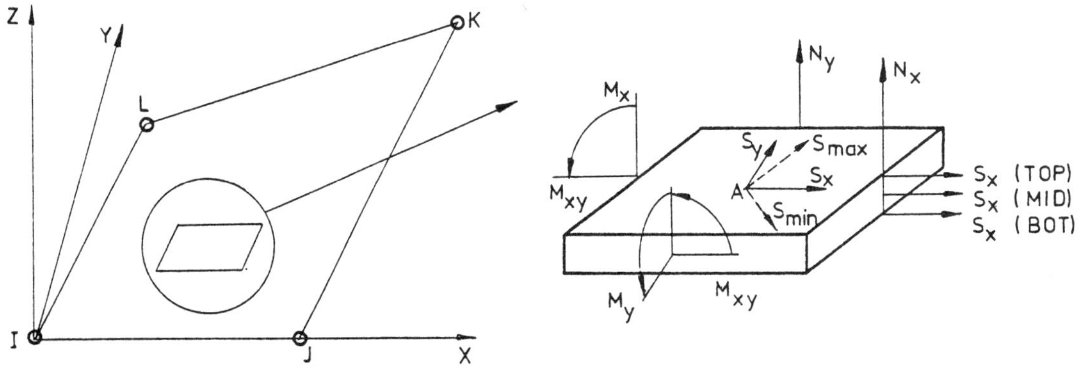

FIGURE 5.32 — Flat quadrilateral shell element output.

NUMERICAL EXAMPLE NO. 1
Approximate Method

Determine stresses and design stiffeners at breech opening of self-supported steel stack to prevent buckling under vertical and wind loading, Figure 5.33.

Assumed

Location: Toronto, Ontario

Material: C.40.21 50 A (equivalent ASTM Steel)
This is weathering steel and although more expensive than 44W, a saving will be made due to lower maintenance costs.

Height Required: 200 ft.

Minimum Discharge Diameter: 10 ft.

Corrosion Allowance: 1/16 inch

Solution

1. Part 1. Determination of Wind Loading

From NBC, Supplement No. 1, 1975, p. 43, the hourly wind pressure for Toronto is 9.9 psf at 1 in 30.

Check $\frac{L}{D}$ ratio: $\frac{L}{D} = \frac{200}{13} = 15.4$

where $D = \frac{D_t + D_b}{2} = \frac{10 + 16}{2} = 13$ ft.

Since $\frac{L}{D} < 100$, Figure B-11, p.77 must be considered.

From Figure B-11, for L/D = 15.4, Cn = 0.65
According to NBC, p. 157

The gust effect factor for structured members is $C_g = 2.0$.

The exposure factor C_e from NBC, Supplement No. 4, p. 56, is

$$C_e = 0.6\left(\frac{H}{60}\right)^{\frac{1}{2}}, \quad C_e > 0.5$$

Total force, Fn, NBC Supplement 4, p. 77 is

$$F_n = C_n q\, C_g\, C_e\, A_c$$

C_e and A_c are functions of H_x, i.e., the height of the stack at a point x from the top as shown in Figure 5.33. Therefore,

$$F_n = 0.65\ (9.9)(0.6)\left[\frac{H_x}{60}\right]^{\frac{1}{2}}(2.0)\left(16 - \frac{6H_x}{200}\right)$$

a. Wind pressure from 0 to 40 ft (exactly 0 to 41′-8″). The exposure factor $C_e = 0.50$

$$F_n = 0.65 \times 9.9 \times 0.5 \times 2.0\left(16 - \frac{6H_x}{200}\right) =$$

$$= 103 - 0.193\ H_x$$

b. Wind pressure from 40 to 200 ft.

$$F_n = 0.65 \times 9.9 \times 0.6\left(\frac{H_x}{60}\right)^{\frac{1}{2}} 2.0\left(16 = \frac{6H_x}{200}\right) =$$

$$= (15.95 - 0.03\ H_x)\sqrt{H_x}$$

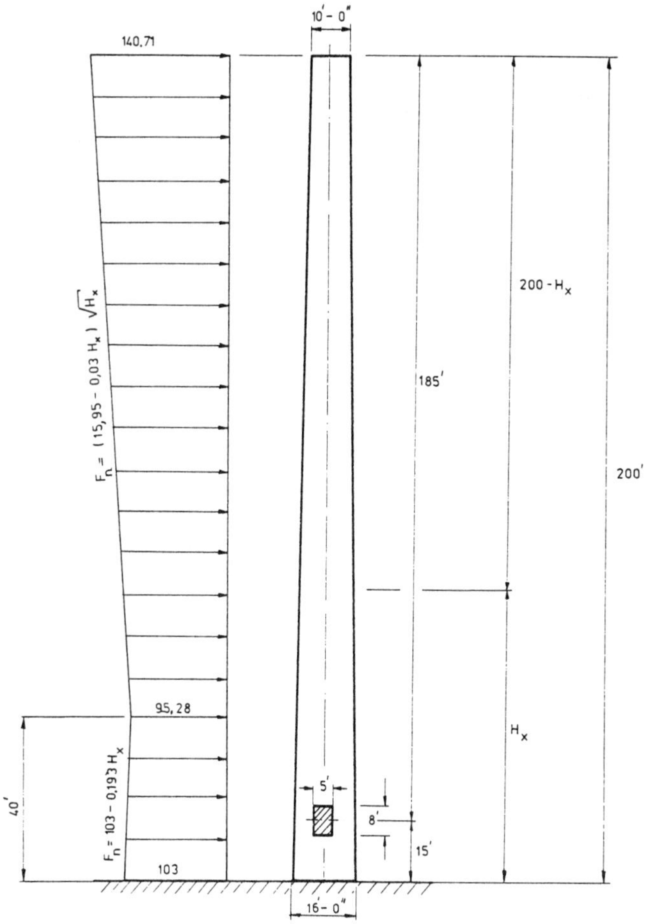

FIGURE 5.33 — Distribution of wind load acting on the stack.

For

$H_x = 0$	$F_n = 103$
$H_x = 40$	$F_n = 95.28$
$H_x = 200$	$F_n = 140.71$

2. Part 2. Bending Moments

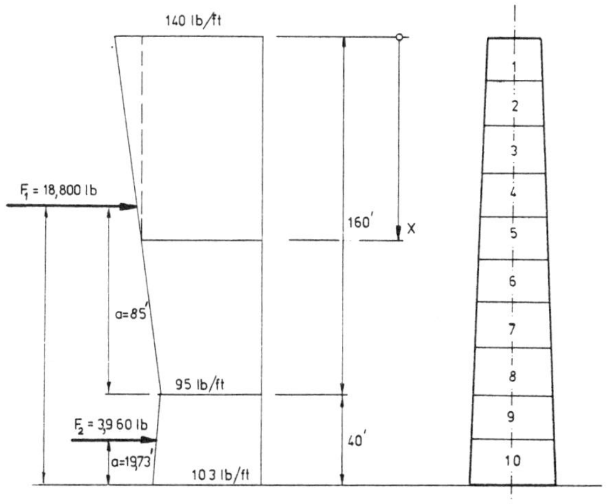

FIGURE 5.34 — Determination of bending moments in stack under wind loading.

Mx for uniform loading if x < 160′-0″

$$M_{xu} = \left[\frac{45(160 - x)}{160} + 95\right]\frac{x^2}{2} =$$
$$= (70 - 0.1406x)x^2$$

Mx for triangular loading if x > 160′-0″

$$Mxt = 0.0938x^3$$

Total moment when x < 160′-0″

$$M_x = (70 - 0.1406x)x^2 + 0.0938x^3 =$$
$$= 70x^2 - 0.0468x^3$$

For the bottom part x > 160′-0″ the exact expression is

$$M_x = \frac{(0.1x^2 + 63x - 12640)(x - 160)(0.2x + 253)}{3\ (0.2\ x + 158)}$$

However, for the bottom 40 ft. the wind loading can be considered to be uniform and set equal to 95 lb/ft, or

$$M_x = \frac{95(x - 160)^2}{2} = 47.5\ (x - 160)^2$$

Moment at the base

$$F_1 = \frac{1}{2}\ (95 + 140)\ 160 - 18800\ \text{lbs}$$

$$a = \frac{160\ (2 \times 140 + 95)}{3(140 + 95)} = \frac{160 \times 375}{3 \times 235}\quad 85\ \text{ft}$$

$$M_{F_1} = 18800 \times (85 + 40) = 2{,}250{,}000\ \text{lb/ft} =$$
$$= 28{,}200{,}000\ \text{lb/in}$$

$$F_2 = \frac{1}{2}\ (95 + 103)\ 40 = 3960\ \text{lbs}$$

$$a_1 = 40 - \frac{40\ (2 \times 103 + 95)}{3(103 + 95)} = 19.73\ \text{ft}$$

$$M_{F_2} = 3960 \times 19.73 = 78130.8\ \text{lb/ft} =$$
$$= 937{,}570\ \text{lb/in}$$

$$M_{tot} = 28{,}200{,}000 + 937{,}570 =$$
$$= 29{,}137{,}570\ \text{lb/in}$$

3. Part 3. Determination of Stack's Wall Thickness

In order to determine the required thickness of stack's wall, the moment due to wind should be determined. The critical buckling stress can be determined from equations:

$$f_s = \frac{W_s}{\pi D_i\ (t-c)}$$

$$f_w = \frac{4M_w}{\pi D_1^{\ 2}(t-c)}$$

Combining both of the above equations and substituting $\sigma_{cr} = f_s + f_w$ yields

$$\sigma_{cr} = \frac{W_s}{\pi D_i(t-c)} + \frac{4M_w}{\pi D_i^2\ (t-c)}$$

setting t′ = t-c, the equation can be rearranged so that t can be optimized. Therefore,

$$t' = \frac{W_s}{\pi\sigma_{cr}D_i} + \frac{4M_w}{\pi\sigma_{cr}D_i^{\ 2}}$$

Assume t′ = 0.50 inch at base

$$\text{Therefore}\quad \frac{D}{t'} = \frac{16 \times 12}{0.5} = 384$$

Since $\frac{D}{t}$ < 400, the Wilson-Newmark equation cannot be used, therefore, the AISI formula applies.

$$\text{Since}\quad \frac{13000}{F_y} = 260 < \frac{D}{t'},$$

$$\sigma_{cr} = \frac{5333}{D/t'} = 13.9\ \text{ksi}$$

To determine the approximate own weight of the stack, assume upper 100 ft thickness of the wall ¼ in. and bottom 100 ft. wall having ½ in. wall thickness

$$W_s = \frac{\pi D \times t \times 490 \times 100}{12}$$

$$W_s = \frac{49000 \times 3.14}{12}\left(\frac{10 + 13}{2} \times \frac{1}{4} + \right.$$
$$\left. + \frac{13 + 16}{2} \times \frac{1}{2}\right) =$$

$$= 12821.66\ (2.875 + 7.25) = 129819.31 \simeq$$
$$\simeq 130{,}000\ \text{lbs}$$

And

$$t_1 = \frac{130{,}000}{13900(16 \times 12)\pi} \times \frac{4 \times 29{,}137{,}570}{13{,}900(16 \times 12)^2\pi} =$$
$$= 0.0155 + 0.0724 = 0.0879\ \text{in}$$

From this preliminary check it can readily be seen that the stress at the bottom of the stack is not critical.

In addition, the deflection of the stack must be considered when determining the required thickness of the stack. According to the British Standard for the design of Steel Chimneys, the maximum allowable deflection at the top of steel stack is

$$\Delta = \ell/200$$

The maximum permissible deflection at the top of this stack should therefore be

$$\Delta = 200/200 = 1\ \text{ft}$$

The actual deflection will be found using the method of virtual work, or

$$1\ \text{lb} \times \Delta = \int_0^L \frac{m\ Mdx}{EI}$$

or

$$\Delta_{tot} = \sum \Delta_i = \sum_{i=1}^{10} \frac{m_i M_i \Delta x}{E I_i}$$

where

m = moment at x due to 1 lb load applied at top of stack

M_i = moment on segment i due to wind loading

I_i = $\pi D_i^3 t/8$ = moment of inertia of segment i at its centroid, which for this example will be considered at ½ the height of the segment

Maximum allowable deflection at top of stack is 12 inches.
Calculated deflection at top is

$$10.794 < 12 \text{ inches}$$

therefore actual deflection is less than allowable.

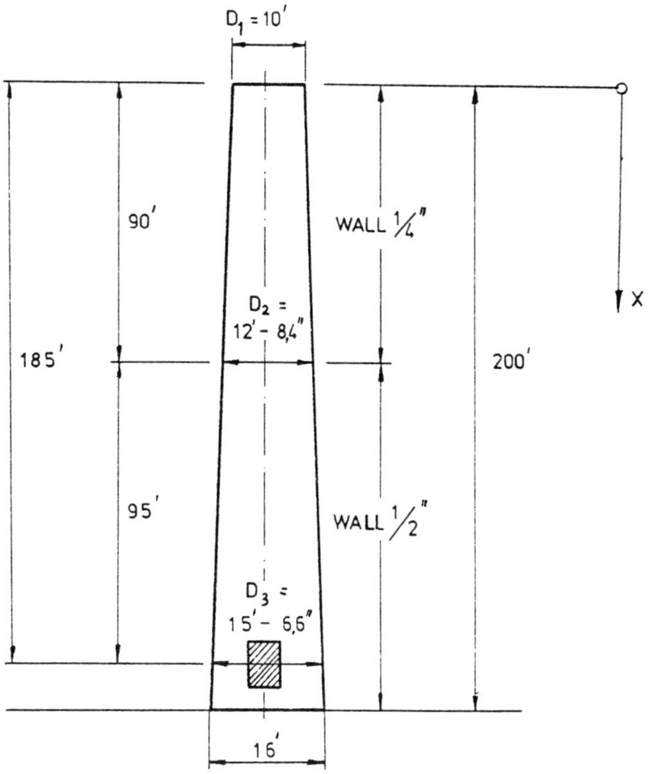

FIGURE 5.35 — Stack's wall dimensions.

4. Part 4. Determination of the Own Weight of Stack and Wind Bending Moment

a. Own Weight of the Stack above Breach Opening

Diameter D_2 = 120 + 0.36x = 120 + 0.36 × 90 = 152.4″ = = 12′-8.4″

Diameter D_3 = 120 + 0.36 × 185 = 186.6″ = 15′-6.6″

$$P_1 = \frac{\pi \times 490 \times 90}{12} \left(\frac{D_1 + D_2}{2} \times t\right) =$$

$$= \frac{3.14 \times 490 \times 90}{12} \left(\frac{10 + 12.7}{2} \times \frac{1}{4}\right) =$$

$$= 23{,}743 \text{ lbs}$$

$$P_2 = \frac{\pi \times 490 \times 90}{12} \left(\frac{12.7 + 15.55}{2} \times \frac{1}{2}\right) =$$

$$= 81{,}498 \text{ lbs}$$

Total P = $P_1 + P_2$ = 114,768 lbs

TABLE 5.6 — Deflections of Stack.

SEC No.	Diameter (in)	x (ft)	m (in-lb)	Mi (in-lb)	m Mix / E	t′ (in)	I =πd³t′/8 (in)	Δtop (in)
(1)	(2)	(3)	(4)	(5)	(6)	(7)	(8)	(9)
1	123.6	10	120	83,438	4	¼	185,375	0.015
2	130.8	30	360	740,832	3,200	¼	219,700	0.015
3	138.0	50	600	2,029,800	24,358	¼	258,000	0.094
4	145.2	70	840	3,923,400	92,278	¼	300,540	0.307
5	152.4	90	1080	6,394,680	248,625	¼	347,500	0.715
6	159.6	110	1320	9,416,520	546,911	½	792,200	0.685
7	166.8	130	1560	12,962,160	1,051,490	½	911,200	1.154
8	174.0	150	1800	17,004,600	1,836,496	½	1,034,400	1.775
9	181.2	170	2040	21,516,840	2,984,816	½	1,178,160	2.555
10	188.4	190	2280	26,472,000	4,587,068	½	1,313,000	3.494
d = 12 + 0.36x, in. where x(ft)					Total deflection = Δtop = 10.794			

b. Determination of the Bending Moment under Wind Load at x = 185 ft., Fig. 5.34

$M_1 = 18{,}800 \times (85 + 25) \times 12 = 24{,}816{,}000$ lb/in

$M_2 = \frac{95}{2} \times (185 - 160)^2 \times 12 = 356{,}250$ lb/in

Total $M_W = 24{,}816{,}000 + 356{,}250 = 25{,}172{,}250$ lb/in

5. Part 5. Determination of Cross-Section Statical Properties

a. Statical Moment of the Compressed Area; Figure 5.36
Assume the position of the neutral axis as shown in Figure 5.36, y_C = in.

According to formula (5.96)

$$S_C = \frac{4}{3}(r^3 - r_1^3)\sin\gamma_1 \sin\gamma_2 + 2r(r^2 - r_1^2)\gamma_1 \sin\beta$$

where $\alpha = 18^\circ 50'$

$$\sin\beta = \frac{36}{93} = 0.3871 \quad ; \quad \beta = 22^\circ 50'$$

$$\gamma_1 = \frac{90 - \alpha + \beta}{2} = \frac{94}{2} = 47^\circ \; ;$$

$$\gamma_1(\text{rad}) = \frac{47}{57.296} = 0.8203$$

$$\gamma_2 = \frac{90 - \alpha - \beta}{2} = \frac{48^\circ 20'}{2} = 24^\circ 10'$$

$$\sin {}_1 = \sin 47^\circ = 0.7314$$

$$\sin {}_2 = \sin 24^\circ 10' = 0.4094$$

$$S_C = \frac{4}{3}(93^3 - 92.5^3) \times 0.7314 \times 0.4094 + 2 \times 93 (93^2 - 92.5^2) \times 0.8203 \times 0.3871 = \frac{4}{3} \times 12904 \times 0.2994 + 186 \times 92.75 \times 0.3175 = 5138 + 5478 = 10{,}616 \text{ in}^3$$

b. Statical Moment of the Tensile Area, Figure 5.36
According to formula (5.103)

$$S_T = \frac{2}{3}(r^3 - r_1^3)\cos^3\beta - \frac{r}{2}(r^2 - r_1^2)\,[2(90-\beta) - \sin 2\beta]\sin\beta$$

where

$$\frac{2(90-\beta)}{57.296} = \frac{2(90-22^\circ 50')}{57.296} = \frac{134.333}{57.296} = 2.3445$$

$$\sin 2\beta = \sin 45^\circ 40' = 0.7153$$

$$\cos\beta = \cos 22^\circ 50' = 0.9216$$

$$S_T = \frac{2}{3}(93^3 - 92.5^3)\,0.9216^3 - \frac{93}{2}(93^2 - 92.5^2) \times [2.3445 - 0.7153] \times 0.3871 = \frac{2}{3} \times 12904 \times 0.7828 - 46.5 \times 9275 \times 1.6293 \times 0.3871 = 6735 - 2720 = 4{,}015 \text{ in}^3$$

c. Moment of Inertia of the Compressed Area According to formula (5.112)

$$I_C = \frac{(r^4 - r_1^4)}{4}\left[2\gamma_1 - \sin(\alpha+\beta)\cos(\alpha-\beta)\right] + 2r(r^3 - r_1^3)\gamma_1 \sin\beta\left(\frac{4\sin\gamma_1 \sin\gamma_2}{3\gamma_1} + \sin\beta\right)$$

where

$$\sin(\alpha+\beta) = \sin(18^\circ 50' + 22^\circ 50') = \sin 41^\circ 40' = 0.6648$$

$$\cos(\alpha-\beta) = \cos(18^\circ 50' = 22^\circ 50') = \cos 4^C = 0.9976$$

$$I_C = \frac{(93^4 - 92.5^4)}{4}[2 \times 0.8203 - 0.6648 \times 0.9976] + 2 \times 93 (93^3 - 92.5^3) \times 0.8203 \times 0.3871 \times \left(\frac{4 \times 0.7314 \times 0.4094}{3 \times 0.8203} + 0.3871\right) =$$

$$= \frac{1595787}{4}[1.6406 - 0.6632] + 186 \times 12904 \times 0.3175 \times (0.4867 + 0.3871) = 389{,}931 + 665{,}876 = 1{,}055{,}816 \text{ in}^4$$

d. Moment of Inertia of the Tensile Area According to formula (5.122)

$$I_C = (r^4 - r_1^4)\left\{\frac{1}{8}\left[2(90-\beta) - \sin 2\beta + 4\cos^3\beta \sin\beta\right] + \frac{8}{9}\,\frac{\cos^6\beta}{2(90-\beta) - \sin 2\beta}\right\} - \frac{4}{3} r (r^3 - r_1^3)\cos^3\beta \sin\beta + \frac{r^2(r^2 - r_1^2)}{2}\left[2(90-\beta) - \sin 2\beta\right]\sin^2\beta$$

where

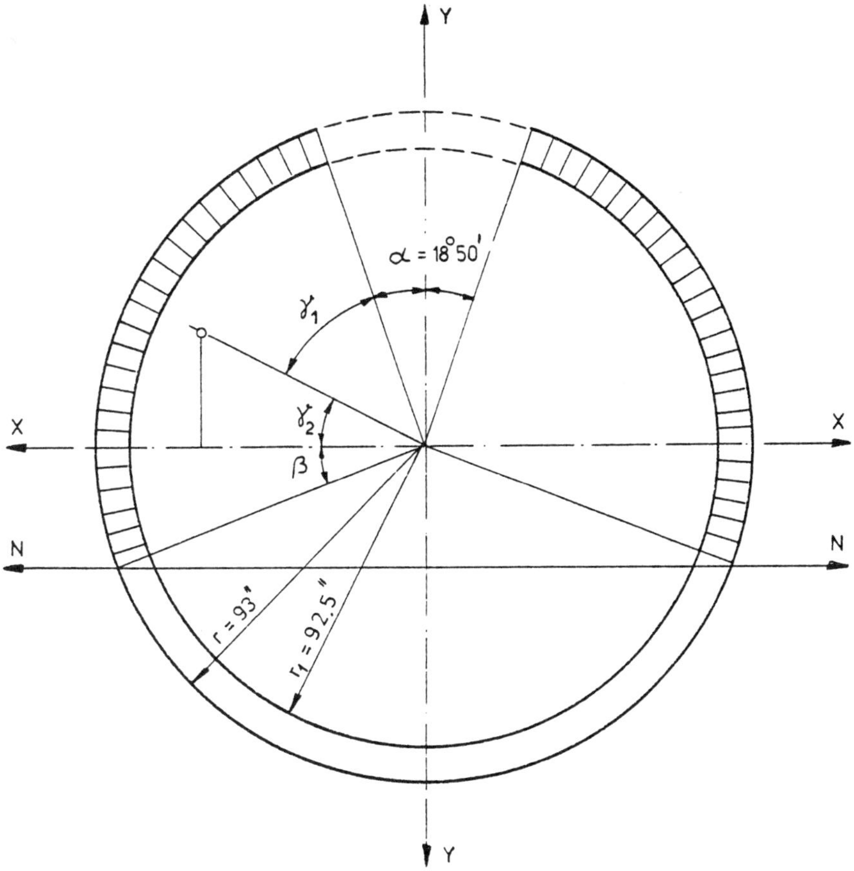

FIGURE 5.36 — Geometric data for statical moments of compressed and tensile areas.

$$\frac{2(90-\beta)}{57.296} = \frac{2(90 - 22^{\circ}50')}{57.296} = \frac{134.333}{57.296} = 2.3445$$

$$\sin 2\beta = \sin 45^{\circ}40' = 0.7153$$

$$\cos^3\beta = \cos^3 22^{\circ}50' = 0.9216^3 = 0.7828$$

$$\sin \beta = \sin 22^{\circ}50' = 0.3871$$

$$\cos^6\beta = \cos^6 22^{\circ}50' = 0.9216^6 = 0.6127$$

$$\sin^2\beta = \sin^2 22^{\circ}50 = 0.3871^2 = 0.1498$$

$$r(r^3 - r_1^{\ 3}) = 93(93^3 - 92.5^3) = 93(804357 - 791453) = 93 \times 12904 = 1,200,072$$

$$r^2(r^2 - r_1^{\ 2}) = 93^2(93^2 - 92.5^2) = 8649\ (8649 - 8556.25) = 8649 \times 92.75 = 802195$$

$$I_T = 1595787 \left\{ \frac{1}{8} \left[2.3445 - 0.7153 + 4 \times 0.7828 \times 0.3871 \right] + \frac{8}{9} \frac{0.6127}{2.3445 - 0.7153} \right\} -$$

$$- \frac{4}{3} \times 1200072 \times 0.7828 \times 0.3871 +$$

$$+ \frac{802195}{2} \left[2.3445 - 0.7153 \right] 0.1498 =$$

$$= 1595787 \left\{ \frac{1}{8} \left[1.6292 + 1.2121 \right] + 0.3345 \right\} -$$

$$- 483,652 + 97890 = 1595787 \times 0.6895 -$$

$$- 386974 = 1,100,235 - 386974 =$$

$$= 713,261 \text{ in}^4$$

6. Part 6. Checking of the Value of Axial Load N. According to formula (5.82) the value of the coefficient ϕ is

$$\phi = \frac{M_W + Px_o}{I_c + I_T} = \frac{25,172,250 + 119,000 \times 36}{1,055,855 + 713,261} =$$

$$= \frac{29,456,250}{1,679,116} = 16.65 \text{ lb/in}^3$$

From formula (5.83) it follows

$$N = \phi\,(S_c - S_T) = 16.65\ (10{,}616 - 4{,}015) =$$

$$= 16.65 \times 6{,}614 = 109{,}907 \text{ lb}$$

$$109{,}907 < 114{,}768$$

Difference is 4.42%

7. Part 7. Determination of Stresses

The average compressive stress in the wall is estimated after formula (5.88)

$$f_{av} = \frac{\phi S_c}{t} \times \frac{180}{\pi(r+r_1)\ (90-\alpha+\beta)}$$

where

$$90 - \alpha + \beta = 90 - 18^o50' + 22^o50' = 94^o$$

$$f_{av} = \frac{16.65 \times 10.616}{0.5} \times \frac{180}{3.14\ (93 + 92.50)\ 94} =$$

$$= 35{,}3513 \times \frac{180}{54{,}752} = 1162 \text{ psi}$$

The maximum stress at the middle of the wall after formula (5.91)

$$f_{max} = \frac{\phi}{2}\left[2x_c + (r + r_1)\cos\alpha\right]$$

where $x_c = 36$ in

$$\cos\alpha = \cos 18^o50' = 0.9465$$

$$f_{max} = \frac{16.65}{2}\left[2 \times 36 + (93 + 92.5) \times 0.9465\right] =$$

$$= \frac{16.65}{2} \times 248 = 2{,}065 \text{ psi}$$

8. Part 8. Determination of Stresses by the Experimental Method

Nondimensional geometric parameter, $\bar{r}$ according to formula (5.125) is

$$\bar{r} = \frac{a + b}{4\sqrt{Rt}}$$

For

a = 5 ft. – 60 in.

b = 8 ft. = 96 in.

R = 93 in.

t = ½ in.

$$\bar{r} = \frac{60 + 96}{4\sqrt{93 \times 0.5}} = \frac{39}{\sqrt{46.5}} = 5.72$$

From Figure 5.23, for $\bar{r}$ = 5.72, we find:

$$\frac{P}{P_{cl}} = 0.22$$

$$\frac{P}{P_{cl}} = 0.6\ E\left(\frac{t^2}{R}\right)$$

Therefore,

$$P = 0.22 \times 0.6\ E\left(\frac{t^2}{R}\right) = 0.132\ E\left(\frac{t^2}{R}\right)$$

or

$$P = 0.132 \times 30 \times 10^6 \times \frac{0.5^2}{93} = 10{,}645 \text{ lb/in}$$

9. Part 9 Design of Stiffeners at Breech Opening

a. The required cross-sectional area of the single vertical stiffener after formula (5.128)

$$A_s = \frac{tR^2\sin\alpha}{d}$$

Assume d = 93cosα + 5 = 93″

$$\sin\alpha = \sin 18^o\ 50' = 0.3228$$

$$A_s = \frac{0.5 \times 93^2 \times 0.3228}{93.00} = 15.00 \text{ in}^2$$

b. Moment of inertia of removed part about a stack center line after formula (5.138) is

$$I_{rem} = R^3\left(\alpha + \frac{\sin 2\alpha}{2}\right)t$$

where $\alpha = 18^o50'$, $\alpha(\text{rad}) = \frac{18.83}{57.296} = 0.3286$

$$\sin 2\alpha = \sin 37^o40' = 0.6111$$

$$I_{rem} = 93^3\left(0.3286 + \frac{0.6111}{2}\right) \times 0.5 =$$

$$= 804357 \times 0.6341 \times 0.5 =$$

$$255{,}021 \text{ in}^4$$

$$2I_{stiff} \geq I_{rem}$$

Therefore, required

$$I_{stiff} = \frac{I_{rem}}{2} = \frac{255021}{2} = 127{,}510 \text{ in}^4$$

Try 10WF54, A = 15.9 in²

$$I_x = 306.0 + 15.9 \times 93^2 = 306.0 + 137{,}578 =$$

$$= 137{,}884 \text{ in}^4$$

c. The axial compressive stress in each stiffener under the weight of the wall above the breech opening according to formula (5.130) is

$$f_G = \frac{C}{A_s} \times \frac{\alpha}{360} = \frac{114{,}768}{15.9} \times \frac{18.88}{360} = 378 \text{ psi}$$

d. The axial compressive stress in each stiffener due to the wind load moment, after formula (5.134)

$$f_w = \frac{M_w d\alpha}{22.5\ t\ D_m^3\ \sin\alpha}$$

$$f_w = \frac{25172250 \times 93 \times 18.83}{22.5 \times 0.5 \times 185.5^3 \times 0.3228} = 1902 \text{ psi}$$

e. The axial compressive stress due to the eccentricity of stiffener, after formula (5.136)

$$f_e = \frac{(G + P_w)e}{S_{stiff}}$$

where

$$G = 114{,}768 \times \frac{18.83}{360} = 6003 \text{ lbs}$$

$$P_w = \frac{M_w \alpha}{90 D_m} = \frac{25172250 \times 18.83}{90 \times 185.5} = 28391 \text{ lbs}$$

$$e = 5 \text{ in.}$$

$$S_{stiff} = 60.5 \text{ in}^3$$

$$f_e = \frac{(6003 + 28391) \times 5}{60.5} = 2842 \text{ psi}$$

f. Total axial stress in vertical stiffener is sum of

$$f_{tot} = f_G + f_w + f_e$$

$$f_{tot} = 378 + 1902 + 2842 = 5122 \text{ psi}$$

g. Buckling Stability of Vertical Stiffeners

Vertical stiffeners under action of axial load and bending moments due to the eccentricity of axial loads at both ends, may be considered as beam-columns. The maximum buckling stress is estimated using formula (5.148)

$$\sigma_{max} = \frac{P}{A} + \frac{M_o c}{I} \sec \frac{h}{2} \sqrt{\frac{P}{EI}}$$

where P is a part of axial load acting on vertical stiffener

$$P = 114{,}768 \times \frac{18.83}{360} = 6003 \text{ lbs}$$

Section properties for stiffener cross-section

A = 15.9 in^2

I = 306.0 in^4

S = 60.5 in^3

Bending moment acting on stiffener due to the wind

$$M_w = f_w S = 1902 \times 60.5 = 115{,}071 \text{ lb in}$$

Bending moment acting on stiffener due to the eccentricity of axial load

$$M_e = 114{,}768 \times \frac{18.83}{360} \times 5 = 30{,}015 \text{ lb in}$$

Total moment

$$M_o = M_w + M_c = 115{,}071 + 30{,}015 =$$

$$= 145{,}086 \text{ lb in}$$

$$\sigma = \frac{6003}{15.90} + \frac{145{,}086 \times 5}{306.0}$$

$$\sec \frac{96}{2} \sqrt{\frac{6224}{30 \times 10^4 \times 306.0}} =$$

$$= 378 + 2370 \sec 0.0395 =$$

$$= 378 + 2370 \sec 2^{\circ}16' =$$

$$= 378 + 2370 \times 1.00078 = 2751 \text{ psi}$$

h. Ring Girder

The total distributed load using formula (5.152)

$$w = \frac{G}{2\pi R} + \frac{M_w}{\pi R^2}$$

$$w = \frac{114{,}768}{2 \times 3.14 \times 92.75} + \frac{25172250}{3.14 \times 92.75^2} =$$

$$= 197 + 932 = 1{,}129 \text{ psi}$$

The bending moment in the girder

$$M = \frac{w\ell^2}{12} = \frac{1129 \times 60^2}{12} = 338{,}700 \text{ lb in}$$

Assuming $\sigma_{all} = 20{,}000$ psi

$$\text{Required SM} = \frac{338700}{20000} = 16.94 \text{ in}^3$$

Use 10 x 5-3/4 21#

$\text{SM} = 21.5 \text{ in}^3 > 16.94 \text{ in}^3$

NUMERICAL EXAMPLE NO. 2 Rigorous Analysis. Application of Finite Element Method

1. INTRODUCTION

The purpose of this investigation is to study the stress distribution and stress concentrations around the breech opening for the individual and combined effects of dead and wind loads. The magnitude and the distribution of stresses in the vicinity of the opening will determine whether it should be reinforced with additional steel or not. These concepts, along with the ones presented in Section 5.8, have been applied to the stress-analysis of a stack-opening as presented in the following pages.

2. DESCRIPTION OF STACK

The stack analyzed in this example, shown in Figure 5.37, has the following characteristic parameters:

Base diameter = 30.0 ft.; Top diameter = 15.0 ft.

Height = 500 ft.; Shell-plate thickness = 1.25 in.

The Young's modulus for the stack material is 29,000 ksi and its yield strength is 40 ksi. The breech, with dimensions 15.0 ft. × 10.0 ft. is located 8 ft. above the ground level.

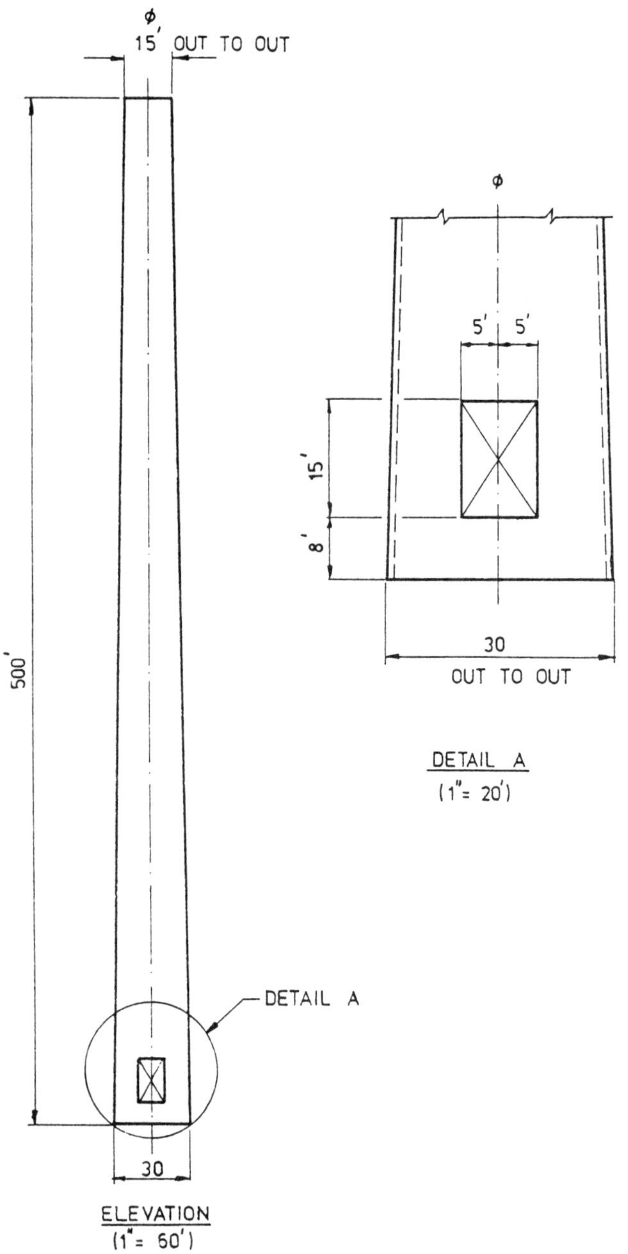

FIGURE 5.37 — Stack and breech dimensions.

3. DESCRIPTION OF FINITE ELEMENT MODELS

In order to study the stress distribution around the breech, two different finite-element-models were related. The first one (or Model I) was 50 ft. high from the base of the stack, while the second one (or Model II) was 200 ft. For a perspective view of Model II, see Figure 5.38 shown on this Figure are also the two major directions, along which the lateral wind has been assumed to act for stress-analysis purposes.

Having reviewed the computer output of Models I and II separately for identical loading conditions, it was concluded that results corresponding to the analysis of Model II should be presented in this report, since its discretization scheme had a much larger number of finite elements as compared to Model I and consequently could predict better results in the vicinity of the stack-opening. In addition, computer results also showed (for Model II analysis) that wind blowing from East to West produced greater stresses around the opening as contrasted to the wind from North to South. Therefore, these values were selected for all stress-calculations in the subsequent pages. Figures 5.39 and 5.40 represent the discretization scheme to the left and right of the opening, respectively, showing its element and node numbers. All element nodes were fed into the computer program in cylindrical coordinates.

Figures 5.41 to 5.46 show the plan view of the discretization scheme at elevations 0.0 ft., 50.0 ft., 80.0 ft., 120.0 ft., 160.0 ft., and 200.0 ft. The node numbers of elements are shown around the outer periphery of the diagram, while an angle subtended at the center of the circle by two adjacent nodes is indicated by an angle-figure shown between the two radii joining the nodes with the center. The change in the stack-radius from 179.375 in., Figure 5.41 to 143.375 in., Figure 5.46 is indicative of the vertical taper that the stack-geometry has a part of its configuration.

4. LATERAL WIND LOAD ON STACK

The lateral wind load on the stack has been calculated according to the National Building Code of Canada. The general equation for the lateral wind load, according to this code, on an exposed surface is as follows:

$$p = q \cdot C_e \cdot C_g \cdot C_p \cdot C_n$$

where

- q = the uniform lateral wind pressure in lbs/ft^2;
- C_e = the exposure factor that varies according to the height of the structure;
- C_g = the gust factor;
- C_p = shape factor; and
- C_n = the additional shape factor for chimneys.

Typical values of the above factors were used as shown below

- q = 9.2 lbs/ft (For Montreal Area);
- C_e = varies from 1.0 for a height of 40 ft. to 2.0 for a height of 1200 ft.
- C_g = 2.0; C_p = 1.6 and C_n = 0.7

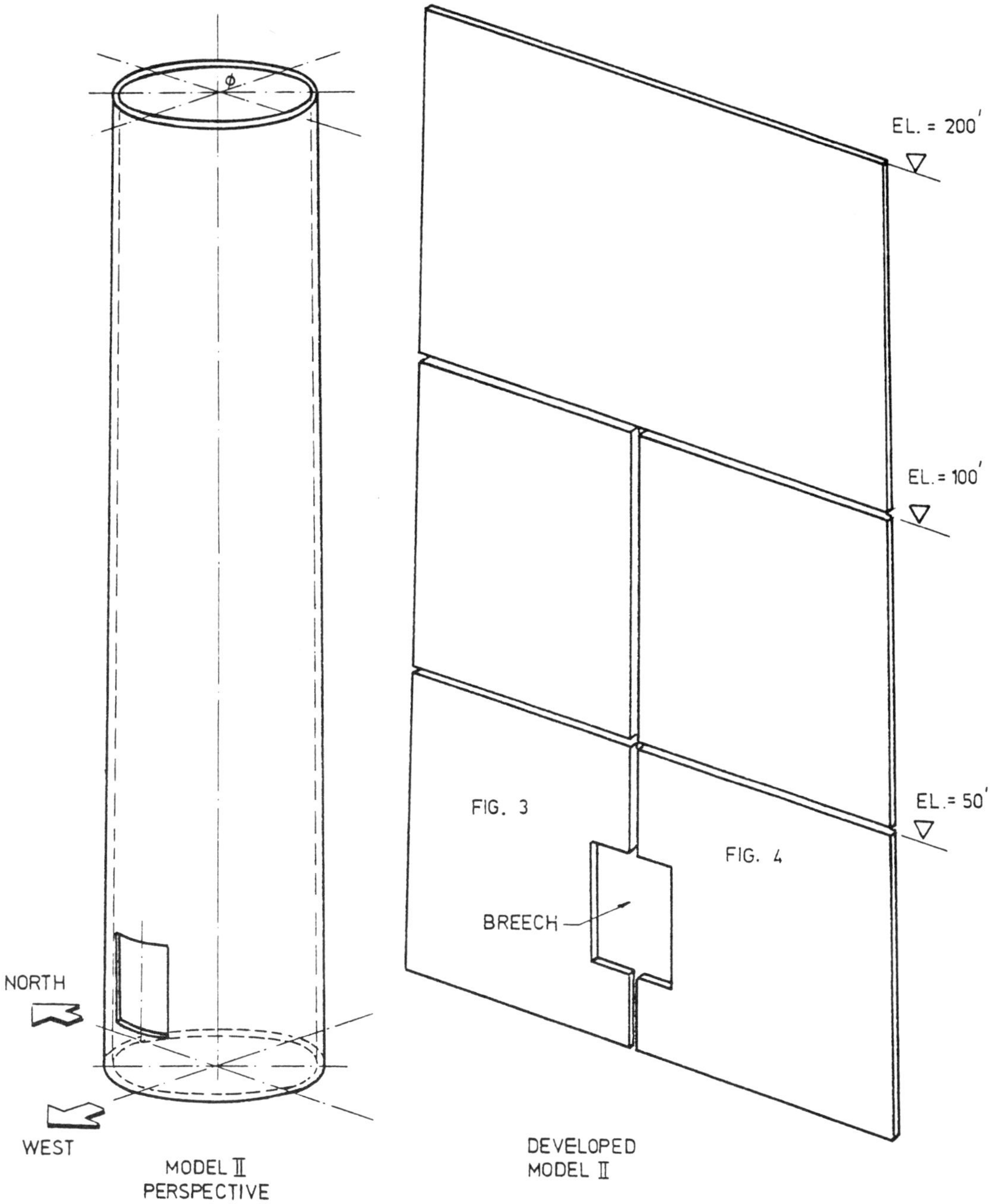

FIGURE 5.38 — Perspective of stack — Model II.

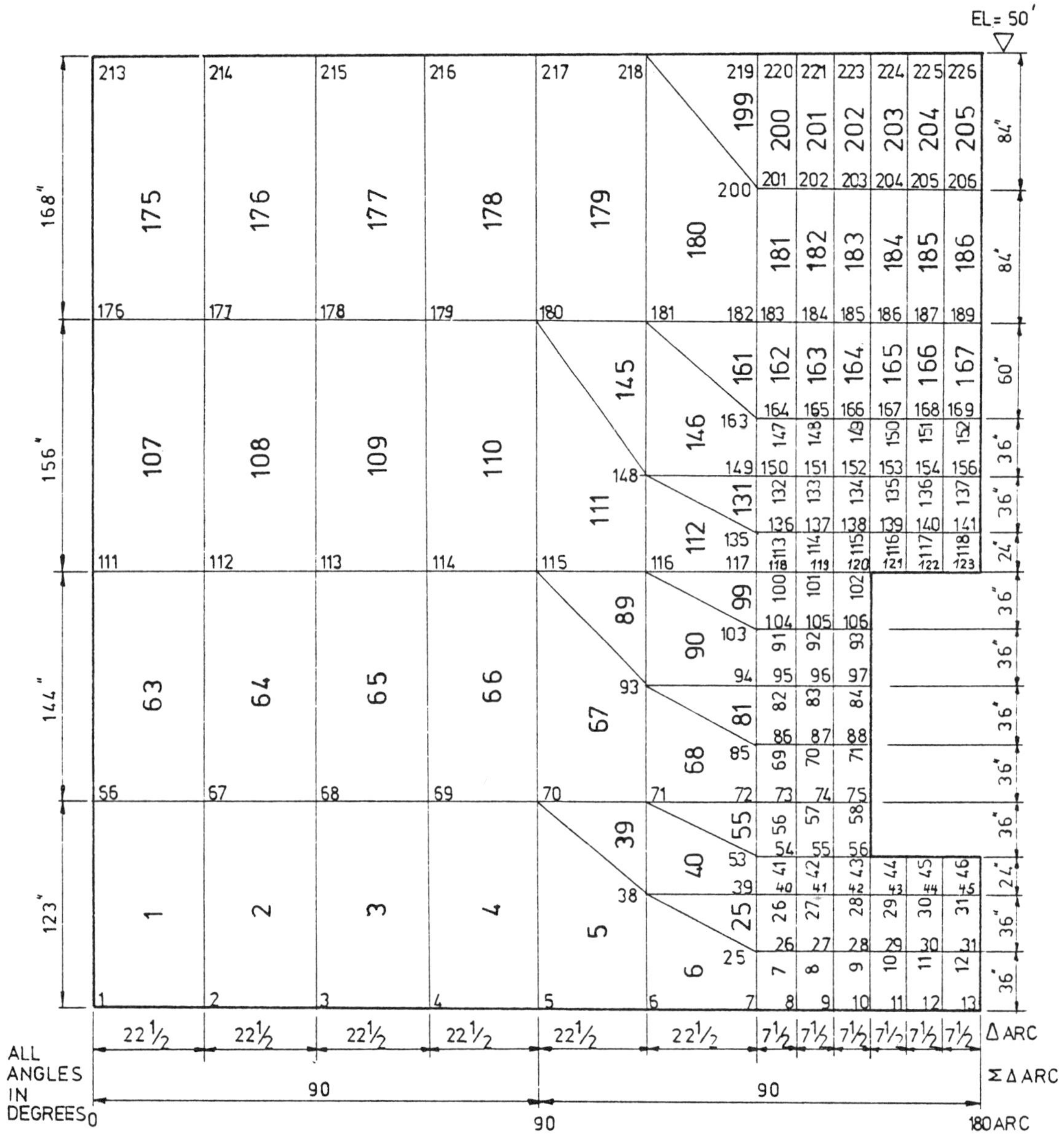

NOTE: The discretization scheme presented on Figure 5.39 is a developed view of the left half (with respect to the opening) of the truncated cone between stack elevations 0.0 and 50.0. The regular-shaped elements *i.e.* rectangles and triangles, and shown, are in fact distorted with vertical lines slightly inclining to the right and the horizontal ones with some curvature.

FIGURE 5.39 — Discretization scheme around opening.

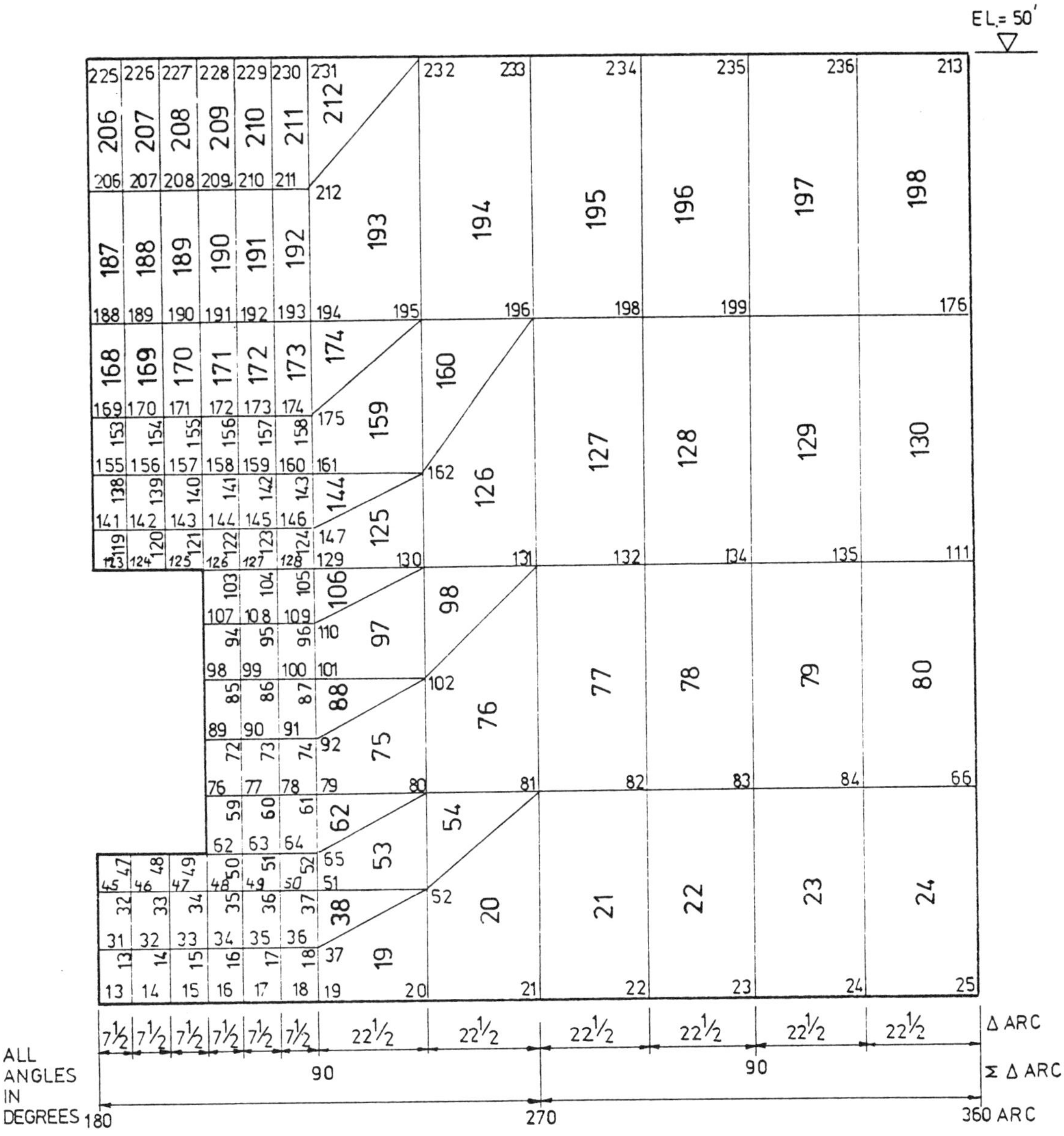

NOTE: The discretization scheme presented on Figure 5.40 is a developed view of the right half (with respect to the opening) of the truncated cone between stack elevations 0.0 and 50.0. The regular shaped elements *i.e.* rectangles and triangles, as shown, are in fact distored with vertical lines slightly inclining to the left and the horizontal ones with some curvature.

FIGURE 5.40 — Discretization scheme around opening.

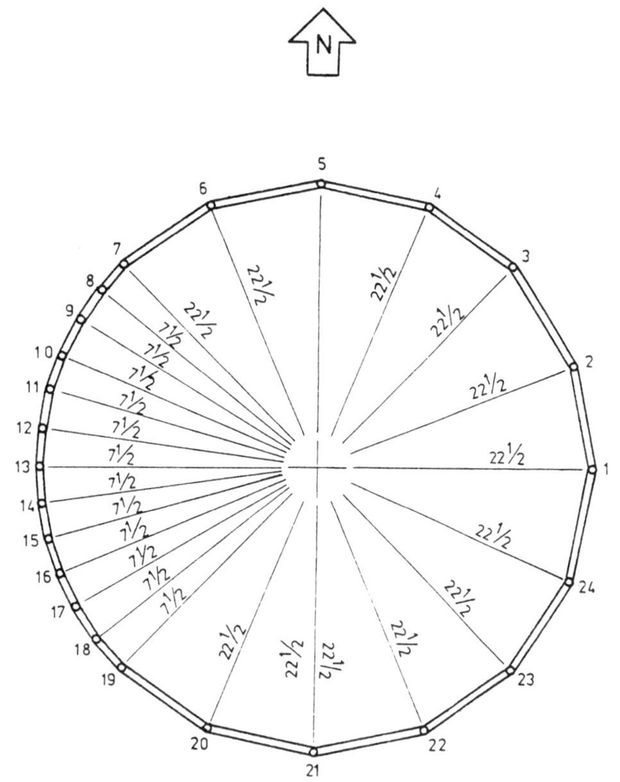

FIGURE 5.41 — Plan cross-section at El. = 0.0′ cross-sectional radius = 179.375″.

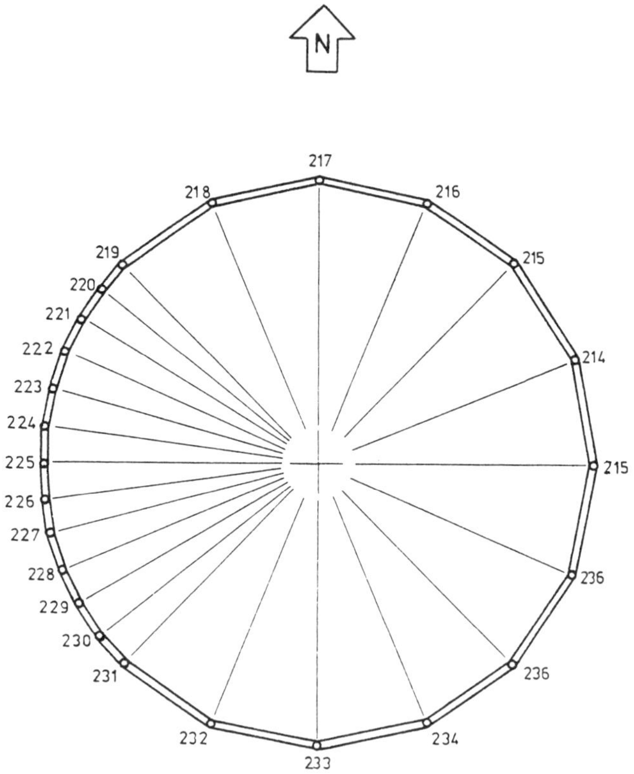

FIGURE 5.42 — Plan cross-section at El. = 50′ cross-sectional radius = 170.375″.

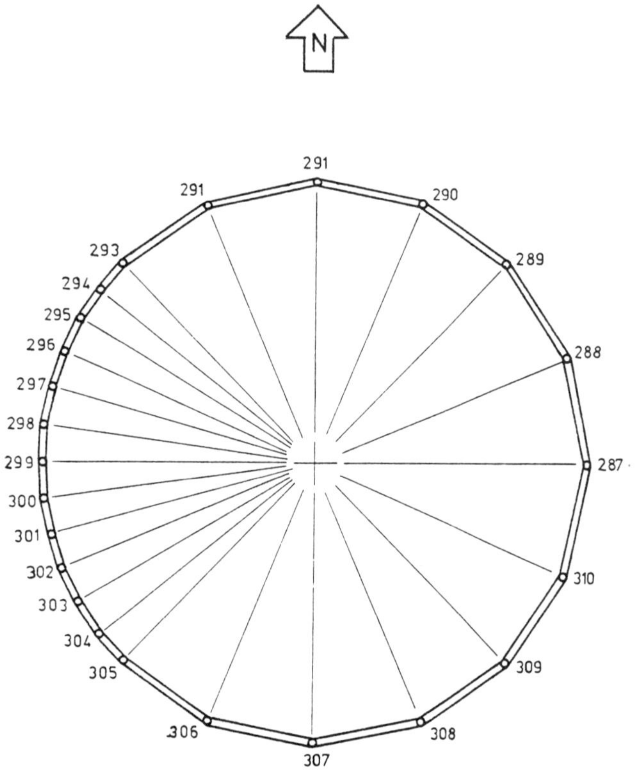

FIGURE 5.43 — Plan cross-section at El. = 80′ cross-sectional radius = 164.975″.

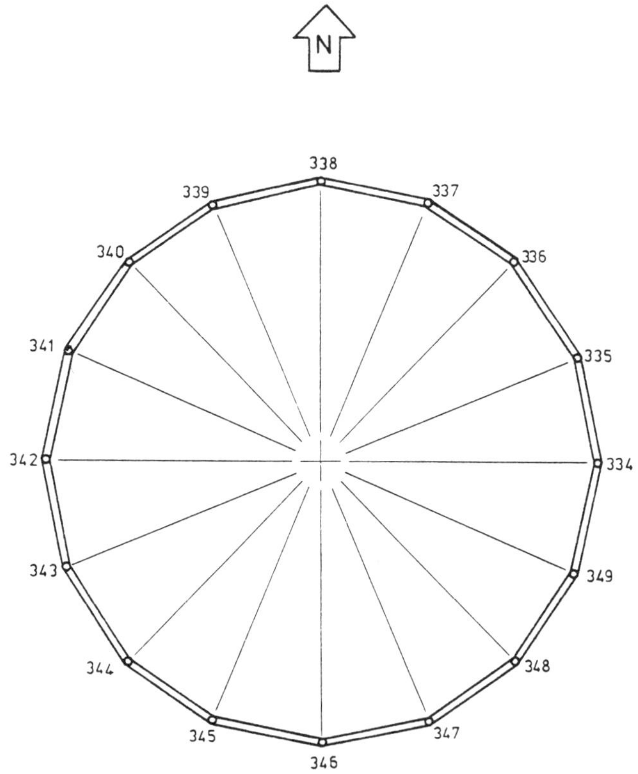

FIGURE 5.44 — Plan cross-section at El. = 180′ cross-sectional radius = 157.775″.

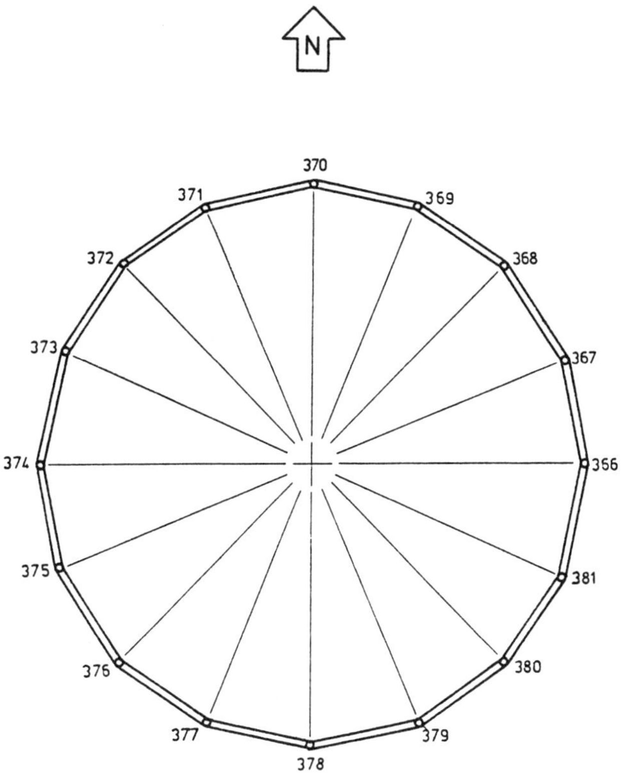

FIGURE 5.45 — Plan cross-section at El. = 160′ cross-sectional radius = 150.575″.

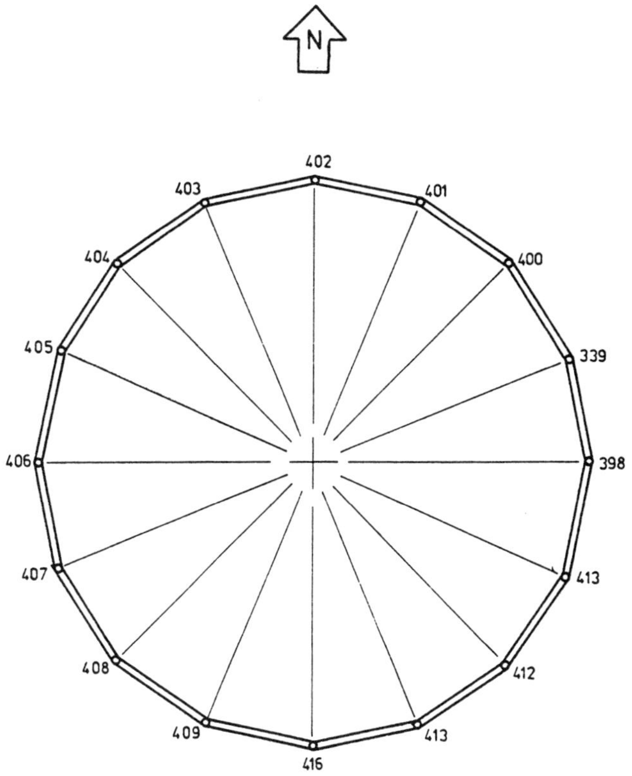

FIGURE 5.46 — Plan cross-section at El. = 200′ cross-sectional radius = 143.375″.

The diagrams for the lateral load appear on Figs. 5.47 and 5.48, for Models I and II respectively, where the wind load appears as a stepped function of the stack-height.

5. TRANSMISSION OF APPLIED FORCES ONTO STACK-MODEL-II

The forces acting on stack-model II are:

(i) Self-weight of the model (height — 200 ft.);

(ii) Dead load of the stack on top of the model acting vertically down; and

(iii) Lateral wind acting on the entire stack.

Figure 5.49 is a graphic display of the application of vertical line loads acting on the top edge of stack-model II due to lateral wind and self-weight of the stack. Moreover, Figure 5.50 shows how the transverse shear due to wind load can be applied to the same model by transforming it into equivalent tangential shear first. Such a conversion has been found very advantageous in ANSYS computer program and leads to substantial savings in computer time used and the cost incurred.

6. THEORETICAL ANALYSIS

The stress analysis of the area adjoining the stack-opening has been performed using the finite element method of analysis for Model I and Model II. However, for reasons given in Section 3 of Example No. 2, only results corresponding to Model II have been used in the following pages. Triangular and rectangular finite elements were chosen for the discretization scheme. ANSYS computer program was used for the stress analysis with loading conditions as dead load, wind load and their combination. The output gave results in terms of bending and membrane stresses, bending and torsional moments and axial loads as well as displacements of elements and nodes. (For detailed results, refer to the computer output).

7. TABULATION OF COMPUTER RESULTS

In order to design the reinforcement around the opening, the most critical combination of external loads on the stack was selected i.e., dead load + lateral E. to W. wind load, Figure 5.48. Moments and axial forces corresponding to this loading condition have been tabulated in Tables 5.8 to 5.11 for elements in the immediate vicinity of the opening, Figures 5.39 and 5.40.

8. DESIGN OF VERTICAL STIFFENERS

It will be assumed that the stiffeners will act integrally with the stack and as such a certain length of the stack plate will have a significant stiffening effect on the stiffeners. In order to determine this length, reference is made to the AISC Specifications for stiffeners of plate girder webs. Notwithstanding the slight disparity between the behavior of a steel stack plate and a plate girder web, we believe we can make use of the following specification with some reservation [5.75].

MODEL I WIND LOADS

The wind moments and shears for Model I were determined tabularly, Table 5.7 in accordance with Fig. 5.47.

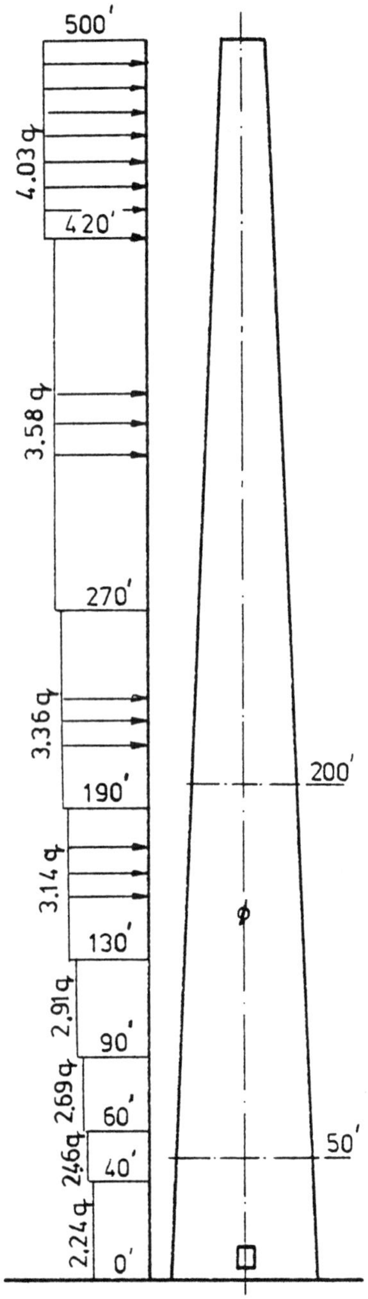

FIGURE 5.47 — Elevation of wind intensity against stack wrt to stack El. in gpsf.

TABLE 5.7

STACK ELEV FT	ELEV DIFF FT	AVG DIAM FT	AVG AREA FT^2	LAT. WIND qPSF	LAT. SHEAR q #	MOM ARM ABT BASE FT	MOMENT ABT BASE q#- FT	MOM ARM ABT. EL=50' FT	MOMENT ABT EL=50' q# - FT
0									
	40	29.4	1176	2.24	2634	20	52680	0	0
40					701				
	20	28.5	570	2.46	----	50	70100	5	3500
60					701				
	30	27.8	833	2.69	2239	75	167925	25	55975
90									
	40	26.7	1068	2.91	3108	110	341880	60	186480
130									
	60	25.2	1512	3.14	4748	160	759680	110	522280
190									
	80	23.1	1848	3.36	6209	230	1428070	180	1117620
270									
	150	19.7	2948	3.58	10552	345	3640440	295	3112840
420									
	80	16.2	1296	4.03	5223	460	2402580	410	2141430
500									
TOTALS AT									
BASE					36115q		8863355 q		
EL.= 50'					32780q				7140125 q

TABLE 5.7 — Table of base and top of Model I lateral wind shears and overturning moments.

Note: The values shwon are for q = 1.0 PSF. For absolute values, they must be multiplied by chosen value of q. For Model I the analysis stresses due to the unit wind were multiplied by q = 9.2, the 1/100 wind for Montreal.

MODEL II LATERAL WIND

Wind forces for Model II, determined with respect to the wind distribution shown in Figure 5.48 and q = 9.2 psf, were as shown.

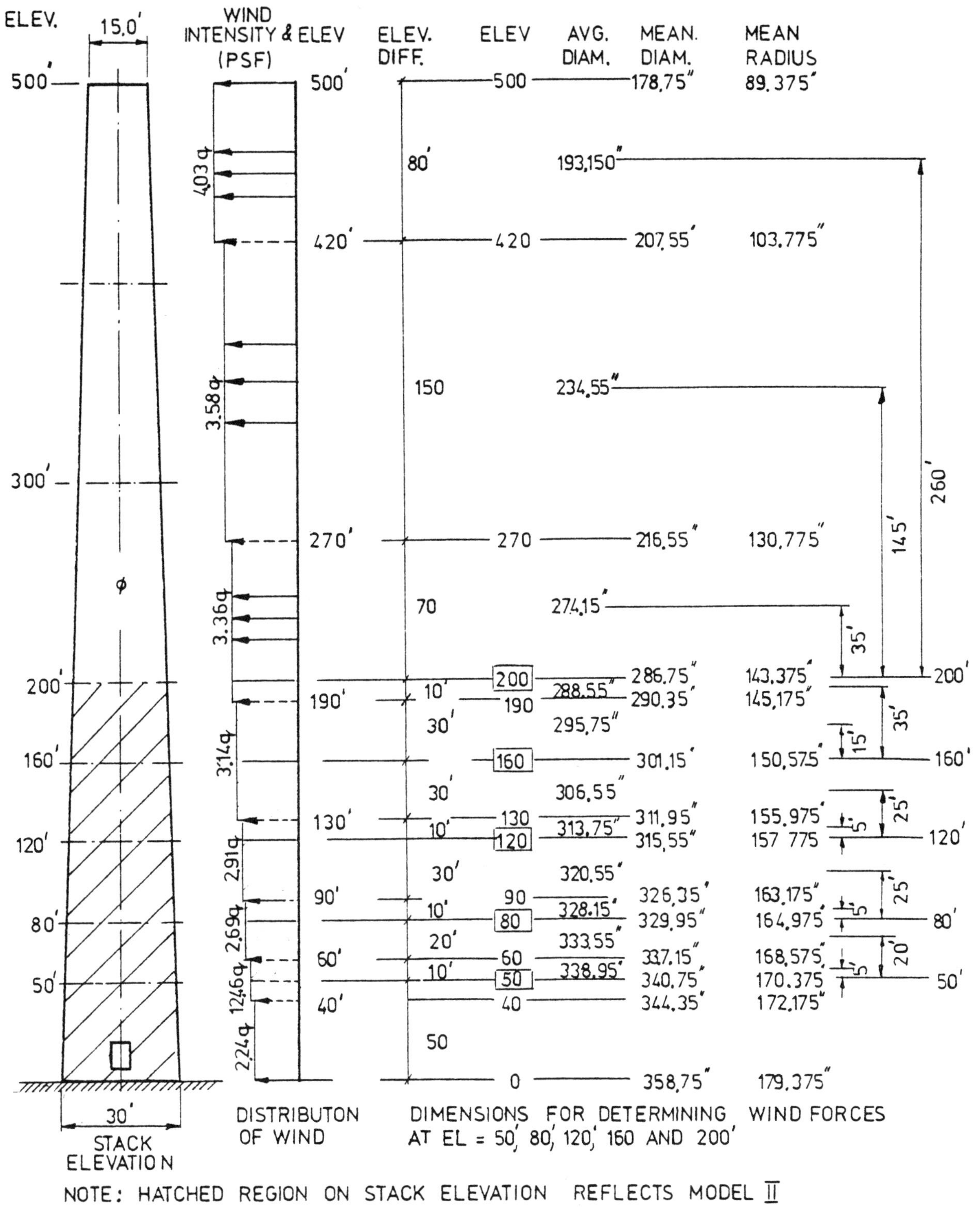

FIGURE 5.48 — Model II region and dimensions required to determine wind forces at node elevations shown.

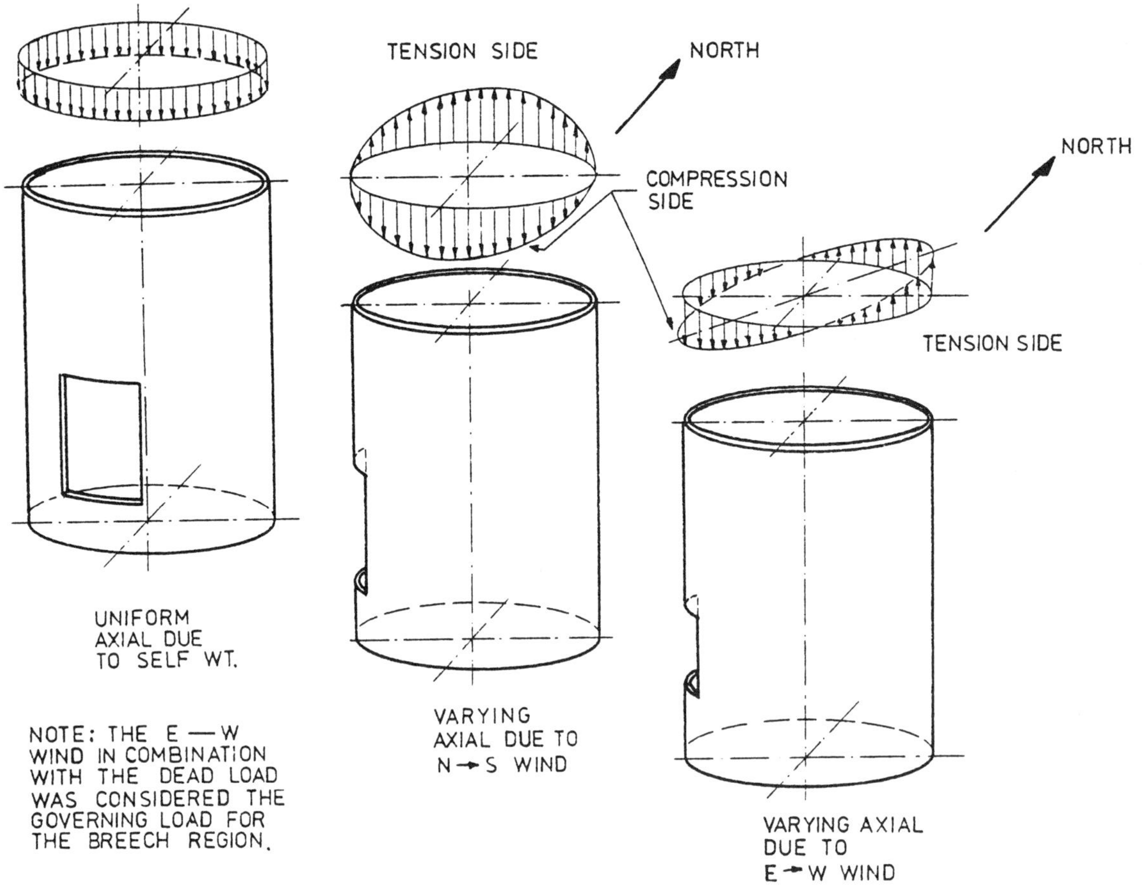

FIGURE 5.49 — Model II top edge vertical line loads due to lateral wind and self-weight dead load.

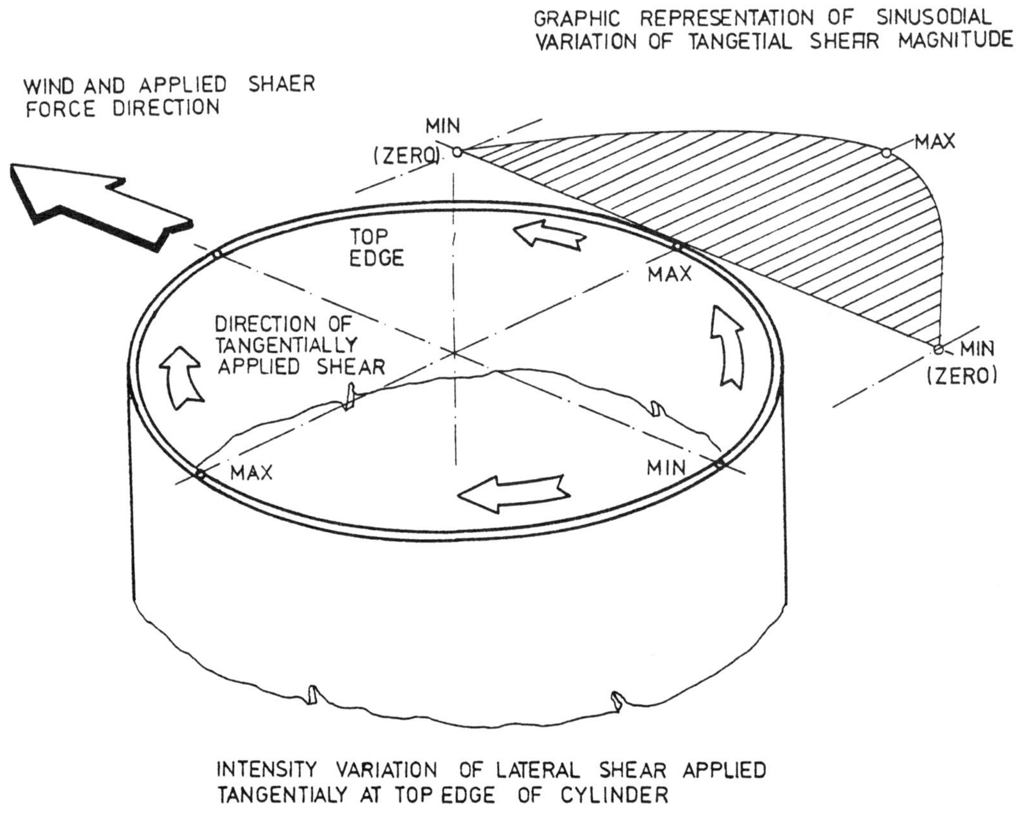

FIGURE 5.50 — Lateral shear along cylindrical plan cross-section.

TABLE 5.8 — Moments and Axial Forces in the Vertical Row of Elements.

(To the left of opening)

Note: Moments are in in-lbs/in whereas axial forces are in lbs/in.

El. No.	M_x	M_y	Nx	Ny
28	70	13	1	-4
43	244	-493	-14	-9
58	385	-119	-24	12
71	216	332	-21	-12
84	49	199	-12	0
93	339	694	-42	25
102	1221	653	-11	22
115	1450	27	28	-32
143	190	67	-23	5

Note: For an isometric plot of Moments M_x and M_y around opening, see Figures 5.51 and 5.52.

TABLE 5.9 — Moments and Axial Forces in the Vertical Row of Elements.

(To the right of opening)

Note: Moments are in in-lbs/in whereas axial forces are in lbs/in.

El. No.	M_x	M_y	Nx	Ny
35	70	13	-1	-4
50	244	-493	14	-9
59	385	-119	24	12
72	216	332	21	-12
85	49	199	12	0
94	339	694	42	25
103	1221	653	11	22
122	1450	27	-28	-32
141	698	-27	-8	-20

Note: For an isometric plot of moments M_x and M_y around opening, see Figures 5.51 and 5.52.

TABLE 5.10 — Moments and Axial Forces in the Horizontal Row of Elements.

(At the bottom of opening)

Note: Moments are in in-lbs/in whereas axial forces are in lbs/in.

El. No.	M_x	M_y	N_x	Ny
42	321	-169	-5	32
43	244	-493	-14	-9
44	-26	-75	-80	6
45	-505	7	19	-19
46	-158	2	20	4
47	-158	2	-20	4
48	-505	7	-19	-19
49	-26	-75	80	6
50	244	-493	14	-9
51	321	-169	5	32

Note: For an isometric plot of moments M_x and M_y around opening, see Figures 5.51 and 5.52.

TABLE 5.11 — Moments and Axial Forces in the Horizontal Row of Elements.

(At the top of opening)

Note: Moments are in in-lbs/in whereas axial forces are in lbs/in.

El. No.	M_x	M_y	N_x	Ny
114	774	265	55	-32
115	1450	27	28	-32
116	1565	146	-223	-21
117	-918	-7	-99	32
118	-1826	-242	-23	-9
119	-1826	-242	23	-9
120	-918	-7	99	32
121	1565	146	223	-21
122	1450	27	-28	-32
123	774	265	-55	-32

Note: For an isometric plot of moments M_x and M_y around opening, see Figures 5.51 and 5.52.

AISC Specification 1.10.5.1:

> Bearing stiffeners shall be placed in pairs at unframed ends on the webs of plate girders and where required at points of concentrated loads. They shall be designed as columns subject to the provisions of Section 1.5.1, assuming the column section to comprise the pair of stiffeners and centrally located strip of the web whose width is equal to not more than 25 times its thickness at interior stiffeners or *a width equal to not more than 12 times its thickness when the stiffeners are located at the end of the web.*

NOTE: Since the output of moments and forces is given in per inch of width, it will be assumed that the contributing width for the forces acting on vertical stiffeners would be 24 in. This is a matter of pure judgement. By scanning the computer output for the two consecutive rows of elements immediately around the opening, it is obvious that moments, shears and axial forces are significantly lower in the second row as compared to their counterparts in the first row (i.e. right next to the opening). Therefore, it is felt that the design of stiffeners should be based on the forces contributed by the entire width of the first row of element via 24 in.

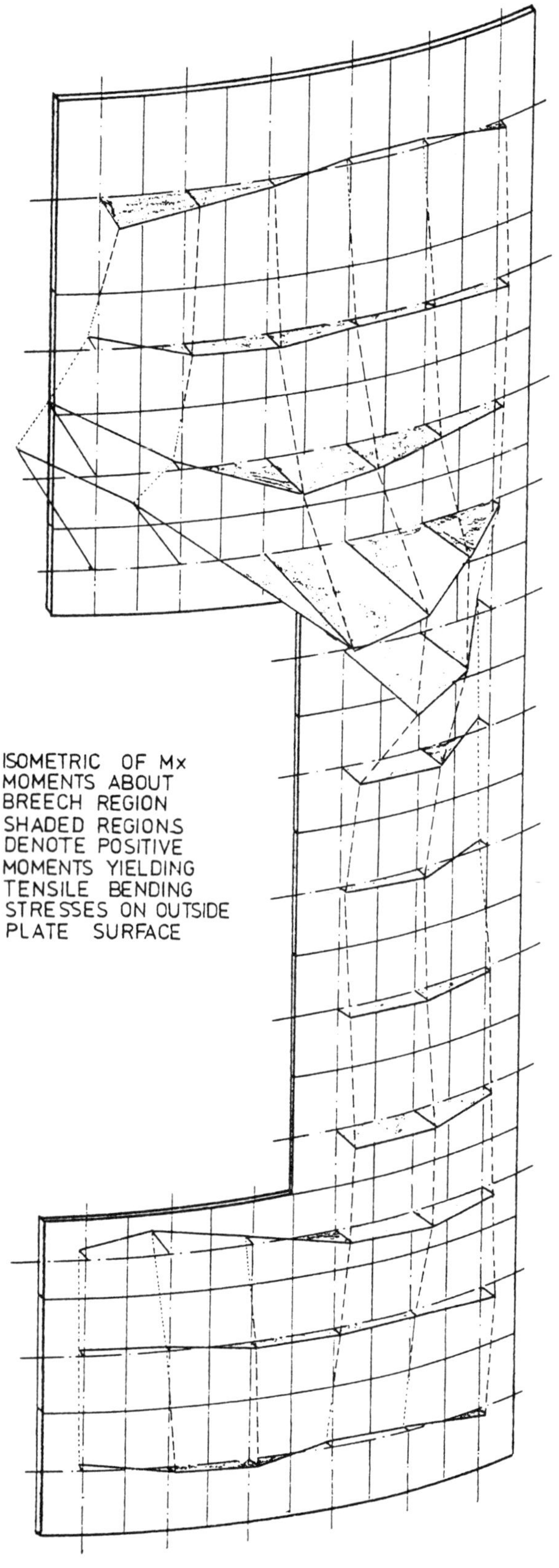

FIGURE 5.51 — Isometric plot of moment, Mx around opening.

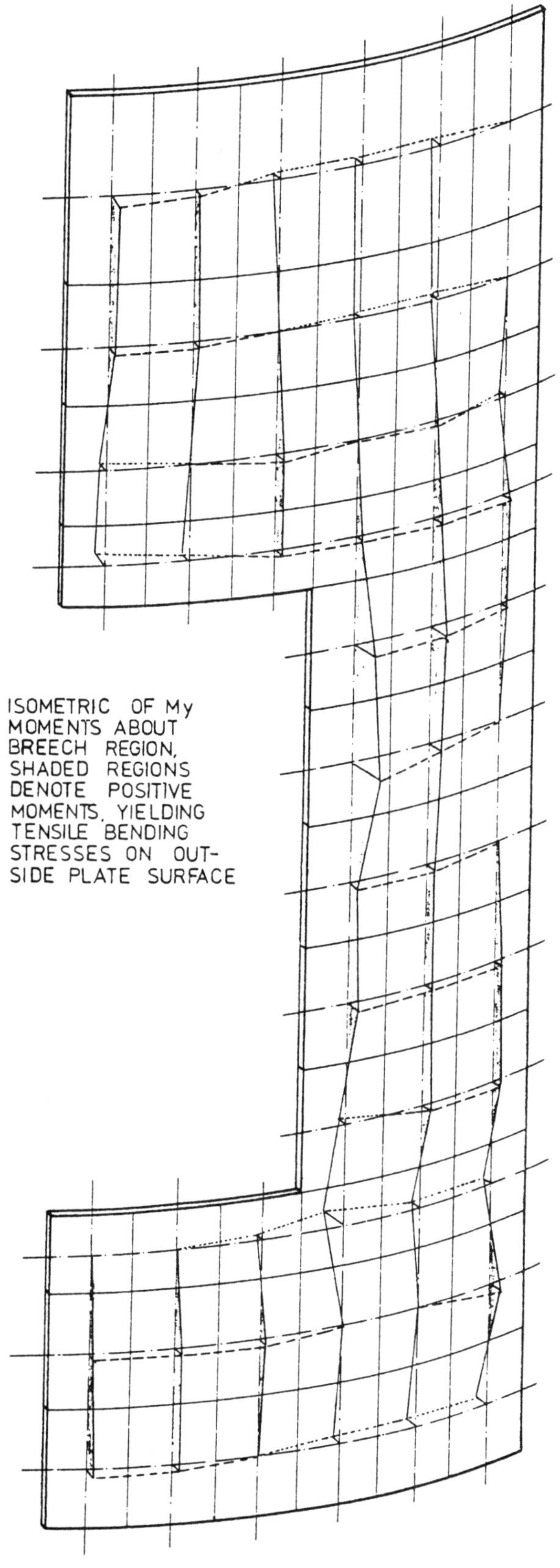

FIGURE 5.52 — Isometric plot of moment, My around opening.

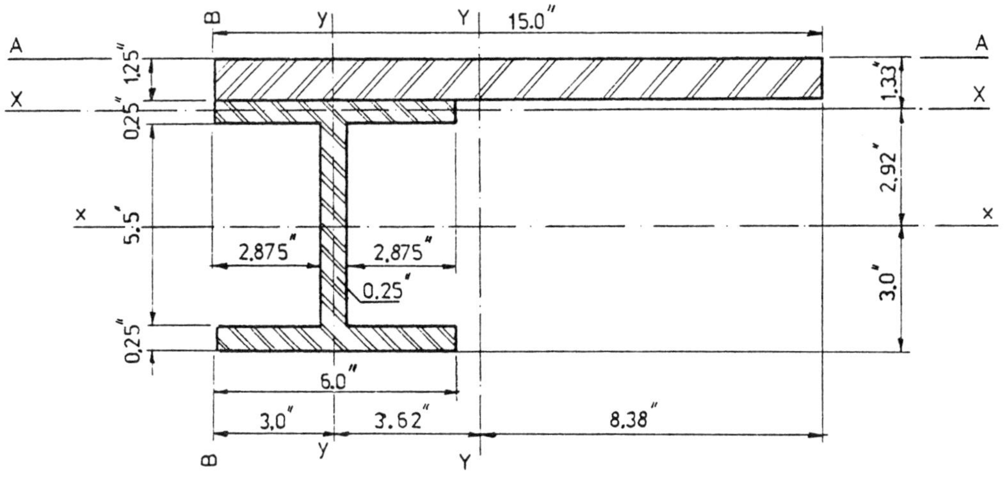

FIGURE 5-53 — Composite vertical stiffener and stack plate section.

Select a WF 6 shape @ 15.5 with the following properties:

A = 4.56 in^2

Ix = 30.1 in^4

Sx = 10.0 in^3

rx = 2.57 in

and

Iy = 9.67 in^4

Sy = 2.23 in^3

ry = 1.46 in

Flange thickness t = ½ in and web thickness = ¼ in

Stack plate thickness = 1.25 in

Stiffening width of stack plate = 12 × 1.25 = 15 in.

8.1 Determination of Neutral Axes of Composite Section

Taking moments about axis A-A we have:
Σ Area moments about axis

$$A\text{-}A = 4.56\,(3 + 1.25) + 15 \times 1.25 \times \frac{1.25}{2} = 19.38 + 11.72 =$$

$$= 31.10 \text{ in}^3$$

Taking moments about axis B-B we have
Σ Area moments about axis

$$B\text{-}B = 4.56 \times 3.0 + 15 \times 1.25 \times \frac{15}{2} = 13.68 + 140.63 =$$

$$= 154.31 \text{ in}^3$$

Total cross-sectional area = 4.56 + 15 × 1.25 = 23.31 in^3

The neutral axis x-x is located at 31.10/23.31 in from axis A-A = 1.33 in, and;

The neutral axis Y-Y is located at 154.31/23.31 in from axis B-B = 6.62 in.

8.2 Determination of I_{xx} and I_{yy}

$$I_{xx} = 30.1 + 4.56\,(2.92)^2 + \frac{15 \times 1.25^3}{12} +$$

$$+ (15 \times 1.25) \times (1.33 - .63)^2 =$$

$$= 30.1 + 38.88 + 2.44 + 9.19 = 80.61 \text{ in}^4$$

$$r_{xx} = \left(\frac{80.61}{23.31}\right)^{\frac{1}{2}} = 1.86 \text{ in.}$$

$$S_{xx} = \frac{80.61}{1.32} = 60.61 \text{ in}^3$$

$$I_{yy} = 9.67 + 4.56\,(3.62)^2 + \frac{1.25(15)^3}{12} +$$

$$+ (15 \times 1.25) \times (8.38 - 7.5)^2 = 9.67 +$$

$$+ 59.76 + 351.56 + 14.52 = 435.51 \text{ in}^4$$

$$r_{yy} = \left(\frac{435.51}{23.31}\right)^{\frac{1}{2}} = 4.32 \text{ in. } S_{yy} = \frac{435.51}{6.62} =$$

$$65.79 \text{ in}^3$$

The opening is 10 ft. wide and 15 ft. high. It is assumed that the stiffeners will be welded to the stack plate around the opening.

Assumed length = 18.0 ft. End conditions assumed as simply-supported. ∴ K = From Tables 5.8 and 5.9.

Max. M_y = 694 × 24 = 16,656 in lbs.

Max. N_y = 25 × 24 = 600 lbs.

$$\frac{K\ell}{r_{xx}} = \frac{18 \times 12}{1.86} = 116 \;; \text{ and}$$

$$\frac{K\ell}{r_{yy}} = \frac{18 \times 12}{4.32} = 50$$

$$C_c = \left(\frac{2\pi^2 E}{F_y}\right)^{\frac{1}{2}} = \left(\frac{2 \times 3.14^2 \times 29{,}000}{40}\right)^{\frac{1}{2}} = 120$$

Since $\frac{K\ell}{r_{xx}} < C_c$

$$F.S. = \frac{5}{3} + \frac{3\left(\frac{K\ell}{r_{xx}}\right)}{8C_c} - \frac{\left(\frac{K\ell}{r_{xx}}\right)^3}{8C_c^{\,3}} =$$

$$= \frac{5}{3} + \frac{3 \times 116}{8 \times 120} - \frac{116^3}{8 \times 120^3} =$$

$$= 1.67 + 0.36 - 0.11 = 1.92$$

$$F_a = \frac{\left[1 - \frac{\left(\frac{K\ell}{r_{xx}}\right)^2}{2C_c^{\,2}}\right] F_y}{F.S.} = \frac{\left[1 - \frac{116^2}{2 \times 120^2}\right] 40}{1.92} =$$

$$= 11.04 \text{ ksi}$$

According to AISC Specification 1.5.1.4.4.

$$F_b = 0.6\ F_y = 0.6 \times 40 = 24 \text{ ksi}$$

$$f_a = \frac{600}{23.31} = 25.74 \text{ psi; and}$$

$$f_b = \frac{16{,}656}{60.61} = 274.81 \text{ psi;}$$

$$\frac{f_a}{F_a} = \frac{25.74}{11040} = 0.0023\ ;$$

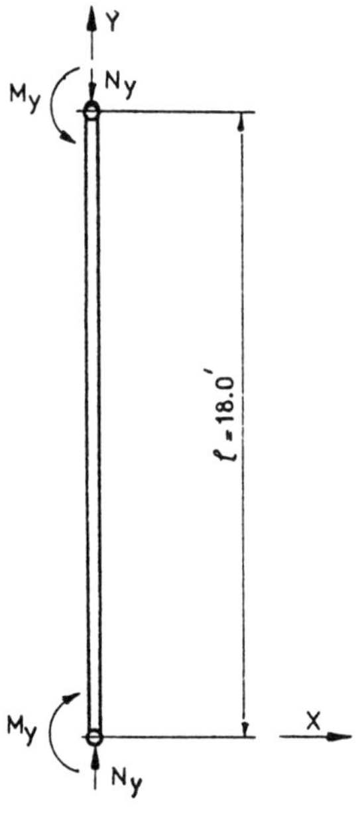

FIGURE 5.54 — Vertical Stiffener.

Since $\frac{f_a}{F_a} < 0.15$, use the criterion $\frac{f_a}{F_a} + \frac{f_b}{F_b} \leq 1.0$; hence

$$\frac{f_a}{F_a} + \frac{f_b}{F_b} = \frac{25.74}{11040} + \frac{274.81}{24{,}000} = 0.0023 + 0.01145$$

$$= 0.01375 \leq 1.0 \text{ Hence quite safe}$$

8.3 Check for Stability

The maximum buckling stress of a vertical stiffener can be estimated using the following expression [5.76].

$$\sigma_{max} = \frac{N_y}{A} + \frac{M_y \cdot c}{I_{xx}} \sec \frac{\ell}{2} \sqrt{\frac{N_y}{EI_{xx}}}$$

where c is the distance between axes x-x and A-A on Figure 5.53. All other notations being the same as defined in the preceding pages.

$$\sigma_{max} = \frac{600}{23.31} + \frac{16{,}656 \times 1.33}{80.61} \sec \frac{18 \times 12}{2} \times$$

$$\times \sqrt{\frac{600}{30 \times 10^6 \times 80.61}}$$

$$= 25.74 + 274.81 =$$

$$= 301 \text{ psi.} \quad \text{Hence quite safe}$$

8.4 Design of Horizontal Stiffeners

The most critical value of bending moment appears in elements 118 and 119 i.e. 1826 lbs-in/in of the plate, Tables 5.10 and 5.11.

Similar to the design of vertical girders, using 24 in. as the contributing width of the stack plate,

Total Moment = 1826 × 24 = 43,823 lbs in

Using σ_{all} = 20,000 psi; the Section modulus required is

$$= \frac{43{,}824}{20{,}000} = 2.19 \text{ in}^3$$

The section modulus provided by the composite section in the y-direction is 65.79 in³. Hence adequately safe.

Therefore, for both vertical stiffeners and horizontal girders around the opening WF 6 shape @ 15.5 should be satisfactory.

REFERENCES — CHAPTER 5

[5.1] Handbook of Industrial Loss Prevent, Prepared by the Staff of the Factory Mutual Engineering Division. General Specifications for the Construction of New Self-Supporting and Guyed Stacks, McGraw-Hill Book Co., N.Y., 1968, pp. 71-7 to 71-11.

[5.2] Ketchum, M.S., Structural Engineers' Handbook, Chapter XI A, Self-Suporting Steel Stacks, McGraw-Hill Book Co., N.Y., 1924, pp. 471-492.

[5.3] British Standards Institution, Specification for Steel Chimney, B.S. 4076: 1966, London, pp. 1-26.

[5.4] British Standard Institution, Code of Basic Data for the Design of Buildings, Chapter V. Loading, Part 1, Dead and Imposed Loads, CP3: 1967, p.6.

[5.5] Steel Stacks. Computation and Construction Design, German Standards DIN 4133, Project, November, 1970, pp. 20. (in German).

[5.6] ACI Committee 307: Proposed Revision of ACI 505-54: Specification for the Design and Construction of Reinforced Concrete Chimneys, ACI Journal, Sept. 1968, pp. 689-712.

[5.7] Abbett, R.W., American Civil Engineering Practice, Vol. III, Reinforced-Concrete Chimneys, John Wiley & Sons, Inc., N.Y., 1957, pp. 32-02 to 32.24.

[5.8] National Research Council of Canada, Canadian Structural Design Manual. Supplement No. 4, to the National Building Code of Canada, 1977, pp. 16, 31, 33.

[5.9] American National Standard Building Code Requirements for Minimum Design Loads in Buildings and Other Structures, 1972, pp. 24, 25.

[5.10] Omori, F., "Wind Pressure on Tall Chimney", Engineering, Vol. 106, 1918.

[5.11] Pagon, W.W., "Vibration Problem in Tall Stacks Solved by Aerodynamics", Eng. News-Record, Vol. 113, No. 2, July 12th, 1934, pp. 41-43.

[5.12] Scruton, C., "Wind-Excited Oscillations of Tall Stacks", The Engineer, June 10th, 1955, pp. 805-808.

[5.13] Ozker, M.S., and Smith, J.O., "Factors Influencing the Dynamic Behavior of Tall Stacks Under the Action of Wind", Trans. of the ASME, Vol. 78, No. 6, August 1956, pp. 1381-1391.

[5.14] Dockstader, E.A., Swiger, W.F., and Ireland, E., "Resonant Vibration of Steel Stacks", Trans. ASCE, Vol. 121, Paper No. 2832, pp. 1088-1112, 1956.

[5.15] Dickey, W.L., and Woodruff, G.B., "The Vibrations of Steel Stacks", Trans. ASCE, Vol. 121, Paper No. 2831, pp. 1054-1087, 1956.

[5.16] Penzien, J., "Wind Induced Vibration of Circular Cylinders", ASCE, Eng. Mech. Div. Proc. Paper 1141, Jan. 1957.

[5.17] Fujino et al, "The Dynamic Behavior of Tall Stacks Under the Action of the Wind", Proc. 7th Nat. Congr. Appl. Mech., Tokyo, 1957, pp. 387-392.

[5.18] Cohen, E., and Perrin, H., "Design of Multilevel Guyed Towers", Papers 1355, 1356, Journ. of St. Div. Proc. ASCE, Vol. 83, St. 5, Sept. 1957.

[5.19] Scruton, C., and Flint, A.R., "Wind-Excited Oscillations of Structures", Proc. Inst. of Civ. Engrs., No. 677, April, 1964.

[5.20] Krupka, V., Design Analysis of Thin-Walled Cylindrical Structures, Praha 1967, Chapter 7 pp. 219-233, (in Czechoslovakian).

[5.21] Nakao, Y., Yamashita, M., Ogata, Y., and Matsumoto, T., "Structural Study on Assembled Four-Cylinder Type Stacks", Technical Review, Mitsubishi Heavy Industries Ltd., January, 1968, pp. 1-10.

[5.22] Vandeghen, A., and Alexandre, M., "Vibrations of the Tall Steel Chimneys Under Wind Action", Publications Internat. Assoc. for Bridge and Struct, Eng., Vol. 29-1, 1969, pp. 95-132, (in French).

[5.23] Wootton, L.R., "The Oscillations of Large Circular Stacks in Wind", Proc. Inst. of Civ. Engrs., Aug. 1969, pp. 573-598.

[5.24] Vellozzi, J., and Cohen, E., "Dynamic Response of Tall Flexible Structures to Wind Loading, Wind Loads on Buildings and Structures", U.S. Dept. of Commerce Publication, Building Science Series 30, 1970, pp. 115-128.

[5.25] Uzsoy, S.Z., "Approximate Analysis of Multiple Circular Cylindrical Shell Wind Shields for High Factory Chimneys", Bull. Int. Assoc. for Shell Structures, No. 41, March, 1970, pp. 43-52, and No. 42, June, 1970, pp. 3-28.

[5.26] Rumman, W.S., "Basic Structural Design of Concrete Chimneys", J. of the Power Div., Proceedings ASCE, June, 1970, pp. 309-318.

[5.27] Anonymous, "Steel Chimney in Trouble", Metal Construction, Vol. 3, No. 4, April, 1971, pp. 145-146.

[5.28] Irish, K., and Cochrane, R.G., "Wind-Induced Oscillation of Circular Chimneys and Stacks", The Structural Engineer, No. 6, Vol. 49, June, 1971, pp. 255-259.

[5.29] Wills, L.G., "Periodic Oscillation of Chimneys", Concrete, May, 1972, p. 35.

[5.30] Irish, K., and Cochrane, R.G., "Wind Vibration of Chimneys", ACI Journal, Sept., 1972, pp. 589-596.

[5.31] Dickey, W.L., "The Design of Two Steam Electric Plants", Trans. ASCE, 1954, Paper No. 2900, pp. 253-272.

[5.32] Dockstader, E.A., Swiga, W.F. and Ireland, E., "Resonant Vibration of Steel Stacks", Proc. ASCE, No. 541, Nov. 1954.

[5.33] Den Hartog, J.P., Mechanical Vibrations, McGraw-Hill Book Co., N.Y., 1947.

[5.34] Dickey, W.L., and Woodruff, G.B., "Vibration of Steel Stacks", Proc. ASCE, No. 540, Nov. 1954.

[5.35] Dockstader, E.A., Swiger, W.F., and Ireland, E., "Resonant Vibration of Steel Stacks", Proc. ASCE, No. 541, November, 1954.

[5.36] Farquharson, F.B., "Wind Forces on Structures: Structures Subject to Oscillation", J. Struct. Div. Proc. ASCE, July, 1958, Paper 1712, pp. 1712-1 and 1712-13.

[5.37] Scrunton, C., "Structures Subject to Oscillation, Discussion of Paper by Farguharson, F.W.", J. Struct. Div., Proc. ASCE, Vol. 85, March, 1959, p. 187.

[5.38] Blume, J.A., "Structural Dynamics in Earthquake Resistant Design", Trans. ASCE, Vol. 125, pp. 1088-1139.

[5.39] Blume, J.A., "A Structural-Dynamic Analysis of Steel Plant Structures Subjected to the May 1960 Chilean Earthquakes", Bulletin of the Seism. Soc. of America, Vol. 53, No. 2, Feb. 1963, pp. 439-48.

[5.40] Kuwano, K., Nakao, Y., Ogato, Y., and Oba, K., "Study on the Seismic Responses of Cylindrical Stacks", Technical Review, Mitsubishi Heavy Industries, Ltd., May 1966, pp. 123-126.

[5.41] Maugh, L.C., and Rumman, W.C., "Dynamic Design of Reinforced Concrete Chimneys", ACI Journal, Proceedings, Vol. 64, No. 9, sept. 1967, pp. 558-567.

[5.42] American Concrete Institute, Specification for the Design and Construction of Reinforced Concrete Chimneys, Standard prepared by ACI Committee 307 (Formerly 503), 1968.

[5.43] Nakao, Y., Yamashita, M., Ogata, Y., and Matsumoto, T., "Structural Study on Assembled Four-Cylinder Type Stacks", Technical Review, Mitsubishi Heavy Industries, Ltd., Jan. 1968, pp. 1-10.

[5.44] Ledwon, J., and Gill, B., "Dynamics of Industrial Chimneys Under Seismic Loads", Intern. Civil Engin, No. 1, Jan. 1969, pp. 10-20.

[5.45] Design Essentials in Earthquake Resistant Buildings, Part 16.3, "Chimneys", Elsevier Publishing Company, N.Y., 1970, pp. 285-287.

[5.46] American National Standard Building Code Requirements for Minimum Design Loads in Buildings and Other Structures, 1972, pp. 25-34.

[5.47] Structural Engineers Association of California, Seismology Committee, Recommended Lateral Force Requirements and Commentary, 1967.

[5.48] Mitchell, W.W., "Determination of the Period of Vibration of Multi-Diameter Columns by the Method Based on Rayleigh's Principle", an unpublished work prepared for the Engineering Department of the Standard Oil Co. of Calif., San Francisco, 1962.

[5.49] Rinne, J.E., "Design of Earthquake-Resistant Structures: Towers and Chimneys", Earthquake Engineering, R.L. Wiegel-Editor, Chapt. 20, Prentice-Hall, Inc. Englewood Cliffs, N.J., 1970, pp. 495-505.

[5.50] Troitsky, M.S., "Design Recommendations Covering Combined Loading Cases and Allowable Stresses for Thin-Walled Cylinders of Large Diameter", CSICC Project #724, 1st Progress Report: Steel Chimneys, Sir George Williams University, January 1973, pp. 1-108.

[5.51] Troitsky, M.S., "Design Recommendations Covering Combined Loading Cases and Allowable Stresses for Thin-Walled Cylinders of Large Diameter", CSICC Project #724, 4th Progress Report: Design Guidelines for Steel Tubular Thin-Walled Structures, Sir George Williams University, January 1974, pp. 1-196.

[5.52] Taylor, F.W., Thompson, S.E., and Smulski, E., Concrete Plain and Reinforced, Vol. 1, John Wiley & Sons, Inc., New York, 4th ed., 1925, pp. 812-819.

[5.53] Timoshenko, S., and Woinowsky-Krieger, S., Theory of Plates and Shells, McGraw-Hill Book Co., Inc., New York, 1959, p. 210.

[5.54] Brownell, L.E., and Young, E.H., Process Equipment Design-Vessel Design, John Wiley & Sons, Inc., New York, 1968, pp. 183-197.

[5.55] Troitsky, M.S., "Design Recommendations Covering Combined Loading Cases and Allowable Stresses for Thin-Walled Cylinders of Large Diameter", CSICC Project #724, 5th Progress Report: Breech Openings at Steel Stacks, Guidelines for Analysis and Design, Concordia University, January 1978, pp. 1-131.

[5.56] Brogan, F. and Almroth, B.O., "Buckling of Cylinders With Cutouts", AIAA Journal, Vol. 8, No. 2, Feb. 1970, pp. 236-241.

[5.57] Almroth, B.O. and Homes, A.M.C., "Buckling of Shells With Cutouts, Experiment and Analysis", Int. J. Solids Structures, Vol. 8, 1972, pp. 1057-1071.

[5.58] Structural Stability Research Council, Guide to Stability Design Criteria for Metal Structures, 3rd ed., Ed. by B.G. Johnston, John Wiley & Sons, New York, U.S.A., 1976, p. 193.

[5.59] Inglis, C.E., "Stresses in a Plate due to the Presence of Cracks and Sharp Corners", Trans. Inst. Nav. Arch., Vol. 55, 1913, pp. 219-230.

[5.60] Savin, G.N., Stress Concentration Around Holes, Pergamon Press, New York, 1961.

[5.61] Greenspan, M., "Effect of a Small Hole on the Stresses in a Uniformly Loaded Plate", Quart. Appl. Mech., Vol. 2, No. 1, 1944, pp. 60-71.

[5.62] Brock, J.S., "Analytical Determination of the Stresses Around Square Holes With Rounded Corners", DTMB Report 1149, Nov. 1957.

[5.63] Muskhelishvili, N.I., Some Basic Problems of the Mathematical Theory of Elasticity, (translation by J.R.M. Radok), P. Noordhoff Ltd., Groningen, Holland, 1953.

[5.64] Brogan, F.A. and Almroth, B.O., "Buckling of Cylinders With Cutouts", AIAA Jnl. 8, pp. 236-241, (1970).

[5.65] Almroth, B.O., Brogan, F.A. and Marlowe, M.B., "Collapse Analysis for Elliptic Cones", AIAA Jnl. 9, pp. 32-36, (1971).

[5.66] Almroth, B.O. and Brogan, F.A., "Collapse Analysis for Shells of General Shape", Vol. I AFFDL-TR-71-8, Wright-Patterson Air Force Base, Ohio, (1971).

[5.67] Holmes, A.M.C. and Almroth, B.O., "An Experimental Study of the Strength and Stability of Thin Monocoque Shells With Reinforced and Unreinforced Rectangular Cutouts". Final Report Under Contract NAS9-10372, prepared for NASA/Manned Spacecraft Centre, Houston, Texas.

[5.68] Almroth, B.O., Brogan, F.A. and Zele, F., "Collapse Analysis for Shells of General Shape", Vol. 11, User's Manual for STAGS, AFFDL-TR-71-8, Wright-Patterson Air Force Base, Ohio, (1971).

[5.69] Turner, M.J., Clough, R.W., Martin, H.C. and Topp, L.J., "Stiffness and Deflection Analysis of Complex Structures", J. Aero, S., Vol. 23, No. 9, 1956, pp. 805-823.

[5.70] Argyris, J.H. and Kelsey, S., Energy Theorems and Structural Analysis, Butterworths, London, 1960 (Collection of papers published in Aircraft Engineering in 1954 and 1955).

[5.71] Bathe, K.J., Wilson, E.L. and Peterson, F.E., SAP IV; "A Structural Analysis Program for Static and Dynamic Response of Linear Systems", Report EERC-73-11, Earthquake Engineering Research Centre, University of California, Berkeley, June 1973 (PB-221-967/3, N.T.I.S.).

[5.72] Zienkiewicz, O.C., The Finite Element Method in Engineering Science, McGraw-Hill Company, London, 1971.

[5.73] Wilson, E.L., Taylor, R.L., Doherty, W.P. and Ghaboussi, J., Incompatible Displacement Models, Numerical and Computer Methods in Structural Mechanics, edited by S.J. Fenves, et al., Academic Press, Inc. N.Y. and London, 1973.

[5.74] DeSalvo, G.J. and Swanson, J.A., ANSYS — Engineering Analysis System: User's Manual, Swanson Analysis Systems, Inc., 879 Pine View Drive, Elizabeth, Pennsylvania, 15037.

[5.75] American Institute of Steel Construction, Manual of Steel Construction, 6th ed., 1967, New York.

[5.76] Johnston, B.G., Guide to Stability Design Criteria for Metal Structures, 3rd ed., John Wiley & Sons, New York, 1967, pp. 26-77.

multilevel guyed stacks

6.1 Parameters Used in the Analysis of a Guyed Stack

6.1.1 Introduction

This part treats a method used to find certain parameters which are necessary to analyze guyed stacks having a uniform cross-section with constant wall thickness. The stack cross-section, thus, has a constant moment of inertia. This approach is essentially similar to that of Kolousek [6.1]. The method of analysis consists in first determining the flexibility of stack-supports at cable connections making use of the fundamental concepts of elasticity and considering the deflected shape of the guys. Having achieved this, the stack is analyzed as a continuous beam having elastic supports [6.2].

6.1.2 Guyed Stack Structural System

Figure 6.1 shows the elevation and plan of a guyed stack. The stack oh is assumed fixed at the bottom point 0. Four guy cables at right-angles to each other are attached to the stack at each of the anchor points e, f and g. The guys are anchored in the ground at points a, b, c and d. For convenience in all subsequent calculations a coordinate system is selected for the stack having its origin at point 0 and the three coordinate axes as shown in Figure 6.1. In addition to its self-weight, the guyed stack is subjected to a lateral uniform wind load acting at an angle β with respect to the x-axis, as shown.

6.1.3 Support Flexibility

The set of guys attached to the stack are provided to prevent the excessive deflection of the stack-supports. Under its self-weight, the guy assumes a catenary shape which does not differ much from the parabolic shape. However, for all calculation purposes, the shape of the guy is assumed as parabolic since the guy, in addition to its self-weight, carries extra loads of isolators and tenders. Under the action of dead loads, the guy is stretched by the axial force F_0 while it deflects transversely because of its self-weight, the maximum displacement being δ_0, Fig. 6.2 (b). In additon, due to the wind action, an axial tension F is developed in the guy with the transverse load component being Q and the maximum displacement, δ, Fig. 6.2 (c).

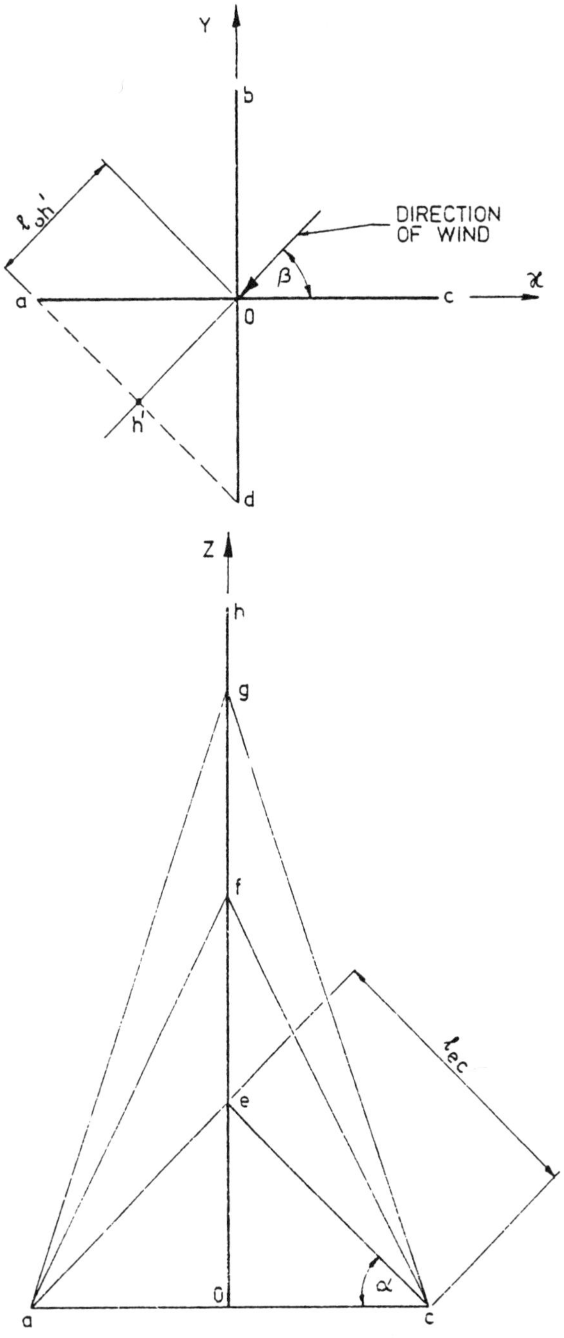

FIGURE 6.1 — General arrangement of a guyed stack.

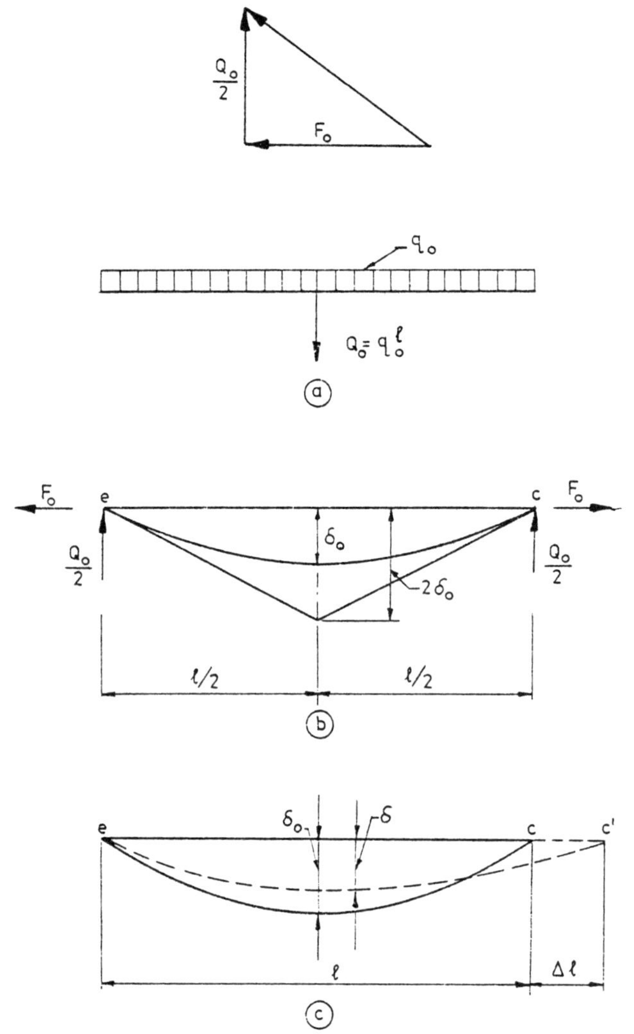

FIGURE 6.2 — Uniform loads on guys.

The change in the length of cord can be expressed by the following equation:

$$\Delta\ell = -\frac{8}{3}\frac{\delta^2}{\ell} + \frac{F\ell}{AE} + \frac{8\delta_0^2}{3\ell} - \frac{F_0\ell}{AE} \qquad (6.1)$$

In equation (6.1), the first and third terms express the difference in the length of the parabolic arch and that of the cord; while the second and fourth terms give the elastic elongation under force, F. A is the cross-sectional area of the guy while E is its modulus of elasticity.

Let us denote the horizontal displacement of any guy-support by y and the angle between the wind direction and the horizontal axis joining two opposite guys, Fig. 6.1, by β and since

$$\frac{2\delta}{(\frac{\ell}{2})} = \frac{Q}{2F} \qquad (6.2)$$

we get

$$\Delta\ell = \frac{Q^2\ell}{24F^2} + \frac{F\ell}{AE} + \frac{Q_0^2\ell}{24F_0} = \frac{F_0\ell}{AE} \qquad (6.3)$$

Relationship between the support displacement y_e and $\Delta\ell_{ec}$ may be found by considering the deformed position of the stack and guy cables under the horizontal wind pressure, Fig. 6.3.

From the triangle ee_1e_2, we have

$$\Delta\ell_{ec} = y_e\cos\alpha$$

However, since the displacement y_e is proportional to the component of the wind load acting along OC, Figure 6.3, it follows that

$$\Delta\ell_{ec} = y_e\ \cos\alpha.\cos\beta;$$

and

$$y_e = \frac{\Delta\ell_{ec}}{\cos\sigma.\cos\beta} \qquad (6.4)$$

Using Equations (6.3) and (6.4), the displacement, y_e can be expressed in terms of the axial force, F_{ec}, as

$$y_e = \frac{A_1}{F_{ec}^2} + A_2\ F_{ec} + A_3 \qquad (6.5)$$

where A_1, A_2 and A_3 are given by

$$A_1 = -\frac{Q^2\ell_{ec}}{24\cos\alpha\ \cos\beta} \qquad (6.6)$$

and

$$A_2 = \frac{\ell_{ec}}{AE\ \cos\alpha\ \cos\beta} \qquad (6.7)$$

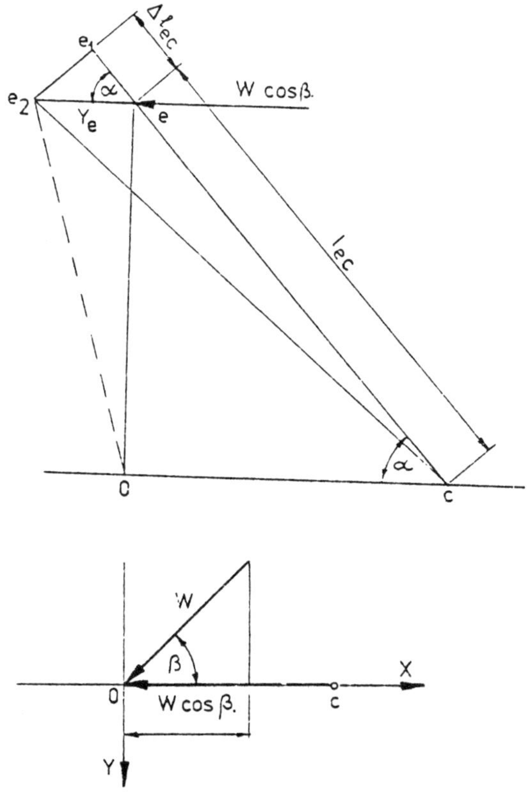

FIGURE 6.3 — Displacement of guy connection e.

$$A_3 = \left(\frac{Q_0^2 \ell_{ec}}{24F_0} - \frac{F_0 \ell_{ec}}{AE}\right) \frac{1}{\cos\alpha \, \cos\beta} \qquad (6.8)$$

At any support of the guyed stack, the reaction of the beam is given by the horizontal components of the resultant forces of the guyed-cables connected to it. For example, for support e, the reaction V_e is found as follows:

Assuming the tension in cable ec to be F_{ec}, its horizontal component, as shown in Figure 6.4, is

$$F_{ec}\cos\alpha = F_{ec} \cdot \frac{\ell_{oc}}{\ell_{ec}} = F_{ec} \cdot \frac{\ell_{oh}\sqrt{2}}{\ell_{ec}} \qquad (6.9)$$

By resolving it into two components — one in the wind direction and one perpendicular to it, we obtain

$$2F_{w_{ec}}^2 = \left[F_{ec} \frac{\ell_{oh}\sqrt{2}}{\ell_{ec}}\right]^2$$

or

$$F_{w_{ec}} = F_{ec} \frac{\ell_{oh}\sqrt{2}}{\ell_{eh}} \qquad (6.10)$$

Because of the symmetry of the structures, a force equal to $F_{w_{ec}}$ is contributed by member eb in the direction of the wind. Therefore, we have

$$2F_{w_{ec}} = 2F_{ec} \frac{\ell_{oh}}{\ell_{ec}} \qquad (6.11)$$

Under the wind direction shown, the tension in cables ec and eb will increase while the initial tension is cables ea and ed will slacken. Assuming tension in cable ed to be F_{ed}, its horizontal component is

$$-F_{ed}\cos\alpha = -F_{ed} \frac{\ell_{oh}\sqrt{2}}{\ell_{ec}} \qquad (6.12)$$

By resolving this force into two components — one in the wind direction and the other perpendicular to it, along with an additional contribution from member ea, we obtain:

$$2F_{w_{ed}} = -F_{ed} \frac{\ell_{oh}}{\ell_{ec}} \qquad (6.13)$$

Therefore, the total horizontal force V_e, at support e is:

$$V_e = 2[F_{ec} - F_{ed}] \frac{\ell_{oh}}{\ell_{ec}} \qquad (6.14)$$

where ℓ_{ec} is the length of member ec and ℓ_{oh} is the horizontal distance along the wind direction as shown in Figure 6.4. Equations (6.5) and (6.14) establish a relationship between the horizontal displacement y_e and reaction V_e at the stack-support e. An examination of the two equations reveals that this relationship is nonlinear. However, a simplified linear relationship between these two parameters of the type shown in Eq. (6.15), is more practical for design purposes

$$y_e = K_e + \xi_e V_e \qquad (6.15)$$

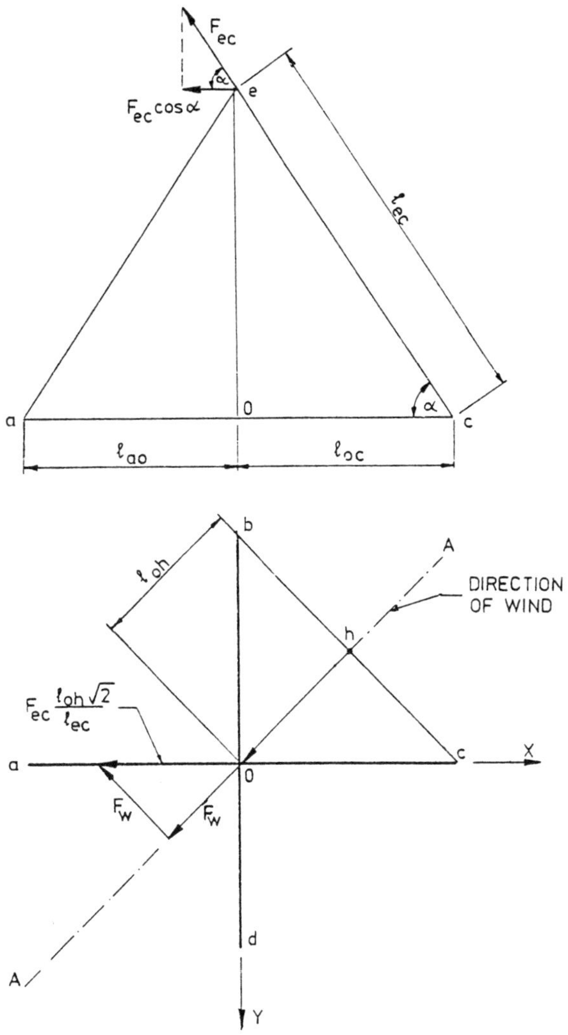

FIGURE 6.4 — Determination of horizontal guy reactions.

where K_e and ξ_e are constants to be determined from the plots of y_e, F_{ec} and V_e as explained below.

The horizontal displacement, y_e, of the support e as given by Eq. (6.5) can be graphically represented for a particular stack, i.e., where all the unknown parameters of Eqs. (6.6), (6.7) and (6.8) are given. A plot of y_e versus the axial force F_{ec} and F_{ed} in cables ec and ed, respectively, is shown in Fig. 6.5.

At this point in our analysis, it will be assumed that under any particular loading condition, e.g., wind, the guyed stack has been stress-analyzed by any of the conventional methods of structural analysis and, as such, the axial forces in cables ec and ed are known. In some cases, if the initial tension in the cables due to pretensioning is known, then that value would be used as the axial force, F, in the guy-cables.

The variation of reaction, V_e, at support e with respect to the horizontal displacement, y_e, at the same point, is shown in Figure 6.6. This diagram has been made with the help of Figure 6.3 and Eq. (6.14) by arbitrarily selecting several values of (F_{ec}-F_{ed}) from Figure 6.3, substituting them into Eq. (6.14) and calculating a corresponding value of V_e, and then making a plot of y_e versus

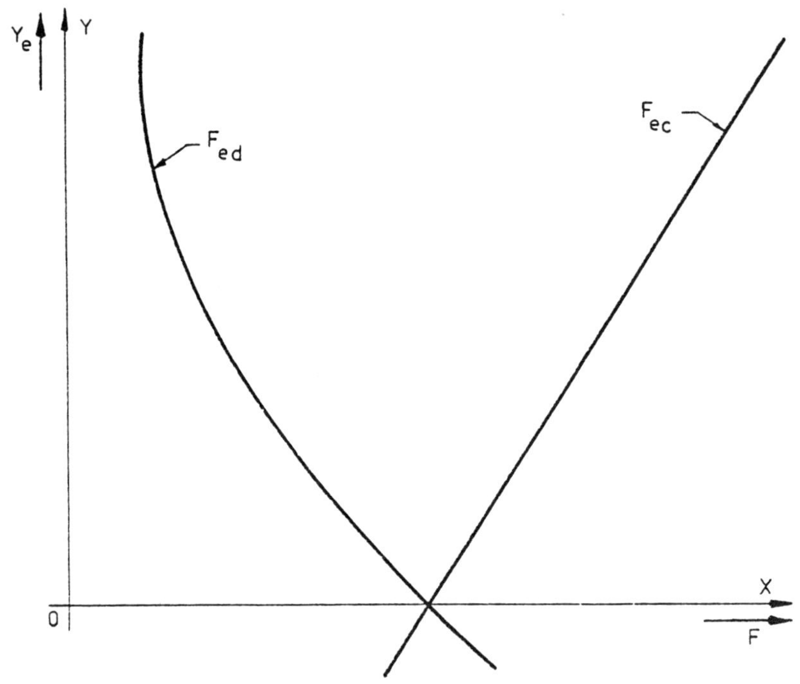

FIGURE 6.5 — Diagram of the horizontal displacement, y_e, of support e with respect to the axial forces F_{ec} and F_{ed} in guy-cables ec and ed, respectively.

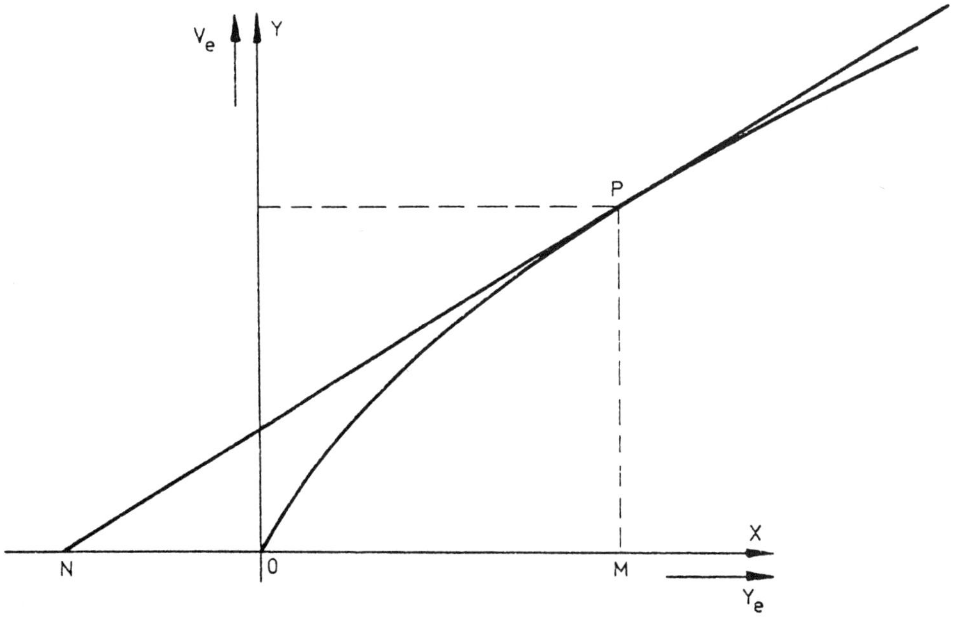

FIGURE 6.6 — Plot of the horizontal displacement, y_e, and horizontal shear, V_e, of support e of the guyed-stack.

V_e. Corresponding to the given axial force in any of the two cables ec or ed, a value of y_e is determined from Figure 6.3 which is transferred to Figure 6.4, as indicated by point P. A tangent to the curve is drawn at this point and extended down to intersect the horizontal axis ox at point N. The distance ON gives the value of K in Equation (6.15); whereas parameter ξ is given by the ratio NM/PM.

By following a similar procedure for all the three supports e, f and g of the guyed-stack, the following set of equations is obtained:

$$y_e = K_e + \xi_e V_e$$

$$y_f = K_f + \xi_f V_f$$

$$y_g = K_g + \xi_g V_g \qquad (6.16)$$

where support-displacements — y_e, y_f and y_g; and support-lateral-shears V_e, V_f and V_g are unknown

6.2 Analysis of the Stack by the Method of Deformation

For the analysis of the stack, Figure 6.7, the method of deformation is used. More specifically, it will be used for determining the transverse displacements ye, yf and y_g of the stack-supports e, f and g, respectively. In addition, this technique gives the corresponding rotation γ_e, γ_f and γ_g of these supports as well. The stack-mast is analyzed as a continuous beam supported on points o, e, f and g. Support 0 is assumed position and direction fixed, while points e, f and g are assumed to be elastic supports experiencing both rotation and lateral displacement. The stack is of an invariable cross-section and each one of its spans is subjected to a uniform but different transverse wind load.

Equations of elastic equilibrium for individual spans of the mast, e.g., fe, eo, etc., will be developed corresponding to their end-support conditions.

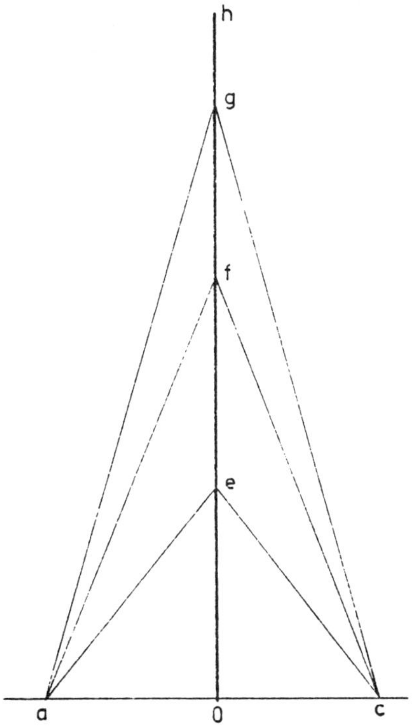

FIGURE 6.7 — Elevation of the guyed-stack.

6.2.1 Equilibrium Equations for Span e-f

Figure 6.8 shows the elastically-deformed shape of the span ef of the stack oh which deforms to e′f′ due to external wind load giving rise to moments M_{fe}, M_{ef} and M_{fg} and shear forces V_{ef}, V_{fe} and V_{fg} acting on the span ef as shown. I is the moment of inertia of the stack cross-section and E is the modulus of elasticity of the material of the stack. MF_{fe} is the negative moment induced at support f because of the continuity of the stack over supports e, f and g; and B_{fe} is the transverse shear at support f corresponding to this moment. V_{fp} is the contribution to the transverse shear at support f from the guy cables.

Since the stack is in equilibrium, the following conditions hold:

$$M_{fe} + M_{fg} = 0$$

$$V_{fe} + V_{fg} + V_{fp} = 0 \qquad (6.17)$$

where M_{fe}, M_{fg} and V_{fp} are given by the expressions

$$M_{fe} = \frac{2EI}{L_{fe}}\left[2\gamma_f + \gamma_e - 3\frac{y_f - y_e}{L_{fe}}\right] + MF_{fe}$$

$$M_{ef} = \frac{2EI}{L_{ef}}\left[2\gamma_e + \gamma_f - 3\frac{y_e - y_f}{L_{ef}}\right] + MF_{ef}$$

$$V_{fe} = \frac{6EI}{L_{fe}^2}\left[-\gamma_f - \gamma_e + 2\frac{y_f - y_e}{L_{fe}}\right] + B_{fe} \qquad (6.18)$$

$$V_{ef} = \frac{6EI}{L_{ef}^2}\left[-\gamma_e - \gamma_f + 2\frac{y_e - y_f}{L_{ef}}\right] + B_{ef}$$

$$B_{ep} = \frac{y_e}{\xi_e} - \frac{K_e}{\xi_e}$$

where L_{ef} (= L_{fe}) is the length of span ef.

6.2.2 Equilibrium Equations for Spans e-o and f-g

A set of equilibrium equations for spans eo (point 0 position as well as direction fixed) and span fg (hinge at point g) are given below. Symbols for moments, shears, translations and rotations used in this section carry the same connotations as explained for the symbols used in equilibrium equations corresponding to span ef.

6.2.2.1 Span e-o

$$M_{eo} = \frac{2EI}{L_{eo}}\left[2\gamma_e + \gamma_o - 3\frac{y_e - y_o}{L_{eo}}\right] + MF_{eo}$$

$$V_{eo} = \frac{6EI}{L_{eo}^2}\left[-\gamma_e - \gamma_o + 2\frac{y_e - y_o}{L_{eo}}\right] + B_{eo} \qquad (6.19)$$

NOTE: Since point 0 is position as well as direction-fixed, both γ_o and y_o are zero.

6.2.2.2 Span f-g

$$M_{fg} = \frac{3EI}{L_{fg}}\left[\gamma_f - \frac{y_f - y_g}{L_{fg}}\right] + MF_{fg}$$

$$V_{fg} = \frac{3EI}{L_{fg}^2}\left[-\gamma_f + \frac{y_f - y_g}{L_{fg}}\right] + B_{fg}$$

$$B_{fp} = \frac{y_f}{\xi_f} - \frac{K_f}{\xi_f} \text{ ; and}$$

$$B_{gp} = \frac{y_g}{\xi_g} - \frac{K_g}{\xi_g} \qquad (6.20)$$

Successive application of Equations (6.17) to (6.20) to supports e, f and g results in five simultaneous equations involving five unknowns viz. y_e, y_f and y_g; and γ_e, γ_f (γ_g usually not required in analysis) — whose values, once determined by the solution of the five simultaneous equations and substituted into Equation (6.16) give the support shearing forces V_e, V_f and V_g. The set of simultaneous equations is as follows:

$$\left[\frac{4EI}{L_{eo}} + \frac{4EI}{L_{ef}}\right]\gamma_e - \left[\frac{6EI}{L_{eo}^2} + \frac{6EI}{L_{ef}^2}\right]y_e + \left[\frac{2EI}{L_{ef}}\right]\gamma_f + \left[\frac{6EI}{L_{ef}^2}\right]y_f = -MF_{eo} - MF_{ef}$$

$$-\left[\frac{2EI}{L_{eo}^2} + \frac{2EI}{L_{ef}^2}\right]\gamma_e + \left[\frac{4EI}{L_{eo}^3} + \frac{4EI}{L_{ef}^3} + \frac{1}{\xi_e}\right]y_e - \left[\frac{2EI}{L_{ef}^2}\right]\gamma_f - \left[\frac{4EI}{L_{ef}^3}\right]y_f = \frac{K_e}{\xi_e} - B_{eo} - B_{ef}$$

$$\left[\frac{4EI}{L_{fe}} + \frac{3EI}{L_{fg}}\right]\gamma_f + \left[\frac{2EI}{L_{fe}}\right]\gamma_e - \left[\frac{6EI}{L_{fe}^2} + \frac{3EI}{L_{fg}^2}\right]y_f + \left[\frac{2EI}{L_{fe}^2}\right]y_e + \left[\frac{3EI}{L_{fg}^2}\right]y_g = -MF_{fe} - MF_{fg}$$

$$-\left[\frac{2EI}{L_{fe}^2} + \frac{3EI}{L_{fg}^2}\right]\gamma_f + \left[\frac{4EI}{L_{fe}^3} + \frac{3EI}{L_{fg}^2} + \frac{1}{\xi_f}\right]y_f - \left[\frac{2EI}{L_{fe}^2}\right]\gamma_e - \left[\frac{4EI}{L_{fe}^3}\right]y_e - \left[\frac{3EI}{L_{fg}^2}\right]y_g = \frac{K_f}{\xi_f} - B_{fg} - B_{fe}$$

$$-\left[\frac{3EI}{L_{fg}^2}\right]\gamma_f + \left[\frac{3EI}{L_{fg}^3}\right]y_f - \left[\frac{3EI}{L_{fg}^3} + \frac{1}{\xi_g}\right]y_g = \frac{K_g}{\xi_g} - B_{fg} \qquad (6.21)$$

6.2.3 Additional Formulas Used in the Static Analysis of the Guyed Stack

Some other pertinent formulas used in the static analysis of a guyed stack, suggested by Max Zar [6.3], are given below. They have been used in this analysis along with those of Kolousek's.

a. The relation between the horizontal component H_e of the guy tension and sag s of the guy is given by:

$$s = \frac{w\,\ell_{ao}^2}{8H_e} \qquad (6.22)$$

where

- w = average weight of cable per unit of horizontal length; and
- ℓ_{ao} = horizontal distance from the center of the stack to the guy anchor.

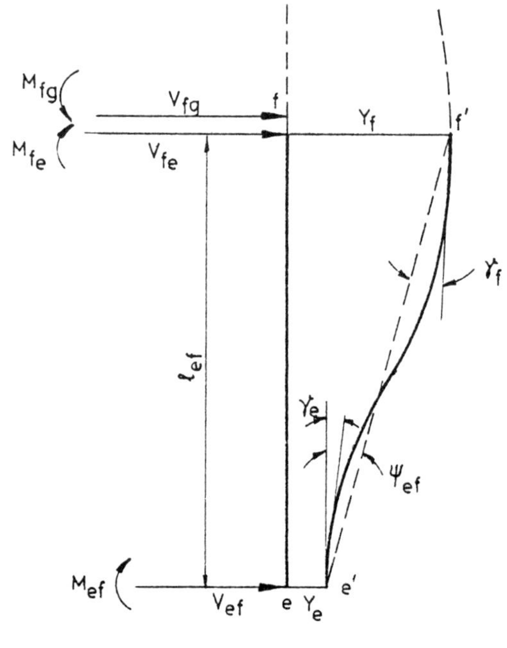

FIGURE 6.8 — Elastic deformation of span e-f.

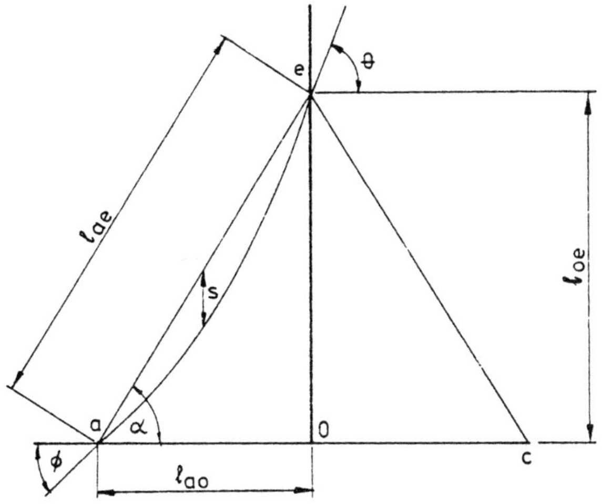

FIGURE 6.9 — Elevation of the guyed stack (showing lower set of cables only).

b. Approximate deflection of the stack at its connection point with guy cables viz., e is given by:

$$y_e = \frac{5\ S_e\ \ell_{ae}}{3\ A_c E_c} \qquad (6.23)$$

where

S_e = the shear force at point e of the stack (contribution from cables only), kips.

ℓ_{ae} = chord dimension of cable, ft.

A_c = cross-sectional area of cable, in^2; and

E_c = modulus of elasticity of cable, psi.

c. The length L of cable stretched between points a and e is given approximately by

$$L = \ell_{ao}\left(\frac{1}{\cos\alpha} + \frac{8s^2\ \cos^3\alpha}{3\ell^2_{ao}}\right) \qquad (6.24)$$

d. Another expression often used for estimating the sag, s of a slack cable on the leeward side of the mast is based on the reduction in its initial tension due to wind and is given by

$$s^2 = \frac{3(\ell_{ao}-y_e)}{8\ \cos^4\alpha}\ [L\ \cos\alpha\ -\ (\ell_{ao}-y_e)] \qquad (6.25)$$

The change in L resulting from change in stress in the slack cable is taken into account in using this equation.

e. Angle θ as shown in Figure 6.9 is determined from:

$$\tan\theta = \frac{w\ell_{ao}}{2S_e} + \frac{\ell_{oe}}{\ell_{ao}} \qquad (6.26)$$

f. Guys at various levels of the stack are so designed that the deflections at their upper ends are proportional to their heights above ground. The required chord length for a cable to maintain a given deflection y_e is given by

$$\ell_{ae} = \left[\left(\frac{R}{2p}\right)^2 + \frac{3A_cE_cy_e}{5p}\right]^{\frac{1}{2}} - \frac{R}{2p} \qquad (6.27)$$

where R is the reaction at point e of the stack due to wind; and

p = load due to wind per foot of cable parallel to the ground and perpendicular to a vertical plane

It should be noted that this formula is applicable only where the guys are anchored at different points on the ground.

6.3 Buckling Analysis of a Guyed Stack

6.3.1 Introduction

The buckling analysis of guyed stacks is a relatively complicated problem for two main reasons:

1. It is difficult to express the elasticity of stack-supports in a linear relationship (which in actuality is a nonlinear one) and;
2. The lateral displacements of stack-supports produce changes in cable tensions which in turn affect the axial forces of the stack. These axial forces acting as functions of the support-displacements cannot accurately predict the buckling load of the stack.

This difficulty is further complicated by the nonlinearity of the relationship between the elastic behavior of the stack-supports and their lateral displacements. These factors are illustrated by studying the buckling behavior of continuous beams subject to the settlement of support in addition to being transversely loaded. The magnitude of axial forces induced in various beam spans due to the lateral settlement of its supports is considered trivial and neglected in its buckling analysis — usually performed by the method of deformations [6.4, 6.5].

Analogically, the stack is considered a continuous beam over elastic supports subject to lateral displacements. Notwithstanding the fact that this assumption introduces a certain amount of approximation into the buckling analysis of a guyed stack, it is nevertheless considered to be reasonable for all practical purposes.

6.3.2 Buckling Load for a Guyed Stack

Figure 6.10 shows a guyed stack subjected to lateral wind. Under the action of this wind, axial forces are induced in the stack. These forces combined with the self-weight of the stack comprise the loads used in the determination of the buckling load for the stack. In this analysis, expressions are developed for determining the buckling loads of individual spans to begin with, and combined later on into composite expressions for predicting the overall buckling of the stack in terms of an arbitrarily-assumed numerical value of a safety factor.

The stack oh is analyzed as a continuous beam supported on elastic supports, o, e, f and g subjected to different axial loads in various spans e.g., force N_{ef} in span ef.

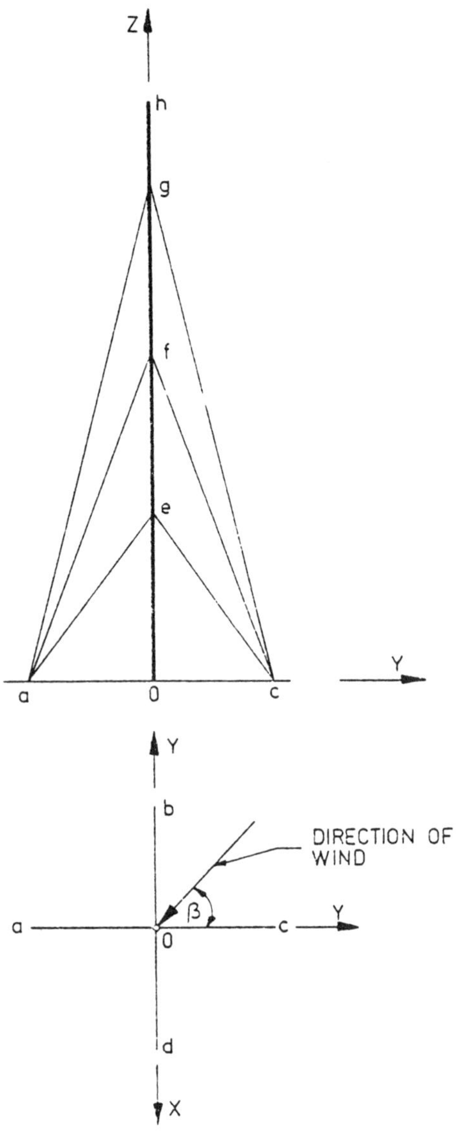

FIGURE 6.10 — Guyed stack with four anchor points.

In addition, the stack supports e, f and g undergo translations and rotations y_e and γ_e; y_f and γ_f; and y_g and γ_g respectively. Support 0 is assumed both position and direction fixed. These joint translations and rotations give rise to moments and shear forces whose magnitude is indirectly affected by the axial forces in various spans of the stack oh.

It is assumed that the buckled system, since it is in equilibrium, satisfies the equilibrium equations of moments and shears at various stack-supports. For example, the set of equilibrium equations at support e is

$$M_{eo} + M_{ef} = 0$$

$$V_{eo} + V_{ef} + B_{ep} = 0 \qquad (6.28)$$

where M_{eo}, M_{ef}, V_{eo} and V_{ef} carry their usual connotation as explained in Part 6.1; whereas B_{ep} is the contribution to the transverse shear at support e from the guy cables and is given by

$$B_{ep} = \frac{y_e}{\xi_e}$$

Buckling equations for various spans of the stack are developed separately. As an illustration, the buckling equation for span e-f is derived in this section by considering its free-body diagram as shown in Figure 6.11.

Span e-f is subjected to these external forces:

1. Axial force N_{ef} and N_{fe} at supports e and f respectively;
2. Bending moments M_{ef} and M_{fe} at supports e and f respectively;
3. Shear forces V_{ef} and V_{fe} at supports e and f respectively;
4. Transverse displacements y_e and y_f respectively; and
5. Rotations γ_e and γ_f at supports e and f respectively.

These external loads acting on span e-f of the stack, assumed to have a constant cross-section in this analysis, give rise to the following equation:

$$\frac{d^4y}{dz^4} - \mu\frac{N_{ef}}{EI_{ef}}\frac{d^2y}{dz^2} = 0 \qquad (6.29)$$

where

μ = an arbitrarily-assumed factor of safety

E = the modulus of elasticity of the stack-material

I_{ef} = the moment of inertia of the stack cross-section

Integration of Equation (6.29) yields the following solution for displacement y of the stack

$$y = A_1 + A_2 z + A_3 \cos\lambda(\tfrac{z}{\ell}) + A_4 \sin\lambda(\tfrac{z}{\ell}) \qquad (6.30)$$

where

$$\lambda = \ell_{ef}\sqrt{\frac{\mu N_{ef}}{EI_{ef}}} \qquad (6.31)$$

and

A_1, A_2, A_3 and A_4 = the constants of integration to be determined from the following boundary conditions:

$$\text{At} \quad z = 0;\ y = y_e \text{ and } \gamma = \gamma_e$$
$$\text{At} \quad z = \ell_{ef};\ y = y_f \text{ and } \gamma = \gamma_f \qquad (6.32)$$

resulting in

$$A_1 = y_e;$$

$$A_2 = \gamma_f\left[\frac{\sin\lambda}{\sin\lambda - \lambda\cos\lambda} - \frac{y_f - y_e}{\ell_{ef}}\right]\frac{\lambda\cos\lambda}{(\sin\lambda - \lambda\cos\lambda)};$$

$$A_3 = 0; \text{ and } A_4 = \frac{-\ell_{ef}\gamma_f + y_f - y_e}{\sin\lambda - \lambda\cos\lambda}$$

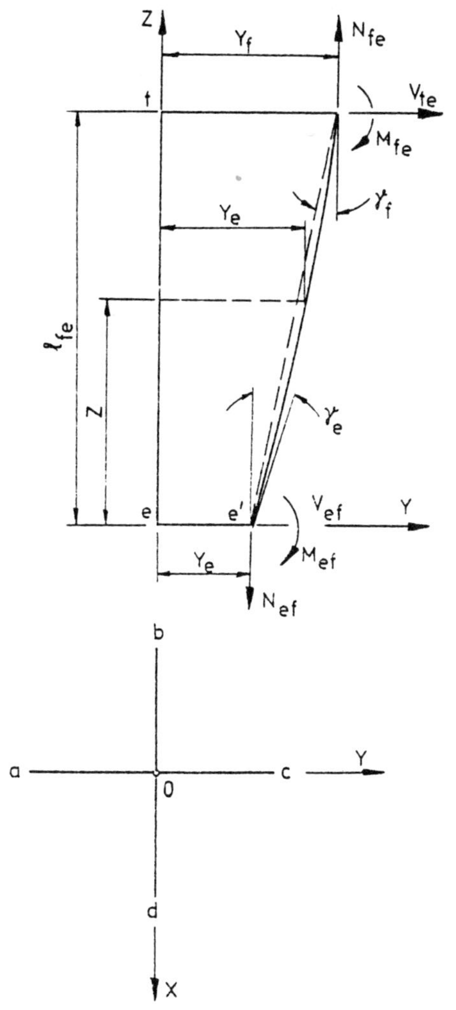

FIGURE 6.11 — Free body diagram of span e-f.

Substituting these values of the unknown coefficients in Equation (6.30), the following expression results

$$y = y_e + \psi z + (\gamma_f - \psi)\frac{\sin\lambda}{\sin\lambda - \lambda\cos\lambda}\left[z - \frac{\ell_{ef}}{\sin\lambda} - \sin\lambda\frac{z}{\ell_{ef}}\right] \qquad (6.33)$$

where

$$\psi = \left[\frac{y_f - y_e}{\ell_{ef}}\right]$$

Equation (6.33) is a generalized equation which determines the shear forces and bending moments in a stack on the assumption that the end displacements of its various spans are known. Equations similar to Eq. (6.33) are derived for spans oe and fg taking into account their respective boundary conditions, e.g., in this particular case, point 0 is both position and direction fixed and point g is free. Successive differentiation of Equation (6.33) gives us bending moments and shear forces at any particular support of the stack o-h, which when substituted into Equation (6.28), gives us a system of equations for the buckling analysis of a continuous beam.

Expressions for moments and shears at support e obtained by using the above approach, are as follows

$$M_{ef} = \gamma_e\left[\frac{\lambda^2_{ef}\sin\lambda_{ef}}{\ell'_{ef}(\sin\lambda_{ef} - \lambda_{ef}\cos\lambda_{ef})}\right] + y_f\left[\frac{\lambda^2_{ef}\sin\lambda_{ef}}{\ell_{ef}\ell'_{ef}(\sin\lambda_{ef} - \lambda_{ef}\cos\lambda_{ef})}\right] - \left[\frac{\lambda^2_{ef}\sin\lambda_{ef}}{\ell_{ef}\ell'_{ef}(\sin\lambda_{ef} - \lambda_{ef}\cos\lambda_{ef})}\right]$$

$$V_{fe} = \gamma_e\left[\frac{\lambda^2_{ef}\sin\lambda_{ef}}{\ell_{ef}\ell'_{ef}(\sin\lambda_{ef} - \lambda_{ef}\cos\lambda_{ef})}\right] + y_f\left[\frac{\lambda^3_{ef}\cos\lambda_{ef}}{\ell^2_{ef}\ell'_{ef}(\sin\lambda_{ef} - \lambda_{ef}\cos\lambda_{ef})}\right] - y_e\left[\frac{\lambda^3_{ef}\cos\lambda_{ef}}{\ell^2_{ef}\ell'_{ef}(\sin\lambda_{ef} - \lambda_{ef}\cos\lambda_{ef})}\right]$$

$$V_{ef} = \gamma_e\left[\frac{\lambda^2_{ef}\cos\lambda_{ef}}{\ell_{ef}\ell'_{ef}(\sin\lambda_{ef} - \lambda_{ef}\cos\lambda_{ef})}\right] - y_f\left[\frac{\lambda^3_{ef}\cos\lambda_{ef}}{\ell^2_{ef}\ell_{ef}(\sin\lambda_{ef} - \lambda_{ef}\cos\lambda_{ef})}\right] + y_e\left[\frac{\lambda^3_{ef}\cos\lambda_{ef}}{\ell^2_{ef}\ell'_{ef}(\sin\lambda_{ef} - \lambda_{ef}\cos\lambda_{ef})}\right] \qquad (6.34)$$

where

$$\frac{1}{\ell'_{ef}} = \frac{EI_{ef}}{\ell_{ef}}$$

Sets of equations for supports, o, f and g, similar to Equation (6.34) when written and combined with it result in the following system of equations for the buckling analysis of the stack

$$\left[\frac{\lambda^2_{oe}\sin\lambda_{oe}}{\ell'_{oe}(\sin\lambda_{oe}-\lambda_{oe}\cos\lambda_{oe})} + \frac{\lambda^2_{ef}\sin\lambda_{ef}}{\ell'_{ef}(\sin\lambda_{ef}-\lambda_{ef}\cos\lambda_{ef})}\right]\gamma_e + \left[\frac{-\lambda^2_{oe}\sin\lambda_{oe}}{\ell_{oe}\ell'_{oe}(\sin\lambda_{oe}-\lambda_{oe}\cos\lambda_{oe})} + \frac{\lambda^2_{ef}\sin\lambda_{ef}}{\ell_{ef}\ell'_{ef}(\sin\lambda_{ef}-\lambda_{ef}\cos\lambda_{ef})}\right]y_e + \left[\frac{-\lambda^2_{ef}\sin\lambda_{ef}}{\ell_{ef}\ell'_{ef}(\sin\lambda_{ef}-\lambda_{ef}\cos\lambda_{ef})}\right]y_f = 0 \quad (6.35)$$

$$\left[\frac{-\lambda^2_{oe}\sin\lambda_{oe}}{\ell_{oe}\ell'_{oe}(\sin\lambda_{oe}-\lambda_{oe}\cos\lambda_{oe})} + \frac{\lambda^2_{ef}\sin\lambda_{ef}}{\ell_{ef}\ell'_{ef}(\sin\lambda_{ef}-\lambda_{ef}\cos\lambda_{ef})}\right]\gamma_e + \left[\frac{\lambda^3_{oe}\cos\lambda_{oe}}{\ell^2_{oe}\ell'_{oe}(\sin\lambda_{oe}-\lambda_{oe}\cos\lambda_{oe})} + \frac{\lambda^3_{ef}\cos\lambda_{ef}}{\ell^2_{ef}\ell'_{ef}(\sin\lambda_{ef}-\lambda_{ef}\cos\lambda_{ef})} + \frac{1}{\xi_e}\right]y_e + \left[\frac{-\lambda^3_{ef}\cos\lambda_{ef}}{\ell^2_{ef}\ell'_{ef}(\sin\lambda_{ef}-\lambda_{ef}\cos\lambda_{ef})}\right]y_f = 0 \quad (6.36)$$

$$\left[\frac{-\lambda^2_{ef}\sin\lambda_{ef}}{\ell_{ef}\ell'_{ef}(\sin\lambda_{ef}-\lambda_{ef}\cos\lambda_{ef})}\right]\gamma_e + \left[\frac{-\lambda^3_{ef}\cos\lambda_{ef}}{\ell^2_{ef}\ell'_{ef}(\sin\lambda_{ef}-\lambda_{ef}\cos\lambda_{ef})}\right]y_e + \left[\frac{\lambda^3_{ef}\cos\lambda_{ef}}{\ell^2_{ef}\ell'_{ef}(\sin\lambda_{ef}-\lambda_{ef}\cos\lambda_{ef})} + \frac{1}{\xi_f}\right]y_f = 0 \quad (6.37)$$

$$\left[\frac{-\lambda_{ef}\sin\lambda_{ef}}{\ell_{ef}\ell'_{ef}(\sin\lambda_{ef}-\lambda_{ef}\cos\lambda_{ef})} + \frac{\lambda_{fg}\sin\lambda_{fg}}{\ell_{fg}\ell'_{fg}(\sin\lambda_{fg}-\lambda_{fg}\cos\lambda_{fg})}\right]\gamma_f + \left[\frac{\lambda^3_{ef}\cos\lambda_{ef}}{\ell^2_{ef}\ell'_{ef}(\sin\lambda_{ef}-\lambda_{ef}\cos\lambda_{ef})} + \frac{\lambda^3_{fg}\cos\lambda_{fg}}{\ell^2_{fg}\ell'_{fg}(\sin\lambda_{fg}-\lambda_{fg}\cos\lambda_{fg})} + \frac{1}{\xi_f}\right]y_f + \left[\frac{-\lambda^3_{fg}\cos\lambda_{fg}}{\ell^2_{fg}\ell'_{fg}(\sin\lambda_{fg}-\lambda_{fg}\cos\lambda_{fg})}\right]y_g = 0 \quad (6.38)$$

$$\left[\frac{\lambda_{fg}\sin\lambda_{fg}}{\ell_{fg}\ell'_{fg}(\sin\lambda_{fg}-\lambda_{fg}\cos\lambda_{fg})}\right] + \left[\frac{-\lambda^3_{fg}\cos\lambda_{fg}}{\ell^2_{fg}\ell'_{fg}(\sin\lambda_{fg}-\lambda_{fg}\cos\lambda_{fg})}\right] + \left[\frac{\lambda^3_{fg}\cos\lambda_{fg}}{\ell^2_{fg}\ell'_{fg}(\sin\lambda_{fg}-\lambda_{fg}\cos\lambda_{fg})} + \frac{1}{\xi_g}\right]y_g = 0 \quad (6.39)$$

The values of λ_{oe}, λ_{ef} and λ_{fg} are calculated by arbitrarily assuming several values of the safety factor μ starting with unity. These values of λs are then substituted into the system of Equations (6.35) to (6.39). The only way to have a non-trivial solution of this system is to have the determinant of the coefficients of γ_e, γ_f, y_e, y_f and y_g equal to zero.

A curve is plotted between the values of this determinant Δ along the ordinate and the safety factor μ along the abscissa, as shown in Figure (6.12). The values of μ corresponding to the point on the abscissa where the (Δ-μ) curve intersects it for the first time represents the minimum factor of safety against buckling. Values of λ_{oe}, λ_{ef} and λ_{fg} are calculated corresponding to this μ and the buckling load for each span is determined separately.

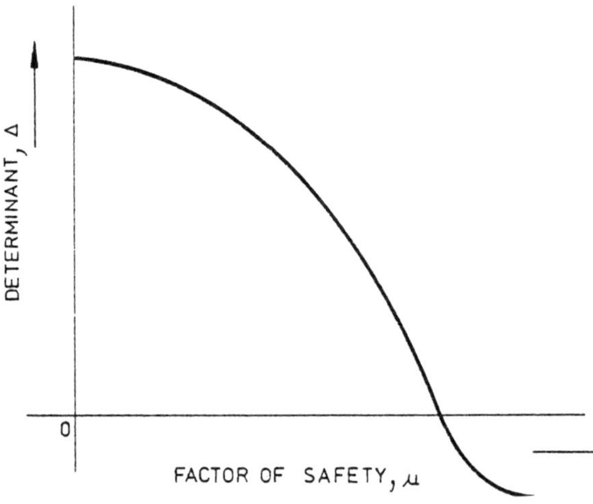

FIGURE 6.12 — Variation of determinant Δ as a function of the safety factor, μ.

6.4 Dynamic Analysis of Guyed Stacks

6.4.1 Introduction

In addition to the static analysis of guyed stacks, as presented in Parts 6.1 to 6.3, its dynamic counterpart is just as much and even more important. In the dynamic study, the external loads considered acting on the stack are: (1) wind pressure, steady or gusting; (2) self-weight of the stack; and (3) the initial tension in the guyed cables. Under these externally applied loads, the stack gyrates giving rise to oscillations which, in general, follow a pattern as postulated in Von Karman's vortex theory and, in particular, can be classified into two broad sets: (a) stack vibrations induced in the direction of the wind produced by a wind load which fluctuates more or less regularly between two extreme values of its pressure, thus giving rise to periodic, impulsive loads acting on the stack consistently for short periods of time; (b) stack experiencing vibrations in a direction perpendicular to the direction of the wind, mainly because of its tall, slender and cylindrical shape. Because of it, slight vibrations set off by a relatively slow wind and partly self-provoked by the stack itself, are sometimes magnified into vibrations with comparatively higher amplitudes of oscillation. If and when the period of the variable wind force acting on the stack coincides with the natural period of its vibration, dangerous and, in extreme circumstances, catastrophic stack-oscillations result. This could happen suddenly under heavy gusts of wind striking the stack for short periods of time, or under a steady wind pressure which builds up gradually over a relatively longer duration. A very important point stressed by Von Karman and later substantiated experimentally by other researchers in this field is that considering all the pertinent parameters involved in a particular study and up to a certain wind velocity, no vibrations of any significance result. Nevertheless, with increasing wind, the amplitudes of stack-vibrations also amplify, and in many cases, are magnified by the stack-configuration. These two loading conditions will form the basis for an analytical study of the dynamic behavior of guyed stacks. The mathematical development presented in this chapter will be illustrated by a numerical example which appears in Appendix A.

A very important factor, worth mentioning at this point and having a significant effect on the amplitude of stack-vibrations, is the slackening of cables under the influence of wind. Incorporating this effect analytically into the dynamic analysis of stacks is very complicated and, at best, can be approximated to a certain degree of accuracy. The insufficiency of available experimental work on this effect, though the only alternative reliable approach of ascertaining it, further complicates the problem.

Another area which defies accurate analysis is the behavior of guyed stacks under forced vibrations, mainly because the oscillations generated in this case are not harmonic.

The effect of self-weight of stack and the initial tension in its cables has not been incorporated in the analysis which follows, apparently for the sake of simplification in calculations. The author felt that the two factors could be neglected because of their comparatively insignificant contribution to the overall dynamic behavior of the guyed stacks. His contribution, nevertheless, can be determined and included in this study, if deemed necessary. [6.7], [6.8], [6.9] and [6.10].

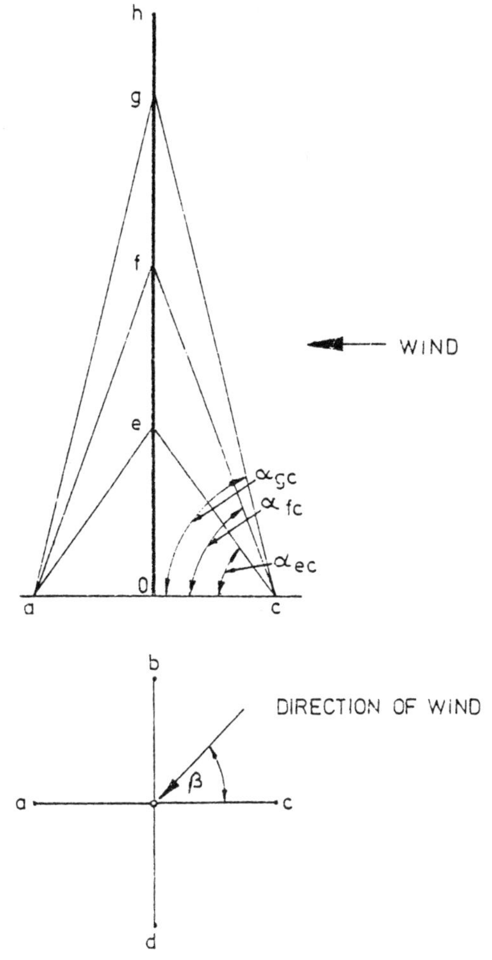

FIGURE 6.13 — Four-cable guyed stack.

6.4.2 Approximate Determination of the Natural Vibrations of a Guyed Stack by the Method of Deformations

Figure 6.13 shows a four-cable guyed stack subjected to lateral wind. The stack is anchored at point 0 and is supported elastically at points e, f and g — these are the points where the cables, anchored at points a, b, c and d on the ground, are connected to the stack as well.

Presented in this chapter is a relatively simplified version of the dynamic analysis of a guyed stack which, in addition to some basic, general assumptions outlined in Part 6.4.1 and applicable to the dynamic behavior of stacks in a broad spectrum, assumes that the amplitude of stack-oscillations under the steady, transverse wind is small. As a result, the guyed stack is, essentially, reduced to a continuous beam elastically supported at several points along its height.

The transverse vibrations of the stack are given by the fundamental differential equation:

$$EI \frac{d^4y}{dx^4} - m\omega^2 y = 0 \tag{6.40}$$

where:

E = Modulus of elasticity of the stack material, psi;

I = Moment of inertia of the uniform stack cross-section, in^4;

m = mass per unit height of the stack, slugs/in; and

ω = angular velocity of vibrations, rad./sec.

The general solution of Eq. (6.40) is given by

$$y = B_1 \cosh \frac{\eta}{\ell} x + B_2 \sinh \frac{\eta}{\ell} x + B_3 \cos \frac{\eta}{\ell} x + B_4 \sin \frac{\eta}{\ell} x \tag{6.41}$$

Where:

ℓ = arbitrary span length

$$\eta = \ell \sqrt[4]{\frac{m\omega^2}{EI}} \tag{6.42}$$

B_1, B_2, B_3, B_4 = the constants of integration to be determined from the boundary conditions of the individual spans of the stack.

Equation (6.41) will be used to determine moments and shears at the ends of each span of the stack. As an illustration, span ef is selected and it is assumed that for this span:

a. γ_e and γ_f are the rotations at joints e and f respectively;
b. Y_e and Y_f are the translations at jointes e and f respectively;
c. V_{ef} and V_{fe} are the shearing forces, (given by the expression: $V = -EI \frac{d^3y}{dx^3}$) at joints e and f respectively; and
d. M_{ef} and M_{fe} are the bending moments, (given by the expression: $M = -EI \frac{d^2y}{dx^2}$) at joints e and f respectively.

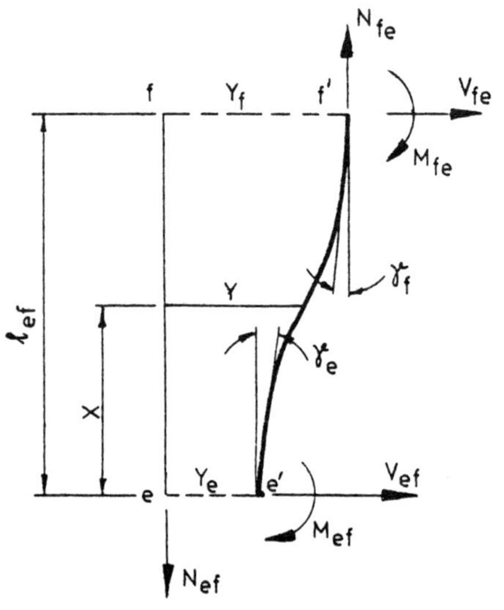

FIGURE 6.14 — Free body diagram of span ℓ_{ef}.

The resulting moments and shears for span ef are as follows:

$$M_{fe} = \frac{1}{\ell'_{fe}} \Pi_2(\eta_{fe})\gamma_f + \frac{1}{\ell'_{fe}} \Pi_1(\eta_{fe})\gamma_e - \frac{1}{\ell_{fe}\ell'_{fe}} \Pi_4(\eta_{fe}) Y_f - \frac{1}{\ell_{fe}\ell'_{fe}} \Pi_3(\eta_{fe}) Y_e$$

$$M_{ef} = \frac{1}{\ell'_{ef}} \Pi_1(\eta_{ef})\gamma_f + \frac{1}{\ell'_{ef}} \Pi_2(\eta_{ef})\gamma_e + \frac{1}{\ell_{ef}\ell'_{ef}} \Pi_3(\eta_{ef}) Y_f + \frac{1}{\ell_{ef}\ell'_{ef}} \Pi_4(\eta_{ef}) Y_e$$

$$V_{fe} = -\frac{1}{\ell_{fe}\ell'_{fe}} \Pi_4(\eta_{fe})\gamma_f + \frac{1}{\ell_{fe}\ell'_{fe}} \Pi_3(\eta_{fe})\gamma_e + \frac{1}{\ell^2_{fe}\ell'_{fe}} \Pi_6(\eta_{fe})\gamma_f + \frac{1}{\ell^2_{fe}\ell'_{fe}} \Pi_5(\eta_{fe})\gamma_e$$

$$V_{ef} = -\frac{1}{\ell_{ef}\ell'_{ef}} \Pi_3(\eta_{ef})\gamma_f + \frac{1}{\ell_{ef}\ell'_{ef}} \Pi_4(\eta_{ef})\gamma_e + \frac{1}{\ell^2_{ef}\ell'_{ef}} \Pi_5(\eta_{ef}) Y_f + \frac{1}{\ell^2_{ef}\ell'_{ef}} \Pi_6(\eta_{ef}) Y_e \tag{6.43}$$

Where:

$\ell_{ef} = \ell_{fe}$ = length of span fe

$\frac{1}{\ell'_{fe}} = \frac{EI_{fe}}{\ell_{fe}}$ where $I_{ef} = I_{fe}$ is the moment of inertia of the uniform stack cross-section of span ef; and,

$$\eta_{ef} = \eta_{fe} = \ell_{ef} \sqrt[4]{\frac{m_{ef}\omega}{EI_{ef}}}$$

where m_{ef} is the mass per unit height of span ef.

$$\Pi_1(\eta_{ef}) = -\eta_{ef} \frac{\sinh \eta_{ef} - \sin \eta_{ef}}{\cosh \eta_{ef} \cos \eta_{ef} - 1}$$

$$\Pi_2(\eta_{ef}) = -\eta_{ef} \frac{\cosh \eta_{ef} \sin \eta_{ef} - \sinh \eta_{ef} \cos \eta_{ef}}{\cosh \eta_{ef} \cos \eta_{ef} - 1}$$

$$\Pi_3(\eta_{ef}) = -\eta^2_{ef} \frac{\cosh \eta_{ef} - \cos \eta_{ef}}{\cosh \eta_{ef} \cos \eta_{ef} - 1} \tag{6.44}$$

$$\Pi_4\left(\eta_{ef}\right) = \eta_{ef}^2 \frac{\sinh \eta_{ef} \sin \eta_{ef}}{\cosh \eta_{ef} \cos \eta_{ef} - 1}$$

$$\Pi_5\left(\eta_{ef}\right) = \eta_{ef}^3 \frac{\sinh \eta_{ef} + \sin \eta_{ef}}{\cosh \eta_{ef} \cos \eta_{ef} - 1}$$

$$\Pi_6\left(\eta_{ef}\right) = -\eta_{ef}^3 \frac{\cosh \eta_{ef} \sin \eta_{ef} + \sinh \eta_{ef} \cos \eta_{ef}}{\cosh \eta_{ef} \cos \eta_{ef} - 1} \tag{6.45}$$

6.4.2.1 *Moments and Shears for Span f — g*

Support g of span f — g is assumed hinged. As such, the moment M_{fg} is zero and the rotation γ_g of support g is given by:

$$\gamma_g = \frac{\Pi_1\left(\eta_{fg}\right)}{\Pi_2\left(\eta_{fg}\right)} + \frac{\Pi_4\left(\eta_{fg}\right)}{\Pi_2\left(\eta_{fg}\right)\ell_{fg}} y_g + \frac{\Pi_3\left(\eta_{fg}\right)}{\Pi_2\left(\eta_{fg}\right)\ell_{fg}} \tag{6.46}$$

with corresponding moments and shears as follows:

$$M_{fg} = \frac{1}{\ell'_{fg}} \Pi_7\left(\eta_{fg}\right)\gamma_f + \frac{1}{\ell_{fg}\ell'_{fg}} \Pi_8\left(\eta_{fg}\right) y_g + \frac{1}{\ell_{fg}\ell'_{fg}} \Pi_9\left(\eta_{fg}\right) y_f$$

$$V_{gf} = \frac{1}{\ell_{gf}\ell'_{gf}} \Pi_8\left(\eta_{gf}\right)\gamma_f + \frac{1}{\ell^2_{gf}\ell'_{gf}} \Pi_{12}\left(\eta_{gf}\right) y_g + \frac{1}{\ell^2_{gf}\ell'_{gf}} \Pi_{10}\left(\eta_{gf}\right) y_f$$

$$V_{fg} = \frac{1}{\ell_{fg}\ell'_{fg}} \Pi_9\left(\eta_{fg}\right)\gamma_f + \frac{1}{\ell^2_{fg}\ell'_{fg}} \Pi_{10}\left(\eta_{fg}\right) y_g + \frac{1}{\ell^2_{fg}\ell'_{fg}} \Pi_{11}\left(\eta_{fg}\right) y_f \tag{6.47}$$

Where:

$\ell_{fg} = \ell_{gf}$ = length of span fg

$\frac{1}{\ell'_{fg}} = \frac{EI_{fg}}{\ell_{fg}}$ where $I_{fg} = I_{gf}$ is the moment of inertia of the uniform stack cross-section of span f — g;

$$\eta_{fg} = \eta_{gf} = \ell_{gf} \sqrt[4]{\frac{m_{gf}\omega}{EI_{gf}}}$$

where m_{gf} is the mass per unit height of gf and,

$$\Pi_7\left(\eta_{fg}\right) = \Pi_2\left(\eta_{fg}\right) - \frac{\Pi_1^2\left(\eta_{fg}\right)}{\Pi_2\left(\eta_{fg}\right)} = \eta_{fg}\left[\frac{2 \sinh \eta_{fg} \sin \eta_{fg}}{\cosh \eta_{fg} \sin \eta_{fg} - \sinh \eta_{fg} \cos \eta_{fg}}\right]$$

$$\Pi_8\left(\eta_{fg}\right) = \Pi_3\left(\eta_{fg}\right) + \frac{\Pi_1\left(\eta_{fg}\right)\Pi_4\left(\eta_{fg}\right)}{\Pi_2\left(\eta_{fg}\right)} = \eta_{fg}^2\left[\frac{\sinh \eta_{fg} + \sin \eta_{fg}}{\cosh \eta_{fg} \sin \eta_{fg} - \sinh \eta_{fg} \cos \eta_{fg}}\right]$$

$$\Pi_9\left(\eta_{fg}\right) = \Pi_4\left(\eta_{fg}\right) + \frac{\Pi_1\left(\eta_{fg}\right)\Pi_3\left(\eta_{fg}\right)}{\Pi_2\left(\eta_{fg}\right)} = -\eta_{fg}^2\left[\frac{\cosh \eta_{fg} \sin \eta_{fg} + \sinh \eta_{fg} \cos \eta_{fg}}{\cosh \eta_{fg} \sin \eta_{fg} - \sinh \eta_{fg} \cos \eta_{fg}}\right]$$

$$\Pi_{10}\left(\eta_{fg}\right) = \Pi_5\left(\eta_{fg}\right) + \frac{\Pi_4\left(\eta_{fg}\right)\Pi_3\left(\eta_{fg}\right)}{\Pi_2\left(\eta_{fg}\right)} = -\eta_{fg}^3\left[\frac{\cosh \eta_{fg} + \cos \eta_{fg}}{\cosh \eta_{fg} \sin \eta_{fg} - \sinh \eta_{fg} \cos \eta_{fg}}\right]$$

$$\Pi_{11}\left(\eta_{fg}\right) = \Pi_6\left(\eta_{fg}\right) - \frac{\Pi_3^2\left(\eta_{fg}\right)}{\Pi_2\left(\eta_{fg}\right)} = \eta_{fg}^3\left[\frac{2 \cosh \eta_{fg} \cos \eta_{fg}}{\cosh \eta_{fg} \sin \eta_{fg} - \sinh \eta_{fg} \cos \eta_{fg}}\right]$$

$$\Pi_{12}\left(\eta_{fg}\right) = \Pi_6\left(\eta_{fg}\right) - \frac{\Pi_4^2\left(\eta_{fg}\right)}{\Pi_2\left(\eta_{fg}\right)} = \eta_{fg}^3\left[\frac{\cosh \eta_{fg} \cos \eta_{fg} + 1}{\cosh \eta_{fg} \sin \eta_{fg} - \sinh \eta_{fg} \cos \eta_{fg}}\right] \tag{6.48}$$

6.4.2.2 *Moments and Shears for Span e — o*

Support o of span e — o is assumed fixed. Consequently, the rotation, γ_o and translation y_o of this support are equal to zero, resulting in the following moments and shears for this span:

$$M_{eo} = \frac{1}{\ell'_{eo}} \Pi_7\left(n_{eo}\right) Y_e + \frac{1}{\ell_{eo}\,\ell'_{eo}} \Pi_9\left(n_{eo}\right) Y_e$$

$$M_{oe} = \frac{1}{\ell'_{eo}} \Pi_8\left(n_{eo}\right) Y_e + \frac{1}{\ell_{eo}\,\ell'_{eo}} \Pi_{10}\left(n_{eo}\right) Y_e$$

$$V_{eo} = \frac{1}{\ell_{eo}\,\ell'_{eo}} \Pi_8\left(n_{eo}\right) Y_e + \frac{1}{\ell^2_{eo}\,\ell'_{eo}} \Pi_{10}\left(n_{eo}\right) ye$$

$$V_{oe} = \frac{1}{\ell_{eo}\,\ell'_{eo}} \Pi_9\left(n_{oe}\right) Y_e + \frac{1}{\ell^2_{eo}\,\ell'_{eo}} \Pi_{11}\left(n_{eo}\right) Y_e \qquad (6.49)$$

Where:

ℓ_{eo} = length of span eo

$\frac{1}{\ell'_{eo}} = \frac{EI_{eo}}{\ell_{eo}}$ where I_{eo} is the moment of inertia of the uniform stack cross-section of span eo and,

$$n_{eo} = \ell_{eo} \sqrt[4]{\frac{M_{eo}\,\omega}{EI_{eo}}}$$

where M_{eo} is the mass per unit height of span eo.

6.4.2.3 Transverse Shear at Stack-Supports from the Cables

The contribution to the transverse shearing forces developed at stack-supports from its cables under the action of wind is determined, first by selecting a typical cable ec for illustration purposes, and then making use of Equations (6.50) and (6.51) given below. Equation (6.50) relates the transverse displacement of the cable ec with its forces and certain other parameters, whereas Equation (6.51) determines the resultant shear force at point e, as shown in Figure 6.13. These two equations have been explained in detail in Part 6.1.

$$y_{ec} = \frac{-Q^2\,\ell_{ec}}{24\,F^2_{ec}\cos\alpha_{ec}\cos\beta} + \frac{\ell_{ec}\,F_{ec}}{A_cE_c\cos\alpha_{ec}\cos\beta} + \frac{Q_0^2\,\ell_{ec}}{24\,F_0^2\cos\alpha_{ec}\cos\beta} - \frac{F_0\,\ell_{ec}}{A_cE_c\cos\alpha_{ec}\cos\beta} \qquad (6.50)$$

and,

$$V_e = 2(F_{ec} - F_{ea})\cos\alpha_{ec} \qquad (6.51)$$

Differentiating y_{ec} with respect to F_{ec} and assuming $\beta = 0$ for the maximum wind pressure on cable ec, we get:

$$\frac{dy_{ec}}{dF_{ec}} = \frac{2Q^2\,\ell_{ec}}{24\,F^3_{ec}\cos\alpha_{ec}} + \frac{\ell_{ec}}{A_cE_c\cos\alpha_{ec}} \qquad (6.52)$$

and,

on differentiating V_e with respect to F_{ec}, the following is obtained:

$$\frac{dV_e}{dF_{ec}} = 2\cos\alpha_{ec} \qquad (6.53)$$

Therefore,

$$\xi^e_o = \frac{dy_{ec}}{dV_e} = \frac{dy_{ec}}{dF_{ec}} \cdot \frac{dF_{ec}}{dV_e} = \frac{Q^2\,\ell_{ec}}{24\,F^3_{ec}\cos^2\alpha_{ec}} + \frac{\ell_{ec}}{2\,A_cE_c\cos^2\alpha_{ec}} \qquad (6.54)$$

Expanding the expression in Eq. (6.54) into a Taylor series, neglecting the higher order terms in it and on writing its inverse, we obtain:

$$\frac{1}{\xi^e_o} = \frac{2\,A_cE_c\cos^2\alpha_{ec}}{\ell_{ec}} - \frac{\left(2\,A_cE_c\cos^2\alpha_{ec}\right)^2}{\ell_{ec}} \cdot \frac{Q^2\,\ell_{ec}}{24\,F^3_{ec}\cos^2\alpha_{ec}} \qquad (6.55)$$

where $\frac{1}{\xi^e_o}$ is a measure of the shear at support e contributed by the cables and is usually expressed in force per unit of horizontal support displacement, e.g., k/in.

Similarly, the shearing forces at supports f and g can be written as:

$$\frac{1}{\xi^f_o} = \frac{2\,A_cE_c\cos^2\alpha_{ec}}{\ell_{fc}} - \frac{\left(2\,A_cE_c\cos^2\alpha_{fc}\right)^2}{\ell^2_{fc}} \cdot \frac{Q^2\,\ell_{fc}}{24\,F^3_{fc}\cos^2\alpha_{fc}} \qquad (6.56)$$

and,

$$\frac{1}{\xi^g_o} = \frac{2\,A_cE_c\cos^2\alpha_{gc}}{\ell_{gc}} - \frac{2\,A_cE_c\cos^2\alpha_{gc}}{\ell^2_{gc}} \cdot \frac{Q^2\,\ell_{gc}}{24\,F^3_{gc}\cos^2\alpha_{gc}}$$

The transverse shears, V_{ep}, V_{fp} and V_{gp} contributed at various stack supports by the cables are

a. For support e: $V_{ep} = \dfrac{y_e}{\xi_o^e}$

b. For support f: $V_{fp} = \dfrac{y_f}{\xi_o^f}$

c. For support g: $V_{gp} = \dfrac{y_g}{\xi_o^g}$ (6.57)

Since the stack is in equilibrium under the dynamic action of wind, conditions of equilibrium at each support are given as follows

a. For support e: $\Sigma M = M_{eo} + M_{ef} = 0$

$\Sigma V = V_{eo} + V_{ef} + V_{ep} = 0$

b. For support f: $\Sigma M = M_{fe} + M_{fg} = 0$

$\Sigma V = V_{fe} + V_{fg} + V_{fp} = 0$

c. For support g: $\Sigma V = V_{gf} + V_{gh} + V_{gp} = 0$

(6.58)

where y_e, y_f and y_g are the lateral displacements of supports e, f and g, respectively.

Substituting Equations (6.43), (6.47), (6.49), (6.55), (6.56) and (6.57) into Equation (6.58) and simplifying, the following set of equations, involving the unknowns γ_e, y_e, γ_f, y_f and y_g is obtained:

$$\left[\frac{1}{\ell'_{oe}}\Pi_7\left(\eta_{oe}\right) + \frac{1}{\ell'_{ef}}\Pi_7\left(\eta_{ef}\right)\right]\gamma_e + \left[\frac{1}{\ell_{oe}\ell'_{oe}}\Pi_9\left(\eta_{oe}\right) - \frac{1}{\ell_{ef}\ell'_{ef}}\Pi_9\left(\eta_{ef}\right)\right]y_e + \left[-\frac{1}{\ell_{ef}\ell'_{ef}}\Pi_8\left(\eta_{ef}\right)\right]\gamma_f = 0$$

$$\left[\frac{1}{\ell_{oe}\ell'_{oe}}\Pi_9(\eta_{oe}) - \frac{1}{\ell_{ef}\ell'_{ef}}\Pi_9(\eta_{ef})\right]\gamma_e + \left[\frac{1}{\ell^2_{oe}\ell'_{oe}}\Pi_{11}(\eta_{oe}) + \frac{1}{\ell^2_{ef}\ell'_{ef}}\Pi_{11}(\eta_{ef}) + \frac{1}{\xi^e_o}\right]y_e + \left[\frac{1}{\ell^2_{ef}\ell'_{ef}}\Pi_{10}(\eta_{ef})\right]y_f = 0$$

$$\left[\frac{1}{\ell'_{ef}}\Pi_7\left(\eta_{ef}\right) + \frac{1}{\ell'_{fg}}\Pi_7(\eta_{fg})\right]\gamma_f + \left[\frac{1}{\ell_{ef}\ell'_{ef}}\Pi_9(\eta_{ef}) - \frac{1}{\ell_{fg}\ell'_{fg}}(\eta_{fg})\right]y_f + \left[-\frac{1}{f_g f'_g}\Pi_8(\eta_{fg})\right]y_g = 0$$

$$\left[\frac{1}{\ell_{ef}\ell'_{ef}}\Pi_9(\eta_{ef}) - \frac{1}{\ell_{fg}\ell'_{fg}}\Pi_9(\eta_{fg})\right]\gamma_f + \left[\frac{1}{\ell^2_{ef}\ell'_{ef}}\Pi_{11}(\eta_{ef}) + \frac{1}{\ell^2_{fg}\ell'_{fg}}\Pi_{11}(\eta_{fg}) + \frac{1}{\xi^f_o}\right]y_f + \left[\frac{1}{\ell^2_{fg}\ell'_{fg}}\Pi_{10}(\eta_{ef})\right]y_g = 0$$

$$-\left[\frac{1}{\ell_{fg}\ell'_{fg}}\Pi_8(\eta_{fg})\right]\gamma_f + \left[\frac{1}{\ell^2_{fg}\ell'_{fg}}\Pi_{10}(\eta_{fg})\right]y_f + \left[\frac{1}{\ell^2_{fg}\ell'_{fg}}\Pi_{12}(\eta_{fg}) + \frac{1}{\xi^g_o}\right]y_g = 0 \quad (6.59)$$

The only method to have a non-trivial solution of the above set of equations is to equate the determinant of the coefficients, Δ of unknown terms to zero. Several values of Δ are calculated corresponding to a number of arbitrarily-assumed values of the natural, angular frequency, ω of the guyed stack. For each separate value of ω, a set of values for the unknown parameters ηs and Πs in Equation (6.59) is obtained, since η and Π are functions of ω as shown in Equations (6.42), (6.43), (6.47) and (6.48).

Values of ω and Δ are plotted along the abscissa and the ordinate of a graph respectively and a curve joining them drawn. The point where the (ω — Δ) curve intersects the abscissa (i.e., the ω — line) for the first time corresponds to the natural frequency of the guyed stack, as shown in Figure 6.15.

6.5 Forced Vibrations of a Guyed Stack

6.5.1 Introduction

An adequate assessment of the characteristic behavior of tall, cylindrical guyed stacks subjected to strong gusty

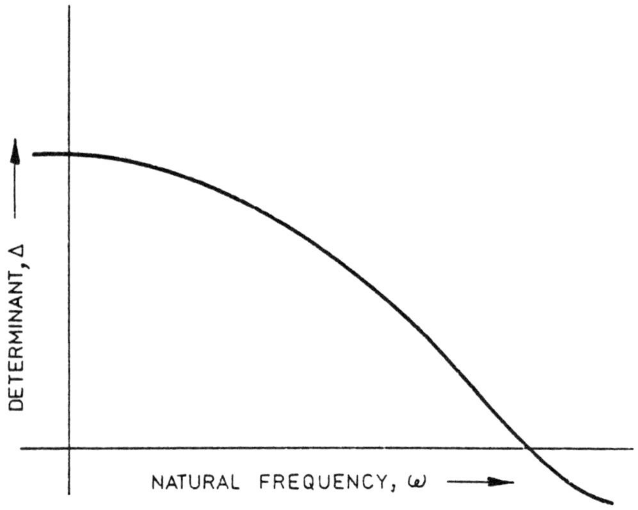

FIGURE 6.15 — Variation of the determinant, Δ as a function of angular natural frequency, ω of the guyed stack.

winds is of great importance in their analysis and design. The wind load imposed on stacks, in such cases, while constituting a significant portion of the entire load, induces vibratory motion in them. This motion is made up of a mean or static component and fluctuating or dynamic component. Whereas the mean response can be easily obtained by means of simplified conventional techniques, as present in Part 6.4, the dynamic response is comparatively difficult to predict mainly because of its non-deterministic character. For design purposes, the mean load is normally multiplied by a gust factor and the stack is then analyzed statically. This approach is adequate for all practical purposes provided that the gust factor used reflects accurately the dynamics of the individual stack.

Notwithstanding the fact that the accurate response of a guyed stack under gusting wind action defies accurate prediction, it is possible to approximate its behavior to a certain degree of accuracy by using a simplified approach — which, to begin with, assumes that the gusts can be approximated to a sinusoidal distribution of load on the stack. It is further assumed that such sinusoidal distribution of load will produce sinusoidal stack-deflections. There is a definite advantage implied on it and that is that by means of harmonic analysis, any loading regardless of how irregular and discontinuous it may be, can be represented by an infinite series. Although the series may be infinite, it is invariably so convergent that usually the first two terms will yield a reasonably accurate solution. In terms of boundary and support conditions, the guyed stack is assumed to be a continuous beam supported on elastic supports while fixed at its base. [6.11], [6.12] and [6.13].

6.5.2 Forced Vibrations from a Harmonic Disturbing Force

The forced vibrations of a guyed stack follow from the analysis pertaining to its natural vibrations presented in Part 6.4. The externally applied wind load is expanded into an infinite series i.e., [$\psi_I(x)$, $\psi_{II}(x)$, $\psi_{III}(x)$, ...] where the individual terms in it correspond to the first, second and third modes of natural vibrations viz., $\nu_I(t)$, $\nu_{II}(t)$, $\nu_{III}(t)$ ... etc. respectively. Thus the amplitude of vibration at any support of the guyed stack is given by:

$$y(x,t) = \nu_I(t)\psi_I(x) + \nu_{II}(t)\psi_{II}(x) + \nu_{III}(t)\psi_{III}(x) + \ldots \tag{6.60}$$

If p(x, t) is assumed to be the externally applied load on the stack, this can also be expanded into an infinite series as follows:

$$p(x,t) = m[p_I(t)\psi_I(x) + p_{II}(t)\psi_{II}(x) + p_{III}(t)\psi_{III}(x) + \ldots] \tag{6.61}$$

where m = mass of the stack per unit height; and $p_I(x)$, $p_{II}(x)$, $p_{III}(x)$, ... = loads corresponding to the first, second and third modes of natural vibrations, respectively.

Applying Equations (6.60) and (6.61) successively to spans oe and ef of the guyed stack, shown in Figure 6.16, we obtain:

a. *For Span oe*

$$p_I(t) = \frac{\int_o^e p(x,t)\psi_I(x)_e dx}{\int_o^e m_{oe}\psi_I^2(x)_e dx}$$

(For first mode of vibration)

$$p_{II}(t) = \frac{\int_o^e p(x,t)\psi_{II}(x)_e dx}{\int_o^e m_{oe}\psi_{II}^2(x)_e dx}$$

(For second mode of vibration)

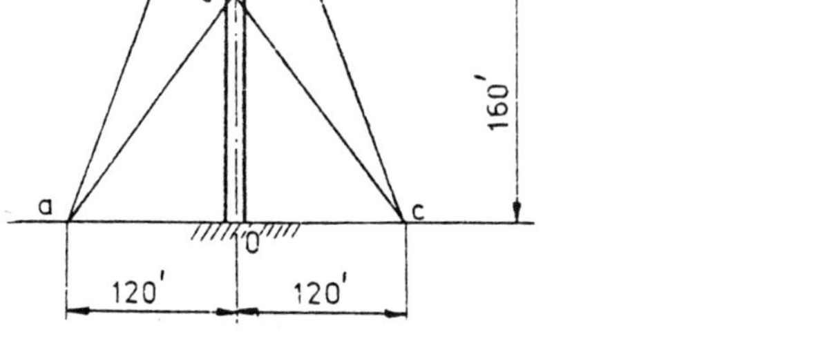

FIGURE 6.16 — Four-cable guyed stack.

b. *For Span e-f*

$$p_I(t) = \frac{\int_e^f p(x,t)\psi_I(x)_f dx}{\int_e^f m_{ef}\psi_I^2(x)_f dx}$$

(For first mode of vibration)

$$p_{II}(t) = \frac{\int_e^f p(x,t)\psi_{II}(x)_f dx}{\int_e^f m_{ef}\psi_{II}^2(x)_f dx}$$

(For second mode of vibration) (6.62)

where m_{oe} and m_{ef} are the masses per unit length of spans oe and ef, respectively.

If the actual load acting on the stack produces resonance with the first mode of natural vibration, it is sufficient to consider only the first terms in the series given in Equations (6.60) and (6.61).

The damped harmonic vibration of the guyed stack is determined from the following equation:

$$\frac{d^2 \nu_I(t)}{dt^2} + 2\zeta \frac{d\nu_I(t)}{dt} + \omega_I^2 \nu_I(t) = p_I(t)\sin\omega_I(t) \qquad (6.63)$$

where

ζ = damping coefficient.

ω_I = the first natural frequency of the guyed stack.

$p_I(t)\sin\omega_I(t)$ = the sinusoidally distributed external load acting on the stack.

In this simplified approach, the horizontal displacement of supports e and f are determined on the basis of the first term in the infinite series, i.e. $\psi_I(x)_e$, given in Equation (5.6) below. Solution of Equation (5.4) gives the displacement-factor, $\nu_I(t)_e$, of the support e, as follows:

$$\nu_I(t)_e = \frac{p_I(t)}{2\zeta\omega_I} \qquad (6.64)$$

Consequently, the horizontal displacement of support e with respect to the first mode of natural vibration is given by:

$$\bar{y}_e = \psi_I(x)_e . \nu_I(t)_e \qquad (6.65)$$

where $\psi_I(x)_e = \dfrac{y_e}{y_f} = \dfrac{\Delta(y_e,1)}{\Delta(y_f,1)}$ and $\Delta(y_e,1)$ & $\Delta(y_f,1)$ are the sub-determinants of the coefficient matrix in the following set of equations (6.66) involving the unknowns γ_e (the rotation of point e); y_e (the translation of point e); and y_f (the translation of point f). Incidentally, this set of equations has already been used in Part 6.4 for predicting the first natural frequency of the guyed stack (See Equation 6.59). For illustration purposes, the set of equations in a matrix from is given below:

$$\begin{matrix} 1 \\ 2 \\ 3 \end{matrix} \begin{bmatrix} xx & xx & xx \\ xx & xx & xx \\ xx & xx & xx \end{bmatrix} \begin{Bmatrix} \gamma_e \\ y_e \\ y_f \end{Bmatrix} = \begin{bmatrix} xx \\ xx \\ xx \end{bmatrix} \qquad (6.66)$$

where the rows of the coefficient matrix have been designated 1, 2 and 3, as shown; and xxs are the numerical values of the individual terms in the actual matrix. The sub-determinants $\Delta(y_e,1)$ and $\Delta(y_f,1)$ are by definition as follows:

$\Delta(y_e,1)$ = sub-determinant of the coefficient matrix in Equation (6.66) obtained by eliminating the column containing the values of y_e and the second row of the matrix.

$\Delta(y_f,1)$ = sub-determinant of the coefficient matrix in Equation (6.66) obtained by eliminating the column containing the values of y_f and the third row of the matrix.

Several values of the sub-determinants $\Delta(y_e,1)$ and $\Delta(y_f,1)$ are obtained based on several arbitrarily-selected values of the natural frequency of the guyed stack. However, in order to obtain consistent plots, it is advantageous to use the same values of these frequencies as used for determining the first natural frequency of the stack. The values of these sub-determinants along with those of the determinant Δ of the coefficient matrix of Equation (6.66) are plotted on a composite graph as a function of the natural frequency, ω of the guyed stack, shown in figure 6.17. Values of $\Delta(y_e,1)$ and $\Delta(y_f,1)$ corresponding to the first natural frequency, ω_I are measured from this graph. The displacement factor, $\psi_I(x)_e$, is then given by:

$$\psi_I(x)_e = \frac{\Delta(y_e,1)}{\Delta(y_f,1)} \qquad (6.67)$$

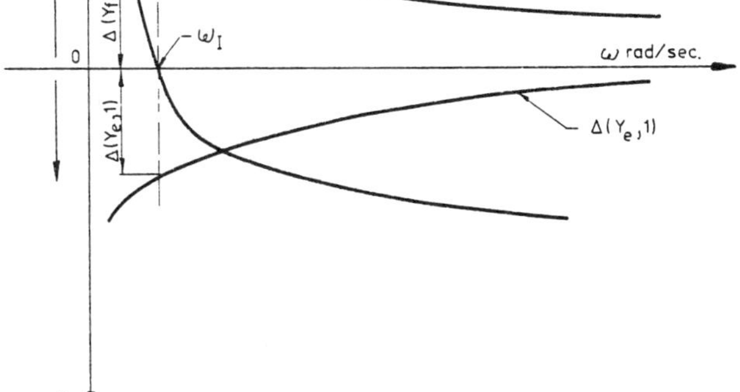

FIGURE 6.17 — Diagram of determinant, Δ of Equation (6.66) and sub-determinants $\Delta(y_e,1)$ and $\Delta(y_f,1)$ as functions of the natural frequency, ω of the guyed stack.

APPENDIX A
Numerical Example

1. DESCRIPTION OF THE PROBLEM

Design a multi-level guyed cylindrical steel stack, having uniform cross-section and an arrangement of guy cables as shown in Figure 6.18. Estimate maximum wind pressure according to the National Building Code of Canada assuming stack location in Montreal. Use a uniform average wind pressure on the stack between the guy support points. Assume prestressed guy cables.

2. PROPERTIES OF THE STACK AND GUY CABLES

The following characteristic parameters have been assumed for the stack and guy cables. The stack will be stress-analyzed for the given loading condition. Based on the stress results, the assumed dimensions and member sizes can be modified.

a. Outside diameter of the stack = 120 in
b. Wall thickness of the circular hollow cross-section of the stack = 1 in
c. Height of the stack above ground level = 400.0 ft
d. Stack-mast is assumed as fixed at its base
e. Modulus of Elasticity

E for the mast material = 30×10^6 psi

f. Guy cables are assumed as ¾ in, bridge strand class A coating for which

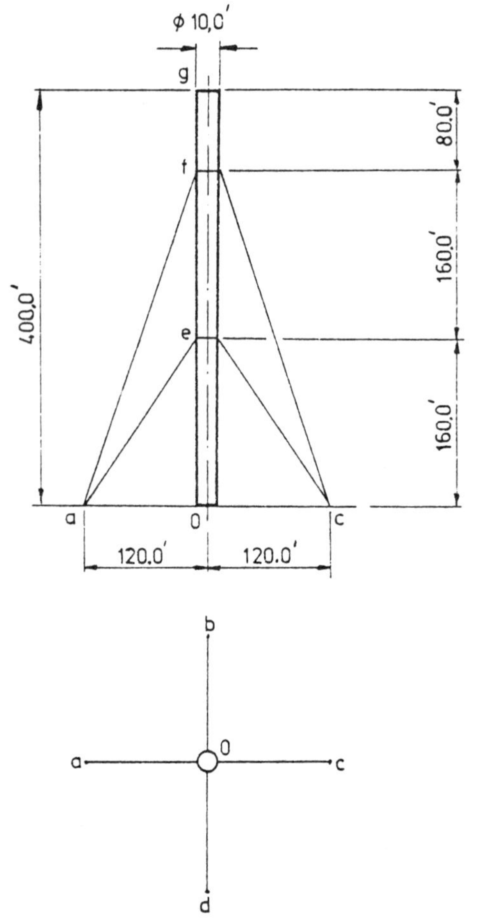

FIGURE 6.18 — General arrangement of the guyed stack.

A_c = 0.338 in^2 = the cross-sectional area of cable.
w = 1.18 lbs/ft = its weight
f_b = 68,000 lbs = breaking strength
E = 24×10^6 psi, = modulus of elasticity of the cable

3. LATERAL WIND LOAD ON THE GUYED STACK

The wind load on the guyed stack has been estimated in accordance with the 1977 National Building Code of Canada.

According to the Code, the lateral pressure on a guyed stack is given by the expression

$$p = q \cdot C_e \cdot C_g \cdot C_p \cdot C_n$$

where

p = the design external pressure acting statically and in a direction normal to the surface of the stack

q = the reference velocity pressure

C_e = exposure factor given in the following Table 6.1 according to the height of the structure

C_g = the gust effect factor normally taken to be 2.0 for structural members

C_p = the shape factor for the structure, which is equal to the algebraic difference of the external pressure coefficients for the windward and leeward sides of the structure

C_p = 1.0 — (−0.8) = 1.8 for the stack in this example

C_n = the force coefficient based on the slenderness ratio of tall structures. C_n was taken as 0.7 for a slenderness ratio of 400/10 = 40 for this stack

TABLE 6.1

Height	40	60	90	130	190	270	420	740
Ce	1.0	1.1	1.2	1.3	1.4	1.5	1.6	1.8

Based on the above factors, lateral wind pressures on the stack as a function of height from the ground level are as follows:

TABLE 6.2

Height ft	p lbs/ft
40	2.52 q
60	2.77 q
90	3.02 q
130	3.28 q
190	3.53 q
270	3.78 q
420	4.03 q
740	4.54 q

where

q = 9.2 lbs/ft for 1:100 wind probability in Montreal area

1. Average pressure on span oe of the stack viz. a span of 160 ft from the ground =
   $= \frac{1}{5}(2.52 + 2.77 + 3.02 + 3.28 + 3.53)q = 3.024\,q$
   $= 27.821$ lbs/ft
2. Average pressure on span of the stack =
   $= \frac{1}{3}(3.53 + 4.78 + 4.03) = 3.780\,q = 34.776$ lbs/ft
3. Average pressure on span fg on the stack =
   $= \frac{1}{2}(4.03 + 4.54) = 4.285\,q = 39.422$ lbs/ft

4. ANALYSIS OF THE STACK AS A CONTINUOUS BEAM

The mast is analyzed as a continuous beam fixed at point 0, elastically supported at points e and f, and subjected to the wind loads, as shown in Figure 6.19. This analysis, performed by the successive application of the three-moment equation to the spans of mast og, will determine the shear forces B_o, B_e and B_f at supports o, e and f respectively. The fixed end condition at point 0 is replaced by an imaginary span oe_1 equivalent in length and load to span oe is shown in Figure 6.20.

Applying the three-moment equation to spans e_1o and oe; and spans oe and ef, the following equations result

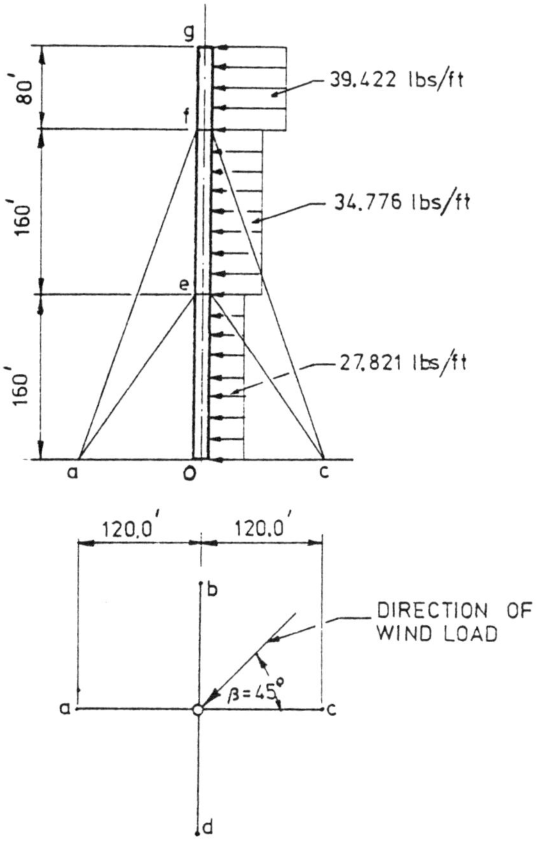

FIGURE 6.19 — Wind load on the guyed stack.

$$M_{e_1}L_{e_1o} + 2M_o(L_{e_1o}+L_{oe}) + M_eL_{oe} = 6\left[\frac{A_{e_1o}x_{e_1}}{L_{e_1o}} + \frac{A_{oe}\,x_e}{L_{oe}}\right]$$

$$M_oL_{oe} + 2M_e(L_{oe}+L_{ef}) + M_fL_{ef} = 6\left[\frac{A_{eo}\,x_e}{L_{oe}} + \frac{A_{ef}\,x_f}{L_{ef}}\right] \qquad (6.68)$$

where

M_{e_1} = $M_f = 0$

M_o and M_e = the negative moments developed at supports 0 and e respectively;

L_{e_1o}, L_{oe}, L_{ef} and L_{fg} = the span lengths of the stack between supports e_1 and 0; 0 and e; e and f; and f and g, respectively;

A_{e_1o} A_{oe} and A_{ef} = the areas of free bending moment diagrams due to wind load on spans e_1o, oe and ef respectively; and

x_{e_1} = the distance between the centroid of area A_{e_1o} and support e_1;

x_{oe} = the distance between the centroid of area A_{oe} and support e;

x_{ef} = the distance between the centroid of area A_{ef} and support f.

w_{e_1o}, w_{oe}, w_{ef} and w_{fg} = the intensities of the uniformly distributed loads on spans L_{e_1o}, L_{oe}, L_{ef} and L_{fg} respectively.

The numerical values of the various parameters are as follows:

$L_{e_1o} = L_{oe} = L_{ef} = 160.0$ ft

$L_{fg} = 80.0$ ft, and

$x_{e_1} = x_{oe} = x_{ef} = 80.0$ ft

$w_{e_1o} = w_{oe} = 278.21$ lbs/ft;

$w_{ef} = 347.76$ lbs/ft; and

$w_{fg} = 394.22$ lbs/ft

$$A_{e_1o} = \left(\frac{w_{e_1o}L_{e_1o}^2}{8}\right) \times \frac{2}{3}L_{e_1o} =$$

$$= \left(\frac{278.21 \times 160^2}{8}\right) \times \frac{2 \times 160}{3} =$$

$$= 94{,}962.35 \text{ k ft}^2$$

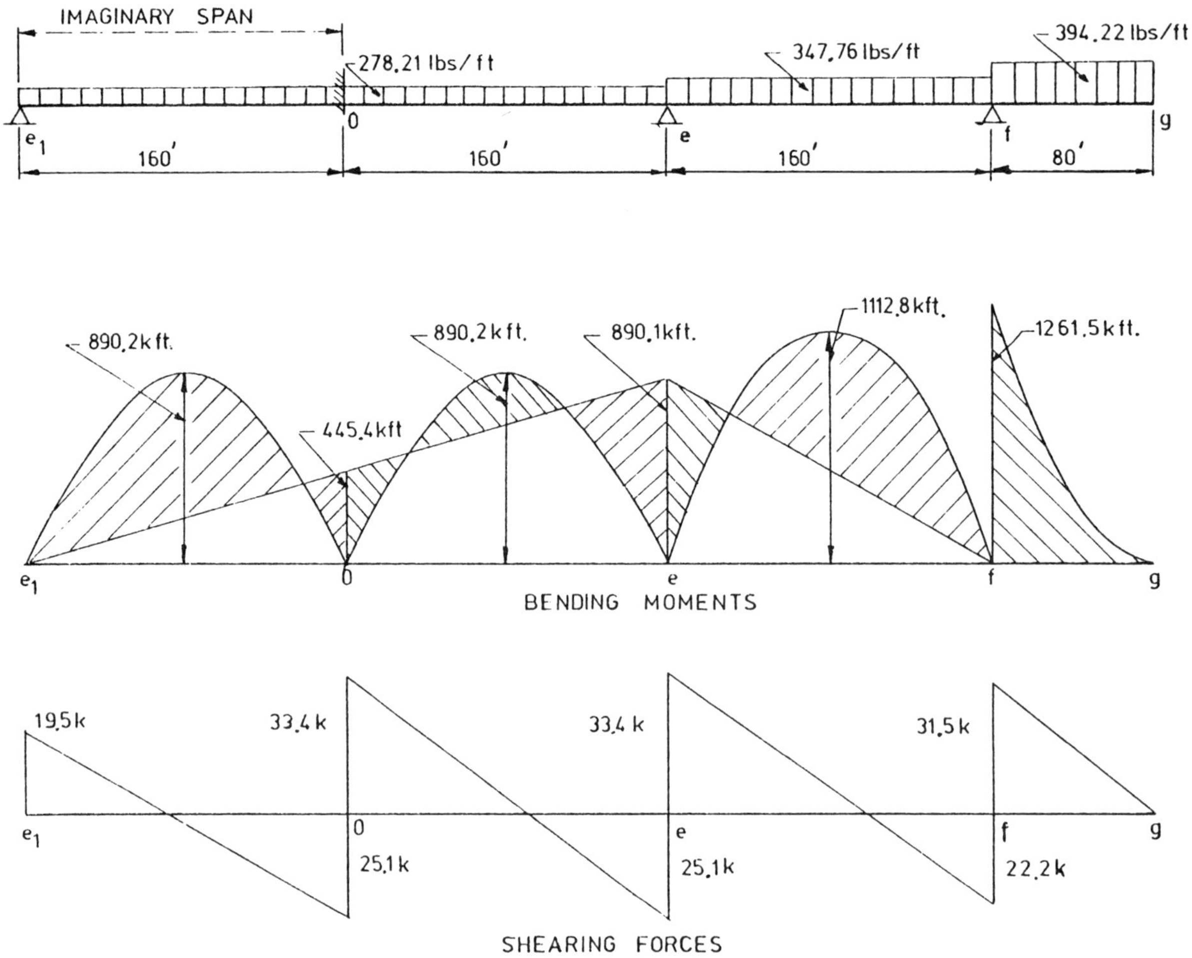

FIGURE 6.20 — Bending moment and shearing force diagrams for the four-cable guyed stack.

$$A_{oe} = \left(\frac{w_{oe}L_{oe}^2}{8}\right) \times \frac{2}{3} L_{oe} = \left(\frac{347.76 \times 160^2}{8}\right) \times$$

$$\times \frac{2 \times 160}{3} = 118{,}702.1 \text{ k ft}^2$$

$$\frac{w_{fg}L_{fg}^3}{6} = \frac{394.22 \times 80^3}{6} = 33{,}640.11 \text{ k ft}^2$$

Substituting the above numerical values into Equation (6.68), the following is obtained:

$$320\, M_e + 640\, M_o = 6\left(\frac{94{,}962.0 \times 80 \times 2}{160}\right) =$$

$$= 569{,}772.0 \text{ k ft}$$

$$160\, M_o + 640\, M_e = 6\left(\frac{94{,}962.0 \times 80}{160} +\right.$$

$$\left. + \frac{118{,}702 \times 80}{160}\right) = 640{,}992 \text{ k ft} \qquad (6.69)$$

Solving simultaneous equations (6.69) M_e and M_o are obtained as

$$M_e = 890.14 \text{ k ft}$$

$$M_o = 445.43 \text{ k ft}$$

Shear force B_o at point

$$0 = \frac{278.21 \times 160}{2000} + \left(\frac{445.43 - 890.14}{160}\right)$$

$$= 19.50 \text{ k}$$

Shear force B_e at point

$$e = [\text{shear force from span oe}]$$

$$= \quad + [\text{shear force from span ef}]$$

$$= [22.30 + 2.8] + \left\{\left[\frac{347.76 \times 160}{2000} + \frac{890.14 - 0.0}{160}\right]\right\}$$

$$= [25.1] + [27.8 + 5.6] = 58.5 \text{ k}$$

Shear force B_f at point

$$f = 27.8 + \frac{394.22 \times 80}{1000} - 5.6 = 53.7 \text{ k}$$

5. ANALYSIS OF STACK CABLES

Figure 6.21 shows the general arrangement of a stack with its sytem of axes ox, oy and oz having their origin at point 0. The direction of the wind is inclined at 45° relative to the direction of ox and parallel to the ground surface. In addition to the wind load shown in Fig. 6.4, each stack cable is assumed as initially tensioned to $\frac{1}{8}$th of its breaking strength, which is 68,000 lbs for each of the cables used in this example. Therefore, the initial tension in each cable is $\frac{1}{8} \times 68{,}000 = 8{,}500$ lbs.

Under the action of wind load, the magnitude of tension in cables ec, eb, fc and fb increases beyond its initial value of 8,500 lbs, whereas cables ea, ed, fa and fd experience a slackening of their initial tension. A detailed analysis of these cables is as follows.

5.1 Analysis of Cable fa

It is assumed that the initial tension of the cable fa is almost fully released, presumably 99%, when the tension in guy fc is maximum. In other words, the force in cable fa is decreased by an amount of 8,425 lbs under the action of the wind. Cable fa will shorten by an amount, $\Delta\ell_{fa}$

$$\Delta\ell_{fa} = \frac{F_{fa} \cdot \ell_{fa}}{A_c E_c} = \frac{8425 \times 341.76}{0.338 \times 24 \times 10^6} = 0.3549 \text{ ft}$$

Therefore the corrected length of fa = 341.76 − 0.3549 = = 341.41 ft

The approximate horizontal deflection of point, f, y_f is given by Equation (6.23)

$$y_f = \frac{5\, S_f\, \ell_{fa}}{3\, A_c E_c}$$

where

S_f = 2.49 + 2.49 = 4.98 kips (see Part 5.2)

ℓ_{fa} = 341.76 ft

A_c = 0.338 in²; and

E_c = 24×10^6 psi

$$y_f = \frac{5 \times 4.98 \times 341.76 \times 10^3}{3 \times 0.338 \times 24 \times 10^6} = 0.3497 \text{ ft} = 4.196 \text{ in}$$

The sag s, of cable f_a is given by Equation (6.25).

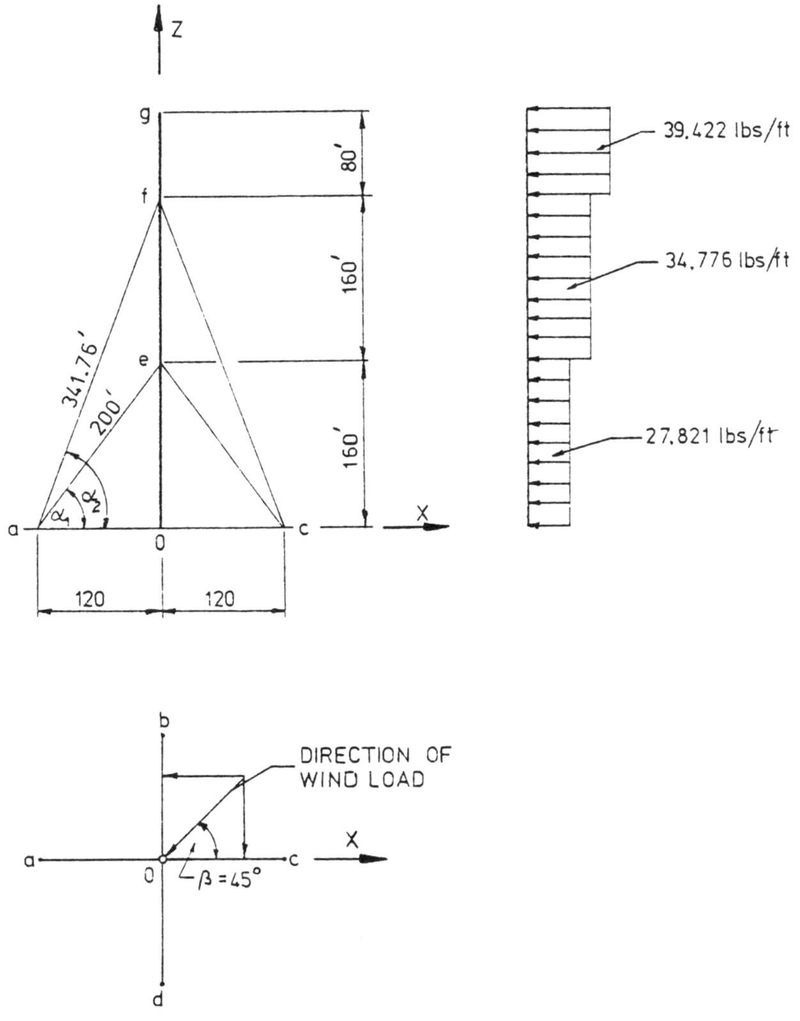

FIGURE 6.21 — Resolution of wind loads onto guy cables.

$$s^2 = \frac{3(\ell_{ao}-y_f)}{8\cos^4\alpha_2}[L\cos\alpha_2 - (\ell_{ao}-y_f)]$$

$$= \frac{3(120.0-0.3497)}{8 \times 0.3511^4}[341.41 \times 0.3511 -$$

$$- 119.6503] = 531.52$$

Therefore,

$$s = (531.52)^{\frac{1}{2}} = 23.05 \text{ ft.}$$

The horizontal component H_f, of the axial force in cable f_a is given by:

$$845 \times 0.3511 = 2958.02 \text{ lbs}$$

And from Equation (6.22),

$$w = \frac{8sH_f}{\ell_{ao}^2} = \frac{8 \times 23.05 \times 2{,}958.02}{120^2} =$$

$$= 37.88 \text{ lbs/ft.}$$

Component η of w, perpendicular to cable fa =

$$= 37.88 \times \cos\alpha_2 = 37.88 \times 0.3511 = 13.30 \text{ lbs/ft}$$

Total transverse load on cable fa =

$$= 13.30 \times 341.76 = 4.54 \text{ kips}$$

Q = 2.27 kips

In addition, cable f_a has the following properties:

$A_c = 0.338$ in²; weight = 1.18 lbs/ft; $E_c = 24 \times 10^6$ psi

Breaking strength = 68,000 lbs

$$q_o = 1.18\cos\alpha_2 = 1.18 \times 0.3511 = $$
$$= 0.414 \text{ lbs/ft length}$$

$$Q_o = \frac{0.414 \times 341.76}{2} = 70.79 \text{ lbs} = 0.071 \text{ K}$$

$$F_o = 8{,}500 \text{ lbs} = 8.5 \text{ K}$$

In order to establish a relationship between the force in cable f_a, i.e., F_{fa} and the displacement at point f, y_f, Equations (6.5), (6.6), (6.7) and (6.8) are used.

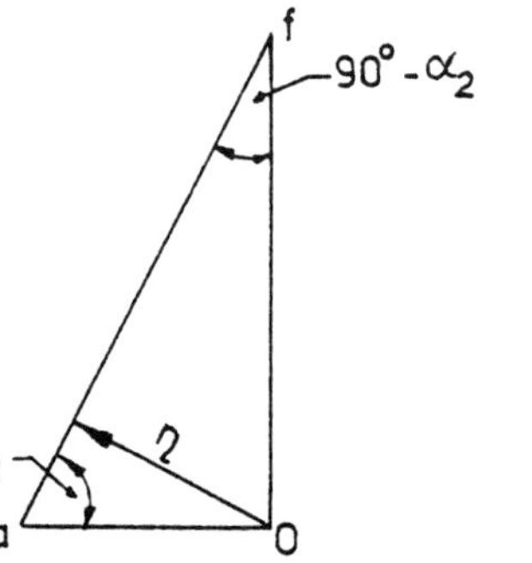

FIGURE 6.22 — Resolution of forces in cable f_a.

$$y_f = -\frac{A}{F_{fa}^2} + A_2 F_{fa} - A_3$$

where

$$A_1 = -\frac{Q^2 \ell_{fa}}{24\cos\alpha_2\cos\beta}$$

$$A_2 = \frac{\ell_{fa}}{A_c E_c \cos\alpha_2 \cos\beta}$$

$$A_3 = \left(\frac{Q_o^2 \ell_{fa}}{24 F_o^2} - \frac{F_o \ell_{fa}}{A_c E_c}\right)\frac{1}{\cos\alpha_2\cos\beta}$$

$$A_1 = -\frac{2.27^2 \text{x} 341.76}{24\text{x}0.3511\text{x}0.707} = -295.60 \text{ k}^2\text{ft}$$

$$A_2 = \frac{341.76\text{x}1000}{0.338\text{x}24\text{x}10^6\text{x}0.3511\text{x}0.707} = 0.169 \text{ ft/k}$$

$$A_3 = \left[\frac{0.071\text{x}341.76}{24\text{x}8.5^2} - \frac{8.5\text{x}341.76\text{x}1000}{0.338\text{x}24\text{x}10^6}\right] \times$$

$$\times \frac{1}{0.3511\text{x}0.707} = -1.4388 \text{ ft}$$

Therefore,

$$y_f = \frac{295.60}{F_{fa}^2} - 0.169 F_{fa} + 1.4388$$

A plot of y_f versus F_{fa} is obtained by assuming various values of F_{fa} and calculating the corresponding values of y_f, as shown in Table 6.3 and Figure 6.23.

TABLE 6.3

F_{fa} K	y_f in
12.0	17.5656
14.0	6.9708
16.0	-1.3224
18.0	-8.2908
20.0	-14.4264
22.0	-22.022

5.2 Analysis of Cable fc

Total wind load on cable fc is

$$= \frac{3.024 \times 9.2 \times 160 + 3.780 \times 9.2 \times 160}{320} =$$

$$= \frac{4451.33 + 5564.16}{320} = 31.29 \text{ lbs/ft}$$

Component parallel to ox = $31.29 \times \frac{1}{\sqrt{2}} = 22.12$ lbs/ft

Component perpendicular to fc =

$= 22.12 \sin\alpha_2 = 22.2 \times 0.9363 = 20.71$ lbs/ft

Actual load on cable $= 20.71 \times \dfrac{0.750}{12} = 1.29$ lbs/ft^2

Total load on cable fc = 1.29 × 341.76 = 440.8704 lbs

Q = 220.4352 lbs = 0.22 K

Total $Q = Q_{wind} + Q_{initial\ tension} =$

= 0.22 K + 2.27 K = 2.49 K

(Approximately the same as in Cable fa.)

In order to establish a relationship between the force in cable fc, i.e., F_{fc} and the displacement at point f, y_f, Equations (6.5), (6.6), (6.7) and (6.8) are used. Additional properties for cable f_c are the same as those for cable fa.

Therefore,

$$A_1 = -\frac{2.49^2 \text{x} 341.76}{24 \text{x} 0.3511 \text{x} 0.707} = -355.6792 \text{ k}^2\text{ft}$$

$$\left.\begin{aligned} A_2 &= 0.169 \text{ ft/K} \\ A_3 &= -1.4388 \text{ ft} \end{aligned}\right\} \text{same as for cable fa}$$

y_f can be written as:

$$Y_f = -\frac{355.68}{F_{fc}^2} + 0.169\, F_{fc} - 1.4388$$

A plot of y_f versus F_{fc} is obtained by assuming various values of F_{fc} anc calculating the corresponding values of y_f, as shown in Table 6.4 and Figure 6.23.

TABLE 6.4

f_{fc} K	y_f in
12.0	-22.5792
14.0	-10.6500
16.0	-1.4904
18.0	6.0648
20.0	12.6240
22.0	18.5316
24.0	23.9964
26.0	29.1480
28.0	35.2620

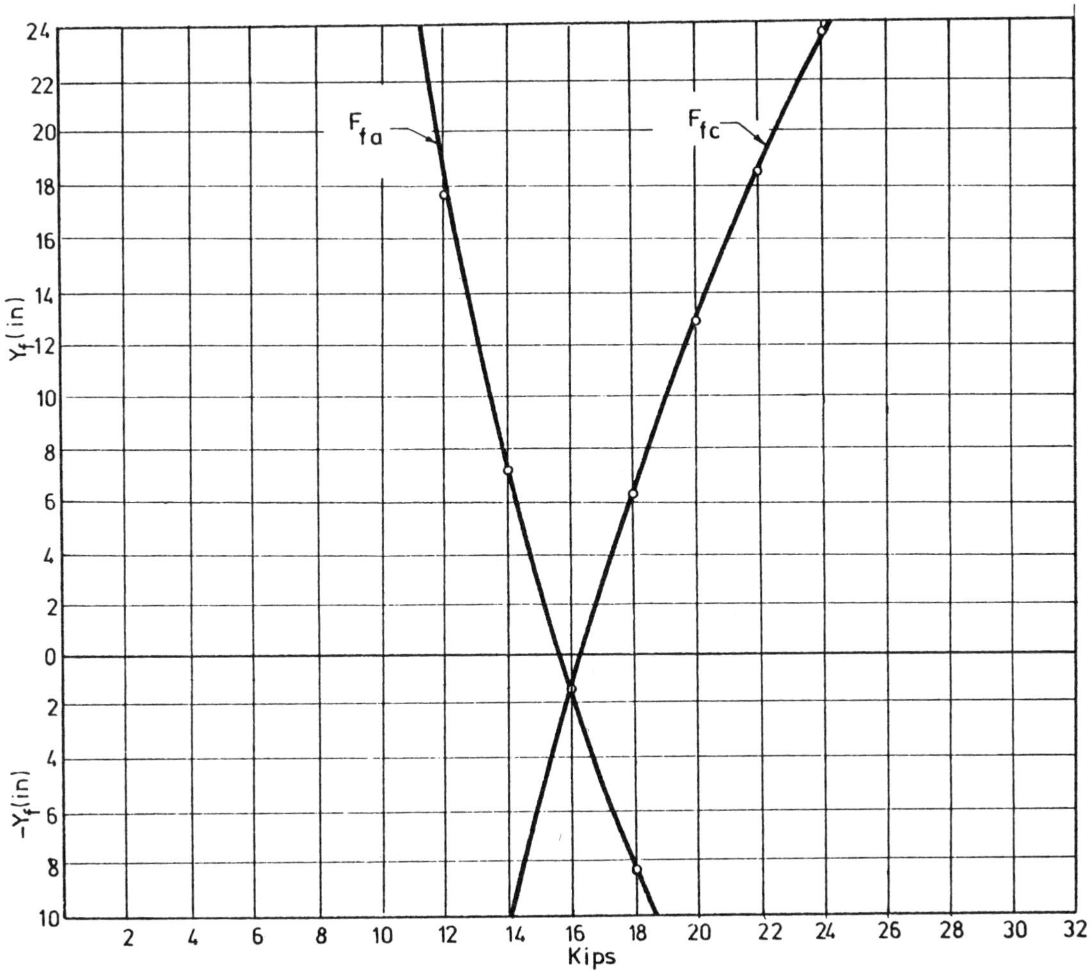

FIGURE 6.23 — Plot of the horizontal displacement y_f, of support f with respect to the axial forces F_{fa} and F_{fc} in guy cables fa and fc, respectively.

5.3 Determination of Shear Force, V_f at Support f

Having established a relationship between y_f and F_{fa}; and, y_f and F_{fc}, a plot between y_f and the total horizontal shear V_f at support f is obtained by making use of Equation (6.14)

$$V_f = 2[F_{fc}-F_{fa}]\frac{\ell_{oh}}{\ell_{fc}}$$

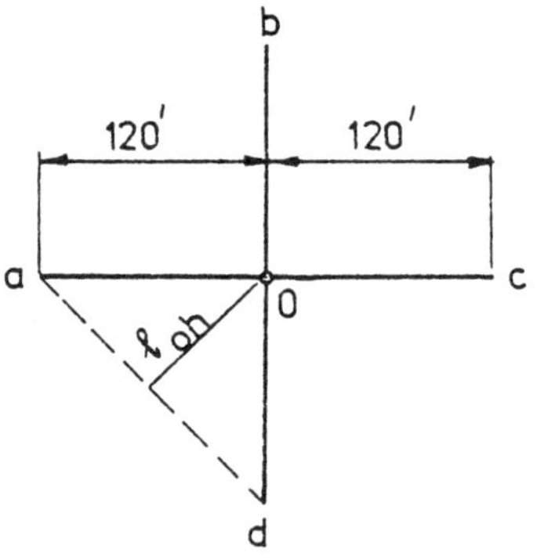

FIGURE 6.24 — Plan of stack.

$$\text{Length } \ell_{oh} = 120 \times \cos 45^{o} =$$

$$= 120 \times \frac{1}{\sqrt{2}} = 84.85 \text{ ft}$$

$$\ell_{fa} \text{ or } \ell_{fc} = 341.76 \text{ ft}$$

$$V_f = \frac{84.85 \times 2}{341.76}[F_{fc}-F_{fa}] = 0.4965\,[F_{fc}-F_{fa}]$$

where the values of $[F_{fc} - F_{fa}]$ are obtained from Figure 6.23 corresponding to the arbitrary values of y_f ranging from 0 to 24 in, as shown in Table 6.5. The values of V_f versus y_f are graphically represented in Figure 6.25, which also determines the parameters K_f and ξ_f of Equation (6.16) for support f as

$$K_f = -1.7 \text{ in; and } \xi_f = 4.06 \text{ in/K}$$

TABLE 6.5

y_f (in)	$(F_{fc} - F_{fa})$	V_f (K)
0	0.8	0.4
1	1.2	0.6
2	1.6	0.8
3	2.2	1.1
4	2.8	1.4
5	3.2	1.6
6	3.6	1.8
7	4.2	2.1
8	4.6	2.3
9	5.2	2.6
10	5.6	2.8
11	6.2	3.1
12	6.6	3.3
13	7.2	3.6
14	7.8	3.9
15	8.4	4.2
16	8.8	4.4
17	9.2	4.6
18	9.8	4.9
19	10.4	5.2
20	11.0	5.5
21	11.4	5.7
22	12.0	6.0
23	12.2	6.1
24	12.8	6.4

5.4 Analysis of Cable ea

It is assumed that the initial tension of the cable ea is released by 85 per cent, when the tension in guy ec is maximum, i.e., the force in cable ea is decreased by an amount of 7225 lbs under the action of wind. Cable ea will shorten by an amount, $\Delta\ell_{ea}$

$$\Delta\ell_{ea} = \frac{F_{ea}.\ell_{ea}}{A_c E_c} = \frac{7225 \times 200}{0.338 \times 24 \times 10^6} = 0.178 \text{ ft}$$

Corrected length ℓ_{ea} = 200.00 − 0.178 = 199.82 ft

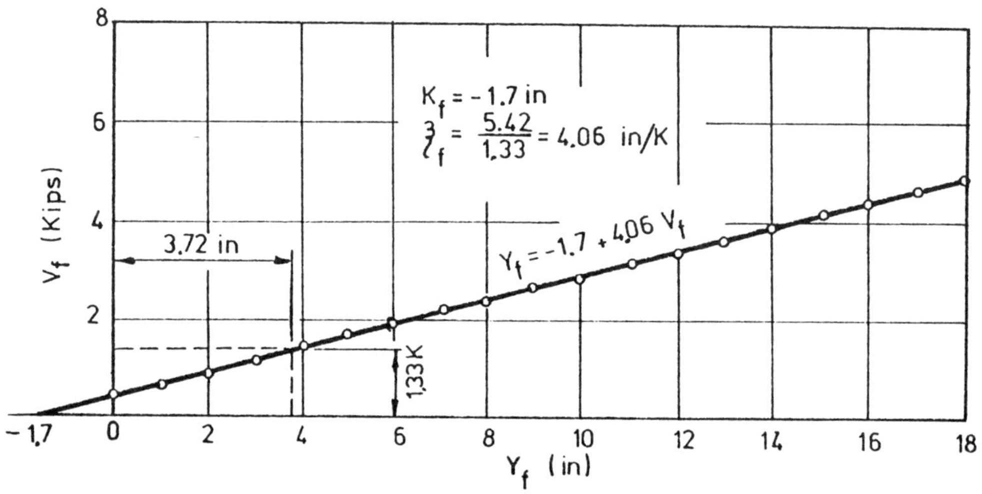

FIGURE 6.25 — Diagram of the horizontal displacement, y_f, and horizontal shear force, V_f of support f of the four-cable guyed stack.

The approximate horizontal deflection y_e of point e is assumed to be one-half that of point f. For a symmetrical structure, e.g., the guyed stack used in this example, displacements y_f and y_e are assumed to be linearly proportional to their respective heights above ground.

Therefore

$$y_e = 0.155 \text{ ft}$$

and the corrected horizontal distance ℓ_{ao} =

$$= 120.0 - 0.155 = 119.85 \text{ ft.}$$

The sag s of cable ea is given by Equation (6.25)

$$s^2 = \frac{3(\ell_{ao} - y_e)}{8 \cos^4 \alpha_1} [L \cos\alpha_1 - (\ell_{ao} - y_e)]$$

$$= \frac{3 \times 119.85}{8 \times 0.6^4} [199.82 \times 0.6 - 119.85]$$

$$= 14.5651$$

Therefore,

$$s = (14.5651)^{\frac{1}{2}} = 3.82 \text{ ft}$$

Axial force in cable ea = 7,225 lbs

Horizontal component H_e of axial force =

$$= 7{,}225 \times 0.6 = 4{,}335 \text{ lbs}$$

From Equation (6.22)

$$w = \frac{8 s H_e}{\ell_{ao}^2} = \frac{8 \times 3.82 \times 4335}{120^2} = 9.20 \text{ lbs/ft}$$

Component η of w, perpendicular to cable ea = 9.20 × cos α_1

$$\eta = 9.20 \times 0.6 = 5.52 \text{ lbs/ft}$$

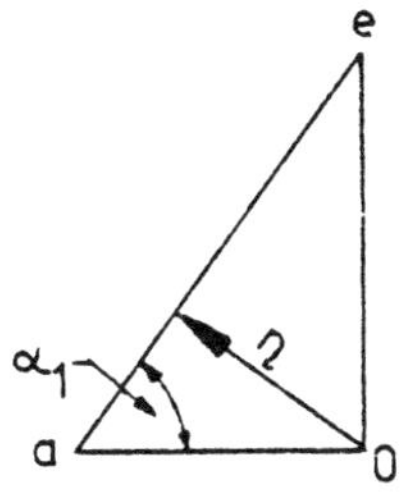

FIGURE 6.26 — Resolution of forces in cable ea.

Total transverse load on cable ea = 5.52 × 200 = 1.104 kips. Hence,

$$Q = 0.552 \text{ K}$$

In addition, cable ea has the following properties,

A_c = 0.338 in²; weight = 1.18 lbs/ft; E_c = 24 × 10⁶ psi

Breaking strength = 68,000 lbs

$$q_o = 1.18 \cos \alpha_1 = 1.18 \times 0.6 = 0.78 \text{ lbs/ft}$$

$$Q_o = \frac{0.78 \times 200}{2} = 78 \text{ lbs} = 0.078 \text{ K}$$

$$F_o = 8{,}500 \text{ lbs} = 8.5 \text{ K}$$

In order to establish a relationship between the force in cable ea, i.e., F_{ea} and displacement at point e, y_e, Equations (6.5), (6.6), (6.7) and (6.8) are used.

Substituting the values of various parameters in the above equations, we obtain:

$$A_1 = \frac{0.552^2 x 200}{24 x 0.6 x 0.707} = 5.99 \text{ K}^2\text{ft}$$

$$A_2 = \frac{-200 x 1.4142 x 1000}{0.338 x 24 x 10^6 x 0.6} = - 0.0581 \text{ ft/K}$$

$$A_3 = \left[\frac{0.078^2 x 200}{24 x 8.5^2} - \frac{8.5 x 200 x 1000}{0.338 x 24 x 10^6}\right] \frac{1.4142}{0.6} =$$

$$= 0.4923 \text{ ft}$$

Therefore,

$$y_e = \frac{5.99}{F_{ea}^2} - 0.0581 F_{ea} + 0.4923$$

A plot of y_e versus F_{ea} is obtained by assuming various values of F_{ea} and calculating the corresponding values of y_e, as shown in Table 6.6 and Figure 6.27.

TABLE 6.6

F_{ea} (K)	y_e (in)
2	22.4832
3	11.6556
4	7.6116
6	3.7212
8	1.4532
10	-0.3444
12	-1.9596
14	-3.4860
16	-4.9668
18	-6.4200
20	-7.8564
22	-9.2820
24	-10.7004
26	-12.1128
28	-13.5228
30	-14.9292
32	-16.3332

5.5 Analysis of Cable ec

Wind load on cable ec = 27.821 lbs/ft

Component parallel to ox = $27.821 \times \frac{1}{\sqrt{2}} = 19.67$ lbs/ft

Component perpendicular to cable ec =

= $19.67 \times \frac{4}{5} = 15.74$ lbs/ft

Actual load on cable = $15.74 \times \frac{0.75}{12} = 0.98375$ lbs/ft

Q = 98.375 lbs = 0.098 K

Total Q =

= $Q_{wind} + Q_{initial\ tension} = 0.098 + 0.552 = 0.65$ K

(Approximately the same as in cable ea.)

In order to establish a relationship between the force in cable ec, i.e., Fec and the displacement at point e, y_e, Equations (6.5), (6.6), (6.7) and (6.8) are used. Additional properties for cable ec are the same as those for cable ea. Therefore,

$$A_1 = \frac{-0.65^2 \times 200}{24 \times 0.6 \times 0.707} = -8.30 \text{ k}^2\text{ft}$$

$A_2 = 0.0581$ ft/k
$A_3 = -0.4923$ ft
} same as for cable ea, and

y_e can be written as:

$$y_e = -\frac{8.30}{F_{ec}^2} + 0.0581\, F_{ec} - 0.4923$$

A plot of y_e versus Fec is obtained by assuming various values of F_{ec} and calculating the corresponding values of y_e, as shown in Table 6.7 and Figure 6.27.

TABLE 6.7

Fec (K)	Ye (in)
2	-29.4132
4	-9.3444
6	-4.4916
8	-1.8864
10	-0.0684
12	1.7676
14	3.3456
16	4.8588
18	6.3348
20	7.7869
22	9.6372
24	10.6524
26	12.0720
28	13.4868
30	14.8980
32	16.3056

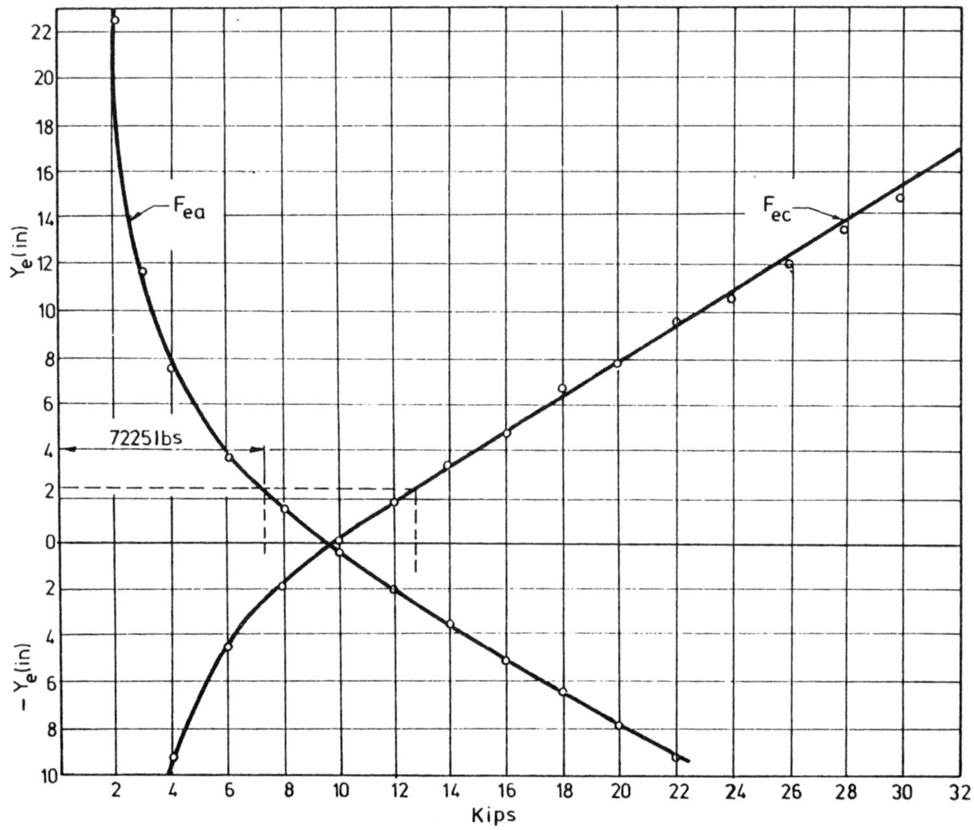

FIGURE 6.27 — Diagram of the horizontal displacement, y_e, of support e with respect to the axial forces F_{ec} and F_{ea} in guy cables ec and ea, respectively.

5.6 Determination of Shear Force, V_e at Support e

Having established a relationship between y_e and F_{ea}; and y_e and F_{ec}, a plot between y_e and the total horizontal shear force V_e at support e is obtained by making use of Equation (6.14)

$$V_e = 2[F_{ec}-F_{ea}]\,\frac{\ell_{oh}}{\ell_{ec}}$$

where ℓ_{oh} is shown in Figure 6.24 and ℓ_{ec} = 200.0 ft.

$$V_e = \frac{84.85 \times 2}{200.0}[F_{ec}-F_{ea}] = 0.8485[F_{ec}-F_{ec}]$$

where the values of $[F_{ec}-F_{ea}]$ are obtained from Figure 6.27, corresponding to the arbitrary values of y_e ranging from 0 to 17 in., as shown in Table 6.8.

The values of V_e versus y_e are graphically represented in Figure 6.28, which also determines the parameters K_e and ξ_e of Equation (6.16) for support e, as:

$$K_e = -0.4 \text{ in; and} \qquad \xi_e = 0.5833 \text{ in/K}$$

TABLE 6.8

Ye (in)	Fec - Fea	Ve (K)
0	0	0
1	2.6	2.21
2	5.0	4.24
3	7.0	5.94
4	8.8	7.47
5	11.2	9.50
6	12.8	10.86
7	14.6	12.39
8	16.2	13.75
9	17.8	15.10
10	19.4	16.46
11	20.8	17.65
12	22.8	19.35
13	24.2	20.54
14	25.4	21.55
15	27.2	23.10
16	28.6	24.27
17	29.8	25.3

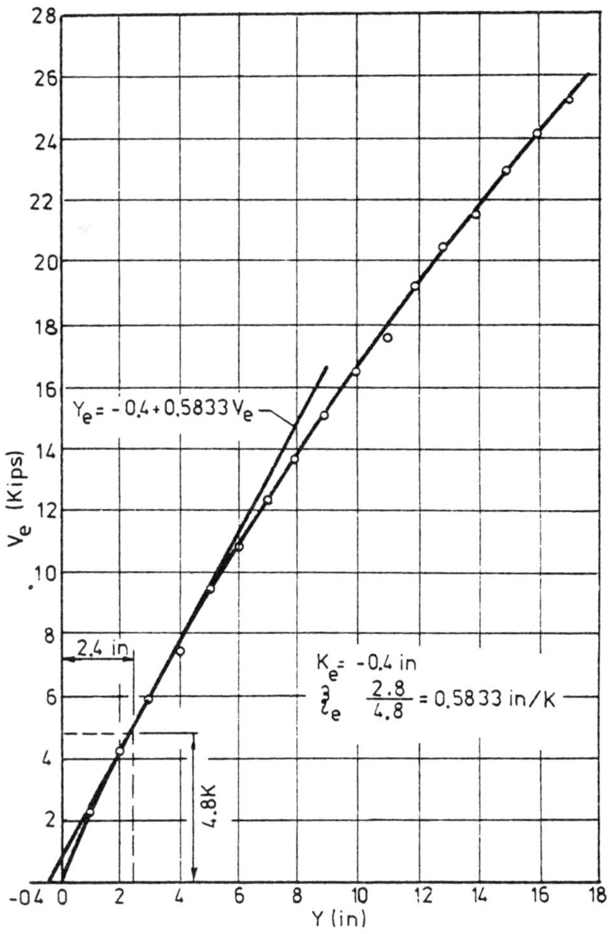

FIGURE 6.28 — Diagram of the horizontal displacement, y_e, and the horizontal shear force, V_e, or support e of the four-cable guyed stack.

5.7 Analysis of the Stack-Mast: Joint Displacements and Support Shears

Joint translations y_e and y_f of the supports e and f respectively and the rotation γ_e of joint e is determined by making use of Equation (6.21) along with the following properties of the stack.

Outside diameter of the stack = 120.0 in

Inside diameter of the stack = 118.0 in

Plate thickness of the stack wall = 1 in

$$I = \frac{\Pi(60^4-59^4)}{4} = 661471.6 \text{ in}^4$$

$$E = 30 \times 10^6 \text{ psi}$$

Applying Equation (6.21) to the stack, the following is obtained

$$\begin{bmatrix} [\frac{4EI}{L_{eo}} + \frac{3EI}{L_{ef}}] & [\frac{-6EI}{L_{eo}^2} + \frac{3EI}{L_{ef}^2}] & \frac{-3EI}{L_{ef}^2} \\ [\frac{-6EI}{L_{eo}^2} + \frac{3EI}{L_{ef}^2}] & [\frac{12EI}{L_{eo}^3} + \frac{3EI}{L_{ef}^3} + \frac{1}{\xi_f}] & \frac{-3EI}{L_{ef}^3} \\ \frac{-3EI}{L_{fe}^2} & \frac{-3EI}{L_{fe}^3} & [\frac{3EI}{L_{fe}^3} + \frac{1}{\xi_f}] \end{bmatrix}$$

$$\begin{Bmatrix} \gamma_e \\ y_e \\ y_f \end{Bmatrix} = \begin{bmatrix} -MF_{eo} - MF_{fe} \\ \frac{K_e}{\xi_e} - (B_{eo} + B_{ef}) \\ \frac{K_f}{\xi_e} - (B_{fe} + B_{fg}) \end{bmatrix}$$

where the individual terms are as follows:

$$\frac{4EI}{L_{eo}} + \frac{3EI}{L_{ef}} =$$

$$= \frac{4x30x10^6x0.66x10^6}{1920} + \frac{3x30x10^6x0.66x10^6}{1920} =$$

$$= 0.07219x10^{12} \quad \text{lbs/in}$$

$$\frac{-6EI}{L^2_{eo}} + \frac{3EI}{L^2_{ef}} =$$

$$= \frac{-6x30x10^6x0.66x10^6}{1920^2} + \frac{3x30x10^6x0.66x10^6}{1920^2} =$$

$$= -16.11 \text{ x } 10^6 \quad \text{lbs}$$

$$\frac{-3EI}{L^2_{ef}} = -16.11 \text{ x } 10^6 \quad \text{lbs}$$

$$[\frac{-6EI}{L^2_{eo}} + \frac{3EI}{L^2_{ef}}] =$$

$$= [\frac{-6x30x10^6x0.66x10^6}{1920^2} + \frac{3x30x10^6x0.66x10^6}{1920^2}] =$$

$$= -16.11 \text{ x } 10^6 \quad \text{lbs}$$

$$\frac{12EI}{L^3_{eo}} + \frac{3EI}{L^3_{ef}} + \frac{1}{\xi_e} =$$

$$= [\frac{12x30x10^6x0.66x10^6}{1920^3} + \frac{3x30x10^6x0.66x10^6}{1920^3} +$$

$$+ 0.001714x10^6] = 0.043674x10^6 \quad \text{lbs/in}$$

$$\frac{-3EI}{L^3_{ef}} = \frac{-3x30x10^6x0.66x10^6}{1920^3} =$$

$$= -0.00839x10^6 \quad \text{lbs/in}$$

$$\frac{3EI}{L^3_{fe}} + \frac{1}{\xi_f} = 0.00839x10^6 + 0.000246x10^6 =$$

$$= 0.008636x10^6 \quad \text{lbs/in}$$

$$\frac{K_e}{\xi_e} = 0.000686x10^6 \text{ lbs;} \quad \frac{K_f}{\xi_f} = 0.000419x10^6 \text{ lbs}$$

$$M_{eo} = \frac{9\omega\ell^2}{128} = \frac{9x9.2x3.024x10x160^2}{128} =$$

$$= 6.009x10^6 \quad \text{lbs in}$$

$$M_{ef} = \frac{\omega\ell^2}{8} = \frac{9.2x3.78x10x160^2}{8} =$$

$$= 13.354x10^6 \quad \text{lbs in}$$

Substituting the above values in Equation (6.21), we obtain

$$\begin{bmatrix} 72190x10^6 & -16.11x10^6 & -16.11x10^6 \\ -16.11x10^6 & 0.043674x10^6 & -0.00839x10^6 \\ -0.00839x10^6 & -0.00839x10^6 & 0.008636x10^6 \end{bmatrix}$$

$$\begin{Bmatrix} \gamma_e \\ y_e \\ y_f \end{Bmatrix} = \begin{bmatrix} -19.363x10^6 \\ 0.0578x10^6 \\ 0.05328x10^6 \end{bmatrix}$$

Solving the above equations, γ_e, y_e and y_f are obtained as follows:

γ_e = 0.003199 radians

y_e = 4.4829 in

y_f = 10.5281 in

6. BUCKLING ANALYSIS OF A 4-CABLE GUYED STACK

6.1 Introduction

The analytical treatment of buckling for a guyed stack, as developed in Part 6.3, will be applied to an actual example, i.e., the 4-cable guyed stack shown in Figure 6.29. It is the same stack which has been analyzed statically in Sections 1 to 5.7.

In order to proceed with the buckling analysis of this stack, it is necessary to determine the magnitude of axial forces induced in the stack from the guy cables under the action of wind. It is assumed that all cables are pretentioned to a magnitude of tension approximately equal to ⅛th the breaking strength of the cables. Under the action of wind, windward cables experience an increase in their tension whereas those on the leeward side have most of their initial tensions released. A reasonable approximation to this effect is made in buckling calculations. Figure

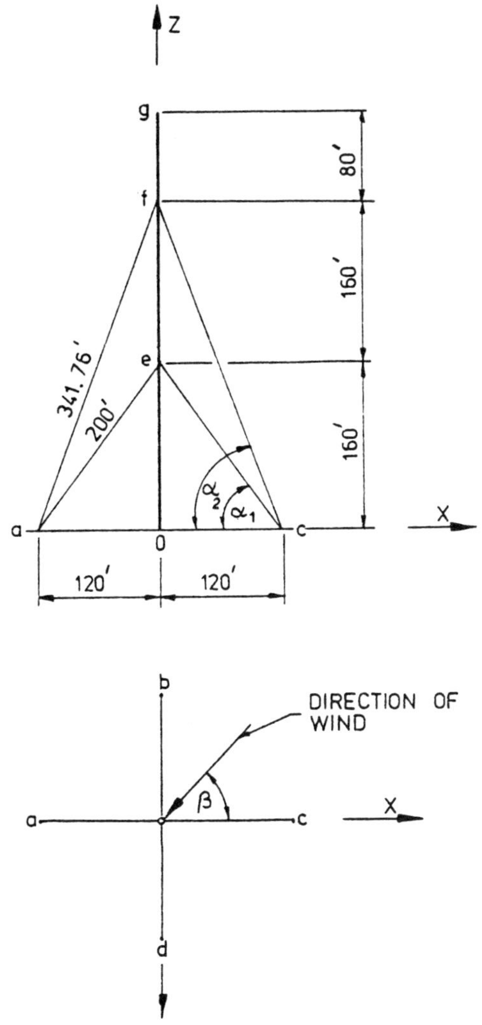

FIGURE 6.29 — Four-cable guyed stack.

6.30 shows the resolution of cable tensions and their contribution to the axial loads of the stack. Stack dimensions and cable forces used in this section are the same as those shown in Section 2. Hence, in order to maintain a continuity of the analytical-numerical presentation, it is suggested that Sections 6 and 2 should be cross-referenced.

From Figure 6.30

Σ Vertical forces at point e =

$$= \{F_{ec}+F_{eb}+F_{ea}+F_{ed}\}\ \sin\alpha_1 =$$

$$= \{F_{ec}+F_{eb}+F_{ea}+F_{ed}\}\ \frac{\ell_{eo}}{\ell_{ec}}\ ;$$

and where F_{ec}, F_{eb}, F_{ea} and F_{ed} are the cable tensions in cables ec, eb, ea and ed, respectively.

Σ Vertical forces at point f =

$$= \{F_{fc}+F_{fb}+F_{fa}+F_{fd}\}\ \frac{\ell_{fo}}{\ell_{fc}}$$

where F_{fc}, F_{fb}, F_{fa} and F_{fd} are the cable tensions in cables fe, fb, fa and fd, respectively.

6.2 Axial Forces of the Stack

The contribution of the cable tensions to the axial forces of the stack is determined by using the following data from Section 5.

$F_{ec} = F_{eb} = 8.5k$; $F_{ea} = F_{ed} = (8.5 - 7.225) = 1.275k$

$F_{fc} = F_{fb} = 9.2k$; $F_{fa} = F_{fd} = (8.5 - 8.425) = 0.075k$

$\ell_{eo} = 160.0'$; $\ell_{fo} = 320.0'$; $\ell_{ec} = 200.0'$; and

$\ell_{fc} = 341.76'$.

$$\Sigma F_f = [2x9.2+2x0.075]\ \frac{320}{241.76} = 17.37\ k$$

$$\Sigma F_e = [2x8.5+2x1.275]\ \frac{160}{200}\quad 15.64\ k$$

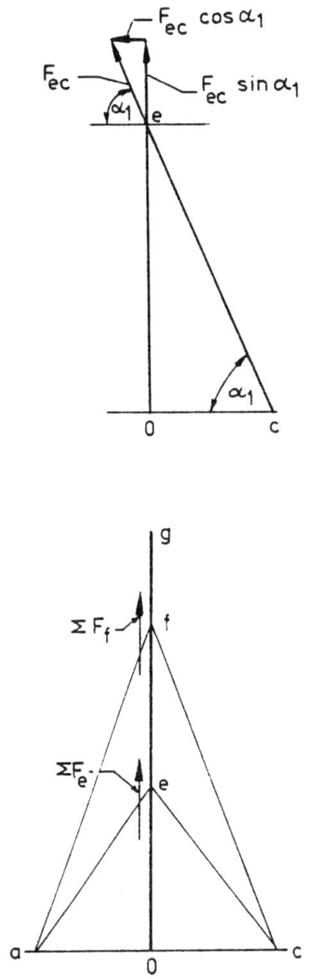

FIGURE 6.30 — Resolution of cable tensions.

Self-Weight of Stack

Cross-sectional area of stack,

$$A = \frac{\pi d^2}{4} - \frac{\pi d_1^2}{4} = \frac{\pi}{4}[120^2 - 118^2] = 373.66 \text{ in}^2$$

where d and d_1 are the outside and inside diameters of the stack, respectively.

$$\text{Self-weight of the stack} = \frac{373.66}{144} \times 0.49 = 2.5949 \text{ k/ft}$$

Contribution of self-weight of the stack at point, f:

$$= \tfrac{1}{2}[80+160] \times 1.27 = 152.4 \text{ k.}$$

Contribution of self-weight of the stack at point, e:

$$= \tfrac{1}{2}[160] \times 1.27 = 101.6 \text{ k.}$$

The total axial force, N_{ef} at point f =

$$= 17.37 + 152.40 = 169.77 \text{ k}$$

The total axial force, N_{eo} at point e =

$$= 15.64 + 101.60 = 117.24 \text{ k}$$

$$\frac{1}{\ell'_{oe}} = \frac{EI_{oe}}{\ell_{oe}} = $$

$$= \frac{30\text{x}10^6 \text{x} 0.66\text{x}10^6}{1920\text{x}10^3} = 10.31 \times 10^6 \text{ k/in}$$

$$\frac{1}{\ell_{oe}\ell'_{oe}} = \frac{10.31\text{x}10^6}{1920} = 0.00537 \times 10^6 \text{ k}$$

$$\frac{1}{\ell^2_{oe}\ell'_{oe}} = 2.7968 \text{ k/in.}$$

Also,

$$\frac{1}{\ell'_{oe}} = \frac{1}{\ell'_{ef}}; \quad \frac{1}{\ell_{oe}\ell'_{oe}} = \frac{1}{\ell_{ef}\ell'_{ef}}; \text{ and}$$

$$\frac{1}{\ell^2_{oe}\ell'_{oe}} = \frac{1}{\ell^2_{ef}\ell'_{ef}}$$

Using Equation (6.31) of Part 6.3, we obtain

$$\lambda_{oe} = 1920 \left[\frac{117.24\text{x}\mu\text{x}10^3}{30\text{x}10^6\text{x}0.66\text{x}10^6}\right]^{\frac{1}{2}} = 0.1477\sqrt{\mu}, \text{ and}$$

$$\lambda_{ef} = 1920 \left[\frac{169.77\text{x}\mu\text{x}10^3}{30\text{x}10^6\text{x}0.66\text{x}10^6}\right]^{\frac{1}{2}} = 0.1778\sqrt{\mu}$$

6.3 System of Equations for the Buckling Analysis of the Stack

Since the stack shown in Figure 6.29 has only three unknowns, i.e., γ_e, y_e and y_f, a simplified version of Eqs. to (6.39) of Part 6.3 is used in this numerical example. The coefficients of these unknowns are calculated by arbitrarily assuming several values of the safety factor, μ, starting with unity. For each value of μ, a system of three equations is obtained and for each such system, the value of determinant, Δ, is computed. A plot of the safety factor, μ versus the corresponding value of the determinant Δ indicates the buckling behavior of the stack, as shown in the following sections:

a. *Safety Factor μ = 1:*
From Section 6.2 of the Appendix, it is known that

$\lambda_{oe} = 0.1477\sqrt{\mu}$; $\lambda_{ef} = 0.1778\sqrt{\mu}$

Therefore, for $\mu = 1$

$\lambda_{oe} = 0.1477$; $\sin \lambda_{oe} = 0.1471$; $\cos \lambda_{oe} = 0.9891$

$\lambda_{ef} = 0.1778$; $\sin \lambda_{ef} = 0.1769$; $\cos \lambda_{ef} = 0.9842$

Substituting these values into Equations (6.35) to (6.39) we obtain

$$\left[\frac{0.1477^2\text{x}0.1471\text{x}10.31\text{x}10^6}{(0.1471-0.1471\text{x}0.9891)} + \right.$$

$$\left. + \frac{0.1778^2\text{x}0.1769\text{x}10.31\text{x}10^6}{(0.1769-0.1778\text{x}0.9842)}\right]\gamma_e +$$

$$+ \left[-\frac{0.1477^2\text{x}0.1471\text{x}0.00537\text{x}10^6}{(0.1471-0.1477\text{x}0.9891)} + \right.$$

$$\left. + \frac{0.1778^2\text{x}0.1769\text{x}0.00537\text{x}10^6}{(0.1769-0.1778\text{x}0.9842)}\right]y_e +$$

$$+ \left[-\frac{0.1778^2\text{x}0.1769\text{x}0.00537\text{x}10^6}{(0.1769-0.1778\text{x}0.9842)}\right]y_f = 0$$

$$\left[-\frac{0.1477^2\text{x}0.1471\text{x}0.00537\text{x}10^6}{(0.1471-0.1471\text{x}0.9891)} + \right.$$

$$\left. + \frac{0.1778^2\text{x}0.1769\text{x}0.00537\text{x}10^6}{(0.1769-0.1778\text{x}0.9842)}\right]\gamma_e +$$

$$+ \left[\frac{0.1477^3\text{x}0.9891\text{x}2.7968}{(0.1471-0.1477\text{x}0.9891)} + \right.$$

$$\left. + \frac{0.1778^3\text{x}0.9842\text{x}2.7968}{(0.1769-0.1778\text{x}0.9842)} + 1.714\right]y_e +$$

$$+ \left[\frac{-0.1778^3\text{x}0.9842\text{x}2.7968}{(0.1769-0.1778\text{x}0.9842)}\right]y_f = 0$$

$$\left[-\frac{0.1778^2\text{x}0.1769\text{x}0.00537\text{x}10^6}{(0.1769-0.1778\text{x}0.9842)}\right]\gamma_e +$$

$$+ \left[-\frac{0.1778^3\text{x}0.9842\text{x}2.7968}{(0.1769-0.1778\text{x}0.9842)}\right]y_e +$$

$$+ \left[\frac{0.1778^3\text{x}0.9842\text{x}2.7968}{(0.1769-0.1778\text{x}0.9842)} + 0.246\right]y_f = 0$$

in which the matrix form is as follows:

$$\begin{bmatrix} 63.4307\text{x}10^6 & -0.0014\text{x}10^6 & -0.0158\text{x}10^6 \\ -0.0014\text{x}10^6 & 0.0000841\text{x}10^6 & -0.00000841\text{x}10^6 \\ -0.0158\text{x}10^6 & -0.00000841\text{x}10^6 & 0.000008659\text{x}10^6 \end{bmatrix}$$

$$\begin{Bmatrix} \gamma_e \\ y_e \\ y_f \end{Bmatrix} = \begin{bmatrix} 0 \\ 0 \\ 0 \end{bmatrix}$$

whose determinant Δ of the coefficient matrix, when evaluated, becomes equal to: 0.000761 k^3/in.

b. *Safety Factor* $\mu = 2$

$\lambda_{oe} = 0.1477\sqrt{2} = 0.2089$; $\lambda_{ef} = 0.1778\sqrt{\mu} = 0.2514$

$\lambda_{oe} = 0.2089$; $\sin \lambda_{oe} = 0.2074$; $\cos \lambda_{oe} = 0.9783$

$\lambda_{ef} = 0.2514$; $\sin \lambda_{ef} = 0.2488$; $\cos \lambda_{ef} = 0.9686$

Substituting these values into Eqs. (6.35) to (6.39), we obtain

$$\left[\frac{0.2089^2\text{x}0.2074\text{x}10.31\text{x}10^6}{(0.2074-0.2089\text{x}0.9783)} + \frac{0.2514^2\text{x}0.2488\text{x}10.31\text{x}10^6}{(0.2488-0.2514\text{x}0.9686)}\right]\gamma_e +$$

$$+ \left[\frac{-0.2089^2\text{x}0.2074\text{x}0.00537\text{x}10^6}{(0.2074-0.2089\text{x}0.9783)} + \frac{0.2514\text{x}0.2488\text{x}0.00537\text{x}10^6}{(0.2488-0.2514\text{x}0.9686)}\right] y_e +$$

$$+ \left[\frac{-0.2514\text{x}0.2488\text{x}0.00537\text{x}10^6}{(0.2488-0.2514\text{x}0.9686)}\right] y_f = 0$$

$$\left[\frac{-0.2089^2\text{x}0.2074\text{x}0.00537\text{x}10^6}{(0.2074-0.2089\text{x}0.9783)} + \frac{0.2514^2\text{x}0.2488\text{x}0.00537\text{x}10^6}{(0.2488-0.2514\text{x}0.9686)}\right]\gamma_e +$$

$$+ \left[\frac{0.2089^2\text{x}0.9783\text{x}0.000002797\text{x}10^6}{(0.2074-0.2089\text{x}0.9783)} + \frac{0.2514^3\text{x}0.9686\text{x}0.000002797\text{x}10^6}{(0.2488-0.2514\text{x}0.9686)} + 1.714\right] y_e$$

$$+ \left[\frac{-0.2514^3\text{x}0.9686\text{x}0.000002797\text{x}10^6}{(0.2488-0.2514\text{x}0.9686)}\right] y_f = 0$$

$$\left[\frac{-0.2514^2\text{x}0.2488\text{x}0.00537\text{x}10^6}{(0.2488-0.2514\text{x}0.9686)}\right]\gamma_e +$$

$$+ \left[\frac{-0.2514\text{x}0.9686\text{x}0.2797\text{x}10^{-5}\text{x}10^6}{(0.2488-0.2514\text{x}0.9686)}\right]y_e +$$

$$+ \left[\frac{0.2514^3\text{x}0.9686\text{x}0.000002797\text{x}10^6}{(0.2488-0.2514\text{x}0.9686)} + 0.246\right]y_f = 0$$

which in a simplified matrix form is as:

$$10^6 \begin{bmatrix} 61.69 & -0.0003 & -0.0159 \\ -0.00027 & 0.0001815 & -0.000008122 \\ -0.0159 & -0.000008122 & 0.000008369 \end{bmatrix}$$

$$\begin{Bmatrix} \gamma_e \\ y_e \\ y_f \end{Bmatrix} = \begin{bmatrix} 0 \\ 0 \\ 0 \end{bmatrix}$$

whose determinant Δ of the coefficient matrix, when evaluated, becomes equal to: 0.000635 k^3/in.

c. *Safety Factor* $\mu = 3$

$\lambda_{oe} = 0.1477\sqrt{3} = 0.2558$; $\lambda_{ef} = 0.1778\sqrt{3} = 0.3080$

$\lambda_{oe} = 0.2558$; $\sin \lambda_{oe} = 0.2530$; $\cos \lambda_{oe} = 0.9675$

$\lambda_{ef} = 0.3080$; $\sin \lambda_{ef} = 0.3032$; $\cos \lambda_{ef} = 0.9529$

Substituting these values into Equations (6.35) to (6.39), we obtain

$$\left[\frac{0.2588^2\text{x}0.2530\text{x}10.31\text{x}10^6}{(0.2530-0.2558\text{x}0.9675)} + \frac{0.3080^2\text{x}0.3032\text{x}10.31\text{x}10^6}{(0.3032-0.3080\text{x}0.9529)}\right]\gamma_e +$$

$$+ \left[\frac{-0.2558^2\text{x}0.2530\text{x}0.00537\text{x}10^6}{(0.2530-0.2558\text{x}0.9675)} + \frac{0.3080^2\text{x}0.3032\text{x}0.00537\text{x}10^6}{(0.3032 - 0.3080\text{x}0.9529)}\right]y_e +$$

$$+ \left[\frac{-0.3080^2\text{x}0.3032\text{x}0.00537\text{x}10^6}{(0.3032-0.3080\text{x}0.9529)}\right] y_f = 0$$

$$\left[\frac{-0.2588^2\times0.2530\times0.00537\times10^6}{(0.2530-0.2558\times0.9675)} + \frac{0.3080^2\times0.3032\times0.00537\times10^6}{(0.3032-0.3080\times0.9529)}\right]\gamma_e +$$

$$+\left[\frac{0.2558^3\times0.9675\times2.7968}{(0.2530-0.2558\times0.9675)} + \frac{0.3080^3\times0.9529\times2.7968}{(0.3032-0.3080\times0.9529)} + 1.714\right]y_e +$$

$$+\left[\frac{-0.3080^3\times0.9529\times2.7968}{(0.3032-0.3080\times0.9529)}\right]y_f = 0$$

$$\left[\frac{-0.3080^2\times0.3032\times0.00537\times10^6}{(0.3032-0.3080\times0.9529)}\right]\gamma_e +$$

$$+\left[\frac{-0.3080^3\times0.9529\times2.7968}{(0.3032-0.3080\times0.9529)}\right]y_e +$$

$$+\left[\frac{0.3080^3\times0.9529\times2.7968}{(0.3032-0.3080\times0.9529)} + 0.246\right]y_f = 0$$

which in a simplified matrix form is:

$$10^6\begin{bmatrix} 61.60 & -0.000246 & -0.0159 \\ -0.00026 & 0.00001797 & -0.0000080277 \\ -0.0159 & -0.0000080277 & 0.0000082737 \end{bmatrix}\begin{Bmatrix}\gamma_e \\ y_e \\ y_f\end{Bmatrix} = \begin{bmatrix}0 \\ 0 \\ 0\end{bmatrix}$$

whose determinant Δ of the coefficient matrix, when evaluated, becomes equal to: 0.0005848 k^3/in.

d. *Safety Factor* $\mu = 4$

$\lambda_{oe} = 0.1477\sqrt{4} = 0.2954$; $\lambda_{ef} = 0.1778\sqrt{4} = 0.3556$

$\lambda_{oe} = 0.2954$; $\sin\lambda_{oe} = 0.2911$; $\cos\lambda_{oe} = 0.9567$

$\lambda_{ef} = 0.3556$; $\sin\lambda_{ef} = 0.3482$; $\cos\lambda_{ef} = 0.9374$

Substituting these values into Equations (6.35) to (6.39), we obtain:

$$\left[\frac{0.2954^2\times0.2911\times10.31\times10^6}{(0.2911-0.2954\times0.9657)} + \frac{0.3556^2\times0.3482\times10.31\times10^6}{(0.3482-0.3556\times0.9374)}\right]\gamma_e +$$

$$+\left[\frac{-0.2954^2\times0.2911\times0.00537\times10^6}{(0.2911-0.2954\times0.9567)} + \frac{0.3556^2\times0.3482\times0.00537\times10^6}{(0.3482-0.3556\times0.9374)}\right]y_e +$$

$$+\left[\frac{-0.3556^2\times0.3482\times10.31\times10^6}{(0.3482-0.3556\times0.9374)}\right]y_f = 0$$

$$\left[\frac{-0.2954^2\times0.2911\times0.00537\times10^6}{(0.2911-0.2954\times0.9567)} + \frac{0.3556^2\times0.3482\times0.00537\times10^6}{(0.3482-0.3556\times0.9374)}\right]\gamma_e +$$

$$+\left[\frac{0.2954^3\times0.9567\times2.7968}{(0.2911-0.2954\times0.9567)} + \frac{0.3556^3\times0.9374\times2.7968}{(0.3482-0.3556\times0.9374)} + 1.714\right]y_e +$$

$$+\left[\frac{-0.3556^3\times0.9374\times2.7968}{(0.3482-0.3556\times0.9374)}\right]y_f = 0$$

$$\left[\frac{-0.3556^2\times0.3482\times0.00537\times10^6}{(0.3482-0.3556\times0.9374)}\right]\gamma_e +$$

$$+\left[\frac{-0.3556\times0.9374\times2.7968}{(0.3482-0.3556\times0.9374)}\right]y_e +$$

$$+\left[\frac{0.3556^3\times0.9374\times2.7968}{(0.3482-0.3556\times0.9374)} + 0.246\right]y_f = 0$$

which in a simplified matrix form is:

$$10^6\begin{bmatrix} 61.28 & -0.00018 & -0.01587 \\ -0.00018 & 0.00001774 & -0.000007911 \\ -0.01587 & -0.000007911 & 0.000008157 \end{bmatrix}\begin{Bmatrix}\gamma_e \\ y_e \\ y_f\end{Bmatrix} = \begin{bmatrix}0 \\ 0 \\ 0\end{bmatrix}$$

whose determinant Δ of the coefficient matrix, when evaluated, becomes equal to: 0.0005145 k^3/in.

e. *Safety Factor* $\mu = 10$

$\lambda_{oe} = 0.1477\sqrt{10} = 0.4671$; $\lambda_{ef} = 0.1778\sqrt{10} = 0.5623$

$\lambda_{oe} = 0.4671$; $\sin\lambda_{oe} = 0.4503$; $\cos\lambda_{oe} = 0.8929$

$\lambda_{ef} = 0.5623$; $\sin\lambda_{ef} = 0.5332$; $\cos\lambda_{ef} = 0.8460$

Substituting these values into Equations (6.35) to (6.39), we obtain:

$$[\frac{0.4671^2 x0.4503x10.31x10^6}{(0.4503-0.4671x0.8929)} + \frac{0.5623^2 x0.5332x10.31x10^6}{(0.5332-0.5623x0.8460)}]\gamma_e + [\frac{-0.4671^2 x0.4503x0.00537x10^6}{(0.4503-0.4671x0.8929)} + \frac{0.5623^2 x0.5332x0.00537x10^6}{(0.5332-0.5623x0.8460)}]y_e + [\frac{-0.5623^2 x0.5332x0.00537x10^6}{(0.5332-0.5623x0.8460)}]y_f = 0$$

$$[\frac{-0.4671^2 x0.4503x0.00537x10^6}{(0.4503-0.4671x0.8929)} + \frac{0.5623^2 x0.5332x0.00537x10^6}{(0.5332-0.5623x0.8460)}]\gamma_e + [\frac{0.4671^3 x0.8929x2.7968}{(0.4503-0.4671x0.8929)} + \frac{0.5623^3 x0.8460x2.7968}{(0.5332-0.5623x0.8460)} + 1.714]y_e + [\frac{-0.5623^3 x0.8460x2.7968}{(0.5332-0.5623x0.8460)}]y_f = 0$$

$$[-\frac{0.5623^2 x0.5332x0.00537x10^6}{(0.5332-0.5623x0.8460)}]\gamma_e + [\frac{0.5623^3 x0.8460x2.7968}{(0.5332-0.5623x0.8460)}]y_e + [\frac{0.5623^3 x0.8460x2.7968}{(0.5332-0.5623x0.8460)} + 0.246]y_f = 0$$

which in a simplified matrix form is:

$$10^6 \begin{bmatrix} 60.65 & -0.0001 & -0.0157 \\ -0.0001 & 0.00002199 & -0.000007316 \\ -0.0157 & 0.000007316 & -0.000007562 \end{bmatrix} \begin{Bmatrix} \gamma_e \\ y_e \\ y_f \end{Bmatrix} = \begin{bmatrix} 0 \\ 0 \\ 0 \end{bmatrix}$$

whose determinant Δ of the coefficient matrix, when evaluated, becomes equal to: 0.00758 k^3/in.

6.4 System of Equations for the Buckling Analysis of the Stack by Using an Axial Load Increased by an Arbitrary Factor of 100

It is apparent from Table 6.9 that the values of Δ are very small and it would take substantially higher values of the safety factor μ before the Δ-μ curve crosses the horizontal axis (i.e., the μ-axis) to determine the values of μ corresponding to the least buckling load of the stack as shown in Part 6.3.

TABLE 6.9 — Values of Determinant, Δ.

μ	$\Delta k^3/in$
1	0.000761
2	0.000635
3	0.000585
4	0.000515
10	0.007580

The conclusion drawn from Table 6.9 and the calculations leading to it, is that the axial load induced in the stack is considerably smaller in comparison to what it is actually capable of supporting before it buckles.

Therefore, in order to establish the upper limit of the stack-buckling-load, the axial forces in its span eo and ef were multiplied by an arbitrary factor of 100. The following two sets of calculations for Δ correspond to this increased axial stack-load.

a. *Safety Factor* $\mu = 1$

$\lambda_{oe} = 1.477\sqrt{\mu}$; $\lambda_{ef} = 1.778\sqrt{\mu}$

$\lambda_{oe} = 1.477$; $\sin\lambda_{oe} = 0.9956$; $\cos\lambda_{oe} = 0.0936$

$\lambda_{ef} = 1.778$; $\sin\lambda_{ef} = 0.9786$; $\cos\lambda_{ef} = -0.2058$

Substituting these values into Equations (6.35) to (6.39), we obtain:

$$[\frac{1.477^2 x0.9956x10.31x10^6}{(0.9956-1.477x0.0936)} + \frac{1.778^2 x0.9786x10.31x10^6}{(0.9786+1.778x0.2058)}]\gamma_e + [\frac{-1.477^2 x0.9956x0.00537x10^6}{(0.9956-1.477x0.0936)} + \frac{1.778^2 x0.9786x0.00537x10^6}{(0.9786+1.778x0.2058)}]y_e + [\frac{-1.778^2 x0.9786x0.00537x10^6}{(0.9786+1.778x0.2058)}]y_f = 0$$

$$[\frac{-1.477^2 x0.9956x0.00537x10^6}{(0.9956-1.477x0.0936)} + \frac{1.778^2 x0.9786x0.00537x10^6}{(0.9786+1.778x0.2058)}]\gamma_e + [\frac{1.477^3 x0.0936x2.7968}{(0.9956-1.477x0.0936)} + - \frac{1.778^3 x0.2058x2.7968}{(0.9786+1.778x0.2058)} + 1.714]y_e + [\frac{1.778^3 x0.2058x2.7968}{(0.9786+1.778x0.2058)}]y_f = 0$$

$$\left[\frac{-1.778^2 x 0.9786 x 0.00537 x 10^6}{(0.9786+1.778 x 0.2058)}\right]\Upsilon_e +$$

$$+\left[\frac{1.778^3 x 0.2058 x 2.7968}{(0.9786+1.778 x 0.2058)}\right]y_e +$$

$$+\left[\frac{-1.778^3 x 0.2058 x 2.7968}{(0.9786+1.778 x 0.2058)} + 0.246\right]y_f = 0$$

which in a simplified version and matrix form is:

$$10^6 \begin{bmatrix} 49.8397 & -0.0012 & -0.0124 \\ -0.0012 & 0.2915 & 2.4063 \\ 0.0124 & 2.4063 & -2.1603 \end{bmatrix} \begin{Bmatrix} \Upsilon_e \\ y_e \\ y_f \end{Bmatrix} = \begin{bmatrix} 0 \\ 0 \\ 0 \end{bmatrix}$$

whose determinant Δ of the coefficient matrix, when evaluated becomes equal to: -272.0394×10^6 k³/in

b. *Safety Factor* $\mu = 2$

$\lambda_{oe} = 1.477\sqrt{2} = 2.0888$; $\lambda_{ef} = 1.778\sqrt{2} = 2.5144$

$\lambda_{oe} = 2.0888$; $\sin\lambda_{oe} = 0.8687$; $\cos\lambda_{oe} = -0.4953$

$\lambda_{ef} = 2.5144$; $\sin\lambda_{ef} = 0.5867$; $\cos\lambda_{ef} = -0.8098$

Substituting these values into Equations (6.35) to (6.39), we obtain:

$$\left[\frac{2.0888^2 x 0.8687 x 10.31 x 10^6}{(0.8687+2.0888 x 0.4953)} +\right.$$

$$\left.+\frac{2.5144^2 x 0.5867 x 10.31 x 10^6}{(0.5867+2.5144 x 0.8098)}\right]\Upsilon_e +$$

$$+\left[\frac{-2.0888^2 x 0.8687 x 0.00537 x 10^6}{(0.8687+2.088 x 0.4953)} +\right.$$

$$\left.+\frac{2.5144^2 x 0.5867 x 0.00537 x 10^6}{(0.5867+2.5144 x 0.8098)}\right]y_e +$$

$$+\left[\frac{-2.5144^2 x 0.5867 x 0.00537 x 10^6}{(0.5867+2.5144 x 0.8098)}\right]y_f = 0.$$

$$\left[\frac{-2.0888^2 x 0.8687 x 0.00537 x 10^6}{(0.8687+2.0888 x 0.4953)} +\right.$$

$$\left.+\frac{2.5144^2 x 0.5867 x 0.00537 x 10^6}{(0.5867+2.5144 x 0.8098)}\right]\Upsilon_e +$$

$$+\left[\frac{-2.0888^3 x 0.4953 x 2.7968}{(0.8687+2.0888 x 0.4953)} +\right.$$

$$\left.+\frac{-2.5144^3 x 0.8098 x 2.7968}{(0.5867+2.5144 x 0.8098)} + 1.714\right]y_e +$$

$$+\left[\frac{2.5144^3 x 0.8098 x 2.7968}{(0.5867+2.5144 x 0.8098)}\right]y_f = 0$$

$$\left[\frac{-2.5144^2 x 0.5867 x 0.00537 x 10^6}{(0.5867+2.5144 x 0.8098)}\right]\Upsilon_e +$$

$$+\left[\frac{2.5144^3 x 0.8098 x 2.7968}{(0.5867+2.5144 x 0.8098)}\right]y_e +$$

$$+\left[\frac{2.5144^3 x 0.8098 x 2.7968}{(0.5867+2.5144 x 0.8098)} + 0.246\right]y_f = 0$$

which in a simplified version and matrix form is

$$\begin{bmatrix} 35.11 x 10^6 & -0.0031 x 10^6 & -0.0076 x 10^6 \\ -0.0031 x 10^6 & -18.6303 & 13.7113 \\ -0.0076 x 10^6 & 13.7265 & 13.9725 \end{bmatrix} \begin{Bmatrix} \Upsilon_e \\ y_e \\ y_f \end{Bmatrix} = \begin{bmatrix} 0 \\ 0 \\ 0 \end{bmatrix}$$

whose determinant Δ of the coefficient matrix, when evaluated becomes equal to: $-14{,}155.538 \times 10^6$ k³/in.

6.5 System of Equations for the Buckling Analysis of the Stack by Using an Axial Load Increased by an Arbitrary Factor of 25

An examination of the values of Δ in Table 6.10 indicates that the (μ-Δ) curve has crossed the horizontal axis, (i.e., the μ-axis). At this point in our investigation, it is interesting to note that the values of Δ in Tables 6.9 and 6.10 correspond to the two extreme loading conditions, i.e., too small and too big axial loads. In order to obtain a reasonably good plot between μ and Δ, it is deemed necessary to multiply the axial loads in spans eo and ef with an arbitrary factor in between 1 and 100. Therefore, the factor selected in third trial is 25.

a. *Safety Factor* $\mu = 1$

$\lambda_{oe} = 0.7235\sqrt{\mu}$; $\lambda_{ef} = 0.8890\sqrt{\mu}$

$\lambda_{oe} = 0.7235$; $\sin\lambda_{oe} = 0.6621$; $\cos\lambda_{oe} = 0.7495$

$\lambda_{ef} = 0.8890$; $\sin\lambda_{ef} = 0.7765$; $\cos\lambda_{ef} = 0.6301$

TABLE 6.10 — Values of Determinant, Δ.
(Axial Loads of Stack Multiplied by 100)

μ	$\Delta_{k^3/in}$
1	-272.04×10^6
2	$-14{,}155.54\times10^6$

Substituting these values into Equations (6.35) and (6.39), we obtain:

$$\left[\frac{0.7235^2\times0.6621\times10.31\times10^6}{(0.6621-0.7235\times0.7495)} + \frac{0.8890^2\times0.7765\times10.31\times10^6}{(0.7765-0.8890\times0.6301)}\right]\gamma_e + \left[\frac{-0.7235^2\times0.6621\times0.00537\times10^6}{(0.6621-0.7235\times0.7495)} + \frac{0.8890^2\times0.7765\times0.00537\times10^6}{(0.7765-0.8890\times0.6301)}\right]y_e + \left[\frac{-0.8890^2\times0.7765\times0.00537\times10^6}{(0.7765-0.8890\times0.6301)}\right]y_f = 0$$

$$\left[\frac{-0.7235^2\times0.6621\times0.00537\times10^6}{(0.6621-0.7235\times0.7495)} + \frac{0.8890^2\times0.7765\times0.00537\times10^6}{(0.7765-0.8890\times0.6301)}\right]\gamma_e + \left[\frac{0.7235^3\times0.7495\times2.7968}{(0.6621-0.7235\times0.7495)} + \frac{0.8890^3\times0.6301\times2.7968}{(0.7765-0.8890\times0.6301)} + 1.714\right]y_e + \left[\frac{-0.8890^3\times0.6301\times2.7968}{(0.7765-0.8890\times0.6301)}\right]y_f = 0$$

$$\left[\frac{-0.8890^2\times0.7765\times0.00537\times10^6}{(0.7765-0.8890\times0.6301)}\right]\gamma_e + \left[\frac{-0.8890^3\times0.6301\times2.7968}{(0.7765-0.8890\times0.6301)}\right]y_e + \left[\frac{0.8890^3\times0.6301\times2.7968}{(0.7765-0.8890\times0.6301)} + 0.246\right]y_f = 0$$

which in a simplified version and matrix form is:

$$\begin{bmatrix} 59.0779\times10^6 & 0.0016\times10^6 & -0.0171\times10^6 \\ -0.0003\times10^6 & 14.0649 & -5.7243 \\ 0.0152\times10^6 & -5.7243 & 5.9703 \end{bmatrix} \begin{Bmatrix} \gamma_e \\ y_e \\ y_f \end{Bmatrix} = \begin{bmatrix} 0 \\ 0 \\ 0 \end{bmatrix}$$

whose determinant Δ of the coefficient matrix, when evaluated becomes equal to: $6{,}544.7\times10^6$ k³/in.

b. *Safety Factor* $\mu = 2$

$\lambda_{oc} = 0.7235\sqrt{2} = 1.0232$; $\lambda_{ef} = 0.8890\sqrt{2} = 1.2572$

$\lambda_{oe} = 1.0232$; $\sin\lambda_{oe} = 0.8538$; $\cos\lambda_{oe} = 0.5206$

$\lambda_{ef} = 1.2572$; $\sin\lambda_{ef} = 0.9513$; $\cos\lambda_{ef} = 0.3084$

Substituting these values into Equations (6.35) to (6.39), we obtain:

$$\left[\frac{1.0232^2\times0.8538\times10.31\times10^6}{(0.8538-1.0232\times0.5206)} + \frac{1.2572\times0.9513\times10.31\times10^6}{(0.9513-1.2572\times0.3084)}\right]\gamma_e + \left[\frac{-1.0232^2\times0.8538\times0.00537\times10^6}{(0.8538-1.0232\times0.5206)} + \frac{1.2572^2\times0.9513\times0.00537\times10^6}{(0.9513-1.2572\times0.3084)}\right]y_e + \left[\frac{-1.2572\times0.9513\times0.00537\times10^6}{(0.9513-1.2572\times0.3084)}\right]y_f = 0$$

$$\left[\frac{-1.0232^2\times0.8538\times0.00537\times10^6}{(0.8538-1.0232\times0.5206)} + \frac{1.2572^2\times0.9513\times0.00537\times10^6}{(0.9513-1.2572\times0.3084)}\right]\gamma_e + \left[\frac{1.0232^3\times0.5206\times2.7968}{(0.8538-1.0232\times0.5206)} + \frac{1.2572^3\times0.3084\times2.7968}{(0.9513-1.2572\times0.3084)} + 1.714\right]y_e + \left[\frac{1.2572^3\times0.3084\times2.7968}{(0.9513-1.2572\times0.3084)}\right]y_f = 0$$

$$\left[\frac{-1.2572^2\times0.9513\times0.00537\times10^6}{(0.9513-1.2572\times0.3084)}\right]\gamma_e + \left[\frac{-1.2572^3\times0.3084\times2.7968}{(0.9513-1.2572\times0.3084)}\right]y_e + \left[\frac{1.2513^3\times0.3084\times2.7968}{(0.9513-1.2572\times0.3084)} + 0.246\right]y_f = 0$$

which in a simplified version and matrix form is:

$$\begin{bmatrix} 62.139\times10^6 & 0.00059\times10^6 & -0.0208\times10^6 \\ 0.0059\times10^6 & 10.9921 & 4.4207 \\ -0.0174\times10^6 & -3.6970 & 3.9865 \end{bmatrix} \begin{Bmatrix} \gamma_e \\ y_e \\ y_f \end{Bmatrix} = \begin{bmatrix} 0 \\ 0 \\ 0 \end{bmatrix}$$

whose determinant Δ of the coefficient matrix, when evaluated becomes equal to: 3,595.7414 k^3/in.

c. *Safety Factor* $\mu = 3$

$\lambda_{oe} = 0.7235\sqrt{3} = 1.2531$; $\lambda_{ef} = 0.8890\sqrt{3} = 1.5398$

$\lambda_{oe} = 1.2531$; $\sin\lambda_{oe} = 0.9500$; $\cos\lambda_{oe} = 0.3123$

$\lambda_{ef} = 1.5398$; $\sin\lambda_{ef} = 0.9995$; $\cos\lambda_{ef} = 0.0309$

Substituting these values into Equations (6.35) to (6.39), we obtain:

$$\Big[\frac{1.2531^2 x0.9500x10.31x10^6}{(0.9500-1.2531x0.3132)} + \frac{1.5398^2 x0.9995x10.31x10^6}{(0.9995-1.5398x0.0309)}\Big]\gamma_e + \Big[\frac{-1.2531^2 x0.9500x0.00537x10^6}{(0.9500-1.2531x0.3123)} + \frac{1.5398^2 x0.9995x0.00537x10^6}{(0.9995-1.5398x0.0309)}\Big]y_e + \Big[\frac{-1.5398^2 x0.9995x0.00537x10^6}{(0.9995-1.5398x0.0309)}\Big]y_f = 0$$

$$\Big[\frac{-1.2531^2 x0.9500x0.00537x10^6}{(0.9500-1.2531x0.3123)} + \frac{1.5398^2 x0.9995x0.0.00537x10^6}{(0.9995-1.5398x0.0309)}\Big]\gamma_e + \Big[\frac{1.2531^3 x0.3123x2.7968}{(0.9500-1.2531x0.3123)} + \frac{1.5398^3 x0.0309x2.7968}{(0.9995-1.5398x0.0309)} + 1.714\Big]y_e + \Big[\frac{-1.5398^3 x0.0309x2.7968}{(0.9995-1.5398x0.0309)}\Big]y_f = 0$$

$$\Big[\frac{-1.5398^2 x0.9995x0.00537x10^6}{(0.9995-1.5398x0.0309)}\Big]\gamma_e + \Big[\frac{-1.5398^3 x0.0309x2.7968}{(0.9995-1.5398x0.0309)}\Big]y_e + \Big[\frac{1.5398^3 x0.0309x2.7968}{(0.9995-1.5398x0.0309)} + 0.246\Big]y_f = 0$$

which in matrix form and simplified version is:

$$\begin{bmatrix} 53.1952x10^6 & -0.0009x10^6 & 0.0134x10^6 \\ -0.0009x10^6 & 5.1214 & -0.3314 \\ -0.0206x10^6 & -0.3314 & 0.5774 \end{bmatrix} \begin{Bmatrix} \gamma_e \\ y_e \\ y_f \end{Bmatrix} = \begin{bmatrix} 0 \\ 0 \\ 0 \end{bmatrix}$$

whose determinant Δ of the coefficient matrix, when evaluated, becomes equal to: $1{,}562.54 \times 10^6$ k^3/in.

d. *Safety Factor* $\mu = 4$

$\lambda_{oe} = 0.7235\sqrt{4} = 1.447$; $\lambda_{ef} = 0.8890\sqrt{4} = 1.7780$

$\lambda_{oe} = 1.447$; $\sin\lambda_{oe} = 0.9924$; $\cos\lambda_{oe} = 0.1234$

$\lambda_{ef} = 1.7780$; $\sin\lambda_{ef} = 0.9786$; $\cos\lambda_{ef} = -0.2059$

Substituting these values into Equations (6.35) to (6.39), we obtain:

$$\Big[\frac{1.4470^2 x0.9924x10.31x10^6}{(0.9924-1.4470x0.1234)} + \frac{1.7780^2 x0.9786x10.31x10^6}{(0.9786+1.7780x0.2059)}\Big]\gamma_e + \Big[\frac{-1.4470^2 x0.9924x0.00537x10^6}{(0.9924-1.4470x0.1234)} + \frac{1.7780^2 x0.9786x0.00537}{(0.9786+1.7780x0.2059)}\Big]y_e + \Big[\frac{-1.7780^2 x0.9786x0.00537x10^6}{(0.9786+1.7780x0.2059)}\Big]y_f = 0$$

$$\Big[\frac{-1.4770^2 x0.9924x0.00537x10^6}{(0.9924-1.4470x0.1234)} + \frac{0.1778^2 x0.9786x0.00537x10^6}{(0.9786+1.7780x0.2059)}\Big]\gamma_e + \Big[\frac{1.4770^3 x0.1234x2.7968}{(0.9924-1.4770x0.1234)} + \frac{-0.1778^3 x0.2059x2.7968}{(0.9786+1.7780x0.2059)} + 1.714\Big]y_e + \Big[\frac{1.7780^3 x0.2059x2.7968}{(0.9786+1.7780x0.2059)}\Big]y_f = 0$$

$$\Big[\frac{-1.7780^2 x0.9786x0.00537x10^6}{(0.9786+1.7780x0.2059)}\Big]\gamma_e + \Big[\frac{1.7780^3 x0.2059x2.7968}{(0.9786+1.7780x0.2059)}\Big]y_e + \Big[\frac{-1.7780^3 x0.2059x2.7968}{(0.9786+1.7780x0.2059)} + 0.246\Big]y_f = 0$$

which in a matrix form and simplified version is:

$$\begin{bmatrix} 50.0441x10^6 & -0.0013x10^6 & -0.0124x10^6 \\ -0.0148x10^6 & 2.6368 & 0.0024 \\ -0.000124x10^6 & 0.0024 & 0.2436 \end{bmatrix} \begin{Bmatrix} \gamma_e \\ y_e \\ y_f \end{Bmatrix} = \begin{bmatrix} 0 \\ 0 \\ 0 \end{bmatrix}$$

whose determinant Δ of the coefficient matrix, when evaluated becomes equal to: -264.4056×10^6 k^3/in.

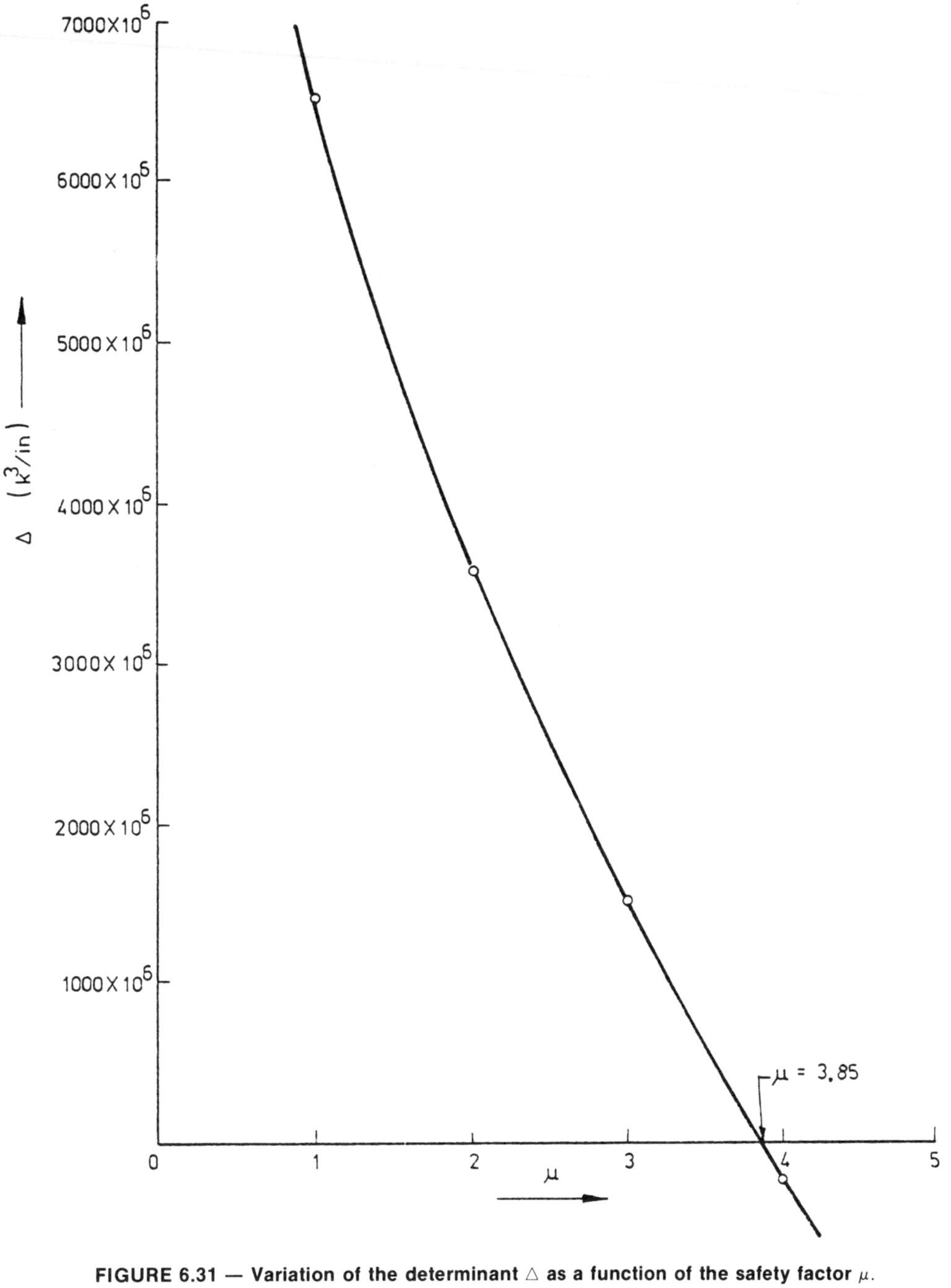

FIGURE 6.31 — Variation of the determinant Δ as a function of the safety factor μ.

e. *Determination of Buckling Loads*
From Figure 6.31, the factor of safety, μ corresponding to the smallest critical load is 3.85. Therefore, the corresponding buckling loads for span eo and ef are:

Critical load$]_{eo}$ =
= $N_{eo} \times 25 \times 3.85 = 117.24 \times 25 \times 3.85 = 11{,}284.35$ k
Critical load$]_{ef}$ =
= $N_{ef} \times 25 \times 3.85 = 169.77 \times 25 \times 3.85 = 16{,}340.36$ k

Since the cross-sectional area of both spans eo and ef is the same, the overall, governing least critical load for the stack is equal to: 11, 284.35 k.

TABLE 6.11 — Values of Determinant, Δ.
(Axial Loads of Stacks Multiplied by 25)

μ	Δk³/in
1	6,544.7x10⁶
2	3,595.7x10⁶
3	1,562.5x10⁶
4	-264.4x10⁶

It is interesting to compare the buckling loads of spans eo and ef with their counterparts of equal spans with end hinges (i.e., assuming the ideal end conditions) as follows:

(i) *Span e-o*

Length ℓ'_{eo} corresponding to the critical load 11,284.3 k is:

$$\ell'_{eo} = \frac{\pi}{2x\sqrt{\frac{N_{eo}}{EI_{eo}}}} =$$

$$= \frac{\pi}{2x\sqrt{\frac{11{,}284.3x10^3}{30x10^6x0.66x10^6}}} = 119.15 \text{ ft}$$

Since the critical load for a span with assumed hinged-end conditions and a reducec length ℓ'_{eo} is the same as that of the span eo (with boundary conditions as stipulated in stack oh), it can be expressed as:

$$P_{cr}]_{eo} = \frac{\pi^2 EI}{(\frac{119.15}{160.0}\ \ell_{eo})^2} = \frac{\pi^2 EI}{(0.7447\ \ell_{eo})^2}$$

(ii) *Span e-f*

Length ℓ'_{ef} corresponding to the critical load 16,340.36 k is:

$$\ell'_{ef} = \frac{\pi}{2x\sqrt{\frac{N_{ef}}{EI_{ef}}}} =$$

$$= \frac{\pi}{2x\sqrt{\frac{16{,}340.30x10^3}{30x10^6x0.66x10^6}}} = 144.08 \text{ ft.}$$

Since the critical load for a span with assumed hinged-end conditions and a reduced length ℓ'_{ef} is the same as that of the span ef (with boundary conditions as stipulated in stack og), it can be expressed as:

$$P_{cr}]_{ef} = \frac{\pi^2 EI}{(\frac{144.08}{160.0}\ \ell_{ef})^2} = \frac{\pi^2 EI}{(0.9005\ell)^2}$$

7. APPROXIMATE DETERMINATION OF THE NATURAL VIBRATIONS OF A FOUR-CABLE GUYED STACK

7.1 Description of the Stack Geometry

The four-cable guyed stack analyzed for its natural vibrations is shown in Figure 6.29. This stack is essentially the same as the one analyzed for the determination of its buckling loads in Section 6 in which its geometrical and mechanical properties are given in detail. In this section, some additional properties necessary for the dynamic study of this stack have been given. For this reason, it is suggested that this section should be read in conjunction with Section 6.

Some important stack properties are as follows:

a. E = Modulus of elasticity of the stack material = 30×10^6 psi

b. I_{oe} = I_{ef} = I_{fg} = Moment of inertia of the invariable stack cross-section = 0.66×10^6 in^4

c. Self-weight of the stack = 1.27 k/ft

d. Acceleration due to gravity = 32.2 ft/sec^2

7.2 Loads Per Unit Height of Various Spans

a. *Span eo*

Self-weight of span eo = 1.27 × 160 = 203.20 k

Longitudinal axial force from the cables at support e .... = 15.64 k

Total load for span eo .... = 218.84 k

Therefore, load per ft. height of span, eo,

$$m_{eo} = \frac{218.84}{160.0} = 1.3678 \text{ k/ft}$$

b. *Span ef*

Total contributing length = 160.0 + 80.0 = 240.0 (Assuming that the mass of span fg is transferred onto span ef.)

Self-weight of span ef = 1.27 × 240 = 304.8 k

Longitudinal axial force from the cables at support f .... = 17.37 k

Total load for span ef .... = 322.17 k

Therefore, load per ft. height of span ef,

$$m_{ef} = \frac{322.17}{160.0} = 2.0136 \text{ k/ft}$$

7.3 Transverse Shear Contributed by Guyed Cables at Stack-Supports

The following cable properties follow from Figure 6.18 and Sections 1 — 6.

a. *Cable ec*

Transverse force Q = 0.65 k; Axial force, F_{ec} = 8.5 k; cos α_{ec} = 0.6; and length, ℓ_{ec} = 200.0 ft

b. *Cable fc*

Transverse force Q = 2.49 k; Axial force F_{fc} = 9.2 k; cos α_{fc} = 0.3511; and length ℓ_{fc} = 341.76 ft

c. *Both Cables ec and fc*

The cross-sectional area of cables, A_c = 0.338 in^2; and the modulus of elasticity of cable material = 24 × 10^6 psi

The transverse shear contributd at point e by the guyed cables is given by Equation (6.55) as:

$$\frac{1}{\xi_0^e} = \frac{2A_cE_c\cos^2\alpha_{ec}}{\ell_{ec}} -$$

$$- \frac{(2A_cE_c\cos^2\alpha_{ec})^2}{\ell_{ec}^2}\ \frac{Q^2\ell_{ec}}{24F_{ec}^3\cos^2\alpha_{ec}} =$$

$$= [\frac{2x0.338x24x10^3}{200x12} x\ 0.6^2] -$$

$$- [(\frac{2x0.338x24x10^3}{200x12} x\ 0.6^2)^2] \times$$

$$\times \left[\frac{0.65^2 \times 200 \times 12}{24 \times 8.5^3 \times 0.6^2}\right] = 2.4336 - 1.1318 =$$

$$= 1.3018 \text{ k/in}$$

Similarly, for point f:

$$\frac{1}{\xi_0^f} = \left[\frac{2 \times 0.338 \times 24 \times 10^3 \times 0.3511^2}{341.76 \times 12}\right]$$

$$- \left[\left(\frac{2 \times 0.338 \times 24 \times 10^3 \times 0.3511^2}{341.76 \times 12}\right)^2\right] \times$$

$$\times \left[\frac{2.49^2 \times 341.76 \times 12}{24 \times 9.2^3 \times 0.3511^2}\right] = 0.4876 - 2.6242 =$$

$$= -2.1366 \text{ k/in}$$

7.4 Determination of the Natural Frequency of the Guyed-Stack

In this analysis, the following additional parameters of the stack-spans will be used:

$$\frac{1}{\ell'_{oe}} = \frac{1}{\ell'_{ef}} = \frac{EI_{oe}}{\ell_{oe}} = \frac{30 \times 10^6 \times 0.66 \times 10^6}{1920 \times 10^3} =$$

$$= 10.31 \times 10^6 \quad \text{k/in}$$

$$\frac{1}{\ell_{oe}\ell'_{oe}} = \frac{1}{\ell_{ef}\ell'_{ef}} = \frac{10.31 \times 10^6}{1920} = 0.00537 \times 10^6 \text{ k}$$

$$\frac{1}{\ell^2_{oe}\ell'_{oe}} = 2.7968 \text{ k/in}$$

Applying Equation (6.42) to spans oe and ef, we obtain

$$\eta_{oe} = \ell_{oe} \sqrt[4]{\frac{m_{oe}\omega^2}{EI_{oe}}} =$$

$$= 160 \times 12 \sqrt[4]{\frac{1.3678}{12 \times 30 \times 10^3 \times 0.66 \times 10^6 \times 32.2 \times 12}} \sqrt{\omega} =$$

$$= 0.6708 \sqrt{\omega}$$

and

$$\eta_{ef} = \ell_{ef} \sqrt[4]{\frac{m_{ef} \cdot \omega^2}{EI_{ef}}}$$

$$= 160 \times 12 \sqrt[4]{\frac{2.0136}{12 \times 30 \times 10^3 \times 0.66 \times 10^6 \times 32.2 \times 12}} \sqrt{\omega} =$$

$$= 0.7389 \sqrt{\omega}$$

Values of η_{oe} and η_{ef} will be calculated for several arbitrarily-assumed values of the radial frequency, ω. The ηs thus obtained, when substituted into Equation (6.48), will determine the corresponding factors Πs for spans oe and ef. These factors along with other pertinent parameters, e.g., $\frac{1}{\ell'_{oe}}, \frac{1}{\ell'_{ef}}, \frac{1}{\xi_{oe}}$, etc., will enable us to calculate the coefficients of unknown deformations, e.g., γ_e, y_e and y_f in Equation (6.59). A determinant, Δ of the matrix of these coefficients, thus obtained, will be plotted as a function of ω — numerically illustrated in the subsequent pages.

1. **Assume ω = 0.1 rad/sec**

Case (a):

$\eta_{oe} = 0.6708\sqrt{0.1} = 0.2121$ rad. $= 12.1548°$

Sinh $\eta_{oe} = 0.2106$; sinh $\eta_{oe} = 0.2137$; cos $\eta_{oe} = 0.9776$; cosh $\eta_{oe} = 1.0226$.

Using Equation (4.8), the following is obtained:

$$\Pi_7(\eta_{oe}) =$$

$$= 0.2121 \left[\frac{2 \times 0.2137 \times 0.2106}{1.0226 \times 0.2106 - 0.2137 \times 0.9776}\right]$$

$$= 0.2121 \left[\frac{2 \times 0.2137 \times 0.2106}{0.0065}\right] = 2.9371.$$

$$\Pi_8(\eta_{oe}) = (0.2121)^2 \left[\frac{0.2106 + 0.2137}{0.0065}\right] = 2.9366$$

$$\Pi_9(\eta_{oe}) = -(0.2121)^2$$

$$\left[\frac{1.0226 \times 0.2106 + 0.2137 \times 0.9776}{0.0065}\right] = -2.9366$$

$$\Pi_{10}(\eta_{oe}) =$$

$$= -(0.2121)^3 \left[\frac{0.9776 + 1.0226}{0.0065}\right] = -2.9362$$

$$\Pi_{11}(\eta_{oe}) = (0.2121)^3 \left[\frac{2 \times 1.0226 \times 0.9776}{0.0065}\right] = 2.9350$$

$$\Pi_{12}(\eta_{oe}) = (0.2121)^3 \left[\frac{0.9776 \times 1.0226 + 1}{0.0065}\right] = 2.9354$$

Case (b):

$\eta_{ef} = 0.7389\sqrt{0.1} = 0.2336$ rad. $= 13.3876°$

Sin $\eta_{ef} = 0.2315$; sinh $\eta_{ef} = 0.2357$; cos $\eta_{ef} = 0.9999$; cosh $\eta_{ef} = 1.0274$

Using Equations (6.48), the following Πs are obtained:

$$\Pi_7(\eta_{ef}) = 0.2336 \left[\frac{2 \times 0.2357 \times 0.2315}{1.0274 \times 0.2315 - 0.2357 \times 0.9999}\right]$$

$$= 0.2336 \left[\frac{2 \times 0.2357 \times 0.2315}{0.0021}\right] = 12.1393$$

$$\Pi_8(\eta_{ef}) = (0.2336)^2 \left[\frac{0.2357 + 0.2315}{0.0021}\right] = 12.1403$$

$$\Pi_9(\eta_{ef}) =$$

$$= -(0.2336)^2 \left[\frac{1.0274 \times 0.2375 + 0.2375 \times 0.9999}{0.0021}\right] =$$

$$= -12.3040$$

$$\Pi_{10}(\eta_{ef}) = -(0.2336)^3\left[\frac{1.0274+0.9999}{0.0021}\right] =$$

$$= -12.4640$$

$$\Pi_{11}(\eta_{ef}) = (0.2336)^3\left[\frac{2x1.0274x0.9999}{0.0021}\right] = 12.4717$$

$$\Pi_{12}(\eta_{ef}) = (0.2336)^3\left[\frac{0.9999x1.0274+1}{0.0021}\right] = 12.3060$$

Substitution of the above values of Πs into Equation (6.59) reduces it to a matrix form as follows:

$$\begin{bmatrix} 10.31x10^6(2.9371+12.1393) & 0.00537x10^6(-2.9366+12.3040) & -0.00537x10^6x12.1403 \\ 0.00537x10^6(-2.9366+12.3040) & 2.7968(2.9350+12.4714)+1.3018 & -2.7968x12.4640 \\ -0.00537x10^6x12.1403 & -2.7968x12.4640 & 2.7968x12.3060-2.1366 \end{bmatrix} \begin{Bmatrix} \gamma_e \\ y_e \\ y_f \end{Bmatrix} = \begin{bmatrix} 0 \\ 0 \\ 0 \end{bmatrix}$$

The determinant, $\triangle$ of the coefficient matrix, when evaluated, is:

$$\triangle = 162{,}083.7995 \times 10^6 \ k^3/in$$

2. **Assume ω = 1/5 = 0.2 rad./sec.**

Case (a):

η_{oe} = $0.6708\sqrt{0.20}$ = 0.2999 rad. = 17.1895°

Sin η_{oe} = 0.2955; sinh η_{oe} = 0.3044; cos η_{oe} = 0.9553; cosh η_{oe} = 1.0453.

$$\Pi_7(\eta_{oe}) = 0.2999\left[\frac{2x0.3044x0.2955}{1.0453x0.2995-0.3044x0.9553}\right]$$

$$= 0.2999\left[\frac{2x0.3044x0.2955}{0.0223}\right] = 2.4195$$

$$\Pi_8(\eta_{oe}) = (0.2999)^2\left[\frac{0.2955+0.3044}{0.0223}\right] = 2.4195$$

$$\Pi_9(\eta_{oe}) = -(0.2999)^2\left[\frac{0.6039}{0.0223}\right] = -2.4356$$

$$\Pi_{10}(\eta_{oe}) = -(0.2999)^3\left[\frac{0.9553+1.0453}{0.0223}\right] =$$

$$= -2.4198$$

$$\Pi_{11}(\eta_{oe}) = (0.2999)^3\left[\frac{2x1.0453x0.9553}{0.0223}\right] =$$

$$= 2.4157$$

$$\Pi_{12}(\eta_{oe}) = (0.2999)^3\left[\frac{1.0453x0.9553+1}{0.0223}\right] =$$

$$= 2.4174$$

Case (b):

$_{ef}$ = $0.7389\sqrt{0.2}$ = 0.3304 rad. = 18.9346°

Sin η_{ef} = 0.3245; sinh η_{ef} = 0.3364; cos η_{ef} = 0.9459; cosh η_{ef} = 1.0551

Using Equation (6.48), the following Πs are obtained:

$$\Pi_7(\eta_{ef}) = 0.3304\left[\frac{2x0.3364x0.3245}{1.0551x0.3245-0.3364x0.9459}\right]$$

$$= 0.3304\left[\frac{2x0.3364x0.3245}{0.0242}\right] = 2.9807$$

$$\Pi_8(\eta_{ef}) = (0.3304)^2\left[\frac{0.3245+0.3364}{0.0242}\right] = 2.9813$$

$$\Pi_9(\eta_{ef}) = -(0.3304)^2$$

$$\left[\frac{1.0551x0.3245+0.3364x0.9459}{0.0242}\right] = -2.9799$$

$$\Pi_{10}(\eta_{ef}) = -(0.3304)^3\left[\frac{0.9459+1.0551}{0.0242}\right] =$$

$$= -2.9823$$

$$\Pi_{11}(\eta_{ef}) = (0.3304)^3\left[\frac{2x1.0551x0.9459}{0.0242}\right] = 2.9749$$

$$\Pi_{12}(\eta_{ef}) = (0.3304)^3\left[\frac{0.9459x1.0551+1}{0.0242}\right] = 2.9779$$

Substitution of the above values of Πs into Equation (6.59) reduces it to a matrix form as follows:

$$\begin{bmatrix} 10.31x10^6(2.4195+2.9807) & 0.00537x10^6(-2.4356+2.9799) & -0.00537x10^6x2.9813 \\ 0.00537x10^6(-2.4356+2.9799) & 2.7968(2.4157+2.9749)+1.3018 & -2.7968x2.9823 \\ -0.00537x10^6x2.9813 & -2.7968x2.9823 & 2.7968x2.9779-2.1336 \end{bmatrix} \begin{Bmatrix} \gamma_e \\ y_e \\ y_f \end{Bmatrix} = \begin{bmatrix} 0 \\ 0 \\ 0 \end{bmatrix}$$

The determinant, Δ of the coefficient matrix, when evaluated is:

$$\Delta = -1{,}702.3859 \times 10^{-6}\ k^3/in$$

3. **Assume ω = 1/4 = 0.25 rad./sec.**

Case (a):

$\eta_{oe} = 0.6708\sqrt{0.25} = 0.3354$ rad. = 19.2184°

Sin η_{oe} = 0.3292; sinh η_{oe} = 0.3417; cos η_{oe} = 0.9443; cosh η_{oe} = 1.0568

Using Equation (6.48), the following is obtained:

$$\Pi_7(\eta_{oe}) = 0.3354\ \left[\frac{2x0.3417x0.3292}{1.0568x0.3417-0.3417x0.9443}\right]$$

$$= 0.3354\ \left[\frac{2x0.3417x0.3292}{0.0384}\right] = 1.9650$$

$$\Pi_8(\eta_{oe}) = (0.3354)^2\left[\frac{0.3417+0.3292}{0.0384}\right] = 1.9654$$

$$\Pi_9(\eta_{oe}) =$$

$$= -\ (0.3354)^2\left[\frac{1.0568x0.3292+0.3417x0.9443}{0.0384}\right]$$

$$= -\ 1.9645$$

$$\Pi_{10}(\eta_{oe}) = -\ (0.3354)^3\left[\frac{0.9443+1.0568}{0.0384}\right] = -\ 1.9662$$

$$\Pi_{11}(\eta_{oe}) = (0.3354)^3\left[\frac{2x1.0568x0.9443}{0.0384}\right] = 1.9611$$

$$\Pi_{12}(\eta_{oe}) = (0.3354)^3\left[\frac{1.0568x0.9443+1}{0.0384}\right] = 1.9631$$

Case (b):

$\eta_{ef} = 0.7389\sqrt{0.25} = 0.3695$ rad. = 21.1695°

Sin η_{ef} = 0.3613; sinh η_{ef} = 0.3780; cos η_{ef} = 0.9326; cosh η_{ef} = 1.0690

$$\Pi_7(\eta_{ef}) = 0.3695\ \left[\frac{2x0.3780x0.3613}{1.0690x0.3613-0.3780x0.9326}\right]$$

$$= 0.3695\ \left[\frac{2x0.3780x0.3613}{0.0337}\right] = 2.9948$$

$$\Pi_8(\eta_{ef}) = (0.3695)^2\left[\frac{0.3613+0.3780}{0.0337}\right] = 2.9952$$

$$\Pi_9(\eta_{ef}) =$$

$$= -\ (0.3695)^2\left[\frac{1.0690x0.3613+0.3780x0.9326}{0.0337}\right] =$$

$$= 2.9927$$

$$\Pi_{10}(\eta_{ef}) = -\ (0.3695)^3\left[\frac{0.9326+1.0690}{0.0337}\right] =$$

$$= -\ 2.9963$$

$$\Pi_{11}(\eta_{ef}) = (0.3695)^3\left[\frac{2x1.0690x0.9326}{0.0337}\right] = 2.9848$$

$$\Pi_{12}(\eta_{ef}) = (0.3695)^3\left[\frac{1.0690x0.9326+1}{0.0337}\right] = 2.9894$$

Substitution of the above values of Πs into Equation (6.59) reduces it to a matrix form as follows:

$$\begin{bmatrix} 10.31x10^6(1.9650+2.9948) & 0.00537x10^6(-1.9645+2.9927) & -\ 0.00537x10^6x2.9952 \\ 0.00537x10^6(-1.9645+2.9927) & 2.7968(1.9611+2.9848)+1.3018 & -\ 2.7968x2.9963 \\ -\ 0.00537x10^6x2.9952 & -\ 2.7968x2.9963 & 2.7968x2.9894-2.1336 \end{bmatrix} \begin{Bmatrix} Y_e \\ y_e \\ y_f \end{Bmatrix} = \begin{bmatrix} 0 \\ 0 \\ 0 \end{bmatrix}$$

The determinant, Δ of the coefficient matrix, when evaluated is:

$$\Delta = -1{,}998.4008 \times 10^{6}\ k^3/in$$

4. **Assume ω = 1/3 = 0.33 rad./sec.**

Case (a):

$\eta_{oe} = 0.6708\sqrt{0.33} = 0.3853$ rad. = 22.0803°

Sin η_{oe} = 0.3759; sinh η_{oe} = 0.3949; cos η_{oe} = 0.9267; cosh η_{oe} = 1.0752.

Using Equation (6.48), the following is obtained:

$$\Pi_7(\eta_{oe}) = 0.3853\ \left[\frac{2x0.3949x0.3759}{1.0752x0.3759-0.3949x0.9267}\right]$$

$$= 0.3853\ \left[\frac{2x0.3949x0.3759}{0.0382}\right] = 2.9945$$

$$\Pi_8(\eta_{oe}) = (0.3853)^2\left[\frac{0.3759+0.3949}{0.0382}\right] = 2.9955$$

$$\Pi_9(\eta_{oe}) =$$

$$= -\ (0.3853)^2\left[\frac{1.0752x0.3759+0.3949x0.9267}{0.0382}\right] =$$

$$= -\ 2.9932$$

$$\Pi_{10}(\eta_{oe}) = -\ (0.3853)^3\left[\frac{0.9267+1.0752}{0.0382}\right] =$$

$$= -\ 2.9976$$

$$\Pi_{11}(\eta_{oe}) = (0.3853)^3\left[\frac{2x1.0752x0.9267}{0.0382}\right] =$$

$$= 2.9840$$

$$\Pi_{12}(\eta_{oe}) = (0.3853)^3\left[\frac{0.9267\times1.0752+1}{0.0382}\right] =$$

$$= 2.9894$$

Case (b):

$\eta_{ef} = 0.7389\sqrt{0.33} = 0.4245$ rad. $= 24.3219°$

Sin $\eta_{ef} = 0.4119$; sinh $\eta_{ef} = 0.4375$; cos $\eta_{ef} = 0.9112$; cosh $\eta_{ef} = 1.0915$.

Using Equation (6.48), the following is obtained:

$$\Pi_7(\eta_{ef}) = 0.4245\left[\frac{2\times0.4375\times0.4119}{1.0915\times0.4119-0.4375\times0.9112}\right]$$

$$= 0.4245\left[\frac{2\times0.4375\times0.4119}{0.0509}\right] = 3.0058$$

$$\Pi_8(\eta_{ef}) = (0.4245)^2\left[\frac{0.4119+0.4375}{0.0509}\right] = 3.0071$$

$$\Pi_9(\eta_{ef}) =$$

$$= -(0.4245)^2\left[\frac{1.0915\times0.4119+0.4375\times0.9112}{0.0509}\right] =$$

$$= -3.0032$$

$$\Pi_{10}(\eta_{ef}) = -(0.4245)^3\left[\frac{0.9112+1.0915}{0.0509}\right]$$

$$= -3.0098$$

$$\Pi_{11}(\eta_{ef}) = (0.4245)^3\left[\frac{2\times1.0915\times0.9112}{0.0509}\right] = 2.9894$$

$$\Pi_{12}(\eta_{ef}) = (0.4245)^3\left[\frac{1.0915\times0.9112+1}{0.0509}\right] = 2.9975$$

Substitution of the above values of Πs into Equation (6.59) reduces it to a matrix form as follows:

$$\begin{bmatrix} 10.31\times10^6(2.9945+3.0058) & -0.00537\times10^6(2.9932+3.0032) & -0.00537\times10^6\times3.0071 \\ -0.00537\times10^6(2.9932+3.0032) & 2.7968(2.9840+2.9894)+1.3018 & -2.7968\times3.0098 \\ -0.00537\times10^6\times3.0071 & -2.7968\times3.0098 & 2.7968\times2.9975-2.1366 \end{bmatrix} \begin{Bmatrix} Y_e \\ y_e \\ y_f \end{Bmatrix} = \begin{bmatrix} 0 \\ 0 \\ 0 \end{bmatrix}$$

The determinant, Δ of the coefficient matrix, when evaluated is:

$$\Delta = -2{,}105.6335 \times 10^6 \text{ k}^3/\text{in}$$

5. **Assume ω = 1/2 = 0.5 rad./sec.**

Case (a):

$\eta_{oe} = 0.6708\sqrt{0.5} = 0.4743$ rad. $= 27.1774°$

Sin $\eta_{oe} = 0.4567$; sinh $\eta_{oe} = 0.4923$; cos $\eta_{oe} = 0.8896$; cosh $\eta_{oe} = 1.1146$.

Using Equation (6.48), the following is obtained:

$$\Pi_7(\eta_{oe}) = 0.4743\left[\frac{2\times0.4923\times0.4567}{1.1146\times0.4567-0.4923\times0.8896}\right]$$

$$= 0.4743\left[\frac{2\times0.4923\times0.4567}{0.071}\right] = 3.0039$$

$$\Pi_8(\eta_{oe}) = (0.4743)^2\left[\frac{0.4567+0.4923}{0.071}\right] = 3.0069$$

$$\Pi_9(\eta_{oe}) =$$

$$= -(0.4743)^2\left[\frac{1.1146\times0.4567+0.4923\times0.8896}{0.071}\right] =$$

$$= -3.0005$$

$$\Pi_{10}(\eta_{oe}) = -(0.4743)^3\left[\frac{1.1146+0.8890}{0.071}\right] =$$

$$= -3.0110$$

$$\Pi_{11}(\eta_{oe}) = (0.4743)^3\left[\frac{2\times1.1146\times0.8896}{0.071}\right] = 2.9802$$

$$\Pi_{12}(\eta_{oe}) = (0.4743)^3\left[\frac{0.8896\times1.1146+1}{0.071}\right] = 2.9929$$

Case (b):

$\eta_{ef} = 0.7389\sqrt{0.5} = 0.5225$ rad. $= 29.9393°$

Sin $\eta_{ef} = 0.4991$; sinh $\eta_{ef} = 0.5466$; cos $\eta_{ef} = 0.8666$; cosh $\eta_{ef} = 1.1396$.

Using Equation (6.48), the following is obtained:

$$\Pi_7(\eta_{ef}) = 0.5225\left[\frac{2\times0.5466\times0.4991}{1.1396\times0.4991-0.5466\times0.8666}\right]$$

$$= 0.5225\left[\frac{2\times0.5466\times0.4991}{0.0951}\right] = 2.9977$$

$$\Pi_8(\eta_{ef}) = (0.5225)^2\left[\frac{0.5466+0.4991}{0.0951}\right] = 3.0019$$

$$\Pi_9(\eta_{ef}) =$$

$$= (0.5225)^2\left[\frac{1.1396\times0.4991+0.5466\times0.8666}{0.0951}\right] =$$

$$= 2.9927$$

$$\Pi_{10}(\eta_{ef}) = -(0.5225)^3\left[\frac{1.1396+0.8666}{0.0951}\right] =$$

$$= -3.0092$$

$$\Pi_{11}(\eta_{ef}) = (0.5225)^3\left[\frac{2\times1.1396\times0.8666}{0.0951}\right] = 2.9606$$

$$\Pi_{12}(\eta_{ef}) = (0.5225)^3\left[\frac{1.1396\text{x}0.8666+1}{0.0951}\right] = 2.9813$$

Substitution of the above values of Πs into Equation (6.59) reduces it to a matrix form as follows:

$$\begin{bmatrix} 10.31\text{x}10^6(3.0039+2.9977) & 0.00537\text{x}10^6(-3.0005+2.9927) & -0.00537\text{x}10^6\text{x}3.0019 \\ 0.00537\text{x}10^6(-3.0005+2.9927) & 2.7968(2.9802+2.9606)+1.3018 & -2.7968\text{x}3.0092 \\ -0.00537\text{x}10^6\text{x}3.0019 & -2.7968\text{x}3.0092 & 2.7968\text{x}2.9813-2.1366 \end{bmatrix} \begin{Bmatrix} Y_e \\ y_e \\ y_f \end{Bmatrix} = \begin{bmatrix} 0 \\ 0 \\ 0 \end{bmatrix}$$

The determinant, Δ of the coefficient matrix, when evaluated, is:

$$\Delta = -2{,}174.6885 \times 10^6 \text{ k}^3/\text{in}$$

6. **Assume ω = 1 rad./sec.**

Case (a):

$\eta_{oe} = 0.6708\sqrt{1} = 0.6708$ rad. $= 38.4368°$

Sin $\eta_{oe} = 0.6216$; sinh $\eta_{oe} = 0.7223$; cos $\eta_{oe} = 0.7833$; cosh $\eta_{oe} = 1.2336$

Using Equation (6.48), the following is obtained:

$$\Pi_7(\eta_{oe}) = 0.6708\left[\frac{2\text{x}0.7223\text{x}0.6216}{1.2336\text{x}0.6216-0.7223\text{x}0.7833}\right]$$

$$= 0.6708\left[\frac{2\text{x}0.7223\text{x}0.6216}{0.2010}\right] = 2.9968$$

$$\Pi_8(\eta_{oe}) = (0.6708)^2\left[\frac{0.7223+0.6216}{0.2010}\right] = 3.0085$$

$$\Pi_9(\eta_{oe}) =$$

$$= -(0.6708)^2\left[\frac{1.2336\text{x}0.6216+0.7223\text{x}0.7833}{0.2010}\right] =$$

$$= -2.9833$$

$$\Pi_{10}(\eta_{oe}) = -(0.6708)^3\left[\frac{1.2336+0.7833}{0.2010}\right] = -3.0288$$

$$\Pi_{11}(\eta_{oe}) = (0.6708)^3\left[\frac{2\text{x}1.2336\text{x}0.7833}{0.2010}\right] = 2.9021$$

$$\Pi_{12}(\eta_{oe}) = (0.6708)^3\left[\frac{1.2336\text{x}0.7833+1}{0.2010}\right] = 2.9528$$

Case (b):

$\eta_{ef} = 0.7389\sqrt{1} = 0.7389$ rad. $= 42.3390°$

Sin $\eta_{ef} = 0.6735$; sinh $\eta_{ef} = 0.8080$; cos $\eta_{ef} = 0.7392$; cosh $\eta_{ef} = 1.2856$

Using Equation (6.48), the following is obtained:

$$\Pi_7(\eta_{ef}) = 0.7389\left[\frac{2\text{x}0.8080\text{x}0.6735}{1.2856\text{x}0.6735-0.8080\text{x}0.7392}\right]$$

$$= 0.7389\left[\frac{2\text{x}0.8080\text{x}0.6735}{0.2686}\right] = 2.9940$$

$$\Pi_8(\eta_{ef}) = (0.7389)^2\left[\frac{0.6735+0.8080}{0.2686}\right] = 3.0111$$

$$\Pi_9(\eta_{ef}) =$$

$$= -(0.7389)^2\left[\frac{1.2856\text{x}0.6735+0.8080\text{x}0.7392}{0.2686}\right] =$$

$$= -2.9742$$

$$\Pi_{10}(\eta_{ef}) = (0.7389)^3\left[\frac{1.2856+0.7392}{0.2686}\right] = -3.0411$$

$$\Pi_{11}(\eta_{ef}) = (0.7389)^3\left[\frac{2\text{x}1.2856\text{x}0.7392}{0.2686}\right] = 2.8546$$

$$\Pi_{12}(\eta_{ef}) = (0.7389)^3\left[\frac{1.2856\text{x}0.7392+1}{0.2686}\right] = 2.9292$$

Substitution of the above values of Πs into Equation (6.59) reduces it to a matrix form as follows:

$$\begin{bmatrix} 10.31\text{x}10^6(2.9968+2.9940) & 0.00537\text{x}10^6(-2.9833+2.9742) & -0.00537\text{x}10^6\text{x}3.0111 \\ 0.00537\text{x}10^6(-2.9833+2.9742) & 2.7968(2.9021+2.8546)+1.3018 & -2.7968\text{x}3.0411 \\ -0.00537\text{x}10^6\text{x}3.0111 & -2.7968 \text{ x } 3.0411 & 2.7968\text{x}2.9292-2.1366 \end{bmatrix} \begin{Bmatrix} Y_e \\ y_e \\ y_f \end{Bmatrix} = \begin{bmatrix} 0 \\ 0 \\ 0 \end{bmatrix}$$

The determinant, Δ of the coefficient matrix, when evaluated is:

$$\Delta = -2{,}669.7910 \times 10^6 \text{ k}^3/\text{in}$$

7. **First Natural Frequency.**

Several values of the determinant, Δ, with respect to the angular frequency, ω of the guyed stack have been tabulated in Table 6.12 and their values are shown in Figure 6.32.

The first natural frequency of stack from Fig. 6.32 = 0.142 rad./sec. Therefore, the natural period of oscillation, $T = \frac{2\pi}{0.142} = 44.2479$ sec.; and the natural frequency,

$$\omega_n = \frac{1}{44.2479} = 0.0226 \text{ cycles/sec.}$$

TABLE 6.12 — Values of Determinant, Δ.

ω rad./sec.	Δ k³/in
1/10 = 0.1	162,083.7995x10⁶
1/5 = 0.2	-1,702.3859x10⁶
1/4 = 0.25	-1,998.4008x10⁶
1/3 = 0.33	-2,105.6335x10⁶
1/2 = 0.50	-2,174.6885x10⁶
1.0	-2,669.7910x10⁶

8. FORCED VIBRATIONS OF A GUYED STACK

8.1 Stack Geometry

The guyed stack analyzed for its behavior under gusting winds is the same stack as studied for its dynamic vibrations under steady wind, presented in Section 7. However, for a quick reference, it is shown in Figure 6.33.

Since the forced vibrations of a guyed stack are an outgrowth of its dynamic oscillations under steady wind, it is suggested that this section should be read in conjunction with Part 6.4 and Section 7. Of special significance is Equation (6.49) of Part 6.4, which will be re-utilized here for the calculation of sub-determinants $\Delta(y_e, 1)$ and $\Delta(y_f, 1)$. Some additional pertinent parameters for this stack are as follows:

a. Self-weight of stack = 1.3678 k/ft.
b. First natural frequency, ω_I = 0.142 rad./sec.
c. Intensity of gusting wind, p(x,t) = 0.0085 sin ωt k/in.
d. Assumed value of the damping coefficient $\zeta = 0.1 \times 2\pi = 0.628$.

8.2 Evaluation of Sub-Determinants $\Delta(y_e, 1)$ and $\Delta(y_f, 1)$

Coefficient matrix of Equation (6.59) of Part 4 is used here for calculating the values of sub-determinants $\Delta(y_e, 1)$ and Δ(yf, 1) as defined in Equation (6.66) of Part 6.5. These sub-determinants will be evaluated for ω = 0.5, 0.33, 0.25, 0.20 and 0.1 rad./sec. successively as shown below:

a. *Natural Frequency ω = 0.5 rad./sec.*

Coefficient matrix of Equation (6.59) =

$$= \begin{array}{c} \\ 1 \\ 2 \\ 3 \end{array} \begin{array}{ccc} Y_e & Y_e & Y_f \\ \end{array}$$
$$\begin{bmatrix} 61.8765\times10^6 & -41.866 & -0.01612\times10^6 \\ -41.866 & 17.9170 & -8.4161 \\ -0.01612\times10^6 & -8.4161 & 6.2015 \end{bmatrix}$$

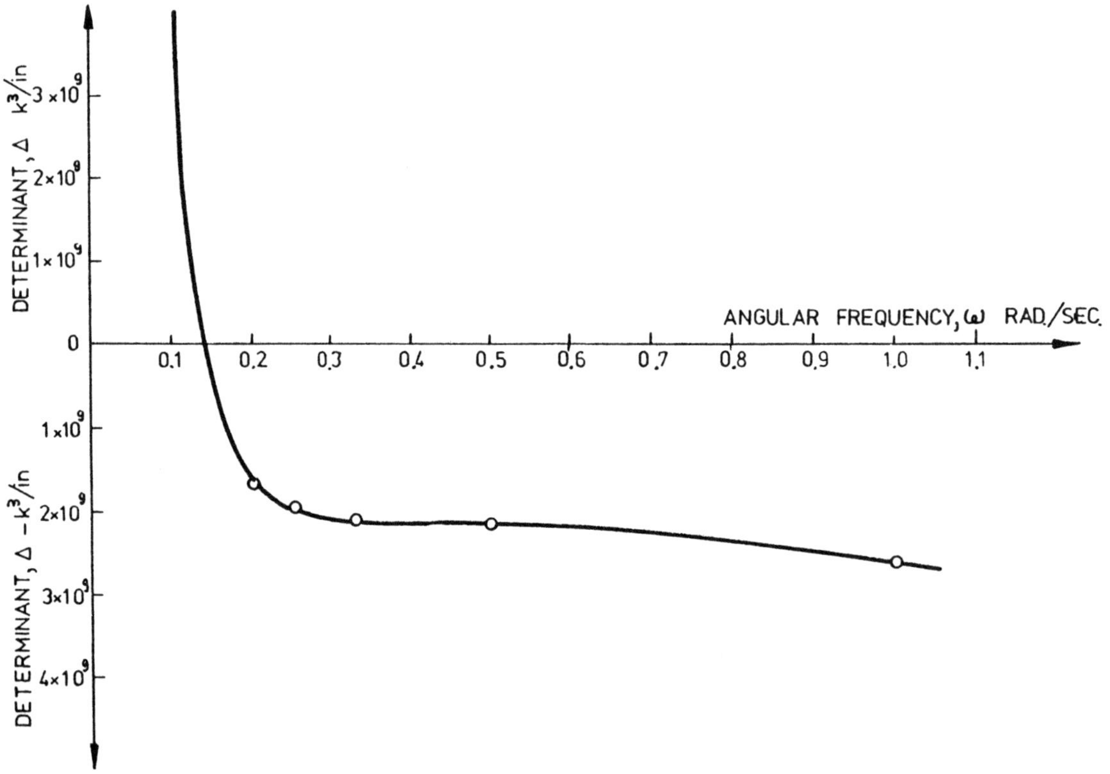

FIGURE 6.32 — Variation of the determinant, Δ as a function of then angular frequency, ω of the guyed stack.

Therefore, $\Delta(y_e,1) =$

$$= \begin{vmatrix} -41.866 & -8.4161 \\ -0.01612\text{x}10^6 & 6.2015 \end{vmatrix} = -0.1358\text{x}10^6 \ \text{k}^2/\text{in}$$

and, $\Delta(y_f,1) =$

$$= \begin{vmatrix} -41.866 & 17.9170 \\ -0.01612\text{x}10^6 & -8.4161 \end{vmatrix} = 0.289\text{x}10^6 \ \text{k}^2/\text{in}$$

b. *Natural Frequency* $\omega = 0.33$ *rad./sec.*

Coefficient matrix of Equation (6.59) =

$$= \begin{matrix} 1 \\ 2 \\ 3 \end{matrix} \begin{bmatrix} Y_e & y_e & y_f \\ 61.8631\text{x}10^6 & 53.7 & -1.6148\text{x}10^{-2}\text{x}10^6 \\ 53.7 & 18.0082 & -8.4178 \\ -1.6148\text{x}10^{-2}\text{x}10^6 & -8.4178 & 6.2468 \end{bmatrix}$$

Therefore, $\Delta(y_e,1) =$

$$= \begin{vmatrix} 53.7 & -8.4178 \\ -1.6148\text{x}10^{-2}\text{x}10^6 & 6.2468 \end{vmatrix} = -0.1356\text{x}10^6 \ \text{k}^2/\text{in}$$

and, $\Delta(y_f,1) =$

$$= \begin{vmatrix} 53.7 & 18.0082 \\ -1.6148\text{x}10^{-2}\text{x}10^6 & -8.4178 \end{vmatrix} = 0.2903\text{x}10^6 \ \text{k}^2/\text{in}$$

c. *Natural Frequency* $\omega = 0.25$ *rad./sec.*

Coefficient matrix of Equation (6.59) =

$$= \begin{matrix} 1 \\ 2 \\ 3 \end{matrix} \begin{bmatrix} Y_e & y_e & y_f \\ 51.1356\text{x}10^6 & 5.5214\text{x}10^{-3}\text{x}10^6 & -1.6084\text{x}10^{-2}\text{x}10^6 \\ 5.5214\text{x}10^{-3}\text{x}10^6 & 15.1345 & -8.3801 \\ -1.6084\text{x}10^{-2}\text{x}10^6 & -8.3801 & 6.2272 \end{bmatrix}$$

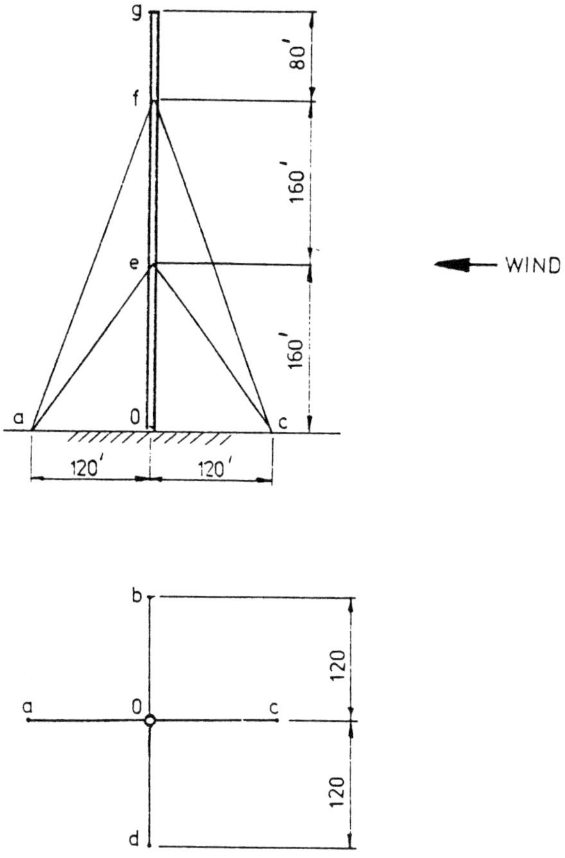

FIGURE 6.33 — Four-cable guyed stack.

Therefore, $\Delta(y_e,1) =$

$$= \begin{vmatrix} 5.5214\text{x}10^{-3}\text{x}10^6 & -8.3801 \\ -1.6084\text{x}10^{-2}\text{x}10^6 & 6.2272 \end{vmatrix} = 0.10032\text{x}10^6 \ \text{k}^2/\text{in}$$

and. $\Delta(y_f,1) =$

$$= \begin{vmatrix} 5.5214\text{x}10^{-3}\text{x}10^6 & 15.1345 \\ -1.6084\text{x}10^{-2}\text{x}10^6 & -8.3801 \end{vmatrix} = 0.1971\text{x}10^6 \ \text{k}^2/\text{in}$$

d. *Natural Frequency* $\omega = 0.20$ *rad./sec.*

Coefficient matrix of Equation (6.59) =

$$= \begin{matrix} 1 \\ 2 \\ 3 \end{matrix} \begin{bmatrix} Y_e & y_e & y_f \\ 55.6761\text{x}10^6 & 2.9229\text{x}10^{-3}\text{x}10^6 & -1.601\text{x}10^{-2}\text{x}10^6 \\ 2.9229\text{x}10^{-3}\text{x}10^6 & 16.3680 & -8.3490 \\ -1.6010\text{x}10^{-2}\text{x}10^6 & -8.3490 & 6.1950 \end{bmatrix}$$

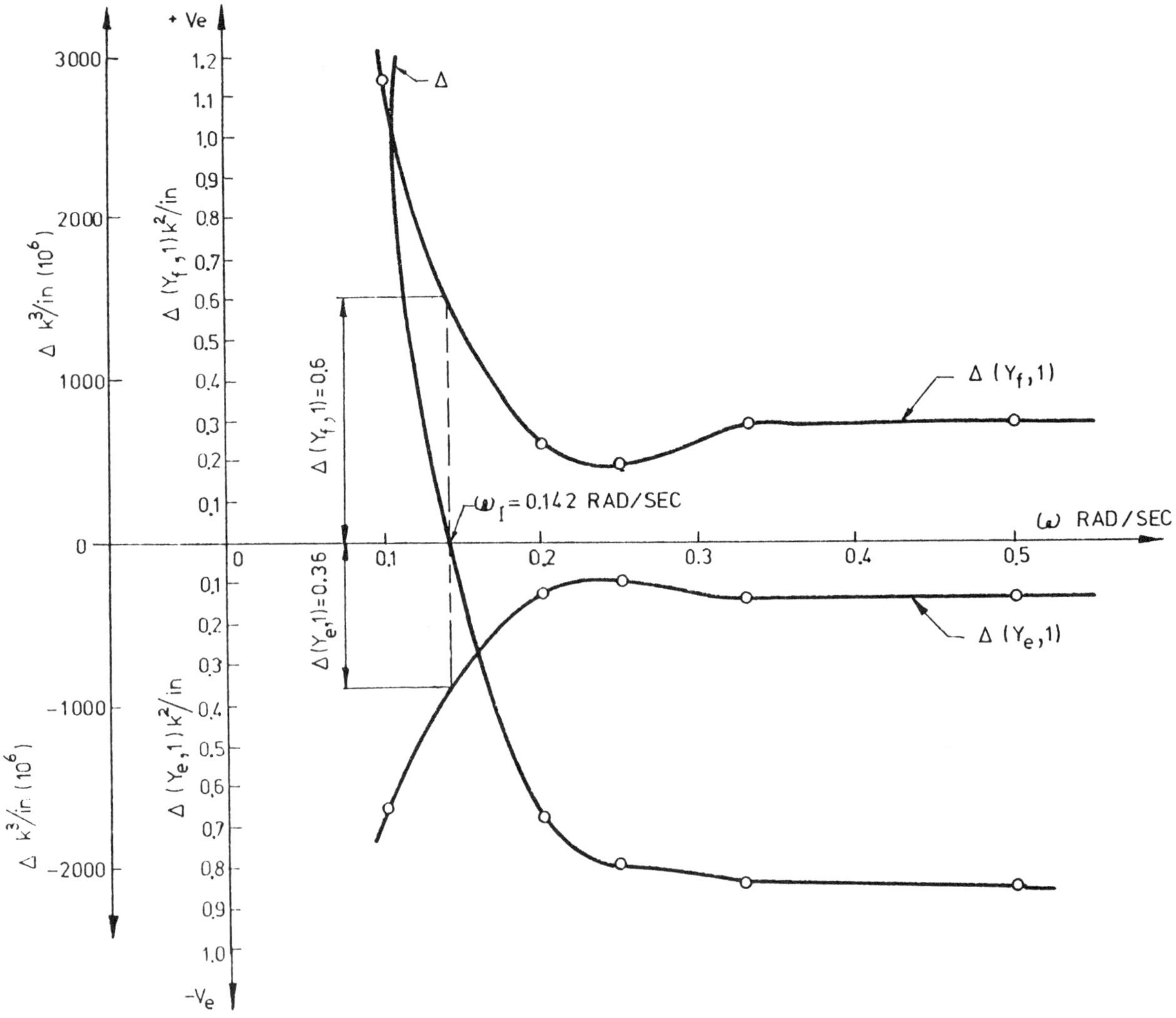

FIGURE 6.34 — Plot of Determinant Δ of Equation (5.7) and sub-determinants $\Delta(y_e,1)$ and $\Delta(y_f,1)$ as functions of the natrual frequency, ω of the guyed stack.

TABLE 6.13 — Values of Determinant, Δ and Sub-Determinants Δ (y_e, 1) & Δ (y_f, 1).

ω rad/sec.	Δ k³/in	Δ (y_e, 1) k²/in	Δ (y_f, 1) k²/in
0.50	-2,174.69x10⁶	-0.1358x10⁶	0.289x10⁶
0.33	-2,105.63x10⁶	-0.1356x10⁶	0.2903x10⁶
0.25	-1,998.40x10⁶	-0.10032x10⁶	0.1971x10⁶
0.20	-1,702.39x10⁶	-0.1156x10⁶	0.2377x10⁶
0.10	162,083.80x10⁶	-0.6500x10⁶	1.1400x10⁶

Therefore, $\Delta(y_e,1) =$

$$= \begin{vmatrix} 2.9229\text{x}10^{-3}\text{x}10^{6} & -8.3490 \\ -1.6010\text{x}10^{-2}\text{x}10^{6} & 6.1950 \end{vmatrix} = -0.1156\text{x}10^{6}\ \text{k}^2/\text{in}$$

and, $\Delta(y_f,1) =$

$$= \begin{vmatrix} 2.9229\text{x}10^{-3}\text{x}10^{5} & 16.3680 \\ -1.6010\text{x}10^{-2}\text{x}10^{6} & -8.3490 \end{vmatrix} = 0.2377\text{x}10\ \text{k}^2/\text{in}$$

e. *Natural Frequency ω = 0.10 rad./sec.*

Coefficient matrix of Equation (6.59) =

$$= \begin{array}{c} \\ 1 \\ 2 \\ 3 \end{array} \begin{bmatrix} Y_e & y_e & y_f \\ 155.4377\text{x}10^{6} & 0.05030\text{x}10^{6} & -0.06519\text{x}10^{6} \\ 0.05030\text{x}10^{6} & 44.3913 & -34.8593 \\ -0.06519\text{x}10^{6} & -34.8593 & 32.2808 \end{bmatrix}$$

Therefore, $\Delta(y_e,1) =$

$$= \begin{vmatrix} 0.05030\text{x}10^{6} & -34.8593 \\ -0.06519\text{x}10^{6} & 32.2808 \end{vmatrix} = -0.65\text{x}10^{6}\ \text{k}^2/\text{in}$$

and,

$$\Delta(y_f,1) = \begin{vmatrix} 0.05030\text{x}10^6 & 44.3913 \\ -0.06519\text{x}10^6 & -34.8593 \end{vmatrix} = 1.14\text{x}10^5 \ \text{k}^2/\text{in}$$

A plot of determinant Δ, and sub-determinants $\Delta(y_e,1)$ and $\Delta(y_f,1)$ as a function of the natural frequency, ω appears in Figure 6.34. From the curves in this figure, the following parameters of the guyed stack corresponding to its first natural frequency of vibration under steady wind, i.e., $\omega_I = 0.142$ rad./sec., are obtained:

$$\psi_I(x)_e = \frac{\Delta(y_e,1)}{\Delta(y_f,1)} = \frac{0.36\text{x}10^5}{0.60\text{x}10^5} = 0.60; \quad \text{and}$$

$$\psi_I(x)_f = \frac{\Delta(y_f,1)}{\Delta(y_f,1)} = \frac{0.60\text{x}10^5}{0.60\text{x}10^5} = 1.0$$

8.3 Determination of Horizontal Displacements of Supports e and f

Applying Equation (6.62) to span o-e of the guyed stack, we obtain:

$$p_I(t) = \frac{\int_o^e p(x,t)\psi_I(x)_e dx}{\int_o^e m_{oe}\psi_I^2(x)_e dx} = \frac{\int_o^{160} 0.0085 \sin \omega t \text{ x } 0.60 \, dx}{\int_o^{160} 1.3678(0.60)^2 dx} = 0.12428 \sin \omega t$$

Using the given first natural frequency of the guyed stack, i.e., $\omega_I = 0.142$ rad./sec. and the damping coefficient, ξ and substituting them into Equation (6.64), it follows that:

$$\nu_I(x)_e = \frac{0.12428}{2\text{x}0.142\text{x}0.628} = 0.6997$$

Therefore, the horizontal displacement of support e with respect to the first mode of natural vibration is given by Equation (6.65) as:

$$\bar{y}_e = \nu_I(t)_e \cdot \psi_I(x)_e = 0.6997\text{x}0.60 = 0.4198 \text{ in}$$

Similarly, the horizontal displacement of support f is:

$$\bar{y}_f = \nu_I(t)_e \cdot \psi_I(x)_f = 0.6997\text{x}1 = 0.6997 \text{ in.}$$

REFERENCES — CHAPTER 6

[6.1] Kolousek, V., "Static and Dynamic Calculations of Stayed Antennae Masts", International Association for Bridge and Structural Engineering Publications, Vol. 8, 1947, pp. 105-138.

[6.2] Troitsky, M.S., "Guyed Steel Stacks. Guidelines for Analysis and Design", Canadian Steel Industries Construction Council, Research Report No. 6, Toronto, September 1978, pp. 1-158.

[6.3] Zar, M., "Towers", Structural Engineering Handbook edited by Gaylord, Jr., E.H., and Gaylord, C.N., McGraw-Hill Book Company, New York, 1968.

[6.4] Cohen, E., and Perrin, H., "Design of Multi-Level Guyed Towers: Wind Loading", J. Struct. Div. ASCE, Vol. 83, Paper No. 1355, September 1957, pp. 1355-1 to 1355-39.

[6.5] Cohen, E., and Perrin, H., "Design of Multi-Level Guyed Towers: Structural Analysis", J. Struct. Div. ASCE, Vol. 83, Paper No. 1356, September 1957, pp. 1356-1 to 1356-29.

[6.6] Rowe, R.S., "Amplified Stress and Displacement in Guyed Towers", Trans. ASCE, Vol. 125, 1960, p. 199.

[6.7] Vandeghen, A., and Alexandre, M., "Wind-Induced Vibrations of Tall Steel Chimney Stacks", International Association for Bridge and Structural Engineering Publications, Vol. 29-1, 1969.

[6.8] Vanasup, J., "Response of Tall Concrete Chimneys Subjected to Fluctuating Wind", Doctoral Dissertation in Civil Engineering, University of Michigan, 1975.

[6.9] Radhakrishnan, R., "Dynamic Wind Analysis of Stacks", Doctoral Dissertation in Civil Engineering, Brigham Young University, 1974.

[6.10] Kolousek, V., "Dynamics in Engineering Structures", translated by Doubravaka Hajsmanovoa, Academia Prague, 1973.

[6.11] Chiu, A.N.L., Sawyer, D.A. and Grinter, L.E., "Vibrations of Towers as Related to Wind Pulses", Journal of the Structural Division, ASCE, October 1964, pp. 137-160.

[6.12] Beitin, K.I., "Dynamic Response of Guyed Towers to Wind Loading", Meeting Preprint 1032; ASCE Annual and Environmental Meeting, Chicago, Illinois, October 13-17, 1969.

[6.13] McCaffrey, R.J., and Hartmann, A.J., "Dynamics of Guyed Towers", Meeting Preprint 1379; ASCE National Structural Engineering Meeting, Baltimore, Maryland, April 19-23, 1971.

CHAPTER 7

bins and bunkers

7.1 Types of Bins

Bins are containers designed for storing and loading granular materials. Considering their size, shape, function, structural analysis and design, there are two main types of bins: *"shallow bins"*, also called *"bunkers"*, and *"deep bins"* called *"bins"*. The important difference between the two is in the behavior of the stored material. This behavior difference is influenced by both-bin geometry and the characteristics of the stored material. Material pressure against the walls and bottom are usually determined by one method for bins and by another for bunkers. Bins and bunkers are made from different structural materials and of different shapes in cross-sections. It should be noted that in this chapter are considered only the analysis and design of bins and bunkers made of structural steel and having either square or rectangular or circular cross-sections in plane.

7.1.1 Methods for Classification Bunkers and Bins

For the proper application of the static pressure equation in the design of a structure, the designer must classify the structure as a *shallow bin or bunker* and a *deep bin or bin*.

For such classification the following two empirical approximations are widely used by designers.

a. *First Method*
By Dishinger, [7.1]

$$H > 1.5\sqrt{A} \tag{7.1}$$

By the Soviet Code, [7.2]

$$H > 1.5a \tag{7.2}$$

where

H = the depth of vertical wall

A = the horizontal cross-section of the inside of the bin

a = the shorter wall of rectangular bin

If the bin in question satisfies either of the above, it is cononsidered a deep bin. If it satisfies neither rule, it is conconsidered to be a shallow bin or bunker.

b. *Second Method*
This method is based on the position of the plan of rupture of the stored material, Figure 7.1.

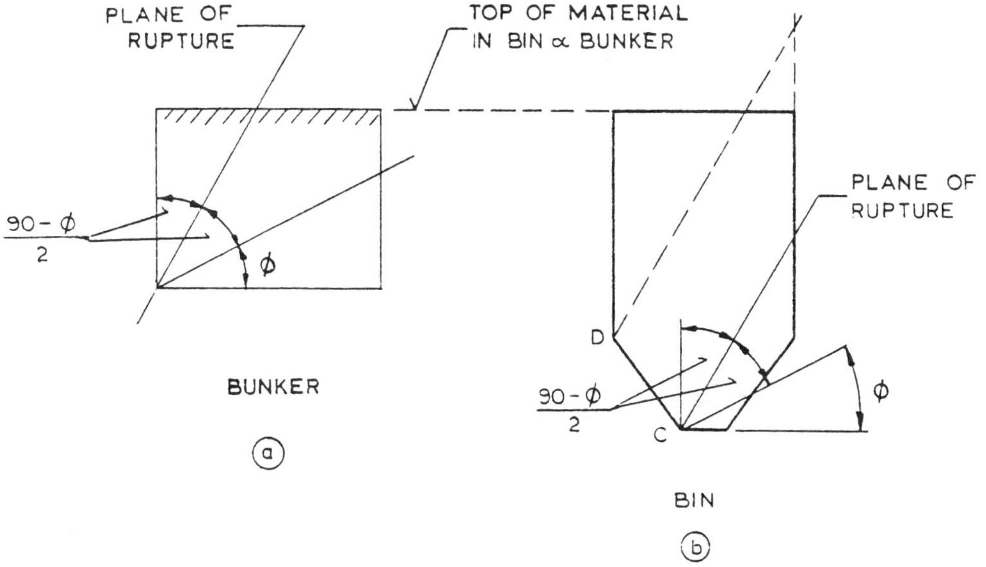

FIGURE 7.1 — Classification of bins using plane of rupture.

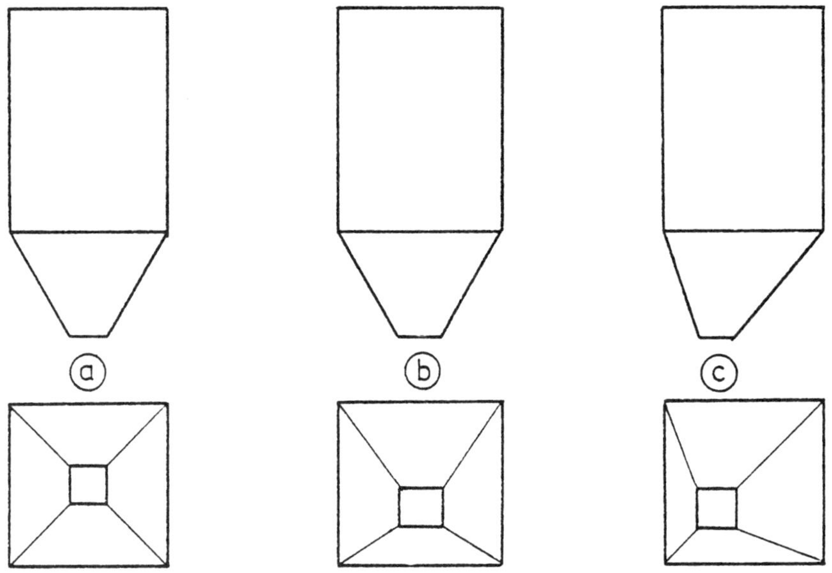

FIGURE 7.2 — Symmetrical and unsymmetrical hoppers.

The plane of rupture is determined by the Coulomb theory. If friction against the wall is neglected, considering the case of a vertical wall and horizontal top surface, then the Coulomb plane of rupture is midway between the angle of repose ϕ and the vertical wall. If the rupture plane intersects the top surface of the stored material, the bin is a bunker, otherwise it is a bin.

However, considering the location of the plane of rupture, there are different approaches. It is considered that the plane of rupture should start either at the bottom of the hopper, point C of Figure 7.1, or at point D. Therefore, by the first interpretation we have the bin, and by the second interpretation it is a bunker.

The classification of the bins on "shallow bins" or bunkers and "deep bins" or bins was introduced to indicate that the pressures of the fills in the bins should be determined by applying different formulas. Thus, in the "shallow bins" or bunkers the vertical force differs only slightly from the weight of the content. However, in "deep bins" or bins this force is substantially reduced because a great part of the weight is carried by the walls of the bin.

7.1.2 Structural Systems of Square and Rectangular Bunkers and Bins

Bunkers and bins may be symmetrical or nonsymmetrical about either, or both, principal axes, according to the location of the discharge opening. Rectangular bunkers and bins usually have pyramidal bottoms with one or more square, circular or elongated discharge openings. Flat bottoms are seldom used since the undesirable dead storage reduces the efficiency of the material flow.

The slopes of the bottom parts of the bunkers and bins must be so inclined that the bunker or bin is self-cleaning, and the line of the least slope must not be excessively flat. This bottom part is called *a hopper* and has pyramidal shape. There are symmetrical, Figure 7.2 (a) and unsymmetrical or eccentrical hoppers, Figures 7.2 (b) and 7.2 (c).

Considering the general arrangements of the bins and bunkers, they may be designed as single or connected in group structures as shown in the following sections of this chapter.

7.1.3 Square and Rectangular Bunkers

Shallow bins or bunkers are designed for storage and loading of granular materials. The volume of the bunkers is defined by the technological requirements. Bunkers are designed either as separate structures or they are constructed as parts of the industrial building. In the first case the bunker consists of the storage part, carrying structure and the gallery above the bunker, Figure 7.3. Between the transverse frames of the bunker, are placed the longitudinal carrying panels which are simultaneously the walls of the bunker. Also in the planes of transverse frames are placed the transverse panels which constitute the walls of the bunker. The longitudinal and transverse panels are supporting all vertical loadings and the horizontal thrust as well as the weight of suspended hoppers.

The spans of the longitudinal panels may be greater than the spans of the transverse panels. In this case the intermediate transverse panels transfer loadings on the longitudinal panels, Figure 7.4.

In the case of a bunker installed in the building, the bunker is supported by the beams of the structure itself, Figure 7.5.

The height of the vertical walls of the bunker is usually not greater than $1\frac{1}{2}$ times that of the maximum dimension of the bunker in plan.

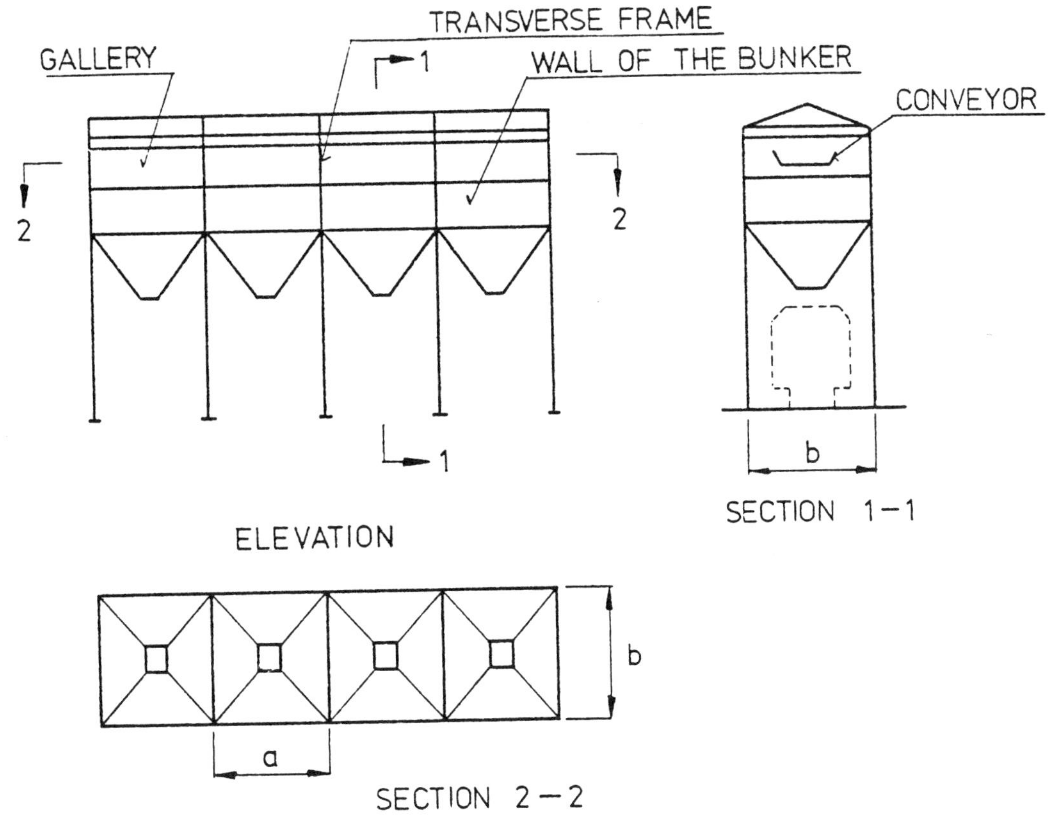

FIGURE 7.3 — Scheme of the bunker.

Depending on the locations of the openings, the bunkers may have one or two axes of symmetry or may be unsymmetrical. It is advisable to design the bunker with a symmetrical disposition of the openings.

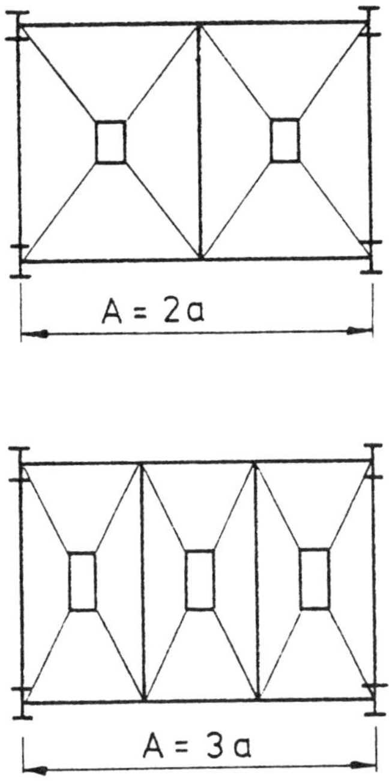

FIGURE 7.4 — Bunkers supported by columns having spacings greater than lengths of the walls.

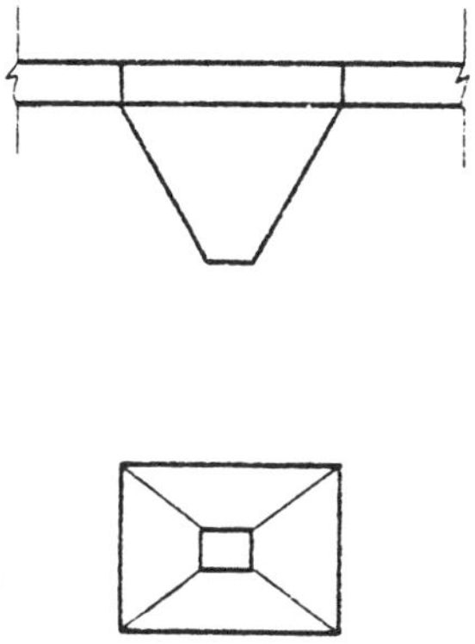

FIGURE 7.5 — Bunker supported by structural frame of the building.

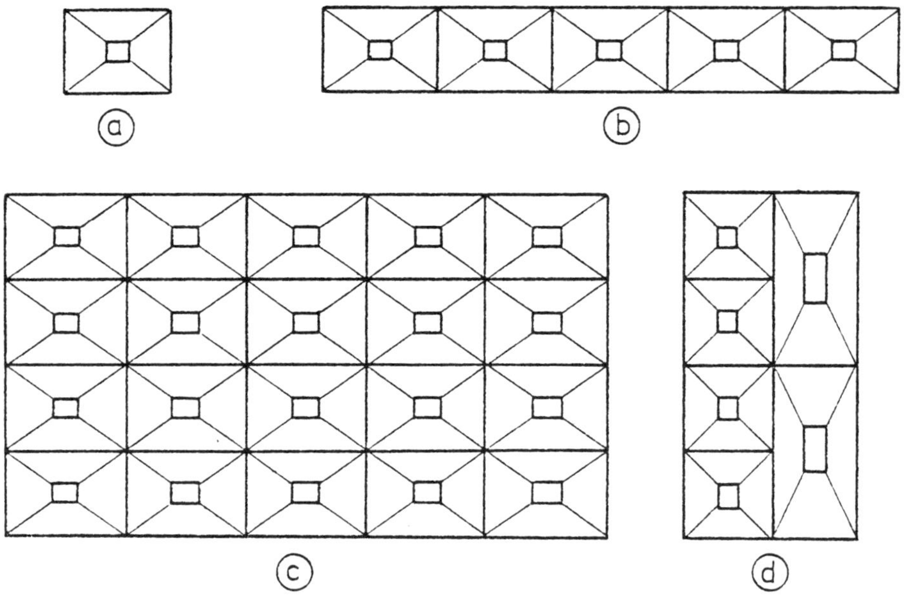

FIGURE 7.6 — Types of bin systems having rectangular cross-sections: (a) Single cell, (b) Single row of cells, (c) Multiple cells, (d) Bin consisting of square and rectangular cells.

7.1.4 Square and Rectangular Bins

Bins of rectangular or square cross-section in plan may be designed either as having a single cell, Figure 7.6 (a) or "single row of cells", Figure 7.6 (b), or "multiple cells" bins, Figure 7.6 (c).

Sometimes it is necessary to design a bin consisting of a combination of square and rectangular cells, Figure 7.6 (d). Dimensions in plan of square or rectangular bins are from 14 ft to 18 ft. In the case of relatively light materials, cells in the plan may be used up to 36 ft.

The walls of the rectangular cells are under tensile forces and bending moments. For a structural analysis of the multicell structure it is necessary to consider the influence of a single loaded cell on the distribution of the bending moments.

Multicell rectangular bins are supported by columns at the corners, their walls act as panels. These panels support the weights of the walls, roof and the load from the hopper.

7.2 Properties of Bulk Solids

The properties of bulk solids such as ore, concentrate, grain, flour, feed, chemicals, rock, sand, powders, refer to the assembly of its particles.

7.2.1 Bulk Density

Bulk density γ is the weight of a bulk material per unit volume. *Loose or aerated bulk density* is determined by weighing a sample in a non-compacted condition.

Packed or compacted bulk density is determined by weighing a sample that has been compacted by vibrating it in a container or by measuring volume change under pressure in a test cell.

Working bulk density is given by

$$\gamma_w = \frac{(\gamma_p - \gamma_a)^2}{\gamma_p} + \gamma_a \tag{7.3}$$

where

γ_a = aerated bulk density

γ_p = packed bulk density

The aerated bulk density should be used to compute bin and hopper capacities.

The packed bulk density gives a conservative estimate of the unconfined compressive strength.

The working bulk density should be used to determine the rate of feed.

7.2.2 Compressibility

Compressibility is a measure of the change in volume of a bulk solid caused by a change in the stress system acting on it.

TABLE 7.1 — Compressibility vs Class of Solid and Type of Flow.

Compressibility Coefficient	Class of Dry Solid	Flow
0.05 to 0.15	Free-flowing granules	Excellent
0.18 to 0.22	Free-flowing; powdered granules	Good
0.18 to 0.22	Flowable, powdered granules	Fair to passable
0.22 to 0.28	Very Fluid powders	Poor-Unstable
0.28 to 0.33	Fluid, cohesive powders	Poor
0.33 to 0.38	Cohesive powders	Very Poor
0.38 to 0.40	Very cohesive powders	Very, Very Poor

The compressibility coefficient of a bulk solid is defined as [7.3]

$$\frac{\gamma_p - \gamma_a}{\gamma_p} = 1 - \frac{\gamma_a}{\gamma_p} \qquad (7.4)$$

This coefficient gives an indication of the flowability of a bulk solid.

The effect of compressibility on the flow of solids is shown in Table 7.1.

7.2.3 Particle Size

Granular materials are generally noncohesive and free-flowing while powders are not. The heavier the material is in bulk density the finer its powder particles will be. Therefore, the powder fraction of a material is defined in terms of its bulk density and a screen mesh size.

According to the classification by Carr [7.3], the powder fraction of material is that portion which passes the mesh size specified in the following Table 7.2, and the granular fraction is the portion retained.

TABLE 7.2

Bulk Density pct	Mesh Size mm
18	100
25 — 55	200
60 — 90	325

7.2.4 Angle of Repose

When an unconsolidated bulk solid falls freely onto a horizontal surface, the particles of the solids roll down the pile. The angle of repose is the slope of the cone formed by dropping material through a funnel onto a horizontal surface.

For free-flowing solids with a narrow range of particle size *the angle of repose* equals the *angle of internal friction* and designated as ϕ and

$$\mu = \tan\phi$$

is the coefficient of friction of filling on filling or the coefficient of internal friction.

7.2.5 External and Internal Angles of Friction

Pressures within a filling of a bin are dependent on the coefficient of friction between the filling and the wall of the bin, given by

$$\mu' = \tan\phi' \qquad (7.5)$$

where ϕ' is the angle of friction of the filling on the bin walls.

When an unconsolidated solid is placed in a bin its weight consolidates it and gives it strength.

Cohesionless solids, for example, gravel and dry sand, gain little strength within the ranges of pressures acting in a bin. Such solids shear when

$$S = W\tan\phi \qquad (7.6)$$

where

S = the minimum force to cause sliding

W = a weight of solid

ϕ = the angle of internal friction

7.2.6 Flow Properties

The relationship between the major and minor principal pressures, p_1 and p_2 respectively, in a bulk solid of semi-infinite mass is given by

$$\frac{p_1}{p_2} \lesseqgtr \frac{1 + \sin\phi}{1 - \sin\phi} + \frac{2c\cos\phi}{p_2(1-\sin\phi)} \qquad (7.7)$$

where

c = cohesion

The inequality represents an elastic state of pressure and the equality a plastic state. Cohesion c varies with the degree of consolidation of the solid.

A particular case of failure called *flow* occurs when pressures are just such that shear occurs within the mass thus destroying the iostropy of the solid.

$$\frac{p_1}{p_2} = \frac{1 + \sin\delta}{1 - \sin\delta} \qquad (7.8)$$

where

δ = the effective angle of internal friction

Equations (7.7) and (7.8) are simultaneously satisfied during the flow, so that c may be expressed in terms of the major pressure p_1 by eliminating p_2 between them, or

$$c = p_1 \frac{\sin\delta - \sin\phi}{(1+\sin\delta)\cos\phi} \qquad (7.9)$$

For cohesionless solids $c = 0$ and from Equations (7.7) and (7.8) it is evident that $\delta = \phi$.

7.2.7 Flow

Some materials will not flow freely through the gate of a bin because of the formation of an *arch* in the solid or a *pipe* through it.

Non-free flow is caused by internal compressive and shearing forces in the solid resulting from cohesion, humidity, time of storage, or electrostatic forces which prevail over gravity forces.

The strength of a given material under given conditions of moisture, temperature, and time of storage at rest is a function of the consolidating pressure which has acted on it. Its strength increases as the consolidating pressure increases.

A solid for which the unconfined yield strength f_c = 0 (dry, round sand, for example) is perfectly free flowing.

The flowability of a solid containing a range of sieve sizes including both fine and coarse particles is invariably governed by the flow properties of the fine particles, because the solid shears across the fine particles.

TABLE 7.3 — Angle of Repose and Flowability.

Angle of Repose deg.	Flowability
25 — 30	Very free flowing granules
30 — 38	Free flowing granules
38 — 45	Powdered granules, fair to passable flow
45 — 55	Cohesive powders, may require special agitation
55 — 70	Very cohesive powders require special agitation

The angle of repose [7.4] is a direct indication of the potential flowability of a material. An evaluation of the results of tests on dry materials is shown in Table 7.3.

7.2.8 Conditions Affecting Flow Properties

Materials which flow freely from a bin under the conditions of continuous flow may gain strength and obstruct the flow after storage at rest. The results of storage time at rest may be due to any one or a combination of the following:

a. During charging, powders entrain the air which escapes with time and results in a more density and cohesive forces.

b. Physical instability due to changes in surface moisture and temperature.

7.2.8.1 Moisture Content

Metallic ores are affected only by surface water. Cohesive forces due to capillary action reach a maximum for a moisture content of between 70 — 90 percent of saturation. In porous solids cohesive forces due to capillary action may be absent.

Water soluble substances and organic materials absorb matter on the surface and produce a saturated solution.

7.2.8.2 Temperature

The effect of temperature changes depend on the type of stored solids.

Properties of *soluble* solids containing moisture under constant temperature do not change. Change of temperatures lead to caking and cyclic changes in temperature lead to severe caking.

Solids are subject to change in hardness with a change in temperature.

Frozen solids at a temperature below 32° F retain their initial properties. The above effects are minimized by keeping the temperature of the stored solid constant.

If all the solids must be kept in motion, the bin must be of the mass-flow type and must have a large hopper in relation to the total volume in order to provide relative motion of the particles.

7.2.8.3 Gradation

An indication of the particle-size content of a solid is determined by passing a sample of the solids through a set of screens of certain standard sizes and recording the percent of each size range passing a given screen size and retained on the next smaller screen.

7.2.8.4 Segregation

Solids with a wide range of particle sizes, and blends containing particles of a wide range of density, size or shape, tend to segregate as they are charged into a bin.

7.2.8.5 Degradation

A degradable solid is one whose lumps of particles may be broken and reduced in size as a result of impact, agitation, or attrition. Degradation usually occurs during the changing although it may occur during discharge.

7.2.8.6 Corrosiveness

Corrosive solids are those which chemically attack the confining surfaces with which they come into contact.

7.2.8.7 Abrasion

Movement of a solid during charging or discharging abrades the walls and bottom of the bins. Funnel-flow bins are advantageous for the storage of hard, abrasive, lumpy solids because there is little wear of the hopper walls.

7.2.9 Flow in Bins

Flow in a bin is called "mass flow" if the flow channel coincides with the bin, or if all of the solids are in motion, and funnel flow if it forms within nonflowing material. A funnel-flow channel is usually shaped like an inverted cone and may intersect the surface of the solid or the wall of the bin, Figure 7.7.

An intersection with the wall is called an effective transition, by analogy with a bin-to-hopper transition. Intersection with the surface usually occurs in shallow bins.

Flow pressures in a mass-flow bin are usually well defined because the flow channel is defined and constant. On the other hand, a funnel-flow channel may expand or contract between fillings and even during flow. Therefore, pressures are more erratic in funnel-flow bins. Another important difference between mass-flow bins and funnel-flow bins has to do with pressure peaks which occur at the boundary between an active-pressure zone and a passive-pressure zone during the flow.

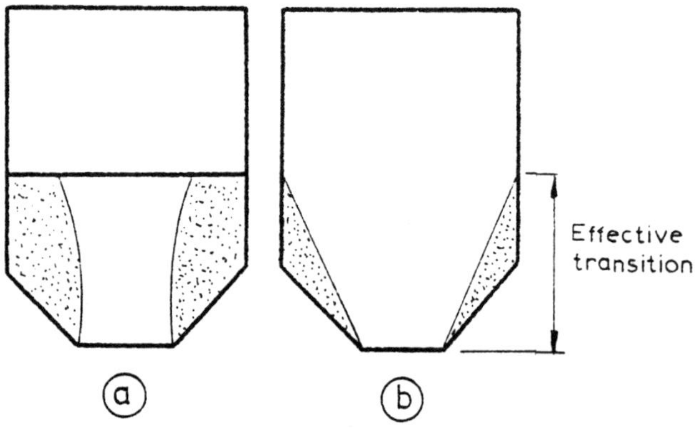

FIGURE 7.7 — Flow in bins.

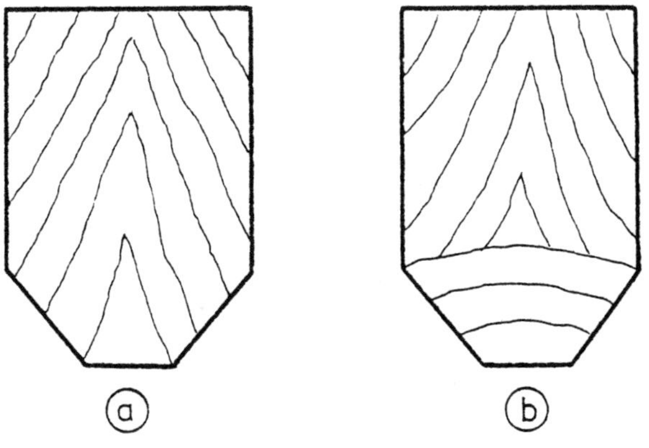

FIGURE 7.8 — Pressure field in bins.

7.2.10 Pressure Field in Bins

When a solid is charged into an empty bin with the gates closed or the feeder is at rest the solid compacts as the initial vertical pressures increse. Therefore, the initial pressure field tends to be active and the major-pressure contours are as shown in Fig. 2.2.

When solid starts flowing out of a bin outlet it expands vertically within the developing flow channel. The channel usually diverges upward, so that the solid also contracts laterally. Therefore, the pressure field tends to be passive, with the major pressure lines arching across the flow channel, Figure 7.8 (b).

7.2.11 Properties of Stored Materials

The properties of the material to be stored affect the intensity of pressure loadings. In addition, they influence the material flow and must be considered in selecting the outlet shape and size and type of unloading system [7.5, 7.6, 7.7].

Table 7.4 shows the properties of commonly stored materials. These values should be used only in the absence of test data for the actual material to be stored. Caution and good judgment must be used in selecting the properties to be used, as pressures are quite sensitive to variations of those properties. The designer must be alert to the possibility of large variation from the tabulated values — for example, the possibility of a material ranging from dry, to damp, to saturated. If the expected condition is not certain, the designer should assume the worst condition for that material.

The designer should be alert to possible varying usage — the storage of different materials in the same bin or bunker.

7.3 Functional Design of Square and Rectangular Bins

7.3.1 Types of Bins

In selecting a type and size of bin the following three types are used in industry according to their pattern of flow:

a. Mass-flow
b. Funnel-flow, and
c. Expanded-flow

The type of flow pattern which develops when solids are withdrawn from a bin significantly affects the reliability and uniformity of flow and, therefore, the bin loads.

7.3.2 Mass-Flow Bins

Mass-flow bins of square and rectangular cross-section usually consist of a vertical cylinder and a mas-flow hopper, Figure 7.9.

When the gate of a mass-flow bin is opened or its feeder started the total volume of solid is in motion.

Mass-flow bins have the following characteristics:

a. The total volume of stored solid is available for process by gravity.
b. Channelling, hang-ups, surging, and flooding do not occur, provided in the case of powder no attempt is

TABLE 7.4 — Typical Design Properties of Granular Materials.

Material	Specific Weight γ lb/ft³	Angle of Repose ϕ	Coefficient of friction μ against steel
Cement, Portland	100	25°	0.30
Cement, Clinker	88	33°	0.3
Peas	50	25°	0.263
Wheat	50	25°	0.414
Beans	46	31.5°	0.366
Barley	39	37°	0.376
Corn	44	27.5°	0.374
Oats	28	28°	0.412
Sugar Granular	62.5	35°	—
Sand Dry	100	35°	0.50
Sand Moist	112.5	40°	0.40
Sand Saturated	125	25°	0.35
Flour	37.5	40°	0.30
Lime, burned (Pebbles)	56.2	35°	0.30
Lime Powder	44	35°	0.30
Coal	50	35°	0.30
Coal, Anthracite	62.5	35°	0.30
Coke	37.5	40°	0.50
Gravel Dry	113	35°	—
Gravel Wet	125	25°	0.75
Manganese Ore	125	40°	—
Iron Ore	165	40°	0.364
Clay Dry	106	40°	0.7
Clay Damp	113	25°	0.4
Clay Wet	138	15°	0.3
Lime Burned, Fine	57	35°	0.3
Lime Burned, Coarse	75	35°	0.3
Gypsum in Lumps Limestone	100	40°	0.3

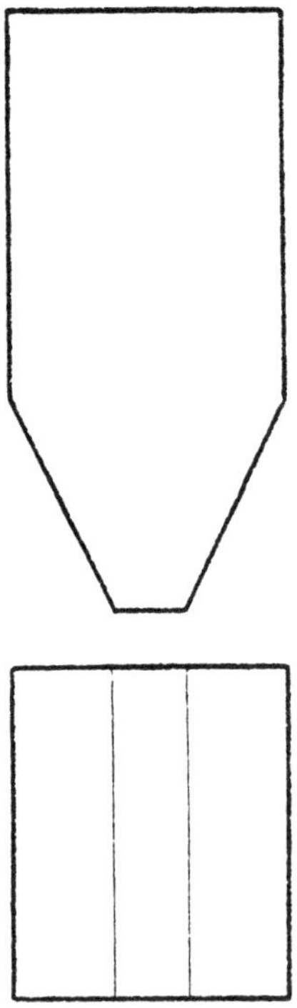

FIGURE 7.9 — Mass-flow bin.

made to withdraw it faster than its inherent rate of discharge.

c. The bulk density of the drawn solid is constant and practically independent of the head of the stored solid. This is advantageous in all cases of controlled flow rate and essential when the rate is controlled volumetrically.

d. A first-in first-out flow pattern, which is desirable for solids that deteriorate during storage and essential in chemical reactors, is readily obtained. This flow pattern is also desirable when segregation must be taken into account.

e. Since no dead regions occur within the container there is a minimum of degradation, spoilage, spontaneous combustion of coal, oxidation of ore, and consolidation at rest.

f. A mixture may be blended by circulating it through and around a suitable mass-flow bin.

g. Pressures throughout the solid and the hopper walls are relatively low, which results in low consolidation and attrition of the solid and moderate or insignificant wear of the walls. Also, the pressures are relatively uniform over a horizontal cross-section of the hopper, which gives uniform consolidation and permeability.

h. Flow is relatively uniform, so that steady-state flow can be closely approached and an analysis based on steady-state flow can be applied with confidence.

i. Low-level controllers and indicators work reliably.

j. Powders degasify and settle out if sufficient storage time is attained.

k. Airlocks (rotary vane feeders) can often be eliminated as long as a minimum level of solid is retained to provide a seal.

7.3.3 Funnel-Flow Bins

Funnel-flow bins generally may consist of a vertical cylinder of rectangular cross-section and a flat-bottom floor or non-mass flow hopper, Figure 7.10.

The solid in a funnel-flow bin flows toward the outlet in a channel formed within the stable solid. The flow channel is typically circular in cross-section and usually assumes a conical shape widening upward from the outlet. The included angle of the cone for a given material depends on its moisture content, temperature, time of storage, and the sequence and rate of charge and draw. In a tall bin the cone may intersect the bin wall. Funnel-flow bins are useful for storing hard, abrasive, lumpy solids because there is little wear of the hopper walls. Funnel flow bins have the following characteristics:

a. Only a small portion of the stored solid is in motion during the draw.

b. A first-in first-out flow sequence prevails. The solid surrounding the flow channel at the bottom of the bin remains at rest until the flow channel is completely empty. This leads to consolidation, caking, deterioration, spontaneous combustion of coal, and oxidation of ores.

c. Any solid which remains at rest under pressure for a long period of time may gain strength and obstruct flow. This may lead to piping (ratholing).

d. Severe segregation usually occurs because no individual particle movement takes place in the hopper.

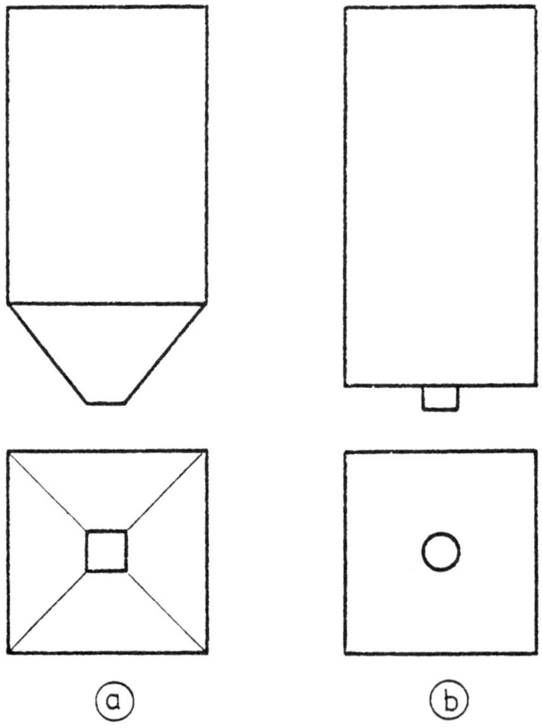

FIGURE 7.10 — Funnel-flow bins: a. Pyramidal; b. Flat-bottom.

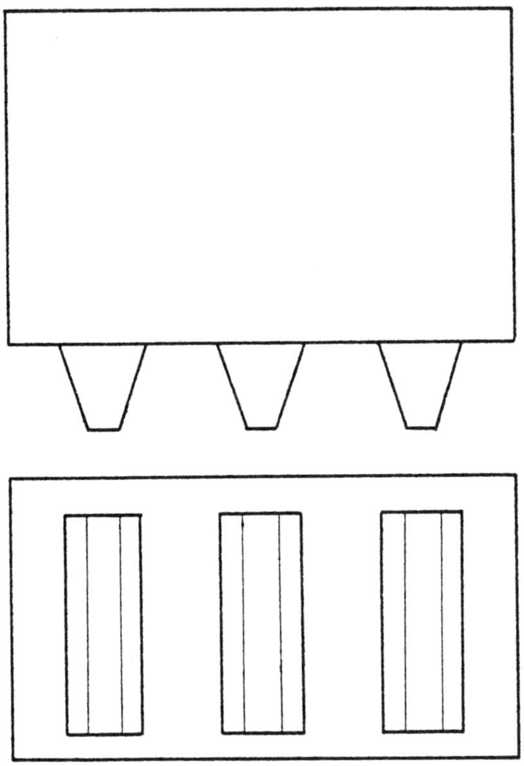

FIGURE 7.11 — Expanded-flow flat-bottom bin.

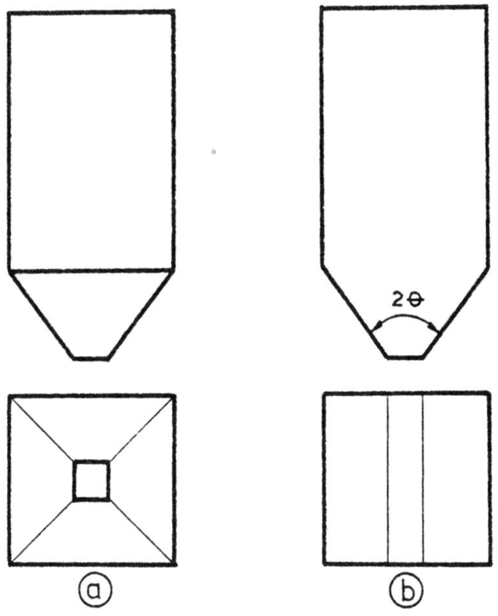

FIGURE 7.12 — Symmetrical Hoppers: a. Pyramidal; b. With two vertical walls.

e. Stress analysis is somewhat unreliable because of the unsteady flow.

7.3.4 Expanded-Flow Bins

Several mass-flow hopper-feeder units can be placed under one large funnel-flow bin, Figure 7.11.

The hopper forces the flow channel to expand to a size large enough to eliminate the possibility of ratholing, to reduce the segregation to an acceptable level, and to ensure de-aeration and smooth flow. This type of bin is useful for the storage of large quantities of non-degrading solids such as ores. A low-level indicator can be placed on the mass-flow hopper.

7.3.5 Hoppers

The success of bin operations depends largely on the design of the hopper. The ease with which the material flows and converges towards the opening depends almost entirely on the shape of the hopper and the smoothness of its walls.

A symmetrical pyramidal hopper may have four walls inclined, or two walls inclined and two vertical, Figure 7.12.

There are also eccentric pyramidal hoppers having single vertical and three inclined walls or two vertical and two inclined walls, Figure 7.13.

Hoppers with one or more vertical walls are preferable. The slotted outlets, Figure 7.12 (b) prevent funneling and assure mass flow, but a feeder over the full length of the slot is necessary to remove the solid.

Square and rectangular bins in which cohesive and non-flowing materials are stored may require an eccentric hopper, Figure 7.13. Stable arches are less likely to form in these hoppers because the solids tend to slide down the vertical wall.

Rectangular gravity-flow bins with flat bottom and multiple circular openings are commonly used to store granular solids. Such bins are low in cost and occupy less vertical space than bins with a single hopper outlet.

To maximize the live storage the openings should be so spaced as to permit the flow patterns above the openings to intersect.

7.3.6 Outlets

The outlet of a hopper must be large enough to assure an unobstructed flow at the required rate. Unless the outlet for granular materials is larger than several particle sizes, the flow may be obstructed by the interlocking of large particles. Furthermore, certain minimum dimensions are necessary to prevent cohesive doming, Figure 7.14 (a), and piping, Figure 7.14 (b).

It is important to distinguish between the *actual area* of an outlet of a hopper, bin, or storage pile and the *effective area,* which is the area through which the solid flows. In the particular case of rectangular outlets, the effective area may be only a small part of the actual area.

7.3.7 Archings

The solid in-a-flow channel may form an arch if the channel is wedge-shaped and a dome if it is conical, Figure 7.15.

The stress S_1 at the abutment must be less than the unconfined compressive strength f_c of the solid if the arch is stable.

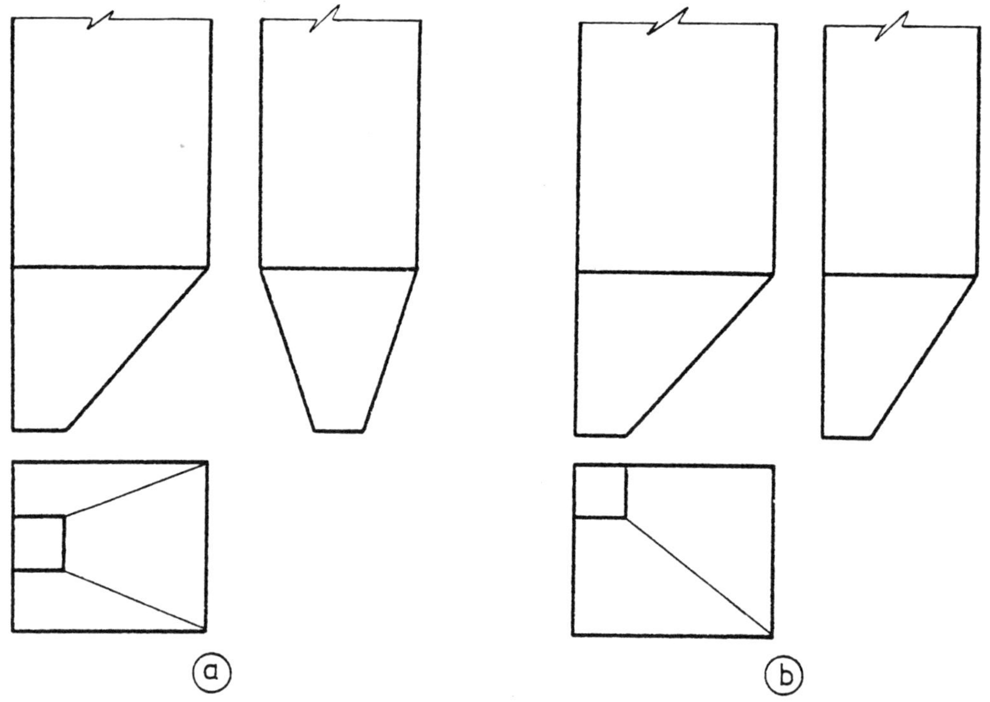

FIGURE 7.13 — Eccentric hopper: a. Single vertical wall; b. Two vertical walls.

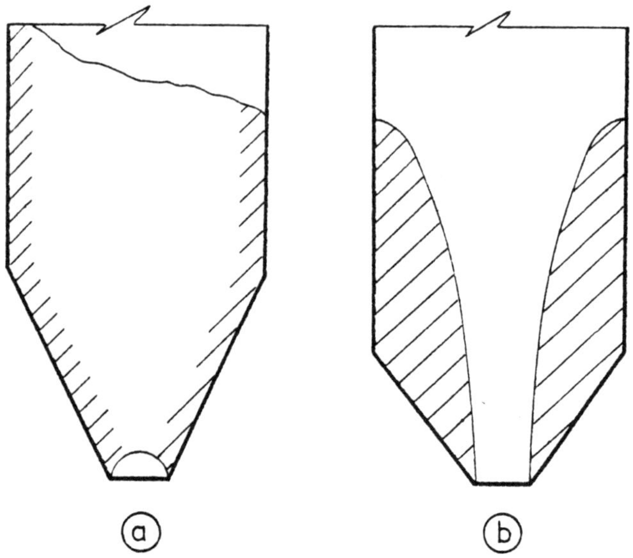

FIGURE 7.14 — A. Doming; b. Piping.

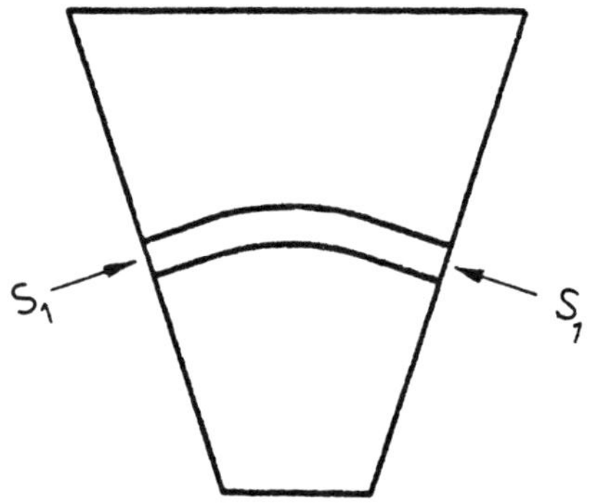

FIGURE 7.15 — Arching effect.

The maximum span of arch which can be self-supporting over this stress system is attained when the shears on the vertical sections at the abutments reach their maximum values. This condition is realized when the major principal stress at the abutments equals the unconfined yield strength f_c of the solid and acts at 45° with the horizontal, Figure 7.16.

The equilibrium of vertical forces for an arch of unit vertical thickness forming over a rectangular opening of

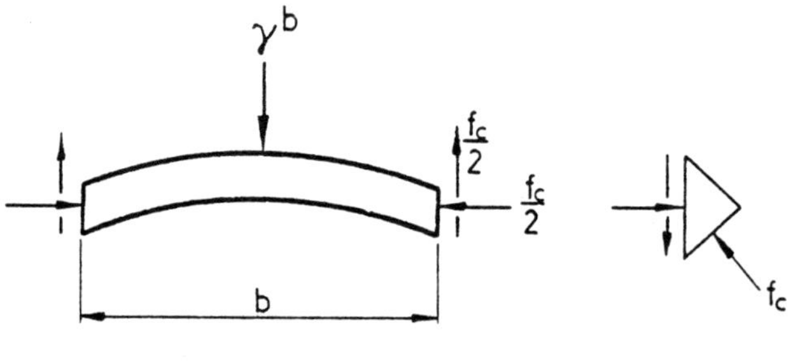

FIGURE 7.16 — Arching effect. Stress components.

width b and length ℓ gives

$$2 \times \tfrac{1}{2} f_c \times 1 \times \ell = \gamma b \ell$$

from which

$$b = \frac{f_c}{\gamma} \qquad (7.10)$$

Similarly, for a dome forming over a circular opening of diameter d

$$\tfrac{1}{2} f_c \times \pi d = 1 \times \gamma \times \pi d^2/4$$

and

$$d = \frac{2f_c}{\gamma} \qquad (7.11)$$

Jenike and Johanson [7.8] suggest the following outlet dimensions:

For circular outlets

$$d \geq \frac{2.2 f_c}{\gamma}$$

and for rectangular outlets

$$b \geq 1.3 \frac{f_c}{\gamma} \qquad (7.12)$$

$$A \geq 3.9 \left(\frac{f_c}{\gamma}\right)^2 \qquad (7.13)$$

where

A = area of the outlet

The corners should be rounded to a radius not less than 0.2b.

7.3.8 Outlets for Mass-Flow Bins

The value of the unconfined compressive strength of the solid f_c to be used to determine the dimensions of the opening for mass-flow bins can be determined first by finding the ratio P_1/S_1 which is called the flow factor of the hopper and is denoted by "ff".

The flow factor is a function of the effective angle of internal friction δ, the angle of friction of the material on the hopper wall φ', and the angle θ of the hopper wall with the vertical.

Values of "ff" are given in Figures 7.27, 7.28 and 7.29.

With "ff" known, the critical value of f_c is determined by drawing the straight line P_1/S_1 = ff on a material properties plot. The intersection of this line the $f_c = f(p_1)$ curve gives the critical value, Figure 7.17.

The evaluation of f_c determined by this procedure is based on the consolidating pressure during steady flow, which is usually the critical value. However, the consolidating pressure caused by the fall of solids into a tall bin may exceed the pressure during flow [7.9].

This consolidating pressure can be estimated by

$$p_1 = 0.705\, v\gamma \sqrt{\frac{h}{g}} \qquad (7.14)$$

where

v = belt velocity, ft/sec

γ = bulk density, lb/ft³

h = height of fall, ft

In determining the bin opening by Equations (7.12) and (7.13) the value of f_c corresponding to pressure p_1 from Equation (7.13) should be used if it exceeds the value of f_c for steady flow.

7.3.9 Hoppers for Mass-Flow Bins

For mass flow to occur the hopper outlet must be large enough to prevent doming of the solid. The walls must be sufficiently steep and smooth to assure sliding of the solid along the wall.

The largest included angle, 2θ, Figure 7.18, of a hopper for which a mass flow will occur depends on the angle of friction φ' between the solid and the wall and on the shape of the horizontal cross-section of the hopper. Since the required values of θ decreases as φ' increases the slope of the hopper walls is usually determined by the value of φ' at the outlet.

For *wedge and chisel hoppers* mass flow will occur if

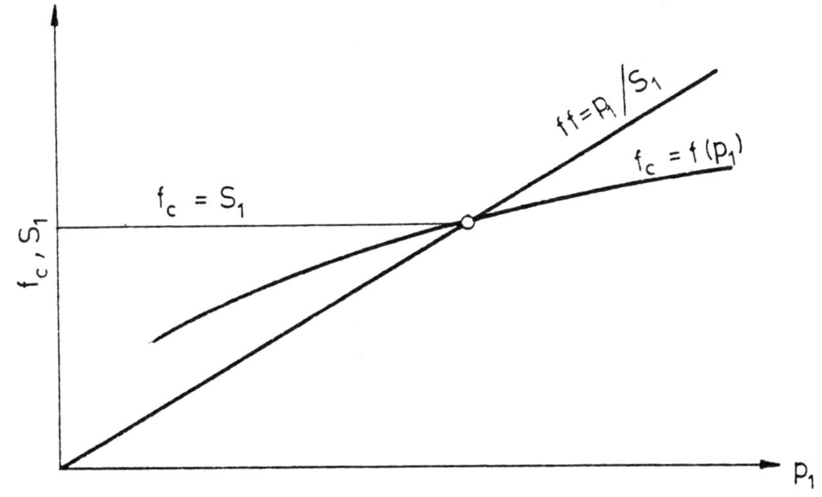

FIGURE 7.17 — Determination of the critical value of f_c.

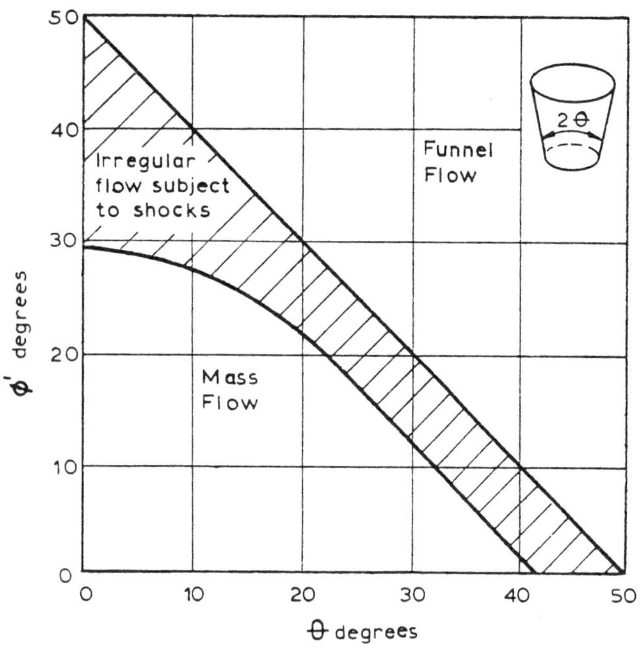

FIGURE 7.18 — Conical hoppers.

$$\theta \leqslant 60° - 1.33\,\phi' \qquad (7.15)$$

$$\phi' \leqslant 0.9\,\delta \qquad (7.16)$$

$$\ell \leqslant 6b \qquad (7.17)$$

And for *transition mass-flow hoppers* with rectangular outlets the end slopes are designed as for conical hoppers, Figure 7.18 and the side slopes as for wedge hoppers. The length-to-width ratio of the outlet should not be less than 3.

It is essential that the completed hopper has a surface which is at least as smooth as the sample of the material used in the determination of the angle of friction ϕ'.

7.3.10 Outlets for Funnel-Flow Bins

The diameter at the bottom of the cone of flowing material in a funnel-flow bin tends to be the same size as the diameter of the outlet. Thus, in a square or rectangular outlet the diameter tends to be that of the diagonal of the square or rectangle, Figure 7.19.

If a pipe forms it is likely to have that same diameter. To make such a pipe unstable so that a stoppage will not develop, the required dimension d, Figure 7.19 is

$$d = \frac{f_c}{\gamma}\,G(\phi) \qquad (7.18)$$

Values of the piping factor $G(\phi)$ are given in Figure 7.20.

The value of f_c is determined as for mass-flow bins, Figure 7.17 but with the flow factor "ff" given by Figure 7.21.

If the outlet is circular or square doming will not occur if the dimension d satisfies Equation (7.18), but in a rectangular outlet the short side b must also be checked for doming by Equation (7.10) or the more conservative Equations (7.12) and (7.13) using the flow factors for flat-bottom bins given in Figure 7.22.

Johanson [7.10] has investigated piping under static filling pressures. The rathole diameter d is given by Equation [7.18] with f_c based on the following flow factors:

For a flat bottom prismatic bin

$$ff = \frac{P_v}{d}\,G(\phi) \qquad (7.19)$$

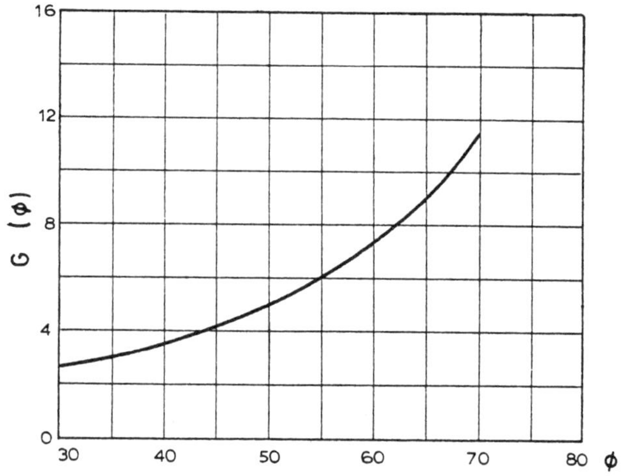

FIGURE 7.20 — Piping factor versus angle of internal friction.

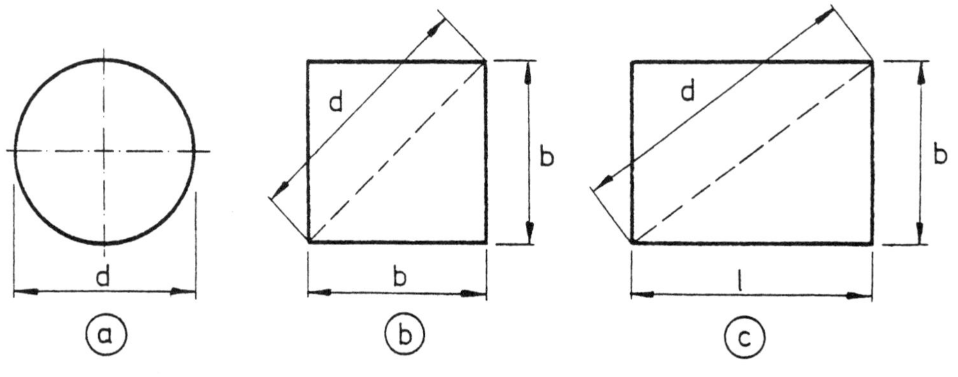

FIGURE 7.19 — Diameter d of outlet in funnel-flow and flat-bottom bins.

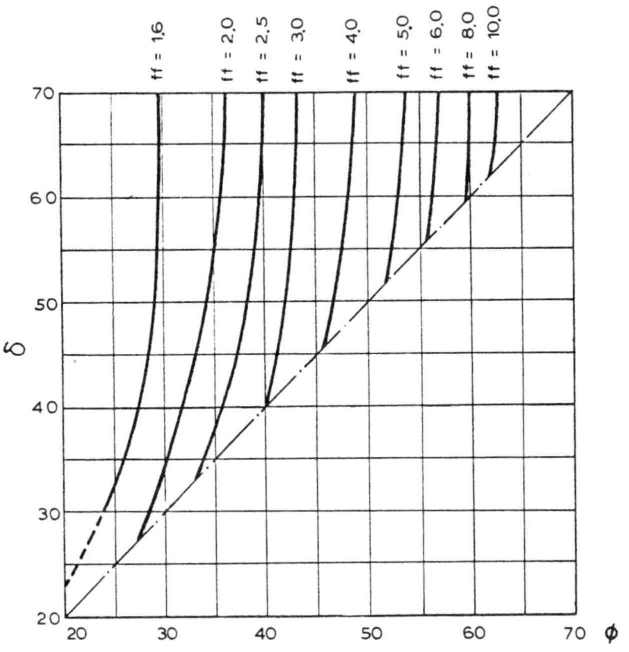

FIGURE 7.21 — Critical flow factor for piping.

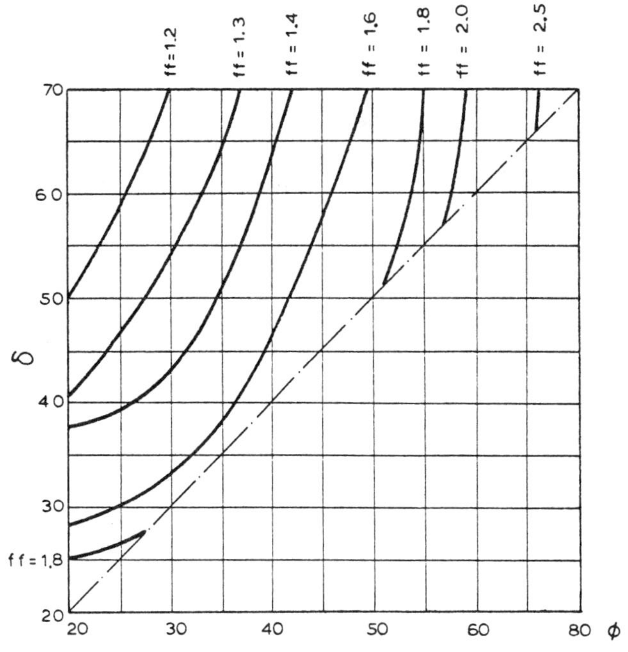

FIGURE 7.22 — Critical flow factor for flat-bottom bin.

where P_V is the vertical filling pressure and $G(\phi)$ is given by Figure 7.20 Johanson suggests that P_V be evaluated by the Janssen formula, Equation (7.32), in deep bins but taken equal to γH,

where

H = height of bin, for shallow bins

or

$$0 < H/B < 1$$

where

B = width of rectangular bin

For a hopper without surcharge

$$ff = FG(\phi) \qquad (7.20)$$

where

F is given by Figures 7.23 and 7.24.

The parameter k upon which F depends is defined in Reference [7.10]. The value 0.8 is typical and is suggested for general use.

For a bin with a hopper the smaller of the values in Equation (7.20) are

$$ff = \left(\frac{P_{vt}}{\gamma d} + \frac{D-d}{2d \tan\theta}\right) G(\phi) \qquad (7.21)$$

where

P_{vt} = vertical pressure in bin at transition

D = width B of rectangular bin

These flow factors are pressures from filling with no outlaw. If material is drawn during the filling, the flow factors for steady flow apply. Flow factors for filling with no outlaw may be several times those for steady flow.

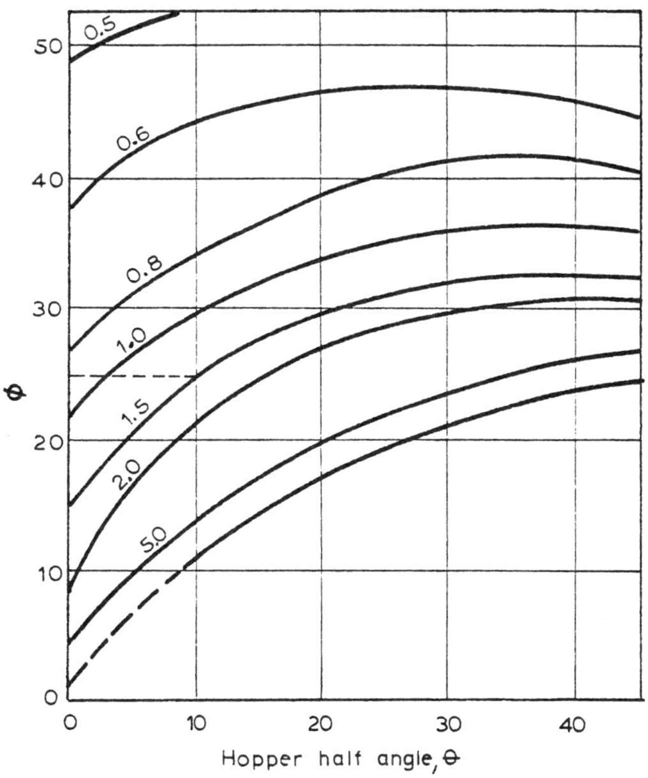

FIGURE 7.23 — Consolidating pressure factor F for conical hoppers, k = 0.8.

7.3.11 Live Storage

An important consideration in the design of funnel-flow bins is the live storage that can be realized. A funnel-flow bin with a steep smooth pyramidal or conical hopper will

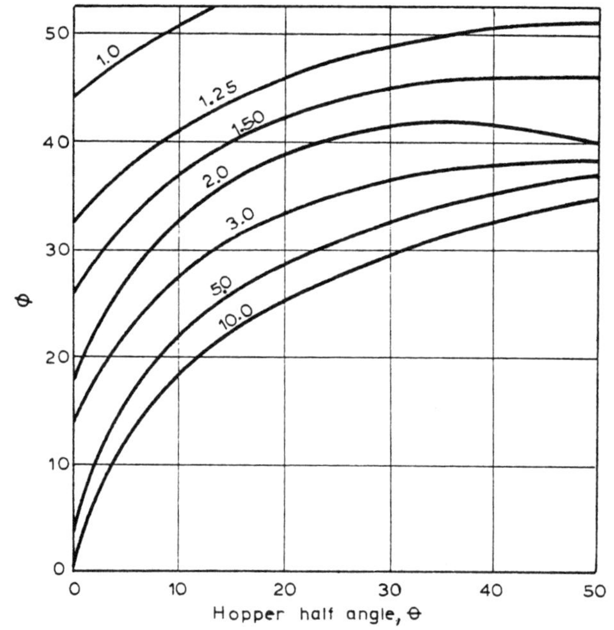

FIGURE 7.24 — Consolidating pressure factor F for wedge-shape hoppers, m = 0, k = 0.8.

clear by gravity. Stainless steel, ceramic tile, and resin coatings can be used to eliminate corrosion during storage so that the hopper will retain its smooth surface. Unless the solids are very free-flowing the slope $0 \leqslant 65° - \phi$ is suggested by Jenike [7.11].

Giunta [7.12] developed formula for predicting the flow boundary in flat-bottomed bins with a round opening where the boundary dimensions, defined in Figure 7.25, are related by

$$d_f = d + \frac{2H - Md}{M + \cot\theta'} \qquad (7.22)$$

where M is a coefficient and θ' is the boundary slope from the outlet, both given in Figure 7.22.

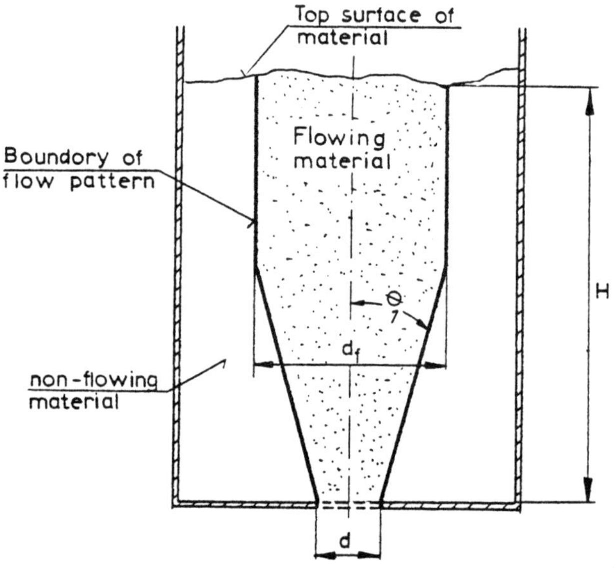

FIGURE 7.25 — Flat-bottom bin with round opening.

This equation is valid only if the opening is large enough to prevent arching and ratholing and if $H > \frac{Md}{2}$.

If $H \leqslant \frac{Md}{2}$ the diameter d_f of the flow boundary equals d.

7.3.12 Rate of Discharge

For calculating the rate of discharge of frictional, cohesive, granular solids it is assumed that the material at the hopper opening forms a continuously failing arch that is in dynamic equilibrium.

The maximum discharge rate Q_m for a wedge hopper is computed by

$$Q_m = \gamma A \sqrt{\frac{bg}{2\tan\theta'}}\sqrt{\left(1 - \frac{f_c}{p_1\, ff}\right)} \qquad (7.23)$$

and for a conical hopper by

$$Q_m = \gamma A \sqrt{\frac{dg}{4\tan\theta'}}\sqrt{\left(1 - \frac{f_c}{p_1\, ff}\right)} \qquad (7.24)$$

The consolidating pressure for a wedge hopper with width b is given by

$$p_1 = \gamma b ff \qquad (7.25)$$

and for a conical hopper with opening diameter d by

$$p_1 = \frac{\gamma d}{2} ff \qquad (7.26)$$

where

ff = flow factor from Figures 7.27, 7.28, 7.29 and 7.22.

The value of P_1, fc is determined from a materials properties curve, Figure 7.30.

This value of Q_m is based on the steady-state velocity of the discharging material. Therefore, the rate of discharge will be overestimated if the gate is open for a short period of time. the average flow rate Q_a for the time the gate is open is given by

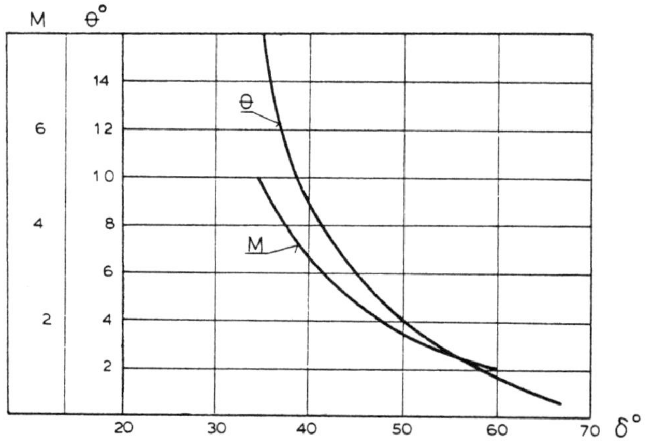

FIGURE 7.26 — Values of M and θ in Equation (7.22).

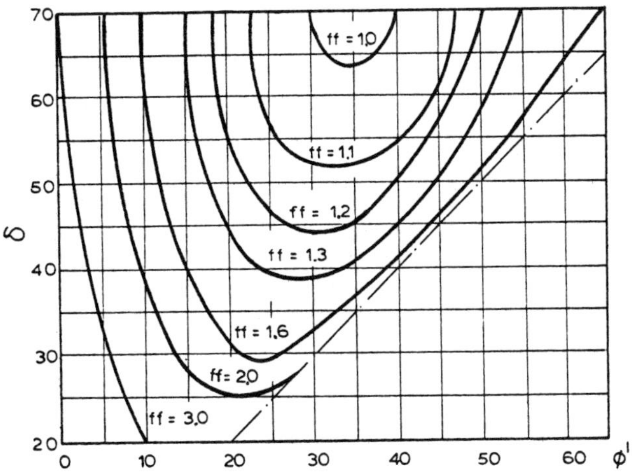

FIGURE 7.27 — Critical flow factor for mass-flow hoppers, θ' = 10°.

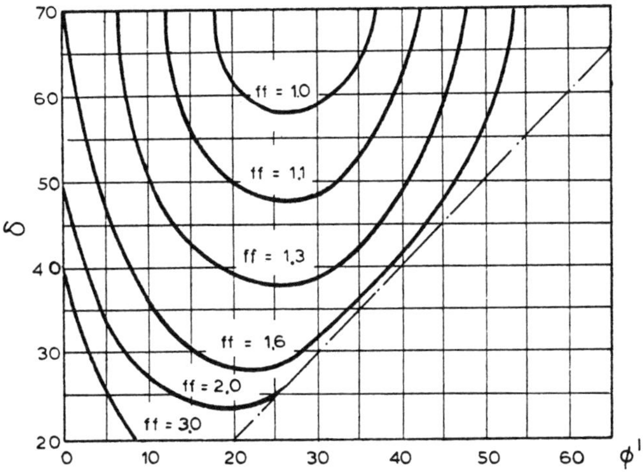

FIGURE 7.28 — Critical flow factor for mass-flow hoppers, θ = 20°.

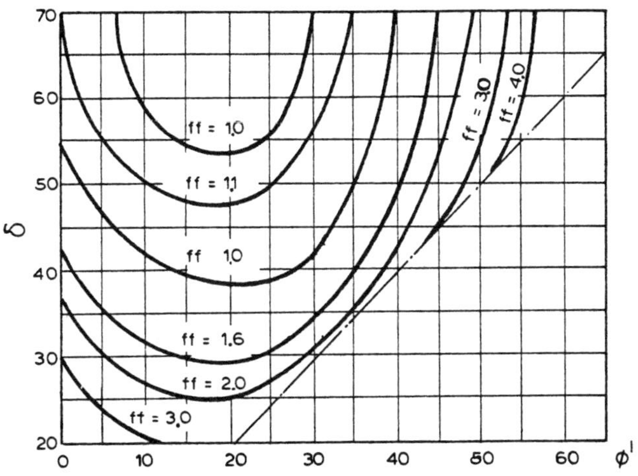

FIGURE 7.29 — Critical flow factor for mass-flow hoppers, θ' = 30°.

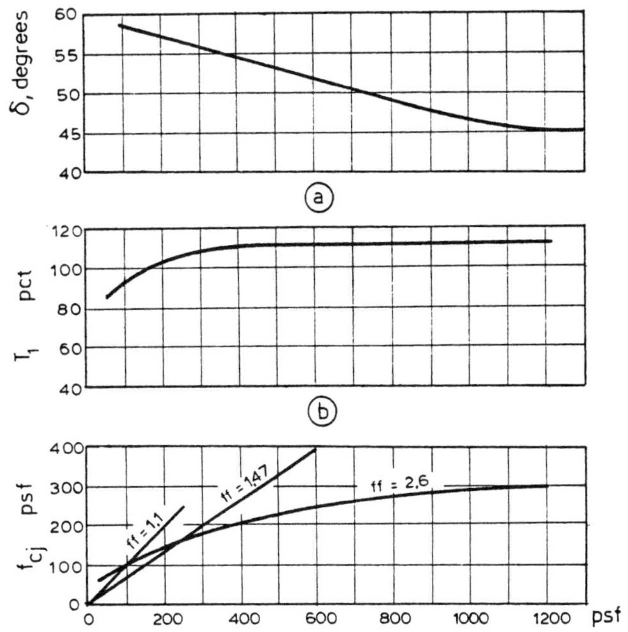

FIGURE 7.30 — Flow properties of dolomite.

$$Q_a = Q_m \left[1 - 2.38 \frac{S}{T}\right] \tag{7.27}$$

where

$$S = \frac{Q_m}{2gA} \cdot \frac{1}{1 - ff(f_c/p_1)} \tag{7.28}$$

and

T = time gate is opened

When T > 4S the error in Equation (7.21) is less than 1%.

7.4 Loads

7.4.1 Design Considerations

In the design of bins and bunkers the following loads should be considered:

1. Dead load of the structure itself and items supported by the structure.
2. Live loads, as follows:
   a. Forces from stored material
   b. Changes in the above due to filling and emptying
   c. Wind
   d. Snow, and
   e. Seismic forces on structure and stored material
3. Thermal stress due to stored material (especially important in long bin groups).

7.4.2 Bin Loading from Stored Material

Material stored in a bin applies lateral forces to the side walls, vertical forces due to friction, to the side walls, vertical forces to horizontal bottoms and both normal and extensional forces to inclined surfaces. The static values of these forces, resulting from materials at rest, are all modified during withdrawal of the material. In general, all forces will increase, so that loads during withdrawal tend to control the design.

Forces applied by stored materials may also be affected by moisture changes, by compaction, and by settling which may accompany alternate expansion and contraction of the walls during daily or seasonal temperature changes.

A rigorous approach in the calculation of bin loads would involve the conditions of material flow during emptying. Until such approach is perfected, equations derived for static forces may be combined with experimental data to approximate the pressure increases occurring during material withdrawal. The procedure involves determining the static pressure or forces and then multiplying these by an "over-pressure" factor, C_d to obtain design pressures or forces.

7.4.3 Pressure on Bin Walls. Janssen's Formula

When a granular mass is deposited in a bin the theory already outlined for bunkers is not directly applicable and the Rankine theory will no longer be satisfactory when calculating vertical and horizontal pressures.

The widely accepted solution of the problem of calculation of the pressure in bins has been proposed by Janssen [7.13].

Figure 7.31 shows the forces acting on a horizontal slice of solid in a bin.

Summation of vertical forces gives

$$dP_v A - \gamma A dy + qL dy = 0 \tag{7.29}$$

where

P_v = vertical pressure in bin

q = frictional force on wall

γ = bulk density of solid

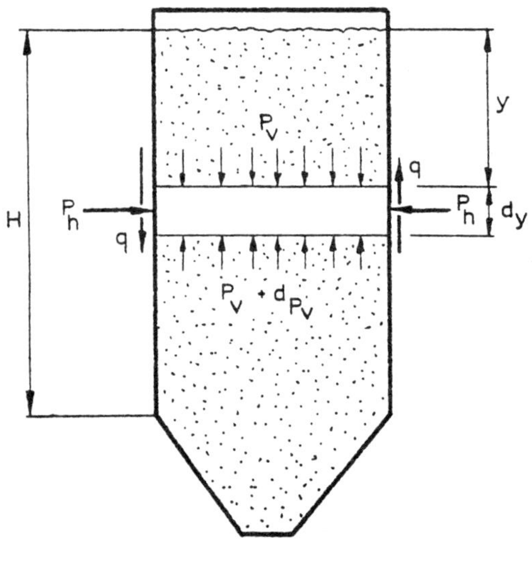

FIGURE 7.31 — Forces acting in a bin.

A = horizontal cross-sectional area of bin

L = perimeter of horizontal cross-section of bin

y = depth of solid above section

The relationship between the friction and the pressure ph on the wall is

$$q = \mu' p_h \tag{7.30}$$

where $\mu' = \tan\phi'$ is the coefficient of friction of the material on the wall. Furthermore, assuming the ratio of the horizontal pressure to the vertical pressure to be constant, we have

$$p_L = K p_v \tag{7.31}$$

where K is the pressure coefficient. Substituting from Equations (7.30) and (7.31) into Equation (7.29) and integrating, we get

$$p_v = \frac{\gamma R}{\mu' K}[1 - e^{-\mu' K y/R}] \tag{7.32}$$

$$p_h = \frac{\gamma R}{\mu'}[1 - e^{-\mu' K y/R}] \tag{7.33}$$

where

p_v = vertical pressure at depth y from top

p_h = horizontal pressure at depth y from top

γ = weight of fill

R = A/L hydraulic radius

L = perimeter of cross-section

e = base natural log

K takes the value by Rankine method or

$$\frac{P_h}{P_v} = K = \frac{1 - \sin\phi}{1 + \sin\phi} \tag{7.34}$$

or more correctly, by experiment. Buisman [7.14] suggested the following value for computing the lateral pressure during filling

$$K = \frac{1 - \sin^2\phi}{1 + \sin^2\phi} \tag{7.35}$$

where ϕ is the angle of repose of the fill.

7.4.4 Computation of Static Pressure — Lateral and Vertical

Two methods for determining static pressures are Janssen's classic method [7.13] and Reimbert's method [7.15]. Figure 7.32 shows the bin dimensions used for each.

Janssen's method is more popular in North America. However, experiments [7.15] show Janssen's method to be unconservative in some cases, wherease Reimbert's method is reported to give pressures agreeing closely with test results.

7.4.4.1 Janssen's Method

Vertical static unit pressure at depth y below the surface according to formula (7.32) is

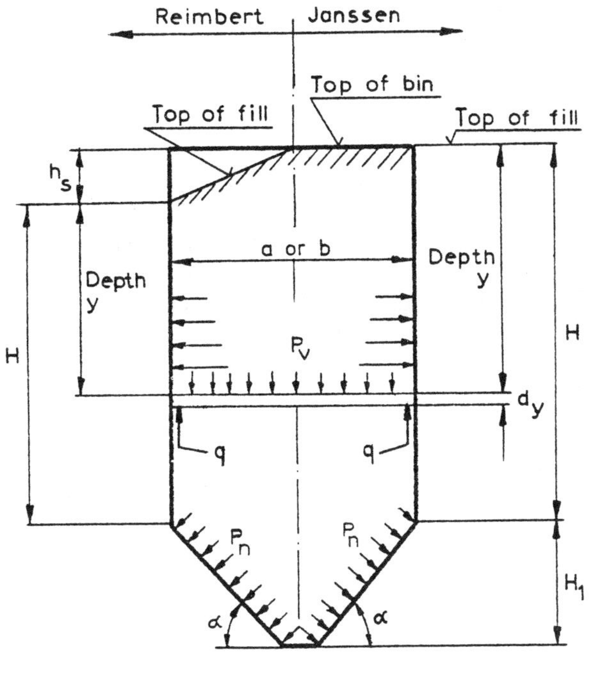

FIGURE 7.32 — Bin dimensions for use in Reimbert's and Janssen's equations.

$$p_v = \frac{\gamma R}{\mu' K}(1 - e^{-\mu' Ky/R})$$

Lateral static unit pressure at depth y according to formula (7.33) is

$$p_h = \frac{\gamma R}{\mu'}(1 - e^{-\mu' Ky/R}) = Kp_v$$

where K is given by formula (7.34)

$$K = \frac{p_h}{p_v} = \frac{1 - \sin\phi}{1 + \sin\phi}$$

In the above, R is the hydraulic radius (area/perimeter) of the horizontal cross-section inside of the bin.

For a square bin or a shorter wall, "a", of a rectangular bin, Figure 7.33, the hydraulic radius is

$$R_a = \frac{a}{4} \quad (7.36)$$

For the long wall, "b", of rectangular bin, Figure 7.33, the hydraulic radius is

$$R_b = \frac{a'}{4} \quad (7.37)$$

where a′, the length of side of an imaginary square bin is [7.15]

$$a' = \frac{2ab - a^2}{b} \quad (7.38)$$

7.4.4.2 Reimbert's Method

The vertical static unit pressure at depth y below the surface is

$$p_v = \gamma[y(\frac{y}{C} + 1)^{-1} + \frac{h_s}{3}] \quad (7.39)$$

Lateral static unit pressure at depth y is

$$p_h = (p_{max})_h\ [1 - (\frac{y}{C} + 1)^{-2}] = z(p_{max})_h \quad (7.40)$$

$(p_{max})_h$ is the maximum lateral unit pressure an C is the characteristic abscissa for use in Equations (7.39) and (7.40), which vary in bin shape, as follows:

For a rectangular bin — on a shorter wall, "a"

$$(p_{max})_a = \frac{\gamma a}{4\tan\phi'} \quad (7.41)$$

$$C_a = \frac{a}{\pi \tan\phi' \tan^2(\frac{\pi}{4} - \frac{\phi}{2})} - \frac{h_s}{3} \quad (7.42)$$

For rectangular bins — on a longer wall, "b"

$$(p_{max})_b = \frac{\gamma a'}{4\tan\phi'} \quad (7.43)$$

$$C_b = \frac{a'}{\pi \tan\phi' \tan^2(45^\circ - \frac{\phi}{2})} - \frac{h_s}{3} \quad (7.44)$$

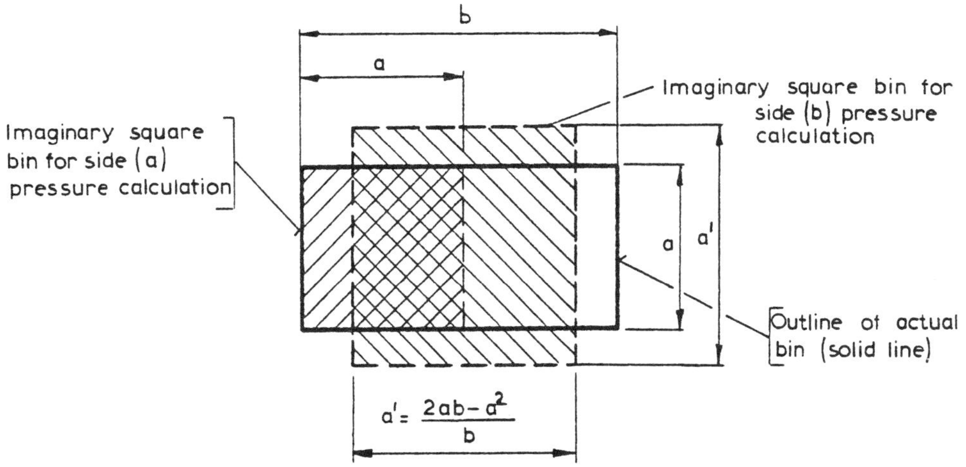

FIGURE 7.33 — Equivalent bin shapes.

where

$$a' = \frac{2ab - a^2}{b} \tag{7.45}$$

For design purposes, the granular material is usually assumed level at the top of the bin, or $h_s = 0$.

7.4.4.3 Static Pressure on Flat Bottoms

Pressure on flat bottoms is given by Equations (7.32) and (7.39). For rectangular bins, however, these equations give different bottom pressures for areas next to the short and long sides. An approximation frequently used to assume the pressure p_a — computed using R after Janssen or C following Reimbert for the short side, "a" — to act on the area A_a, as shown in Figure 7.34. Similarly, pressure p_b computed using R or C for side "b", is assumed to act on area A_b.

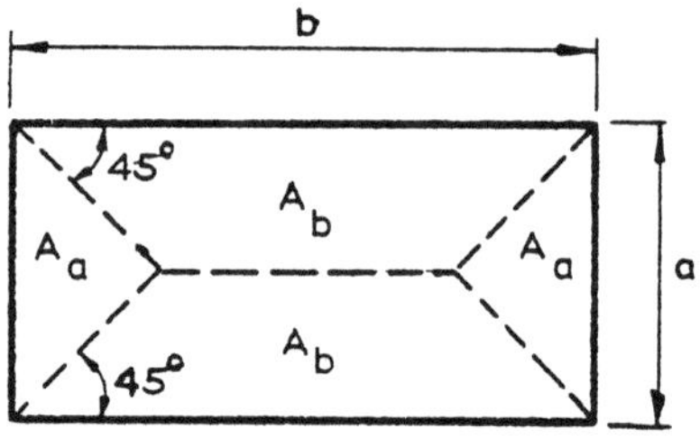

FIGURE 7.34 — Assumed distribution of vertical pressure on horizontal plane of rectangular bin.

7.4.4.4 Static forces — vertical friction

For *square bins*, the total static frictional force per foot-wide vertical strip of wall above depth y is approximately

By Reimbert's method

$$q = (\gamma y - p_v)A/L \tag{7.46}$$

By Janssen's method

$$q = (\gamma y - 0.8\, p_v)A/L \tag{7.47}$$

For *rectangular bins*, lateral pressures differ for the long and short walls, therefore, side friction and vertical pressures also differ. The friction loads may be approximated by Equations (7.46) and (7.47) when terms p_v, A and L, respectively, are substituted by p_a, A_a and a for side "a", and by p_b, A_b, and b for side "b", Figure 7.34.

7.4.5 Overpressure and Overpressure Factors, Cd

Equations (7.32) through (7.47) above, are for static pressures only, due to stored material at rest, before withdrawal is begun. During the withdrawal, these pressures may increase [7.15, 7.16, 7.17, 7.18, 7.19]. The increases are sometimes called dynamic effects, but the term "overpressure" is preferred since the increase includes both static and dynamic effects. Among these are arching of the material, increasing higher wall pressure and vertical friction, and a collapse of the arched material causing an increased vertical pressure due to impact.

Basically, the total pressure may be as much as three to four times [7.20, 7.21] the static pressure computed by the Janssen or Reimbert equations. Bins have been designed bearing overpressure. Many bins so designed have failed and many others not yet showing signs of distress may have dangerously low margins of safety.

Overpressure is not yet well enough understood to consider by rational methods. However, its effect can be approximated using the overpressure factor, C_d, to convert from computed static pressures to design pressures. In general

$$\text{Design pressure} = C_d \times \text{static pressure} \tag{7.48}$$

The following Table 7.5 shows tentative overpressure factors, C_d. The factors for use with Janssen's method are from the Soviet Code [7.22] but with slight modifications. Those for use by Reimbert's method are computed from those for Janssen's method.

To promote the better flow of material, designers sometimes use a flow-guiding insert in the bin or lever

TABLE 7.5 — Values of Overpressure Factor, C_d.

(For funnel flow only. With mass flow, higher C_d factors than shown here are recommended)

Description of Area of Application	Reimbert	Janssen	Note
1. Correction factor C_d for use in calculating horizontal pressure. Square and rectangular single bin and bin groups Upper H_1 portion of bin. Lower 2/3 of bin height	1.10	1.50	For Reimbert's equation C_d for H/a or H/b between 1.5 and 4.5 should be determined by interpolation.
H/(a or b) = 1.5	1.40	1.65	For both methods, if
H/(a or b) $\geqslant$ 4.5	1.65	1.65	H/a or H/b > 5 C_d values 15% higher than shown above are recommended.
$(H_1)a$ = ban $\phi \leqslant$ 1/3 H			
$(H_1)b$ = atan $\phi \leqslant$ 1/3 H			
2. Correction factor C_d for use in calculating bottom pressures in bin. Steel hoppers and ring-beams; steel beams; steel columns.			
a. For flour and bran	1.75	1.50	
b. All types of grain	1.50	1.25	
c. All types of granular material except (a) and (b)	2.00	1.75	

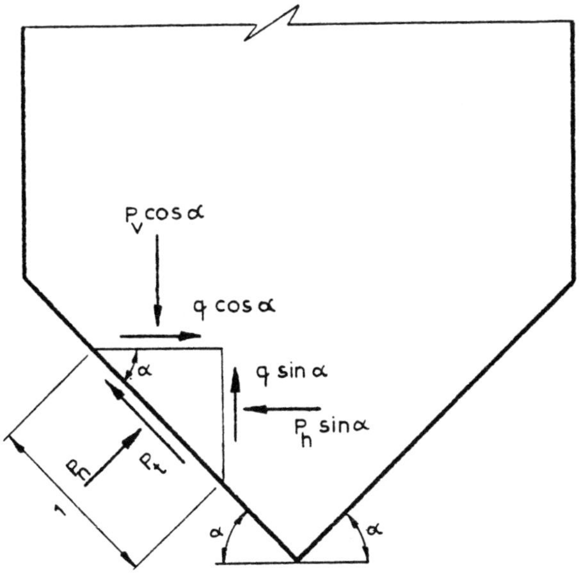

FIGURE 7.35 — Components of forces on hopper wall.

directly above the hopper. Tests show that this instrument may cause large additional local overpressures, beyond the normal overpressures without the insert. When the insert is used, overpressure factors, C_d, for walls at the level of the insert should be 50% higher than that shown in Table 7.5.

7.4.6 Static Pressures in Bin Hoppers

Static horizontal pressures, p_h, and p_v, on inclined hopper walls are calculated by the Janssen or Reimbert formulas. The hydraulic radius, R, may be retained within the hopper depth, but is usually assumed constant and equal to that of the bin.

The forces on a triangular element at the wall of a hopper are shown in Figure 7.35.

Projecting these forces on the lines of action of p_n and p_t gives

$$p_n = p_v \cos^2\alpha + p_h \sin^2\alpha + q \sin 2\alpha \quad (7.49)$$

$$p_t = \tfrac{1}{2}(p_v - p_h)\sin 2\alpha - q \cos 2\alpha \quad (7.50)$$

where

q = $\mu' p_h$ is the frictional force on the wall

If the pressure field is assumed to be a Rankine field, or $q = 0$, the above equations reduce to

$$p_n = p_v \cos^2\alpha + p_h \sin^2\alpha \quad (7.51)$$

$$p_t = \tfrac{1}{2}(p_v - p_h)\sin 2\alpha \quad (7.52)$$

7.4.7 Design Pressure in Bins

For bins with centrally located discharge openings, design pressures due to stored material are

$$p_{v,des} = C_d\, p_v \quad (7.53)$$

$$p_{h,des} = C_d\, p_h \quad (7.54)$$

$$p_{n,des} = p_{h,des} \sin^2\alpha + p_{v,des} \cos^2\alpha \quad (7.55)$$

The maximum wall friction force per unit length of wall is instantaneous with a maximum, or static, p_v at depth y.

Therfore, for bins

$$q_{des} = q \quad (7.56)$$

7.4.8 Pressures on Bin Walls. Eccentric Outlets

Pressures on bins walls during the discharge of granular material through eccentric outlets may differ considerably from flow pressures in bins with concentric outlets. Two methods are considered, as follows.

7.4.8.1 DIN Specifications Method

A procedure for computing pressures in bins with eccentric outlets is given in DIN 1055 Specifications[7.23]. The change in pressure Δp_h from that for a concentric outlet is given for rectangular bins by

$$\Delta p_h = p_{hi} - p_h \quad (7.57)$$

where

p_h = pressure on the wall B at depth y in a bin with sides B and L, Figure 7.36.

p_{hi} = pressure on wall B at depth y computed for a bin with sides B and L + 2e, Figure 7.36.

The pressure increment is to be subtracted from p_h on the wall adjacent to the outlet and added onto the opposite wall. If the outlet is also eccentric in other directions the pressure changes on the other two walls are computed in the same way.

Equation (7.57) reduces to a simple form for the pressure increment at the bottom of the tall bins of a square section. In this case, Janssen's formula gives

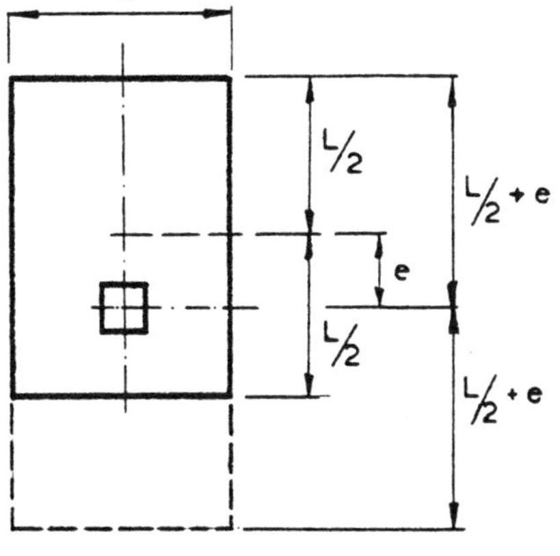

FIGURE 7.36 — Eccentric outlet.

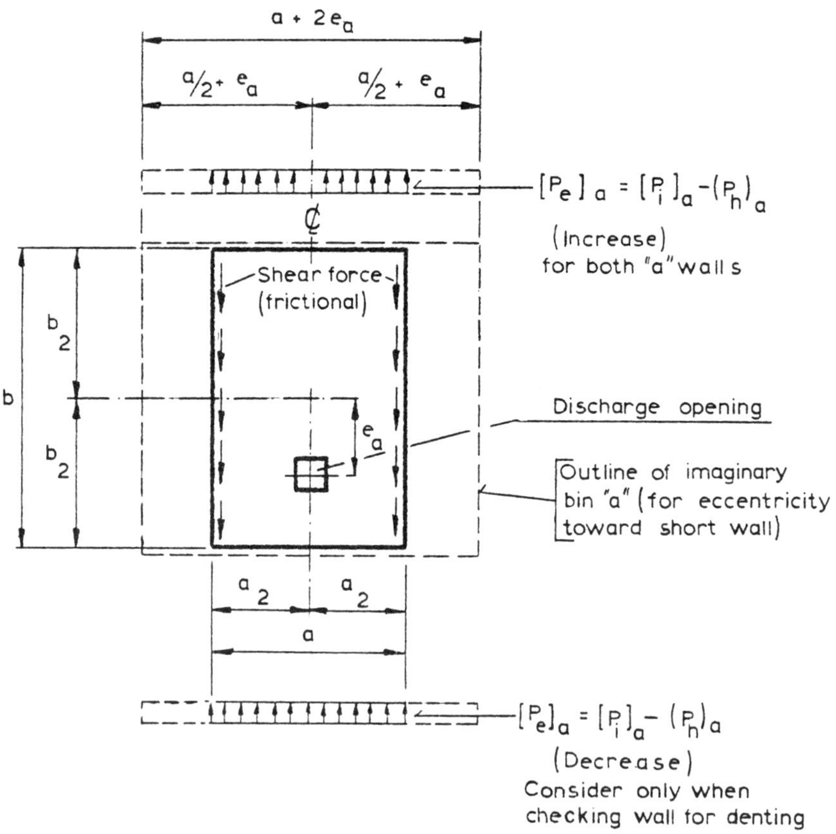

FIGURE 7.37 — Pressure change due to eccentric discharge in rectangular bin.

$$p_h = R/\mu' \qquad (7.58)$$

so that p_h and p_{hi} are directly proportional to the corresponding hydraulic radii. This gives

$$\Delta p_h = \frac{e}{B + e} p_h \qquad (7.59)$$

for square bins where

e = eccentricity of the outlet, Figure 7.36.

B = side of a square bin.

7.4.8.2 Imaginary Bin Method

An eccentric discharge may be considered by adding a pressure p_e corresponding to the lateral design pressure, $p_{h,des}$ computed at y by either the Janssen or Reimbert formula. p_e is considered to vary from zero at the top of the bin to a maximum at depth y = H, Figure 7.32 and to remain at that maximum for the full depth of the hopper. Within a height H, the lateral design pressure at depth y is then

$$p_{h,des} = C_d p_h + \frac{y}{H} (p_e) \qquad (7.60)$$

The correction p_e at depth H is

$$p_e = p_i - p_h \quad (\text{at } y = H) \qquad (7.61)$$

where

p_i = the lateral static pressure at depth H in an imaginary bin

as shown in Figure 7.37.

For a rectangular bin, the imaginary bin is determined as shown by Figure 7.37.

When the opening is displaced toward side "a", correction p_e for sides "a" is computed using an imaginary bin measuring $(a + 2e_a) \times b$. If e_a is larger than a, the imaginary bin should measure 3a × b. Similarly, if the opening is eccentric toward side "b", the imaginary bin will measure $(b + 2e_b) \times a$. If both eccentricities occur, each correction is computed separately, using the first described imaginary bin to determine p_e for sides "a", and the second for sides "b".

Where multiple discharge openings occur, even though the group is centrally located, an eccentric discharge is always possible and should be considered.

7.4.9 Lateral Pressure Design Curve

Table 7.5 shows different C_d values for different depth zones [24] as follows:

Zone 1:

The upper portion of depth H_1 = $D\tan\phi$. This is a zone of negligible "arching" of the stored material, one for which some authorities [7.24, 7.25] consider the Rankine formula valid for computing lateral static pressure provided it is modified for the bin shape.

Zone 2:
This zone, the lower two-thirds has significant arching of the stored material and, consequently, large lateral pressures. In this zone, C_d varies with the type of structure.

Zone 3:
The pressure curve in this zone is assumed as a straight line. It joins the maximum pressure for Zone 1 and the minimum for Zone 2 and is horizontal when $H_1 = H/3$.

Figure 7.38 shows the resulting curve of lateral design pressures. Note that pressure increases, p_e, due to eccentric discharge, if any, are added to the product $C_d p_h$.

7.4.10 Limiting the Height of a Bin Wall

Various formulas defining the limiting height of a shallow bin have been proposed. According to one of the assumptions in Coulomb's theory, the failure surface in the material must intersect the top of the fill, Figure 7.39.

This condition is satisfied if

$$\frac{H}{B} \leq \tan\phi + \sqrt{\frac{\tan\phi(1+\tan^2\phi)}{\tan\phi+\tan\phi'}} \qquad (7.62)$$

This equation gives H/B limits of 1.12 and 2.30 for $\phi = \phi' = 20$ degrees and $\phi = \phi' = 50$ degrees, respectively.

It should be noted that Coulomb's formula does not strictly apply to a bin with a hopper because the failure prism is not wedge-shaped. Where it has been used in this situation, its limits of validity have sometimes been defined as shown in Figure 7.40, where B_1 replaces B in Equation (7.62).

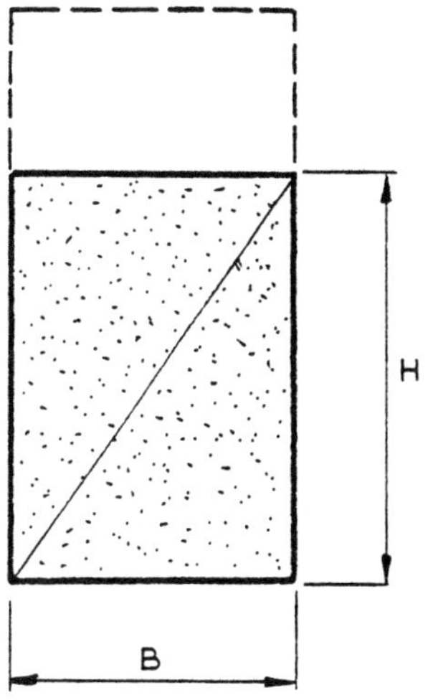

FIGURE 7.39 — Limiting height of a bin.

A second method is to classify the structure as a shallow bin based on the position of the plane of rupture of the stored material.

Shallow bins or bunkers are considered as containers being relatively small in height by comparison with the dimensions in plan, in which the plane of rupture DK, Figure 7.41 cuts the free surface of the fill AB in the bunker.

Analytically, this condition is expressed as follows:

$$m \geq x = \frac{h_1}{\tan\beta} \qquad (7.63)$$

where β is an angle of the plane of rupture. According to Coulomb's theory, the wedge ADK between this plane

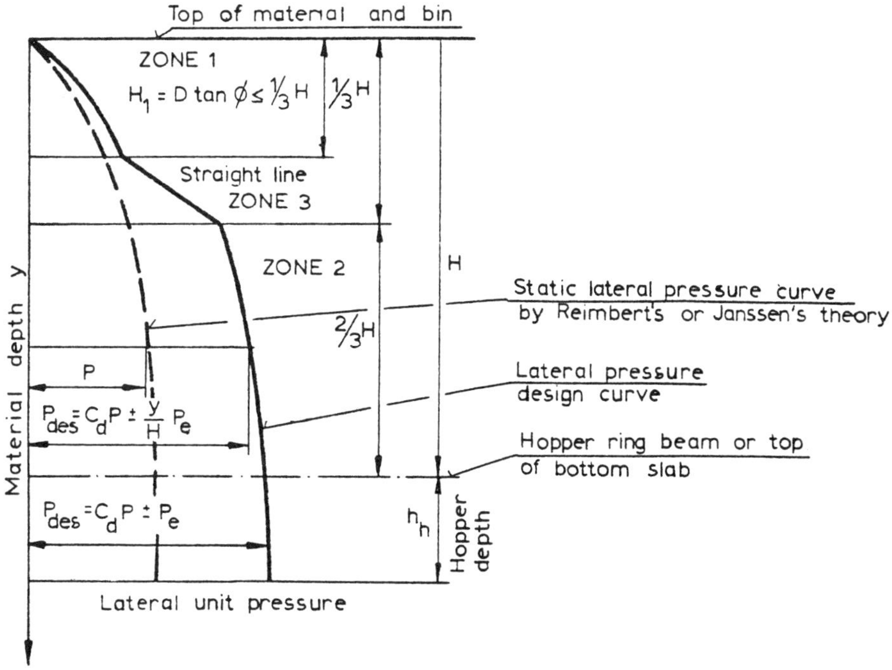

FIGURE 7.38 — Proposed lateral pressure design curve.

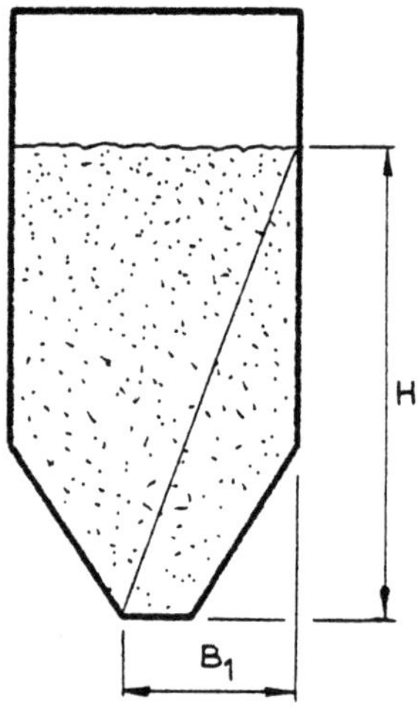

FIGURE 7.40 — Reduced width of failure surface.

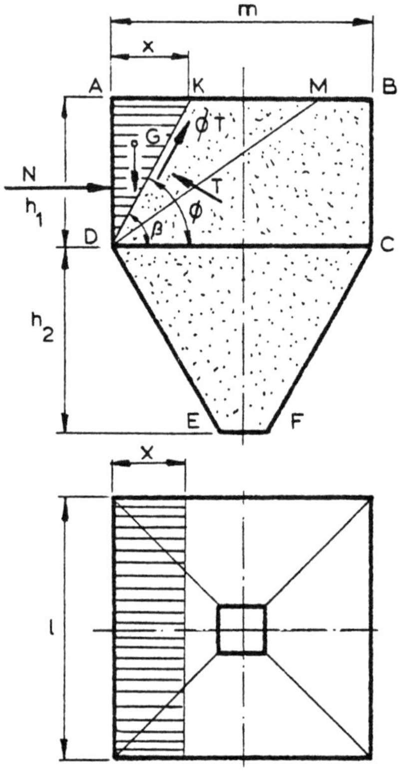

FIGURE 7.41 — Scheme of shallow bin or bunker.

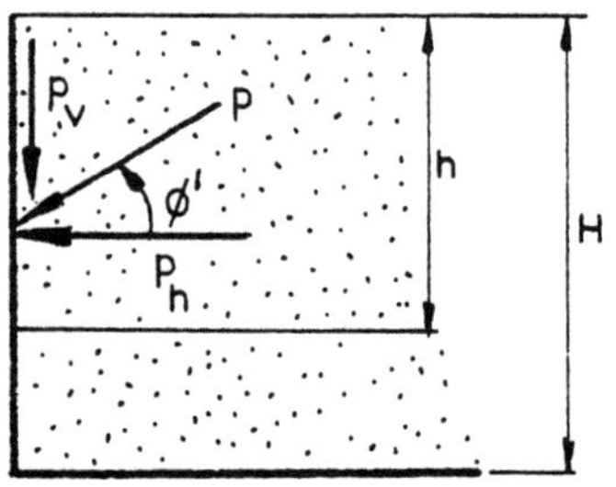

FIGURE 7.42 — Fill pressure on a wall.

and the wall of the bunker produces the maximum pressure. The angle of rupture is

$$\beta = \frac{90 + \phi}{2} \tag{7.64}$$

where ϕ is the angle of repose of the material in the bunker.

Practically speaking, in bunkers, the depth of the upper vertical wall or the prismatic part is seldom greater than the length and width of the bunker in plan.

7.4.11 Bunker Loads and Forces

Bunkers are designed for loads imposed by storing material, their own weight, together with the weight of equipment and platforms carried by the structure. The impact, roof, loads, wind, and earthquake may also need to be considered.

Material pressures on the bunker walls and bottom may be determined by the Janssen or Reimbert equations for bins, or by the Rankine method, which follows. This theory is not entirely accurate since it is based on assumptions not fully met, and also ignores the boundary conditions. Friction forces on the walls are neglected and pressures are assumed to be normal to the surface against which they act.

7.4.11.1 Top Surface of the Stored Material is Horizontal

For a level fill against a vertical surface, Coulomb's theory gives the following values of vertical and horizontal pressures, Figure 7.42 [7.26].

Vertical pressure

$$p_v = \gamma h \frac{\cos^2\phi \tan\phi'}{\left[1 + \sqrt{\frac{\sin(\phi+\phi')\sin\phi}{\cos\phi'}}\right]^2} \tag{7.65}$$

and horizontal pressure

$$p_h = \gamma h \frac{\cos^2\phi}{\left[1 + \sqrt{\frac{\sin(\phi+\phi')\sin\phi}{\cos\phi'}}\right]^2} \tag{7.66}$$

where ϕ' is an angle of friction of the fill on the bunker walls.

If ϕ is assumed equal to ϕ', Equations (7.65) and (7.66) gives

$$p_v = \gamma h \frac{\sin\phi \cos\phi}{[1 + \sqrt{2}\sin\phi]^2} \tag{7.67}$$

$$p_h = \gamma h \left[\frac{\cos\phi}{1 + \sqrt{2}\sin\phi}\right]^2 \tag{7.68}$$

If the wall friction is neglected, $\phi' = 0$, Equations (7.65) and (7.66) reduce to the Rankine active horizontal static unit pressure at depth h, or

$$p_v = 0$$

$$p_h = \gamma h \frac{1 - \sin\phi}{1 + \sin\phi} = K\gamma h \tag{7.69}$$

where

$$K = \frac{1 - \sin\phi}{1 + \sin\phi} \quad (7.70)$$

The total horizontal pressure on the vertical wall is

$$P_H = \int_0^H p_h dh = \gamma \frac{1 - \sin\phi}{1 + \sin\phi} \int_0^H h dh =$$

$$= \tfrac{1}{2}\gamma H^2 \frac{1 - \sin\phi}{1 + \sin\phi} \quad (7.71)$$

Vertical static unit pressures at depth h below the surface is

$$q = \gamma h \quad (7.72)$$

The unit static pressure normal to an inclined surface at depth h is

$$q_\alpha = \gamma h(\cos^2\alpha + K \sin^2\alpha) \quad (7.73)$$

7.4.11.2 Top Surface of the Stored Material Slopes

If the surface of the stored material slopes at the angle of repose which is approximately equal to ϕ, Figure 7.43.

The vertical static unit pressure at depth y is

$$p_v = \gamma(y + a_0 \tan\phi) \quad (7.74)$$

The horizontal static unit pressure at depth y is

$$p_h = \gamma y \cos^2\phi \quad (7.75)$$

The unit static pressure normal to the inclined surface of the hopper wall at depth y is

$$p_n = \gamma\cos^2\alpha(y+a_0 \tan\phi) + \gamma y \cos^2\phi \sin^2\alpha \quad (7.76)$$

For small trapezoidal walls, it is sometimes sufficiently accurate to use average pressures. An average pressure on an inclined trapezoid of upper edge a and lower edge a_1 is

$$p_{n,avg.} = \frac{a(5p_{n,top}+4p_{n,bot}) + a_1(p_{n,top}+2p_{n,bot})}{6(a+a_1)}$$

$p_{h,avg}$ for a vertical trapezoid is computed by the same equation, but with $p_{n,top}$ and $p_{n,bot}$ substituted by p_h — values computed from Equations (7.69) or (7.75).

7.4.12 Bunker Design Pressures

Design pressures are

$$P_{h,des} = C_d P_h \quad (7.77)$$

$$P_{v,des} = C_d P_v \quad (7.78)$$

$$P_{n,des} = C_d P_n \quad (7.79)$$

C_d for bunkers is usually considered to be unity. However, a significant impact may occur when the volume of material suddenly dumped into a bunker is large compared to the bunker capacity. In this case, C_d will be larger. Values [7.27] for use in a bunker design are shown in Table 7.6.

TABLE 7.6 — Values of Dynamic Coefficient C_d for Bunkers.

Ratio of volume dumped in one load to total bunker capacity	1:2	1:3	1:4	1:5	1:6 and less
Dynamic Coefficient C_d	1.4	1.3	1.2	1.1	1.0

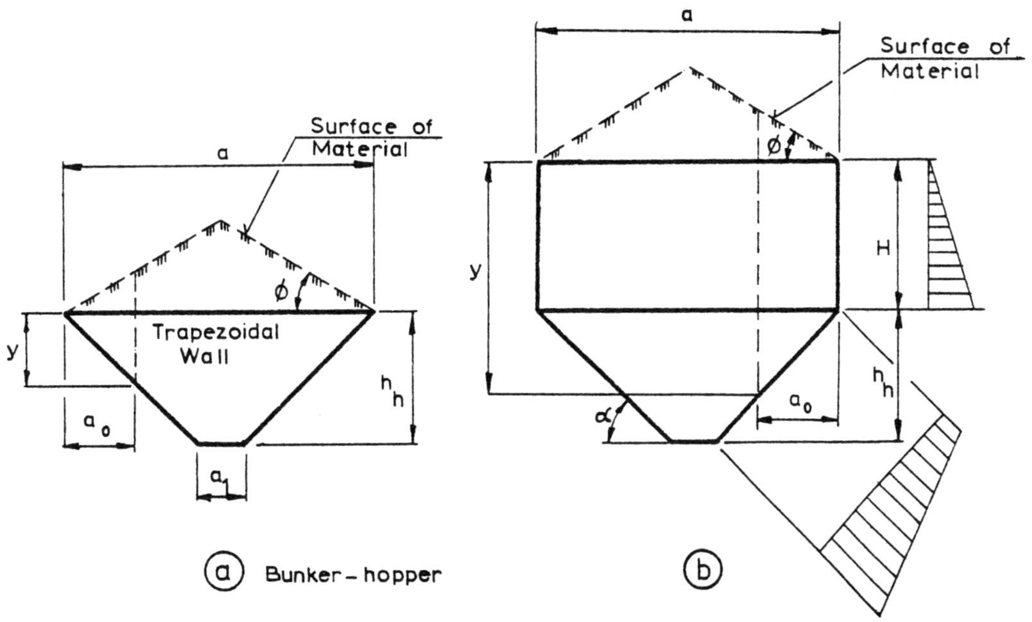

FIGURE 7.43 — Bunker and hopper dimensions.

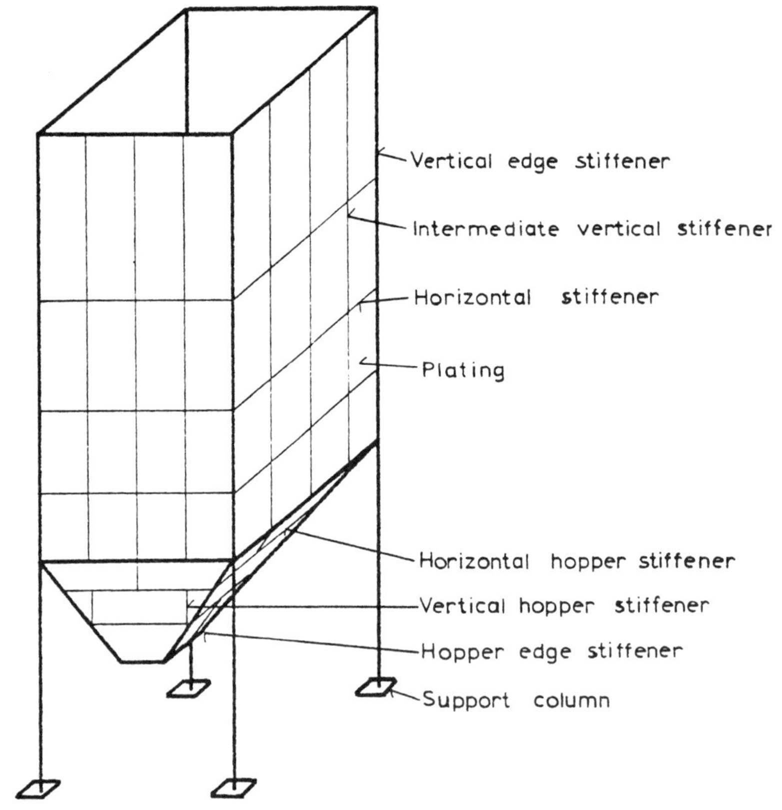

FIGURE 7.44 — Structural system of the bin.

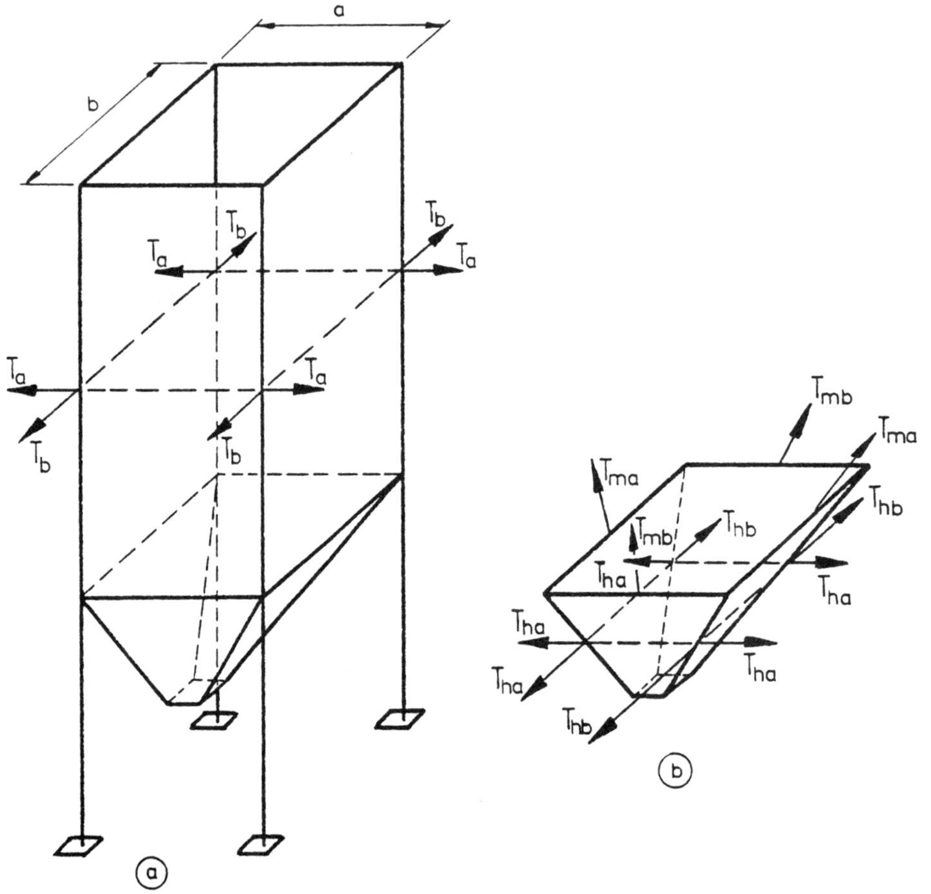

FIGURE 7.45 — Forces acting on the structural system of the bin.

7.5 Structural Analysis and Design of Square and Rectangular Bins

7.5.1 General Concept

Under the action of statical and dynamic loadings, in the walls of a loaded bin, originates complex tensile, compressive, flexural and torsional stresses. Torsional stresses originate from the assymmetric shape of the bin and are due to eccentric loading of the bin by materials.

An exact strength analysis of the bin is extremely complex and for this reason, usually only approximate methods of analysis are used. However, these approximate methods provide practical and safe solutions. The exact methods of analysis are used for the design of complicated and large bins.

A structural system of the typical square or rectangular bin is a rigid space frame consisting of vertical and horizontal stiffening beams supporting plating, Figure 7.44.

7.5.2 Forces and Moments Acting on the Structural System of the Bin

The structural system of the bin is analyzed and designed, considering the action of the tensile and bending forces originating from the pressure of stored material, Figure 7.45.

In the vertical walls of the bin supported by the columns, Figure 7.45 (a), T_a, T_b act as horizontal tensile forces. In addition, the walls under local bending are under the action of the side loading pressure. The forces set up in a square and rectangular bin will induce direct, bending and shear stresses in various parts. It is not practical to accommodate these forces on a skin plate only. Any attempt to do so will result in very thick plates and, as a consequence, a totally uneconomical design.

Usually, the skin plate, which forms the bin wall, is designed to span between suitably positioned stiffeners, which may be either vertical or horizontal, depending upon the design approach. Most often, stiffeners are arranged to act in a horizontal plane and as a result, this layout probably forms large numbers of bins in service. Skin plates may be treated as either simply supported or continuous between the stiffeners, which in turn, also act as beams, but spanning at right angles to the designed direction of the skin plating.

The inclined walls of the hopper are under the action of the above-mentioned loads, as well as under tensile forces acting horizontally and along the slope of the hopper wall (meridional forces).

Also it is necessary to consider vertical compressive forces due to own weight of the hopper and the frictional forces originating between the loads and walls.

The maximum pressure on the skin plate is on the bottom of the hopper wall. The skin plate spans between the successive stiffeners will be subject to bending, some shear, and longitudinal tension. Normally, the plate is designed for bending, the shear being neglected and the longitudinal tension is taken up on the stiffeners, which may include a part of the plate.

In summary, forces which must be accommodated in any bin are due to:

a. the weight of the contained material which exerts both vertical and horizontal forces, and
b. own weight of the bin including the roof.

7.5.3 Vertical Walls, Lateral Pressure

Lateral pressures are usually computed using Janssen's formula (7.33)

$$P_h = \frac{\gamma R}{\mu'}\left(1-e^{-\mu' K_y/R}\right)$$

where, according to (7.34)

$$K = \frac{1-\sin\phi}{1+\sin\phi}$$

For the shorter wall (7.36)

$$R_a = \frac{a}{4}$$

and for the longer wall (7.38) and (7.37)

$$a' = \frac{2ab - a^2}{b}$$

$$R_b = \frac{a'}{4}$$

Using these R-values, the static lateral pressure values are computed at various depths, y, using formula (7.33), and are tabulated. Overpressure factors, Cd, are selected from Table 7.5, and the design pressures computed by Equation (7.48) are added to the tabulated values.

7.5.4 Vertical Walls, Tension and Bending Moments

Vertical walls of rectangular bins are subject to horizontal membrane tensile forces and two-way bending and frequently also to in-plane bending, vertical shear, and vertical membrane forces, depending on the manner of support used.

7.5.4.1 Tensile Forces

In the following, we are considering a closed rectangular frame having unit height and loaded at the short side by the pressure p_a and long side under pressure p_b, Figure 7.46.

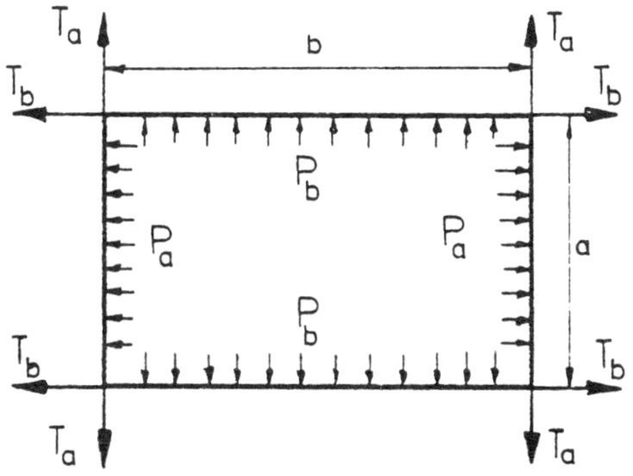

FIGURE 7.46 — Tensile forces.

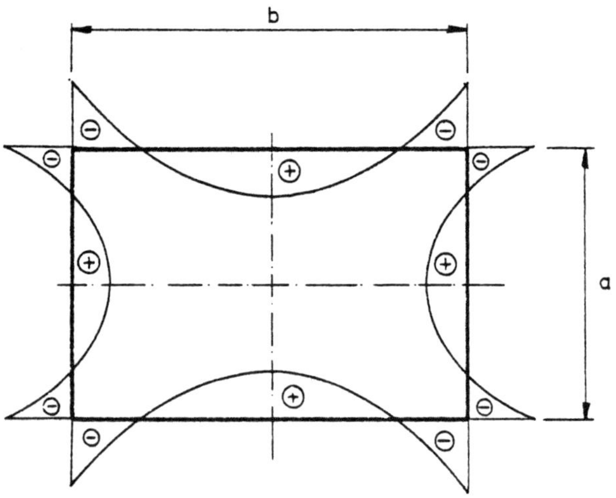

FIGURE 7.47 — Bending moments for rectangular bin.

The tensile forces acting in the walls are

$$\text{Side "a"} \quad T_a = \frac{1}{2} P_b b \tag{7.80}$$

$$\text{Side "b"} \quad T_b = \frac{1}{2} P_a a \tag{7.81}$$

These values of tensile forces are computed for various depths, y, and are tabulated.

7.5.4.2 Bending Moments

By indicating as I_a and I_b the moments of inertia of the sides having lengths a and b, Figure 7.47, the auxiliary values is [7.28]

$$K = \frac{I_b a}{I_a b} \tag{7.82}$$

The corner bending moments are

$$M_c = -\frac{P_a a^2 K + P_b b^2}{12(1+K)} \tag{7.83}$$

The midspan bending moments are

$$M_a = \frac{P_a a^2}{8} + M_c \tag{7.84}$$

$$M_b = \frac{P_b b^2}{8} + M_c \tag{7.85}$$

In particular cases where

$$I_a = I_b \tag{7.86}$$

the corner bending moments are

$$M_c = \frac{P_a a^3 + P_b b^3}{12(a+b)} \tag{7.87}$$

and in the case of square bins, when

$$a = b,\ I_a = I_b,\ K = 1,\ P_a = P_b = P \tag{7.88}$$

$$M_c = -\frac{pa^2}{12} \tag{7.89}$$

$$M_a = M_b = \frac{pa^2}{24} \tag{7.90}$$

The values of the corner and midspan bending moments are computed for various depths, y, and tabulated.

7.5.5 Vertical Walls. Plating

7.5.5.1 Small Deflection Theory

The walls of the bin are constructed from the steel plate spanning between the stiffener beams forming the frame of the bin walls. Normally, plating is designed for bending, the shear being neglected and the longitudinal tension being taken up by the stiffeners, which may include a part of the plating.

Generally speaking, the plate will tend to span in two directions simultaneously, that is at right-angles and parallel to the stiffeners, provided the ratio

$$b/a \leqslant 3 \tag{7.91}$$

with reference to Figure 7.48.

Nonuniform pressure of the fill along the height of the bin is transformed into an average uniformly distributed loading acting on each panel. The introduced approximation certainly differs from the actual loading acting on the plating. However, the final results provide a safe design as proven by experience.

In the following, formulas are given for the design of rectangular plates under uniformly distributed loading, having fixed edges [7.29]. The maximum bending moment acts along the axis x-x in the center of the plate, Figure 7.49.

$$M_{max} = \alpha P_n a^2 b \tag{7.92}$$

where

$$a > b$$

and

P_n = the average pressure

α = the proportionally coefficient given in Table 7.7.

For the plates having simply supported edges, Figure 7.50, the maximum bending moment will also act in the center of the plate and will be defined by the same equation as (7.92) having only different coefficients α_1, as given in Table 7.8.

TABLE 7.7 — The Proportionality Coefficients for Plates Having Fixed Edges.

K = b/a	1.00	1.25	1.50	1.75	2.00	∞
α	0.0513	0.0665	0.0757	0.0817	0.0829	0.0833

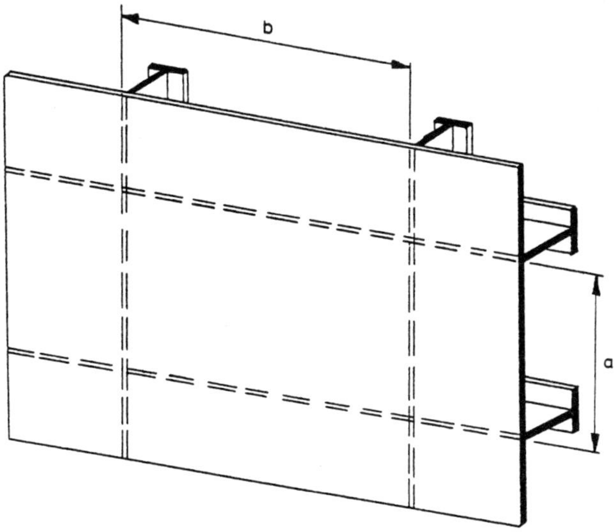

FIGURE 7.48 — Stiffened plating of the vertical wall.

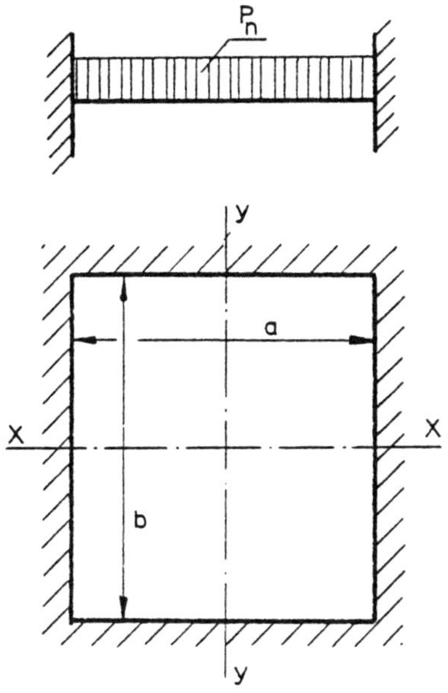

FIGURE 7.49 — Plate having fixed edges.

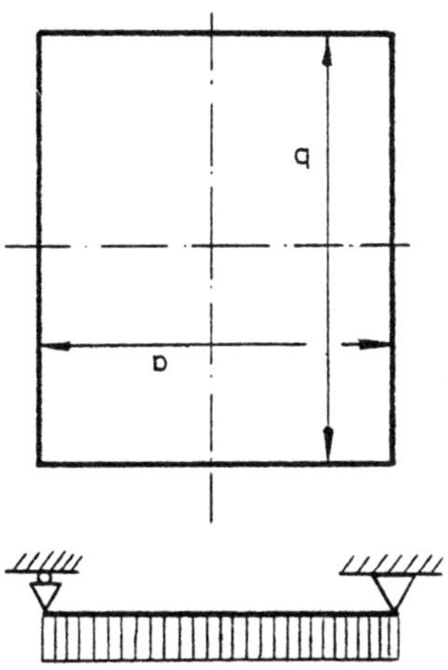

FIGURE 7.50 — Plate having simply supported edges.

TABLE 7.8 — The Proportionality Coefficients for Plates Having Simply Supported Edges.

K = b/a	1.0	1.1	1.2	1.3	1.4	1.5	1.6
α_1	0.0479	0.0553	0.0626	0.0693	0.0753	0.0812	0.0862
K = b/a	1.7	1.8	1.9	2.0	3.0	4.0	∞
α_1	0.0908	0.0948	0.0985	0.1017	0.1189	0.1235	0.1250

$$M_{max} = \alpha_1 P_n a^2 b \tag{7.93}$$

Both formulas (7.92) and (7.93) have been developed under the assumption that the edges were either fixed or free, and may be displaced in the plane of the plate under bending. Poisson's ratio for steel has been used as 0.3.

A comparison of the coefficients α and α_1 indicates a relatively small difference between them up to $b/a \leqslant 1.4$. This condition indicates that the formula (5.14) is correct for an analysis of the bin walls, because some of the sheets are not fully fixed at the edges.

When skin plating is designed for one-way bending only, the maximum bending moment will be:

For simply supported plates

$$M_{max} = \frac{P_n b^2}{8} \tag{7.94}$$

and for continuous plates

$$M_{max} = \frac{P_n b^2}{12} \tag{7.95}$$

All the above expressions are based upon the small deflection theory.

Where large deflections of bin plating are considered acceptable, there may be some advantage in designing such plates in accordance with the large deflection theory to be discussed in the following section.

7.5.5.2 Large Deflection Theory

Plating may be designed as a membrane having a large deflection [7.30].

Let us consider the strip of the plate under uniformly distributed loading, having a unit width and ℓ-span between the stiffening ribs, used as supports, Figure 7.51.

As limiting conditions at supports, they are considered to be fixed hinges.

The bending moment per unit width of the plating is

$$M = \frac{P_n \ell^2}{8} - \frac{4P_n \ell^2}{\pi^3} \cdot \frac{S}{S + S_E} \tag{7.96}$$

or

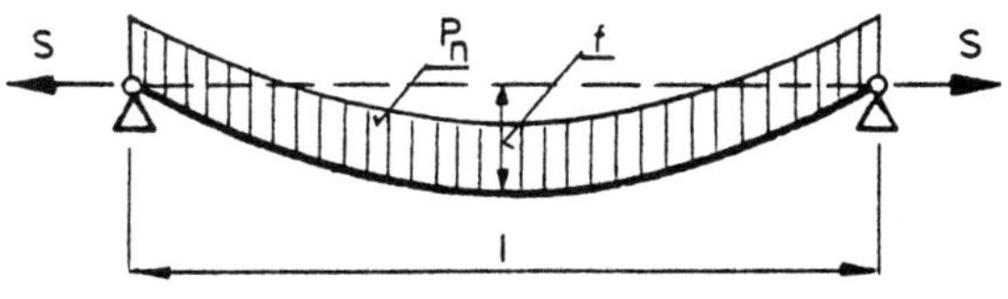

FIGURE 7.51 — Plate as membrane.

$$M = \frac{P_n \ell^2}{8} \left(1 - 1.034 \frac{S}{S + S_E}\right) \qquad (7.97)$$

where

$$S = \frac{1}{\pi} \sqrt[3]{\frac{4(p_n \ell)^2 Et}{\pi(1-\nu^2)}} \qquad (7.98)$$

and

$$S_E = \frac{\pi^2 EI}{\ell^2(1-\nu^2)} \qquad (7.99)$$

where

p_n = uniformly distributed loading per 1 inch width of the plating (lbs/in)

ℓ = span of the plating-spaced between the stiffening ribs (in)

t = thickness of the plate (in)

I = moment of inertia of the plating per unit width (in^4)

E = 29×10^6 — modulus of elasticity of steel (psi)

ν = 0.3 — Poisson's ratio

7.5.5.3 Determination of the Plating Thickness

Plating for side walls of a bin are designed under pressure of the fill, as explained in Sections 7.5.5.1 and 7.5.5.2 considering the following two cases:

a. According to small deflection theory

After determination of the average pressure of the fill on panel, we find M_{max} using either Equation (7.92) or (7.93).

Using a section modulus of the sheet as

$$W = \frac{bt^2}{6} \qquad (7.100)$$

we have

$$M_{max} = Wf_{all} = \frac{bt^2}{6} f_{all} \qquad (7.101)$$

Therefore, the required thickness of the plate, considering also the corrosion, is

$$t = \sqrt{\frac{6M_{max}}{bf_{all}}} + (1/32 \text{ to } 1/16) \text{ in} \qquad (7.102)$$

b. According to large deflection theory

Assuming a thickness of the plate t, the stress in the plate should satisfy the following formula

$$f = \frac{S}{t} + \frac{6M}{t^2} \leq f_{all} \qquad (7.103)$$

where the tensile force S in the plate is used after formula (7.98) and the bending moment M after formula (7.97).

7.5.5.4 The Effective Width of the Plating

Part of the skin plating may be included in the stiffener design. It is suggested that the width of the plate should not exceed [7.31]:

a. When $\ell/t < 40$, the actual width center-to-center of stiffeners ℓ should be used.

b. when $\ell/t > 40$, no more than 40t.

In any event, the distance center-to-center of successive stiffeners should not exceed 80t.

7.5.6 Vertical Walls. Horizontal and Vertical Stiffeners

Square and rectangular bins have side wall stiffeners arranged horizontally and vertically.

When the stiffeners are arranged horizontally, they will be subject to a uniformly distributed load, and may be designed either as simply supported beams trimmed into an extension of the corner support columns, or where this is not possible, as closed rigid frames.

In addition to the bending moments, the horizontal stiffeners will also be subject to direct tension due to pressure on the wall, and at right angles to the wall actually being designed. Side wall stiffeners may be designed to include a part of the skin plate, as given in Section 7.5.5.4.

Vertical stiffeners are designed under horizontal pressure having spans between the horizontal stiffening frames.

The bending moments for simply supported vertical stiffeners are

$$M = \frac{P_h h^2}{8} \qquad (7.104)$$

and for vertical stiffeners having fixed ends

$$M = \frac{P_h h^2}{12} \qquad (7.105)$$

where

P_h = uniformly distributed loading

h = span of the vertical stiffeners

The fixation of the ends of the vertical stiffeners cannot be considered as a full fixation due to possible torsional deformations of the horizontal stiffeners. For this reason, in designing the vertical stiffeners under bending, it is recommended that some average value of the bending moment, or

$$M = \frac{P_h h^2}{10} \qquad (7.106)$$

be used.

7.5.7 Walls Analysis of Multicell Bins

A structural analysis of multicell bins as space-type structures present one of the most complicated problems in statics. For a simplification of the structural analysis, a multicell bin is usually divided into separate units under certain assumptions.

In cases of rectangular cells having shapes which do not substantially differ from square shapes, a single cell may be analyzed without taking into consideration the influence of the adjacent cells. However, the influence of the intermediate connections considering the bending moments produced by the loadings in adjacent cells increases with the deviation of the square cell from the rectangular one.

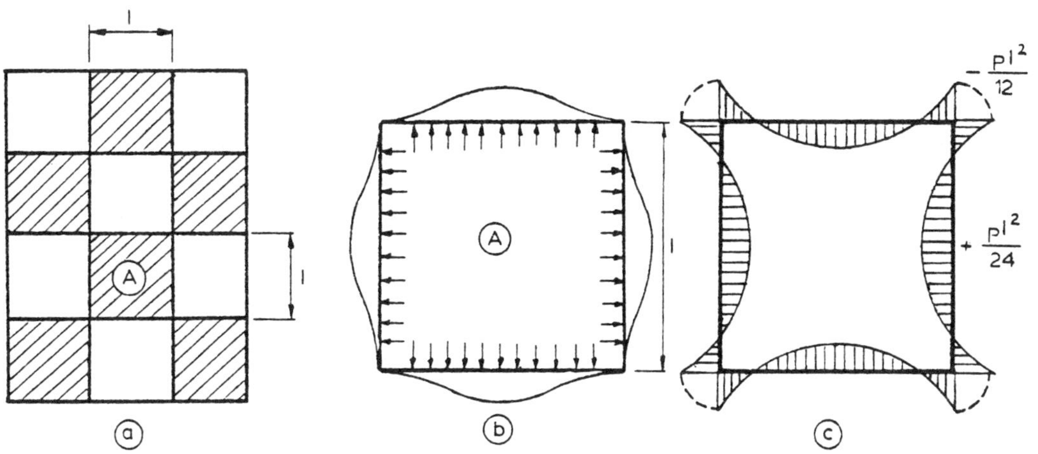

FIGURE 7.52 — Deformation and moment diaphragm of single cell in one multicell bin: a. Multicell bin; b. Deformations, c. Moment diagram.

All walls in one intermediate square cell A, Figure 7.52, indicate equal deflections at their mid-spans.

In this case, the corners of the walls do not undergo any deformations and do not influence the neighboring cells. Form this it follows that by an analysis of the square cell, the values of the corner and midspan moments may be expressed as follows [7.32]:

Moments in corners

$$M = -\frac{p\ell^2}{12} \qquad (7.107)$$

Moments in midspan

$$M = +\frac{p\ell^2}{24} \qquad (7.108)$$

The method used by Cross in comparison to other methods, is more practical in analyzing multicell rectangular bins. However, in this case, it is necessary for each loading condition to perform new iteration. In these calculations, it is necessary to introduce into the analysis the stiffnesses of the members.

It is known from experience that at multicell bins, having more than three cells, the influence of the loaded cells on the neighboring cells is relatively small.

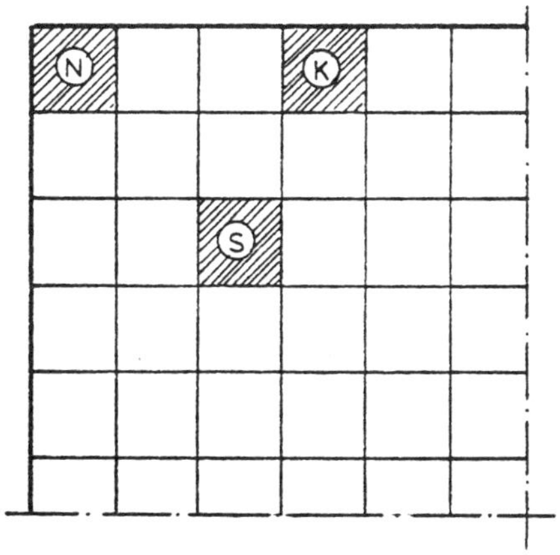

FIGURE 7.53 — Multicell bin.

Let us consider bins consisting of nine cells and introduce the following designations: N-corner cell, K-edge cell and S-inside cell, Figure 7.53.

We may obtain maximum values of the moments by solving the following scheme of a loaded nine-cell system, Figure 7.54.

The problem consists of the determination of four unknowns, using the Cross method.

Considering equal cells having the ratios for the walls

$$I_1/\ell_2 = I_2/\ell_1 \qquad (7.109)$$

where I_1 and I_2 are the moments of inertia of the walls, we may use Table 7.9 for analysis; M_1 and M_2 are span mo-

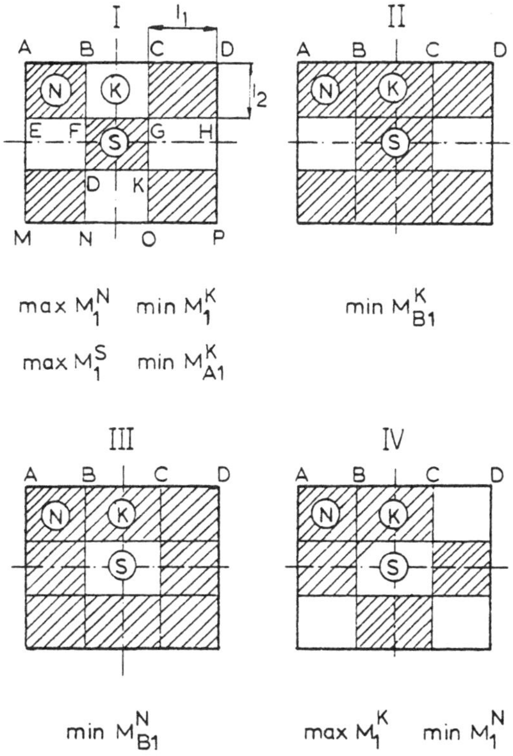

FIGURE 7.54 — Loading scheme for nine-cell bin.

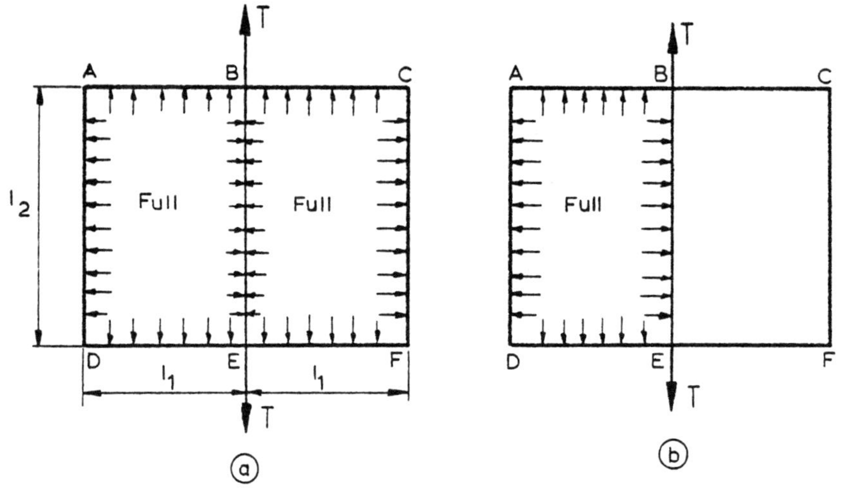

FIGURE 7.55 — Tension in multicell bin.

ments and M_{A_1}, M_{A_2}, and M_{B_1} are corner moments.

In the case of $\ell_1 = \ell_2$ and $I_1 = I_2$, the formulas shown in Table 7.9 are not exact, but may be used without great error if for $\ell_2 < \ell_1$, the condition $0.75 < \ell_2/\ell_1 < 1.0$ is satisfied.

The walls of the bins are not affected only by the bending, but are also stressed by the tension, Figure 7.55.

Under loading shown in Figure 7.55 (a), the tensile load in the wall is

$$T = \ell_1 p \tag{7.110}$$

and under the loading shown in Figure 7.55 (b) the tensile force is

$$T = \frac{\ell_1 p}{2} \tag{7.111}$$

In the first case the bending moment in this wall is equal to zero. The advantage of the analysis of a sqaure

TABLE 7.9 — Maximum Values of the Corner and Span Moments for the Most Dangerous Loadings of Multicell Bins (ℓ_1, $= \ell$, $\alpha = \ell_2/\ell_1$).

Inside Cell	Edge Cell	Corner Cell
	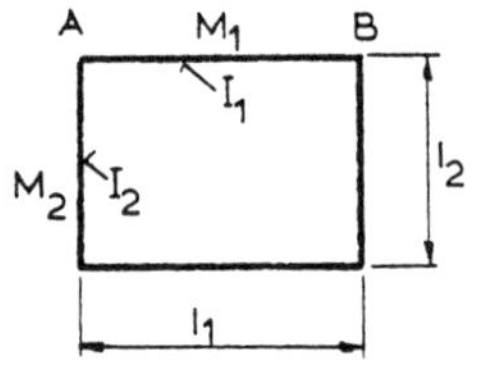	
$\max M_1 = \frac{p\ell^2}{24}\,\frac{48-23\alpha^2}{25}$	$\max M_1 = \frac{p\ell^2}{24}\,\frac{41-16\alpha^2}{25}$ $\min M_1 = \frac{-p\ell^2}{24}\,\frac{19(1-\alpha^2)}{25}$	$\max M_1 = \frac{p\ell^2}{24}\,\frac{91-41\alpha^2}{50}$ $\min M_1 = \frac{-p\ell^2}{24}\,\frac{24(1-\alpha^2)}{25}$
$\min M_{A_1} = \frac{-p\ell^2}{12}\,\frac{12-\alpha^2}{11}$	$\min M_{A_1} = \frac{-p\ell^2}{12}\,\frac{28-3\alpha^2}{25}$	$\min M_{A_1} = \frac{-p\ell^2}{12}\,\frac{1+\alpha^2}{2}$ $\min M_{B_1} = \frac{-p\ell^2}{12}\,\frac{61-11\alpha^2}{50}$
$\max M_2 = \frac{p\ell^2}{24}\,\alpha^2$	$\min M_2$ — as for the inside cell	$\max M_2$ — as for the inside cell $\min M_2 > 0$
$\min M_{A_2} = \frac{-p\ell^2}{12}\,\frac{1+\alpha^2}{2}$	$\min M$ — as for the A_2 inside cell	$\min M$ — as for the A_2 inside cell

bin is that the maximum tensile force does not work simultaneously with the maximum bending moment. The above-stated assumptions for rectangular bins stiffened by frames have actually not agreed with the main stresses.

Let us compare the deformations in horizontal planes for two closed strips of equal depth dh so arranged that the deflections of the top, with the horizontal pressure p loaded strips are shorter than the bottom loaded with the pressure p + dp. Due to this deflection differences appear between the closed strips, transverse forces and moments in a lateral direction.

7.5.8 Hoppers. Determination of Dimensions

In the following we are considering generally, the case of a bin having unsymmetrical hopper, Figure 7.56.

The volume of both the bin and the hopper is given by the following formula

$$V = a_2 b_2 h_2 + \frac{h_1}{6}\left[(2a_2 + a_1)b_2 + (2a_1 + a_2)b_1\right] \qquad (7.112)$$

and the coordinates of the center of gravity of the hopper are

$$x_c = x_0 h_1 \frac{(a_1 + a_2)(b_1 + b_2) + 2a_1 b_1}{12V} \qquad (7.113)$$

$$y_c = y_0 h_1 \frac{(a_1 + a_2)(b_1 + b_2) + 2a_1 b_1}{12V} \qquad (7.114)$$

where

x_0, x_0 = are distances in plan between the centers of gravities of the upper and bottom areas of the hopper.

c_1, c_2, c_3, c_4 = distances between the edges of the hopper and sides of the opening

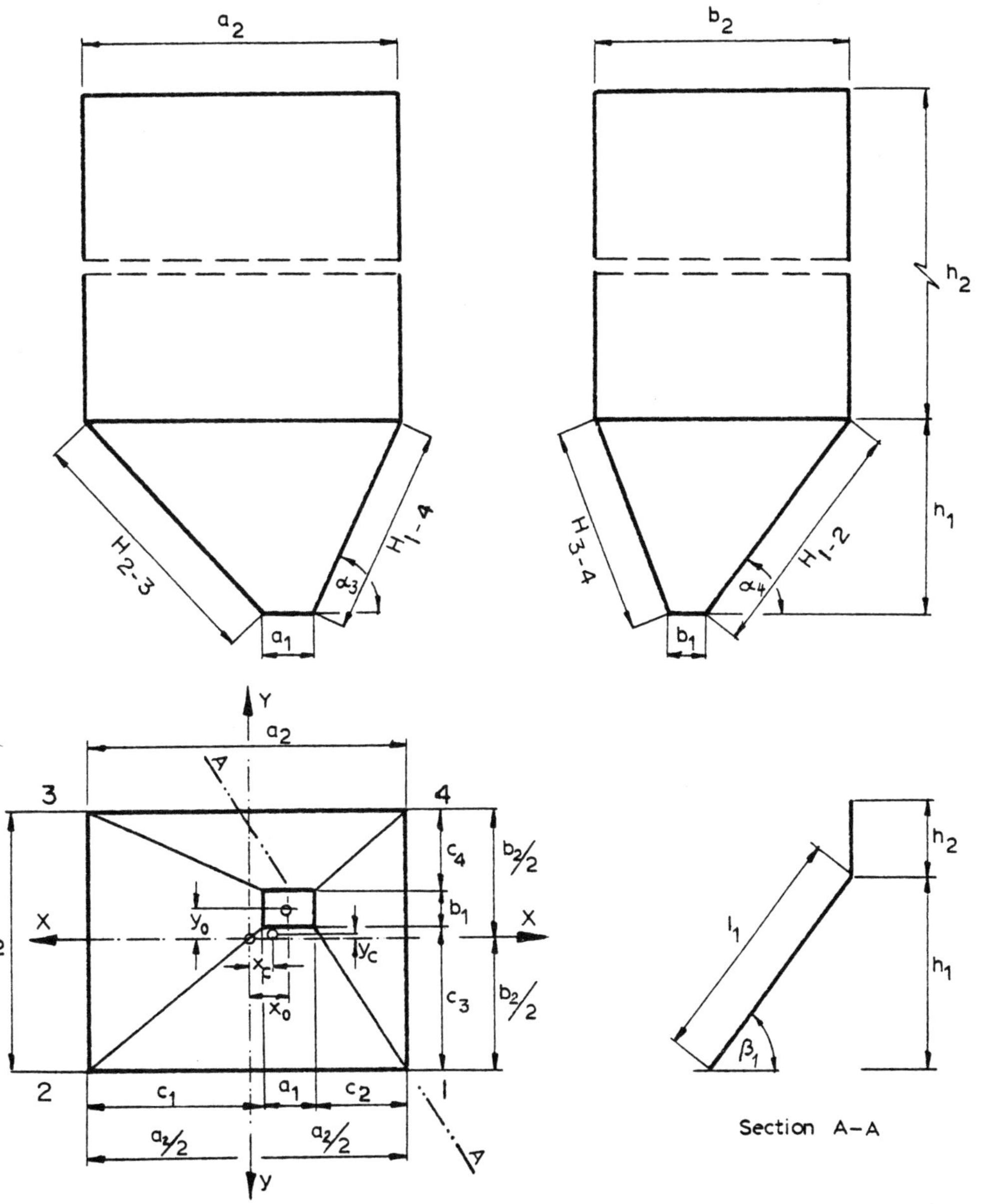

FIGURE 7.56 — Geometry of the bunker and hopper.

The depth H_n and the angle of the inclination of the wall of the hopper α_n are determined from the formulas

$$H_n = \sqrt{h_1^2 + C_n^2} \qquad (7.115)$$

$$\tan \alpha_n = \frac{h_1}{C_n} \qquad (7.116)$$

where the symbol n indicates an arbitrary wall under consideration.

The length of the edge along intersection of two walls n and m as well as the angle of inclination are

$$\ell_{nm} = \sqrt{h_1^2 + C_n^2 + C_m^2} \qquad (7.117)$$

$$\tan \beta_{nm} = \frac{h_1}{\sqrt{C_n^2 + C_m^2}} \qquad (7.118)$$

The angle betwen two intersecting walls n and m is estimated using the following formula

$$\alpha_{nm} = 180^\circ - \beta \qquad (7.119)$$

where angle β may be found from the formula

$$\cos\beta = \cos\alpha_n \cos\alpha_m \qquad (7.120)$$

7.5.9 Forces Acting on Hopper Walls

Pyramidal hopper walls are subject to bending as well as tensile membrane forces. The bending always includes two-way, plate-type bending and may include insignificant in-plane bending.

All of the following is for symmetrical or nearly symmetrical hoppers only. The angles of slope for walls "a" and "b" are α_a and α_b respectively.

The membrane tensile forces in the inclined walls are assumed to consist of horizontal forces, T_h, and forces, T_m, in the plane of the wall and normal to T_h, Figure 7.57.

Forces T_m per unit width may vary in intensity along the width of the wall. For simplicity, however, they are usually assumed to be uniform along any one wall.

Material pressures for each hopper wall are assumed as those computed using the properties of the bin wall to which it is attached, plus a share of the dead load.

7.5.9.1 Design Pressure Normal to the Inclined Walls

The design pressure P_n may be computed from Equation (7.55)

$$P_{n,des} = P_{h,des} \sin^2\alpha + P_{v,des}\cos^2\alpha$$

In computing P_n designers usually use the dimensions at the top of the hopper to obtain R or L for the Janssen (7.32) or Reimbert (7.33) formulas, ignoring the reduction of cross-section within the hopper, Figure 7.57.

When angles of slope are α_a and α_b, the design pressures are

$$P_{na} = P_h \sin^2\alpha_a + P_v \cos^2\alpha_a \qquad (7.121)$$

$$P_{nb} = P_h \sin^2\alpha_b + P_v \cos^2\alpha_b \qquad (7.122)$$

Normal components of the own weight of the walls, Figure 7.58, also produce normal forces acting on the inclined walls.

The normal component due to the own weight is

$$g_n = g \cos\alpha \qquad (7.123)$$

where

g = weight of wall per square unit

α = inclination angle of the wall

7.5.9.2 Horizontal Tensile Forces

The horizontal tensile forces T in the hopper walls are calculated per unit width of the hopper wall, Figure 7.59.

The horizontal tensile forces in the hopper wall are

$$T_a = \frac{1}{2} P_{nb} \sin\alpha_b \qquad (7.124)$$

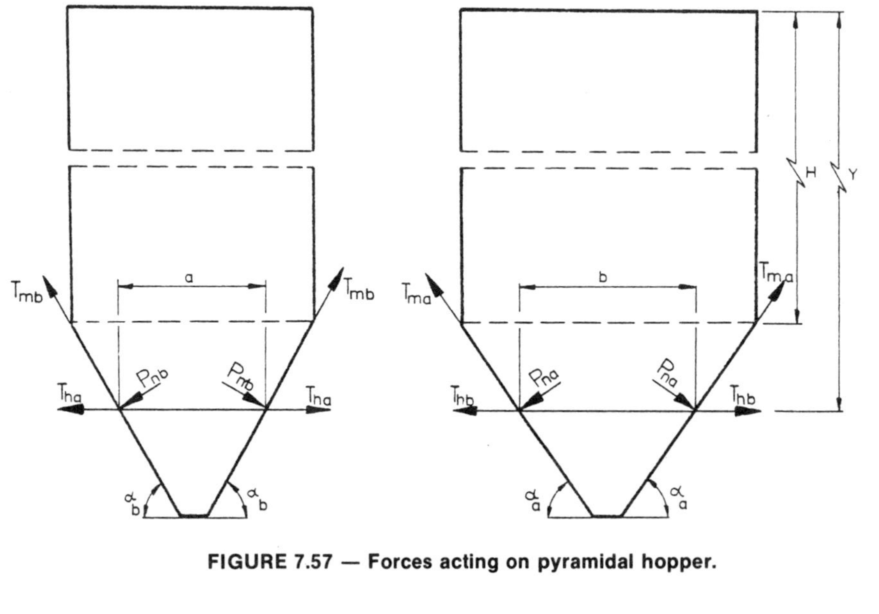

FIGURE 7.57 — Forces acting on pyramidal hopper.

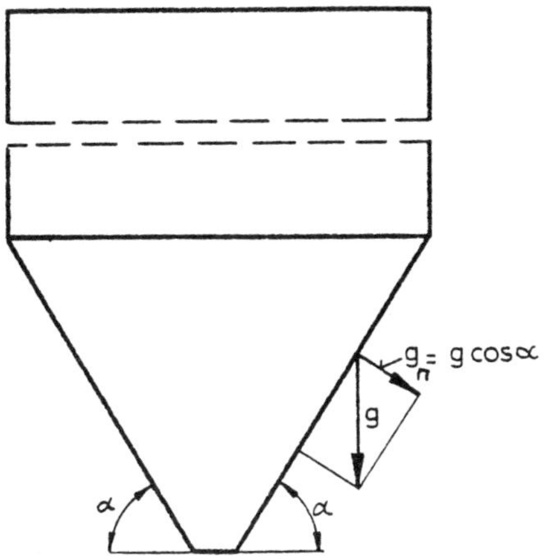

FIGURE 7.58 — Load due to the own weight of the hopper's wall.

$$T_b = \frac{1}{2} P_{na} \sin\alpha_a \qquad (7.125)$$

where

P_{na}, P_{nb} = average normal pressures on the walls "a" and "b"

α_a, α_b = angles of inclination of the walls "a" and "b"

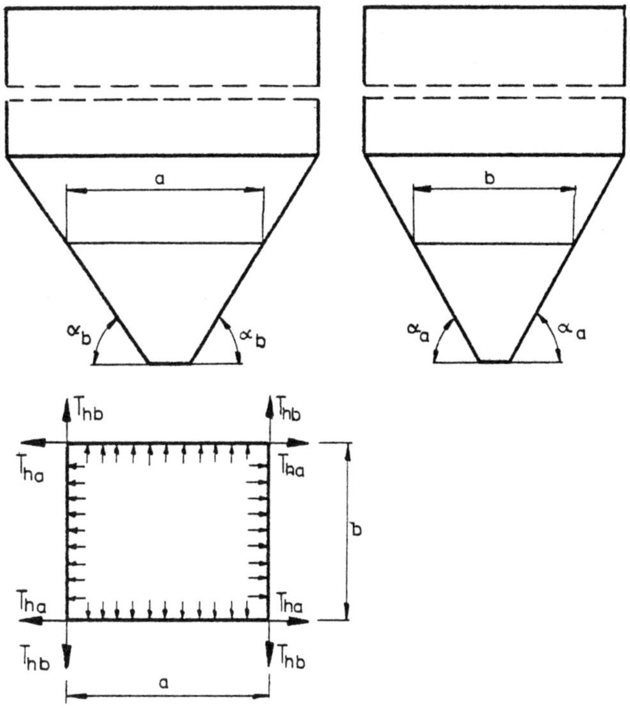

FIGURE 7.59 — Horizontal tensile forces.

7.5.9.3 Meridional Tensile Forces at Symmetrical Hopper

The hopper walls are also under tensile forces along their inclined lengths. These forces tend to separate the hopper from the upper vertical walls of the bin. These tensile forces are determined considering the load action shown in Figure 7.69 [7.33].

In the hopper material pressures applied to area Aa are assumed to affect the value T_{ma} for wall "a", and those applied to area A_b to affect T_{mb} for wall "b".

If the area of the discharge opening is ignored and if some arbitrary factors C_a and C_b are selected to define

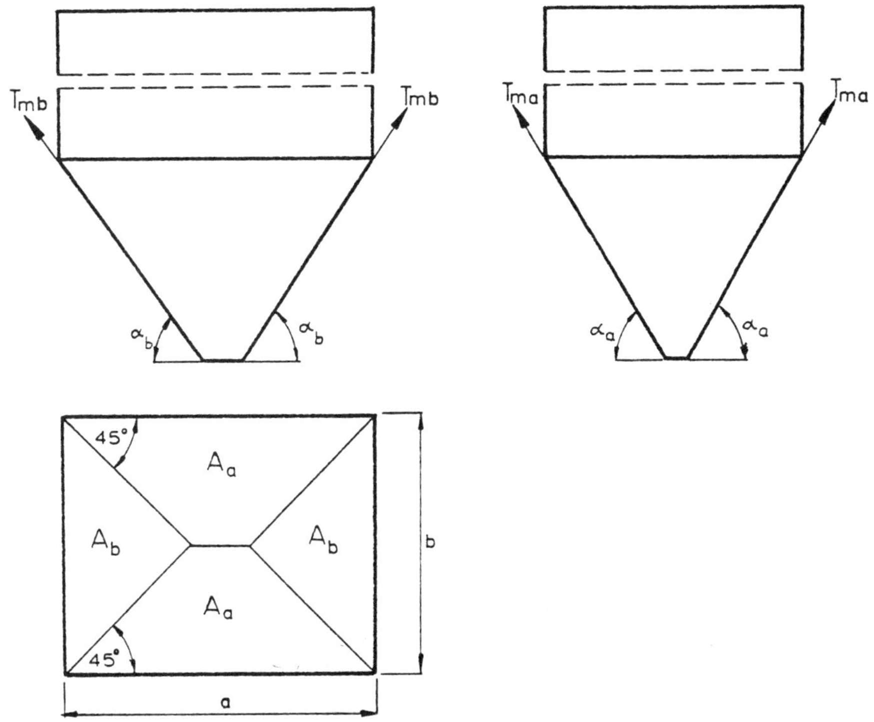

FIGURE 7.60 — Meridional tensile forces at symmetrical hopper.

the divison of the load to areas A_a and A_b such that, for a symmetrical case

$$2C_a + 2C_b = 1.0 \tag{7.126}$$

then the maximum meridional forces per unit width of wall are

$$T_{ma} = \frac{C_a W + A_a P_{a,des} + C_a G}{a \sin\alpha_a} \tag{7.127}$$

$$T_{mb} = \frac{C_b W + A_b P_{b,des} + C_a G}{b \sin\alpha_b} \tag{7.128}$$

In the above formulas, $P_{a,des}$ is computed using properties for the short wall and $P_{b,des}$ using an imaginary square bin. For simplicity, sometimes the average unit for $P_{a,des}$ and $P_{b,des}$ is used for all walls of the hopper.

W = total weight of stored material in a hopper and bin

G = own weight of the hopper

7.5.9.4 *Meridional Tensile Forces at Unsymmetrical Hopper*

With unsymmetrical hoppers, Figure 7.61, the tensile forces are nonuniform along the perimeter due to the influence of the weights of the fill.

For the determination of tensile forces T in the hopper walls the following formulas are used [7.34]

$$T_1 = \frac{W + G}{2(a+b)\sin\alpha_1} t_x t_y \tag{7.129}$$

$$T_2 = \frac{W + G}{2(a+b)\sin\alpha_2} (2-t_x) t_y \tag{7.130}$$

$$T_3 = \frac{W + G}{2(a+b)\sin\alpha_3} (2-t_x)(2-t_y) \tag{7.131}$$

$$T_4 = \frac{W + G}{2(a+b)\sin\alpha_4} t_x (2-t_y) \tag{7.132}$$

where

$\alpha_1, \alpha_2, \alpha_3, \alpha_4$ = angles of inclinations of the walls

t_x, t_y = distribution coefficient given in Table 7.10.

TABLE 7.10 — Values of the Coefficient t_x and t_y.

$\frac{d_a}{a}$	Coefficient t_x			$\frac{d_b}{b}$	Coefficient t_y		
	$\frac{a}{b}$ = 1.0	$\frac{a}{b}$ = 1.5	$\frac{a}{b}$ = 2.0		$\frac{a}{b}$ = 1.0	$\frac{a}{b}$ = 1.5	$\frac{a}{b}$ = 2.0
0.30	1.600	1.666	1.720	0.30	1.600	1.545	1.514
0.35	1.450	1.500	1.540	0.35	1.450	1.409	1.386
0.40	1.300	1.333	1.360	0.40	1.300	1.272	1.257
0.45	1.150	1.166	1.180	0.45	1.150	1.136	1.129
0.50	1.000	1.000	1.000	0.50	1.000	1.000	1.000
0.55	0.850	0.834	0.820	0.55	0.850	0.864	0.871
0.60	0.700	0.667	0.640	0.60	0.700	0.728	0.743
0.65	0.550	0.500	0.460	0.65	0.550	0.591	0.614
0.70	0.400	0.334	0.280	0.70	0.400	0.455	0.485

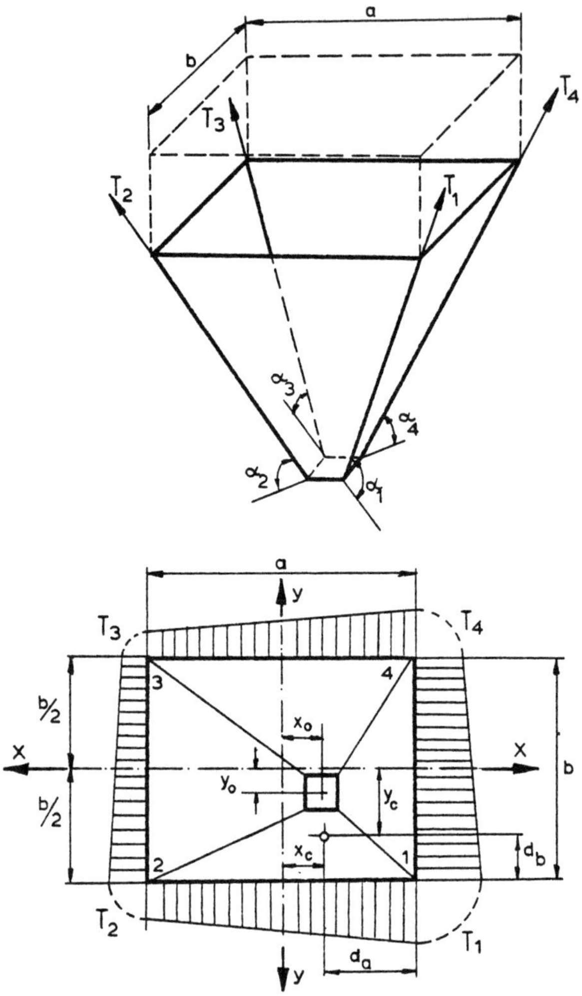

FIGURE 7.61 — Meridional tensile forces at unsymmetrical hopper.

7.5.10 Hoppers. Design of Plating

7.5.10.1 *Average Pressures on Hopper Panels*

The normal pressure P_n acting on some panels of the hopper wall will change along the depth of the hopper. However, for practical analysis it is assumed that the normal pressure along the depth is uniformly distributed. In this case it is possible to use formulas given below for the determination of average normal pressures P_n on the panels [7.29].

a. *Rectangular panel*

The average uniformly distributed equivalent pressure, Figure 7.62, is

$$P_n = \frac{P_n' + P_n''}{2} \tag{7.133}$$

b. *Triangular panel*

The average uniformly distributed equivalent pressure, Figure 7.63, is

$$P_n = \frac{P_n' + 2P_n''}{3} \tag{7.134}$$

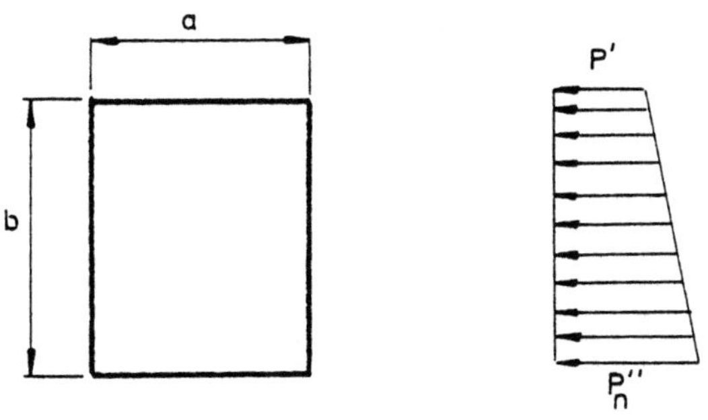

FIGURE 7.62 — Rectangular panel.

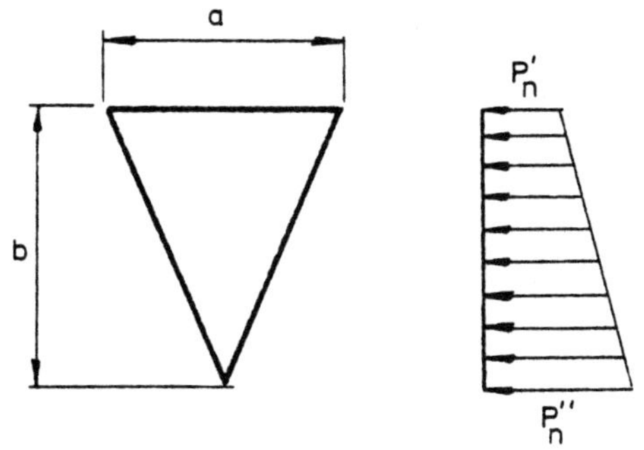

FIGURE 7.63 — Triangular panel.

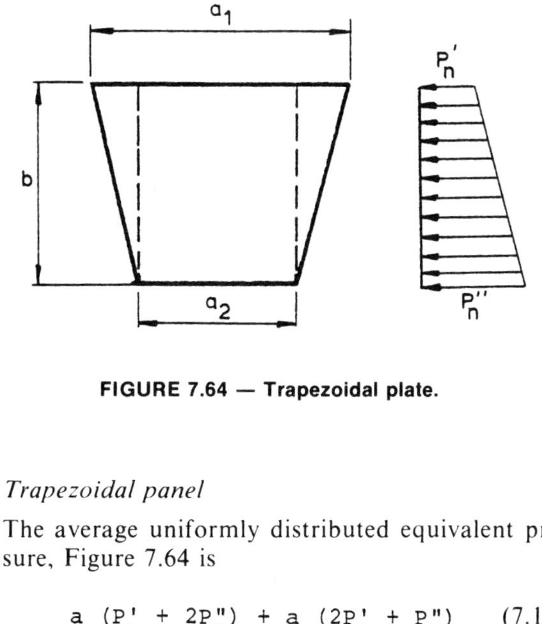

FIGURE 7.64 — Trapezoidal plate.

c. *Trapezoidal panel*

The average uniformly distributed equivalent pressure, Figure 7.64 is

$$P_n = \frac{a_1(P_n' + 2P_n'') + a_2(2P_n' + P_n'')}{3(a_1 + a_2)} \qquad (7.135)$$

7.5.10.2 *Converting of Triangular and Trapezoidal Panels into Rectangular Panels*

Plating of hoppers supported by stiffeners may form panels of triangular or trapezoidal shapes. However, formulas for normal pressures on hopper walls may be used only for rectangular panels. It is possible to use this formula for the approximate design of triangular and trapezoidal panels only after converting such panels into rectangular shape. Examples of such converting are given as follows [7.29].

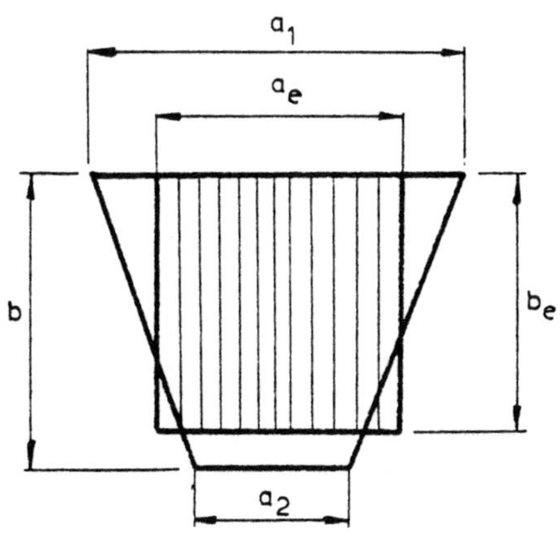

FIGURE 7.65 — Conversion of the triangle into an equivalent rectangle.

a. *Converting of triangular panel into equivalent rectangular, Figure 5.22*

Line EF passes through the center of gravity of the triangular parallel to the side AB. A length of the line EF defines the width of the rectangle, or

$$a_e = \frac{2}{3}a \qquad (7.136)$$

The height of the rectangle is

$$b_e = b - \frac{a}{6} \qquad (7.137)$$

Such transformation is possible only for an equilateral or equivalent triangle. In general cases, the areas of the rectangle, obtained after transformation of the triangle are not equal.

According to formulas (7.101) and (7.102), the ratio of the areas of the triangle and the rectangle is

$$\frac{A_t}{A_r} = \frac{ab/2}{a_e b_e} = \frac{a^2\tan\alpha/4}{a^2\tan\alpha/3 - a^2/9} = \qquad (7.138)$$

$$= \frac{9}{4} \cdot \frac{\tan\alpha}{3\tan\alpha - 1}$$

we obtain

$$\tan\alpha = \frac{4}{3} \text{ and } b = \frac{2}{3}a$$

At an equilateral triangle, this angle is

$$\alpha = 53° \; 5'$$

This angle should be taken as an average and most often used in the rectangular bins.

Assuming some close values of the angle α, then from formula (7.138), the ratios of the triangular and rectangular areas will be

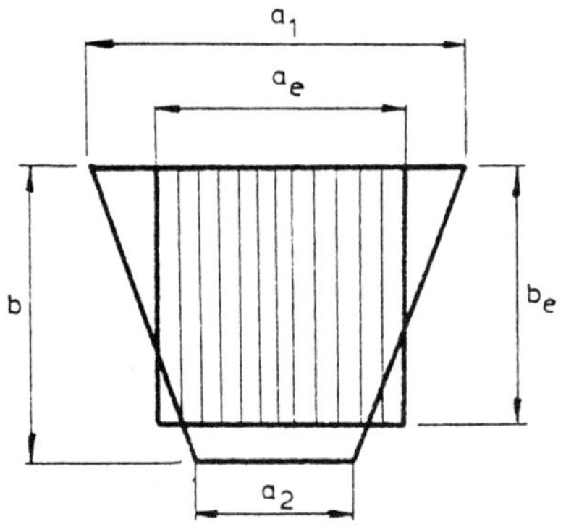

FIGURE 7.66 — Conversion of the trapezoidal panel into an equivalent rectangle.

At $\alpha = 65^\circ$ $\quad \frac{A_t}{A_r} \approx 0.89$

$\alpha = 53^\circ 5'$ $\quad \frac{A_t}{A_r} = 1.0$

$\alpha = 45^\circ$ $\quad \frac{A_t}{A_r} \approx 1.12$

The above ratios indicate a possible application of the areas obtained after transformation.

b. *Conversion of the trapezoidal panel into an equivalent rectangular, Figure 7.66*

The width of the rectangle is

$$a_e = \frac{2}{3} \cdot \frac{(a_1 + 2a_2)a_1}{a_1 + a_2} \tag{7.139}$$

and the height is

$$b_e = b - \frac{1}{6} \cdot \frac{(a_1 - a_2)a_1}{a_1 + a_2} \tag{7.140}$$

Using formulas (7.136) and (7.137) or (7.139) and (7.140) for the determination of the values a_e and b_e, it is possible to design the plating of the hopper as rectangular plates.

7.5.11 Hoppers. Horizontal and Inclined Stiffeners

7.5.11.1 Horizontal Stiffeners

Horizontal stiffeners are subject to uniformly distributed loads and are designed as closed frames. In addition to bending moments, horizontal stiffeners will also be subject to direct tension due to the pressure on the wall at right angles to the wall actually being designed.

The total stress f in the stiffener is determined after conventional formula

$$f = \frac{T}{A} + \frac{M}{W} \tag{7.141}$$

where

T = tensile force

M = bending moment

A = cross-sectional area of the stiffener

W = section-modulus of the stiffener

7.5.11.2 Inclined Stiffeners

Inclined stiffeners, apart from tensile forces, are also designed considering the bending moment pressure of the fill on the walls.

The bending moments for simply supported inclined stiffeners is

$$M = \frac{P_n \ell^2}{8} \tag{7.142}$$

and for stiffeners having fixed ends

$$M = \frac{P_n \ell^2}{12} \tag{7.143}$$

where

P_n = loading normal to the wall

ℓ = span of inclined stiffener

The fixity of the ends of the inclined stiffeners cannot be considered as full fixity, because of the possible torsional deformations of the horizontal stiffeners. For this reason, the design of the inclined stiffeners under bending, it is recommended that the use of some average values of the bending moment be made, or

$$M = \frac{P_n \ell^2}{10} \tag{7.144}$$

7.5.12 Procedure for the Design of Bins

On the basis of detailed structural analyses discussed in this Chapter, the following procedure is recommended for the design of bins:

1. Determine the properties of the material to be stored.
2. Establish the basic bin size and supporting system.
3. Determine the necessary geometry data-angle of the inclined walls, wall sizes, center of gravity of material, etc.
4. Determine the design pressures of granular material on the walls and bottom, considering the dynamic effects, if any. Sketch a pressure diagram.
5. Determine the tensile forces in all vertical and inclined walls.
6. Determine the bending moments in the vertical and inclined walls.
7. Determine the moments and shears due to in-plane bending.
8. Design the bin components:

   a. plating
   b. stiffening frame

7.6 Square and Rectangular Bunkers. Structural Analysis and Design

7.6.1 General Concept

Square and rectangular bunkers may be symmetrical or nonsymmetrical about either, or both, principal axes, according to the location of their discharge opening. Rectangular bunkers usually have pyramidal bottoms with one or more square, circular or elongated discharge openings.

A bunker may consist of a pyramidal hopper only, without the vertical walls. In this case a continuous horizontal supporting beam is located along the upper edge and resists both the vertical and horizontal forces and torsion applied by the hopper shell.

7.6.2 Structural Analysis

The forces set up in a rectangular bunker will induce direct, shear and bending stresses in the various parts. Because of their shape, it is not a practical solution to accommodate these forces on the skin plating only, which may become very thick and, therefore, result in uneconomical design.

Usually, the skin plating, which forms the bunker wall, is designed to span between suitably positioned stiffeners, which may be either *vertical* or *horizontal* depending upon the design approach. Normally these stiffeners are arranged to act in the *vertical plane*. Skin plating may be treated as a beam, either simply supported or continuous between the stiffeners, which in turn also acts as beams, spanning at right-angles to the designed direction of the skin plate [7.35].

7.6.3 Symmetrical Bunkers

7.6.3.1 Forces Acting on Side Walls

Figure 7.67 shows a typical cross-section through a rectangular symmetrical bunkers. For purposes of analysis [7.36, 7.37] only the unit length will be considered. Because the cross-section is symmetrical about the center, this analysis will be based upon one half only.

1. *The vertical forces*

The load in the prismatic section

$$W_1 = \gamma h_1 a \tag{7.145}$$

The load in the hopper section

$$W_2 = \gamma h_2 a/2 \tag{7.146}$$

The load immediately above the bottom plate or grate

$$W_3 = \gamma (h_1 + h_2) b \tag{7.147}$$

2. *The horizontal forces*

$$L_1 = \frac{\gamma h_1^2}{2} K \tag{7.148}$$

$$L_2 = \gamma h_1 h_2 K \tag{7.149}$$

$$L_3 = \frac{\gamma h_2^2}{2} K \tag{7.150}$$

where

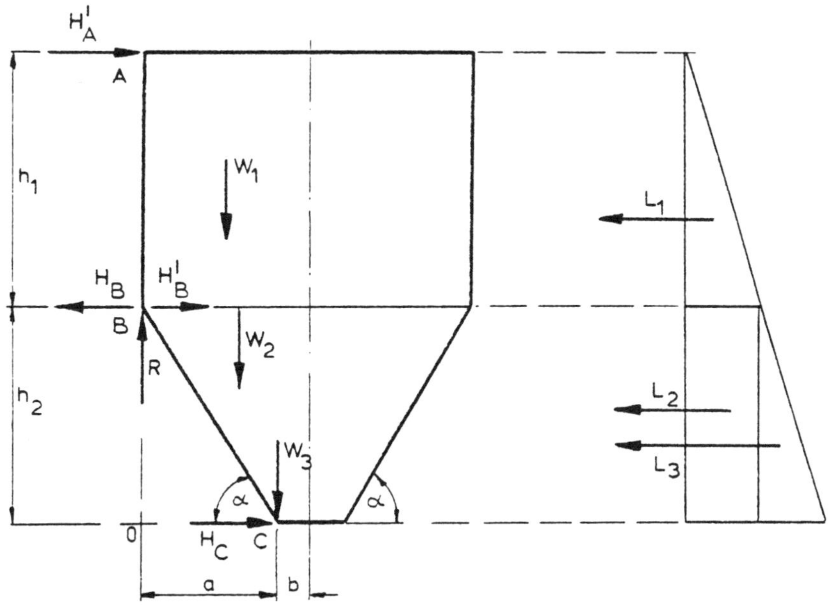

FIGURE 7.67 — Symmetrical bunker. Forces.

$$K = \frac{1-\sin\phi}{1+\sin\phi} \qquad (7.151)$$

is the ratio of lateral-to-vertical pressures usually taken as Rankine's factor

ϕ = angle of repose

γ = specific weight of the fill

Under the action of the vertical forces W_1, W_2 and W_3 and horizontal forces L_2 and L_3 there will be a tendency for the bunker to burst at the level of the discharge opening. This bursting forces has the value of H_C in Figure 7.67. Likewise, the force H_B may be looked upon as a force which is trying to cause collapse across the waist of the bunker.

These two forces may be found as follows:

For H_B taking moments about point C, then

$$H_B = \frac{1}{h_2}\left[W_1\frac{a}{2} + W_2\frac{2a}{3} + W_3 a - L_2\frac{h_2}{2} - L_3\frac{h_3}{3}\right] \qquad (7.152)$$

and for H_C by taking moment about point B

$$H_C = \frac{1}{h_2}\left[W_1\frac{a}{2} + W_2\frac{2a}{3} + W_3 a + L_2\frac{h_2}{2} + L_3\frac{2h_3}{3}\right] \qquad (7.153)$$

Above the waist, the two forces causing bursting action are

$$H'_A = \frac{L_1}{3} \qquad (7.154)$$

$$H'_B = \frac{2L_1}{3} \qquad (7.155)$$

7.6.3.2 Forces Acting on Hopper Walls

The total pressure on the sloping wall of the hopper will be in the area of the pressure diagram between B and C, Figure 7.68.

The total normal load acting on the sloping wall is

$$R_N = \left(\frac{P_{nC} + P_{nB}}{2}\right) h_2 \operatorname{cosec}\alpha \qquad (7.156)$$

and acts through the center of gravity of the pressure diagram as shown in Figure 7.68.

The value of the normal pressure according to formula (7.51) is

$$P_n = P_v \cos^2\alpha + P_h \sin^2\alpha \qquad (7.157)$$

The vertical and horizontal reactions at B and C may be resolved to give the normal and tangential components required to design the hopper stiffeners, shown in Figure 7.69.

The normal reaction or shear at B is

$$R_B = R\cos\alpha - H_B\sin\alpha \qquad (7.158)$$

and the tangential component of tension at B is

$$T_B = R\sin\alpha + H_B\cos\alpha \qquad (7.159)$$

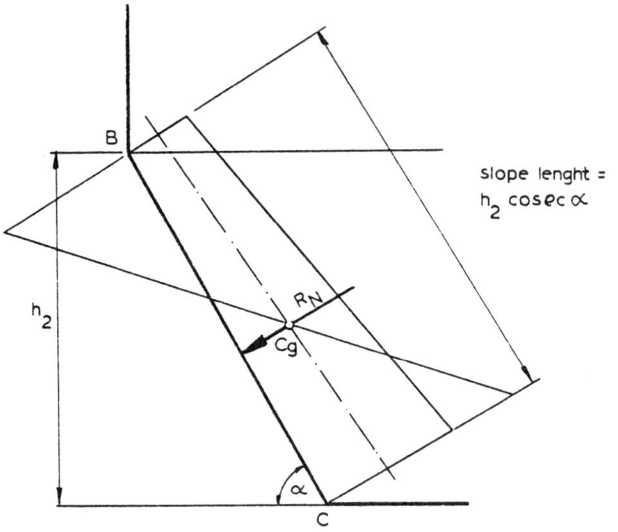

FIGURE 7.68 — Pressure on the sloping wall of the hopper.

Similarly,

$$R_C = H_C\sin\alpha - W_3\cos\alpha \qquad (7.160)$$

$$T_C = H_C\cos\alpha + W_3\sin\alpha \qquad (7.161)$$

The final forces and pressures necessary to enable strength calculation performances are shown in Figure 7.70.

7.6.4 Unsymmetrical Bunkers

7.6.4.1 Forces Acting on Walls

Figure 7.71 shows a typical cross-section through an unsymmetrical rectangular bunker about a vertical center line, passing through the discharge opening.

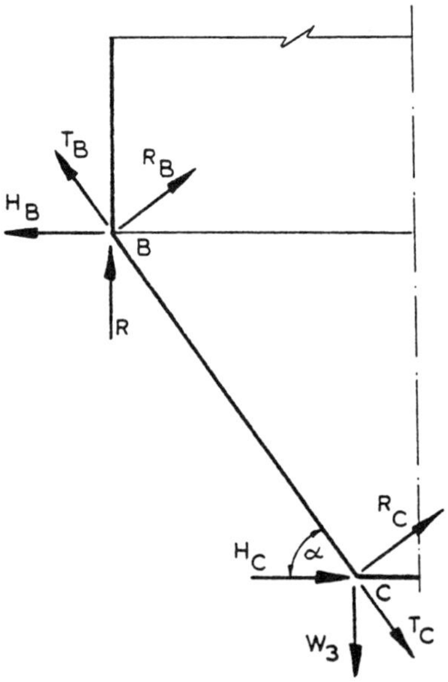

FIGURE 7.69 — Components of reactions at B and C.

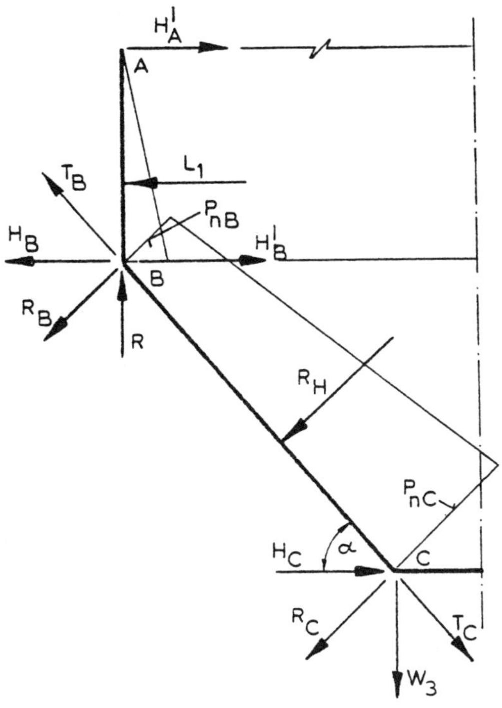

FIGURE 7.70 — Resulting forces acting on the walls of a bunker.

Above the waist BE the horizontal forces will have exactly the same meaning as already outlined for the symmetrical case, that is

$$H'_A = H'_F = \frac{L_1}{3} \tag{7.162}$$

$$H'_B = H'_E = \frac{2L_1}{3} \tag{7.163}$$

where

$$L_1 = \frac{wh_1^2}{2} K_1 \tag{7.164}$$

Below the waist the forces are

$$H_B = \frac{1}{h_2} [W\frac{a}{2} + W_2\frac{2a}{3} + W_3 a - L_2\frac{h_2}{2} - L_3\frac{h_2}{3}] \tag{7.165}$$

$$H_C = \frac{1}{h_2} [W_1\frac{a}{2} + W_2\frac{2a}{3} + W_3 a + L_2\frac{h_2}{2} + L_3\frac{2h_2}{3}] \tag{7.166}$$

Similarly, for the less steep side

$$H_E = \frac{1}{h_2} [W_4\frac{a'}{2} + W_5\frac{2a'}{3} + W_3 a' - L_2\frac{h_2}{2} - L_3\frac{h_2}{3}] \tag{7.167}$$

$$H_D = \frac{1}{h_2} [W_2\frac{a'}{2} + W_3\frac{2a'}{3} - W_3 a' + L_2\frac{h_2}{2} + L_3\frac{2h_2}{3}] \tag{7.168}$$

The vertical reactions are

$$R_{VB} = W_1 + W_2 + W_3 \tag{7.169}$$

$$R_{VE} = W_4 + W_5 + W_3 \tag{7.170}$$

The reactions R_{VB} and R_{VE} are *not* simply supported beam reactions. This is because there is an out-of-balance reaction at the outlet having the value

$$H = H_D - H_C \tag{7.171}$$

Below the waist, BE the horizontal forces, must be in equilibrium, that is

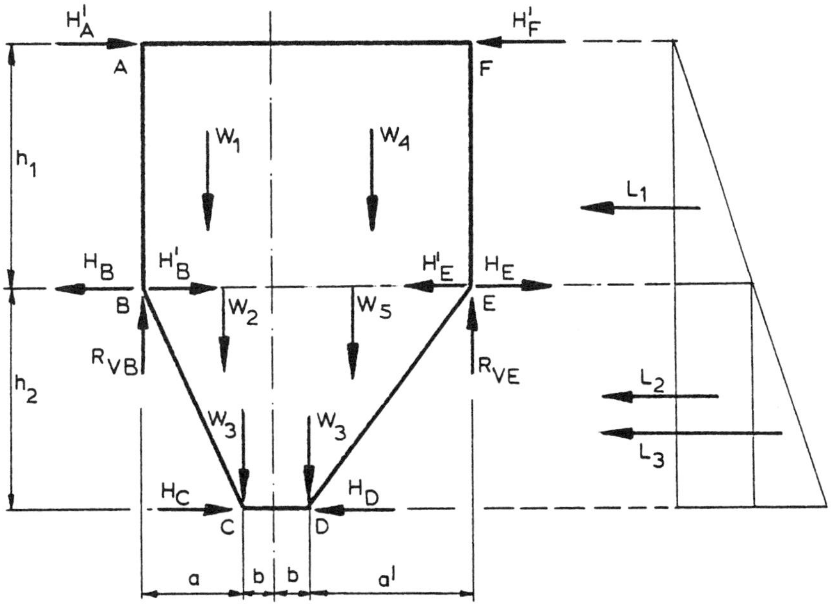

FIGURE 7.71 — Unsymmetrical bunker. Forces.

$$H_B + L_2 + L_3 - H_C = 0$$

$$H_E + L_2 + L_3 - H_D = 0$$

$$H_D - H_C = H_E - H_B \qquad (7.172)$$

The procedure for calculating the normal and tangential pressures and reactions follows the methods given for the symmetrical bunker.

7.6.5 Side Wall Design

The side walls are designed to accommodate the side pressure, horizontal reactions from the hopper and self weight, Figure 7.72.

The walls are designed for side pressure only with the skin stiffeners arranged either vertically or horizontally as best suits the layout. The hopper reactions may then be carried on a beam of reasonable dimension situated immediately at the bottom of the wall. This beam may be part of the wall. Where the side wall is of such proportions that the simple beam theory may be used, the wall is treated as a special form of plate girder.

Stiffeners are usually used, and the vertical stiffeners are subject to a triangular load distribution and must resist a bending moment

$$M_{max} = \frac{1}{16} ph^2 \qquad (7.173)$$

when they extend for the full uninterrupted depth of the wall. If they are broken due to the presence of horizontal stiffeners the bending moment for each separate length will be the appropriate value taken from Table 7.11.

Square or near square bunkers may have stiffeners arranged either vertically or horizontally.

If the stiffeners are arranged horizontally they will be subject to a uniformly distributed load, and may be designed either as simply supported beams trimming into an extension of the corner support column, or where this is not possible as closed frames.

At rectangular bunkers the walls are generally under action of bending and tension. Bending moments in horizontal cross-section of the bunker are determined using formulas from (7.82) to (7.90) and tensile forces are determined using formulas (7.80) and (7.81).

Side wall stiffeners should be designed to include part of the skin plate, as given in Section 7.5.5.4.

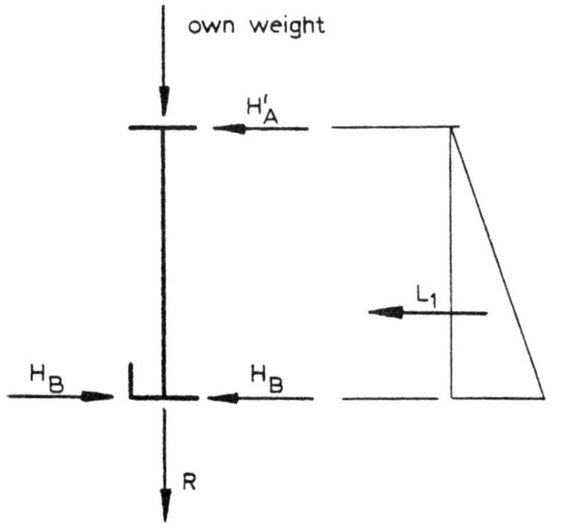

FIGURE 7.72 — Loadings on side wall.

7.6.5.1 Side Wall Plating

Side wall plating is designed in the following procedure given in this Chapter, Section 7.5.5, after chasing the spacing of the vertical stiffeners.

The wall plating should be checked for shear under vertical loading carried by the wall plate, or

$$F_s = \frac{R\ell}{ht} \qquad (7.174)$$

where

- R = vertical reaction at the waist
- ℓ = spacing between vertical frames
- h = depth of the side wall
- t = thickness of the plating

Also, plating should be checked against possible crippling using formula given by Timoshenko [7.38]

$$f_{cr} = \frac{K\pi^2 D}{b^2 t} \qquad (7.175)$$

where

- D = $\frac{Et^3}{12(1 - \nu^2)}$ = the flexural rigidity of the plate
- K = $5.35 - 4(b/h)^2$ = a factor
- b = spacing between vertical stiffeners
- E = 30×10^6 psi = modulus of elasticity of steel plating
- ν = 0.3 = Poisson's ratio

The factor of safety against crippling is

$$F.S = \frac{f_{cr}}{f_s} \qquad (7.176)$$

7.6.5.2 Side Wall Stiffeners

a. *Vertical stiffeners*
After the spacing of the vertical stiffeners has been chosen, the stiffeners are designed including the effective width of the plating 40t, as discussed in Section 7.5.5.4. The bending moment is used according to formula (7.172).

b. *Horizontal stiffeners at level A*
Horizontal stiffeners at level A are designed considering horizontal bursting load due to the horizontal pressure

$$H'_A = L_1/3$$

Figure 7.67.

The bending moment is

$$M = \frac{H'_A \ell^2}{8} \qquad (7.177)$$

where ℓ is spacing between vertical frames supporting the bunker.

c. *Horizontal stiffeners at level B*
Horizontal stiffeners at level B resist the bending moment in the horizontal plane

$$M = \frac{(H_B - H_B')\ell^2}{8} \quad (7.178)$$

Also the vertical flange of the horizontal stiffener should be checked for transverse tension under reaction R.

7.6.6 Hopper Wall Design

7.6.6.1 Hopper Wall Plating

Hopper wall plating is designed considering the maximum normal pressure at C, P_{nC}.

Considering the continuity of the wall, the bending moment between stiffeners at spacing b, is

$$M = \frac{P_{nC} b^2}{12} \quad (7.179)$$

Assuming the permissible bending stress, f, and thickness of the plating, t, the section modulus of the plating is

$$S = \frac{t^2}{6} \quad (7.180)$$

and the required spacing of the inclined stiffeners may be found from the equation

$$\frac{P_{nC} b^2}{12} = \frac{t^2}{6} f \quad (7.181)$$

or

$$b = t\sqrt{2f/P_{nC}} \quad (7.182)$$

Considering the effective width of the plating acting integrally with the stiffener, the second condition to evaluate spacing of the stiffeners is

$$b = 60t \quad (7.183)$$

The final choice of plate thickness will be deferred until the stiffeners have been designed.

7.6.7 Hopper Stiffeners

These stiffeners are required to give support to the hopper plate which is carrying the load. The hopper stiffener is an inclined beam subject to bending, shear and direct tension under uniformly varying pressure and tension, Figure 7.73.

The bending moment at some distance x from B will be

$$M_x = \frac{L}{6}(2p_B + p_C)x - p_B\frac{x^2}{2} - (p_C - p_B)\frac{x^3}{6L} \quad (7.184)$$

Table 7.11 gives values of coefficients of x and M_X expressed in the form

$$x = KL \quad (7.185)$$

$$M = \alpha p_B L^2 \quad (7.186)$$

using

$$\frac{P_C}{P_B} = \beta \quad \text{and} \quad L = 1 \quad (7.187)$$

Together with the bending moment will be a direct tension, having the value

$$T_x = T_B - [p_{tB}x - (p_{tB} - p_{tC})\frac{x^2}{2L}] \quad (7.188)$$

The above bending moment M_X and direct tension T_X will each need to be multiplied by the distance center-to-center of the stiffeners to give the actual design values required to design the stiffeners.

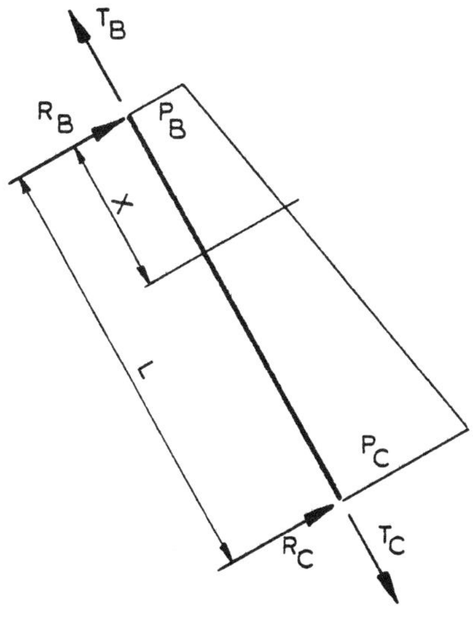

FIGURE 7.73 — Hopper stiffener loading.

TABLE 7.11 — Values of Coefficients.

β	K	α
0	0.4226	0.0641
0.25	0.4516	0.0789
0.50	0.4726	0.0940
0.75	0.4904	0.1094
1.00	0.5000	0.1250
1.25	0.5092	0.1406
1.50	0.5166	0.1564
1.75	0.5228	0.1722
2.00	0.5276	0.1881
2.50	0.5352	0.2199
3.00	0.5408	0.2517
4.00	0.5486	0.3155

7.7 Supports for Square and Rectangular Bins and Bunkers

7.7.1 Types of Supports

Bins are normally supported by means of suitable structures such as columns or skirts, resting on a reinforced-concrete foundation. The skirt type of the support structure between the bin and the foundation may consist of a steel shell. The final choice may depend on the relative cost, service requirements, aesthetics, and customer preference. For relatively low structures, the skirt support may be the most economical and may offer an added advantage of providing a housing for equipment. For high, elevated structures columns are preferable and such type will be discussed in this Chapter.

7.7.2 Loads

This section discusses only the load requirements peculiar to bunker and bin supports. A selection of foundation types and design methods are given in special technical literature and will therefore, not be treated here.

Loads to be considered are:

a. Dead Load — weight of structure, equipment, and supported backfill (if any).
b. Live load — full weight of stored material and snow.
c. Wind
d. Earthquake

Load combinations to be considered in the design of supports and foundations for strength, stability, and reactive pressure include:

1. Dead and live loads, (all cells full)
2. Dead and live loads, (all cells full) and earthquakes
3. Dead and wind loads (all cells empty)
4. Unsymmetrical loading of bins or bunker groups. Assume half of the cells are full and the other half empty.
   a. Dead and live loads
   b. Dead and live and wind or earthquake loads (whichever controls)
5. Dead and live loads in all cells but one: one cell empty. (May control for bending for the foundation slab.)

7.7.3 Wind and Earthquake Loads

Wind may affect the stability of empty bins, particularly tall, narrow bins or bin groups. Foundation pressures and column stresses, however, may be worse with wind acting on the full bin.

Earthquake forces may affect the stability and strength of the structure. Columns and walls supporting the bin may be particularly vulnerable and their foundations may also be affected. In the absence of better codes covering the seismic design of bins, it is suggested that only the earthquake requirements of the latest Uniform Building Code be applied [7.39].

There is experimental evidence [7.40] that only a fraction of stored material weight need be considered when computing earthquake forces. The authors use not less than 80% of the weight of the stored material as an effective live load, W_{eff}, from which to determine seismic forces. Lower percentages have been suggested, but it is considered that more research is needed to substantiate the lower values.

When bin bottoms are on supports independent from the walls, the bin structure and independent bottom support will share the lateral force from any seismic action on the effective weight of the stored materials. Safarian and Harris [7.41] suggest computing the above distribution as follows:

To the independent bottom structure, the *smaller* of the following:

1. A portion of the force due to W_{eff}, divided according to relative stiffness of the two structures.
2. Force due to the effective weight within the hopper (if any) plus the product of coefficient of internal friction times effective weight of the stored material above.

To the bin structure, the *larger* of the following:

1. The total force due to W_{eff}, minus that computed above for the independent bottom structure.
2. 50 percent of the total force due to W_{eff}.

7.7.4 Axial Forces in Columns Under Vertical Loads

The columns supporting single or multicell bins are usually loaded by axial and horizontal forces causing bending of the columns. [7.42]

The axial forces in columns are caused by the own weight of the structure and the weight of the stored material.

Bending is caused by wind action.

Axial forces in the columns will be determined considering own weight of the bin, stored material and the forces due to wind pressure. The axial forces are calculated by acting on the top of the supporting columns, considering the total load distributed in the ratio of the base areas. In general, we are considering a multicell bin having different dimensions in plan, as shown in Figure 7.74.

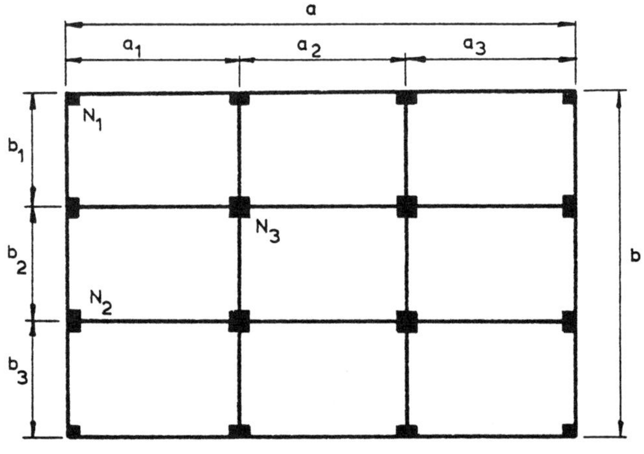

FIGURE 7.74 — Plan of the columns.

Column at corner:

$$N_1 = \frac{q\, a_1 b_1}{4} \qquad (7.189)$$

Column at the edge:

$$N_2 = \frac{q}{4}(b_2+b_3)\,a_1 \qquad (7.190)$$

Column at the middle:

$$N_3 = \frac{q}{4}(b_1+b_2)\,(a_1+a_2) \qquad (7.191)$$

where

q = total uniformly distributed load due to the own weight and stored material

Assuming that all cells have equal dimensions or

$$a_1 = a_2 = a_3 = a$$
$$b_1 = b_2 = b_3 = b \qquad (7.192)$$

we have

$$N_1 = q\frac{ab}{4}\ ;\quad N_2 = q\frac{ab}{2}\ ;\quad N_3 = qab \qquad (7.193)$$

For the determination of axial loads in columns due to wind loads on the bin, it is necessary to evaluate the resulting moment with respect to the center of gravity of all cross-sectional areas of the columns, Figure 7.75.

The eccentricity of the vertical loading, Figure 7.75 is

$$e = \frac{M}{\Sigma N} \qquad (7.194)$$

and the distance of the center of gravity of all cross-sectional areas of the columns from the leeward side will be

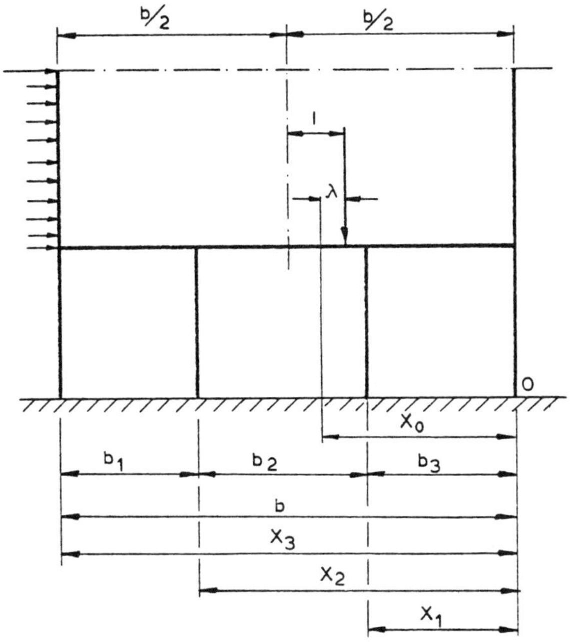

FIGURE 7.75 — Eccentricity of the vertical load due to wind action.

$$x_0 = \frac{\sum_{1}^{n} A_i\, x_i}{\sum_{1}^{n} A_i} \qquad (7.195)$$

where

ΣA_i = summary of all cross-sectional areas of the columns

$x_1, x_2, \ldots \ldots x_n$ = distances of the rows of columns from the edge 0

The moment of the summary of ΣN with respect to the center of gravity of all cross-sectional areas of the columns is

$$M = \Sigma N\lambda = \Sigma N\left(e - \frac{b}{2} + x_0\right) \qquad (7.196)$$

If the moment M is known, then the force in any column may be found from the formula

$$N' = \pm\frac{M_x}{\Sigma x^2} \qquad (7.197)$$

where Σx^2 is the sum of the distances of a single column from the axis passing through the center of gravity of all columns in the square and x is the distance of the row of columns.

The resulting axial force in a column is

$$N = N_n \pm \frac{M_x}{\Sigma x^2} \qquad (7.198)$$

The first term in Equation (7.197) defines the influence of the vertical loading and the second one the influence of the wind loading on the columns.

In cases when the length of all cells are equal, the Equation (7.197) is simplified.

For example, for three cells having $b_1 = b_2 = b_3 = \ell$ considering n columns in each row

$$N = N_n \pm \frac{M_x}{5\ell^2 n} \qquad (7.199)$$

7.7.5 Bending Moments in Columns Under Wind Loading

Wind pressure produces bending moments in columns supporting the bin. The distribution of bending moments in columns depends on the type of supports of the columns. Two basic cases should be considered, namely hinged and fixed supports of the columns.

7.7.5.1 Hinged Supports

In cases of hinged supports the resulting wind pressures are applied at the top of columns, Figure 7.76.

Similarly to rigid frames, the bending moments are distributed on the columns in ratio of the moments of inertia of its cross-sections. From the condition of the equilibrium of the system, it follows

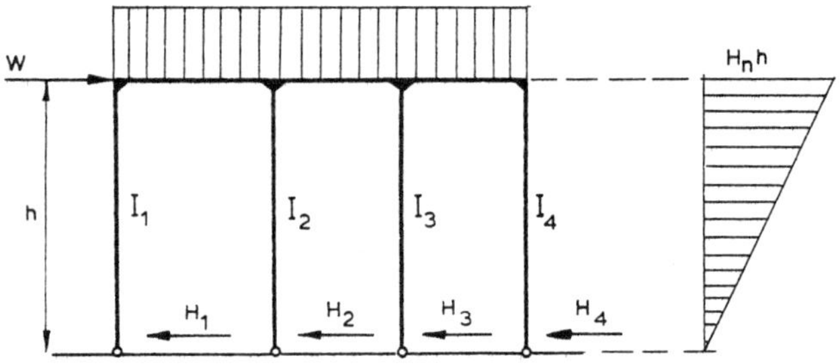

FIGURE 7.76 — Hinged Supports.

$$Wh = H_1 h + H_2 h + H_3 + H_4 h \quad (7.200)$$

or

$$W = H_1 + H_2 + H_3 + H_4$$

and

$$\left.\begin{aligned} M_1 &= Wh\,\frac{I_1}{\Sigma I} \\ M_2 &= Wh\,\frac{I_2}{\Sigma I} \\ M_3 &= Wh\,\frac{I_3}{\Sigma I} \\ M_4 &= Wh\,\frac{I_4}{\Sigma I} \end{aligned}\right\} \quad (7.201)$$

where

$$\Sigma I = I_1 + I_2 + I_3 + I_4 \quad (7.202)$$

If the cross-sections of the columns are known, we calculate from formulas (7.201) their moments of inertia and bending moments.

7.7.5.2 Fixed Supports

In the following only two rows of columns having fixed supports are considered, Figure 7.77.

In cases of columns having fixed supports, it is assumed that the bending moments are zero at the mid-height of the columns.

The resulting fixity moments and axial forces in columns are

$$M_1 = M_2 = \frac{W}{2} \cdot \frac{h_1}{2} = \frac{Wh_1}{4} \quad (7.203)$$

$$N' = \frac{Wh - W\frac{h_1}{2}}{\ell} = \frac{W}{\ell}\left(h - \frac{h_1}{2}\right) \quad (7.204)$$

A resulting axial force acting on the column is obtained as a result of the axial forces N′ and the axial force due to own weight of the structure and weight of the fill in the bin.

A similar analysis may be performed for a number of rows of the columns supporting multicell bins.

7.8 Circular Bins

7.8.1 Vertical and Horizontal Static Pressures on Cylindrical Wall

According to Janssen's method the vertical static unit pressure at depth y below the surface, formula (7.32) is

$$P_v = \frac{\gamma R}{\mu' K}\left(1 - e^{-\mu' K y/R}\right) \quad (7.205)$$

and the lateral static unit pressure, formula (7.33) is

$$P_h = \frac{\gamma R}{\mu'}\left(1 - e^{-\mu' K y/R}\right) \quad (7.206)$$

where K is given by formula (7.23)

$$K = \frac{P_h}{P_v} = \frac{1-\sin\phi}{1+\sin\phi} \quad (7.207)$$

and the hydraulic radius is

$$R = \frac{\pi D^2}{4 \times \pi D} = \frac{D}{4} \quad (7.208)$$

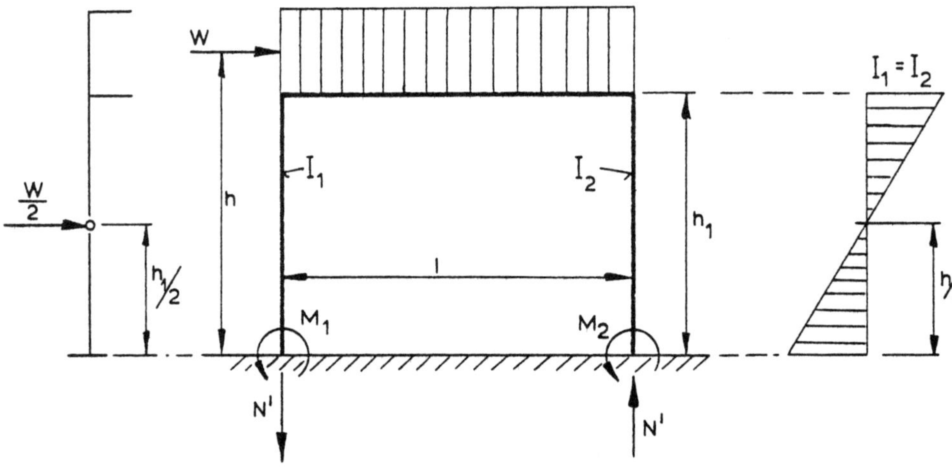

FIGURE 7.77 — Fixed Supports of Columns.

where D is the internal diameter of the bin. Therefore, vertical and horizontal pressures, after substituting into (7.205) and (7.206) the value (7.208), we obtain

$$P_V = \frac{\gamma D}{4\mu' K}(1 - e^{-4\mu' Ky/D}) \quad (7.209)$$

$$P_H = \frac{\gamma D}{4\mu'}(1 - e^{-4\mu' Ky/D}) \quad (7.210)$$

7.8.2 Determination of Forces and Stresses

7.8.2.1 Cylindrical Shell

The cylindrical shell will be subject to hoop tension and meridional compression.

The hoop tension can be derived from Equation (7.33) and Formula (7.210).

$$T_1 = P_H \frac{D}{2} = \frac{\gamma D^2}{8\mu'} (1 - e^{-4\mu' Ky/D}) \text{ lbs/ft} \quad (7.211)$$

or

$$T_1 = \frac{\gamma D^2}{96\mu'} (1 - e^{-4\mu' Ky/D}) \text{ lbs/in}$$

The hoop stress is

$$f_H = \frac{T}{t} = \frac{\gamma D^2}{96\mu' t} (1 - e^{-4\mu' Ky/D}) \text{ psi} \quad (7.212)$$

In a bin the vertical pressure is no longer simply γy because a part of its contents is carried on the wall by friction. This frictional force causing the compression is the difference between the total free vertical pressure and for the level under consideration takes the value

$$T_2 = \frac{1}{\pi D} [\gamma y \frac{\pi D^2}{4} - \frac{\gamma D}{4\mu' K} (1 - e^{-4\mu' Ky/D}) \frac{\pi D^2}{4}] =$$

$$= \frac{\gamma y D}{4} - \frac{\gamma D^2}{16\mu' K} (1 - e^{-4\mu' Ky/D}) \text{ lbs/ft} \quad (7.213)$$

or

$$T_2 = \frac{\gamma y D}{48} - \frac{\gamma D^2}{192\mu' K} (1 - e^{-4\mu' Ky/D}) \quad \frac{\gamma y D}{48} -$$

$$- \frac{T_1}{2K} \text{ lbs/in} \quad (7.214)$$

The meridional compressive stress is

$$f_c = \frac{1}{t} (\frac{\gamma y D}{48} - \frac{T_1}{2K}) \text{ psi} \quad (7.215)$$

To find the maximum stress in the plating at the level in question f_H and f_c may be combined in the form

$$\max f_H = f_H + \nu f_c \quad (7.216)$$

and

$$\max f_c = f_c + \nu f_H \quad (7.217)$$

where ν is Poisson's ratio.

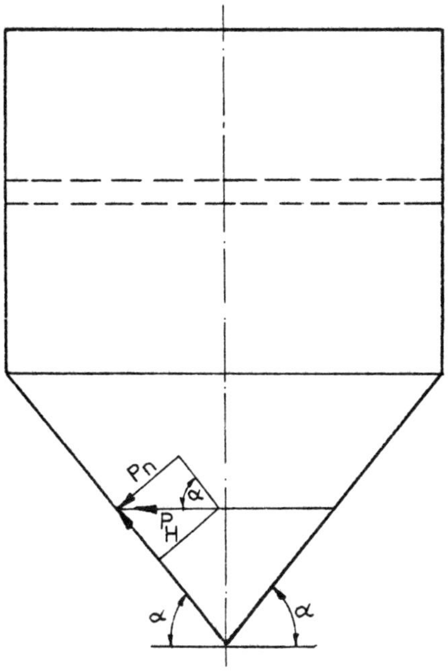

FIGURE 7.78 — Components of forces on conical hopper shell.

7.8.2.2 Conical Hopper Shell

The plates forming the cone will be subjected to both hoop tension and meridional tension. Janssen's expression for both lateral and horizontal pressure may be used, Figure 7.78.

Normal pressure using formula (7.51) is

$$P_n = P_V(\cos^2\alpha + K\sin^2\alpha) = \gamma y(\cos^2\alpha + K\sin^2\alpha) \quad (7.218)$$

and the horizontal pressure

$$P_h = \frac{\gamma y}{\sin\alpha} (\cos^2\alpha + K\sin^2\alpha)$$

giving a hoop tension

$$T_1 = \frac{P_h D}{2} = \frac{\gamma y D}{2} (\cos^2\alpha + K\sin^2\alpha) \quad (7.219)$$

The meridional force is parallel to the generator line of the cone. The meridional force per unit width at depth y is

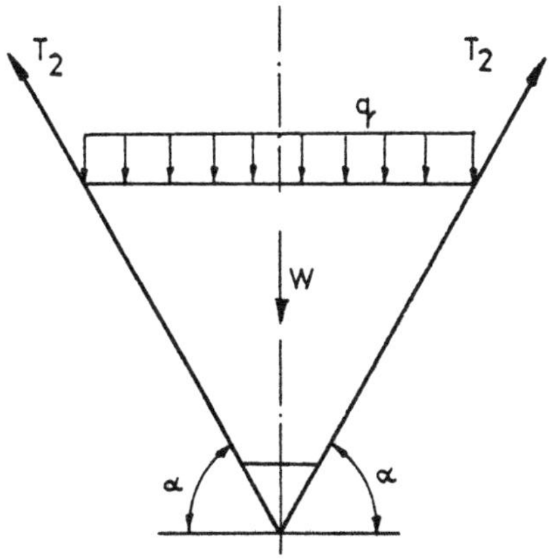

FIGURE 7.79 — Forces on conical hopper.

computed from the equilibrium of the loads on the cone below that depth. These loads, shown in Figure 7.79, are the result of vertical pressures Pv (at depth y) and G, the combined weights of the hopper itself, and material stored below depth y plus any equipment supported by the hopper

$$T_2 = \frac{\gamma y D}{4\sin\alpha} + \frac{W}{\pi D \sin\alpha} + \frac{G}{\pi D \sin\alpha} \qquad (7.220)$$

where

W = total weight of stored material below depth y

G = own weight of the hopper

Both forces are maximum at the upper edge of the hopper, and approach zero at their lower edge.

7.8.3 Circular Girder Supporting Bin

The circular girder supports the weight of the shell, the contents of the bin and own weight of the girder. The above load is uniformly distributed along the circular girder, which is supported by equally spaced columns. A transition from the vertical wall of the bin and conical hopper may be made abruptly. In this case a separate stiffening ring must be provided to resist the horizontal inward pull from the cone.

7.8.3.1 Shear Forces and Bending Moments in Circular Girder

Shear forces and bending moments acting on the circular girder under uniformly distributed vertical loading are based on the analysis by Beyer [7.43], and the following designations are used, Figure 7-80.

W = total load on girder in lbs

R = radius of the girder in inches

N = number of columns

$\alpha = \frac{2\pi}{N}$ = angle at the center subdued by radii through two consecutive columns.

Reaction force at column

$$K = \frac{2\pi RW}{N} \qquad (7.221)$$

Shear force left and right of column are

$$Q_{N,K} = \pm\frac{\pi RW}{N} \qquad (7.222)$$

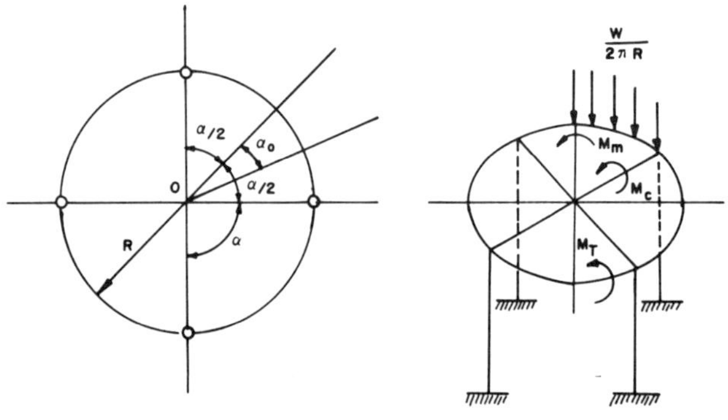

FIGURE 7.80 — Circular girder supported by columns.

Due to the symmetry of the loading the torsional moments at columns and midspan between columns are zero.

The uniformly distributed vertical load $\frac{W}{2\pi R}$ acting on the circular girder causes bending moments perpendicular to the plane of the circular girder as well as the torsional moments.

The bending moment at the column is

$$M_c = -\frac{WR}{2\pi}\left(1 - \frac{\pi}{N}\text{ctg}\frac{180°}{N}\right) \qquad (7.223)$$

The bending moment at any section in the girder defined by the angle α_0 is

$$M_b = \frac{WR}{2\pi}\left(\frac{\pi}{N} \times \frac{\cos\alpha_0}{\sin\frac{\alpha}{2}} - 1\right) \qquad (7.224)$$

The bending moment at a section in the girder midway between columns at $\alpha_0 = 0$ is

$$M_m = \frac{WR}{2\pi}\left(\frac{\pi}{N} \times \frac{1}{\sin\frac{\alpha}{2}} - 1\right) \qquad (7.225)$$

The torsional moment at any section of the girder defined by the angle α_0 is

$$M_t = -\frac{WR}{2}\left(\frac{\sin\alpha_0}{N\cos\frac{\alpha}{2}} - \frac{\alpha_0}{180}\right) \qquad (7.226)$$

The maximum torsional moment is located at the point of contra-flexure of bending moment, or $M_b = 0$, which should be solved for α_0.

By equating to zero the expression (7.224), we have

$$\frac{WR}{2\pi}\left(\frac{\pi}{N} \times \frac{\cos\alpha_0}{\sin\frac{\alpha}{2}} - 1\right) = 0$$

and

$$\cos\alpha_0 = \frac{N}{\pi} \times \sin\frac{\alpha}{2} \qquad (7.227)$$

For proposed number of columns N and angle α we find from the Equation (7.227) the value of the angle α_0. After substitution of this value into Equation (7.226), we obtain the value of the maximum torsional moments.

The stresses in the circular girder are

$$\sigma_t = \frac{M}{S_t}; \qquad S_b = -\frac{M}{S_b} \qquad (7.228)$$

where M indicates corresponding bending moment and

S_t and S_b = section modulus of the circular girder with respect to the top and bottom edges.

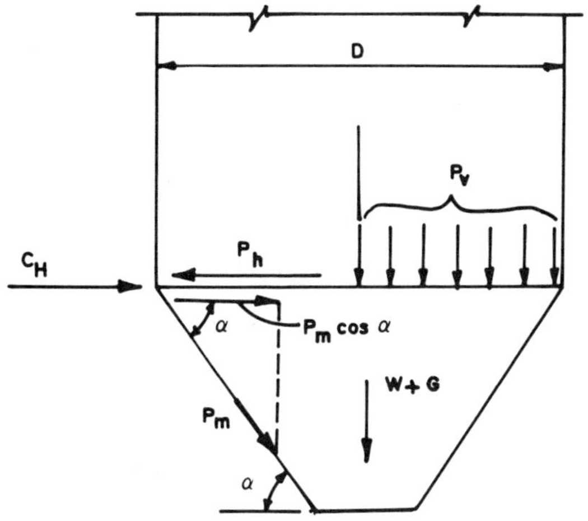

FIGURE 7.81 — Forces at the junction of vertical wall and hopper at circular bin.

Considering the circular girder supported by four columns the values of the moments are for N = 4 and $\alpha = \frac{\pi}{4}$

The bending moment at column, formula (7.223)

$$M_c = -\frac{WR}{2\pi}\left(1 - \frac{\pi}{N}\text{ctg}\frac{180°}{N}\right) =$$

$$= -\frac{WR}{2 \times 3.141}\left(1 - \frac{3.141}{4}\text{ctg}\frac{180°}{4}\right) = -0.0342\,WR \qquad (7.229)$$

The bending moment at a section in the girder midway between columns, formula (7.225)

$$M_m = \frac{WR}{2\tau}\left(\frac{\pi}{N} \times \frac{1}{\sin\frac{\alpha}{2}} - 1\right) =$$

$$= \frac{WR}{2 \times 3.141}\left(\frac{3.141}{4} \times \frac{1}{\sin 450} - 1\right) = 0.0176\,WR \qquad (7.230)$$

To find maximum torsional moment, we first calculate the value of α_0; after formula (7.227)

$$\cos\alpha_0 = \frac{N}{\pi}\sin\frac{\alpha}{2} = \frac{4}{3.141} \times \sin 45° = 0.900$$

$$\alpha_0 = 25°10' \text{ or } 64°56' \qquad (7.231)$$

After substituting this value into the formula (7.226), we have

$$M_t = -\frac{WR}{2}\left(\frac{\sin\alpha_0}{N\cos\frac{\alpha}{2}} - \frac{\alpha_0}{180°}\right) =$$

$$= -\frac{WR}{2}\left(\frac{\sin 25°10'}{4\cos 45°} - \frac{25.166}{180°}\right) = -0.0053\,WR \qquad (7.232)$$

7.8.3.2 Stiffening Ring

At junction of vertical bin wall and inclined wall of the hopper cone originates the resulting horizontal compression under horizontal pressure of the fill material and horizontal component of the meridional force acting in the wall of the hopper, Figure 7.81.

The intensity of the resulting uniformly distributed horizontal inward pull, generally, is

$$C_H = p_h - p_m\cos\alpha \qquad (7.233)$$

7.8.3.3 Checking of the Stiffening Ring for Buckling

According to Timoshenko [7.44] the minimum value of the required moment of inertia of the cross-section of the ring under the critical value of the compression force C_H is

$$I_{min} = \frac{C_H R^2}{3E} \qquad (7.234)$$

where

C_H = uniformly distributed compressive force in lbs/in.

R = centroidal radius of ring, in.

E = modulus of elasticity, psi

I_{min} = minimum moment of inertia in.^{4}

7.8.3.4 Horizontal Bracing System

Horizontal bracing system of the circular girder is resisting either the horizontal wind force or seismic forces. These forces produce shear flow in the girder. Considering horizontal force W acting in the plane of the circular girder, Figure 7.82, the intensity of shear flow according to Roark [7.45] is

$$S = \frac{W\sin\varphi}{\pi R} \qquad (7.235)$$

where

φ = angle from the direction of force W to any cross-section of circular girder

R = centroidal radius of the circular girder.

The bracing system consisting of struts between columns, Figure 7.83, resists shear flow.

Assuming the number of columns as N, the length between columns is $\frac{2\pi R}{N}$. Therefore, the force acting in the

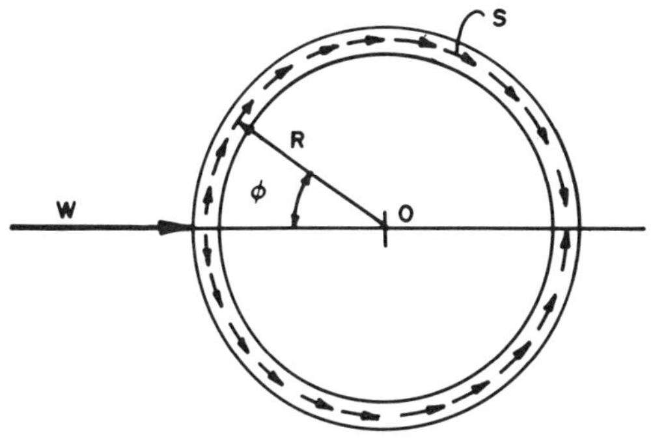

FIGURE 7.82 — Shear Flow in circular girder.

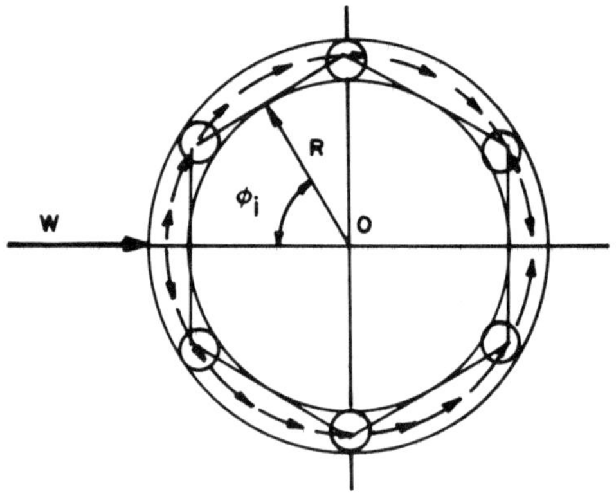

FIGURE 7.83 — Forces in bracing system.

strut of the bracing system under shear flow is

$$F = \frac{2\pi R}{N} \times S = \frac{2\pi R}{N} \times \frac{W \sin\varphi_i}{\pi R} = \frac{2W \sin\varphi_i}{N} \qquad (7.236)$$

where φ_i is the angle between direction of force N and normal to the strut.

7.8.4 Columns

In the design of columns supporting the bin the following loads should be considered:

a. Dead load of structure and contents;
b. Direct load on the bin due to wind or seismic force which may result in the compression and uplift of the columns.

All the columns must be designed for the most severe conditions, since the horizontal forces may act in any direction.

To calculate the column load due to the wind or seismic forces, it is necessary to consider the horizontal forces acting on all areas of the structure in elevation, Figure 7.84.

The total moment due to the horizontal loads is

$$M_H = H_1h_1 + H_2h_2 + H_3h_3 \qquad (7.237)$$

where

H_1, H_2, H_3 = horizontal loads due to wind or seismic forces acting on the bin, hopper and columns, respectively

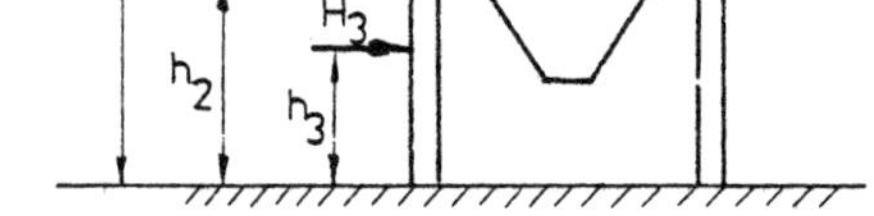

FIGURE 7.84 — Horizontal loads distribution.

To calculate the column load due to horizontal forces we consider the horizontal section through six columns supporting the bin, Figure 7.85.

The maximum compressive load acting on column 1 will occur on the leeward side when the horizontal force acting in the direction 1-1, Figure 7.85.

The moment due to the horizontal forces, MH, about the axis A-B should be equal to resisting moment of the loads in the columns about axis A-B. In a bin with six columns we have

$$M_H = P_1 x 2R + P_2 x 4R \cos\theta_n \qquad (7.238)$$

where

P_1, P_2 = vertical loads acting on the columns 1 and 2, respectively

From the ratio between loads $\frac{P_1}{P_2} = \frac{R}{R\cos\theta_n}$, we obtain $P_2 = P_1\cos\theta_n$ and substituting this value in the moment equation, we have

$$M_H = 2P_1R(1+2\cos^2\theta_n) \qquad (7.239)$$

or

$$P_1 = \pm \frac{M_H}{2R(1+2\cos^2\theta_n)} \qquad (7.240)$$

at

$$\theta = 60^\circ \qquad P_1 = \frac{M_H}{3R}$$

By designating dead load component as V, the total load is

$$P_{tot} = -\frac{V}{6} \pm \frac{M_H}{3R} = -\frac{1}{6}\left(V \pm \frac{2M_H}{R}\right) \qquad (7.241)$$

It is proven that, generally, for the bin supported by the N columns, the maximum compressive load may be expressed by the formula

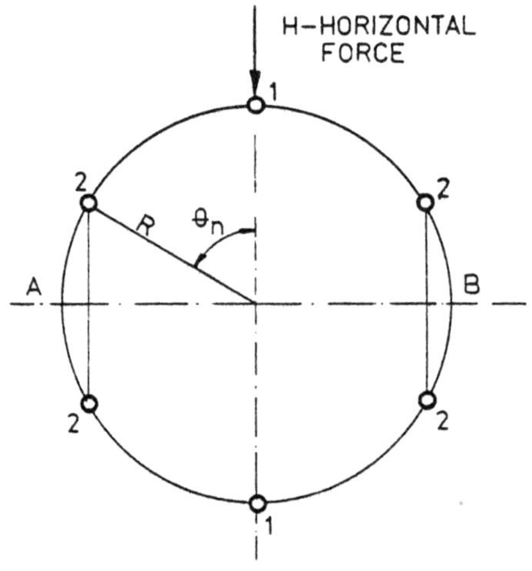

FIGURE 7.85 — Horizontal load acting on bin supported by six columns.

$$P_c = -\frac{1}{N}\left(V + \frac{2M_H}{R}\right) \qquad (7.242)$$

and in the case when

$$\frac{2M_H}{R} > V$$

the uplift load is

$$P_u = -\frac{1}{N}\left(V - \frac{2M_H}{R}\right) \qquad (7.243)$$

Loads acting on the columns should be investigated for the following cases:

a. Bin empty + horizontal force

$$P_{EH} = -\frac{1}{N}\left(V_E \mp \frac{M_H}{R}\right) \qquad (7.244)$$

b. Bin full + horizontal force

$$P_{FH} = -\frac{1}{N}\left(V_F \mp \frac{M_H}{R}\right) \qquad (7.245)$$

c. Bin full, no horizontal force

$$P_F = -\frac{V_F}{N} \qquad (7.246)$$

7.8.5 Roofs for Circular Bins

For circular bins roofs are generally of conical or dome systems.

A conical steel sheet roof supported on a radial framework is the most satisfactory. The sheets should be cut radially and welded. The roof sheets are sometimes detailed developed and the sheets are bent to shape. Typical radial framework for the steel bin is shown in Fiugre 7.86.

Domes have been used to roof circular bins. They are relatively lightweight and can be shaped so that loads induce mainly axial stresses. For a shallow dome, a tension ring usually is provided around the base to resist thrusts. Designers have a choice of a wide variety of structural systems for domes. In general, dome construction may be designed as a single-layer framing, double-layer (truss) framing, or space frame, for greater resistance to buckling, and stressed skin, with the roof deck acting integrally with structural framing.

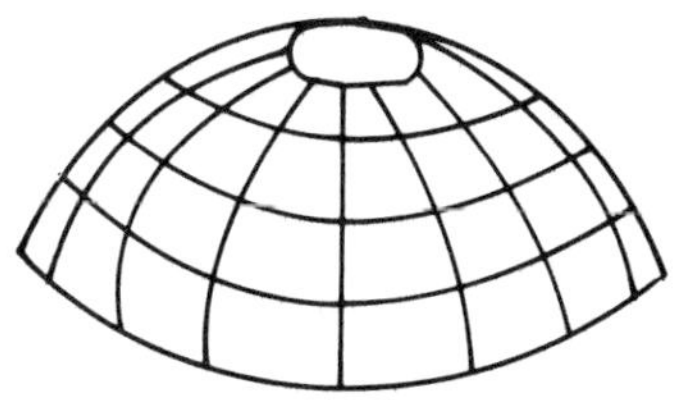

FIGURE 7.87 — Schwedler dome.

In a ribbed dome, its principal components are half arches. They are connected at the crown, but usually, to avoid a complicated joint with numerous members converging there, the ribs at top terminated at the small diameter compression ring.

A Schwedler dome offers more even distribution of the dead load and reduces the unbraced length of the ribs, Figure 7.87.

Principal members are the arch rib and a series of horizontal rings with diameter increasing with distance from the crown. The ribs transmit loads to the base mainly by axial compression, and the ring resists hoop stresses. With simplifying assumptions this system can be considered statically determinate. For spherical domes of this type, an economic rise-span ratio is $\frac{1}{7.5}$ achieved by making the radius of the dome equal to the diameter of its base.

7.8.5.1 Tension Ring

A conical and dome type roof systems rafters are distributed radially and supported by the top edge of the wall of the bin, which is under horizontal thrust and vertical load. To resist this thrust the horizontal tension ring should be provided, which is loaded with N horizontal loads, correspond-

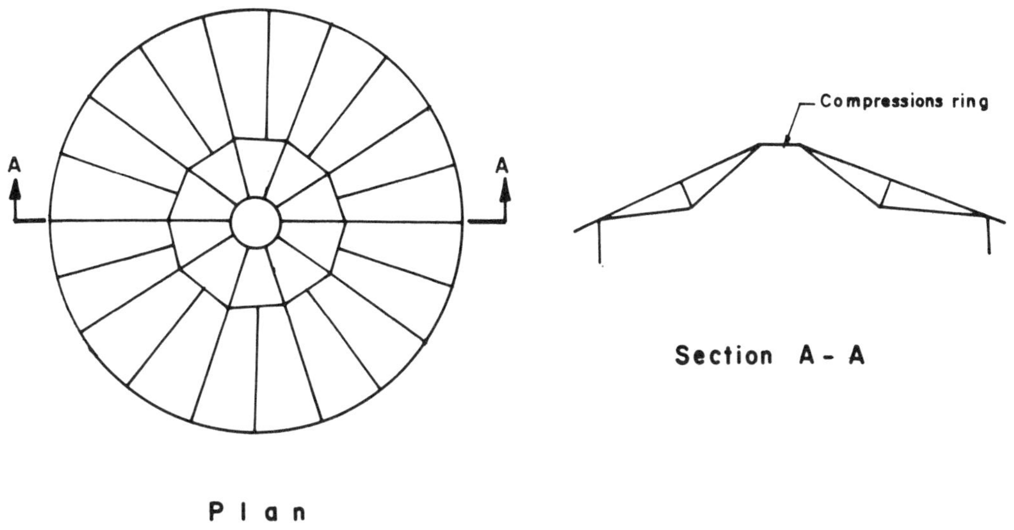

FIGURE 7.86 — Framing for roof of circular bin.

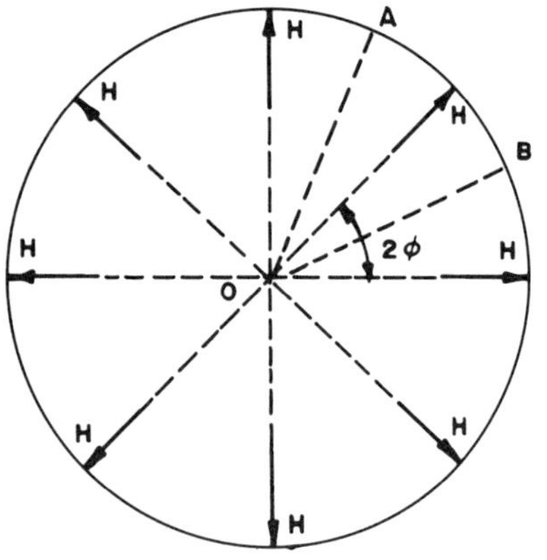

FIGURE 7.88 — Tension ring under radial horizontal loads H.

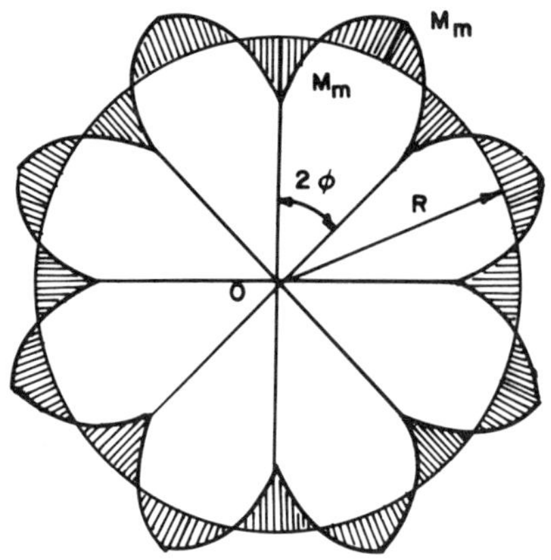

FIGURE 7.89 — Bending moments diagram.

ing to the number of rafters, acting radially and producing tension and bending moments in the ring, Figure 7.88. The ring carries a number of horizontal radial loads H spaced 2φ apart.

From considerations of symmetry it is clear that the resultant actions at corresponding points in each segment such as OAB need to be considered. A and B are midway between loaded points. The following formulas are based on the theory by Pippard [7.46].

1. Shears in the tension ring at each side of the rafters are

$$S = \pm \frac{H}{2} \tag{7.247}$$

and at midspan between loads

$$S = 0 \tag{7.248}$$

2. At radial load H points value of tenion is

$$T_H = \frac{H}{2}\cot\frac{180°}{N} \tag{7.249}$$

and at midpoint between loads H

$$T_H = \frac{H}{2}\csc\frac{180°}{N} \tag{7.250}$$

3. The values of negative bending moments at radial loads H are

$$M_H = \frac{HR}{2}\left(\cos\varphi - \frac{1}{\varphi}\right) = \frac{HR}{2}\left(\cos\frac{180°}{N} - \frac{N}{\pi}\right) =$$

$$= -\frac{HR}{2}\left(\frac{N}{\pi} - \cos\frac{180°}{N}\right) \tag{7.251}$$

and the positive bending moments midway between radial loads H are

$$M_m = \frac{HR}{2}\left(\csc\varphi - \frac{1}{\varphi}\right) = \frac{HR}{2}\left(\csc\frac{180°}{N} - \frac{N}{\pi}\right) \tag{7.252}$$

where

N = number of radial loads H

R = ring radius, ft

$2\varphi = \frac{2\pi R}{N}$ = angle between radial load H

The diagram of bending moments acting on tension ring is shown in Figure 7.89.

Radial forces H produce tension in the ring, as well as the positive bending moments M_o, at the inside edge of the ring.

The resulting stresses in the ring are

$$f = +\frac{H}{A} \pm \frac{M}{W} \tag{7.253}$$

where

A = cross-sectional area of the ring, in.^{2}

W = section modulus of the cross-section of the ring with respect to inner edge, in.^{3}

7.9 Circular Bunkers

7.9.1 Forces in the Cylindrical Section of a Bunker

Circular bunkers consist of upper or cylindrical, and bottom or conical shells, supported by columns, Figure 7.90.

The principles involved in finding the vertical and horizontal pressures in shallow circular bunkers are similar to those given for the rectangular type. However, the horizontal or lateral pressure set-up will be acting against a radial surface, and for this reason the skin plate will be subject to direct stresses only.

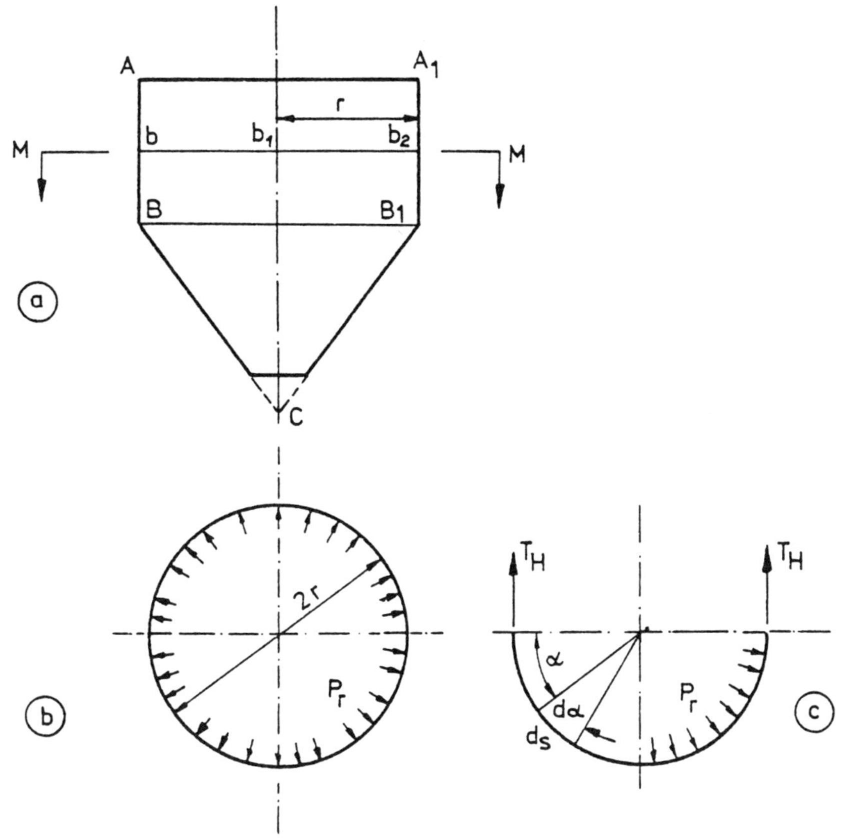

FIGURE 7.91 — Vertical and horizontal cross-sections through a circular bunker.

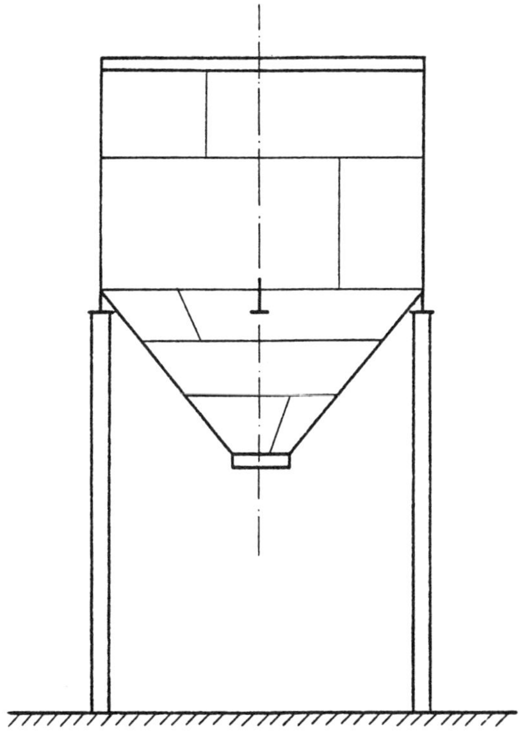

FIGURE 7.90 — General view of a circular bunker.

7.9.2 Pressures in Cylindrical Section

As before, the vertical pressure according to formula (7.72) is

$$q = p_v = \gamma h \tag{7.254}$$

and the lateral pressure after formula (4.43)

$$p_h = K\gamma h = \frac{1-\sin\phi}{1 \sin\phi} h \tag{7.235}$$

Figure 7.91 (a) shows a cross-section through a typical symmetrical circular bunker. Section M-M at Figure 7.91 (b) shows the shell subjected to a uniform radial pressure P_r.

By cutting through the diameter, Figure 7.91 (c), the ring must be in equilibrium under the action of the pressure P_r and thre reaction T_H.

The vertical component of the radial force acting on the arc ds is

$$p_r ds \cos\alpha = pr \cos\alpha \, dx \tag{7.256}$$

Therefore, the total pressure P is

$$P = P_r r \int_{-\pi/2}^{\pi/2} \cos\alpha dx = 2P_r r \tag{7.257}$$

which must equal the sum of the reactions $2T_H$, or

$$T_H = P_r r = \frac{P_r d}{2} \quad (7.258)$$

But $P_r = P_n = K\gamma h$, so the hoop tension is

$$T_H = K\frac{\gamma h d}{2} \quad (7.259)$$

and will have a maximum value at the waist BB_1.

7.9.3 Meridional Tension in Cylindrical Section

Depending upon the location of supports, there may also be a meridional tension in the cylindrical portion.

a. *Bunker supported at AA level*
Let W_c be the weight of the content and W_B the bunker weight, then the meridional tension per unit length of circumference will be

$$T_t = \frac{W_c + W_B}{2\pi r} \quad (7.260)$$

b. *Bunker supported at bb_1b_2 level*
The meridional tension in the shell immediately below this level will be due to the weight of contents W_c plus the self weight of the bunker W_b below bb_1b_2

$$T_t = \frac{W_c + W_b}{2\pi r} \quad (7.261)$$

c. *Bunker supported at BB_1 level*
If the bunker is supported at the waist BB_1 which is to be preferred, this meridional tension may be discounted, or

$$T_t = 0 \quad (7.262)$$

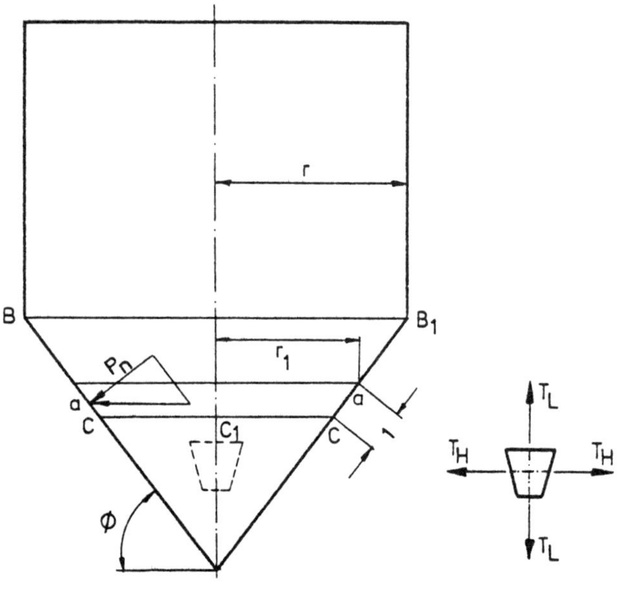

FIGURE 7.92 — Forces in conical bunker bottom.

7.9.4 Forces in the Conical Section

Considering that the supports are never below the waist, there will be both hoop tension and meridional tension.

a. *The hoop tension*
To find the stress in the ring a-a, let us assume a section of unit slant height cut from the cone, Figure 7.92.

Now the normal pressure P_n on the element may be found from the ellipse of stress and is given by the formula [1]

$$P_n = V\frac{\sin^2(\phi+\theta)}{\sin^2\phi\left[1+\frac{\sin\theta}{\sin\phi}\right]^2} \quad (7.263)$$

and the hoop tension due to Pn will be found from the formula

$$T_H = P_n r_1 \operatorname{cosec}\phi \quad (7.264)$$

where

V = the vertical pressure at the level under consideration

ϕ = the slope of the wall

θ = the angle of repose

T_H has a maximum value at BB_1.

b. *Inclined or tangential tension*
Tangential tension will be due to the vertical pressure acting upon the plane cc_1c plus the weight of the material and the weight of the shell immediately below the plane. The total vertical load at this plane is therefore

$$W_1 = V\pi r_1^2 + \frac{w\pi r_1^2 h}{3} + G \quad (7.265)$$

and the circumferential tension due to W_1 is

$$\frac{W_1}{2\pi r_1} \quad (7.266)$$

which, when resolved up the cone gives

$$T_L = \frac{W_1}{2\pi r_1}\operatorname{cosec}\phi \quad (7.267)$$

The maximum value of the meridional tension in the cone will be at BB_1 when W_1 equals the total weight of the contents plus the self weight of the cone, and $r_1 = r$.

A minimum tension will occur at the outlet, and will apply to both the hoop and tangential tension.

7.9.5 Stresses in the Walls

The stresses induced in the shell plates will be due to hoop and meridional tension.

a. *Cylindrical section*
In the cylindrical section the actual hoop stress will be

$$f_H = \frac{T_H}{t} = \frac{whdK_1}{2t} \quad (7.268)$$

and the meridional stress, if applicable

$$f_L = \frac{T_L}{t} = \frac{W}{2\pi rt} \qquad (7.269)$$

where t is the thickness of the plate.

b. *Cone section*
In the cone the hoop stress will be

$$f_H = \frac{T_H}{t} = \frac{P_n r_1}{t} \operatorname{cosec}\phi \qquad (7.270)$$

and the meridional stress is

$$f_L = \frac{T_L}{t} = \frac{W_1}{2\pi r_1 t} \operatorname{cosec}\phi \qquad (7.271)$$

7.9.6 Ring Beam and Supports

Whenever possible a circular bunker should be supported at the waist BB_1, Figure 7.92. In order to provide against heavy local stresses set up due to the presence of supporting brackets, it is common practice to have a ring beam connected to the shell plate. The ring beam will take the load from the bunker uniformly distributed.

APPENDIX A — EXAMPLE NUMBER 1
Design of Rectangular Bin

THE PROBLEM

Design a single cell rectangular steel bin for storing peas. Dimensions are shown in Figure 7.93. The bottom is a symmetrical pyramidal hopper. The bin walls rest on the hopper base, which are supported by four columns.

For peas:

$$\gamma = 50 \text{ lb/ft}^3$$

$$\phi = 25^\circ$$

$$\mu' = 0.296$$

A. Bin Wall Forces

1. LATERAL PRESSURES

Lateral pressures are computed using Janssen's formula (7.33)

$$P_h = \frac{\gamma R}{\mu'} (1 - e^{-\mu' K y/R})$$

where (7.34)

$$K = \frac{1-\sin\phi}{1+\sin\phi} = \frac{1-\sin 25^\circ}{1+\sin 25^\circ} = \frac{1-0.422}{1+0.422} =$$

$$= \frac{0.578}{1.423} = 0.406$$

For the shorter wall, formula (7.36)

$$R_a = \frac{a}{4} = \frac{12.0}{4} = 3.0 \text{ ft}$$

For the longer wall formulas (7.38) and (7.37)

$$a' = \frac{2ab-a^2}{b} = \frac{(2 \times 12 \times 20 - 12 \times 12)}{20}$$

$$= 16.8 \text{ ft}$$

$$R_b = \frac{a'}{4} = \frac{16.8}{4} = 4.2 \text{ ft.}$$

Using these R-values and the above material properties, the static lateral pressure values are computed at various depths, y, using Equation (7.33) and tabulated in Table 7.12.

For wall "a"

$$P_h = \frac{50 \times 3}{0.296} (1 - e^{-\frac{0.296 \times 0.406}{3} y}) =$$

$$= 506.75 (1 - e^{-0.04y})$$

For wall "b"

$$P_h = \frac{50 \times 4.2}{0.296} (1 - e^{-\frac{0.296 \times 0.406}{4.2} y}) =$$

$$= 709.45 (1 - e^{-0.0286y})$$

Overpressure factors, C_d, are selected from Table 7.5, and design pressures computed by Eq. (7.48) are added to Table 7.12.

TABLE 7.12 — Horizontal Pressures on Bin Walls.

Depth y ft	Coefficient C_d from Table 7.5	Wall "a" Static Pressure psf	Wall "a" Design Pressure psf	Wall "b" Static Pressure psf	Wall "b" Design Pressure psf
5	1.50	92	138	93	140
10	1.50	167	250	176	264
20	1.65	279	460	400	660
30	1.65	354	584	409	675
40	1.65	405	668	485	800

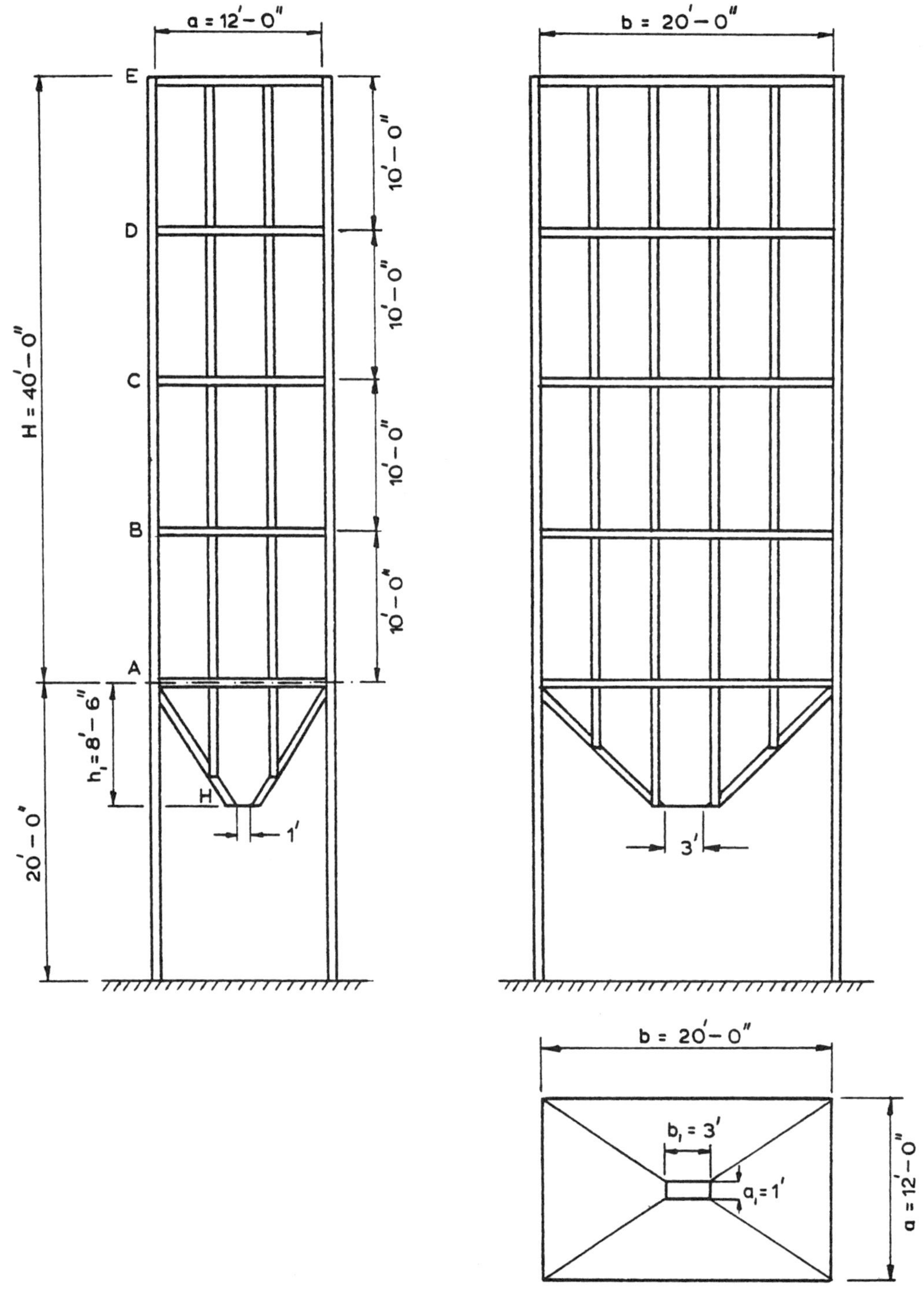

FIGURE 7.93 — Single cell rectangular bin with pyramidal hopper, supported on four columns.

2. VERTICAL WALL LOADS DUE TO FRICTION

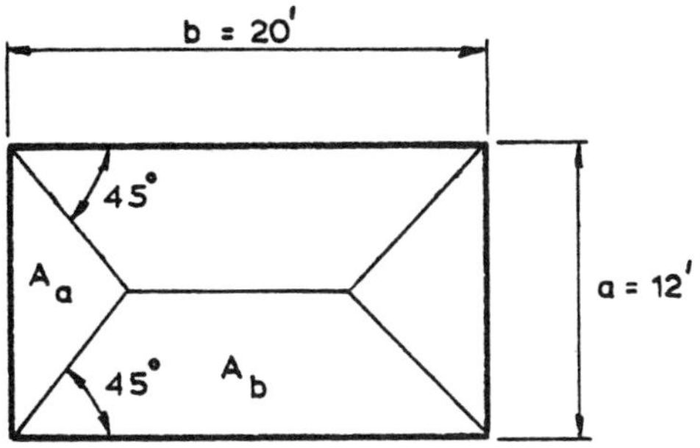

FIGURE 7.94 — Distribution of loads.

$$A_a = \frac{1}{2} 6 \times 12 = 36 \ ft^2$$

$$A_b = \frac{1}{2} (12 \times 20 - 2 \times 36) = 84 \ ft^2$$

At the bottom, y = 40 ft, using formula (7.47), we find for *Wall "A"*

$$q_{a,des} = (\gamma y - \frac{0.8P_h}{K}) \frac{A_a}{a} =$$

$$= (50 \times 40 - 0.8 \times \frac{405}{0.406}) \frac{36}{12} =$$

$$= 3,606 \ lb/ft$$

and for *Wall "B"*

$$q_{b,des} = (\gamma y - 0.8\frac{P_h}{K}) \frac{A_b}{b} =$$

$$= (50 \times 40 - 0.8 \times \frac{485}{0.406}) \frac{84}{20} =$$

$$= 4386 \ lb/ft$$

3. SIDE WALL TENSIONS AND BENDING MOMENTS

The side wall tensions for the walls "a" and "b" are calculated by the formulas (7.80) and (7.81)

$$T_a = \frac{1}{2} p_b b \quad ; \quad T_b = \frac{1}{2} p_a a$$

Wall bending moments are calculated using formula (7.83), assuming $I_a/I_b = 1$ and $k = a/b$. Therefore, corner moment is

$$M_c = \frac{P_a a^3 + p_b b^3}{12(a + b)}$$

and the midspan moments according to formulas (7.84) and (7.85), are

$$M_a = \frac{P_a a^2}{8} - \frac{P_a a^3 + P_b b^3}{12(a + b)}$$

$$M_b = \frac{P_b b^2}{8} - \frac{P_a a^3 + P_b b^3}{12(a + b)}$$

The numerical values are tabulated in Table 7.13.

B. Hopper. Forces

For all depths use R equal to hydraulic radius at the top of the hopper. In this case use the pressure for y = 40 ft. throughout.

1. VERTICAL DESIGN PRESSURE AT y = 40 FT

According to formulas (7.32) and (7.33), and using values from Table 7.12, we find

For Wall "a"

$$P_{va} = \frac{P_{ha}}{K} = \frac{668}{0.406} = 1645 \ lb/ft^2$$

For wall "b"

$$P_{vb} = \frac{P_{hb}}{K} = \frac{800}{0.406} = 1970 \ lb/ft^2$$

2. DESIGN PRESSURES NORMAL TO THE INCLINED WALLS AT y = 40 FT

According to formula (7.51)

$$P_n = P_v \cos^2\alpha + P_h \sin^2\alpha$$

Angles of the slope are

$$\tan\alpha_a = \frac{8.5}{\frac{1}{2}(20-3)} = 1; \quad \alpha_a = 45^\circ$$

$$\tan\alpha_b = \frac{8.5}{\frac{1}{2}(12-1)} = 1.732 \ ; \quad \alpha_b = 60^\circ$$

Wall "a"

$$P_{na} = P_v \cos^2 45^\circ + P_h \sin^2 45^\circ =$$

$$= (668+1645)0.707^2 = 1157 \ lb/ft^2$$

Wall "b"

$$P_{nb} = 800 \times 0.866^2 + 1970 \times 0.5^2 =$$

$$= 1092 \ lb/ft^2$$

TABLE 7.13 — Tensile Forces and Bending Moments.

	Wall "a"			Wall "b"		
	Tension	At Support	At Midspan	Tension	At Support	At Midspan
Depth y ft	T_a k/ft	M_c ftk/ft	M_a ftk/ft	T_b k/ft	M_c ftk/ft	M_b ftk/ft
5	1.40	-3.54	1.05	0.83	-3.54	3.46
10	2.64	-6.62	2.12	1.50	-6.62	6.58
20	6.68	-15.82	2.54	2.76	-15.82	17.18
30	6.75	-16.69	8.16	3.50	-16.69	31.40
40	8.00	-19.67	9.67	4.00	-19.67	37.65

3. HORIZONTAL TENSILE FORCES

According to formulas (7.124) and (7.125) at y = 40 ft.

Wall "a"

$$T_a = (\frac{b}{2}) P_{nb} \sin\alpha_b = (\frac{20}{2}) \times 1092 \times 0.866 =$$

$$= 9460 \text{ lb/ft}$$

Wall "b"

$$T_b = (\frac{a}{a}) P_{na} \sin\alpha_a = (\frac{12}{2}) \times 1157 \times 0.707 =$$

$$= 4900 \text{ lb/ft}$$

4. THE WEIGHT OF THE CONTENT OF THE HOPPER

According to formula (7.112) the volume of the hopper is

$$V = \frac{h_1}{6} [(2a_2 + a_1)b_2 + (2a_1 + a_2)b_1] =$$

$$= \frac{8.5}{6}[(2 \times 12 + 1)20 + (2 \times 1 + 12)3] =$$

$$= 768 \text{ ft}^3$$

The weight of the content is:

$$W = 768 \times 50 = 38 \text{ kips}$$

5. THE WEIGHT OF THE HOPPER

The approximate weight of the hopper is estimated as

$$G \approx 6 \text{ kips}$$

6. MERIDIONAL TENSILE FORCES

Meridional tensile forces are estimated according to formulas (7.127), (7.128) and (5.126), where

$$A_a = A_b = \frac{1}{4} \times 12 \times 20 = 60 \text{ ft}^2$$

For wall "a" at y = 40 ft

$$T_{ma} = \frac{C_a W + A_a P_{va} + C_a G}{b \sin\alpha_b} =$$

$$= \frac{\frac{1}{4} \times 38 + 60 \times 1.645 + \frac{1}{4} \times 6}{12 \times 0.707} =$$

$$= 14.53 \text{ K/ft}$$

For wall "b" at y = 40 ft

$$T_{mb} = \frac{C_b W \quad A_b P_{vb} \quad C_b G}{b \sin\alpha_b} =$$

$$= \frac{\frac{1}{4} \times 38 + 60 \times 1.970 + \frac{1}{4} \times 6}{20 \times 0.866} =$$

$$= 8.24 \text{ K/ft}$$

At y = 44 ft

$$A_a = A_b = \frac{1}{4} \times 6.5 \times 11.5 = 18.68 \text{ ft}^2$$

$$T_{ma} = \frac{\frac{1}{4} \times 38 + 18.68 \times 1.645 + \frac{1}{4} \times 6}{6.5 \times 0.707} =$$

$$= 9.09 \text{ K/ft}$$

$$T_{mb} = \frac{\frac{1}{4} \times 38 + 18.68 \times 1.970 + \frac{1}{4} \times 6}{11.5 \times 0.866} =$$

$$= 4.80 \text{ K/ft}$$

At y = 48 ft

$$A_a = A_b = \frac{1}{4} \times 1.0 \times 3.0 = 0.75 \text{ ft}^2$$

$$T_{ma} = \frac{0.75 \times 1.645}{1.0 \times 0.707} = 1.74 \text{ K/ft}$$

$$T_{mb} = \frac{0.75 \times 1.970}{3.0 \times 0.866} = 0.57 \text{ K/ft}$$

The following Tables 7.14 and 7.15 show the membrane forces at various depths y.

TABLE 7.14 — Design pressures and forces.

Depth y ft	Wall "a"				
	Design pressure psf			Forces k/ft	
	Lateral	Vertical	Normal	Horizontal T_a	Meridional T_{ma}
40	668	1645	1157	9.46	14.53
44	↓	↓	↓	5.44	9.09
48				1.42	1.74

TABLE 7.15 — Design pressures and forces.

Depth y ft	Wall "b"				
	Design pressure psf			Forces k/ft	
	Lateral	Vertical	Normal	Horizontal T_b	Meridional T_{mb}
40	800	1970	1092	4.9	8.24
44	↓	↓	↓	2.66	4.80
48				0.41	0.57

C. Design of Side Wall a = 12 ft

The calculations for the side walls will be based upon the assumption that the skin plate is designed to span between vertical stiffeners. The skin plates will be treated as continuous between the stiffeners. The horizontal forces and bending moments will be taken by horizontal stiffeners.

1. SKIN PLATING ALONG WALL "a"

a. *Small deflection theory*

Use stiffeners at 4 ft centers. The ratio of panel height-to-width is $\frac{10}{4} = 2.5$, therefore the design plate should span in two directions. The average pressure between levels A and B, Figure 7.93, is

$$w = \frac{1}{2}(584 + 668) = 626 \text{ lb/ft}^2$$

Assuming a bending moment of

$$M = \frac{w\ell^2}{10}$$

then

$$M = \frac{626 \times 4^2 \times 12}{10} = 12019 \text{ lb in.}$$

Assume a plate thickness of $\frac{9}{16}$ in. The section modulus for 12 in. width of plate is then

$$S = \frac{bh^2}{6} = \frac{12}{6}\left(\frac{9}{16}\right)^2 = 0.632 \text{ in}^3$$

Stress

$$f = \frac{M}{S} = \frac{12019}{0.632} = 19{,}017 \text{ psi}$$

By applying the large deflection theory, analyzed in Section 7.5.5.2, it is possible to design plating more economically, as follows.

b. *Skin plating along wall "a"*
Large deflection theory

Use a plate having $t = \frac{1}{2}$ in. According to formula (7.98), the tensile force

$$S = \frac{1}{\pi}\sqrt[3]{\frac{4(P_n\ell)^2Et}{\pi(1-\nu^2)}}$$

where

P_n = 626 psf = 4.34 psi

ℓ = 4 × 12 = 48 in

E = 29 × 10^6 psi

ν = 0.3

$$S = \frac{1}{3.14}\sqrt[3]{\frac{4(4.34\text{x}48)^2 \text{ x } 29 \text{ x } 10^6 \text{ x } 0.5}{3.14(1-0.3^2)}} =$$

$$= 3057 \text{ lbs}$$

According to formula (7.99)

$$S_E = \frac{\pi^2EI}{\ell^2(1-\nu^2)}$$

where

$$I = \frac{1}{12}\left(\frac{1}{2}\right)^3 = \frac{1}{96} \text{ in}^4$$

$$S_E = \frac{3.14^2 \text{ x } 29 \text{ x } 10^6}{48^2 \text{ x } (1-0.3)^2 \text{ x } 96} = 1420 \text{ lbs}$$

The bending moment after formula (7.97)

$$M = \frac{P_n\ell^2}{8}\left(1 - 1.034\frac{S}{S + S_E}\right)$$

or

$$M = \frac{4.34 \text{ x } 48^2}{8}\left(1 - 1.034\frac{3057}{3057 + 1420}\right) =$$

$$= 367 \text{ lb in.}$$

The stress in the plate, according to formula (7.103)

$$f = \frac{S}{t} + \frac{6M}{t^2} = \frac{3057}{0.5} + \frac{6 \text{ x } 367}{0.25} = 14{,}933 \text{ psi}$$

Use plating

$$t = \frac{1}{2} \text{ in.}$$

2. VERTICAL STIFFENERS BETWEEN LEVELS A AND B

From Table 7.12, the pressures are

$$P_A = 668 \text{ lb/ft}^2;\ P_B = 584 \text{ lb/ft}^2$$

The bending moment according to Equation (7.186)

$$M = \alpha P_B L^2$$

Since

$$\beta = \frac{P_A}{P_B} = \frac{668}{584} = 1.14$$

Then, the coefficient in Table 7.11 is

$$\alpha = 0.125 + \frac{0.0156 \text{ x } 0.14}{0.25} = 0.133$$

$$M = 0.133 \text{ x } 584 \text{ x } 10^2 \text{ x } 4 \text{ x } 12 =$$

$$= 372{,}825 \text{ lb in.}$$

Try a stiffener structural ST6 WF 15.5#, including

$$40t = 40 \text{ x } \frac{1}{2} = 20 \text{ in}$$

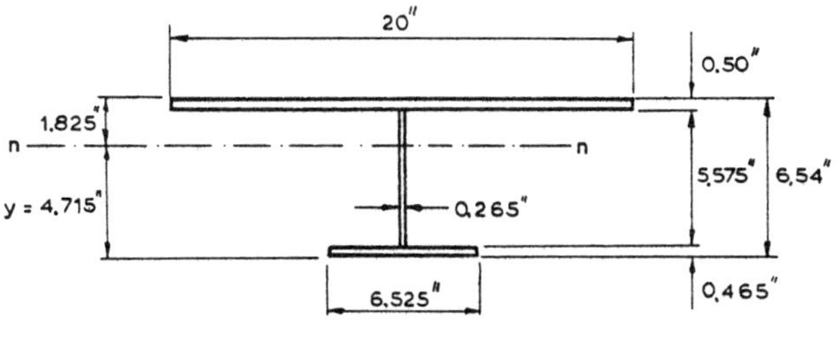

FIGURE 7.95 — Vertical stiffener cross-section.

$$y = \frac{20x0.50x6.29+0.265x5.575x3.252+6.525x0.465x0.233}{20\ 0.5+0.265x5.575+6.525x0.465}$$

$$= \frac{62.90+4.804+0.707}{10+1.477\ 3.034} = \frac{68.411}{14.511} = 4.715 \text{ in}$$

$$I_n = \frac{1}{12} \times 20 \times 0.5^3 + 20 \times 0.5 \times 1.57^2 +$$

$$+ \frac{1}{12} \times 0.265 \times 5.575^3 + 0.265 \times$$

$$\times 5.575 \times 1.463^2 + \frac{1}{12} \times 6.525 \times$$

$$\times 0.465^3 + 6.525 \times 0.465 \times 4.483^2 =$$

$$= 92.878 \text{ in}^4$$

$$S_c = \frac{92.878}{1.825} = 50.892 \text{ in}^3$$

$$S_t = \frac{92.878}{4.715} = 19.698 \text{ in}^3$$

$$f_{comp} = \frac{372,825}{50.892} = 7335 \text{ psi}$$

$$f_{ten} = \frac{372,825}{19.698} = 18,927 \text{ psi}$$

which are satisfactory.

3. HORIZONTAL STIFFENER AT LEVEL B

From Table 7.13 we find:

Tensile force

$$T_a = 6750 \times 10 = 67500 \text{ lbs}$$

Moment at support

$$M_c = 16,690 \times 10 \times 12 = 2,002,800 \text{ lb in}$$

Use WF10 × 10 @ 100#, A = 29.43 in^2

$$S_x = 112.4 \text{ in}^3$$

$$f = \frac{T_a}{A} + \frac{M_c}{S} = \frac{67,500}{29.43} + \frac{2,002,800}{112.4} =$$

$$= 2293 + 17,818 + 20,111 \text{ psi}$$

D. Design of Side Wall b = 20 ft

1. SKIN PLATING ALONG WALL "b"

a. *Small deflection theory*

Use stiffeners at 4 ft centers.

The average pressure between levels A and B, Figure 7.93, is

$$w = \frac{1}{2}(675 + 800) = 738 \text{ lb/ft}^2$$

The bending moment is

$$M = \frac{w\ell^2}{10} = \frac{738 \times 4^2 \times 12}{10} = 14,170 \text{ lb in}$$

Assume a plate thickness of $\frac{5}{8}$ in. The section modulus for 12 in. width of a plate are

$$S = \frac{bh^2}{6} = \frac{12}{6}(\frac{5}{8})^2 = 0.781 \text{ in}^3$$

Stress

$$f = \frac{M}{S} = \frac{14,170}{0.781} = 18,143 \text{ psi}$$

b. *Large deflection theory*

Use a plate

$$t = \frac{1}{2} \text{ in}$$

$$P_n = 738 \text{ psf} = 5.125 \text{ psi}$$

The tensile force

$$S = \frac{1}{3.14} \sqrt[3]{\frac{4(5.125x48)^2 \times 29 \times 10^6 \times 0.5}{3.14(1-03^2)}}$$

$$= 3408 \text{ lbs}$$

$$S_E = 1420 \text{ lbs}$$

The bending moment

$$M = \frac{5.125 \times 48^2}{8} (1-1.034\frac{3408}{3408+1420})$$

$$= 400 \text{ lb in.}$$

The stress in the plate

$$f = \frac{3408}{0.5} + \frac{6 \times 400}{0.25} = 16,416 \text{ psi}$$

Use plating

$$t = \frac{1}{2} \text{ in}$$

2. VERTICAL STIFFENERS BETWEEN LEVELS A AND B

From Table 7.12, the pressures are

$$P_A = 800 \text{ lb/ft}^2 \ ; \quad P_B = 675 \text{ lb/ft}^2$$

Since

$$\beta = \frac{P_A}{P_B} = \frac{800}{675} = 1.185$$

Then, the coefficient in Table 7.11 is

$$\alpha = 0.1250 + \frac{0.185 \times 0.0156}{0.250} = 0.1365$$

$$M = 0.1365 \times 675 \times 10^2 \times 4 \times 12 =$$

$$= 442{,}260 \text{ lb in}$$

Try a stiffener structural ST6 WF 18#.

$$y = \frac{20x0.5x6.37+0.305x5.58x3.33+6.565x0.54x0.27}{20x0.5+0.305x5.58+6.565x0.54}$$

$$= 4.61 \text{ in}$$

$$I_n = \frac{1}{12} \times 20 \times 0.5^3 + 20 \times 0.5 \times 1.76^2$$

$$+ \frac{1}{12} \times 0.305 \times 5.58^3 + 0.305 \times 5.58 \times$$

$$\times 1.28^2 + \frac{1}{12} \times 6.565 \times 0.54^3 +$$

$$+ 6.565 \times 0.54 \times 4.34^2 = 105.248 \text{ in}^4$$

$$S_c = \frac{105.248}{2.01} = 52.36 \text{ in}^3$$

$$S_t = \frac{105.248}{4.61} = 22.83 \text{ in}^3$$

$$f_{comp} = \frac{442{,}260}{52.36} = 8{,}446 \text{ psi}$$

$$f_{ten} = \frac{442{,}260}{22.83} = 19{,}372 \text{ psi}$$

Use a stiffener ST6 WF @ 18#.

3. HORIZONTAL STIFFENER AT LEVEL B

From Table 7.13, we find:
Tensile force

$$T_a = 3500 \times 10 = 35{,}000 \text{ lbs}$$

Moment at support

$$M_c = 16{,}690 \times 10 \times 12 = 2{,}002{,}800 \text{ lb in}$$

Use WF10 × 10 @ 100#, A = 29.43 in²

$$S_x = 112.4 \text{ in}^3$$

$$f = \frac{35{,}000}{29.43} + \frac{2{,}002{,}800}{112.4} = 1189 + 17{,}818 =$$

$$= 19{,}007 \text{ psi}$$

E. Design of Hopper. Wall "a" = 12 ft

1. PLATING OF THE HOPPER

a. *Small deflection theory*

Use stiffeners at 4 ft centers. The pressure on the plating from Table 7.14, is

$$P_{na} = 1157 \text{ psf}$$

The bending moment is

$$M = \frac{1157 \times 4^2 \times 12}{10} = 22{,}214 \text{ lb in}$$

Assume a plate thickness of t = ¾ in. The section modulus for 12 in width of plate is

$$S = \frac{bt^2}{6} = \frac{12}{6}\left(\frac{3}{4}\right)^2 = 1.125 \text{ in}^3$$

$$f = \frac{M}{S} = \frac{22{,}214}{1.125} = 19{,}745 \text{ psi}$$

b. *Large deflection theory*

Use plate having

$$t = \frac{1}{2} \text{ in}$$

$$P_{na} = 1157 \text{ psf} = 8 \text{ psi}$$

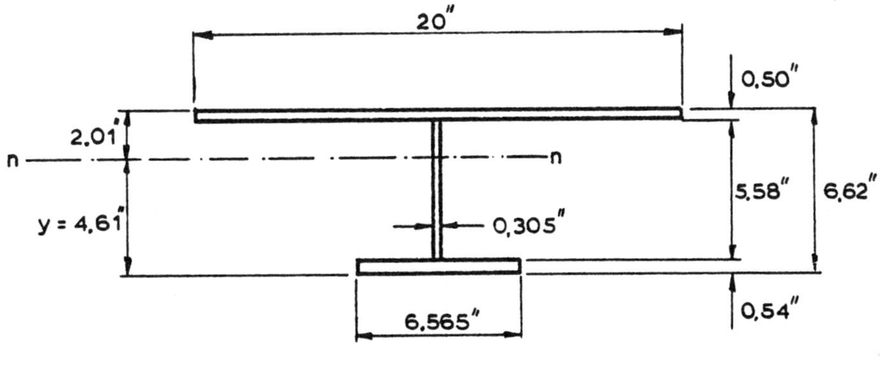

FIGURE 7.96 — Vertical stiffener cross-section.

The tensile force

$$S = \frac{1}{3.14} \sqrt[3]{\frac{4(8x48)^2 x29x10^6 x0.5}{3.14(1-0.3^2)}} = 4585 \text{ lbs}$$

$$S_E = 1420 \text{ lbs}$$

$$M = \frac{8 \times 48^2}{8} (1-1.034\frac{4585}{4585+1420}) = 486 \text{ lb in}$$

$$f = \frac{4585}{0.5} + \frac{6 \times 486}{0.25} = 20,834 \text{ psi}$$

which is satisfactory. Use plate

$$t = \frac{1}{2} \text{ in}$$

2. DESIGN OF HOPPER STIFFENERS

The actual length of hopper stiffener

$$\frac{8.5}{\sin\alpha_a} = \frac{8.5}{0.707} = 12.02 \text{ ft}$$

As A and H the normal pressures from Table 7.14 are:

$$P_{nA} = 1157 \text{ lb/ft}^2$$

$$P_{nH} = 1157 \text{ lb/ft}^2$$

Using formula (7.175)

$$\beta = \frac{1157}{1157} = 1.00$$

From Table 7.11

$$\alpha = 0.125$$

The maximum bending moment per unit length is

$$M_x = \alpha P_{nA} L^2 = 0.125 \times 1157 \times 12.02^2 \times 12 =$$

$$= 250,745 \text{ lb in}$$

Using stiffeners at 4 ft spacings, the total bending moment is

$$M_{tot} = 250,745 \times 4 = 1,002,980 \text{ lb in}$$

and total tensile force from Table 7.14 is

$$T_{ma} = 9090 \times 4 = 36,360 \text{ lbs}$$

Try ST 10 WF @ 31#

$$y = \frac{20x0.5x10.74\ 0.40x9.875x5.552\ 8.24x0.615x0.307}{20x0.5\ 9.875x0.4\ 8.24x0.615}$$

$$= 6.88 \text{ in}$$

$$I_n = \frac{1}{12} \times 20 \times 0.5^3 + 20 \times 0.5 \times 3.86^2 + \frac{1}{12} \times 0.40 \times 9.875^3 + 0.40 \times 9.875 \times 1.33^2 + \frac{1}{12} \times 8.24 \times 0.615^3 + 8.24 \times 0.615 \times 6.573^2 = 407,288 \text{ in}^4$$

$$S_{comp} = \frac{407,288}{4.11} = 99.097 \text{ in}^3$$

$$S_{tens} = \frac{407,288}{6.88} = 59.198 \text{ in}^3$$

$$f_{tens} = \frac{36,360}{19.12} + \frac{1,002,980}{59.198} = 18,843 \text{ psi}$$

Use ST 10 WF @ 31#

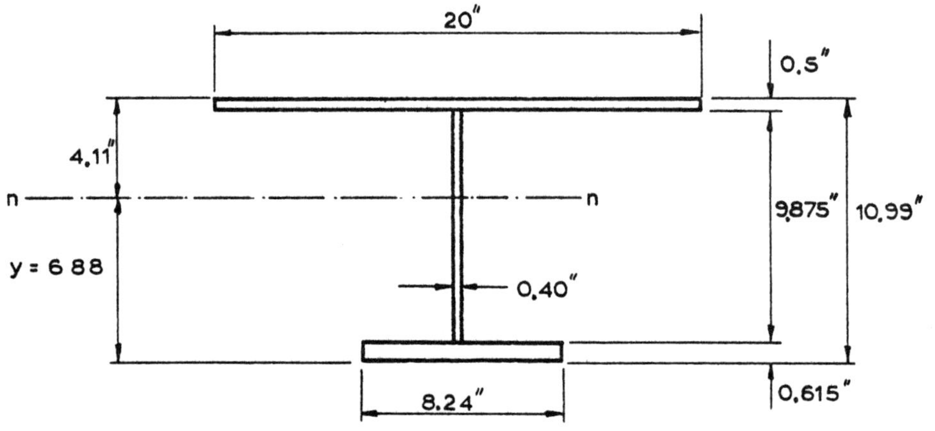

FIGURE 7.97 — Hopper stiffener cross-section.

F. Design of Hopper Wall "b" = 20 ft

1. PLATING OF THE HOPPER

a. *Small deflection theory*

Use stiffeners at 4 ft centers.

The pressure on the plating from Table 7.15 is

$$P_{nb} = 1{,}092 \text{ psf}$$

The bending moment is

$$M = \frac{1092 \times 4^2 \times 12}{10} = 20{,}966 \text{ lb in}$$

Assume a plate thickness of t = ¾ in.

The section modulus for 12 in width of plate is

$$S = \frac{bt^2}{6} = \frac{12}{6}\left(\frac{3}{4}\right)^2 = 1.125 \text{ in}^3$$

$$f = \frac{M}{S} = \frac{20{,}966}{1.125} = 18{,}636 \text{ psi}$$

b. *Large deflection theory*

Use plate having

$$T = \frac{1}{2} \text{ in}$$

$$P_{nb} = 1092 \text{ psf} = 7.58 \text{ psi}$$

The tensile force

$$S = \frac{1}{3.14}\sqrt[3]{\frac{4(7.58 \times 48)^2 \times 29 \times 10^6 \times 0.5}{3.14(1-0.3^2)}} =$$

$$= 4427 \text{ lbs}$$

$$S_E = 1420 \text{ lbs}$$

$$M = \frac{7.58 \times 48^2}{8}\left(1 - 1.034\frac{4427}{4427 + 1420}\right) =$$

$$= 474 \text{ lb in}$$

$$f = \frac{4427}{0.5} + \frac{6 \times 474}{0.25} = 8854 + 11376 =$$

$$= 20{,}230 \text{ psi}$$

which is satisfactory.

Use plate

$$t = \frac{1}{2} \text{ in}$$

2. DESIGN OF HOPPER STIFFENERS

The actual length of hopper stiffener

$$\frac{8.5}{\sin\alpha_b} = \frac{8.5}{0.866} = 9.82 \text{ ft}$$

At A and H the normal pressures from Table 7.15 are:

$$P_{nA} = 1092 \text{ lb/ft}^2$$

$$P_{nH} = 1092 \text{ lb/ft}^2$$

Using formula (7.175)

$$\beta = \frac{1092}{1092} = 1.00$$

From Table 7.11

$$\alpha = 0.125$$

The maximum bending moment per unit length is

$$M_x = \alpha P_{nA} L^2 = 0.125 \times 1092 \times 9.82^2 \times 12 =$$

$$= 157{,}956 \text{ lb in}$$

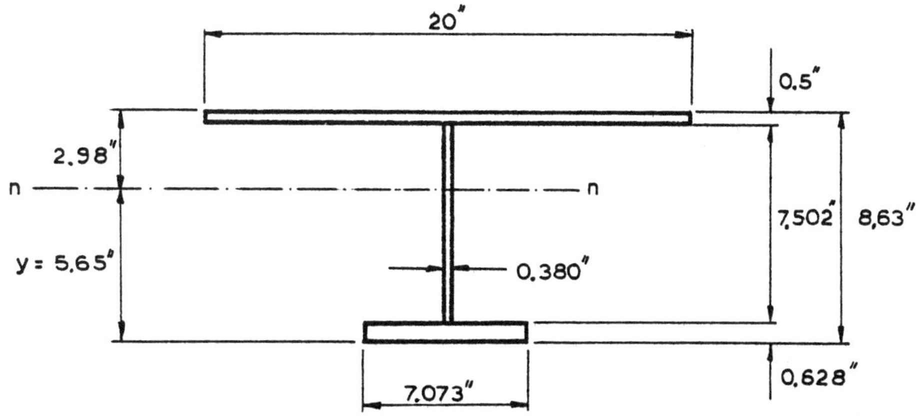

FIGURE 7.98 — Stiffener cross-section.

Using stiffeners at 4 ft spacings, the total bending moment is

$$M_{tot} = 157{,}956 \times 4 = 631{,}824 \text{ lb in}$$

and total tensile force from Table 7.18 is

$$T_{mb} = 4800 \times 4 = 19{,}200 \text{ lbs}$$

Try ST 8 WF @ 25#

$$y = \frac{20\text{x}0.5\text{x}8.38+0.38\text{x}7.502\text{x}4.378+7.073\text{x}0.628\text{x}0.314}{20\text{x}0.5+0.38\text{x}7.502+7.073\text{x}0.628}$$

$$= 5.65 \text{ in}$$

$$I_n = \frac{1}{12} \times 20 \times 0.5^3 + 20 \times 0.5 \times 2.73^3$$

$$+ \frac{1}{12} \times 0.38 \times 7.502^3 + 0.38 \times 7.502 \times$$

$$\times 1.272^2 + \frac{1}{12} \times 7.073 \times 0.628^3 +$$

$$+ 7.073 \times 0.628 \times 5.336^2 = 219.34 \quad \text{in}^4$$

$$S_{comp} = \frac{219.34}{2.98} = 73.60 \text{ in}^3$$

$$S_{ten} = \frac{219.34}{5.65} = 38.82 \text{ in}^3$$

$$f = \frac{19{,}200}{17.35} + \frac{631{,}824}{38.82} = 17{,}382 \text{ psi}$$

APPENDIX B — EXAMPLE NUMBER 2
Design of Rectangular Bunker

THE PROBLEM

Design a single cell rectangular steel bunker for storing coal. The dimensions are shown in Figure 7.99. The bottom is a symmetrical pyramidal hopper. The bunker walls rest on the hopper base, which is supported by four columns.

For coal:

$$\gamma = 32 \text{ lb/ft}^3$$

$$\phi = 30^{\circ}$$

$$K = \frac{1-\sin 30}{1+\sin 30} = \frac{1-0.5}{1.5} = \frac{0.5}{1.5} = \frac{1}{3}$$

A. Vertical and Horizontal Forces, Figure 7.99

Both forces are calculated for the unit length of the bunker.

1. VERTICAL FORCES

Weight of coke

$$W_1 = 32 \times 12 \times 8.5 = 3264 \text{ lbs}$$

$$W_2 = \frac{1}{2} \times 32 \times 12 \times 8.5 = 1632 \text{ lbs}$$

$$W_3 = 32 \times (12 + 12) \times 1.5 = 1152 \text{ lbs}$$

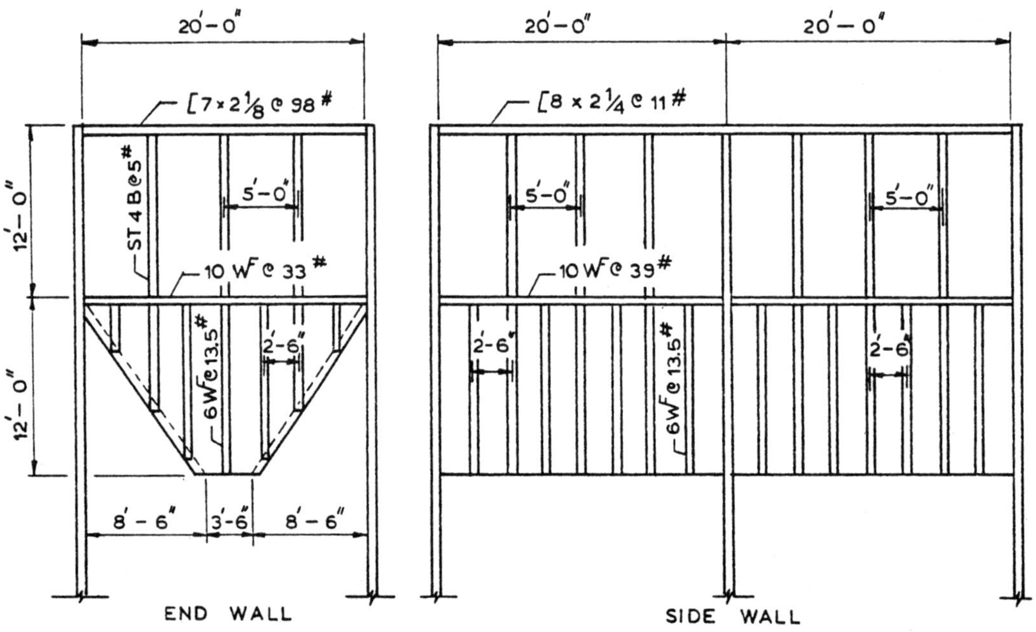

FIGURE 7.99 — Single rectangular bunker with pyramidal hopper.

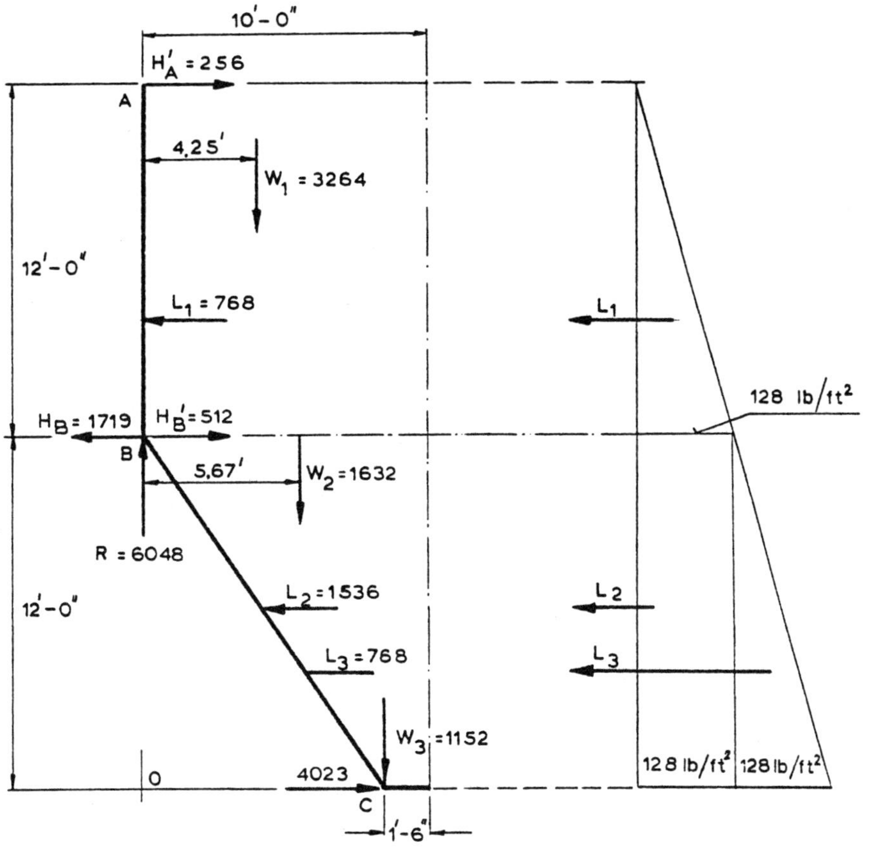

FIGURE 7.100 — Horizontal and vertical forces.

2. HORIZONTAL FORCES

At B $\quad K\gamma h = \frac{1}{3} \times 32 \times 12 = 128 \text{ lb/ft}^2$

At C $\quad K\gamma h = \frac{1}{3} \times 32(12 + 12) = 256 \text{ lb/ft}^2$

The later pressure is separated into two components: a constant lateral pressure between B and C of 128 lb/ft² plus a varying pressure, maximum at C of

$$\frac{1}{3} \times 32 \times 12 = 128 \text{ lb/ft}^2$$

Then

$$L_1 = \frac{1}{2} \times 128 \times 12 = 768 \text{ lb/ft}$$

$$L_2 = 128 \times 12 = 1536 \text{ lb/ft}$$

$$L_3 = \frac{1}{2} \times 128 \times 12 = 768 \text{ lb/ft}$$

3. HORIZONTAL EXTERNAL REACTIONS

Above the waist, the horizontal reactions at A and B are

$$H'_A = \frac{768}{3} = 256 \text{ lb/ft}$$

$$H'_B = \frac{2}{3} \times 768 = 512 \text{ lb/ft}$$

Below the waist, the reactions are: Horizontal reaction at B. By taking moments about the point 0 we obtain:

$$H_B = \frac{1}{12}(3264 \times 4.25 + 1632 \times 5.67 + 1152 \times$$

$$\times 8.5 - 1536 \times 6.0 - 768 \times 4.0) =$$

$$= \frac{1}{12} \times 20629 = 1719 \text{ lbs}$$

4. The vertical reaction below the waist is then

$$R = 3264 + 1632 + 1152 = 6048 \text{ lbs}$$

And checking to show that $\Sigma H = 0$

Forces	Reactions
-768 -1536 -768 —	+ 256 + 512 -1719 +4023
-3072	+3072

B. Pressure on Hopper Wall

At B the vertical pressure is 32 × 12 = 384 lb ft² and the horizontal pressure 128 lb/ft². Similarly, at C these pressures are 32 × (12 + 12) = 768 lb/ft² and 256 lb/ft² respectively. Using formula (7.51)

$$P_n = P_v \cos^2\alpha + P_h \sin^2\alpha$$

where

$$\cos\alpha = \frac{8.5}{\sqrt{8.5^2 + 12^2}} = \frac{8.5}{14.70} = 0.578$$

$$\sin\alpha = \frac{12}{12.70} = 0.816$$

The normal pressure at B is

$$P_{nB} = 384 \times 0.578^2 + 128 \times 0.816^2 =$$

$$= 128.28 + 85.23 = 213.51 \text{ lb/ft}$$

and the tangential pressure, formula (7.52)

$$P_{tB} = (P_v - P_h) \sin\alpha \cos\alpha$$

$$P_{tB} = (384 - 128) \times 0.816 \times 0.578 =$$

$$= 120.74 \text{ lb/ft}^2$$

Similarly at C the normal pressure is

$$P_{nC} = 768 \times 0.578^2 + 256 \times 0.816^2 =$$

$$= 256.57 + 170.46 = 427.03 \text{ lb/ft}$$

and the tangential pressure

$$P_{tC} = (768 - 256) \times 0.816 \times 0.578 =$$

$$= 241.48 \text{ lb/ft}$$

The normal load supported on the hopper wall excluding self weight, according to formula (7.156) is

$$R_N = \left(\frac{P_{nC} + P_{nB}}{2}\right) h_2 \operatorname{cosec}\alpha$$

$$R_N = \frac{(427.03 + 213.51)}{2} \times 12 \times \frac{14.70}{12} =$$

$$= 4708 \text{ lbs}$$

By resolving the vertical and horizontal reactions at B and C this will serve as a check on the proceeding calculations.

1. At b, the normal reation is

$$6048 \cos\alpha - 0.719 \times \sin\alpha = 6048 \times 0.578 - 1719 \times 0.816 =$$

$$= 3496 + 1403 = 2093 \text{ lbs}$$

2. The tangential component is

$$6048 \sin\alpha + 1719 \cos\alpha = 6048 \times 0.816 + 1719 \times 0.578 =$$

$$= 4935 + 994 = 5929 \text{ lbs}$$

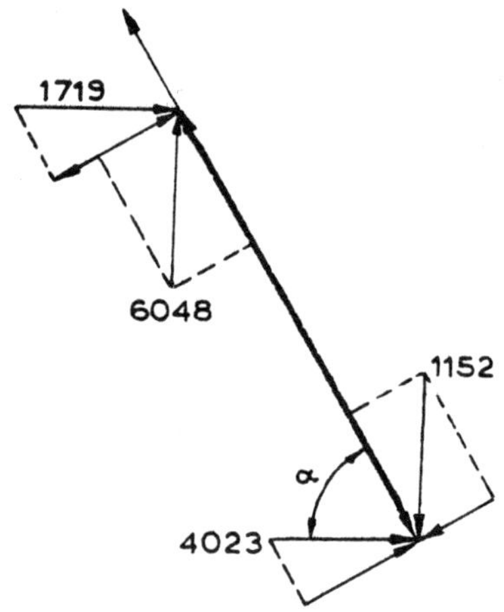

FIGURE 7.101 — Determination of normal and tangential components at B and C.

3. At C, the normal component will be

$$4023 \sin\alpha - 1152 \cos\alpha = 4023 \times 0.816 - 1152 \times 0.578 =$$

$$= 3283 - 666 = 2617 \text{ lbs}$$

4. The tangetial component is

$$4023 \cos\alpha + 1152 \sin\alpha = 4023 \times 0.578 + 1152 \times 0.816 =$$

$$= 2325 + 940 = 3265 \text{ lbs}$$

Therefore, the total normal load supported on the hopper wall will be

$$2093 + 2617 = 4710 \text{ lbs}$$

C. Design of Side Walls

The bunker will be carried by three pairs of supports arranged to give two 20 ft. by 20 ft. grids. The corner supports will coincide with the junction of the side and end walls, while one pair of intermediate supports will be treated as a portal frame spanning 20 ft.

The side walls will span 20 ft. Calculations for the side walls will be based upon the assumption that as a plate stiffened against horizontal pressure, it will be strong enough in its total depth to support vertical loads without the provision of a separate beam. The horizontal forces, at A and B, will be taken by horizontal beams. The skin plate will have vertical intermediate stiffeners.

1. SKIN PLATES

Stiffeners at 5 ft centers are used. The ratio of panel height to width is $\frac{12}{5} = 2.4$, therefore the design plate should span in one direction.

The pressure at B is 128 lb/ft^2. Assume a bending moment of

$$M = \frac{w\ell^2}{10}$$

then

$$M = \frac{128 \times 5^2 \times 12}{10} = 3840.0 \text{ lb in}$$

Assume a plate thickness of 5/16 in. The section modulus for 12 in. width of the plate is then

$$S = \frac{bh^2}{6} = \frac{12}{6} \times \left(\frac{5}{16}\right)^2 = 0.195 \text{ in}^3$$

stress

$$f = \frac{M}{S} = \frac{3840}{0.195} = 19{,}692 \text{ psi}$$

which is satisfactory.

The vertical load carried by the side wall is

$$W = 20 \times 6048 = 120{,}960 \text{ lbs}$$

giving a shear stress in the wall plate of

$$f_s = \frac{120{,}960 \times 16}{2 \times 12 \times 12 \times 5} = 1344 \text{ psi}$$

The shear stress is very small. Checking critical stresses according to the formula

$$f_{cr} = \frac{K\pi^2 D}{b^2 t}$$

where

$$D = \frac{Et^3}{12(1-\nu^2)} = \text{the flexural rigidity of the plate}$$

$$K = 5.35 - 4(b/h)^2 = \text{a factor}$$

For $b = 5$ ft, $h = 12$ ft, $b/h = 0.416$

$$K = 5.35 - 4\left(\frac{5}{12}\right)^2 = 5.35 - 0.69 = 4.66$$

$$D = \frac{30 \times 10^6\ t^3}{12(1-0.3^2)} = 274 \times 10^4\ t^3$$

$$f_{cr} = \frac{4.66 \times 3.14^2 \times 274 \times 10^4}{60^2} \times \left(\frac{5}{16}\right)^2 =$$

$$= 4098 \text{ psi}$$

Therefore, the safety factor against crippling is

$$\frac{f_{cr}}{f} = \frac{4098}{1344} = 3.05$$

The plate is therefore safe.

2. VERTICAL STIFFENERS

The total triangular load to be supported is

$$W = 768 \times 5 = 3840 \text{ lbs}$$

Using the coefficients in Table 7.11, since $\beta = 0$, $\alpha = 0.0641$, then, according to Equation (7.186)

$$M = \alpha P_h \ell^2$$

or

$$M = 0.0641 \times 128 \times 12^2 \times 60 = 70{,}889 \text{ lb in}$$

Try a stiffener structural ST4B @ 5#, including

$$40t = 40 \times \frac{5}{16} \quad 12.5 \text{ in}$$

$$y = \frac{12.5 \times 0.312 \times 4.106 + 0.17 \times 3.746 \times 2.077 + 3.94 \times 0.204 \times 0.102}{12.5 \times 0.312 + 0.17 \times 3.746 + 3.94 \times 0.204}$$

$$= \frac{16.000 + 1.322 + 0.081}{3.90 + 0.64 + 0.80} = \frac{17.40}{5.33} = 3.264 \text{ in}$$

$$I_n = \frac{1}{12} \times 12.5 \times 0.312^3 + 12.5 \times 0.312 \times$$

$$\times 0.842^2 + \frac{1}{12} \times 0.17 \times 3.746^3 + 0.17 \times$$

$$\times 3.746 \times 1.187^2 + \frac{1}{12} \times 3.94 \times$$

$$0.204^3 + 3.94 \times 0.204 \times 3.162^2$$

$$= 0.031 + 2.764 + 0.745 + 0.897 + 0.027 +$$

$$+ 8.036 = 12.475 \text{ in}^4$$

$$S_c = \frac{12.475}{0.998} = 12{,}500 \text{ in}^3$$

$$S_t = \frac{12.475}{3{,}264} = 3.821 \text{ in}^3$$

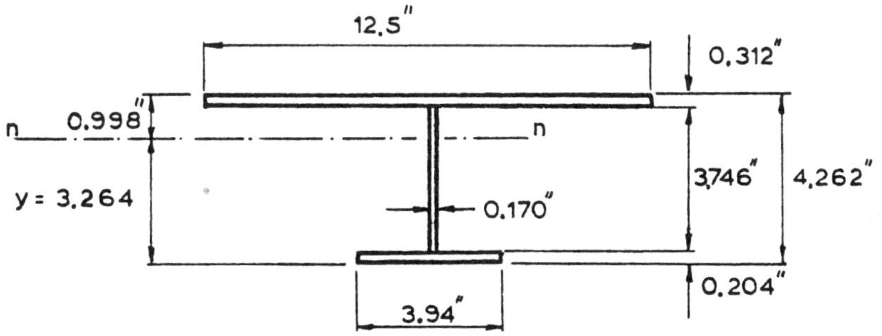

FIGURE 7.102 — Vertical stiffener. Cross-section.

$$f_{comp} = \frac{70,889}{12.500} = 5,671 \text{ psi}$$

$$f_{ten} = \frac{70,889}{3.821} = 18,552 \text{ psi}$$

which are satisfactory.

3. HORIZONTAL STIFFENER AT A

The horizontal force at this level is 256 lbs/ft and the bending moment is

$$M = \frac{256 \times 20^2 \times 12}{8} = 153,600 \text{ lb in}$$

using a stiffener from an [8 × 2¼ @ 11.5#] S_{XX} = 8.1 in³, the stress is

$$f = \frac{153,600}{8.1} = 18,962 \text{ psi}$$

the compression flange is fully restrained.

4. HORIZONTAL STIFFENER AT B

The net horizontal forces at this level is

$$1719 - 512 = 1207 \text{ lbs}$$

The bending moment is

$$M = \frac{1207 \times 20^2 \times 12}{8} = 724,200 \text{ lb in}$$

using a stiffener WF 10 × 8 @ 39#, S_X = 42.2 in³

$$f = \frac{724,200}{42.2} = 17,161 \text{ psi}$$

The vertical force is 6048 lb/ft which must be resisted in transverse tension in the flange. This force gives a stress of

$$f_t = \frac{6048}{12 \times 7.99 \times 0.528} = 119.46 \text{ psi}$$

D. End Walls

The end walls serve the same purpose as the side walls and will be subject to similar horizontal pressures. The values given for L_1, L_2, and L_3 in Figure 7.99, are appropriate.

At the center of the end wall the horizontal reactions, called levels A, B and C, will take the values

$$H_A = \frac{L}{3} = \frac{768}{3}$$

$$H_B = \frac{2L}{3} + \frac{L}{2} + \frac{L}{3} = \frac{2 \times 768}{3} + \frac{1536}{2} + $$

$$+ \frac{768}{3} = 512 + 768 + 256 = 1536 \text{ lb ft}$$

$$H_C = \frac{L}{2} + \frac{2L}{3} = \frac{1536}{2} + \frac{2 \times 768}{3} =$$

$$= 768 + 512 = 128.0 \text{ lb ft}$$

Above the waist, the vertical stiffeners will be positioned at 5 ft centers. Their size will be identical to those of the side wall stiffeners.

Below the waist the vertical stiffeners in the middle section of the end plate will be placed at 2.5 ft centers and opened out to 5 ft centers toward the ends. A calculation will show that these stiffeners can be made identical with the side wall stiffeners.

The horizontal stiffeners at levels A and B will be designed assuming that the lateral pressure is shared between the side and end walls.

a. *Horizontal stiffener at level A*

Figure 7.106 shows the loading arangement assumed for the horizontal stiffener at level A.

$$P_1 = \frac{1}{2} \times 256 \times 8.5 = 1088 \text{ lbs}$$

$$P_2 = 256 \times 3.0 = 768 \text{ lbs}$$

Reactions

$$R = 1088 + \frac{1}{2} \times 768 = 1472 \text{ lbs}$$

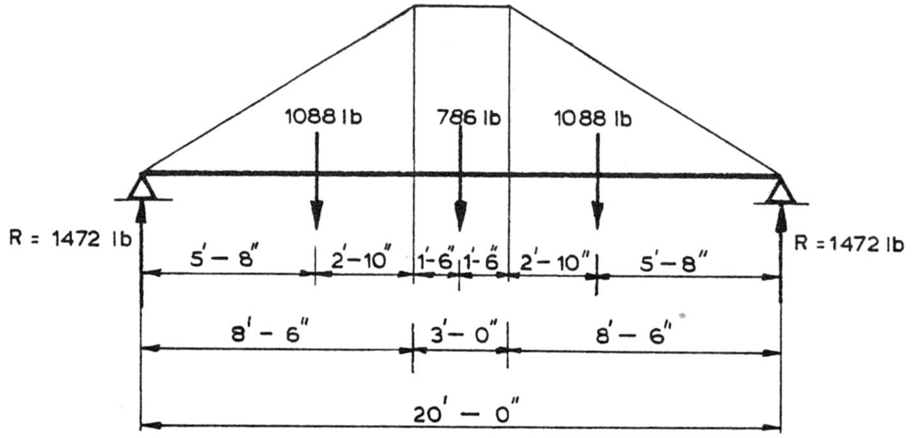

FIGURE 7.103 — Diagram of loading at level A.

Maximum bending moment

$$M_{max} = 12[1472 \times 10 - 1088 \times 4.33 - 384 \times 0.75] = 116{,}652 \text{ lb/in}$$

Using a stiffener from a [7 × 2⅛ @ 9.8#], $S_X = 6 \text{ in}^3$, the stress is

$$f = \frac{116{,}652}{6} = 19{,}442 \text{ psi}$$

b. *Horizontal stiffener at level B*

Figure 7.104 shows the loading arrangement assumed for the horizontal stiffener at level B.

$$P_1 = \frac{1}{2} \times 1536 \times 8.5 = 6528 \text{ lbs}$$

$$P_2 = 1536 \times 3.0 = 4608 \text{ lbs}$$

Reaction

$$R = 6528 + \frac{1}{2} \times 4608 = 8832 \text{ lbs}$$

Maximum bending moment

$$M_{max} = 12[8832 \times 10 - 6528 \times 4.33 - 2304 \times 0.75] = 699{,}912 \text{ lb in}$$

Using a stiffener WF 10 × 8 @ #33, $S_x = 35 \text{ in}^3$

$$f = \frac{699{,}912}{35} = 19{,}997 \text{ psi}$$

E. Plating of the Hopper

The maximum normal pressure at C is

$$P_{nC} = 427.0 \text{ lb/ft}$$

The tangential pressure at C is

$$P_{tC} = 241.48 \text{ lb/ft}$$

Considering the continuity of the wall, assume

$$M = \frac{P_{nC} b^2}{12}$$

The permissible bending stress

$$f = 20{,}000 \text{ psi}$$

With 5/16 in plate, the section modulus is

$$\frac{12}{6} \times \left(\frac{5}{16}\right)^2 = 0.195 \text{ in}^3$$

The spacing of the stiffeners is

$$\frac{427.0 \times b^2 \times 12}{12} = 0.195 \times 20{,}000$$

$$b = \sqrt{\frac{0.195 \times 20{,}000}{427.0}} = 3.02 \text{ ft} = 36.26 \text{ in}$$

or

$$b = 60\,t = 60 \times \frac{5}{16} = 18.75 \text{ in}$$

The final choice of plate thickness will be deferred until the stiffeners have been designed.

F. Design of Hopper Stiffeners

The actual length of hopper stiffeners is

$$\frac{12}{\sin\alpha} = \frac{12}{0.816} = 14.70 \text{ ft}$$

At B the normal pressure is

$$P_{nB} = 213.51 \text{ lb/ft}$$

and the tangential pressure is

$$P_{tB} = 120.74 \text{ lb/ft}$$

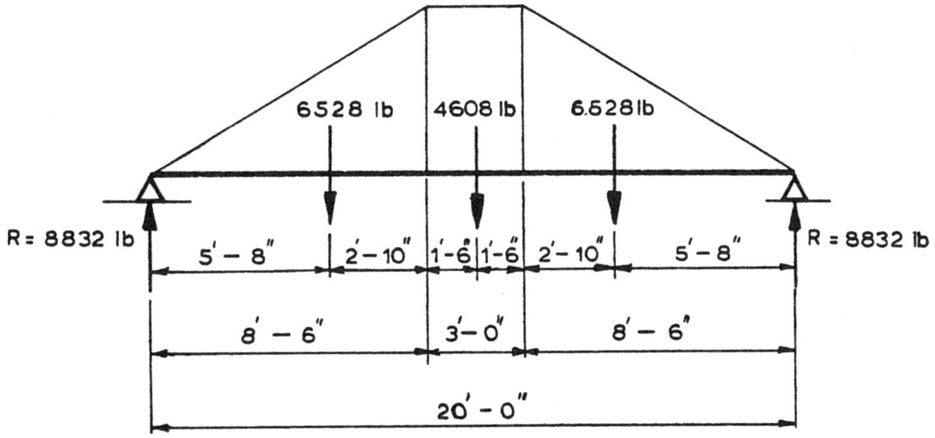

FIGURE 7.104 — Diagram of loading at level B.

Similarly, at C the normal pressure is

$$P_{nC} = 427.03 \text{ lb/ft}$$

and the tangential pressure is

$$P_{tC} = 241.48 \text{ lb/ft}$$

Using formula (6.187)

$$\beta = \frac{P_C}{P_B} = \frac{427.03}{213.51} = 2.00$$

From Table 7.11

$$x = 0.5276 \text{ L and } \alpha = 0.1881$$

Therefore, the maximum bending moment per unit length is

$$M_x = \alpha P_B L^2 = 0.1881 \times 213.51 \times 14.70^2 =$$

$$= 8678.44 \text{ lb ft}$$

The tension to be resisted per unit length, according to the expression (7.188)

$$T_x = T_B - [P_{tB}x - (P_{tB}-P_{tC})\frac{x^2}{2L}]$$

$$x = 0.5276 \times 14.70 = 7.75 \text{ ft}$$

$$T_x = 5929 - [120.74 \times 7.75 -$$

$$- (120.74-241.48)\frac{7.75^2}{2 \times 14.70}] =$$

$$= 5929 - [935.73 + 246.65] = 4747 \text{ lbs}$$

Try a stiffener arrangement at 30 in, bending and tension carried by a built-up section consisting of a T-section and part of the hopper plate, Figure 7.105.

Try a stiffener structural ST 6 WF @ 13.5#

$$y = \frac{12.5\text{x}0.312\text{x}6.136+0.24\text{x}5.58\text{x}3.19+6.5\text{x}0.4\text{x}0.2}{12.5\text{x}0.312+0.24\text{x}5.58+6.5\text{x}0.4}$$

$$= \frac{23.93+4.27+0.52}{3.900+1.339+2.6} = \frac{28.72}{7.889} = 3.66 \text{ in}$$

$$I_n = \frac{1}{12} \times 12.5 \times 0.312^3 + 12.5 \times 0.312 \times 2.476^2 + \frac{1}{12} \times 0.24 \times 5.58^3 + 0.24 \times 5.58 \times 0.47^2 + \frac{1}{12} \times 6.5 \times 0.4^3 + 6.5 \times 0.4 \times 3.46^2 = 0.032 + 23.91 + 3.474 + 0.295 + 0.035 + 31.126 =$$

$$= 58.87 \text{ in}^4$$

$$S_c = \frac{58.87}{2.632} = 22.36 \text{ in}^3$$

$$S_t = \frac{58.87}{3.66} = 16.08 \text{ in}^3$$

$$f_{comp} = \frac{8678.44\text{x}2.5\text{x}12}{22.36} = 11{,}643 \text{ psi}$$

$$f_{tens} = \frac{8678.44\text{x}2.5\text{x}12}{16.08} = 16{,}191 \text{ psi}$$

$$f_{tens} = \frac{4747\text{x}2.5}{} = 1{,}506 \text{ psi}$$

Total tensile stress

$$f_{tot} = 16{,}191 + 1{,}506 = 17{,}697 \text{ psi}$$

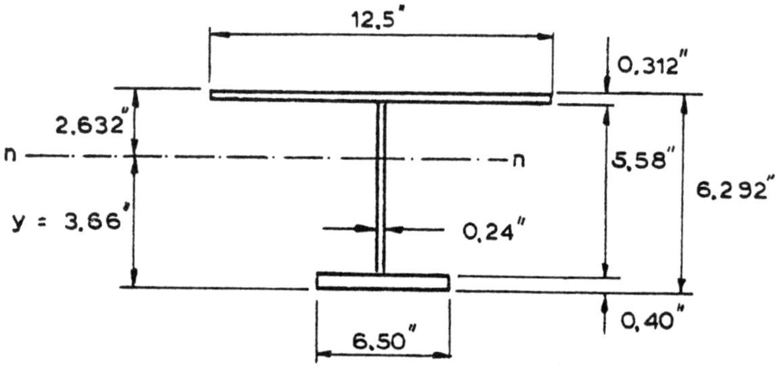

FIGURE 7.105 — Hopper stiffener. Cross-section.

REFERENCES — CHAPTER 7

[7.1] Fischer, W., "Silos and Bunkers in Stahlbau", VEB Verlag fur Bauwesen, Berlin, 1966 (in German).

[7.2] Soviet Concrete and Reinforced Concrete Code (Stroitelnie Normi i Pravila, Tshast II, Razdel B, Glava 1, Betonnie i Zelezobetonnie Konstruktsii SNKP) II-B.1-62. Publication of the Committee of Soviet Ministers USSR of Construction Works, Moscow 1962 (in Russian).

[7.3] Carr, R.L., Jr., "Evaluating Flow Properties of Solids", Chemical Engineering, Jan. 18th, 1965.

[7.4] Carr, R.L., Jr., "Particle Behavior, Storage and Flow", ASME Publication No. 68 — MH-6, Oct. 1968.

[7.5] Jenike, A.W., "Flow of Bulks Solids", Bulletin 64, Utah Engineering Experimental Station, University of Utah, March, 1954.

[7.6] Theimer, O.F., "Ablauf fordernde Trichterkonstructionen von Silozellen", Aufberechnungs — Technik No. 10, Oct. 1969, pp. 547-556.

[7.7] Kvapil R., and Taubmann, H.J., "Flow and Extraction of Solids from Bins", Publication 68-MH-32, ASME, New York, 1968.

[7.8] Jenike, A.W., and Johanson, J.R., "Annual Report," Project No. 126 — Fourth Part, "Mass-Flow Bins" and Project 126A — First Part, "Funnel-Flow Bins," American Iron and Steel Institute, 1971.

[7.9] Johanson, J.R., and Colijn, H., "New Design Criteria for Hoppers and Bins," Iron and Steel Engineer, Vol. XLI, October 1964, pp. 85-104.

[7.10] Johanson, J.R., "Effect of Initial Pressures on Flowability of Bins," Journal of Engineering for Industry, May 1965, pp. 395-399.

[7.11] Jenike, A.W., "Storage and Flow of Solids", Bul. 123, Utah Engineering Experiment Station, University of Utah, Nov. 1964.

[7.12] Giunta, J.S., "Flow Patterns of Granular Materials in Flat-Bottom Bins", Trans. ASME, J. of Engr. for Ind., May 1969.

[7.13] Janssen, H.A., "Versuch uber Getreidedruck in Silozellen", Zeitschrift verein Deutscher Ingenieure, Vol. 39, August 1895, pp. 1045-1049.

[7.14] Buisman, A.S.K., "Grondmechanica", Ed. Waltman, Delft, 1940, pp. 205-209.

[7.15] Reimbert, M., and Reimbert, A., "Silos-Traite′ Theoretique et Pratique", Editions Eyrolles, Paris, 1961.

[7.16] Petrov, B.A., "Experimental Determination of Cement Pressure on Reinforced Concrete Silo Walls", Cement No. 2, 1958, pp. 21-25 (in Russian).

[7.17] Pieper, K., and Wenzel, F., "Druckverhaltnisse in Silozellen", Verlag von Wilhelm Ernst and Sohn, Berlin, Munich, 1968.

[7.18] Pieper, K., "Investigation of Silo Loads in Measuring Models", Publication 68-MH-30, ASME, New York, 1968.

[7.19] Pieper, K., and Wagner, K., "Der Einfluss Verschiedner Auslaufarten auf die Seitendrucke in Silozellen", Aufbereitungs Technik No. 10, Oct. 1969, pp. 542-546.

[7.20] Deutsch, G.P., and Schmidt, L.C., "Pressures on Silo Walls," Publication 68-MH-24, ASME, New York, 1968.

[7.21] Takahashi, K., "Silo Design", Concrete Journal, Japanese National Council on Concrete, Vol. 5, No. 9, Aug. 1969 (in Japanese).

[7.22] Soviet Code CH-302-65, "Instructions for Design of Silos for Granular Materials," Moscow, USSR, 1965 (in Russian).

[7.23] Deutsche Normen, DIN-1055, Blatt 6, Lasten in Silozellen, November 1964, pp. 1-6 (in German).

[7.24] Plátonov, P.N., and Kovtun, A.P., "The Pressure of Grain on Silo Walls", Mukomolno Elevatornaia Promyschlennost, Moscow, Vol. 25, Dec. 1959, pp. 22-24 (in Russian).

[7.25] Leonhardt, F., "Zur Frage der Sicheren Bemessung von Zement-Silos", Beton-und Stahlbetonbau, Berlin, Vol. 55, No. 3, March 1960, pp. 48-58.

[7.26] Terzaghi, K., "Theoretical Soil Mechanics", 5th Ed., John Wiley & Sons Inc., New York, 1948, pp. 77-81.

[7.27] Lipnitski, M.E., and Abramovitsch, S.P., "Reinforced Concrete Bunkers and Silos," Izdatelstvo Literaturi Po Stroitelstvu, Leningrad, 1967 (in Russian).

[7.28] Reimbert, M., and Reimbert, A., "Silos Berechnung, Beitrieb und Ausfuhrung", Bauverlag GMBH, Berlin, 1975, p. 123.

[7.29] Alferov, K.V., "Bunkers, Gates and Conveyors", Mashgiz, Moskow, USSR, 1946 (in Russian).

[7.30] Lichtarnikov, J.M., Klykov, V.M., and Ladyzhensky, D.V., "Design of Steel Structures", Budivelnik, Kiev, USSR, 1976 (in Russian).

[7.31] Lambert, F.W., "The Theory and Practical Design of Bunkers", The British Constructional Steelwork Association Ltd., London, 1968.

[7.32] Gieselski, R., et al., "Behalter, Bunker, Silos, Shornsteine, Fernsehturme and Freileitungsmaste", Verlag von Wilhelm Ernst und Sohn, Berlin, 1970.

[7.33] Safarian, S., and Harris, E.C., "Silos and Bunkers", Handbook of Concrete Engineering, Ed. by M. Fintel, Van Nostrand Reinhold Company, New York, 1974.

[7.34] Lipnitski, M.E., and Abramovitsch, S.P., "Reinforced Concrete Bunkers and Silos," Izdatelstvo Literaturi Po Stroitelstvu, Leningrad, 1960 (in Russian).

[7.35] Ketchum, M.S., "Walls, Bins and Grain Elevators", McGraw-Hill, New York, 3rd Edition, 1919.

[7.36] Lambert, F.W., "The Theory and Practical Design of Bunkers", The British Constructional Steelwork Association Ltd., London, 1968.

[7.37] Rogers, P., "Design of Large Coal Bunkers", Transactions ASCE, Vol. 117, Paper No. 2507, 1952.

[7.38] Timoshenko, S., and Woinowsky-Krieger, S., "Theory of Plates and Shells", 2nd Edition, McGraw-Hill Company, Inc., New York, 1959.

[7.39] Uniform Building Code, Vol. 1, International Conference of Building Officials, Pasadena, California, 1970.

[7.40] Chandrasekaran, A.R., and Jain, P.C., "Effective Live Load of Storage Materials Under Dynamic Conditions", Indian Concrete Journal, pp. 369-385, Sept. 1968.

[7.41] Safarian, S.S., and Harris, E.C., "Silos and Bunkers", Handbook Concrete Engineering, Van Nostrand Reynolds Ltd., New York, p. 497, 1974.

[7.42] Gieselski, R., et al., "Behalter, Bunker, Silos, Shornsteine, Fernsehturme und Freileitungsmaste", Verlag von Wilhelm Ernst und Sohn, Berlin, 1970.

[7.43] Beyer, K., Die Statik im Stahlbetonbau, 2 Auflage; Springer-Verlag, Berlin, 1956, pp. 31, 621-622.

[7.44] Timoshenko, S.P. and Gere, J.M., Theory of Elastic Stability, 2nd Ed., McGraw-Hill Book Company, Inc., New York, 1961, p. 289.

[7.45] Roark, R.J. and Young, W.C., Formulas for Stress and Strain, 5th Ed., McGraw-Hill Book Company, New York, 1975, p. 236.

[7.46] Pippard, A.J.S., Studies in Elastic Structures, Edward Arnold & Co., London, 1952, pp. 7-8.

Similarly, at C the normal pressure is

$$P_{nC} = 427.03 \text{ lb/ft}$$

and the tangential pressure is

$$P_{tC} = 241.48 \text{ lb/ft}$$

Using formula (6.187)

$$\beta = \frac{P_C}{P_B} = \frac{427.03}{213.51} = 2.00$$

From Table 7.11

$$x = 0.5276 \text{ L and } \alpha = 0.1881$$

Therefore, the maximum bending moment per unit length is

$$M_x = \alpha P_B L^2 = 0.1881 \times 213.51 \times 14.70^2 =$$

$$= 8678.44 \text{ lb ft}$$

The tension to be resisted per unit length, according to the expression (7.188)

$$T_x = T_B - [P_{tB}x - (P_{tB}-P_{tC})\frac{x^2}{2L}]$$

$$x = 0.5276 \times 14.70 = 7.75 \text{ ft}$$

$$T_x = 5929 - [120.74 \times 7.75 -$$

$$- (120.74-241.48)\frac{7.75^2}{2 \times 14.70}] =$$

$$= 5929 - [935.73 + 246.65] = 4747 \text{ lbs}$$

Try a stiffener arrangement at 30 in, bending and tension carried by a built-up section consisting of a T-section and part of the hopper plate, Figure 7.108.

Try a stiffener structural ST 6 WF @ 13.5#

$$y = \frac{12.5 \text{x} 0.312 \text{x} 6.136 + 0.24 \text{x} 5.58 \text{x} 3.19 + 6.5 \text{x} 0.4 \text{x} 0.2}{12.5 \text{x} 0.312 + 0.24 \text{x} 5.58 + 6.5 \text{x} 0.4}$$

$$= \frac{23.93+4.27+0.52}{3.900+1.339+2.6} = \frac{28.72}{7.889} = 3.66 \text{ in}$$

$$I_n = \frac{1}{12} \times 12.5 \times 0.312^3 + 12.5 \times 0.312$$

$$\times 2.476^2 + \frac{1}{12} \times 0.24 \times 5.58^3 + 0.24$$

$$\times 5.58 \times 0.47^2 + \frac{1}{12} \times 6.5 \times 0.4^3 +$$

$$+ 6.5 \times 0.4 \times 3.46^2 = 0.032 + 23.91 +$$

$$+ 3.474 + 0.295 + 0.035 + 31.126 =$$

$$= 58.87 \text{ in}^4$$

$$S_c = \frac{58.87}{2.632} = 22.36 \text{ in}^3$$

$$S_t = \frac{58.87}{3.66} = 16.08 \text{ in}^3$$

$$f_{comp} = \frac{8678.44 \text{x} 2.5 \text{x} 12}{22.36} = 11{,}643 \text{ psi}$$

$$f_{tens} = \frac{8678.44 \text{x} 2.5 \text{x} 12}{16.08} = 16{,}191 \text{ psi}$$

$$f_{tens} = \frac{4747 \text{x} 2.5}{} = 1{,}506 \text{ psi}$$

Total tensile stress

$$f_{tot} = 16{,}191 + 1{,}506 = 17{,}697 \text{ psi}$$

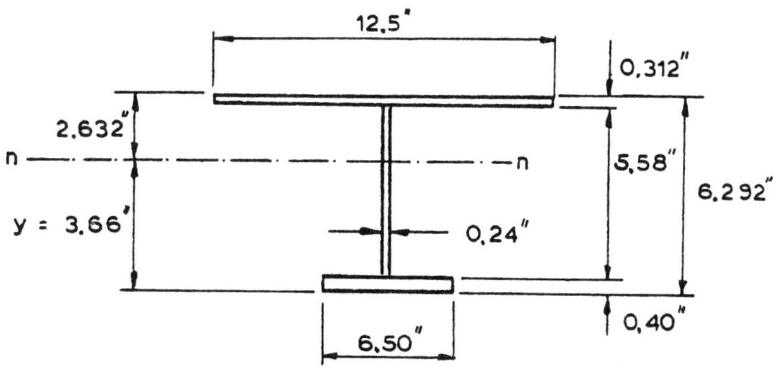

FIGURE 7.108 — Hopper stiffener. Cross-section.

REFERENCES — CHAPTER 7

[7.1] Fischer, W., "Silos and Bunkers in Stahlbau", VEB Verlag fur Bauwesen, Berlin, 1966 (in German).

[7.2] Soviet Concrete and Reinforced Concrete Code (Stroitelnie Normi i Pravila, Tshast II, Razdel B, Glava 1, Betonnie i Zelezobetonnie Konstruktsii SNKP) II-B.1-62. Publication of the Committee of Soviet Ministers USSR of Construction Works, Moscow 1962 (in Russian).

[7.3] Carr, R.L., Jr., "Evaluating Flow Properties of Solids", Chemical Engineering, Jan. 18th, 1965.

[7.4] Carr, R.L., Jr., "Particle Behavior, Storage and Flow", ASME Publication No. 68 — MH-6, Oct. 1968.

[7.5] Jenike, A.W., "Flow of Bulks Solids", Bulletin 64, Utah Engineering Experimental Station, University of Utah, March, 1954.

[7.6] Theimer, O.F., "Ablauf fordernde Trichterkonstructionen von Silozellen", Aufberechnungs — Technik No. 10, Oct. 1969, pp. 547-556.

[7.7] Kvapil R., and Taubmann, H.J., "Flow and Extraction of Solids from Bins", Publication 68-MH-32, ASME, New York, 1968.

[7.8] Jenike, A.W., and Johanson, J.R., "Annual Report," Project No. 126 — Fourth Part, "Mass-Flow Bins" and Project 126A — First Part, "Funnel-Flow Bins," American Iron and Steel Institute, 1971.

[7.9] Johanson, J.R., and Colijn, H., "New Design Criteria for Hoppers and Bins," Iron and Steel Engineer, Vol. XLI, October 1964, pp. 85-104.

[7.10] Johanson, J.R., "Effect of Initial Pressures on Flowability of Bins," Journal of Engineering for Industry, May 1965, pp. 395-399.

[7.11] Jenike, A.W., "Storage and Flow of Solids", Bul. 123, Utah Engineering Experiment Station, University of Utah, Nov. 1964.

[7.12] Giunta, J.S., "Flow Patterns of Granular Materials in Flat-Bottom Bins", Trans. ASME, J. of Engr. for Ind., May 1969.

[7.13] Janssen, H.A., "Versuch uber Getreidedruck in Silozellen", Zeitschrift verein Deutscher Ingenieure, Vol. 39, August 1895, pp. 1045-1049.

[7.14] Buisman, A.S.K., "Grondmechanica", Ed. Waltman, Delft, 1940, pp. 205-209.

[7.15] Reimbert, M., and Reimbert, A., "Silos-Traite′ Theoretique et Pratique", Editions Eyrolles, Paris, 1961.

[7.16] Petrov, B.A., "Experimental Determination of Cement Pressure on Reinforced Concrete Silo Walls", Cement No. 2, 1958, pp. 21-25 (in Russian).

[7.17] Pieper, K., and Wenzel, F., "Druckverhaltnisse in Silozellen", Verlag von Wilhelm Ernst and Sohn, Berlin, Munich, 1968.

[7.18] Pieper, K., "Investigation of Silo Loads in Measuring Models", Publication 68-MH-30, ASME, New York, 1968.

[7.19] Pieper, K., and Wagner, K., "Der Einfluss Verschiedner Auslaufarten auf die Seitendrucke in Silozellen", Aufbereitungs Technik No. 10, Oct. 1969, pp. 542-546.

[7.20] Deutsch, G.P., and Schmidt, L.C., "Pressures on Silo Walls," Publication 68-MH-24, ASME, New York, 1968.

[7.21] Takahashi, K., "Silo Design", Concrete Journal, Japanese National Council on Concrete, Vol. 5, No. 9, Aug. 1969 (in Japanese).

[7.22] Soviet Code CH-302-65, "Instructions for Design of Silos for Granular Materials," Moscow, USSR, 1965 (in Russian).

[7.23] Deutsche Normen, DIN-1055, Blatt 6, Lasten in Silozellen, November 1964, pp. 1-6 (in German).

[7.24] Platonov, P.N., and Kovtun, A.P., "The Pressure of Grain on Silo Walls", Mukomolno Elevatornaia Promyschlennost, Moscow, Vol. 25, Dec. 1959, pp. 22-24 (in Russian).

[7.25] Leonhardt, F., "Zur Frage der Sicheren Bemessung von Zement-Silos", Beton-und Stahlbetonbau, Berlin, Vol. 55, No. 3, March 1960, pp. 48-58.

[7.26] Terzaghi, K., "Theoretical Soil Mechanics", 5th Ed., John Wiley & Sons Inc., New York, 1948, pp. 77-81.

[7.27] Lipnitski, M.E., and Abramovitsch, S.P., "Reinforced Concrete Bunkers and Silos," Izdatelstvo Literaturi Po Stroitelstvu, Leningrad, 1967 (in Russian).

[7.28] Reimbert, M., and Reimbert, A., "Silos Berechnung, Beitrieb und Ausfuhrung", Bauverlag GMBH, Berlin, 1975, p. 123.

[7.29] Alferov, K.V., "Bunkers, Gates and Conveyors", Mashgiz, Moskow, USSR, 1946 (in Russian).

[7.30] Lichtarnikov, J.M., Klykov, V.M., and Ladyzhensky, D.V., "Design of Steel Structures", Budivelnik, Kiev, USSR, 1976 (in Russian).

[7.31] Lambert, F.W., "The Theory and Practical Design of Bunkers", The British Constructional Steelwork Association Ltd., London, 1968.

[7.32] Gieselski, R., et al., "Behalter, Bunker, Silos, Shornsteine, Fernsehturme and Freileitungsmaste", Verlag von Wilhelm Ernst und Sohn, Berlin, 1970.

[7.33] Safarian, S., and Harris, E.C., "Silos and Bunkers", Handbook of Concrete Engineering, Ed. by M. Fintel, Van Nostrand Reinhold Company, New York, 1974.

[7.34] Lipnitski, M.E., and Abramovitsch, S.P., "Reinforced Concrete Bunkers and Silos," Izdatelstvo Literaturi Po Stroitelstvu, Leningrad, 1960 (in Russian).

[7.35] Ketchum, M.S., "Walls, Bins and Grain Elevators", McGraw-Hill, New York, 3rd Edition, 1919.

[7.36] Lambert, F.W., "The Theory and Practical Design of Bunkers", The British Constructional Steelwork Association Ltd., London, 1968.

[7.37] Rogers, P., "Design of Large Coal Bunkers", Transactions ASCE, Vol. 117, Paper No. 2507, 1952.

[7.38] Timoshenko, S., and Woinowsky-Krieger, S., "Theory of Plates and Shells", 2nd Edition, McGraw-Hill Company, Inc., New York, 1959.

[7.39] Uniform Building Code, Vol. 1, International Conference of Building Officials, Pasadena, California, 1970.

[7.40] Chandrasekaran, A.R., and Jain, P.C., "Effective Live Load of Storage Materials Under Dynamic Conditions", Indian Concrete Journal, pp. 369-385, Sept. 1968.

[7.41] Safarian, S.S., and Harris, E.C., "Silos and Bunkers", Handbook Concrete Engineering, Van Nostrand Reynolds Ltd., New York, p. 497, 1974.

[7.42] Gieselski, R., et al., "Behalter, Bunker, Silos, Shornsteine, Fernsehturme und Freileitungsmaste", Verlag von Wilhelm Ernst und Sohn, Berlin, 1970.

[7.43] Timoshenko, S.P., and Gere, J.M., "Theory of Elastic Stability", 2nd Ed., McGraw-Hill Book Co., Inc., New York, 1961, p. 289.

[7.44] Brownell, L.E., and Young, E.H., "Process Equipment Design-Vessel Design", John Wiley & Sons, Inc., New York, 1968, pp. 198, 199.

[7.45] Seide, P., Weingarten, V.I., and Morgan, E.J., "Final Report on the Development of Design Criteria for Elastic Stability of Thin Shell Structures", Space Technology Laboratories, Inc., Los Angeles, 1960.

[7.46] Weingarten, V.I., Morgan, E.J., and Seide, P., "Elastic Stability of Thin-Walled Shells Under Internal Pressure and Axial Compression", J. Am. Inst. Aero and Astro, June 1965.

[7.47] API Standard 650, "Welded Steel Tanks for Oil Storage", American Petroleum Institute, Washington, D.C.

[7.48] Timoshenko, S.P., and Gere, J.M., "Theory of Elastic Stability", 2nd Ed., McGraw-Hill Book Company, New York, 1961.

[7.49] Kloppel, K., and Roos, E., "Buckling of Thin-Walled Stiffened and Unstiffened Spherical Shells Under Full and One-Sided Loading", Der Stahlbau, March 1956.

[7.50] DIN4119 Deutsche Normen, "Above ground Cylindrical Steel Tank Construction".

[7.51] Kloppel, K. and Jungbluth, O., "Buckling Problems of Thin-Walled Spherical Shells", Der Stahlbau, May 1953.

horizontal storage tanks

8.1 Introduction

This chapter is based on the available technical literature on horizontal steel storage tanks. The available information related to the loads and stress analysis are brought together so as to simplify the designer's task of designing such type of tanks. Emphasis has been placed on the local buckling stress analysis since this is usually the governing criteria in the design of these structures.

Figure 8.1 presents the major classification of tanks which can be applied depending on their purpose and location.

These classifications conform to two major types of tanks, namely unstiffened and stiffened tanks. Of these main types there are sub-types depending on end design and method of support. Figure 8.2 shows the types of end construction commonly in use today.

8.2 Principal Design Codes

a. *Low pressure tanks*

Practically all tanks for the storage of liquids in the United States are constructed in accordance with either the American Petroleum Institute Standard 650 [8.1] covering welded steel tanks for oil storage and other liquids [8.2, 8.3, 8.4], or American Water Works Association (AWWA) Standard D100, covering steel tanks for water storage [8.5, 8.6]. It should be noted that there are basic differences between storing liquids other than water or oil, however, the designer should consider which type of tank best fits his circumferences. In either case, the design standards provide minimum requirements for safe con-

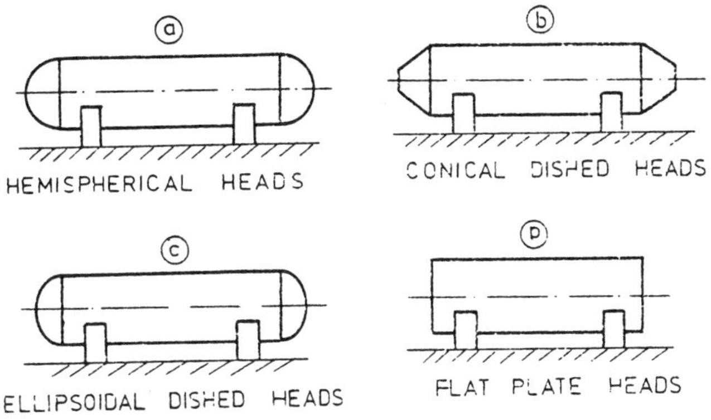

FIGURE 8.2 — Head types for horizontal storage tanks.

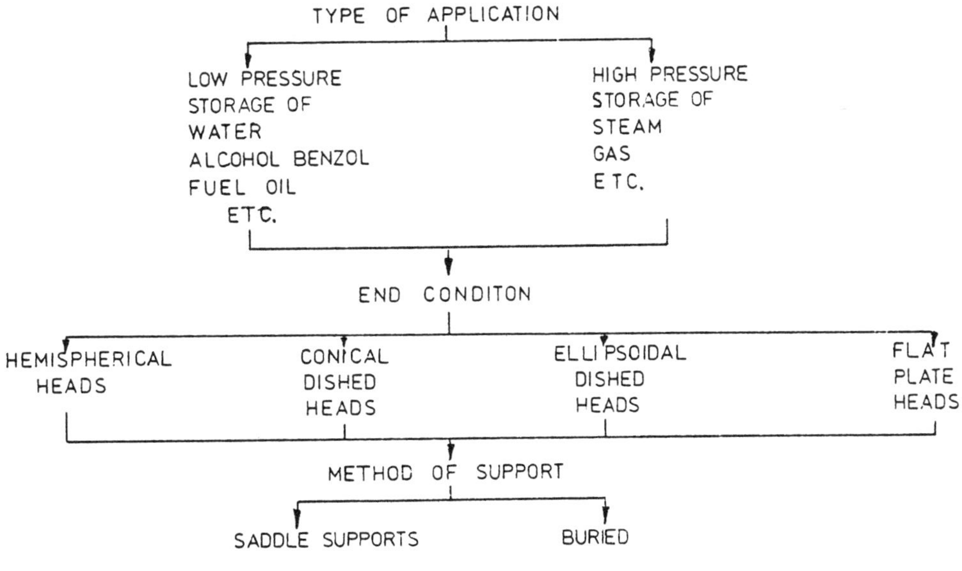

FIGURE 8.1 — Tank classification.

struction and should not be covering all possible service conditions.

In addition, there are specifications for gravity water tanks by the Association of Factory Mutual Fire Insurance Companies [8.7, 8.8], by the American Railway Engineering Association [8.9, 8.10] and by the American Welding Society [8.11].

b. *High pressure tanks*

In most countries (Australia, Austria, Canada, Finland, Germany, Holland, India, Italy, New Zealand, Switzerland, U.S.A.) the National Codes have the force of law and strict adherence to their rules is mandatory. In the United States the American Society of Mechanical Engineers (ASME) rules are the most widely used. The ASME Boiler and Pressure Vessel Code is the best known.

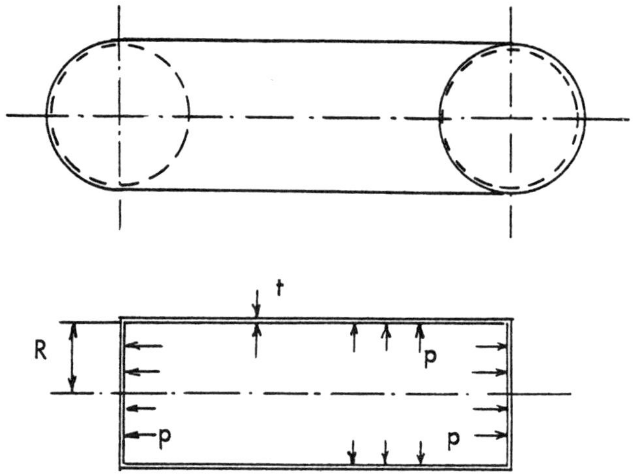

FIGURE 8.3 — Cylindrical shell under internal pressure.

8.3 Loads and Design Assumptions

a. *Loads*

In the design of horizontal storage tanks the following dead and live loads should be considered. [8.12]

1. Dead loads should consist of the unit weight of the tank shell and the unit weight of attached platforms, stairway brackets, and connecting piping.
2. Live loads should consist of the weight of the product stored in the tank and the maximum pressure exerted on the tank by its contents.
3. Overturning by the wind, when the tank is empty, should be checked.
4. Temperature differences should be checked to see if thermal stresses influence the design of the tank.
5. Snow and ice loads should be investigated if climatic conditions warrant.
6. Loading due to tank supports, both local and general, should be investigated, including settlement effects if foundation conditions warrant.
7. Earthquake forces should be considered if geographic conditions require it. (Tanks should be checked for overturning by seismic forces when they are full.)

b. *Design assumptions*

Theoretical stress analysis makes the following simplifying assumptions:

a. The material is homogeneous and isotropic.
b. The shell has a perfect geometric shape.
c. An elastic analysis is used to maintain the linear relationship between stress and strain.
d. All structures are practically may be subject to residual stresses at zero load, which thermal stress relief does not remove completely. Elastic analysis only calculates the change in stress due to a change in load.

8.4 Tank Under Internal Pressure

Suppose a long circular cylindrical tank is subjected to an internal pressure p, which may be due to a fluid or gas enclosed within the tank, Figure 8.3.

The following equations, based on membrane theory were derived to determine the longitudinal and circumferential stresses in the wall of the tank as a result of internal pressure.

The longitudinal stress

$$\sigma_1 = \frac{pR}{2t} \qquad (8.1)$$

and the circumferential or the "hoop" stress

$$\sigma_2 = \frac{pR}{t} \qquad (8.2)$$

where

R = the internal radius

t = the thickness of the wall

p = the internal pressure

Figure 8.4 shows the distribution of the longitudinal and circumferential stresses producing shearing stresses in the wall of the tank.

The maximum shearing stress in the plane of σ_1 and σ_2 is

$$\tau_{max} = \frac{pR}{4t} \qquad (8.3)$$

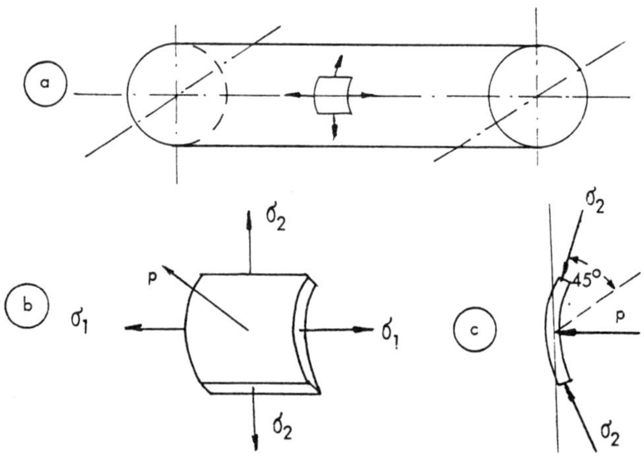

FIGURE 8.4 — Stresses acting on an element of the wall of a tank under internal pressure.

This is not, however, the maximum shearing stress in the wall of the tank, because in the plane of σ_2 and p, the maximum shearing stress is

$$\tau_{max} = \frac{pR}{2t} \tag{8.4}$$

Again, in the plane of σ_1 and p, the maximum shearing stress is

$$\tau_{max} = \frac{pR}{4t} \tag{8.5}$$

The greatest of these maximum shearing stresses is given by Eq. (8.4). It occurs on the plane at 45 degrees to the tangent and parallel to the longitudinal axis of the cylinder, Figure 8.4 (c).

A comparison of Eq. (8.1) with Eq. (8.2) indicates that for a fixed diameter, wall thickness, and given pressure, the circumferential stress is twice that of the longitudinal stress.

Therefore, the stress as determined by Eq. (8.2) is "controlling" and is commonly used "thin-walled equation" referred to in the various codes for tanks. This equation does not recognize the fact that welded seams or joints may cause weakness.

8.5 Heads of the Tanks Under Internal Pressure

The sharp discontinuity in the shape existing at the junction of a cylindrical tank and a head results in localized stress concentration at the junction.

The nature of the stress concentration is complex in that bending moments, shear, and stress reversals must be considered in addition to the membrane stresses resulting from internal pressure.

8.5.1 Hemispherical Heads [8.13]

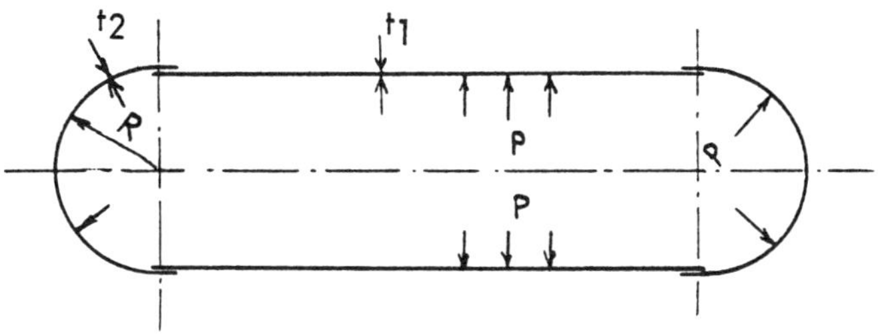

FIGURE 8.5 — Cylindrical tank with hemispherical heads.

In the case where the thicknesses of the heads and sides are so proportioned that there is no local bending stresses near the junction of the heads with the body of the shell, then

$$t_2 = \frac{t_1}{2.4} \tag{8.6}$$

The hemispherical head is therefore thinner than the cylindrical section.

8.5.2 Conical Dished Heads [8.14]

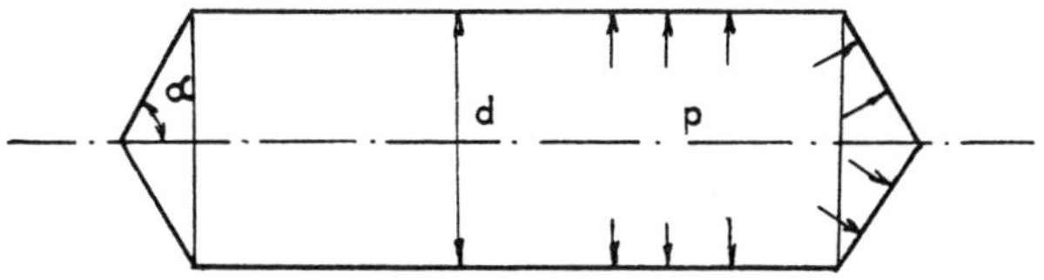

FIGURE 8.6 — Cylindrical tank with conical heads.

For a conical head with a half-open angle, α, not greater than 30°

$$t = \frac{pd}{2\cos\alpha(fE - 0.6p)} \tag{8.7}$$

where

f = the allowable stress

E = modulus of elasticity

The discontinuity stresses at the junction of the conical closure with the shell may cause excessive deformation. This may be prevented by the addition of a compression ring at the junction. When α exceeds Δ, as determined from Table 8.1, conical heads without a knuckle will require a compression ring at the section where the cone joins the shell.

TABLE 8.1

p/fE	0.01	0.002	0.003	0.004	0.005	0.006
Δ, deg.	13	18	22	25	28	31

The required cross-sectional area of the compression ring in square inches is given by

$$A = \frac{p}{fE}\left(\frac{d^2\tan\alpha}{8}\right)\left(1 - \frac{\Delta\ deg}{\alpha\ deg}\right) \tag{8.8}$$

where Δ = the critical value for Table 8.1.

8.5.3 Ellipsoidal Dished Heads [8.15]

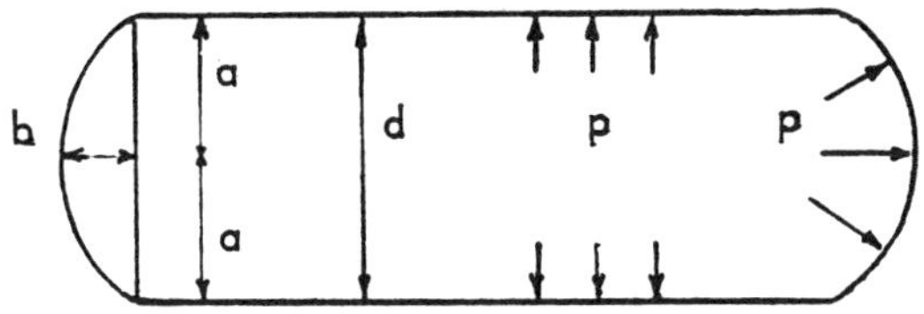

FIGURE 8.7 — Cylindrical tank with ellipsoidal heads.

The most common types of closure for tanks under internal pressure is the ellipsoidal head with a major-to-minor-axis ratio equal to 2.0:1.0. In this case, the following relationship applies:

$$t = \frac{pd}{2fE - 0.2p} \tag{8.9}$$

For ellipsoidal dished heads which have a major-to-minor axis ratio other 2.0:1.0, the following formulas are used:

$$V = \frac{1}{6}(2 + K^2) \tag{8.10}$$

$$t = \frac{pdV}{2fE - 0.2p} + c \tag{8.11}$$

where

K $= \frac{a}{b}$ = major-to-minor axis ratio

d = inside diameter of the head skirt

V = stress-intensification factor

E = welded-joint efficiency

f = allowable stress

c = corrosion allowance

Basic allowable efficiencies for welded joints are shown in the following Table 8.2.

TABLE 8.2

Type of Joint	Basic Joint Efficiency, Per Cent
1 Double-welded butt joint. Single-welded butt joint with backing strip	80
2 Single-welded butt joint without backing strip	70
3 Double fulfilled lap joint	65
4 Single fulfilled lap joint with plug welds	60
5 Single fulfilled lap joint without plug welds	50

8.5.4 Flat-Plate Heads [8.16]

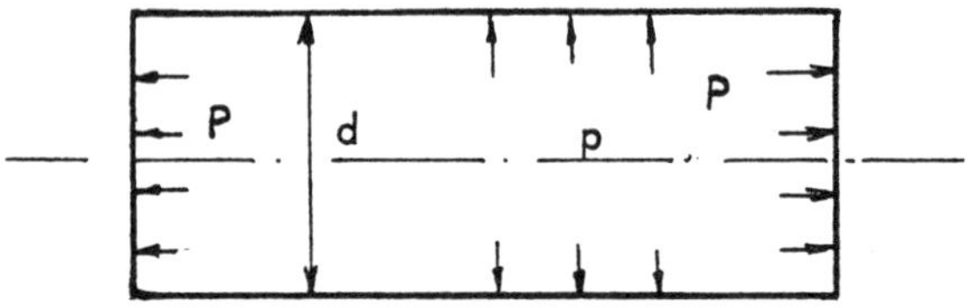

FIGURE 8.8 — Cylindrical tank with flat-plate heads.

The minimum thickness of the flat-plate head is determined as

$$t = d\sqrt{\frac{3}{16} \times \frac{p}{f}} \tag{8.12}$$

or by the formula

$$t = d\sqrt{C(p/f)} \tag{8.13}$$

where the constant C is given by the ASME Boiler and Pressure Vessel Code, Unfired Pressure Vessels for different types of head connections to the shell.

8.6 Stability Under Axial Compression

The design-allowable buckling stress for a thin-walled circular cylinder subjected to axial compression is given by the equation

$$\sigma_{cr} = K_c \frac{\pi^2 E}{12(1-\mu^2)} \left(\frac{t}{L}\right) \tag{8.14}$$

where

E = Young's modulus of material

μ = Poisson's ratio

t = the thickness of the shell

L = length of the cylinder

For short cylinders

$$\gamma Z < \pi^2 K_{co}/2\sqrt{3} \tag{8.15}$$

The buckling coefficient may be expressed approximately by the equation

$$K_c = K_{co} + \frac{12}{\pi^4}\frac{\gamma^2 Z^2}{K_{co}} \tag{8.16}$$

and for moderately long cylinders

$$\gamma Z > \pi^2 K_{co}/2\sqrt{3} \tag{8.17}$$

$$K_c = \frac{4\sqrt{3}}{\pi^2}\gamma Z \tag{8.18}$$

where

$$Z = \frac{L^2}{Rt}\sqrt{1-\mu^2} \tag{8.19}$$

K_{co} = 1 for simply supported edges.

K_{co} = 4 for clamped edges, γ may be obtained from Figure 8.9.

The correlation factor γ has been included to account for the difference between theoretical and experimental results. The classical theoretical value for the buckling coefficients is $\gamma = 1$. The majority of the tests data available for cylinders subjected to axial compression are presented in [8.17, 8.18]. Very few data are available in the short-cylinders range.

The formula for the buckling stress of moderately long cylinders may be rewritten in a more useful form

$$\sigma_{cr} = \gamma\, C_c E\left(\frac{t}{R}\right) \tag{8.20}$$

where

$$C_c = \frac{1}{\sqrt{3(1-\mu^2)}} \approx 0.6 \tag{8.21}$$

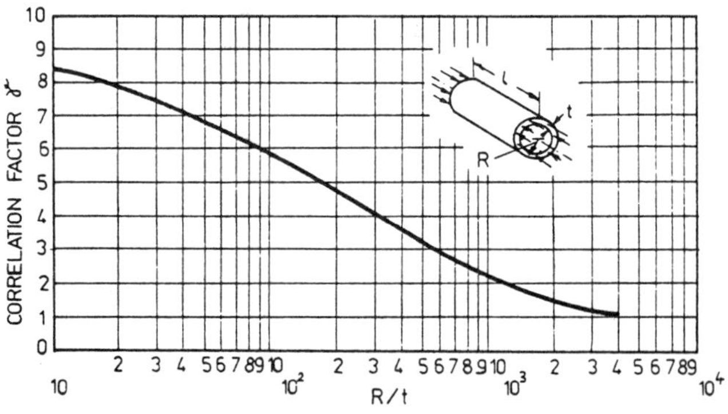

FIGURE 8.9 — Correlation factors for unstiffened circular cylinders subjected to axial compression.

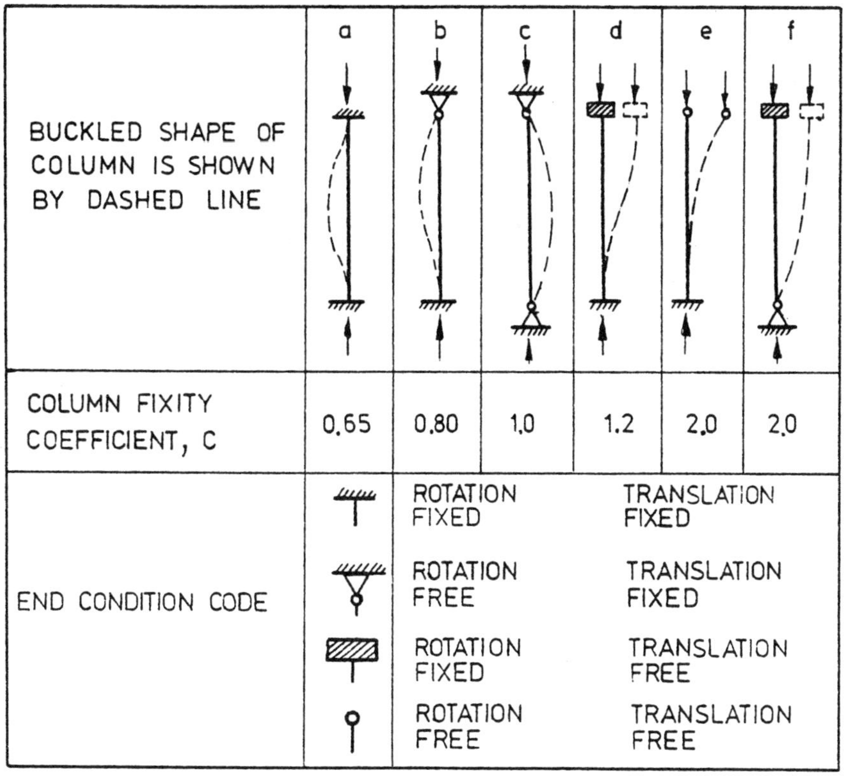

FIGURE 8.10 — Column fixity coefficients.

Very long cylinders must be checked for buckling as a Euler column, by the equation

$$\sigma_{cr} = \frac{\pi^2}{2} CE\left(\frac{R}{L}\right)^2 \tag{8.22}$$

where the column fixity coefficient C for a simple support is equal to 1. For any other support conditions C may be obtained from Figure 8.10.

8.7 Bending Stability

Relatively little theoretical work has been done on the problem of the buckling of thin cylinders under pure bending. The smallness of the buckling waves found for axial loading immediately suggested that buckling will take place on the compression side of the bending specimen in practically the same way as it does in an axially loaded specimen, and that any results found for axial loading would apply also to pure bending, with some factors to allow for the fact that the stress varies from zero to a maximum, instead of being constant around the circumference. Numerous bending tests on specimens similar to those tested in axial compression completely confirmed this opinion. Buckling occurred over the compression side of the specimens in the same wave form, with approximately the same wave-lengths, as in the axially loaded specimens. Comparison of axial compression and pure-bending tests results show exactly the same decrease of axial load P with an increase of the ratio R/t, as shown by the axially-loaded specimens.

If the compressive load is applied with some eccentricity, we obtain a combined compression and bending of a cylindrical shell.

The analysis performed by Timoshenko [8.19] indicates that in a circumferential direction we have no longer simple sinusoidal waves, and that a more complicated form of buckling occurs. The calculation of the critical value of the maximum compressive stress becomes much more complicated than before. This calculation made by Flugge [8.20] shows that we can get a satisfactory approximation and are on the safe side by assuming, for any ratio P/P_1, that buckling occurs when the maximum compressive stress becomes equal to the critical stress calculated for symmetrical buckling, therefore

$$\frac{1}{t}(P_o + P_1)_{cr} = \frac{E}{\sqrt{3(1-\mu^2)}} \cdot \left(\frac{t}{R}\right) \tag{8.23}$$

where

P_o/t = the uniform compressive stress

P_1/t = the maximum compressive stress due to bending

For pure bending ($P_o = 0$) the exact solution gives for the critical compressive stress a value which is about 30 percent higher than that obtained from Eq. (8.23).

An interesting case of instability of a thin cylindrical shell in bending has been discussed by Brazier [8.21] and Chwalla [8.22], considering flattening of the cross-sections of thin curved tubes. They found that for thin metal tubes with built-in edges, failure always occurred as a result of buckling in small waves of the same kind as in the case of uniformly compressed cylindrical shells. The results show exactly the same decrease of σ_{cr} with the increase of R/t as are shown by the axially loaded cylinders.

The values found for σ_{cr} are about 1.4 times those found in axial-compression tests for all values of R/t.

For many years, the theoretical maximum bending stress that will cause buckling of a circular cylindrical shell has been commonly accepted as being equal to 1.3 times the compressive buckling stress.

However, the investigation performed by Dinnik [8.23], indicated that the critical compressive stress

$$\sigma_{cr} = 0.334E\left(\frac{t}{R}\right) \tag{8.24}$$

is about 55 percent smaller than that obtained from Eq. (2.1).

The investigation by Seide and Weingarten [8.24] in 1961, by means of Batdorf's [8.25] modified Donnell's equation for buckling and using the Galerkin method, was undertaken to try to settle the problem, at least from the viewpoint of small-deflection theory. The results indicate that the ratio of bending and compressive stresses can vary widely with wave length, but that when we minimize with respect to wave length, the maximum critical bending stress is, for all practical purposes, equal to the critical compressive stress.

The design-allowable buckling stress for a thin-walled circular cylinder subjected to bending may be obtained from the equations presented for axial compression if γ is obtained from Figure 8.11. The theoretical buckling stress is

$$\sigma_{cr} = \frac{E}{\sqrt{3(1-\mu^2)}}\left(\frac{t}{R}\right) \tag{8.25}$$

The majority of the test data available for [8.17], cylinders subjected to bending are presented in References [8.17], [8.26]. Very few data are available in the short-cylinder range.

The formula for the buckling stress of moderately long cylinders subjected to bending may be written in the more useful form

$$\sigma_{cr} = \gamma\, C_b E\left(\frac{t}{R}\right) \tag{8.26}$$

where

$$C_b = \frac{1}{\sqrt{3(1-\mu^2)}} \tag{8.27}$$

σ_{cr} is the maximum stress due to the bending moment (e.g., the outer fiber stress).

If the stresses are elastic, the allowable moment is

$$M_{cr} = \pi R^2 \sigma_{cr} t \tag{8.28}$$

8.8 External Pressure Effect

A wide variety of chemical and petrochemical processes require equipment operating under partial vacuum. Such tanks are under external pressure from the atmosphere.

A cylindrical vessel under external pressure has an induced circumferential compressive stress equal to twice the longitudinal compressive stress because of external-pressure effects alone. Under such a condition, the vessel is apt to collapse because of elastic instability caused by the circumferential compressive stress. The collapsing strength of such vessels may be increased by the use of uniformly spaced, internal or external circumferential stiffening rings. From the standpoint of elastic stability, such stiffeners have the effect of subdividing the length of the shell into subsections equal in length to the center-to-center spacing of the stiffeners.

Long, thin cylinders without stiffeners or with stiffeners spaced beyond a "critical length" will buckle at stresses below the yield point of the material. The corresponding critical pressure at which buckling occurs is a function only of the L/D ratio and the modulus of elasticity, E, of the material. If the length of the shell with closures L, or the distance between circumferential stiffeners, L, as the case may be, is less than the critical length, the critical pressure at which collapse occurs is a function of the L/D ratio, as well as of the t/D ratio and the modulus of elasticity, E.

8.8.1 The Relationship for the Conditions Beyond the Critical Length

Considering buckling of circular rings under uniform external pressure, Timoshenko [8.27] pointed out that the formula for the critical load obtained for a ring can be applied also in the case of cylindrical shells with free edges submitted to a uniform lateral pressure

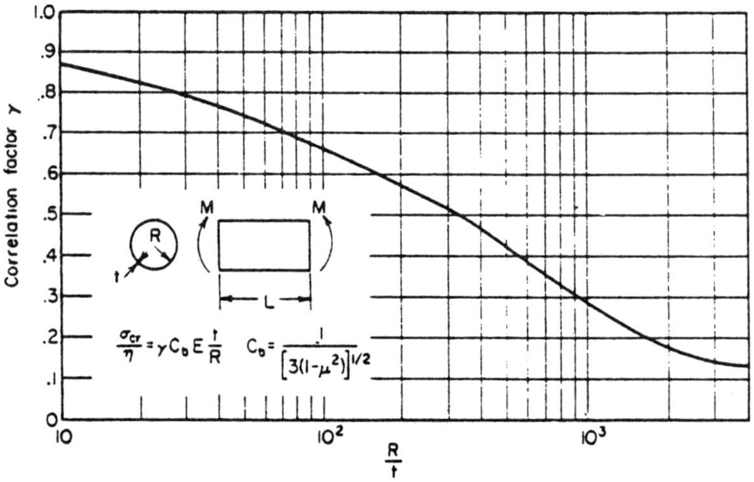

FIGURE 8.11 — Correlation factors for circular cylinder under bending.

$$q_{cr} = \frac{E}{4(1-\mu^2)} \left(\frac{t}{R}\right)^3 \tag{8.29}$$

and the critical buckling stress

$$\sigma_{cr} = \frac{E}{4(1-\mu^2)} \left(\frac{t}{R}\right)^2 \tag{8.30}$$

The same formula can be applied also in the case of a shell with some constraint at the edges if the length of the shell is so large that the stiffening effect of any constraint at the edges can be neglected.

Substituting for Poisson's ratio $\mu = 0.3$ into formula (8.29) gives

$$q_{cr} = 0.27\, E\left(\frac{t}{R}\right)^3 \tag{8.31}$$

Eq. (8.31) gives the theoretical "critical" external pressure.

Stewart [8.28] in a number of tests using commercial tubing and pipe investigated the applicability of Eq. (8.31) and found that collapse occurred at a critical pressure of 27 percent less than the theoretically predicted pressure. For the design of long, thin, cylindrical vessels operating under external pressure, a factor of safety of 4 may be applied by Eq. (8.31) giving

$$q_{all} = 0.07 \left(\frac{t}{R}\right)^3 \tag{8.32}$$

8.8.2 Instability of Cylinders Shorter Than Critical Length

If the length of a cylinder is not very large in comparison with its diameter, we can no longer disregard the end conditions, and in calculating the intensity of lateral pressure at which buckling occurs, we must consider the general equations of deformation of a cylindrical shell. The first investigations of this kind were made by Lorenz [2.1] and Southwell [2.3]. A more accurate formula for the critical load was developed by Von Mises [8.29]

$$q_{cr} = \frac{Et}{R(1-\mu^2)} \left[\frac{1-\mu^2}{(n^2-1)\left(1+\frac{n^2}{\lambda^2}\right)^2} + \frac{t^2}{12R^2}\left(n^2-1+\frac{2n^2-1-\mu}{1+\frac{n^2}{\lambda^2}}\right) \right] \tag{8.33}$$

and critical stress

$$\sigma_{cr} = \frac{E}{1-\mu^2} \left[\frac{1-\mu^2}{(n^2-1)\left(1+\frac{n^2}{\lambda^2}\right)^2} + \frac{t^2}{12R^2}\left(n^2-1+\frac{2n^2-1-\mu}{1+\frac{n^2}{\lambda^2}}\right) \right] \tag{8.34}$$

where

n = number of half-waves

$\lambda = \frac{\pi R}{\ell}$ = geometrical characteristic of the shell

ℓ = length of the shell or the spacing of the stiffening rings.

R = radius of the shell

μ = Poisson's ratio

When the shell is very long, $\ell/R > 8$ is a large number; neglecting, in Eq. (8.34), the terms containing the square of the ratio t/R, we obtain

$$\sigma_{cr} = \frac{Et^2(n^2-1)}{12R^2(1-\mu^2)} \tag{8.35}$$

which, with $n = 2$, coincides with the previous result given by Eq. (8.30).

$$\sigma_{cr} = \frac{E}{4(1-\mu^2)} \left(\frac{t}{R}\right)^2$$

In calculating the critical load for shorter cylinders, the values of n should be so chosen that the critical buckling stress defined by the formula (8.34) will be minimum.

In the following Table 8.3, are given the values of n in function of the ratio R/t and ℓ/R at which the critical stress is minimum.

TABLE 8.3

ℓ/R \ R/t	25	50	200	250	300
∞	2	2	2	2	2
10	2	2	3	4	4
5	3	3	4	5	5
2	4	5	6	8	9

8.8.3 The Design-Allowable Curve for External Pressure

If a cylindrical shell with simply supported edges is subjected to uniform external pressure, p, the design-allowable buckling stress in the circumferential direction is [8.31, 8.3].

$$\frac{\sigma_{cr}}{\eta} = K_p \frac{\pi^2 E}{12(1-\mu^2)} \left(\frac{t}{L}\right)^2 \tag{8.36}$$

The buckling coefficient K_p and a definition of the geometrical parameters are given in Figure 8.12. For elastic buckling, $\eta = 1$ is used.

For long cylinders, when

$$L^2/R^2 \geqslant R/t$$

the design-allowable buckling stress is

$$\frac{\sigma_{cr}}{\eta} = \frac{\gamma E}{4(1-\mu^2)} \left(\frac{t}{R}\right)^2 \tag{8.37}$$

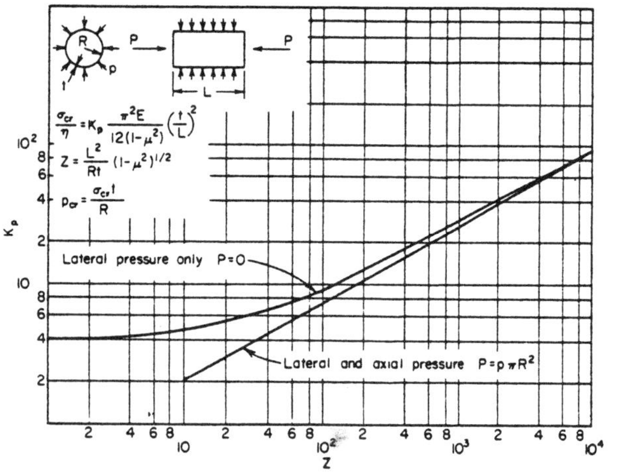

FIGURE 8.12 — Buckling coefficient for circular cylinder subjected to external pressure.

The factor γ was introduced to reduce the theory to a design value. It is recommended to use $\gamma = 0.9$ [8.32]. The majority of test data for cylinders subjected to external pressure are presented in Refs. [8.17, 8.31, 8.33].

The design-allowable pressure for any length of cylinder may be obtained from the formula

$$P_{cr} = \frac{\sigma_{cr} t}{R} \tag{8.38}$$

The pressure P_{cr} is the design-allowable pressure for complete buckling of the shell (e.g., when buckles have formed all the way around the cylinder). For some values of the parameters (large R / t and / or large initial imperfections), single buckles will occur at pressures less than P_{cr}, but complete buckling will occur at higher pressures. Therefore, for some applications these results should be used with caution.

8.9 Stability of Unstiffened Tanks. Combined Loadings

The criterion for structural failure of a member under combined loading is frequently expressed in terms of a stress-ratio equation

$$R_1^x + R_2^y = 1 \tag{8.39}$$

In general, the stress ratio R is the ratio of the allowable value of the stress caused by a particular kind of load in a combined-loading condition to the allowable stress for the same kind of load when it is acting alone. The subscript denotes the stress due to a particular kind of loading (compression, shear, etc.), and the exponents (usually empirical) express the general relationship of the quantities for failure of the member. The stress ratio is most easily understood if it is defined first of a particular loading condition. A curve drawn from a stress-ratio equation is termed a stress-ratio interaction curve. In simple loadings, the term "stress-ratio" is used to denote the ratio applied to allowable stress.

8.9.1 Internal Pressure and Axial Compression

The buckling stress of moderately long cylinders subjected to internal pressure and axial compression may be determined by using Figure 8.9 in conjunction with Figure 8.13. Figure 8.13 presents a curve that allows the calculation of the increase in buckling stress as a function of pressure and geometry only.

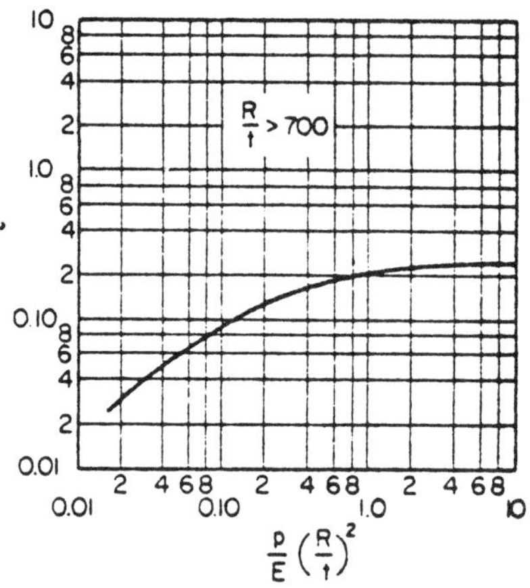

FIGURE 8.13 — Increase in axial-compressive buckling — stress coefficient for cylinders due to internal pressure.

The design-allowable buckling stress is

$$\frac{\sigma_{cr}}{\eta} = (\gamma C_c + \Delta C_c) E \left(\frac{t}{R}\right) \tag{8.40}$$

where γ is obtained from Figure 8.9 and ΔC_c is obtained from Figure 8.13. The cylinder under internal pressure is capable of resisting a total compressive load P_{cr}, which may be obtained from the equation

$$P_{cr} = 2\pi R t \sigma_{cr} + \pi R^2 p \tag{8.41}$$

It should be noted that the design curve in Figure 8.13 is valid only for moderately long cylinders, where $1.72\sqrt{t/R} < \ell / r < 2.85\sqrt{R/t}$. Very long cylinders must be checked for buckling as Euler columns.

8.9.2 Internal Pressure and Bending

The buckling stress of moderately long cylinders subjected to internal pressure and bending may be determined by using Figure 8.14 in conjunction with Figure 8.11.

Figure 8.14 presents curves that allow the calculation of the increase in critical stress as a function of pressure and geometry only. The design-allowable buckling stress is

$$\sigma_{cr} \doteq (\gamma C_b + \Delta C_b) E \left(\frac{t}{R}\right) \tag{8.42}$$

where γ is obtained from Figure 3.12 and ΔC_b is obtained from Figure 8.14.

Two curves for calculating the increment in critical stress caused by internal pressure are presented in Figure 8.14. The curve labeled "No external axial load" should be used to calculate the critical stress of a cylinder subjected to bending and internal pressure only. The curve

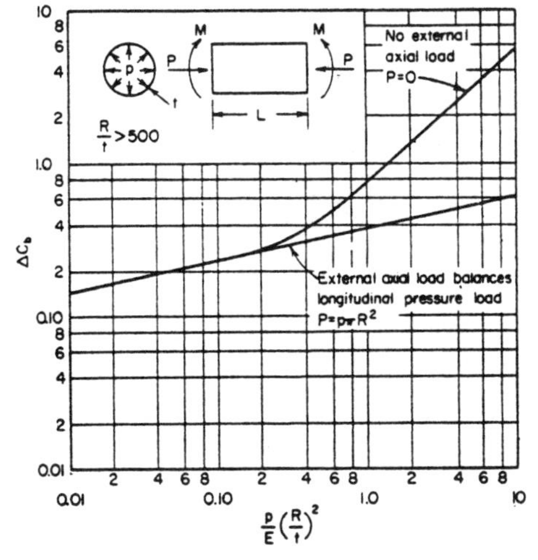

FIGURE 8.14 — Increase in bending buckling-stress coefficient for cylinder due to internal pressure.

labeled "External axial load balances longitudinal pressure load" should be used to calculate the critical stress of a cylinder subjected to bending and internal pressure plus an external axial compression load equal to the internal pressure load $\pi R^2 p$ acting on the heads of the cylinder.

If the curve for no axial load is used and the stresses are elastic, the design-allowable moment is

$$M_{cr} = \pi R^2 (\sigma_{cr} t + \frac{pR}{2}) \qquad (8.43)$$

It should be noted that the design curves for internal pressure in Figure 8.14 are valid only for moderately long cylinders.

8.9.3 Axial Compression and Bending

The test data presented in Refs. [8.17, 8.26] indicate that the linear interaction for the case of cylinders under combined axial compression and loading shown in Figure 8.15 may be used.

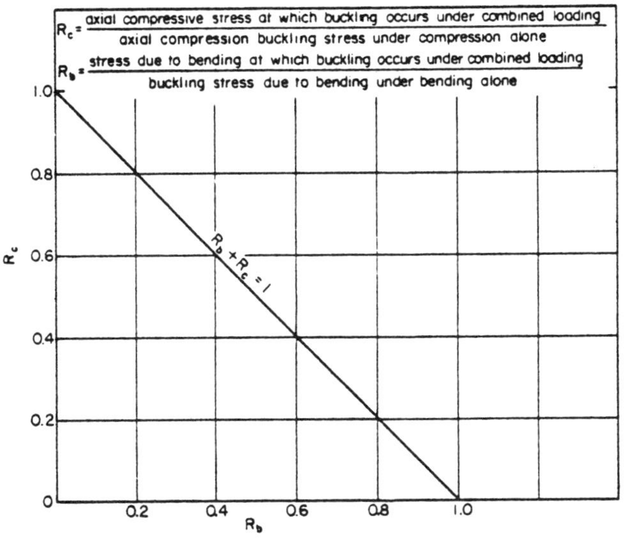

FIGURE 8.15 — Buckling-stress interaction curve for circular cylinder under combined axial compression and bending.

The buckling stress due to bending alone may be found from Figure 8.11 and the buckling stress under axial compression alone may be found using Figure 8.9.

8.9.4 Axial Compression and External Pressure

The limited data from Ref. [8.17] for cylinders subjected to axial compression and external lateral and axial pressure indicate that the linear interaction curve presented in Figure 8.16 may be used for design. σ_{cr} is found using Figure 8.9 and P_{cr} using Figure 8.16.

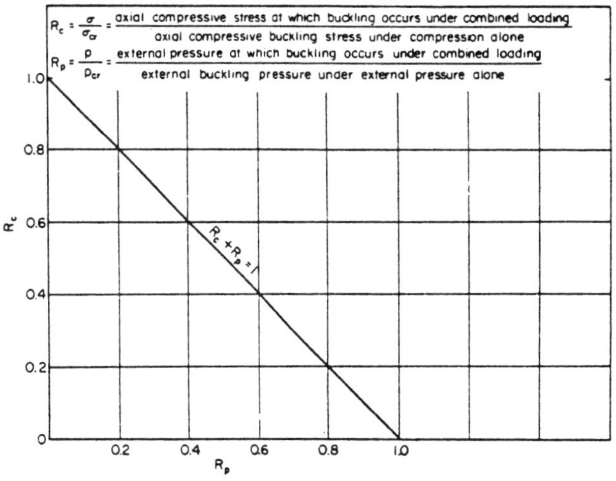

FIGURE 8.16 — Buckling-stress interaction curve for circular cylinder under combined external pressure and axial compression.

8.10 Approximate Stress Analysis of the Tanks

The following tank stress analysis is based on Vlasov's Theory [8.34].

For the analysis of the cylindrical shells Vlasov introduced the bending moment theory based on the following assumptions:

a. The longitudinal bending moments and twisting moments are neglected.
b. In the cross-section the shell is due to take without deformations the axial forces (normal and tangential)

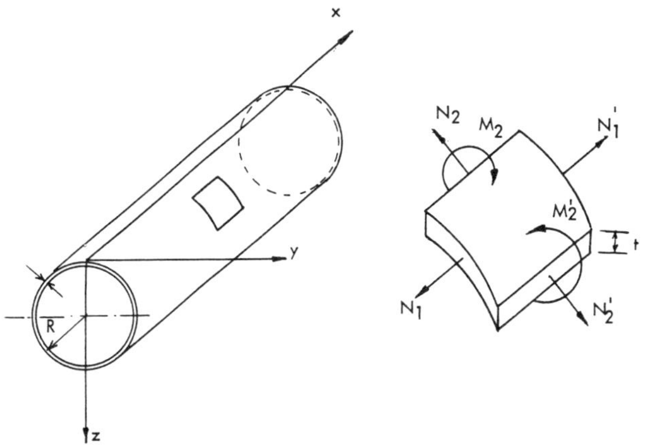

FIGURE 8.17 — Element of the shell with stress resultant.

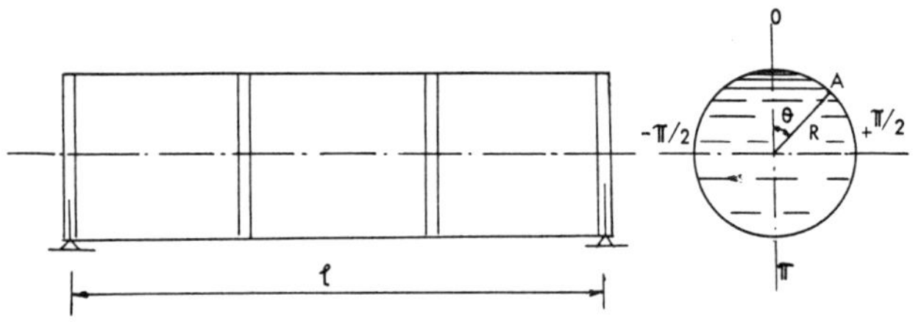

FIGURE 8.18 — Tank under the weight of the liquid.

and also the transverse bending moments.

c. The shear deformations are neglected.

The internal forces and bending moments considered in the stress analysis are shown in Figure 8.17.

8.10.1 Stresses in the Tank Under the Weight of Liquid

In the following, horizontal cylindrical tank having intermediate flexible stiffening rings at equal spacings and rigid end rings freely supported, Figure 8.18 is considered.

For the filled up tank, we obtain [8.35]

a. Longitudinal normal force in the middle cross-section of the shell

$$N_1 = -\frac{4\gamma\ell^2}{\pi^3}\cos\theta \qquad (8.44)$$

b. The tangential force at the support of the shell

$$T = \frac{\gamma R\ell}{2}\sin\theta \qquad (8.45)$$

c. The circumferential normal force in the middle cross-section of the shell

$$N_2 = \gamma R^2(1 - \cos\theta) \qquad (8.46)$$

where γ is the specific weight of the liquid.

8.10.2 Stresses in the Tank Under Its Own Weight

In the limits $0 \leqslant \theta \leqslant \pi$ the components of the load are

$$Z = \frac{4}{\pi} g \cos\theta \qquad (8.47)$$

$$Y = \frac{4}{\pi} g \sin\theta \qquad (8.48)$$

where g represents the weight of the unit cross-sectional area of the wall of the shell.

8.10.3 Tank Having Rigid Stiffening End Rings, Either Freely Supported or Fixed Under Weight of the Liquid and Its Own Weight

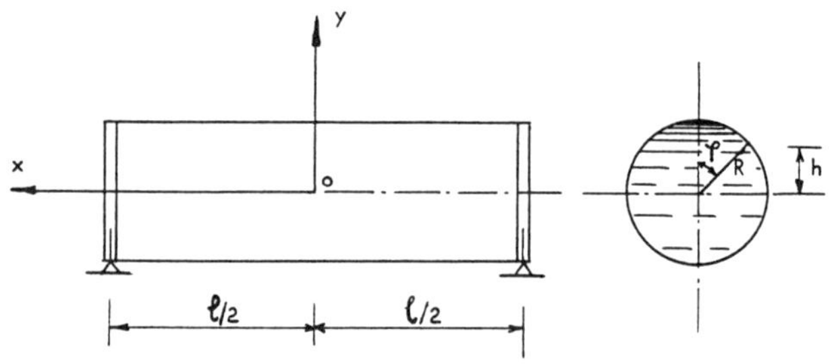

FIGURE 8.19 — Tank having rigid stiffening end rings.

The stresses in the shell of this tank due to the full liquid load and its own weight, considering different types of end supports, are shown in the folowing Table 8.4 [8.36].

The functions $f(\psi)$ and $f_1(\psi)$ depend on the boundary conditions and may be determined from the following Table 8.5.

TABLE 8.4

Loading	Stresses		
	Longitudinal Normal Stress, N_1	Circumferential Normal Stress, N_2	Tangential Stress, T
Filled with Liquid	$\sigma_1 = \frac{\gamma x^2}{2t}\cos\psi + f(\psi)$	$\sigma_2 = \frac{\gamma R}{t}(h - R\cos\psi)$	$\tau = \frac{\gamma x R \sin\psi}{t}$
Own Weight of the Shell	$\sigma_1 \frac{\gamma_1 x^2}{R}\cos\rho + f_1(\psi)$	$\sigma_2 = -\gamma_1 R\cos\psi$	$\tau = -2\gamma_1 x \sin\psi$

TABLE 8.5

Case 1	Case 2	Case 3
Rigid end rings are freely supported.	Rigid end rings are fixed at supports.	Rigid end rings cannot rotate, but may move freely.
	Full by Liquid	
$f(\psi) = \frac{\gamma\ell^2}{8t}\cos\psi$	$f(\psi) = -\frac{\gamma}{t} x$ $x\left[\frac{\ell^2}{24} + R^2\mu\right) x$ $x \cos\psi - hR\mu]$	$f(\psi) = -\frac{\gamma}{t} x$ $x\left(\frac{\ell^2}{24} + R^2\mu\right)\cos\psi$
	Own Weight	
$f_1(\psi) = -\frac{\gamma_1\ell^2}{4R}\cos\psi$	$f_1(\psi) = -\frac{\gamma_1}{R} x$ $x\left(\frac{\ell^2}{12} - R^2\mu\right)\cos\psi$	$f_1(\psi) = -\frac{\gamma_1}{R} x$ $x\left(\frac{\ell^2}{12} - R^2\mu\right)\cos\psi$

8.11 Stability of Unstiffened Tanks

8.11.1 Critical Length Between Stiffeners

Equations (8.31)

$$q_{cr} = 0.27E\ (\frac{t}{R})^3$$

and (8.32)

$$q_{all} = 0.07\ (\frac{t}{R})^3$$

apply to long, thin cylinders under external pressure without circumferential stiffening rings or with the stiffening rings spaced at or beyond the "critical length". To make allowance for the added restraint offered by stiffeners spaced at less than the critical length, the critical length may first be evaluated.

The expression for the critical length was first developed by Southwell [8.37]. The relationship resulting from the Southwell analysis is given by

$$\ell_{cr} = \frac{16\pi\sqrt{3}}{27}\ (\sqrt[4]{(1-\mu^2)})\ \ (R\sqrt{R/t}) \qquad (8.49)$$

Substituting $\mu = 0.3$, gives

$$\ell_{cr} = 3.13\ R\sqrt{R/t} \qquad (8.50)$$

where

ℓ_{cr} = critical length

R = radius of cylinder

t = wall thickness

8.11.2 Collapsing Pressure on Tank with Circumferential Stiffeners

For tanks in which circumferential stiffeners are spaced at less than the critical length, the coefficient of Eq. (8.31) must be modified according to the proportions of the tank [8.38, 8.39], or

$$q_{cr} = KE\ (\frac{t}{D})^3 \qquad (8.51)$$

Applying a factor of safety of 4 gives

$$q_{all} = \frac{K}{4}\ E\ (\frac{t}{D})^3 \qquad (8.52)$$

where

K = coefficient according to the properties of the tank, as indicated in Figure 8.20 [8.40].

Substituting Eq. (8.52) into an equation expressing the circumferential stress

$$f = \frac{qD}{2t} \qquad (8.53)$$

we find that the circumferential compressive stress from external pressure at which the collapse occurs is

$$f = \frac{DKE}{2t}\ (\frac{t}{D})^3 \qquad (8.54)$$

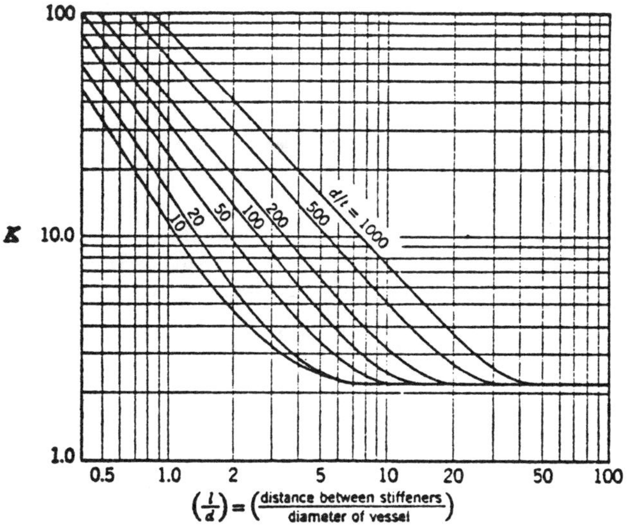

FIGURE 8.20 — Collapse coefficients for cylindrical shells under external pressure.

Using a design factor of safety of 4, in which the allowable pressure is considered to be ¼ of the theoretical pressure at which collapse occurs, we obtain

$$q_{cr} = \frac{2ft}{D} = 4q_{all} \qquad (8.55)$$

where

q_{cr} = theoretical external pressure at which collapse occurs.

q_{all} = allowable external pressure.

8.11.3 Design of Circumferential Stiffeners

In designing circumferential stiffening rings for tanks under external pressure each stiffener is considered to resist the external load for an $\ell/2$ distance on either side of the ring, where ℓ is the spacing between rings. Thus, the load per unit length on the ring at collapse is equal to

$$\ell x q_{cr}$$

Considering the action of uniform pressure acting on the circular ring, Figure 8.21, the critical value of the compressive force is [8.41].

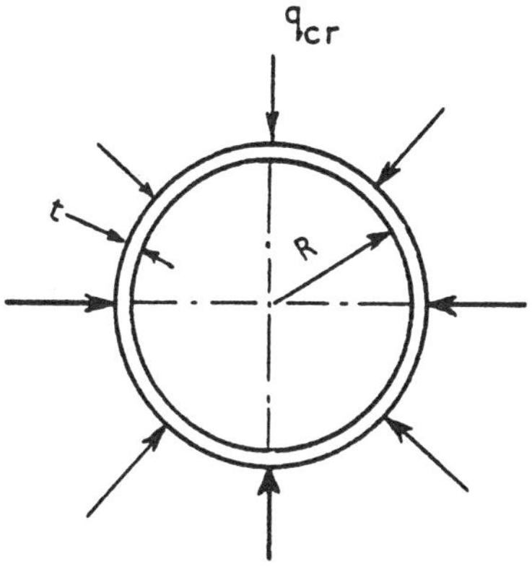

FIGURE 8.21 — Stability of circular ring.

$$q_{cr} = \frac{3EI}{R^3} = \frac{24EI}{D^3} \tag{8.56}$$

This equation expresses the theoretical or critical load per unit circumferential length of unit width of circumference.

For a strip of unit width, the critical load is the pressure at which buckling theoretically occurs.

We may rewrite Eq. (8.56), noting that in this equation the term ℓ is taken as unity. Therefore,

$$P = q_{cr} x \ell = \frac{24EI}{D^3} \tag{8.57}$$

where P is the load on combined shell and stiffener in psi of circumferential length, or

$$I = \frac{q_{cr} D^3 \ell}{24EI}$$

Multiplying by t/t and rearranging, gives

$$I = \left(\frac{D^2 \ell t}{12EI}\right)\left(\frac{q_{cr} D}{2t}\right)$$

Substituting Eq. (3.53)

$$f = \frac{qD}{2t}$$

and formula $\epsilon = f/E$, gives

$$I = \frac{D^2 \ell t \varepsilon}{12} \tag{8.58}$$

The moments of inertia of the stiffening ring and the shell act together to resist collapse of the tank under external pressure. Timoshenko [8.41] has shown that the combined moment of inertia of the shell and stiffener may be considered as equivalent to that of a thicker shell, or

$$t_y = t + \frac{A_y}{d_y} = t + \frac{A_y}{\ell} \tag{8.59}$$

where

t_y = equivalent thickness of shell

A_y = cross-sectional area of one circumferential stiffener

ℓ = d_y = distance between circumferential stiffeners

Substituting Eq. (8.59) into Eq. (8.58) gives

$$I = \frac{D^2 \ell}{12}\left(t + \frac{A_y}{\ell}\right)\varepsilon \tag{8.60}$$

where I is the required moment of inertia of stiffening ring.

8.11.4 Short, Ring-Stiffened Tanks

If a cylindrical shell with rings (circumferential stiffeners) is subjected to lateral and axial external pressure, it may fail in three distinct ways. The types of failure will be classified as material failure, buckling between rings, and general instability. A brief discussion of these failure modes is presented here. A more detailed discussion of ring-stiffened cylinders subjected to external pressure is presented in [8.42].

8.11.4.1 Material Failure

For the purpose of analysis, the stress and deflection distributions of a cylinder prior to buckling can be assumed to be axisymmetric. Therefore, the analysis methods of previous sections can be used to determine the stresses in the cylinder. A more detailed procedure is discussed in [8.43]. The actual stresses must be compared with the material's allowable stresses. In addition, the compressive stress in the frame must be compared with the local buckling stresses of the ring.

8.11.4.2 Buckling Between Rings

This failure will occur in a cylinder having relatively heavy rings. The sheet will buckle between rings and the rings will remain circular in cross-section. The allowable buckling stress for this mode of failure can be obtained from Chapter 4, using a cylinder length equal to the ring spacing. A more detailed method of analyzing this mode of failure is presented in [8.44].

8.11.4.3 General Instability

General instability will occur when the ring buckles with the sheet at the critical load. The design-allowable general instability pressure P_{cr} for a ring-stiffened cylinder subjected to lateral and axial external pressure can be obtained from

$$P_{cr} = \frac{\gamma E}{30 \times 10^6}(P_s + P_f) \tag{8.61}$$

where E = Young's modulus of the material (the ring and sheet must be made from the same material).

P_s can be obtained from Figure 8.22 if

$$\left(\frac{I_e}{dR^3}\right) \times 10^6 < 10 \tag{8.62}$$

and from Figure 8.23 if

$$\left(\frac{I_e}{dR^3}\right) \times 10^6 > 10 \tag{8.63}$$

P_f can be obtained from Figures 8.24 and 8.25.

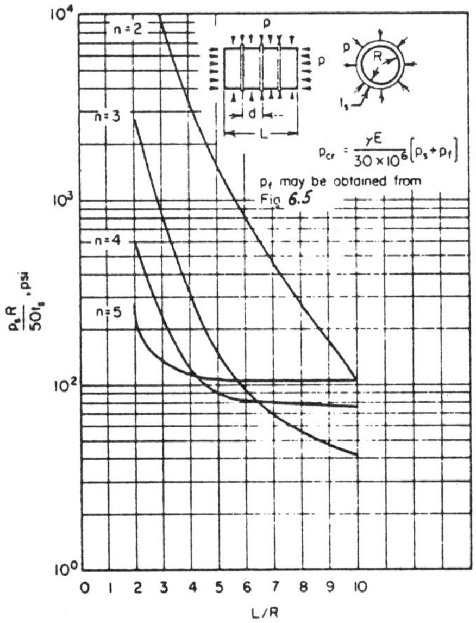

FIGURE 8.22 — Shell pressure factor P_s as a function of ring spacing L/R for $(L_s/dr^3) \times 10 < 10$.

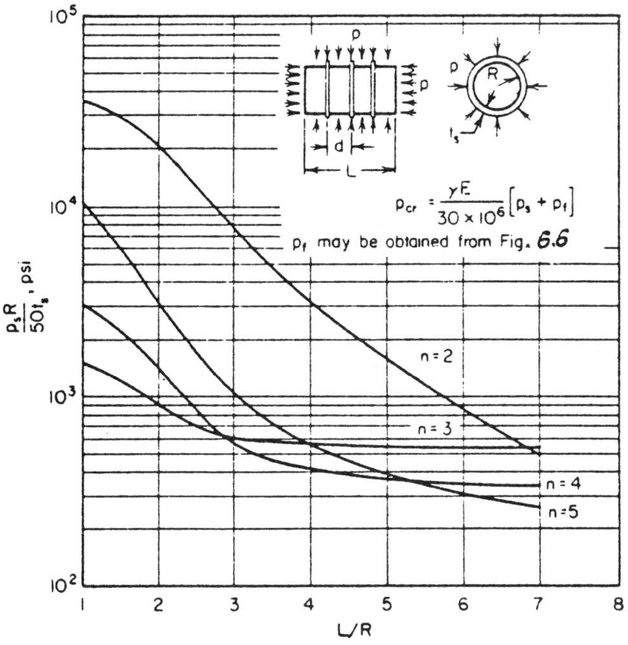

FIGURE 8.23 — Shell pressure factor P_s as a function of ring spacing L/R for $(I_e/dR^3) \times 10^6 \leq 10$.

The quantity of P_{cr} must be computed for n = 2, 3, 4 and 5; its lowest value is the critical allowable buckling pressure. This graphical method for determining P_s and P_f was obtained from [8.45, 8.46], which are based on [8.47]. Reference [8.47] analyses simply supported cylinders. The test results of [8.48] indicate an appreciable increase in buckling pressure due to edge fixity, but a method of including these effects for design purposes has not been developed.

The parameter γ is introduced to reduce the theoretical results of [8.47] to a value which may be used for design purposes. It is a function of many variables such as initial out of roundness of the cylinder. For the test data given in [8.48, 8.49], γ = 1 is adequate.

However, the cylinders in [8.48, 8.49] were machined to very close tolerances. Presently, it is not known what value of γ to use, but for

$$\sqrt{(12I_e/dt_s)^3}\,(L^2/t_s) > 4 \times 10^3 \qquad (8.64)$$

γ = 0.9 is probably reasonable. The value of I_e required in Figures 8.24 or 8.25 can be calculated from

$$I_e = \frac{A_f e^2}{1 + A_f/d_e t_s} + I_f + \frac{d_e t_s^3}{12} \qquad (8.65)$$

where

A_f, I_f = the area of ring and moment of inertia of ring about its own neutral axis, respectively

e = distance from middle surface of sheet to centroid of ring

$d_e = (d - t_w)\,F_1 + t_w \qquad (8.66)$

t_w = ring web thickness

The value of F_1 can be obtained from Figure 8.26.

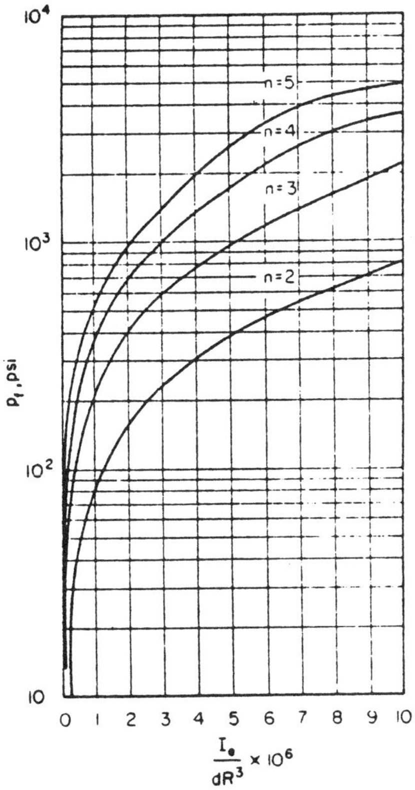

FIGURE 8.24 — Ring pressure factor P_f as a function of ring stiffness I_e/dR^3.

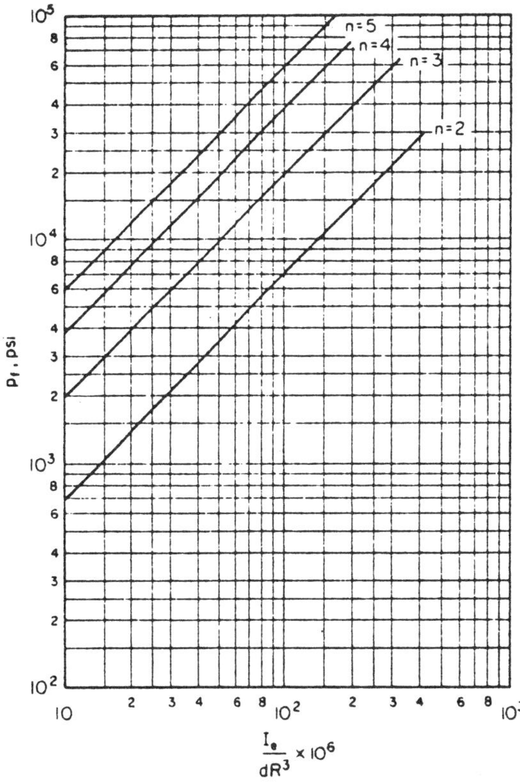

FIGURE 8.25 — Ring pressure factor P_f as a function of ring stiffness I_e/dR^3.

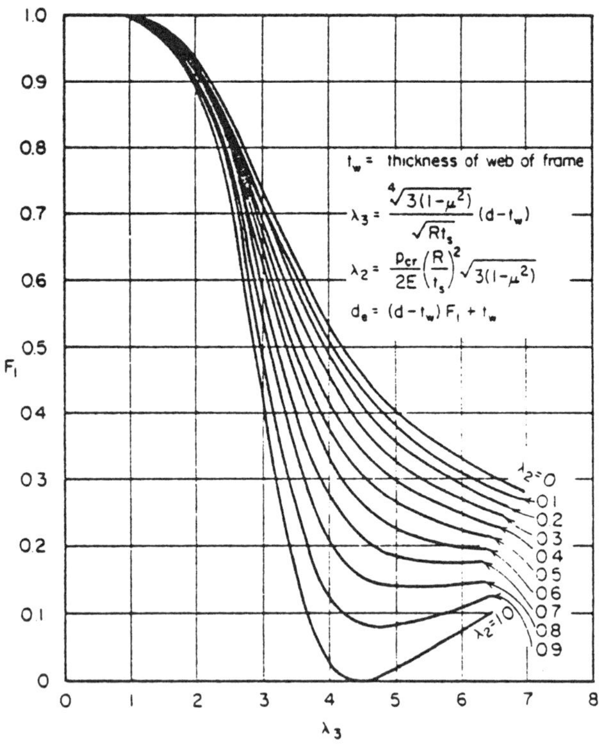

FIGURE 8.26 — Stress function for determining effective length.

Parameter λ_2 in Figure 8.26 is a function of P_{cr}, which is unknown. A good approximation can be obtained by using $\lambda_2 = 0$. For cases in which $\lambda_2 < 2$, this approximation gives results within 5 percent accuracy.

For the loading case of external pressure [8.50] states that a cylinder with rings on the inside will be stronger than a cylinder with rings on the outside if $Z > 500$, where

$$Z = \sqrt{(1-\mu^2)}\ \frac{L^2}{Rt_s} \qquad (8.67)$$

If the rings are on the inside and $Z > 500$, the results will be slightly conservative because the curves presented are for external rings. Therefore, the curves presented will probably be conservative if the rings are located on the outside because most stiffened cylinders fall in the range $Z > 500$.

If the parameters for a particular design do not fail in the range of parameters presented in Figures 8.24 and 8.25, the following formula [8.51] can be used to estimate the design-allowable external pressure

$$P_{cr} = \frac{5.5\gamma E\ \sqrt[4]{(I_e/d)^3 t_s}}{L\sqrt{R^3}} \qquad (8.68)$$

provided

$$\sqrt{\left(\frac{12I_e}{dt_s}\right)^3 \frac{1}{t_s^{\,4}}} \cdot \frac{L^2}{R} > 4 \times 10^3 \qquad (8.69)$$

The theoretical results of [8.47] were reduced 10 percent for design purposes, and it has been found that the theory of [8.47] predicts buckling pressures that are, in some cases, 80 percent of the theoretical results given in [8.51]; therefore, $\gamma = 0.8 \times 0.9 = 0.72$ can be used in the preceding formula to obtain the design-allowable buckling pressure.

8.12 Stability of Heads Under Uniform External Pressure

8.12.1 Spherical Head

The buckling of a spherical head under uniform external pressure, Figure 8.27, has been treated extensively.

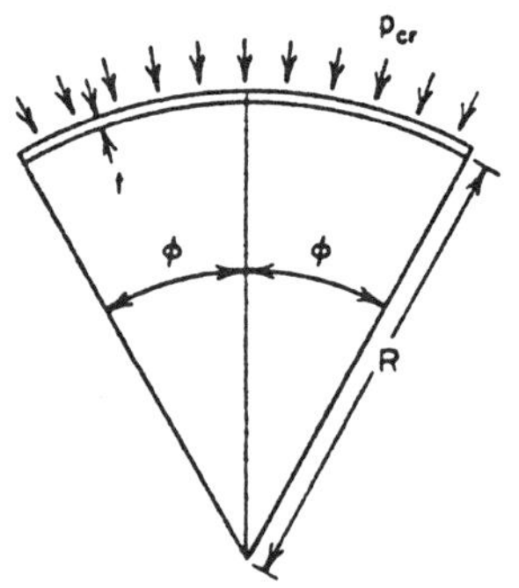

FIGURE 8.27 — Geometry of spherical cap under uniform external pressure.

Theoretical results are presented in [8.52, 8.53] for axisymmetric snap-through of shallow spherical shells with edges that are restrained against translation but are either free to rotate or are clamped. Results for asymmetric buckling are given in [8.54, 8.55] for the same boundary conditions. The results reported in these references are presented as the ratio of the buckling pressure P_{cr} for the spherical cap and the classical pressure P_{cl} for a complete spherical shell as a function of geometry parameter λ

$$\frac{P_{cr}}{P_{cl}} = f(\lambda) \qquad (8.70)$$

with

$$P_{cl} = \frac{2E}{\sqrt{3(1-\mu^2)}}\left(\frac{t}{R}\right)^2 \qquad (8.71)$$

$$\lambda = \sqrt[4]{12(1-\mu^2)}\ \sqrt{\frac{R}{t}}\ 2\sin\frac{\phi}{2} \qquad (8.72)$$

where ϕ is half the included angle of the spherical cap, Figure 8.27.

The function $f(\lambda)$ depends on the boundary conditions imposed on the shell.

Most of the available test data apply to spherical shells, and values are lower than theoretically predicted buckling pressures. The discrepancy between theory and experiment can be largely attributed to initial deviations from the ideal spherical shape, [8.53, 8.56, 8.57] and to differences between actual and assumed edge conditions [8.58, 8.59].

Most of the available data are summarized in [8.60];

some other test results are given in [8.56, 8.61]. A lower bound to the data for clamped shells is given by

$$\frac{P_{cr}}{P_{cl}} = 0.14 \quad \frac{3.2}{\lambda^2} \qquad \lambda > 2 \tag{8.73}$$

which is plotted in Figure 8.28.

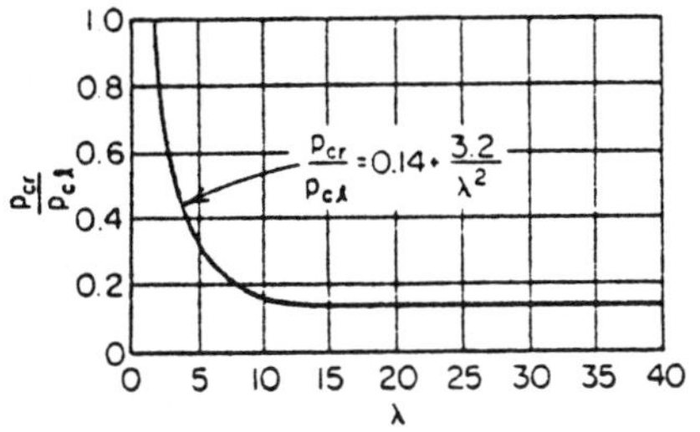

FIGURE 8.28 — Recommended design buckling pressure of spherical cap.

While the λ parameter is used in shallow-shell analysis, Figure 8.28 may be applied to deep shells, as well as to shallow shells.

8.12.2 Hemispherical and Torispherical Heads

The elastic stability of a thin sphere under external pressure is given by the formula (8.71)

$$P_{cl} = \frac{2E}{\sqrt{3(1-\mu^2)}} \left(\frac{t}{R}\right)^2$$

If a spherical shell is submitted to uniform external pressure, it may retain its spherical form and undergo only a uniform compression. The magnitude of the uniform compressive stress in this case is [8.62].

$$\sigma = \frac{pR}{2t} \tag{8.74}$$

where

- p = the pressure per unit area of the middle surface
- R = the radius of the sphere
- t = the thickness of the shell

By substituting (8.74) into (8.71), we obtain

$$\sigma_{cr} = \frac{E}{\sqrt{3(1-\mu^2)}} \left(\frac{t}{R}\right) \tag{8.75}$$

This stress has the same magnitude as the critical stress for an axially compressed cylindrical shell of radius R and of thickness t.

Hemispherical and torispherical heads, Figure 8.29 under external pressure are subject to failure by elastic instability, as are spherical shells.

Eq. (8.75) applies in the case of hemispherical or torispherical heads and gives the theoretical pressure at which collapse would occur because of elastic instability.

8.12.3 Ellipsoidal Head

The problem of elastic stability of the ellipsoidal shell, which is under the action of the external uniformly distributed pressure q is similar with the problem of the spherical shell: in both cases the local buckling should be considered.

The radius of curvature of an ellipsoidal head changes about the meridian of the head, Figure 8.30.

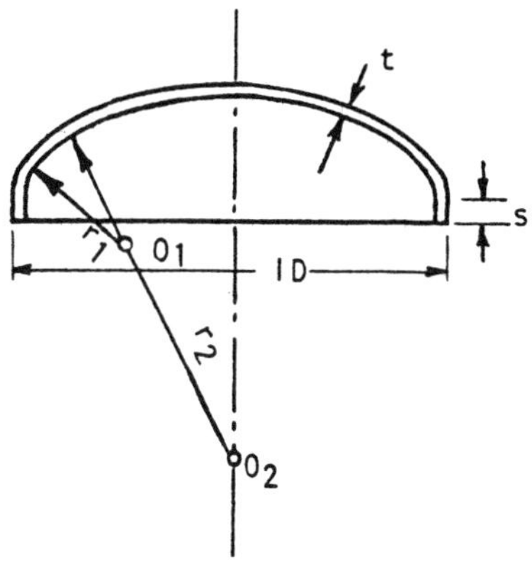

FIGURE 8.30 — Meridional section of element of an ellipsoidal head.

To use the previous relationships for ellipsoidal head, an equivalent radius of curvature must be used. The radius of curvature of an ellipsoidal head is maximum at the center of the head and at this point, is equal to twice the radius of the shell for a head having a major-to-minor axis ratio of 2.0.

Design of the head based on this minimum radius of curvature would result in considerable overdesign because the radius of curvature decreases as the point under consideration is moved away from the center toward the junction with the shell. This decrease in the radius results

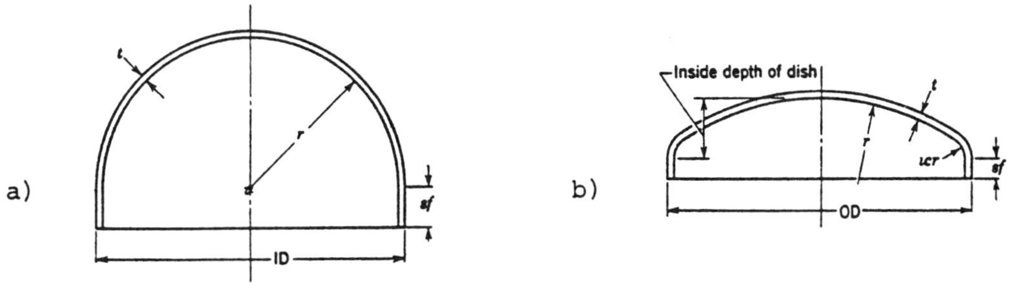

FIGURE 8.29 — (a) Hemispherical and (b) Torispherical Heads.

in an increase in rigidity and greater elastic stability. Thus, an ellipsoidal dished head has greater elastic stability than a torispherical dished head having the same diameter, thickness, and radius of curvature at the center of the head. As the radius of curvature of an ellipsoidal dished head varies along the meridian, an *average* radius may be used. However, the average must not be taken too far from the center of the head, which is the least stable point on the head.

8.12.4 Conical Head

Conical heads under external pressure, Figure 8.31 can be classified in three groups.

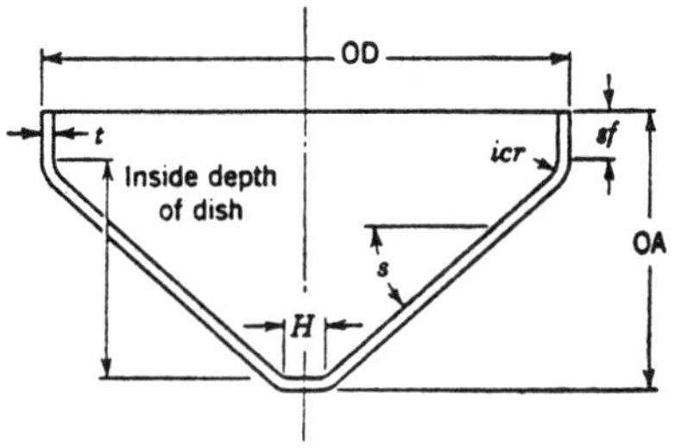

FIGURE 8.31 — Typical conical head.

If the apex angle is small (45° or less) the conical head is considered to behave as a cylindrical shell having the same diameter as the large end of the cone and a length equal to the axial length of the cone, provided the cone has no stiffening rings. If circumferential stiffening rings are used, the metal thickness may be decreased in each successive section as the apex is approached. In this case, each section is designed by using the greatest diameter of the section as the equivalent shell diameter, D, and the axial length between stiffeners (center-to-center) as the equivalent shell length between stiffeners, L. For conical heads having an *intermediate* apex angle (45° to 120°) the same procedure is followed except that the diameter at the large end of the cone is taken as the length of the equivalent cylinder if no circumferential stiffeners are used. If circumferential stiffeners are used, the procedure is the same as for stiffened cones with apex angles of less than 45° described above.

For flat cones having apex angles greater than 120°, the conical head is designed as a flat plate having a diameter equal to the largest diameter of the cone.

8.13 Stability of Horizontal Tank with Saddle Supports

The cylindrical shell acts as a beam over the two supports to resist by bending the uniform load of the tank and its contents, Figure 8.32.

An analysis of the stresses induced in the shell by the supports was reported by Zick [8.63] who developed formulas for the stresses.

A cylindrical tank with dished closures at the ends may be treated as an equivalent cylinder having a length equal to (L + 4/3 H) where L is the distance between the

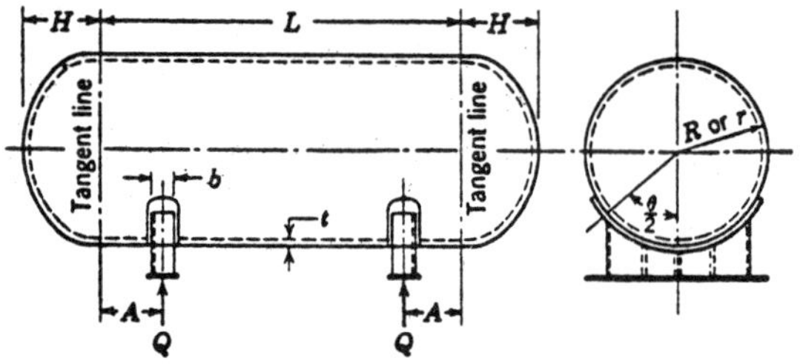

FIGURE 8.32 — Location and type of support for horizontal storage tank on two supports.

tangent lines of the tank and H is the depth of a dished closure. This approximation assumes that the weight of the head and the fluid contained in the head is equal to 2/3 of the weight of a cylinder of length H and the fluid contained in it. This approximation is valid for hemispherical heads and elliptical dished heads.

The maximum axial compressive stress in the shell f_s at the saddle will be

$$f_s = -\frac{3K_1QL}{\pi R^2 t} \qquad (8.76)$$

where

Q = load on one saddle, lb

L = tangent length of the tank, ft

A = distance from center line of saddle to tangent line, ft

H = depth of head, ft

R = radius of cylindrical shell, ft

t = thickness of cylindrical shell, in

r = radius of cylindrical shell, in

K_1 = dimensionless constant for various support conditions

$$K_1 = \left[\frac{4A}{L}\left(1 - \frac{1 - \frac{A}{L} + \frac{R^2 - H^2}{2AL}}{1 + \frac{4H}{3L}}\right) \left(\frac{\pi\left(\frac{\sin\Delta}{\Delta} - \cos\Delta\right)}{\left(\Delta + \sin\Delta\cos\Delta - 2\frac{\sin^2\Delta}{\Delta}\right)}\right)\right] \qquad (8.77)$$

$$\Delta = \frac{\pi}{180}\left(\frac{\theta}{2} + \frac{\beta}{6}\right) \qquad (8.78)$$

where

2Δ = arc, in radians, of unstiffened shell in plane of saddle effective against bending

θ = angle of contact of saddle with shell, degrees

β = $(180 - \frac{\theta}{2})$ = central angle from vertical to

The above values are illustrated by Figure 8.33.

The maximum axial compressive stress in the shell f_m at the midspan will be

$$f_m = \frac{3K_2QL}{\pi R^2 t} \qquad (8.79)$$

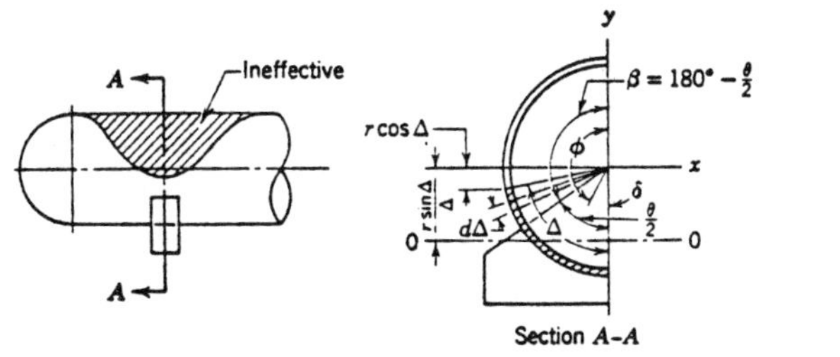

FIGURE 8.33 — The effective area of shell under beam action.

where

$$K_2 = \left[\frac{1 + 2\left(\frac{R^2 - H^2}{L^2}\right)}{1 + \frac{4H}{3L}} - \frac{4A}{L}\right] \quad (8.80)$$

The compression stresses f_s and f_m should not exceed ½ of the compression yield point of the material or the value given by

$$f \leq \left(\frac{E}{29}\right)\left(\frac{t}{R}\right)\left[2 - \left(\frac{2}{3}\right)(100)\left(\frac{t}{R}\right)\right] \quad (8.81)$$

which is based upon the accepted formula for buckling of short steel cylindrical columns.

REFERENCES — CHAPTER 8

[8.1] American Petroleum Institute, "Welded Steel tanks for Oil Storage", API Standard 650.

[8.2] APT, "Recommended Rules for the Design and Construction of Large, Welded, Low-Pressure Storage Tanks, APT Standard 620.

[8.3] APT, "Standard Tank with Riveted Shells", APT Standard 12-A, "Bolted Tanks", APT Standard 12-B; "All-Welded Oil Storage Tanks", APT Standard 12-C; "All-Welded Production Tanks", APT Standard 12-D.

[8.4] American Standard Institute, "Gasoline Tanks", B.90.1, 1963.

[8.5] American Water Work Association, "Standard for Steel Tanks, Standpipes, Reservoirs and Elevated Tanks, for Water Storage", AWWA Standard D-100-59.

[8.6] AWWA, "Standard Specifications for Elevated Steel Water Tanks, Standpipes and Reservoirs", AWWA-AWS.

[8.7] Association of Factory Mutual Fire Insurance Companies, "Specifications for Gravity Water-Tanks and Steel Towers".

[8.8] Natural Board of Fire Underwriters, "Standards for the Construction and Installation of Tanks, Gravity and Pressure, Towers, etc."

[8.9] American Railway Engineering Association, "Specifications for Riveted Steel Water and Oil Tanks".

[8.10] American Railway Engineering Association, "Specifications for Welded Steel Tanks for Railway Water Services".

[8.11] American Welding Society, "Standard Rules for Field-Welding of Steel Storage Tanks, D.S.1".

[8.12] American Iron and Steel Institute, "Steel Tanks for Liquid Storage", (No date), p. 18.

[8.13] Brownell, L.E. and Young, E.H., "Process Equipment Design", John Wiley & Sons, Inc., New York, 1968, pp. 138, 139, 258.

[8.14] Ibid., pp. 113-118.

[8.15] Ibid., pp. 120-140.

[8.16] Ibid., pp. 110-112, 261.

[8.17] Weingarten, V., Morgan, E., and Seide, P., "Final Report on Development of Design Criteria for Elastic Stability of Thin Shell Structures", Space Technology Laboratories, Inc., STL/TR-60-000-19425, 1960.

[8.18] Harris, L.A., Suer, H.S., Skene, W.T., and Benjamin, R.J., "The Stability of Thin Walled Unstiffened Cylinders Under Axial Compression Including the Effects of Internal Pressure", J. Aerospace Sci., Aug. 1957.

[8.19] Timoshenko, S.P., and Gere, J.M., "Theory of Elastic Stability", Second Ed., McGraw-Hill Book Co., N.Y., 1961, pp. 482-485.

[8.20] Flugge, W., "Die Stabilitat der Kreiszylinderschale", Ingenieur-Archiv, Vol. 3, No. 5, 1932, p.463, (in German).

[8.21] Brazier, L.G., "On the Flexure of Thin Cylindrical Shells and Other Thin Sections", Proceedings of the Royal Soc. of London, Ser. A., Vol. 116, 1927, pp. 104-114.

[8.22] Chwalla, E., "Reine Beigung Schlanker, dunwandiger Rohr mit gerader Achse", Z. Angen. Math. u. Mech. Vol. 13, No. 1, 1933, pp. 48-58. (in German).

[8.23] Dinnik, A.N., "Collected Works", Vol. III, Academy of Sciences, Ukrainian SSR, Kiev, 1956, pp. 88-90. (in Russian).

[8.24] Seide, P., and Weingarten, V.I., "On the Buckling of Circular Cylindrical Shells Under Pure Bending", Trans. ASME, Vol. 83, Series E., March, 1961, pp. 112-116.

[8.25] Batdorf, S.B., "A Simplified Method of Elastic-Stability Analysis for Thin Cylindrical Shells, II — Modified Equilibrium Equation", NACA TN 1342, June, 1947.

[8.26] Gerard, G., and Becker, H., "Handbook of Structural Stability", Part III, "Buckling of Curved Plates and Shells", NACA TN 3783, 1957.

[8.27] Timoshenko, S.P., and Gere, J.M., "Theory of Elastic Stability", McGraw-Hill Book Company, Inc., New York, 1961, p. 289.

[8.28] Stewart, R.T., "Collapsing Pressure of Bessemer Steel Lap-Welded Tubes, Three to Ten Inches in Diameter", Trans. Am. Soc. Mech. Engrs., 27, 1906, p. 780.

[8.29] Von Mises, R., "Der Kritische Aussendruck Zylindrischer Rohrer", Zeit, V.D.I., Vol. 58, 1914, pp. 750-755. (in German).

[8.30] Batdorf, S.B., "A Simplified Method of Elastic-Stability Analysis for Thin Cylindrical Shells", NACA Rept. 874, March, 1947.

[8.31] Gerard, G., and Becker, H., "Handbook of Structural Stability. Part III — Buckling of Curved Plates and Shells", NACA TN 3783, 1957.

[8.32] NASA, "Buckling of Thin-Walled Circular Cylinders", NASA SP 8007, 1968.

[8.33] Gerard, G., "Handbook of Structural Stability", Supplement to Part III — "Buckling of Curved Plates and Shells", NASA TN D-163, September, 1959.

[8.34] Vlasov, V.Z., "General Theory of Shells and Its Application in Engineering", Translation — NASA, Department of Commerce, Washington, D.C., 1964.

[8.35] Lessig, E.N., Lileyev, A.F., and Sokolov, A.G., "Steel Plate Structures", Moscow, (in Russian), 1956, pp. 343-345.

[8.36] Ibid., p. 352.

[8.37] Southwell, R.V., "On the General Theory of Elastic Stability," Philos. Trans. Royal Soc., London, Series A, Vol. 123, 1913, pp. 187-244.

[8.38] Windenberg, D.F., and Trilling, C., "Collapse by Instability of Thin Cylindrical Shells Under External Pressure," Trans. Am. Soc. Mech. Engrs., 56, 1934, p. 819.

[8.39] Cook, G., "The Collapse of Short Thin Tubes by External Pressure", Phil. Mag., July, 1914, pp. 51-56.

[8.40] Strum, R.G., and O'Brien, H.I., "Computing Strength of Vessels Subjected to External Pressure", Trans. Am. Soc. Mech. Engrs., 69, 1947, p. 353.

[8.41] Timoshenko, S.P., and Gere, J.M., "Theory of Elastic Stability", Sec. Ed., McGraw-Hill Book Co., Inc., New York, 1961, p. 289.

[8.42] Pulos, J., "Structural Analysis and Design Considerations for Cylindrical Pressure Hulls", David Taylor Model Basin Report, 1639, April, 1963.

[8.43] Pulos, J., and Salerno, V., "Axisymmetric Elastic Deformations and Stresses in Ring-Stiffened Perfectly Circular Shell Under External Hydrostatic Pressure", David Taylor Model Basin Report 1497, September, 1961.

[8.44] Reynolds, T.E., "Elastic Lobar Buckling of Ring-Supported Cylindrical Shells Under Hydrostatic Pressure", David Taylor Model Basin Report 1614, September, 1962.

[8.45] Reynolds, T.E., "Graphical Method for Determining the General-Instability Strength of Stiffened Cylindrical Shells", David Taylor Model Basin Report 1106, September, 1957.

[8.46] Ball, W., "Formulas and Curves for Determining the Elastic General-Instability Pressures of Ring-Stiffened Cylinders", David Taylor Model Basin Report 1570, January, 1962.

[8.47] Kendrick, S., "The Buckling Under External Pressure of Circular Cylindrical Shells With Evenly-Spaced Equal Strength Circular Ring-Frames", Part III, Naval Construction Research Establishment Report R-244, 1953.

[8.48] Reynolds, T., and Blumenberg, W., "General Instability of Ring-Stiffened Cylindrical Shells Subjected to External Hydrostatic Pressure", David Taylor Model Basin Report 1324.

[8.49] Galletly, G., et al, "General Instability of Ring-Stiffened Cylindrical Shells Subjected to External Hydrostatic Pressure — A Comparison of Theory and Experiment", J. Appl. Mech., Vol. 25, Trans. ASME, Vol. 80, June, 1958.

[8.50] Singer, J., Baruch, M., and Harari, O., "Further Remarks on the Effects of Eccentricity of Stiffeners on the General Instability of Stiffened Cylindrical Shells", Technion-Israel Institute of Technology, TAE Report 42, August, 1965.

[8.51] Becker, H., "Handbook of Structural Stability", Part VI, "Strength of Stiffened Curved Plates and Shells", NASA TN 3786, 1956.

[8.52] Weinitschke, H., "On the Stability Problem for Shallow Spherical Shells", J. Math. Phys. Vol. 38, No. 4, December, 1960, pp. 209-231.

[8.53] Budiansky, B., "Buckling of Clamped Shallow Spherical Shells", Proc. UITAM Symp. Theory of Thin Elastic Shells, North-Holland Publishing Company, Amsterdam, 1960, pp. 64-94.

[8.54] Huang, N.C., "Unsymmetrical Buckling of Thin Shallow Spherical Shells", J. Appl. Mech., Vol. 31, No. 3, September, 1964, pp. 447-457.

[8.55] Weinitschke, H., "On Asymmetric Buckling of Shallow Spherical Shells", J. Math. Phys., Vol. 44, No. 2, June, 1965, pp. 141-163.

[8.56] Thurston, G.A., and Penning, F.A., "Effect of Axisymmetric Imperfections on the Buckling of Spherical Caps Under Uniform Pressure", J. Am. Inst. Aeron. Astronautics, Vol. 4, No. 2, February, 1966, p. 319.

[8.57] Bushnell, D., "Nonlinear Axisymmetric Behavior of Shells of Revolution", J. Am. Inst. Aeron, Astronautics, Vol. 5, No. 3, March, 1967, pp. 432-439.

[8.58] Wang, L.R.L., "Effects of Edge Restraint on the Stability of Spherical Caps", J. Am. Inst. Aeron. Astron. Vol. 4, No. 4, April, 1966, pp. 718-719.

[8.59] Bushnell, D., "Buckling of Spherical Shells Ring-Supported at the Edges", J. Am. Inst. Aeron. Astron. Vol. 5, No. 11, November, 1967, pp. 2041-2046.

[8.60] Wang, L.R.L., "Discrepancy of Experimental Buckling Pressures of Spherical Shells", J. Am. Inst. Aeron. Astron., Vol. 5, No. 2, February, 1967, pp. 357-359.

[8.61] McComb, H.G., Jr., and Fitcher, W.B., "Buckling of a Sphere of Extremely High Radius-Thickness Ratio, Collected Papers on Instability of Shell Structures", NASA TN D-1510, 1962, pp. 561-570.

[8.62] Timoshenko, S.P., and Gere, J.M., "Theory of Elastic Stability", Sec. Ed., McGraw-Hill Book Company, Inc., N.Y., 1961, pp. 517-518.

[8.63] Zick, L.P., "Stresses in Large Horizontal Pressure Vessels on Two Saddle Supports", Welding Journ. Res. Suppl. 30, 1951, p. 435 s.

CHAPTER 9

above ground pipelines

9.1 Introduction

This chapter is concerned with the stress analysis of above ground pipelines suitable for use in conveying liquids and gases.

In North America, practically all oil pipelines are constructed in accordance with the American Petroleum Institute (API) Standard Specifications. [9.1] Water-supply pipelines use the American Water Works Association Standard Specifications [9.2] and gas pipelines — following gas — the American Standard Code for Pressure Piping [9.3].

In the above-mentioned codes, it has become common to publish formulas giving pipe wall thicknesses in inches as a function of certain coefficients. Therefore, the design stresses and conversion from feet to inches, are all hidden in the numerous coefficients. Now that different design stresses are permitted under varying conditions and materials, it appears more correct to revert to more basic formulas in which the designer is aware of the stresses being used. It is this approach that this chapter is based upon.

9.2 Design Loads and Forces

In the design of the pipelines the following dead loads and imposed live loads and forces should be considered:

1. *Main Loads*
   a. Dead Load
   b. Live Load
   c. Wind Forces
   d. Icing
   e. Temperature
2. *Special Loads*
   f. Earthquake Forces

9.2.1 Dead Load

Dead loads are computed on the basis of the unit weight of the materials. Dead load consists of the weight of the pipe and reinforcing rings, if any.

9.2.2 Live Load

Live load constitutes the weight of the liquid or the maximum gas pressure.

9.2.3 Wind Forces

Surface pipelines should be checked for wind pressure according to the National Standard Building Code [9.4].

9.2.4 Icing

In locations where the strongest winds and icing may occur simultaneously, pipelines must be calculated assuming and ice covering based on climatic conditions and local experience, according to the National Standard Building Code.

9.2.5 Temperature Influence

Temperature influences are considered where extreme differences in temperature warrant them.

9.2.6 Earthquake Forces

Earthquake forces are applicable where geographic conditions require them, according to the National Standard Building Code.

9.3 Pipe Stress Analysis

The American Standard Association Code for Oil Transportation Piping [9.5] considers the following stress-producing forces:

1. *Maximum Working Pressure*

   The maximum pressure expected at any point in the line under normal working conditions, including static pressure and pressure required to overcome friction losses.

2. *Internal Design Pressure*

   The maximum internal liquid pressure which shall not be less than the maximum working pressure plus allowance for surge pressure if anticipated.

3. *External Design Pressure*

   The effect of possible vacuum within the pipe shall be recognized and the pipe wall selected shall provide adequate strength to prevent collapse.

4. *Working and Design Temperature*

   It is not necessary to vary the design stress for metal temperature between 250° F and -20° F.

5. *Fluid Expansion Effects*

   Pressure shall be considered for exposed sections of pipelines, valves and fittings, and pressure relieving devices shall be installed when required.

6. *Dynamic Effects*

   Unusual loadings such as those caused by impact, wind, vibrations, and resonance shall be considered and provided for in accordance with sound engineering practice.

7. *Weight Effects*

   The weight of pipe, piping components, liquid contents, covering and other sustained weights shall be considered when piping is exposed and not supported continuously.

8. *Thermal Expansion and Contraction Loads*

   When substantial temperature variations are anticipated, the effects of thermal expansion and contraction shall be considered.

The basic criteria for pressure design of a piping system including valves, flanges, fittings, headers and special assemblies are their suitability to withstand internal pressure and other specified loadings.

In the following, we analyze basic conditions of the overland pipelines under internal pressure but subjected to different external loadings.

9.3.1 Straight Pipe Under Internal Pressure

9.3.1.1 Circumferential or Hoop Stresses

Suppose a long pipe is subjected to an internal pressure p, which may be due to a fluid or gas enclosed within the pipe.

The average unit tensile stress in the pipe is

$$f_c = \frac{pr}{t} = \frac{pd}{2t} \tag{9.1}$$

which is called the circumferential or hoop stress, where

p = the uniform internal pressure, psi

r = the radius of the mean circumference of the pipe, in.

t = thickness of the wall, in.

This relation is independent of the length of the pipe, provided the section is far enough removed from the flanges or other reinforcements to receive no support from them. Equation (9.1) is generally known as the "common formula for bursting pressure" and is frequently used in calculations of bursting strength for thin-walled pipes without taking into account the effect of lateral contraction. In Eq. (9.1), it is assumed that the hoop stress is uniformly distributed across the cylinder wall. This condition does not hold, except in the case of pipes having walls of infinitesimal thickness.

9.3.1.2 Barlow's Formula

An empirical formula for internal fluid pressure which gives results on the side of safety for all practical thickness ratios is that known as Barlow's formula [9.6]. This formula is similar to the common formula except that the outside diameter of the pipe is used instead of the inside. Barlow's formula is

$$f_c = \frac{pD}{2t} \tag{9.2}$$

While Barlow's formula is widely used because of its convenience of solution, it was not generally considered to have any theoretical justification until formulae based on the maximum-energy-of-distortion theory showed that for thin-walled pipe with no axial tension, Barlow's formula actually is theoretically correct. Since most commercially important pipes have a ratio of wall thickness to outside diameter less than 0.10, Barlow's formula for thin-walled pipes is of great significance. Comprehensive bursting tests on commercial steel pipe have demonstrated that the formula predicts the pressure at which the pipe will rupture with an accuracy well within the limits of uniformity of commercial pipe thickness. In general, failure occurred at a pressure about three percent higher than predicted.

Barlow's formula has been employed in the ASA Standards for Wrought Iron and Wrought Steel Pipe and the ASA Code for Pressure Piping.

9.3.1.3 The Allowable Unit Stress f_c

The allowable unit stress f_e is obtained by taking the unit stress of the pipe material at the yield point, and reducing it:

1. by a suitable factor of safety
2. by a factor for the efficiency of the welded longitudinal joint; and
3. by a factor to allow for the effect of extreme temperatures.

   The derived thickness is then increased to care for:

   a. the effect of corrosion, and
   b. the manufacturing tolerances of the material

9.3.1.4 Longitudinal Stresses

The longitudinal unit tension stess in the wall of the pipe is

$$f_\ell = \frac{pr}{2t} = \frac{pd}{4t} \tag{9.3}$$

or half the intensity of the circumferential stress.

If the continuity of the pipe is interrupted by expansion joints, Dressler couplings, bell and spigot joints, or other joints incapable of carrying tension, then this end force must be otherwise provided for.

9.3.2 Stresses in Circular Bends [9.7]

Let us consider a circular bend, with dimensions as shown in Figure 9.1.

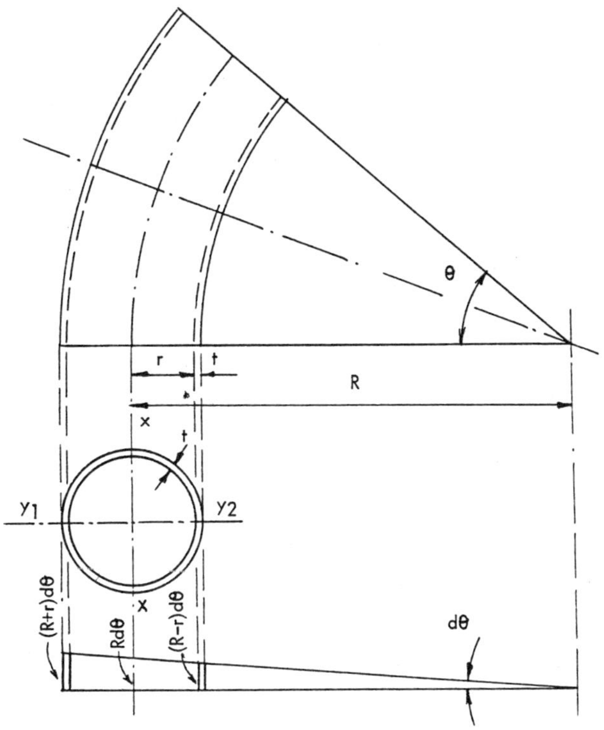

FIGURE 9.1 — Stresses in circular bend.

9.3.2.1 Circumferential or Hoop Stresses

On section x-x, the unit tension stress is

$$f_x = \frac{pr}{t}\left(1 - \frac{\pi r}{4R}\right) \qquad (9.4)$$

which is smaller than in the straight pipe, formula (9.1).

On section y-y the unit stress at y_1 is

$$f_{y_1} = \frac{pr}{t}\left(\frac{R+\frac{r}{3}}{R+r}\right) \qquad (9.5)$$

which is smaller than in the straight pipe.

On section y-y the unit stress at y_2 is

$$f_{y_2} = \frac{pr}{t}\left(\frac{R-\frac{R}{3}}{R-r}\right) \qquad (9.6)$$

which is greater than in the straight pipe.

For a commonly used bend radius of five pipe diameters R = 10r, the overstress is 7½ percent. For a standard long radius pipe fitting R = 3r, the overstress is 33 percent, which could be serious in a highly stressed line.

9.3.3 Bending Stresses [9.8]

In general, the ordinary theory of flexure applies when a circular pipe is supported at intervals, is held circular at and between the supports, and is completely filled. If the pipe is only partially filled and the cross-section at points between supports become out-of-round, the maximum fibre stress is considerably greater than indicated by the ordinary flexure formula, and is highest for the half-filled condition.

The maximum bending moment in continuous pipe supported on equidistant multiple supports, due to the weight of the pipe g and its contained liquid w is

$$M = \pm \frac{q\ell^2}{12} \qquad (9.7)$$

where

$$q = g + w \qquad (9.8)$$

The maximum compressive longitudinal fibre-stress due to the bending is

$$f_e = \frac{q\ell^2}{12} \cdot \frac{1}{S} \qquad (9.9)$$

where

$$S = \frac{\pi}{64D/2}(D^4 - d^4) = \frac{\pi}{32D}(D^4 - d^4) \qquad (9.10)$$

is the section modulus of the pipe.

If both ends of the pipeline are closed, longitudinal stresses will originate and the total stress is

$$f_{tot} = \frac{q\ell^2}{12S} + \frac{pD}{4t} \qquad (9.11)$$

Excessive deflection should be avoided when pipe acts as a beam. A maximum permissible deflection of $\frac{1}{360}$ of the span is suggested as good practice.

The maximum theoretical deflection can be determined using

$$\Delta = \frac{1}{184} \cdot \frac{q\ell^4}{EI} \qquad (9.12)$$

where

- Δ = maximum deflection at center of span in inches
- q = total load on span in pounds per inch
- ℓ = length of the span in inches
- E = modulus of elasticity, 30 × 10^6 psi for steel pipe
- I = moment of inertia of pipe in in^4

9.4 Buckling Stability of Multispan Straight Pipe Under Own Weight and Liquid Pressure

In general, the ordinary theory of flexure applies when a circular pipe is supported at intervals, is held circular at and between the supports, and is completely filled. If the pipe is only partially filled and the cross-section at points between supports become out-of-round, the maximum fibre stress is considerably greater than indicated by the ordinary flexure formula, and is highest for the half-filled condition.

The maximum bending moment in continuous pipe supported on equidistant multiple supports, Figure 9.2, due to the weight of the pipe g and its contained liquid w is

$$M = \pm \frac{q\ell^2}{12} \qquad (9.13)$$

where

$$q = g + w \qquad (9.14)$$

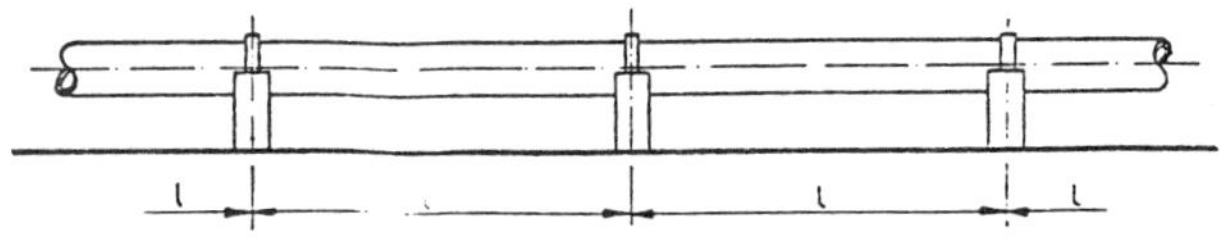

FIGURE 9.2 — Multispan pipe.

The maximum compressive longitudinal fibre-stress due to the bending is

$$\sigma_q = \pm \frac{q\ell^2}{12S} \tag{9.15}$$

where

$$S = \frac{\pi D_m^{\ 2} t}{4} \tag{9.16}$$

is the section modulus of the pipe, Figure 9.3. Therefore

$$\sigma_q = \pm \frac{q\ell^2}{3\pi D_m^{\ 2} t} \tag{9.17}$$

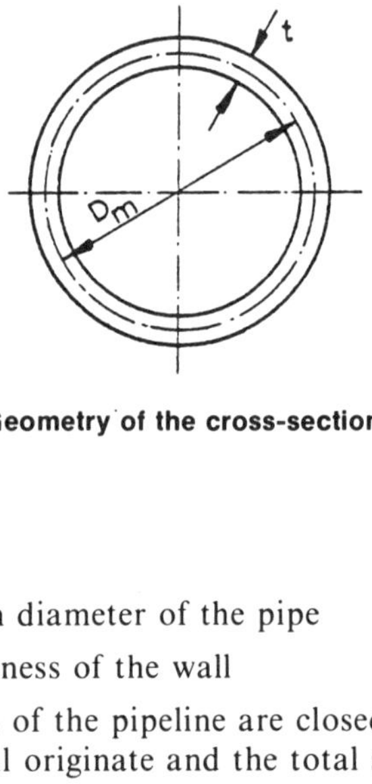

FIGURE 9.3 — Geometry of the cross-section of the pipe.

where

D_m = mean diameter of the pipe

t = thickness of the wall

If both ends of the pipeline are closed, the longitudinal stresses will originate and the total stress is

$$\sigma_{tot} = \pm \frac{q\ell^2}{3\pi D_m t} + \frac{pD_m}{4t} < \sigma_{cr} \tag{9.18}$$

where

p = the internal pressure

and the permissible critical buckling stress σ_{cr} may be determined after the diagram in Chapter 1.2.

The permissible span considering possible buckling is then

$$\ell = D_m \sqrt{3\pi t/q\ (\sigma_{cr} - PD_m/4t)} \tag{9.19}$$

9.5 Multispan Pipe Under Different Loadings

In the following we analyze the multispan pipe under the influence of different asymmetric loadings such as the partial filling of the pipe with liquid, and the outside ice and wind action. The asymmetric loading is defined with respect to the longitudinal axis of the pipeline.

In such cases the pipe should be considered as a space type structure and, therefore, analyzed as space type structure or a cylindrical shell. Formulas for the determination of critical buckling forces in such shell are based on the analysis developed by Lessig [9.9].

For the multispan pipe, shown in Figure 9.4, formulas for the axial compressive forces are as follows:

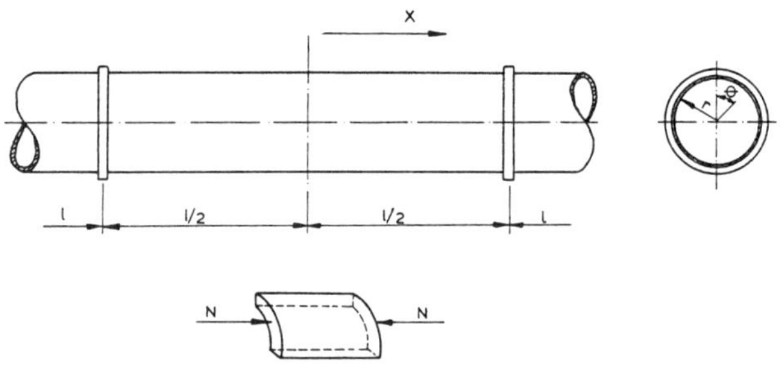

FIGURE 9.4 — Multispan pipe.

Expansions for axial compressive forces are given as a Fourier series. For the midspan of a multispan pipe, the compressive force is

$$N_m = \frac{m}{3} \sum_1^n A_n \cos n\phi \tag{9.20}$$

and at the support

$$N_s = \frac{2}{3} m \sum_1^n A_n \cos n\phi \tag{9.21}$$

where the values of the coefficients are:

$$m = \frac{2\sqrt{3}\, r}{t} \tag{9.22}$$

$$A_n = -\frac{rmK^2 n^2}{(mK^2)^2 + n^8} \left(a_n + \frac{b_n}{n}\right) \tag{9.23}$$

$$K = \frac{\pi r}{\ell} \tag{9.24}$$

where

m and K = the geometric characteristics of the shell

r = mean radius of the wall

t = thickness of the wall

As experience indicates the expressions (9.20) and (9.21) converge very rapidly, and sufficient accuracy is obtained by taking only two terms, or using n = 1, 2.

After substituting expressions (9.22) and (9.24) into (9.23), we then obtain

$$A_n = \frac{2\sqrt{3}\,\pi^2 r^4 t\, l^2 n^2}{12\pi^4 r^6 + t^2 l^4 n^8} \left(a_n + \frac{b_n}{n}\right) \tag{9.25}$$

and the final expressions for the compressive forces are

$$N_m = 4\tfrac{2}{\pi} r^5 \; {}^2\sum_1^n \frac{(a_n + b_n/n)\, n^2 \cos n}{12\pi^4 r^6 + t^2 \; {}^4 n^8} \tag{9.26}$$

and

$$N_s = 8\tfrac{2}{\pi} r^5 \; {}^2\sum_1^n \frac{(a_n + b_n/n)\, n^2 \cos n}{12\pi^4 r^6 + t^2 \; {}^4 n^8} \tag{9.27}$$

In the following Table 9.1 the expressions for the Fourier coefficients are given for the different transverse loadings. In this Table, the following designations are used:

- g = own weight of the pipe per unit length
- h = depth of the ice on bottom surface of the pipe
- γ_ℓ = the specific weight of the liquid
- γ_i = the specific weight of the ice
- w = velocity wind pressure

The axial stress is

$$\sigma = \frac{N}{t} \tag{9.28}$$

TABLE 9.1 — Fourier's Coefficients.

		a_1	a_2	a_n n > 2	b_1	b_2	b_n n > 2
1	**Own Weight** 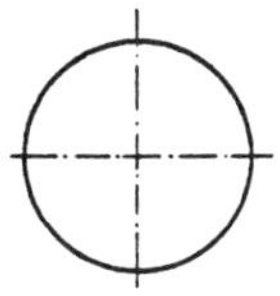	$\frac{4g}{\pi}$	0	0	$\frac{4g}{\pi}$	0	0
2	**Pressure of Liquid** 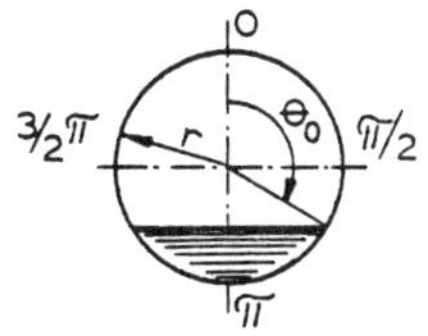	$\frac{4\gamma r}{\pi^2}(\pi - \theta_0 + \frac{\sin 2\theta_0}{2}) - \frac{8\gamma r}{3\pi^2}\sin^3\theta_0$		$\frac{8\gamma r}{\pi} \times \frac{n \sin\theta_0 \cos n\theta_0}{n(n^2-1)} \times \frac{\cos\theta_0 \sin n\theta_0}{}$	0	0	0
3	**Outside Ice** 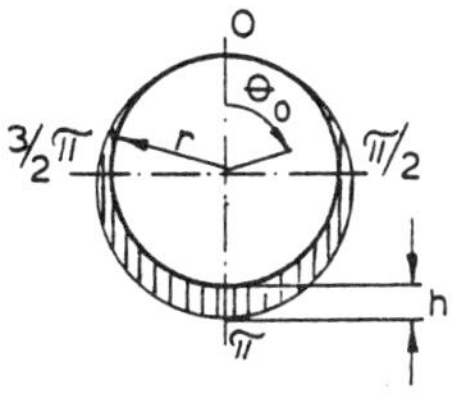	$\frac{2\gamma h}{\pi}$	$\frac{\gamma h}{\pi}$	0	$\frac{2\gamma h}{\pi}$	$-\frac{\gamma h}{\pi}$	0
4	**Wind** 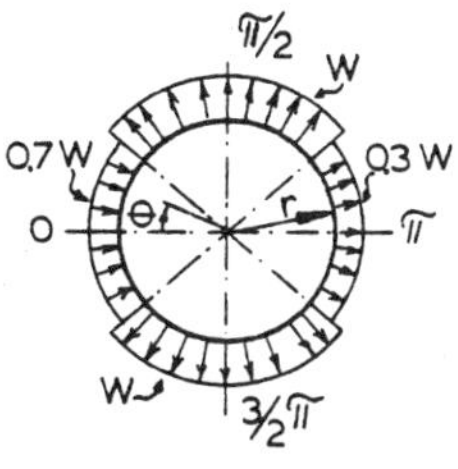	$\frac{5.66w}{\pi^2}$	$\frac{9.60w}{\pi^2}$	$\frac{w}{n\pi^2} \times (13.6 \sin\frac{n\pi}{4} - 5.6 \sin\frac{3n\pi}{4})$	0	0	0

and the conditions of the stability of the pipe against buckling are:

In the span:

$$\sigma_m = \frac{N_m}{t} = \frac{4\pi^2 r^5 l^2}{t} \sum_{1}^{n} \frac{(a_n + b_n/n)\, n^2 \cos n\phi}{12\pi^4 r^6 + t^2 l^4 n^8} \leq \sigma_{cr} \tag{9.29}$$

and at the support:

$$\sigma_s = \frac{N_m}{t} = \frac{8\pi^2 r^5 l^2}{t} \sum_{1}^{n} \frac{(a_n + b_n/n)\, n^2 \cos n\phi}{12\pi^4 r^6 + t^2 l^4 n^8} \leq \sigma_{cr} \tag{9.30}$$

where the permissible critical buckling stress should be determined by the diagram in Section 2.6.

9.6 Buckling of the Pipe Due to Ovalling Effect

The elastic stability of the pipes under bending may be affected by a relatively large displacement caused by the flattening or ovalling of the cross-section of thin wall pipes, known as Karman's effect.

Brazier [9.10] investigated this problem and proved that the increase of the bending moment due to the ovalling of the pipe results in the substantial reduction of the pipe stiffness and consequently to a loss of stability. For a solution of this problem Brazier assumed that a long pipe is bent into a circle of large radius, having the same transverse distortions of every cross-section along the pipe, Figure 9.5.

After bending of the pipe, the relation between the bending moment transmitted by the pipe at any section and the curvature of the pipe are considered. The longitudinal compression on the inner side of the pipe and the longitudinal tension on the outer side, both have a component directed towards the center of the pipe that tends to flatten it. These pressures actually flatten the circular cross-section into the quasi-oval form shown in Figure 9.5 (b).

Brazier obtained the following expression for the critical bending moment

$$M_{cr} = \frac{2\sqrt{2}}{9} \cdot \frac{E\pi R_m t^2}{\sqrt{1 - \mu^2}} \tag{9.31}$$

Using section modulus $S = \pi R_m{}^2 t$ and Poisson's ratio $\mu = 0.3$, we obtain the buckling stress

$$M_{cr} = \frac{2\sqrt{2}\, E\pi R_m t^2}{9\pi t R_m^2 \sqrt{1 - 0.3^2}} = 0.33E\left(\frac{t}{R_m}\right) \tag{9.32}$$

and the maximum displacement in a radial direction.

$$w = -0.22\, R_m \tag{9.33}$$

Therefore, considering the ovalling deformation of the pipe, the critical buckling stress given by Equation (9.18) shall be smaller than the stress given by expression (9.32) or

$$\frac{q\ell^2}{3\pi D_m^2 t} + \frac{PD_m}{4t} \leq 0.33\, E\left(\frac{t}{R_m}\right) \tag{9.34}$$

to prevent the possible ovalling deformation of the pipe.

9.7 Buckling of the Pipe as Beam-Column

Internal pressure loading can lead to the instability of a pipe as a beam-column. This kind of instability arises in piping systems in which the axial pressure load is restrained by some structure other than the pipe itself. The

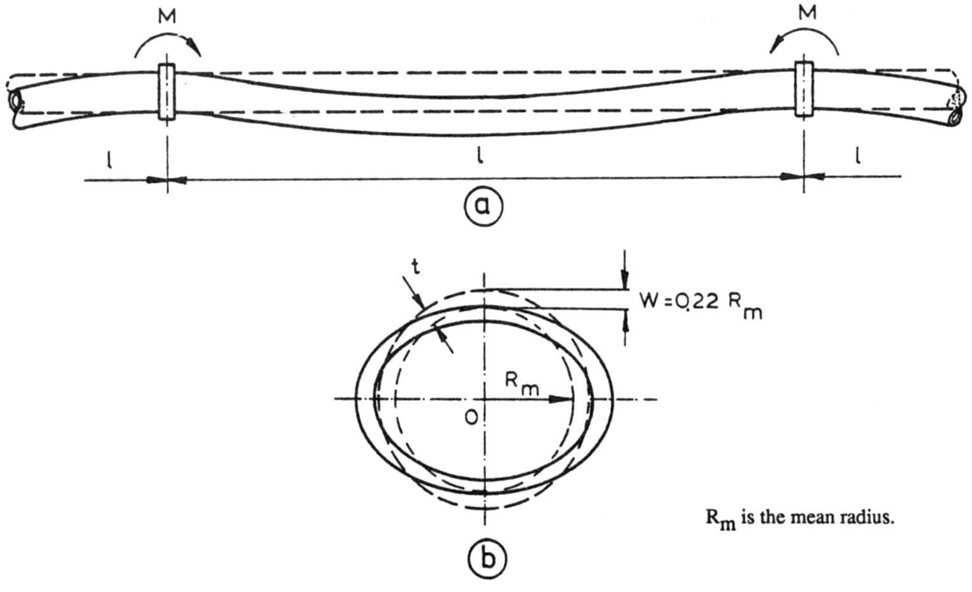

FIGURE 9.5 — Deformation of thin pipe immediately before buckling.

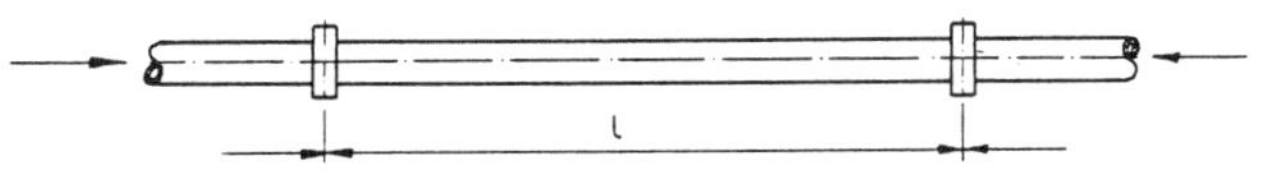

FIGURE 9.6 — Pipe under buckling pressure.

most common examples occur in piping systems using either bellows expansion joints, or packed slip-joints.

Flugge [9.11] and Haringx [9.12] presented a theory for this type of instability which is essentially that of a column loaded in axial compression by the pressure-end force, namely $\pi R^2 P$.

We are considering a straight uniform pipeline, attached to a rigid anchor block at each end, and initially unstressed, Figure 9.6.

When an internal pressure is applied, it may cause the pipe to buckle sideways.

The longitudinal stress set up in a straight thin-walled axially constrained pipe by an internal pressure, p, is

$$\mu pD/2t \tag{9.35}$$

where μ is Poisson's ration, D = diameter and t = wall thickness of the pipe.

The resultant force over a complete cross-section perpendicular to the pipe axis is

$$\frac{\pi D^2}{4}\ p\ (1-2\mu) \tag{9.36}$$

it is compressive since $\mu < 1/2$.

The difference between the compressive force

$$\frac{\pi D^2}{t}\ p \tag{9.37}$$

carried by the fluid within the pipe and the tensile force

$$\mu\frac{PD}{2t}\ (\pi Dt) \tag{9.38}$$

carried by the wall of the pipe. The pipe can be thought of as a column with clamped ends carrying this resultant compressive force, and buckling can be expected to occur when the force reaches the Euler buckling load.

The flexural rigidity of the pipe is

$$EI = \frac{E\pi D^3 t}{8} \tag{9.39}$$

and the Euler buckling load is

$$P_{cr} = \frac{4\pi^2 EI}{\ell^2} = \frac{\pi^3 ED^3 t}{2\ell^2} \tag{9.40}$$

Buckling will occur when

$$\frac{\pi D^2}{4}\ p(1-2\mu) = \frac{\pi^3 ED^3 t}{2\ell^2} \tag{9.41}$$

and the buckling pressure is

$$p_b = \frac{2\pi^2 EDt}{(1-2\mu)\ell^2} \tag{9.42}$$

The experimental investigations confirmed buckling of the pipe under certain critical internal pressure determined by the formula (9.42), [9.13].

9.8 Buckling Stability of Curved Pipes

9.8.1 Basic Parameters of the Curved Pipe

The term "curved pipe" as used herein is intended to cover both shop or field bent pipes and welding elbows manufactured to meet butt-welded fitting specifications.

Analytically, such components may be considered as a section of an annular torus. They are basically defined by a bend radius R, a cross-section radius r, wall thickness t, and arc angle α, as shown in Figure 9.7, where

M_i = in-plane bending moment

M_o = out-of-plane bending moment

It is normal practice to absorb the expansion of a straight pipeline by an expansion bend or curved pipe. Two conditions have to be fulfilled:

a. The curved pipe must not be overstressed, and
b. The force needed to deflect the bend by the required amount must not be greater than the pipeline or than its anchorages can safely sustain.

It is known that the forces and deflections in a curved pipe are not those predicted by the simple Bernoulli-Euler bending theory, as was proven by Bantlin [9.14] and Von Karman [9.15]. Also, we have to differentiate between the structural behavior of the so-called "long radius bend" pipes where $R \leqslant 10r$ and a "short radius bend" having $R \leqslant 3r$.

9.8.2 Effect of Flattening of the Cross-Section During Bending

It has been established that a curved pipe subjected to bending is more flexible and has higher stresses than would be indicated by the elementary theory of bending. That a curved pipe subjected to a moment loading behaves differently than a curved solid bar was noted experimentally by Bantlin [9.14] in 1910. Because of the ability of the pipe cross-section to deform and to flatten, a curved pipe is more flexible than a curved bar of the same moment of inertia. For the same reason high bending stresses can develop in the hoop direction.

The theory of flattening of the circular cross-section of a curved pipe during flexure has been developed by a number of investigators in explanation of the lack of agreement between ordinary curved-bar formulas and test results on curved pipes.

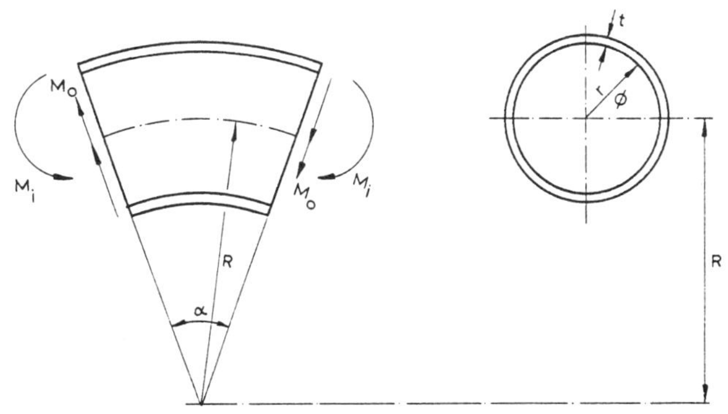

FIGURE 9.7 — Nomenclature for curved pipe.

The characteristics of curved pipes are recognized in pipe-system flexibility calculations by the use of "flexibility factors" and "stress-intensification factors" which are simply the ratios of actual flexibility and stress to those predicted by the elementary bending theory.

Theoretical studies of the flexibility and stress-intensification factors of curved pipe began some 65 years ago with the work of Von Karman [9.15] who developed theoretical formulas for in-plane bending of curved pipe *without internal pressure.*

In 1947 Barthelemy [9.16] published his development of the theory for in-plane bending *with internal pressure.* This paper apparently escaped attention on this continent. In 1956, Kafka and Dunn [9.17] directly developed the theory for in-plane bending with internal pressure.

During this period from 1911 to 1943 numerous tests were run on pipe bends with both in-plane and out-of-plane bending. It was observed that curved pipe flexibility for out-of-plane bending was also higher than anticipated by curved beam theory. However, it was not until 1943 that Vigness [9.18] gave the development of the theory for *out-of-plane moments.*

In 1957, Rodabaugh and George [9.19] developed the theory for both in-plane and out-of-plane bending and also developed comparatively simple equations with adequate accuracy for purposes of practical piping problems.

9.8.3 Longitudinal and Transverse Stresses Due to Bending

Considering stresses in curved pipes, the following definitions are introduced:

a. Longitudinal stresses — in the direction of the length of pipe bend parallel to the axis of the bend, and
b. Transverse stresses — around the pipe bore in the hoop direction.

The stresses that exist in curved pipes occur as a result of the changes in their free lengths in relation to their points of anchorages. The maximum stress in the pipe walls at the bends is usually several orders of magnitude higher than that which would be calculated by treating the pipe as a simple beam. The origin and mathematical nature of these stresses have been quite thoroughly demonstrated by the above-mentioned investigators.

Figure 9.8 (a) shows how the cross-section of a pipe becomes elliptical under the influence of bending moments tending to decrease the radius of curvature, R, called an "in-plane bending moment."

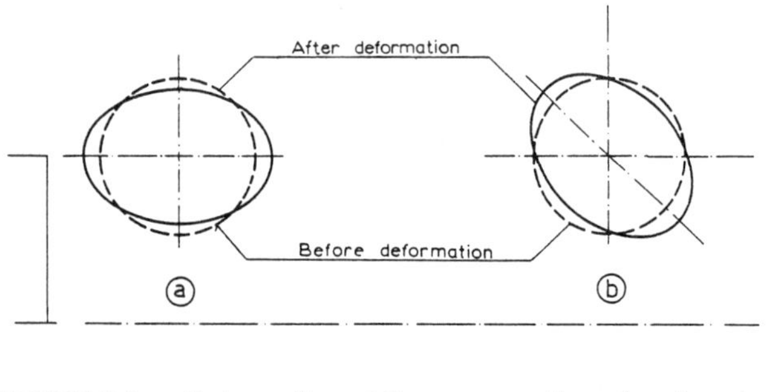

FIGURE 9.8 — Deformation of the cross-section of a pipe at a bend of radius R. (a) In-plane bending, (b) Out-of-plane bending.

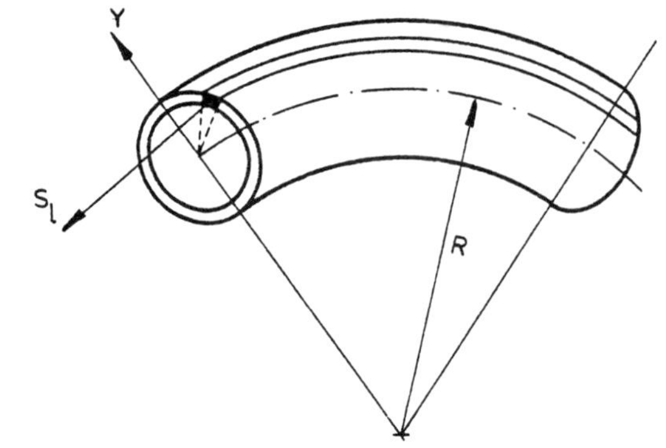

FIGURE 9.9 — Pipe bend showing longitudinal stress S_ℓ.

An "out-of-plane bending moment" acting alone causes the angle of the major axis of the ellipse to be inclined at approximately 45 degrees to the horizontal as illustrated in Figure 9.8 (b).

At any cross-section of a pipe bend the following stresses are known to exist:

1. Longitudinal stresses which act in a direction perpendicular to the plane of the circular cross-section of the pipe. These stresses occur primarily as the result of bending moments being applied to the pipe bend, but may be augmented by the end loads causing direct tensile or compressive stresses. This stress component is shown in Figure 9.9.

2. Transverse stresses occurring in the plane of the circular cross-section of the pipe tangent to the pipe wall. These stresses are comparable to hoop stresses. Part of this transverse stress is due to the transverse bending of the pipe wall in becoming elliptical in shape. At the point of the sharpest curvature of the pipe wall, this stress is tensile at the outer surface of the pipe wall and compressive at the inner surface. In addition, a uniform transverse stress has also been shown to exist by Gross [9.20] which, when superposed upon the one due to wall distortion, has the effect of increasing the compressive stress at the inner wall and decreasing the tensile stress at the outer wall at the point of sharpest curvature, as shown in Figure 9.10.

There is considerable evidence by Hovgaard [9.21] and Cross [9.22] supported by current practical experience to justify the conclusion that the maximum longitudinal stresses in the pipe wall determine the point of breakdown of the pipe bends. The maximum combined or effective stress in a pipe may locally exceed the yield strength of the material by a substantial amount without causing the pipe to suffer an observable permanent set.

9.8.4 Bending of the Pipe In-Plane of Curvature

In 1911, Karman [9.15] published a theoretical analysis of the characteristics of a curved pipe subjected to "in-plane" bending. He did not consider the internal pressure and assumed Poisson's ratio μ = 0. Karman showed that the apparent decrease in stiffness of a curved pipe under the action of a bending moment, is due to an ovalization of the pipe in the curved section, Figure 9.11.

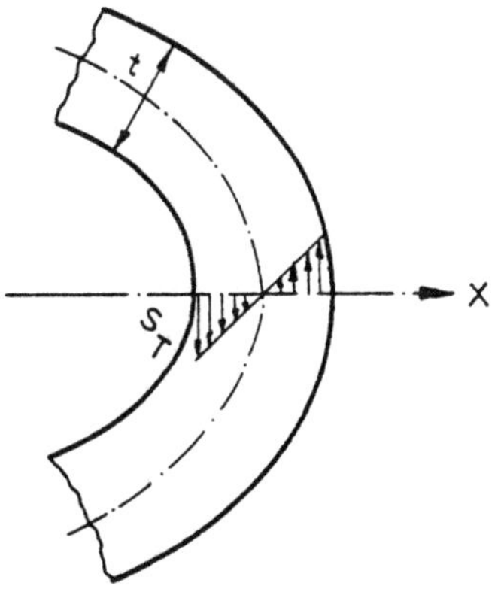

FIGURE 9.10 — Pipe wall cross-section showing transverse stresses, S_T, due to in-plane bending.

The effect of the ovalization is to reduce the stiffness parameter EI of the curved pipe to a reduced value

$$(EI)_{red} = \frac{EI}{K} \tag{9.43}$$

where the flexibility factor

$$K = \frac{12h^2 + 10}{12h^2 + 1} > 1 \tag{9.44}$$

and

$$h = tR/r^2 \tag{9.45}$$

where

R = bend radius

r = radius of pipe cross-section

I = moment of inertia of pipe cross-section

E = modulus of elasticity

This ovalization also yields a stress distribution somewhat different from that computed by the simple formula $\sigma = \frac{My}{I}$ of beam theory.

The first term approximation by Karman is sufficiently accurate for relatively heavy-wall pipe bends with large bend radii. However, with the increasing use of welding elbows having relatively thin walls, it has become more apparent that the first-term approximations given by Equation (9.44), grossly underestimate the flexibility present in curved pipes or welding elbows with small values of the parameter $h = tR/r^2$.

Beskin [9.23] has extended Karman's analysis of in-plane bending, to include sufficient terms in the series solution. He obtained his results as a function of h, thereby showing that for values of h less than about 0.3, the value of the flexibility factor is given by the simple formula

$$K = \frac{1.65}{h^{2/3}} \tag{9.46}$$

In the formulas for K given by Karman and subsequent investigators the influence of internal pressure on the ovalization has not been considered. Kafka and Dunn [9.17] extended Karman's work by including into the analysis of curved pipe the influence of internal pressure. According to their theory the maximum bending stresses, which may result in buckling, are

$$\sigma_{max} = \frac{Mr_m}{I} K_1 \left(1 - \frac{2B}{2r_m}\right) \leq \sigma_{cr}, \quad \text{if } \frac{6B}{A} < r_m \tag{9.47}$$

and

$$\sigma_{max} = \frac{2}{3} \frac{Mr_m}{I} K_1 \sqrt{\frac{r_m A}{6B}} \leq \sigma_{cr}, \quad \text{if } \frac{6B}{A} > r_m \tag{9.48}$$

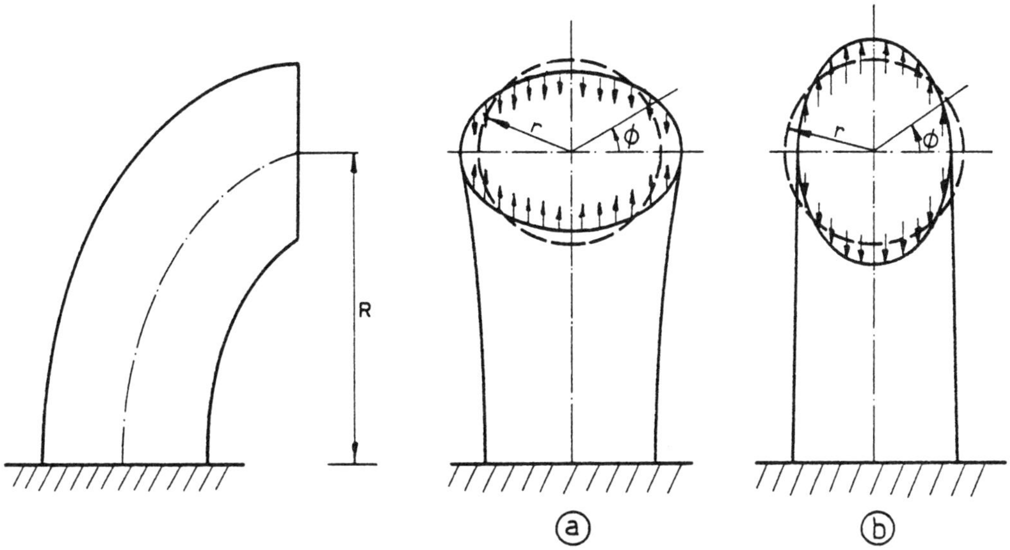

FIGURE 9.11 — Deformation of curved pipe cross-section under in-plane-of-curvature bending moment: (a) Moment tending to decrease radius of curvature of bend, (b) Moment tending to increase radius of curvature of bend.

where the flexibility factor is

$$K_1 = \cfrac{1}{1 - \cfrac{(1-\mu^2)rB^2}{Er_m^3\ tA}} \tag{9.49}$$

and

$$A = \frac{5}{2}\,\frac{Etr_m}{R^2(1-\mu^2)} + \frac{36D}{r_m^{\ 2}}\left(\frac{1}{r_m} + \frac{4\mu}{15R\pi}\right) + 12p \tag{9.50}$$

$$B = \frac{3}{2}\,\frac{Etr_m^{\ 2}}{R^2(1-\mu^2)} + \frac{4D\mu}{Rr_m} \tag{9.51}$$

where

- R = radius of pipe bend
- r_m = mean radius of pipe cross-section
- D = $\dfrac{Et^3}{12(1-\mu^2)}$ = plate stiffness
- t = pipe wall thickness
- I = moment of inertia of pipe cross-section
- p = internal pressure
- M = applied bending moment

σ_{cr} is estimated from the diagram shown in Section 2.6.

In 1957, Rodabaugh and George [9.19] further extended the theory, considering large diameters and relatively thin-walled pipes, including the effect of internal pressure. They expressed the longitudinal stresses in-plane bending by the following formula

$$\sigma = \frac{K_pM_ir}{I(1-\mu^2)}\left\{(1 + \frac{3d_1}{2})\sin\phi + \tfrac{1}{2}\sum_{n=1}^{\infty}[d_n(1-2n) + d_{n+1}(2n+3)]\sin(2n+1)\phi \pm \pm \frac{\mu\lambda}{2}\sum_{n=1}^{\infty} d_n(2n-8n\ )\cos 2n\phi\right. \tag{9.52}$$

where

$$K_p = \frac{5 + 6\lambda^2 + 24\psi}{0.5 + 6\lambda^2 + 24\psi} \tag{9.53}$$

$$\lambda = \frac{tR}{r}\sqrt{1-\mu^2} = \frac{t}{\sqrt{1-\mu^2}} \tag{9.54}$$

is the flexibility characteristic

$$\psi = \frac{pR^2}{Ert} = \text{parameter related to pressure} \tag{9.55}$$

$$d_1 = \frac{3}{5 + 6\lambda^2 + 24\psi} \tag{9.56}$$

$$d_n = \frac{C_n}{r\eta} \tag{9.57}$$

The values of the coefficient Cn are determined from a set of n linear equations with (n+1) unknown C's as follows

$$3r\eta = (5+6\lambda^2+24\psi)C_1 - \frac{5}{2}C_2 \ldots = 0$$

$$-\frac{5}{2}C_1 + (17+600\lambda^2+480\psi)C_2 - \frac{21}{2}C_2 \ldots = 0$$

$$-C_{n-1}\frac{(2n-3)(2n+1)}{2} + C_n\{(4n^2+1) + (8n^3-2n)^2\frac{\lambda^2}{6} + [8n^2(4n^2-1)]\psi\} - C_{n+1}\frac{(2n-1)(2n+3)}{2} \ldots = 0 \tag{9.58}$$

By assuming that $C_{n+1} = 0$, all constants may be evaluated. Since the C's all contain a factor of $r\eta$, it is convenient to introduce the relation

$$d_n = \frac{C_n}{r\eta} \tag{9.59}$$

$$\eta = \frac{\Delta\alpha}{\alpha} \tag{9.60}$$

α and ϕ indicates pipe bend arcs and circumferential location angles, as shown in Figure 9.12.

Considering first approximation from Equation (9.52) at n = 1, we obtain

$$\sigma = \frac{K_pM_ir}{I(1-\mu^2)}\{(1 + \frac{3d_1}{2})\sin\phi - \frac{d_1}{2}\sin\phi - \frac{d_1}{2}\sin 3\phi \pm 3d_1\mu\lambda\cos 2\phi\} \tag{9.61}$$

and the maximum value of σ at $\phi = \frac{\pi}{2}$, (9.62)

or

$$\sigma_{max} = \frac{K_pM_ir}{I(1-\mu^2)}\{1 + 2d_1 \pm 3d_1\mu\lambda\} \leq \sigma_{cr} \tag{9.63}$$

where σ_{cr} is estimated from the diagram shown in Section 2.6.

9.8.5 Bending of the Pipe Out-of-Plane of Curvature

During the period from 1911 to 1943 numerous tests were run on pipe bends with out-of-plane bending, Figure 9.13.

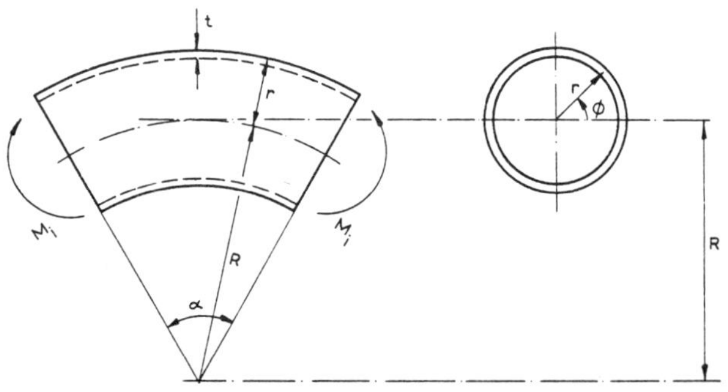

FIGURE 9.12 — Pipe bend arc and circumferential location angle.

In 1943, Vigness [9.18] established a theory of flexibility of pipe bends perpendicular to the plane of the bend. Experimental results have verified the developed bending equations and have shown the pipes to be more flexible than expected from the application of "rod" theory. He obtained a flexibility factor which is identical with that found for the bending of pipe in the plane of the bend. The increased flexibility is caused by a distortion of the cross-section of the pipe and the K-factor, for the first-term approximation is the same as that given in Equation (9.44).

Figure 9.13 shows the manner in which the cross-section of a pipe at a bend will distort when acted upon by a moment which is applied perpendicular to the plane of the bend so that compressional longitudinal stresses occur on the sides BA'B', Figure 9.13 (a) and BAB', Figure 9.13 (b). Tensional stresses occur on the opposite sides and a center line BB' is contained in the neutral plane. The cross-section is circular before the application of stress, and after the application of stress it assumes the shape as represented by the dotted oval, the major axes of which are 45 degrees to the plane containing the pipe bend. Longitudinal stresses are concentrated at their point of maximum value and are of greater magnitude than expected from ordinary theory. The maximum value of longitudinal buckling stresses for bending perpendicular to the plane of the bend is given by the equation

$$\sigma_{max} = \frac{M_r}{I}\left(\frac{12h^2+4}{12h^2+1}\right) = B\frac{M_r}{I} \leq \sigma_{cr} \quad (9.64)$$

where B is called the longitudinal-stress multiplication factor and is equal to

$$B = \frac{12h^2+4}{12h^2+1} \quad (9.65)$$

where

$$h = \frac{tR}{r^2} \quad (9.66)$$

and σ_{cr} is estimated from the diagram shown in Chapter 2.

Rodabaugh and George [9.19] developed a theory establishing the flexibility and stresses due to out-of-plane bending including the effect of internal pressure and proved its adequacy by means of carefully conducted tests.

The longitudinal stress for out-of-plane bending is expressed by the following formula

$$\sigma = \frac{K_pM_or}{I(1-\mu^2)}\{(1+\frac{3d}{2})\cos\phi + \frac{1}{2}\sum_{n=1}^{\infty} d_n(1-2n) + d_{n+1}(2n+3)]\cos(2n+1)\phi \pm \quad (9.67)$$

$$\pm \frac{\mu\lambda}{2}\sum_{n=1}^{\infty} d_n(8n^3-2n)\sin 2n\phi$$

where the values of parameters are given by the expressions from Eqs. (9.53) to (9.60).

Considering the first approximation from Equation (9.67) at n = 1, we obtain

$$\sigma = \frac{K_pM_or}{I(1-\mu^2)}\{(1+\frac{3d_1}{2})\cos\phi - \frac{d_1}{2}\cos 3\phi \pm \pm \mu\lambda d_1\sin 2\phi\} \quad (9.68)$$

and the maximum value of σ at $\phi = \frac{\pi}{4}$, or

$$\sigma_{max} = \frac{K_pM_or}{I(1-\mu^2)}\{0.707 + 1.414d_1 \pm 3\mu\lambda d_1 \pm \leq \sigma_{cr} \quad (9.69)$$

where σ_{cr} is estimated from the diagram shown in Chapter 2.

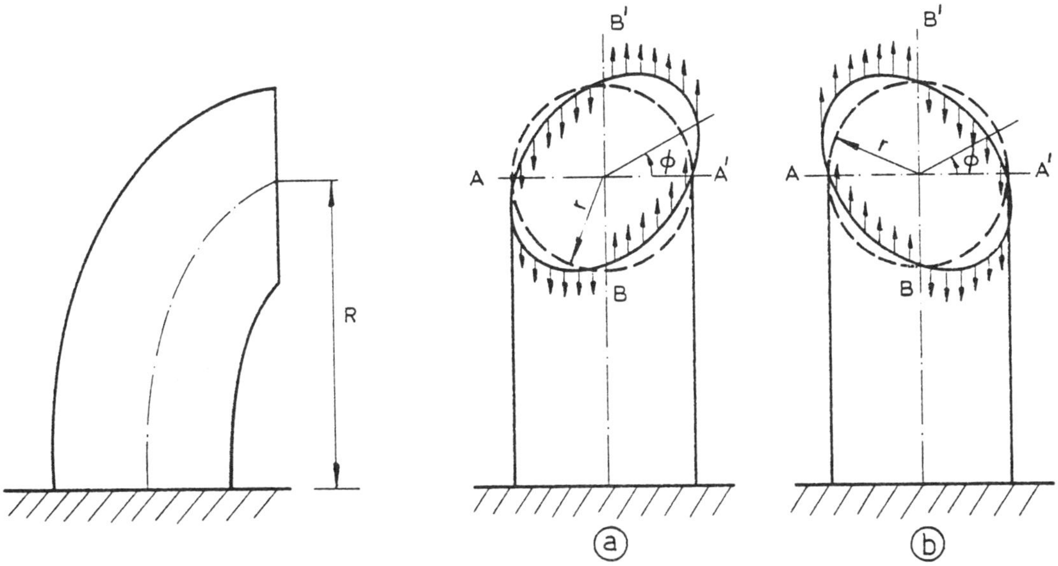

FIGURE 9.13 — Deformation of curved pipe cross-section under out-of-plane bending moment acting perpendicular to plane of bend.

9.9 Buckling Stability of Stiffened Pipe

9.9.1 Method of Solution

In the following we are considering the stability of multi-span pipes stiffened by rings and subjected to different kinds of bending loadings.

Until a few years ago, the customary approach to the stability analysis of a stiffened pipe was to replace it by an equivalent orthotropic cylindrical shell. This approach, however, did not permit taking into account the eccentricity of the stiffeners which were found to be of importance in stiffened pipes. A second approach also assumes the stiffness to be "smeared" over the whole surface of the shell, but considers the "distributed" stiffness of the stiffeners separately, which permits the inclusion of the effect of eccentricity.

The "smearing" of stiffeners appears reasonable for pipes with many closely spaced stiffeners, but it is doubtful when their number is small. Hence, the buckling of a stiffened pipe is analyzed by a third approach in which, instead of "smearing" the stiffeners, the "discrete" analysis, based on the Vlasov's theory of shells [9.24] is used. A "discrete" analysis includes a consideration of the local instability between the stiffeners, whereas the first two approaches deals with the general instability only.

According to Vlasov's theory, the longitudinal normal stress component N, acting in the wall of the pipe, may cause buckling, Figure 9.14.

9.9.2 Normal Stress Components for Different Loading

In Table 9-2 are shown the values of the normal stress components for different cases of loadings, using the following designations:

- h = depth of the ice on bottom surface of the pipe
- t = thickness of wall of the pipe
- R = radius of the cross-section of the pipe
- ℓ = span between support
- g = own weight of the pipe per unit length
- γ_ℓ = the specific weight of the liquid
- γ_i = the specific weight of the ice
- w = velocity wind pressure

TABLE 9.2 — Pipeline Under Different Loadings.

OWN WEIGHT	PRESSURE OF LIQUID
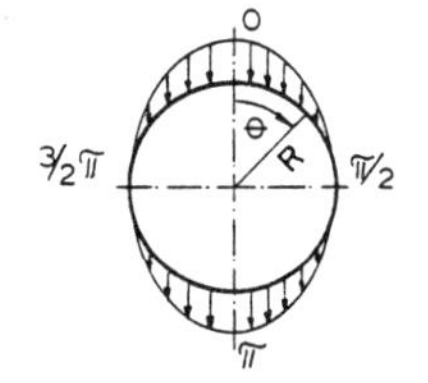 $N_g = \phi_1 K_1 \frac{gR^2}{t} \cos\frac{\pi X}{l}$	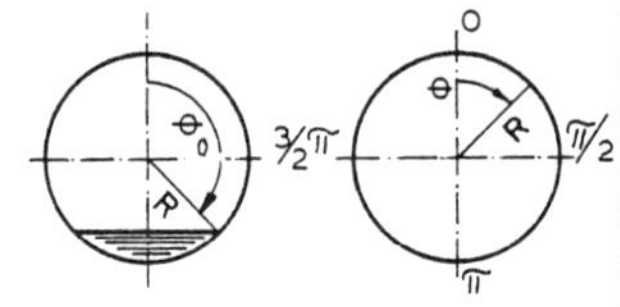 $N_l = \phi_2 K_2 \gamma_l l^2 \cos\frac{\pi X}{l}$
OUTSIDE ICE	**WIND**
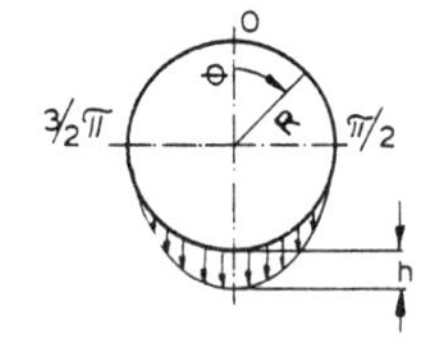 $N_i = \phi_3 K_3 \frac{\gamma_i h R^2}{t} \cos\frac{\pi X}{l}$	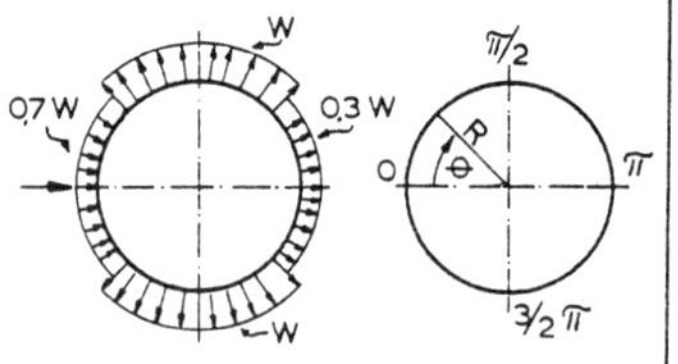 $N_w = \phi_4 K_4 \frac{wR^2}{t} \cos\frac{\pi X}{l}$

The parameter ϕi depends on the support conditions. At the supports, when the pipe may freely rotate, $\phi_i = 1$. If the support conditions are such that the pipe cannot rotate, then at supports $\phi_i = -2/3$, and at the mid-span $\phi_i = 1/3$. The values of the parameter Ki are given in Table 9.3, where

- $\epsilon = \frac{R^3}{t^2}$ = the geometric characteristic of the pipe
- $c = \frac{1}{t}\sqrt[3]{\frac{12I}{b}}$ = the equivalent thickness of the shell in transverse direction
- I = moment of inertia of the stiffening ring including the wall of the pipe in width = b.
- b = spacing between stiffening rings
- θ_0 = central angle, defining the level of liquid in the pipe

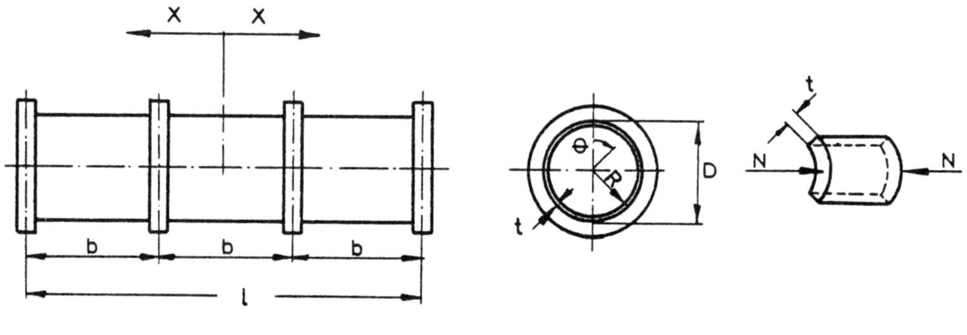

FIGURE 9.14 — Multispan pipe reinforced by stiffening ring.

TABLE 9.3 — Values of the Parameter K_i.

$K_1 = -0.258 \frac{1}{\varepsilon}\cos\theta$
$K_2 = -0.041(\pi-\theta_0+\frac{\sin 2\theta_0}{2})\cos\theta+\frac{128\varepsilon^2\sin\theta_0\cos 2\theta}{1168.92\varepsilon^2+256c^3}$
$K_3 = -0.129 \frac{1}{\varepsilon}\cos\theta+\frac{226.2\varepsilon\cos 2\theta}{1168.92\varepsilon^2+256c^3}$
$K_4 = -0.0581 \frac{1}{\varepsilon}\cos\theta-\frac{460.8\varepsilon\cos 2\theta}{1168.92\varepsilon^2+256c^3}$

The above formulas for the analysis of the multi-span pipes of large diameter permit the determination of stresses in the wall between stiffening rings. The length of the pipe wall acting together with the stiffening ring, may be determined using the following formula [9.25]

$$\ell_s = \delta + 1.664\sqrt{Rt} + \frac{w_r t_r}{(t + t_r)} \qquad (9.70)$$

where

δ = the thickness of the web of the stiffening ring

R = radius of the cross-section of the pipe

w_r = width of the flange of the stiffening ring

t = thickness of wall of the pipe

t_r = the average depth of the flange of the stiffening ring

It was assumed that the intermediate stiffening rings are spaced at equal distances between the supports.

9.9.3 Determination of the Critical Buckling Stresses

The critical buckling stress in the wall of the pipe is determined by the formula

$$f_{cr} = \frac{N_i}{t} \qquad (9.71)$$

where N_i is determined for the different cases of loading of the pipe, as shown in Table 9.3, and using the corresponding values of the parameter K_i as shown in Table 9.4.

The stability of the compressed zone of the pipe against buckling should be checked according to the condition

$$f_{cr} \leqslant \sigma_{cr} \qquad (9.72)$$

where σ_{cr} is found from the diagram of the permissible local buckling stresses given in Chapter 2, Figure 2.6.

9.10 Temperature Effect Upon a Restrained Pipe

9.10.1 Free Expansion or Contraction of the Pipe

If a pipe of length ℓ undergoes a uniform temperature change $(t - t_o)$, its length will alter by an amount

$$\Delta L = \pm\ell(t - t_o)\alpha \qquad (9.73)$$

where α is the coefficient of thermal expansion.

The diameter will change by

$$\Delta D = \pm\alpha D(t - t_o) \qquad (9.74)$$

where D denotes mean diameter of the pipe.

Extensions and tensile stresses will be denoted by a positive sign, and contractions and compressive stresses by a negative sign.

Since the pipe is free to change its dimensions, these strains will not produce any stresses in the material of the walls.

9.10.2 Stresses Produced by Restraints at Uniform Change of Temperature

If the free strains already determined are restricted in any way, additional stresses are produced.

Let us consider a straight length of pipe, AB = ℓ, which has a liquid flowing through it from A to B. At A the pressure is p_A and at B it is p_B, Figure 9.15.

The variation of pressure from A to B is assumed to be linear. If the whole length were under uniform pressure, then

$$p_m = \frac{1}{2}(p_A + p_B) \qquad (9.75)$$

The free longitudinal increase in length owing to this pressure is

$$\frac{p_m D\ell}{4Et}(1 - 2\mu) \qquad (9.76)$$

where

E = modulus of elasticity for the pipe metal

t = wall-thickness of the pipe

μ = Poisson's ratio

If the initial temperature of the pipe is t_o and its final temperature is t, it will increase in length by

$$\alpha\ell(t - t_o)$$

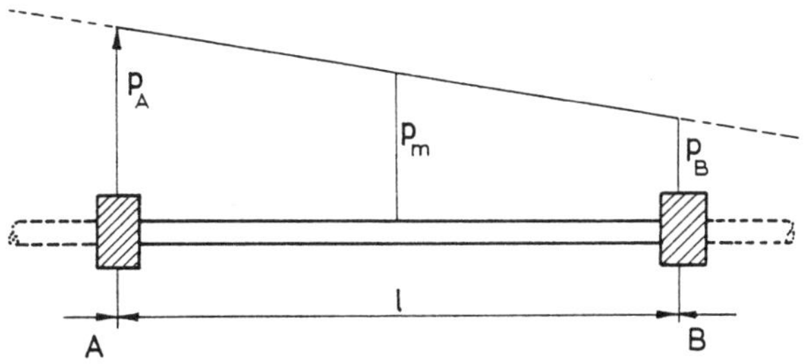

FIGURE 9.15 — Restrained pipe.

The total increase in length owing to pressure and temperature change will therefore be

$$\frac{p_m D \ell}{4Et}(1 - 2\mu) + \alpha \ell (t - t_o) \qquad (9.77)$$

If these movements are prevented by anchor blocks at A and B, an additional longitudinal stress f_ℓ will be induced in the pipe. The longitudinal strain produced by it will be

$$\frac{f_\ell}{E}$$

and the alteration in length

$$\frac{f_\ell \ell}{E} \qquad (9.78)$$

Since the total alteration in length ℓ due to restrain is zero

$$\ell\left[\frac{p_m D}{4Et}(1 - 2\mu) + \alpha(t - t_o) + \frac{f_e}{E}\right] = 0 \qquad (9.79)$$

Therefore

$$f_\ell = -\frac{p_m D}{4t}(1 - 2\mu) - \alpha E(t - t_o) \qquad (9.80)$$

Adding this stress to that due to pressure, the total longitudinal stress at any point where the pressure is p is

$$\Sigma f_\ell = -\frac{p_m D}{4t}(1 - 2\mu) - \alpha E(t - t_o) + \frac{pD}{4t} \qquad (9.81)$$

The first term is always negative since, for steel, μ is approximately = 0.3. The second term may be either negative or positive.

The maximum negative compressive stress will therefore occur when t has a maximum value t_1, and at the point where p is a minimum, i.e., at the outlet and where it is p_B.

The maximum tensile stress will occur when t has a minimum value, t_2, and at the point where p is a maximum, i.e., at the inlet end where it is p_A.

The values of these maximum stresses are, respectively compression

$$\Sigma f_\ell = -\left[\frac{p_m D}{4t}(1 - 2\mu) + \alpha E(t_1 - t_o) - \frac{p_B D}{4t}\right] \qquad (9.82)$$

and tension

$$\Sigma f_\ell = -\left[\frac{p_m D}{4t}(1 - 2\mu) + \alpha E(t_o - t_2) + \frac{p_A D}{4t}\right] \qquad (9.83)$$

When the pipe is empty, the pressure terms vanish and stresses are caused by temperature changes only.

If the pipe is subjected to the same temperature whether it is full or empty, the maximum compressive stress when it is under pressure will be greater than, equal to, or less than the maximum compressive stress when it is empty, as follows

$$\frac{p_m D}{4t}(1 - 2\mu) - \frac{p_B D}{4t} \gtreqless 0 \qquad (9.84)$$

Substituting ½ (p_A p_B) for p_m, this condition becomes

$$\frac{p_A}{p_B} \gtreqless \frac{1 + 2\mu}{1 - 2\mu} \qquad (9.85)$$

In conditions of temperature differential the cross-country surface-laid sections of the pipeline take care of expansion and contraction by alternating curves.

In rocky areas, vertical curves may allow for movement from temperature change.

However, in cases where such conditions do not exist, the buckling stability of the pipe between the anchors should be checked, namely the value of the maximum compressive stress Eq. (9.80) should be smaller than the permissible buckling stress used in Fig. 2.6, Chapter 2, for the ratio D/t.

9.10.3 Stresses Produced by Restraints. Non-Uniform Change of Temperature

If a pipe receives the direct rays of the sun, its temperature will not be uniform throughout the cross-section. The metal which is exposed will be hotter than that on the other side which will be in the shade.

If the two temperatures are, respectively, t_1' and t_2', the rise above the initial conditions is

$$(t_1' - t_o), \text{ on the hotter side}$$

and

$$(t_2' - t_o), \text{ on the cooler side}$$

It will be assumed that the temperature gradient across the pipe is linear.

The effect is the same as if the pipe were subjected to a uniform rise of temperature

$$[\tfrac{1}{2}(t_1' + t_2') - t_o]$$

superimposed on a rise which varies linearly from

$$\tfrac{1}{2}(t_1' - t_2') \quad \text{to} \quad -\tfrac{1}{2}(t_1' - t_2')$$

across the pipe, as illustrated in Figure 9.16.

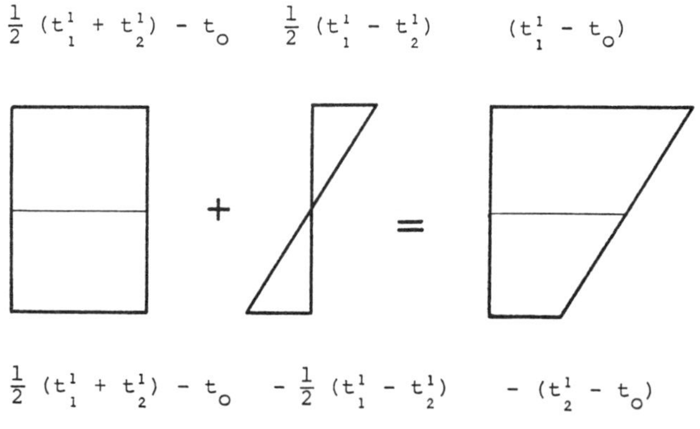

FIGURE 9.16 — Differential heating.

The uniform rise will give only a longitudinal strain

$$\varepsilon_\ell = \alpha\left[\frac{1}{2}(t_1^1 + t_2^1) - t_o\right] \quad (9.86)$$

The varying temperature acting alone will produce a positive strain on the hot side and an equal negative strain on the cool side, and the pipe will curve in a circular arc away from the sun. The center-line will be unchanged in length.

If, as in Figure 9.17, the radius of the neutral axis is R, the difference in length between the arcs AB and CE is

$$(R + \frac{D}{2})\phi - (R - \frac{D}{2})\phi = \alpha\ell(t_1^1 - t_2^1)$$

Therefore

$$\phi = \frac{\ell}{R} = \frac{\alpha\ell(t_1^1 - t_2^1)}{D}$$

and curvature

$$\frac{1}{R} = \frac{\alpha(t_1^1 - t_2^1)}{D} \quad (9.87)$$

If a restrained pipe is heated differentially, a uniform longitudinal stress

$$-\alpha E\left[\frac{1}{2}(t_1^1 + t_2^1) - t_o\right] \quad (9.88)$$

will be induced owing to the prevention of an extension along the center-line of the pipe. In addition, the restraints will prevent the rotation of the ends of the pipe by applying moments, constant throughout the section between anchorages, to counteract the tendency of the free pipe to assume a uniform curvature. The restrained pipe will therefore remain straight and will be subjected at every section to a bending moment

$$M = \frac{EI}{R} = \frac{\pi D^2 \alpha E(t_1^1 - t_2^1)t}{8} \quad (9.89)$$

The maximum stresses due to this will be

$$\pm \frac{\alpha E}{2}(t_1^1 - t_2^1) \quad (9.90)$$

The total fibre-stresses will be the sum of those due to the bending moment Eq. (9.90) and the uniform stress due

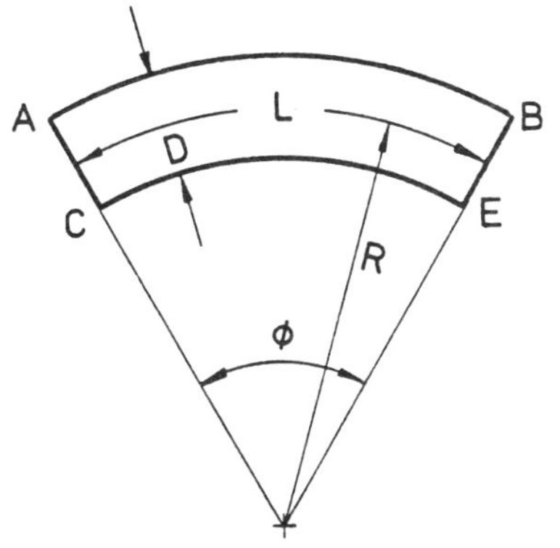

FIGURE 9.17 — Curvature geometry.

to the strain along the neutral axis, Eq. (9.88), or maximum compressive stresses at

$$\text{top} = -\alpha E(t_1^1 - t_o)$$

$$\text{bottom} = -\alpha E(t_2^1 - t_o) \quad (9.91)$$

If the pipe is under pressure, corresponding stresses must be added to those given above, or

maximum compressive stress =

$$= -\left[\frac{p_m D}{4t}(1-2\mu) + \alpha E(t_1^1 - t_o) - \frac{p_B D}{4t}\right]$$

and

minimum compressive stress =

$$= -\left[\frac{p_m D}{4t}(1-2\mu) + \alpha E(t_2^1 - t_o) - \frac{p_A D}{4t}\right]$$

$$(9.92)$$

The above defined compressive stresses may result in a buckling of the pipe. To prevent buckling, the first Eq. (9.92) should be smaller than the values given in Chapter 2, Fig. 2.6, considering the ratio D/t.

9.11 Stresses Produced by Bending and Temperature Under Restrain

9.11.1 Bending Stresses Under Weight of Pipe and Liquid

The maximum compressive longitudinal fibre-stress due to bending is given by the expression

$$f_q = \frac{q\ell^2}{12S} \quad (9.93)$$

where

- q = g + w, and
- g = weight of the pipe per unit length
- w = weight of the liquid in the pipe per unit length
- S = $\frac{\pi D^2 t}{4}$ = section modulus of the pipe

or

$$f_q = \frac{q\ell^2}{3\pi D^2 t} \quad (9.94)$$

9.11.2 Bending Stresses Due to the Eccentricity of the Axial Restrain Load

Bending due to the eccentricity of an axial load can occur in any plane and in extreme cases, the effects are therefore additive.

In currently accepted theories of axially loaded members, the imperfections of manufacture, material, and loading are represented either by an equivalent initial curvature of the member of an equivalent intital eccentricity of loading. These curvatures and eccentricities are empirical values, obtained to adjust formulas derived from mathematical analysis with observed tests. The following formula expressing the equivalent eccentricities of the pipe, [9.26] may be suggested:

$$\delta = \frac{\ell}{120} + \frac{D}{8} \quad (9.95)$$

The axial load in any length of pipe between successive anchor-blocks is, from Eq. (9.80)

$$P = -\pi Dt \left[\frac{p_m D}{4t}(1-2\mu) + \alpha E(t-t_o)\right] \quad (9.96)$$

and the bending moment at any point due to the equivalent eccentricity of loading, is

$$M_e = P\delta$$

and the compressive fibre stress is

$$f_e = \frac{P\delta}{S} = \frac{4\delta}{D}\left[\frac{p_m D}{4t}(1-2\mu) + \alpha E(t-t_o)\right] \quad (9.97)$$

9.11.3 Axial Compressive Stress

Equation (9.81) contains a term involving p, the actual pressure at a point in the pipe, but since the pressure-drop between two anchor blocks is unlikely to be great it simplifies the work if p_m is used instead of p. Then, we have

$$\alpha E(t_1-t_o) - \frac{pD\mu}{2t} \quad (9.98)$$

The combined compressive stress is obtained as the sum of the Eqs. (9.94), (9.97) and (9.98).

$$f_1 = -\left\{\alpha E(t_1-t_o) - \frac{pD\mu}{2t} + \frac{4\delta}{D}\left[\alpha E(t_1-t_o) + \frac{pD}{4t}(1-2\mu)\right] + \frac{q\ell^2}{3\pi D^2 t}\right\} \quad (9.99)$$

To prevent buckling, the stress given by Eq. (9.99) should be smaller than the values given in Chapter 2, Fig. 2.6, considering the ratio D/t.

At the same point, there will be a circumferential tensile stress

$$f_2 = \frac{pD}{2t} \quad (9.100)$$

and f_1 should be restricted to such a value that f_1 and f_2, acting together, do not over-stress the metal.

9.12 Criteria of Safety in Complex Stress System

Design criteria by Mises, Huber, and Hencky have been suggested for the maximum safe stress in material under complex stress. They are based on the assumption that a material yields when the energy of deformation per unit volume reaches a certain value, irrespective of the stress system imposed.

If the normal stresses along three mutually perpendicular axes are f_1, f_2 and f_3, respectively, and f_y is the yield-stress in direction tension, the Mises-Hencky criteria are

$$(f_1-f_2)^2 + (f_2-f_3)^2 + (f_3-f_1)^2 = 2f_y^2 \quad (9.101)$$

The stresses should receive the appropriate signs. In the present case, f_3 is zero and the criteria becomes

$$f_1^2 + f_2^2 - f_1 f_2 = f_y^2 \quad (9.102)$$

Substituting the value $f_2 = \frac{pD}{2t}$, the maximum value which f_1 may reach, without yielding in the metal, is found to be

$$f_1 = \frac{pD}{4t} - \sqrt{f_y^2 - \frac{3p^2D^2}{16t^2}} \quad (9.103)$$

It is, however, undesirable to permit stresses to occur which cause yielding under normal working conditions, since no margin is left for an emergency such as, for instance, a surge in pressure. A load factor, n, is therefore necessary and is most conveniently introduced by using $\frac{f_y}{n}$ instead of f_y in Equation (9.102).

If expressions (9.99) and (9.103) are equated, with $\mu = 0.3$, the spacing of ring supports is found to be, with measured value of δ

$$\ell = \frac{\pi Dt}{40q}\left\{\sqrt{K^2 - \frac{1.528q}{t}\frac{3}{4}K - \frac{f_y^2}{n} - \frac{3p^2D^2}{16t^2}}^{1/2} - K\right\} \quad (9.104)$$

where

$$K = 2\,e(t_1-t_o) + \frac{pD}{5t} \quad (9.105)$$

This equation provides a criterion as to whether the pipeline is overstressed or not. More information is required to establish if the eccentricity value adopted is reasonable, but the form of the equation will be unaltered. For specified values of the essential data, ℓ can be calculated. If it is positive, the pipe may be carried on supports not farther apart than the determined value of ℓ, without violation of the Mises-Hencky criterion.

If ℓ is zero, the pipe requires continuous support, and if it is negative, the criterion is violated and the pipe cannot safely withstand the pressure and temperature specified.

When the pipe is empty, the equation becomes

$$\ell = \frac{\pi Dt}{40g}\left[\sqrt{K^2 - \frac{1.528g}{t}\cdot\frac{3}{4}K - \frac{f_y}{n}} - K\right] \quad (9.106)$$

where

$$K = 2\alpha E(t_1-t_o) \quad (9.107)$$

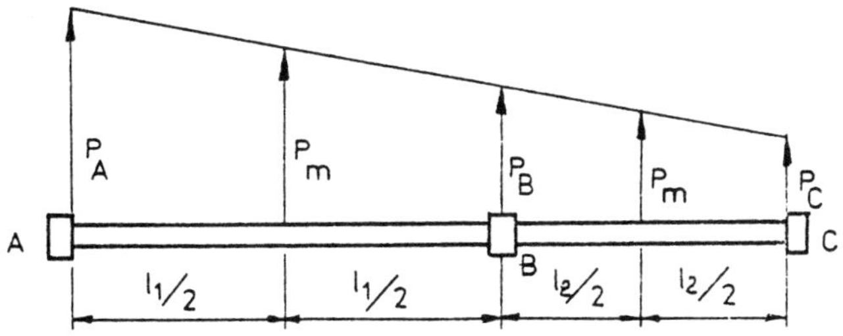

FIGURE 9.18 — Pipeline on anchor blocks.

9.13 Loads on Anchor and Buttress Blocks

9.13.1 Anchor Blocks

The anchor blocks must be designed to resist the resultant forces from the length of pipe which they support. The conditions at end anchorages differ from those at intermediate blocks and may vary considerably in different arrangements. As typical example, we are considering the pipeline shown in Figure 9.18 which is anchored at the ends A and C and at an intermediate point B.

The hydraulic gradient is as shown and p_m and p_m are the mean pressures in the spans AB and BC respectively.

The longitudinal stress required to prevent movement of AB is, from Equation (9.80)

$$f_\ell = -\frac{P_m D}{4t}(1-2\mu) - \alpha E(t-t_o)$$

and the force which will be exerted on the end block A is therefore

$$P_A = -\pi Dt\left[\frac{P_m D}{4t}(1-2\mu) - E(t-t_o)\right] \quad (9.108)$$

Similarly, the force exerted on the end block C is given by the same expression with p_m substituted for P_m.

The force to be resisted by block B will be the resultant of the forces required to restrain the spans AB, and BC, and so far an intermediate block

$$P = -\frac{\pi D^2}{4}(P_m - p_m')(1-2\mu) \quad (9.109)$$

Due to differential heating of the pipe an end anchorage may, in addition to the direct thrust, be subjected to a moment

$$M = \frac{\pi D^2 \alpha E(t_1' - t_2')t}{8} \quad (9.110)$$

9.13.2 Buttress Blocks

At any point where the pipeline changes direction, whether in the horizontal or vertical plane, the tendency to lateral movement must be prevented by a buttress block. This block will also be an anchor block and the forces acting upon it may be calculated as shown in the previous section.

Suppose the pipeline is deflected at B through an angle ϕ and the forces acting on the block B thus calculated at P_1 and P_2, Figure 9.19.

Then the resultant thrust on the buttress is

$$Q = \sqrt{P_1^2 + P_2^2 - 2P_1P_2\cos\phi} \quad (9.111)$$

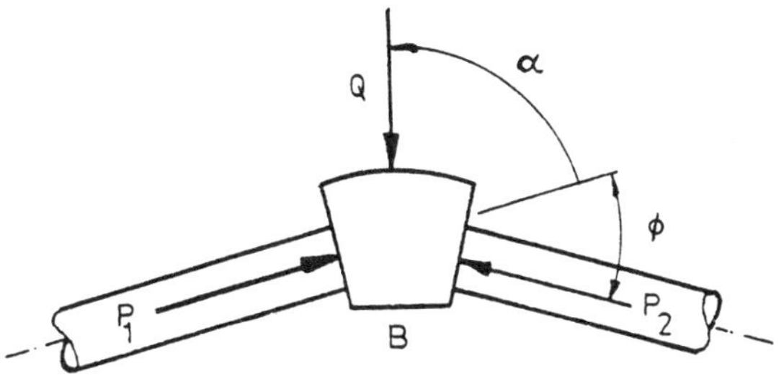

FIGURE 9.19 — Forces acting on buttress block B.

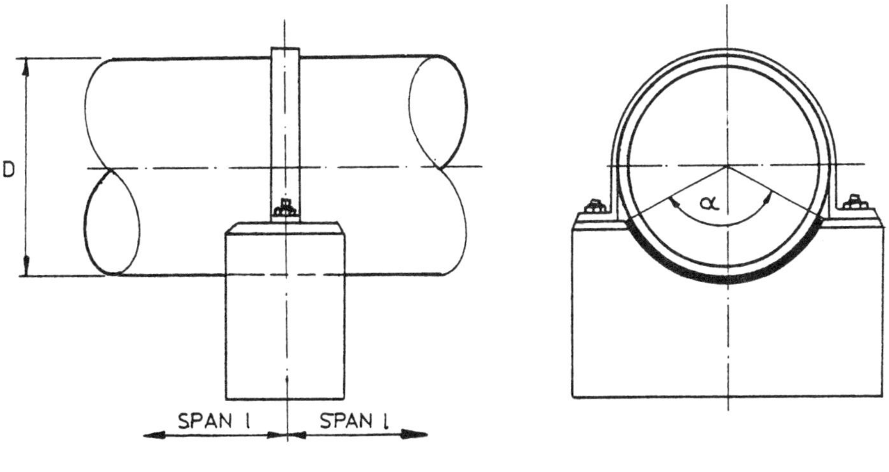

FIGURE 9.20 — Concrete saddle.

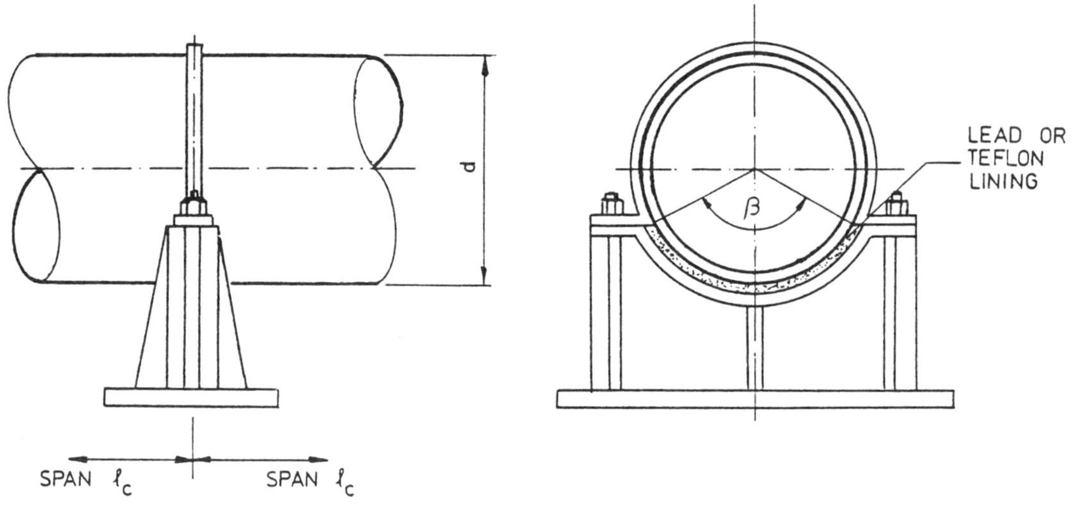

FIGURE 9.21 — Steel saddle.

which acts at an angle α to the direction of P_1, and α being given by

$$\tan\alpha = \frac{P_2 \sin\phi}{P_1 - P_2 \cos\phi} \quad (9.112)$$

9.14 Buckling of the Pipe at Supports

9.14.1 Saddle Supports

Pipes may be supported in various ways depending upon size, field conditions and economics [9.27]. There are two basic types of support: saddle and ring girder. Figure 9.20 shows a detail of a concrete saddle and Figure 9.21 a steel saddle.

Pipe costs are usually lower when saddles can be used, since this permits greater flexibility in field erection. Ring girder construction usually adds to fabrication costs but allows for greater clear span.

The contact angle α varies from 90 to 120 degrees. When pipe lengths resting on saddles are joined by flanges or mechanical couplings, the strength and position of the joints must be such that they will safely resist the bending and shear forces while remaining tight. Ordinarily, it is advisable to place joints at, or as near as practicable to, the point of zero bending moment in the span or spans.

Saddle supports cause high local stresses both longitudinally and circumferentially in unstiffened, comparatively thin wall pipe at the tips and edges of the supports. Their intensity is practically independent of the width of the saddle.

Because saddle supports cause critical points of stress in the metal adjacent to the saddle edges, it is frequently more economical to increase the wall thickness of the pipe when it is overstressed than to provide stiffening rings.

9.14.2 Stress Effect Due to Saddle Support

For single or multiple spans of uniform thickness the maximum stress at the saddle is:

a. for the pipe with *unrestrained* ends

$$\sigma_{tot} = \sigma_b + \sigma_\ell \quad (9.113)$$

b. for the pipe with *restrained* ends

$$\sigma_{tot} = \sigma_b + 0.25\sigma_p + \sigma_\ell \quad (9.114)$$

where

σ_b = bending stress in span with unrestrained ends

σ_p = circumferential stress due to internal liquid pressure

Following experimental tests [9.28, 9.29], Roark [9.30] stated that the maximum of the local stresses adjacent to the tips of rigid saddles would probably not exceed that given by the formula

$$\sigma_\ell = K \frac{P}{t^2} \log_e \frac{R}{t} \quad (9.115)$$

where

P = total saddle reaction in pounds

R = pipe radius in inches

t = pipe wall thickness in inches

K = 0.02 − 0.00012 (α − 90), in which α is in degrees, Figure 9.20

He also stated that the maximum saddle reaction a pipe can stand is about twice the value of P when σ_ℓ equals the yield point of the steel used. Formula (9.115) does not account for temperature stresses.

Presumably the maximum saddle stress given by the formula (9.115) would occur in the circumferential direction although Roark did not go beyond stating that the high local stresses were both longitudinal and circumferential directions.

However, design practice showed that the formula (9.115) is too conservative [9.31].

On the basis of the detailed analytical and experimental investigations, considering rigid saddles, Stokes [9.32] proposed to introduce the reduction factor (R.F.) into the formula (9.115), as follows

$$R.F = \frac{\tanh A}{A} \quad (9.116)$$

where

$$A = K\frac{R}{t}\sqrt{\frac{f_c}{E}} \qquad (9.117)$$

K = 1.1, a constant

R = mean radius of pipe

t = thickness of wall

f_c = the circumferential stress

E = 30×10^6 psi, modulus of elasticity for steel

The R.F.'s calculated from the maximum circumferential bending stresses, uniform pressure saddle, having α = 150 degrees, have been plotted in Figure 9.22.

Stokes suggested that more theoretical and experimental data should be obtained before using the above diagram for saddles with considerably different contact angles.

To this local stress at the saddle must be added the maximum longitudinal bending stress at the support for the pipe acting as a continuous beam, unless a joint occurs at the support.

Therefore, for the pipe with unrestrained ends, the total buckling stress according to formula (9.113) is the summary of expressions (9.17) and (9.115)

$$\sigma_{tot} = \sigma_b + \sigma_\ell = \frac{q\ell^2}{3\pi D_m^2 t} + \frac{pD_m}{4t} + K\frac{P}{t^2}\log_e\frac{R}{t} \leq \sigma_{cr} \qquad (9.118)$$

where the permissible critical buckling stress σ_{cr} must be determined after the diagram in Figure 2.6, Chapter 2.

For the pipe with restrained ends, the total buckling stress according to formula (9.114) is the summary of expressions (9.115), (9.118) and the value of the circumferential stress due to internal liquid pressure, or

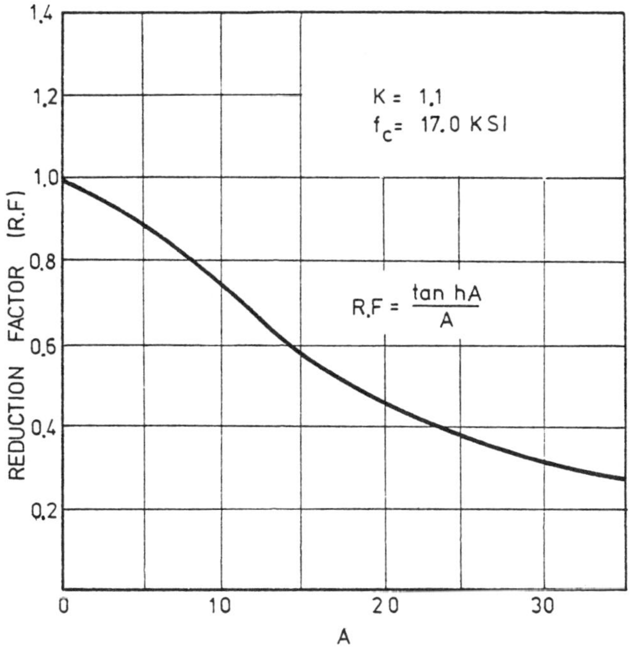

FIGURE 9.22 — Reduction factor for circumferential stress when internal pressure acts.

$$\sigma_{tot} = \sigma_b + 0.25\sigma_p + \sigma_\ell = \frac{q\ell^2}{3\pi D_m^2 t} + \frac{pD_m}{4t} + 0.25\frac{pD_m}{2t} + K\frac{P}{t^2}\log_e\frac{R}{t} \leq \sigma_{cr} \qquad (9.119)$$

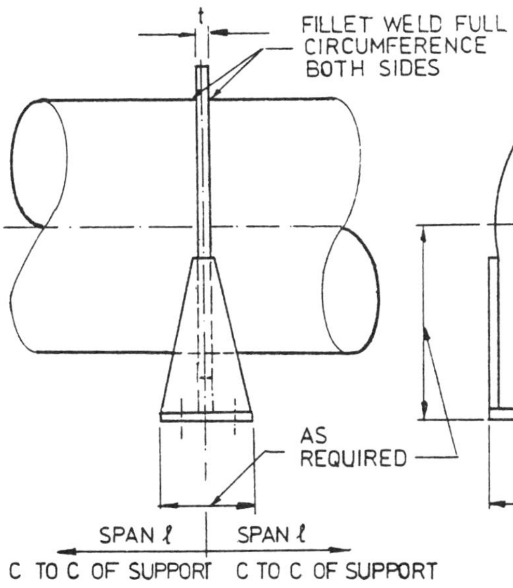

FIGURE 9.23 — Details of ring girder support for small pipes.

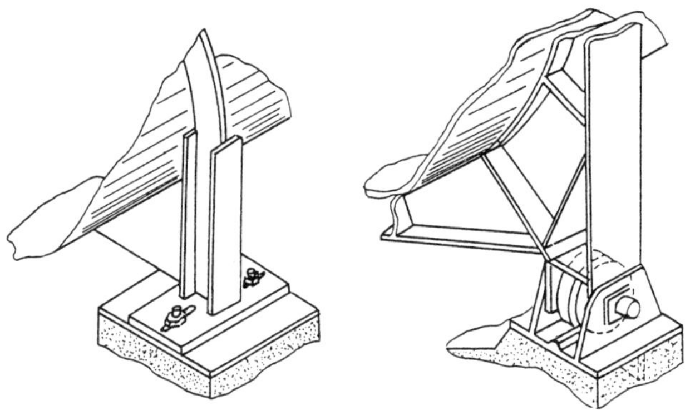

FIGURE 9.24 — Details of ring girder support for large pipes (a) simple type (b) roller support.

9.14.3 Ring Girder Supports

When a large-diameter steel pipe is laid either above ground or streams, rigid ring girders, spaced at relatively long intervals, have been found to be very effective supports. Buckling of the pipe and stresses in these circumstances may be more severe than under the ground, as there is no sidefill support and the pipe saddles cause stress concentrations. Rigid ring girders at each support are useful in these circumstances, because these girders prevent the distortion of the pipe at the points of support and thus maintain its ability to act as a beam. Generally, practical considerations limit spans to a minimum of 40 and a maximum of 100 feet. Details of ring girder supports for small pipes are shown in Figure 9.23 and for large pipes in Figure 9.24.

The rings may be plain and rectangular in cross-section, of T or H-shaped, with the back fixed all around the pipe.

A design for ring girder construction based on elastic theory was developed by Schorer [9.33] and the nomenclature, interpreted by Figure 9.25, is used in the design equations.

9.14.3.1 Stress in Pipe Shell

If the pipe is under pressure, corresponding to a head, h, above the bottom of the pipe, then the maximum combined ring stress is

$$f_r = \frac{D_m}{2t}(w + gh) \qquad (9.120)$$

The combined maximum longitudinal stress for free end pipe condition is

$$f_\ell = \pm \frac{\ell^2}{4t}\left(\frac{2w}{D_m} + \frac{q}{2}\right) \qquad (9.121)$$

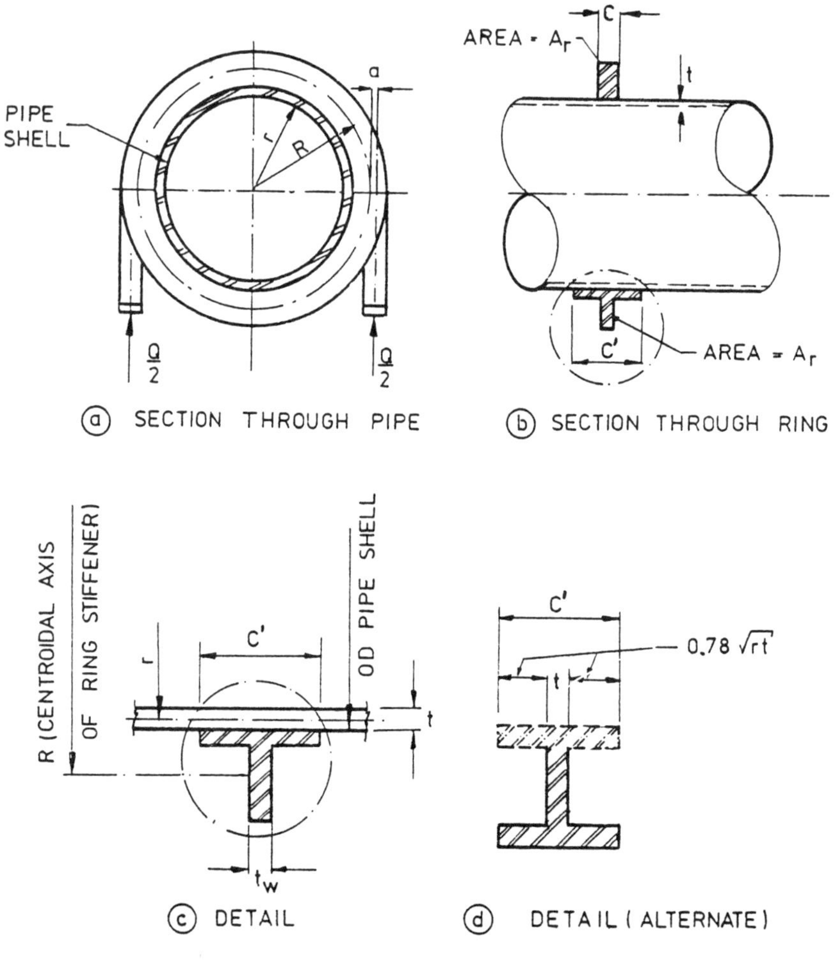

FIGURE 9.25 — Pipe and ring girder support.

In the case of continuous pipelines the direct longitudinal stress is equal to 2/3 of the stress given by the formula (9.121), or

$$f_\ell = \frac{\ell^2}{6t}\left(\frac{2w}{D_m} + \frac{q}{2}\right) \qquad (9.122)$$

where

D_m = mean diameter of a pipe

h = head

ℓ = length of span from center to center of ring-girder supports

t = thickness of shell

q = unit weight of the fluid flowing in a pipe

w = weight of pipe shell per unit of area

For unequal spans and other end conditions Equation (9.121) should be multiplied by moment coefficients for those conditions, with due regard given to the position and sign of maximum moment.

If the pipe ends are fixed, the longitudinal stresses due to temperature change must be added to f_ℓ in Equation (9.121). When expansion joints and bearings of low frictional resistance are provided, however, the temperature stresses may be practically eliminated. Additional thrust or pull due to installation on a slope also must be considered.

9.14.3.2 Rim Stresses

In practical cases the pipe is built as a continuous structure over more than two supports providing the intermediate ring disks members. Figure 9.26 shows the state of deformation of pipe shell, which is under a uniform inside pressure, p, and is rigidly connected to a circular ring of constant cross-sectional area, A_r.

The rim deformation due to internal radial pressure was investigated by Pasternak [9.34], and the maximum rim bending stress in the shell is

$$f_{bo} = \frac{1.82(A_r - ct)}{(A_r + 1.56t\sqrt{rt})} \times \frac{pr}{t} \qquad (9.123)$$

where

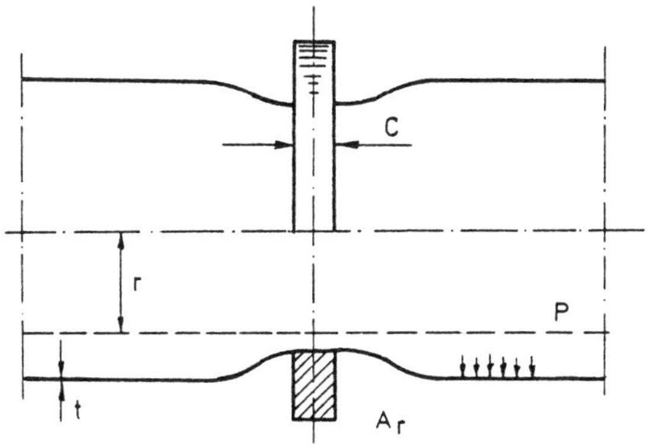

FIGURE 9.26 — Deformation of a pipe shell at location of rim.

A_r = area of supporting ring in a plane along the axis of the pipe

c = width of circular girder ring

This equation was developed on the assumption that the stiffener ring is integral with the pipe shell and that the rim load is symmetrical.

As the rim load is not symmetrical, because of the weight of the liquid, a good approximation of the maximum value of f_{bo} is obtained by substituting the value of f_r from Equation (9.120) in place of $\frac{pr}{t}$.

If the girder ring is fitted to the pipe in a loose manner, the rim bending stress due to the reaction at the supports should be taken into account, because the load will be transmitted mostly by direct bearing on the lower half of the ring rather than through shear distributed around the pipe.

The total combined longitudinal shell stress, f, is

$$f = f_\ell + f_{bo} \qquad (9.124)$$

In Equation (9.123), the factor, $\frac{pr}{t}$, is identical with the ring stress in the unrestricted shell, also called the circumferential stress. The rim bending stress is, therefore, a function of the circumferential stress in the more distant shell portions. If A_r becomes very large, comparable to an absolutely unyielding disk, the rim bending stress approaches an absolute maximum value for which Equation (9.123) gives

$$f_{bm} = 1.82\,\frac{pr}{t} \qquad (9.125)$$

In this extreme case the rim bending stress exceeds the circumferential stress by 82 percent. The smaller the area of the ring disk, the smaller the rim bending stress will be. The bending stresses in the rim zone cannot be neglected in the design, because in continuously supported pipelines the longitudinal stresses due to beam action are combined with the longitudinal rim bending stresses.

Equation (9.123) also shows the influence of the width, c, on the magnitude of the rim bending stress. The wider the ring disk, the smaller will be the bending stress for a constant area, A_r.

9.14.3.3 Stress in Ring Girder

Figure 9.27 shows the circular ring disk, symmetrically loaded at the inner circumference by the shearing forces, S, acting as tangent to the circle at each point.

The ring is held in equilibrium by two vertical reactions, $\frac{Q}{2}$, applied at the distance, a, from the neutral axis of the ring. The forces, s and $\frac{Q}{2}$, then represent the external load system, which is symmetrical to the vertical diameter.

The resulting moment at the top of the ring is

$$M_o = \frac{QR}{\pi}\left[\frac{3}{4} - \frac{\pi}{4} + \frac{a}{R}\left(1 - \frac{\pi}{4}\right)\right] \qquad (9.126)$$

and the maximum intermediate moment is given as

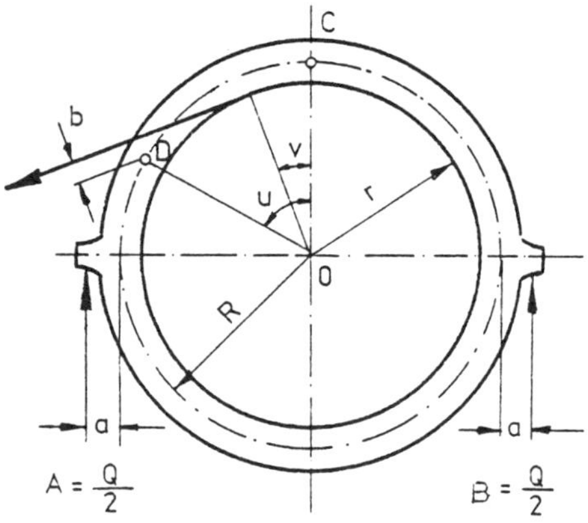

FIGURE 9.27 — Loading on the ring girder.

$$M_a = -\frac{QR}{4} \times \frac{a}{R} \quad (9.127)$$

The maximum possible ring moment is obtained for a = 0.004R outside the neutral axis, in which case

$$M_i = M_a = 0.01\ QR \quad (9.128)$$

Figure 9.28 shows the absolute values of M_o, M_i and M_a plotted in the function of the ratio, $\frac{a}{R}$.

The maximum bending stress, f_1, is

$$f_1 = \frac{My}{I} \quad (9.129)$$

where I is the moment of inertia of the supporting ring with respect to the neutral axis and y is the distance from neutral axis to extreme fiber.

The maximum ring stress, f_2, due to shear forces

$$f_2 = \frac{Q}{4A_r} \quad (9.130)$$

The ring stress, f_3, due to tensile force is

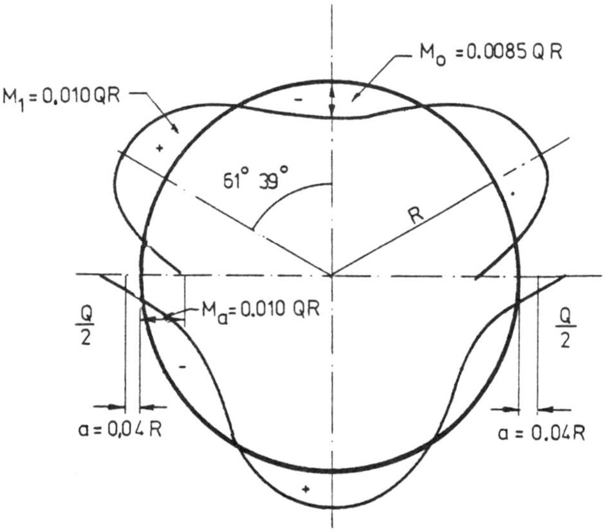

FIGURE 9.28 — Bending moment in supporting ring for a = 0.04R.

$$f_3 = \frac{pr}{A_r}\left[c + 1.56\sqrt{rt}\ \frac{(A_r - ct)}{(A_r\ 1.56t\ \sqrt{rt})}\right] \quad (9.131)$$

As all of these stresses are combined at the horizontal diameter, the total maximum stress, f_T, in the ring girder is

$$f_T = f_1 + f_2 + f_3 \quad (9.132)$$

The maximum allowable stress in the ring girder or the pipe shell when the pipe is fully loaded is usually 10,000 psi, or 18,000 psi if half loaded.

To support the ring girder, a short column on each side of the pipe is attached to the girder and supported on a pier either by direct bearing or by a roller device, rocker assembly, or pin connection. In any event, the design must permit longitudinal movement of the pipe as well as afford adequate support.

Some advantages of ring girder support for pipelines are that it reduces flood and highway hazards, avoids expensive substructures required by other types of construction, and affords a practical method of crossing streams, swamps, and marshes.

9.15 Vibration of Pipelines

Vibration of pipeline may start under dynamic forces at insufficient stiffness of the system or in the case when the natural frequencies of the structure are approaching the frequencies of the induced forces.

Wind action on the pipeline having outside diameter D_o depends on the value of the Raynolds number R_e. If $R_e > 50$, the Karman's vertices are generated. The frequencies of these vertices are

$$n = S\frac{V}{D_o} = 0.20\ \frac{V}{D_o} \quad (9.133)$$

and the angular velocity is

$$\omega = 2\pi n = 1.26\ \frac{V}{D_o}\ \text{rad/sec} \quad (9.134)$$

where

v = wind velocity

D_o = outside diameter of a pipe

In the case when the natural frequencies of vibration of the pipe coincide with the frequencies of the induced forces, the pipeline will be in the state of resonance and undergoes substantial amplitudes. When we find the natural frequencies of vibration of the pipeline it is necessary to compare them with the frequency due to the wind force, corresponding to the velocity of the wind at given location. The natural frequency of the pipeline should differ not less than 20-30 percent from the induced frequencies.

The natural frequency of the vibration of the pipeline at multispan system, may be determined using the following formulas [9.35]

$$P_{min} = \sqrt[4]{\frac{\sum_{i=1}^{N} (f_i)^4}{N}} \tag{9.135}$$

and

$$P_{max} = \sqrt[4]{\frac{\sum_{i=1}^{N} (\rho_i)^4}{N}} \tag{9.136}$$

where

N = number of spans

$f_1, f_2, \ldots, f_N$ = natural frequencies of the single spans of the pipeline, obtained from the given N-spans system by dividing it at all intermediate supports

$\rho_1, \rho_2, \ldots, \rho_N$ = natural frequencies of vibration of the single span of the pipeline at full fixity at the intermediate supports

Natural frequencies of the flexural vibration of the single span of the pipeline are determined using the following formula

$$f = \frac{a^2}{\ell^2} \sqrt{\frac{EI}{m}} \text{ rad/sec} \tag{9.137}$$

or

$$\nu = \frac{f}{2\pi} \approx 0.159 f \text{ cycles/sec} \tag{9.138}$$

where

ℓ = span of the pipeline, in

E = modulus of elasticity of material, psi

I = moment of inertia of the cross-section of the pipe, in^4

m = $\frac{q}{g}$ = mass per unit length

q = total vertical uniformly distributed load, lbs/in

g = acceleration of the gravity, in/sec^2

a = coefficients depending on the type of fixity at supports, given in the following Table 9.4.

9.16 Influence of Imperfections

9.16.1 Introduction

The following factors may affect the buckling strength of the pipe and are to be an ideal structure after the fabrication and installation.

1. The effect on the buckling behavior of some initial out-of-roundness of the pipe, or so-called "geometric imperfections".
2. Variation of the wall thickness.

TABLE 9.4 — Values of the Coefficients a.

	Type of Support	a
1	Hinged supports	3.14
2	Fixed supports	4.73
3	One support-hinge, other — fixed	3.93
4	One support fixed, other — free	1.88

3. Variation of the modulus of elasticity of the pipe material.
4. Variation of the modulus of failure of the pipe material.
5. Initial stresses, or residual stresses introduced after welding.
6. Accidental lateral loading.

To establish the factors from (2) to (4) it is necessary to conduct a detailed analysis of the material used for the pipe by laboratory methods.

Factor (5) is checked by the known stress-relieving methods.

In the following the factor (1) or geometric imperfections of the pipe which may occur during fabrication or installation will be investigated.

9.16.2 Stability of the Pipe Having Local Eccentricity

In the following Figure 9.29 is shown the geometry of the local eccentricity of multispan pipeline.

An analysis of the problem [9.36] indicates that the critical bending moment when the pipe approaches its instability condition is

$$M_{cr} = \phi\frac{EI}{\gamma R} \tag{9.139}$$

where ϕ is a non-dimensional coefficient and is defined by:

$$\phi = 1.02 - \frac{1}{\sqrt{1 + \lambda/3}} \frac{r^2}{Rt} + \frac{0.33}{1 + \lambda/3} \left(\frac{r^2}{Rt}\right)^2 \tag{9.140}$$

where

$$\phi < 2$$

$$\gamma = \sqrt{\frac{12(1-\mu^2)}{1 + \lambda/3}} \cdot \frac{r^2}{Rt} \tag{9.141}$$

$$\lambda = 12(1-\mu^2) \frac{pr^3}{Et^3} \tag{9.142}$$

$$R = \frac{a^2}{8f} \tag{9.143}$$

and

r = the radius of the cross-section of the pipe

p = the internal pressure

R = the radius of the curvature of the pipe

f = the maximum value of the eccentricity

a = the length of the chord

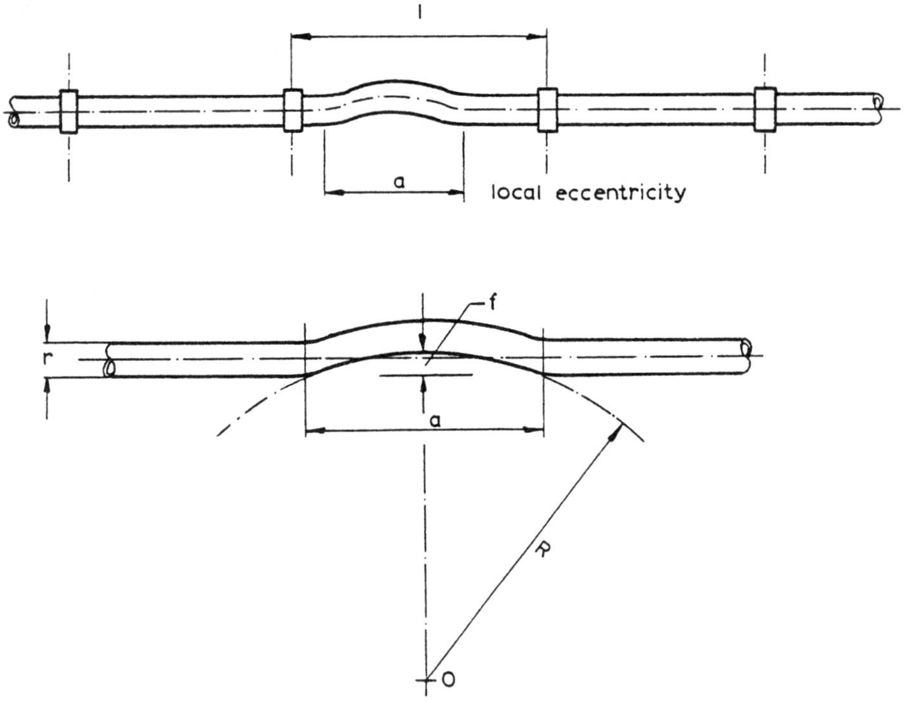

FIGURE 9.29 — Local eccentricity.

For straight pipe

$$1/R = 0 \text{ and } \phi = 1.02.$$

9.16.3 Out-of-Round Pipe

An anlaysis of stresses in out-of-round pipes was published by Haigh [9.37] in 1936. A similar analysis is given by Schmidt [9.38].

In these theories, the initial cross-sectional shape is assumed to be described by the equation, Figure 9.30.

$$R = R_m + \Sigma U_n \cos n\theta \qquad (9.144)$$

where

- R = initial pipe radius
- R_m = mean pipe radius
- U_n = radial displacement

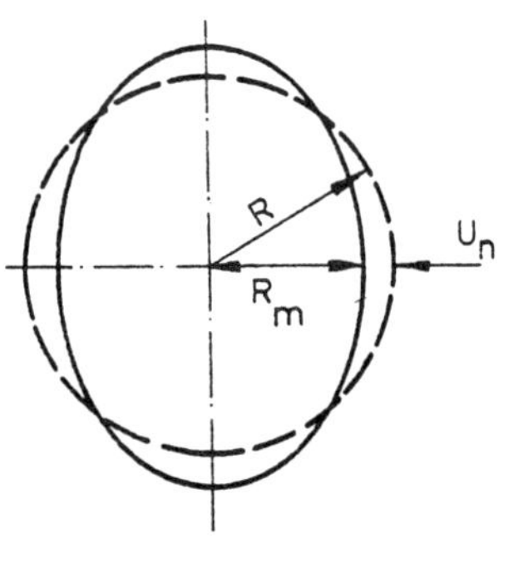

FIGURE 9.30 — Displacement of the pipe cross-section.

Equation (9.144) can, of course, be used to describe any cross-sectional shape of the pipe. It is assumed that this shape persists for a long distance along the pipe axis. If the out-of-roundness is small

$$U_n/R \leqslant 1$$

the membrane stress is still essentially pR/t. The bending stress is given by

$$\sigma_b = \frac{pR}{t}\left(\frac{U_n}{t}\right)\left[\frac{6}{1 + \dfrac{12pR^3(1-\mu^2)}{Et^3(n^2-1)}}\right]\cos n\theta \qquad (9.145)$$

For a simple ovality of the cross-section, n is equal to 2, the maximum buckling stress at $\theta = 0, \pm \pi/2$ due to bending as given by Equation (9.145) is essentially the same as that given in the ASME Code [9.39].

$$\sigma_b = \frac{pR}{t}\left(\frac{U_n}{t}\right)\left[\frac{6}{1 + \dfrac{4pR^3(1-\mu^2)}{3Et^3}}\right] \leq \sigma_{cr} \qquad (9.146)$$

where the permissible critical buckling stress σ_{cr} may be determinted after the diagram in Figure 2.6. It can be seen that Equation (9.145) is nonlinear with pressure.

Figure 9.31 shows an example of thin nonlinear effect for the particular case of n = 2, R/t = 40, U_2 = 0.01.

Figure 9.32 shows how the total stress, or bending and membrane stresses, compares with the membrane stress for a range of values of R/t and for $U_2/R = 0.01$ or $U_2/R = 0.003$.

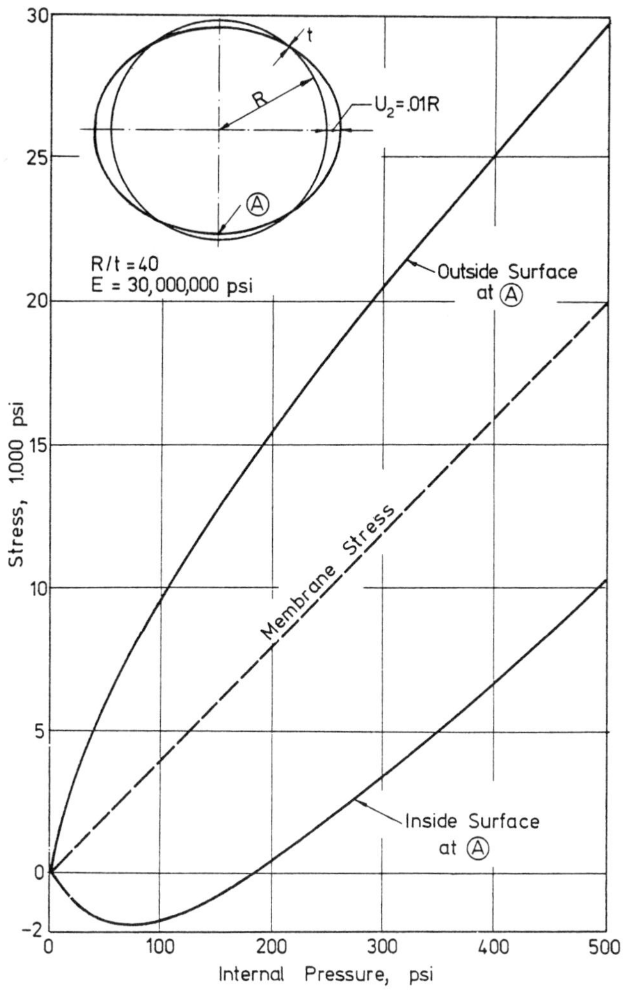

FIGURE 9.31 — Example of pressure vs. stress for an out-of-round pipe, n = 2.

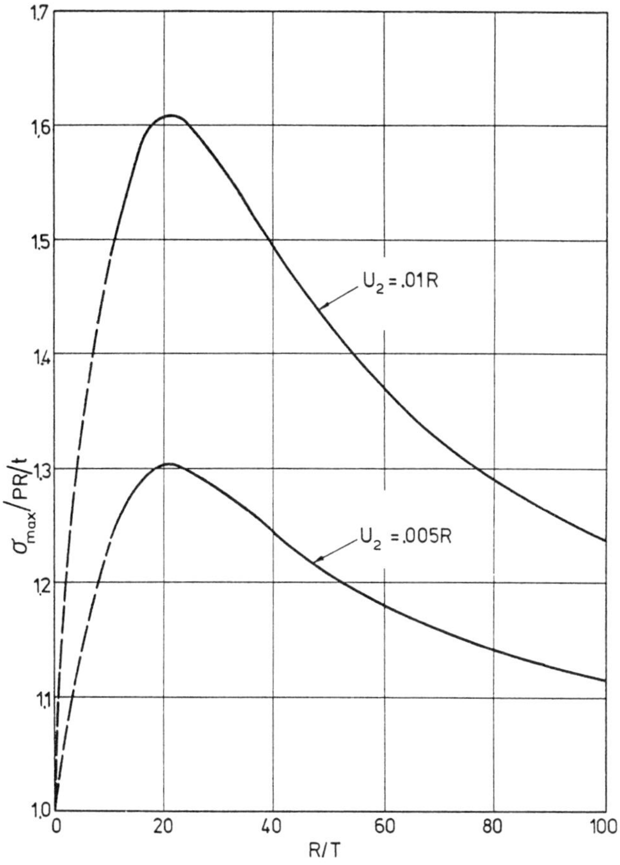

FIGURE 9.32 — Maximum stress as a function of R/t, out-of-round pipe, n = 2.

It might be remarked that the nonlinear effect is quite significant for large values of R/t. For example, at R/t = 60, U_2/R = 0.01, a linear analysis would give

$$\frac{\sigma_{max}}{pR/t} = 4.6$$

as compared to the nonlinear analysis result of

$$\frac{\sigma_{max}}{pR/t} = 1.37$$

for p such that

$$pR/t = 20{,}000 \text{ psi}$$

Pipes made from rolled-and-welded plates may have a local out-of-roundness at the longitudinal weld because of either under-rolling or over-rolling of the abutting plate edges. A description of this shape by Equation (9.144) involves higher values of n. It can be seen in Equation (9.145) that as n increases, the nonlinear effect decreases.

At this time, no calculations for local out-of-roundness using Equations (9.144) and (9.145) are available. However, a theory and sample calculation given by Schmidt [9.40] are pertinent. Schmidt developed the linear theory for a deformed shape as shown in Figure 9.33.

In a specific example in which R/t = 45.5, and U is such that the out-of-roundness is 2% of the diameter, Schmidt found that the maximum bending stress is 11.1 times the nominal membrane stress pR/t.

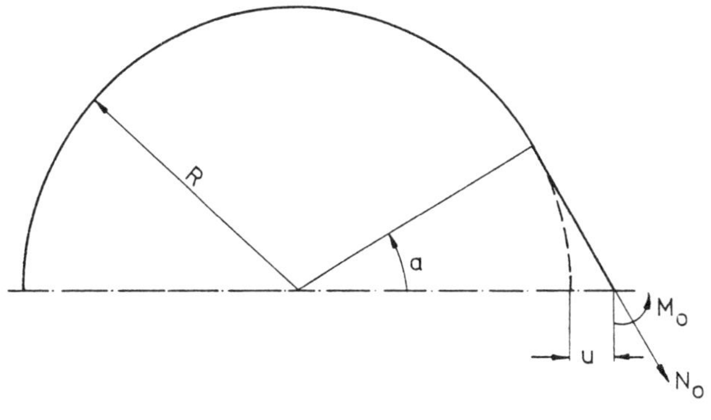

FIGURE 9.33 — Cross-section of pipe with an angular buckle, analysis by Schmidt.

$$u = 2(\sec\alpha - 1)(\sec\alpha + 3) \qquad (9.147)$$

$$M_o = PR^2\, g(\alpha) \qquad N_o = PR\, H(\alpha) \qquad (9.148)$$

$$G(\alpha) = [3(\pi-\alpha)^2 \tan^2\alpha + (\pi-\alpha)\tan^3\alpha(3+\sin^2\alpha + 0.5\tan^2\alpha) + \tan^6\alpha(1\ 29\ \cos^2\alpha)/12]/I(\alpha) \qquad (9.149)$$

$$H(\alpha) = [6(\pi-\alpha)^2\sec^2\alpha + (\pi-\alpha)\sin\alpha(12+16\tan^2\alpha + 2.5\tan^4\alpha) + \tan\alpha\sin\alpha(6+10+\tan^2\alpha + 0.5\tan^4\alpha)]/I(\alpha) \qquad (9.150)$$

$$I(\alpha) = 6(\pi-\alpha)^2 + (\pi-\alpha)\sin 2\alpha(6+9\tan^2\alpha + 2\tan^4\alpha) + \sin^2\alpha(6+12+\tan^2\alpha+\tan^4\alpha) \qquad (9.151)$$

Such high bending stresses are presumably relieved by yielding, however, cyclic pressure tests on pipe with longitudinal welds indicate that the combination of out-of-roundness and weld irregularities may constitute a significant weakness in pipes subjected to cyclic internal pressure.

9.16.4 Out-Of-Round Curved Pipe

When internal pressure is applied to curved pipe, there will be a rotation of one end of the curved pipe with respect to the other end if one or both ends are free.

If both ends are fixed, a moment will develop at the fixed ends. For values of ν greater than about 10 and for $b/c \approx 1$, the value of the moment is given in Reference [9.41] as

$$M_p = \pi\left(1 - \frac{b^2}{c^2}\right)\left(1 - \frac{2}{\nu}\right) Rrt\left(\frac{pr}{t}\right) \qquad (9.152)$$

where

M_p = end moment due to internal pressure with the ends fixed

ν = $\sqrt{12(1-\mu^2)}\,\dfrac{bc}{at}$ = a parameter

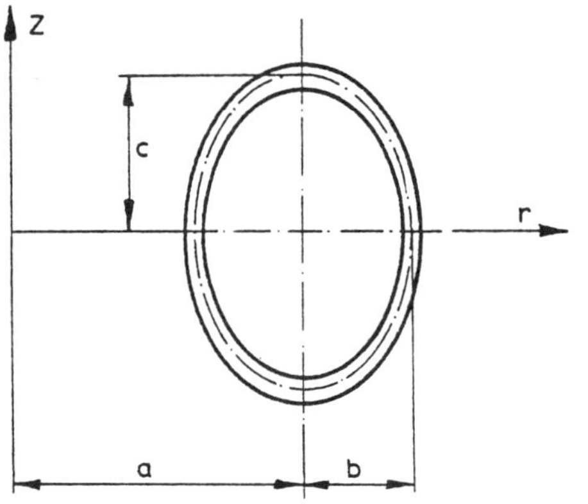

FIGURE 9.34 — Cross-section of toroidal pipe.

b,c = the semiaxes of the elliptical section

a = the distance of the center of the section from the axis of revolution

t = thickness of the wall

All of the above values are shown in Figure 9.34.

The value of M_p can be express in terms of nominal buckling stress in straight pipe due to bending as

$$\sigma_p = \frac{M_p}{S} = \frac{M_p}{\pi r^2 t} = \left(1 - \frac{b^2}{c^2}\right)\left(\frac{R}{r}\right)\left(1 - \frac{2}{\nu}\right)\left(\frac{pr}{t}\right) \leq \sigma_{cr} \qquad (9.153)$$

where the permissible critical buckling stress σ_{cr} may be determined after the diagram in Figure 2.6, Chapter 2.

Expressed in the form of Equation (9.153) it may be seen that for curved pipe not more than ±1% out-of-round, the equivalent bending stress σ_p will not be more than about 10 ro 12% of the nominal hoop stress, $\frac{pr}{t}$. However, for a large, closely coupled piping system attached to load-sensitive equipment, the moment produced by internal pressure may not be negligible.

REFERENCES — CHAPTER 9

[9.1] American Petroleum Institute. Standard Specificatons.

[9.2] American Water Works Association. Standard Specifications.

[9.3] Gas-American Standard Code for Pressure Piping.

[9.4] American National Standard Building Code Requirements for Minimum Design Loads in Buildings and Other Structures, ANSI A58.1-1972, Revision of A58.1-1965.

[9.5] American Standards Association. Code for Oil Transportation Piping.

[9.6] Lester, C.B., "Hydraulics for Pipelines", Oldom Publishing Co., Bayonne, New Jersey, 1958, p. 61.

[9.7] Wilbur, W.E., "Analyzing Pipe Line Stresses", Pipe Line Industry, February 1963, p. 26.

[9.8] American Water Works Association, "Steel Pipe Design and Installation", Manual M11, 1964.

[9.9] Lessig, E.N., "On the Problem of Design of Above-Ground Multi-Span Pipelines of Large Diameter Under Transverse Loading Conditions", Stroitelstvo i Architectura (in Russian) No. 1, 1959, pp. 48-58.

[9.10] Brazier, L.G., "On the Flexure of Thin Cylindrical Shells and Other 'Thin' Sections", Proceedings of the Royal Society of London, Ser. A., Vol. CXVI, November, 1927, pp. 104-114.

[9.11] Flugge, W., "Stresses in Shells", Springer-Verlag, Berlin, 1960, pp. 434-436.

[9.12] Haringx, J.A., "Instability of Thin-Walled Cylinders Subjected to Internal Pressure", Philips Res. Rep., Vol. 7, 1952, pp. 112-118.

[9.13] Palmer, A.C., and Baldry, J.A.S., "Lateral Buckling of Axially Constrained Pipeline", Journal of Petroleum Technology, Nov. 1974, pp. 1283-1284.

[9.14] Bantlin, A., "Formanderung und Beanspruchung federnder Ausgleichsrohren", Z. Ver. deut. Ing., Vol. 54, 1910, pp. 43-49.

[9.15] Karman, Th. von, "Uber die Formanderung dunwandiger Rohre, insbesondere federnder Ausgleichsrohr", Z. Ver. deut. Ing., Vol. 55, 1911, pp. 1889-1895.

[9.16] Bathe'lemy, J., "Etude de la deformation et des tensions internes des Tuyaux a Ligne Moyenne Plane, Soumis a des efforts exterieurs et a une pression interne", Bulletin de l'Association Technique Maritime, 1947.

[9.17] Kafka, P.G., and Dunn, M.B., "Stiffness of Curved Circular Tubes With Internal Pressure", Journal of Appl. Mechanics, Trans. ASME, Vol. 78, pp. 247-254.

[9.18] Vigness, I., "Elastic Properties of Curved Tubes", Trans. ASME, Vol. 65, 1943, pp. 105-120.

[9.19] Radabaugh, E.C., and George, H.H., "Effect of Internal Pressure on Flexibility and Stress-Intensification Factors of Curved Pipe or Welding Elbows", Trans. ASME, Vol. 79, 1957, pp. 939-948.

[9.20] Gross, N., "Experiments on Short-Radius Pipe-Bends", Proceedings. The Institution of Mechanical Engineers (B), Vol. 1B, 1952-1953, p. 465.

[9.21] Hovgaard, W., "Further Research on Pipe Bends", Journal of Mathematics and Physics, Vol. 7, 1917-1928, pp. 239-297.

[9.22] Cross, N., and Ford H., "The Flexibility of Short-Radius Pipe-Bends", Proceedings. The Institution of Mechanical Engineers (b), Vol. 1B, 1952-1953, p. 480.

[9.23] Beskin, L., "Bending of Curved Thin Tubes", Journal of Applied Mechanics, March 1945, pp. A-1 to A-7.

[9.24] Vlasov, V.Z., "General Theory of Shells and Its Application in Engineering", translation — NASA, Department of Commerce, Washington, D.C., 1964.

[9.25] Klykov, V.M., "Investigation of the Supporting Rings at Pipeline of Large Diameter", Journal — Construction and Architecture, (in Russian), No. 8, 1958.

[9.26] Pippard, A.J.S., "Stresses in Restrained Pipe-Line", Journal of the Institution of Civil Engineers, No. 2., Dec. 1950.

[9.27] American Water Works Association, Steel Pipe Design and Installation, Manual M11, 1964.

[9.28] Wilson, W.M., and Olson, E.D., "Tests of Cylindrical Shells", University of Illinois Eng. Exper. Station Bulletin, No. 331, Sept. 23, 1941, p. 129.

[9.29] Hartenberg, R.S., "The Strength and Stiffness of Thin Cylindrical Shells on Saddle Supports", Dissertation (Ph.D.), University of Wisconsin, 1941.

[9.30] Roark, R.J., "Formulas for Stress and Strain", 3rd Edition, McGraw-Hill, New York, 1954, p. 282.

[9.31] Nitschke, J.E., "Investigation of Supports for Mannum-Adelaide Pipeline, South Australia", Eng. and Water Supply Department, Report, 1950.

[9.32] Stokes, R.D., "Stress in Steel Pipeline at Saddle Supports", The Institution of Engineers, Australia, Civil Engineering Transactions, October 1965, pp. 119-133.

[9.33] Schorer, H., "Design of Large Pipelines" Transactions ASCE, 98:101, 1933.

[9.34] Pasternak, P., "Die praktische Berechnung biegefester Kugelschalen, Kreisrunder Fundamentplatten auf elastischer Bettung und Kreiszylindrischer Wandungen in gegenseitiger monolither Verbindung", Zeitschrift fur angewandte Mathematik und Mechanik, Vol. 6, 1926, p. 1.

[9.35] Spiridonov, V.V., "Vibration of the Above Ground Pipelines Systems", Construction of the Pipelines, No. 11, 1961 (in Russian), pp. 11-14.

[9.36] Axelrod, E.L., "Analysis of the Permissible Span of the Pipeline Under Stability Condition", Structural Mechanics and Design of Constructions, (in Russian), No. 6, 1963, pp. 34-37.

[9.37] Haigh, B.P., "An Estimate of the Bending Stresses Induced by Pressure in a Tube that is not Initially Quite Circular", Proc. Instit. Mechan. Engrs., Vol. 133, 1936, pp. 96-98.

[9.38] Schmidt, K., "Stress in Out-Of-Round Pressure Vessel", V.D.I., Vol. 102, No. 1, 1960, pp. 11-15.

[9.39] ASME Boiler and Pressure Vessel Code, Section VIII, Par. UF-27, 1965 Edition, published by the American Society of Mechanical Engineers, New York.

[9.40] Schmidt, K., "Calculation of Stress for an Out-Of-Round Tube Under Internal Pressure", V.D.I., Vol. 98, No. 4, 1956, pp. 121-125.

[9.41] Clark, R.A., Gilroy, T.I., and Reissner, E., "Stresses and Deformations of Toroidal Shells of Elliptical Cross-Section", Journal of Appl. Mechs., Vol.74, 1952, pp. 37-48.

CHAPTER 10

underground pipelines

10.1 Introduction

In the following we are considering flexible pipelines, whose cross-sectional shapes can be distorted sufficiently to change their vertical or horizontal dimensions more than 3 percent before causing damage.

Steel pipes having either light or heavy walls because of its physical characteristics can always function as a flexible conduit within the limitations imposed by the relative rigidity of coatings of the pipe.

Although the maximum load-carrying capacity of flexible pipe depends to some extent on the wall thickness and its section modulus, the pipe by deflecting, is able to make full use of the load-carrying ability of the soil surrounding it. As the pipe may change its slope without failure, it transfers part of the vertical load into a horizontal or radial thrust which is resisted by the passive pressure of the soil at its sides as these move outward. The flexible pipe actually divides the load with the soil enclosing it.

10.2 Types of Loads

Earth loads on pipe are of two kinds. The one most known is practice is trench loading, the pipe being laid in an excavated trench and backfilled. The second kind occurs when a pipe is laid on a graded or prepared ground surface and a fill of earth is placed around and over the pipe. As the load carried by the pipe may differ in the two cases, trench loading should be distinguished from fill loading. Further, the reaction of flexible pipe to either fill or trench load is different from the reaction of rigid pipe to these loads.

When installation is made in a trench, the load may be calculated in accordance with Marston's formula for ditch conduit [10.1] or for steel pipe, the load again may be taken as the weight of the soil prism above the pipe. The Marston ditch conduit formula is

$$W_c = C_d w B_d^2 \tag{10.1}$$

where

W_c = load on the conduit, lb/ft

C_d = a calculation coefficient, Figure 10.1

w = unit weight of the ditch fill, lb/ft^3

B_d = width of ditch at top of pipe, ft

This formula gives the total vertical load in the ditch or trench at the level of the top of the pipe. Marston and others have shown by experiment and observation of actual structures that practically all of this load is carried by the conduit itself when the conduit is rigid.

If, however, the conduit is relatively flexible, such as a thin-walled steel pipe, it will deflect vertically and, by so doing, transmit a portion of the load to the trench walls by friction and to the trench bottom by settlement.

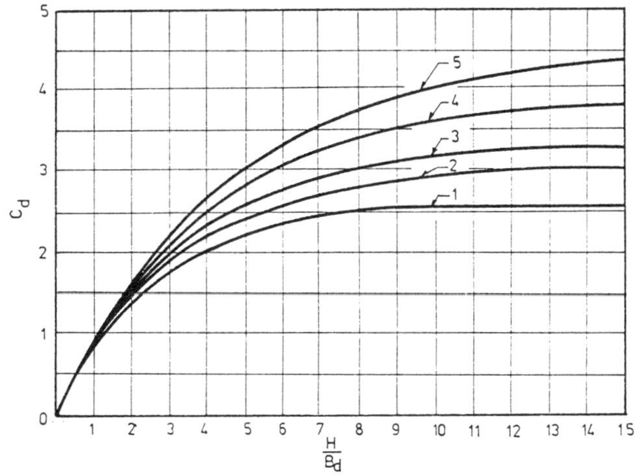

FIGURE 10.1 — Computation diagram for loads on pipe completely buried in ditches.

C_d refers to the calculation coefficient; H, to the height (ft) of fill over the top of the conduit; and B_d to the width (ft) of the ditch at the top of the pipe. The numbers on the curves represent: [1] minimum for granular materials without cohesion; [2] maximum for sand and gravel; [3] maximum for saturated topsoil; [4] ordinary maximum for clay; and [5] maximum for saturated clay.

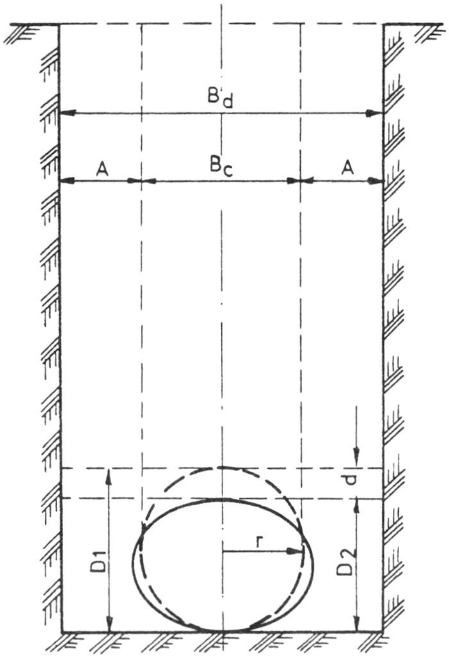

FIGURE 10.2 — Deflection of steel pipe.

In Figure 10.2, the trench fill is shown divided into three prisms, where

- B_d = the trench width at the top of the pipe
- B_c = D_1 = diameter of installed pipe
- A = the difference between B_d and B_c
- d = pipe deflection
- D_2 = diameter of deflected pipe
- r = radius of installed pipe

It is suggested that, as a minimum loading condition, if the pipe deflection d equals the settlement of side fill below the top of the pipe, the pipe and the side fills may each carry the same amount of the load per unit of width. Under these conditions, the load on the flexible pipe may be determined by multiplying Marston's load expression, Equation [10.1] by the ratio of the width of the pipe to the width of the trench, giving

$$W_c = c_d\, w\, B_d^2 \left(\frac{B_c}{B_d}\right) \qquad (10.2)$$

where

- B_c = the width of the pipe

It should be pointed out that, for steel pipe, Equation (10.1) represents the maximum loading conditions and Equation (10.2) represents the probable minimum loading condition which may develop on a pipe in a trench. The actual load in a given case may lie somewhat between these limits, depending upon the soil properties and the relative rigidity of the conduit compared with the side fills. For more rigid pipe, the actual load will approach that given by Equation (10.1), whole, for quite flexible pipe, the actual load will approach, or equal, that given by Equation (10.2).

TABLE 10.1 — Values of Bedding Constant

Bedding Angle α Degrees	Bedding Constant K_e
0	0.110
15	0.108
22½	0.105
30	0.102
45	0.096
60	0.090
90	0.083

10.3 Determination of Load

According to Spangler [10.2] when the pipe deflection is limited to not more than R/10 (5 percent of diameter)

$$\Delta_x = \Delta_y = \frac{K_e W_e R^3}{12EI} \qquad (10.3)$$

where

- E = modulus of elasticity of the material of the pipe, psi
- I = moment of inertia of pipe wall per inch of length, in^4
- K_e = the bedding or deflection constant
- R = external radius of a smooth steel pipe, in.
- W_e = the total effective external load, lb/ft
- Δx = deflection of the horizontal diameter of a flexible pipe, in
- Δy = deflection of the vertical diameter of a flexible pipe, in

Values of the bedding constant K_e for various forms of the bedding angle are shown in Table 10.1 [10.3].

The bedding angle α is defined as one-half the angle subtended by the arc of the pipe ring which is in contact with the pipe bedding, as shown in Figure 10.3.

But in the presence of an external fluid pressure P_e the deflection caused by the fill load will be increased to

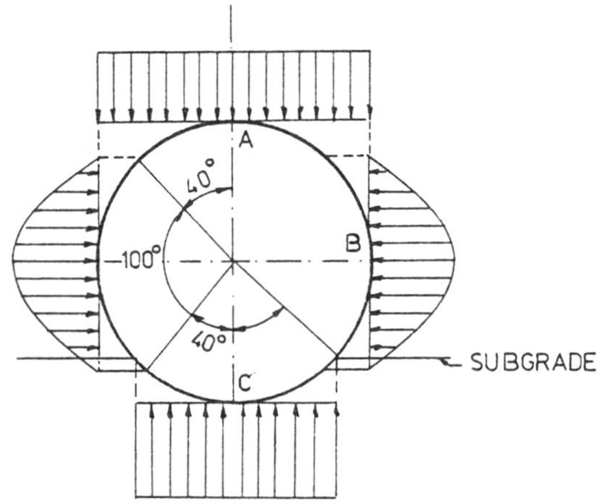

FIGURE 10.3 — Distribution of load and thrust.

The diagram gives an idea of the distribution of the horizontal thrust of the pipe side against the soil. A, B, and C are reference points; α is one-half the bedding contact angle.

Δ_x by the excess of the vertical pressure load over the horizontal pressure load, caused by the difference in the vertical and horizontal projected areas of the pipe [10.2], or the difference between the vertical and horizontal loads

$$P_e(2R+\Delta'_x) - P_e(2R-\Delta'_x) = 2P_e\Delta'_x =$$

$$= \text{horizontal load} \qquad (10.4)$$

Then the total effective vertical load is

$$\frac{W_e}{12} + 2P_e\Delta'_x = \frac{\Delta'_x}{\Delta_x} \cdot \frac{W_e}{12} \quad \text{lb/in} \qquad (10.5)$$

since from Equation (10.3), the ratio of the load to the deflection is constant. After substitution the value Δ_x from Equation (10.3), we obtain

$$\Delta'_x = \frac{K_e W_e R^3}{12(EI-2P_e K_e R^3)} \qquad (10.6)$$

For smooth-walled pipes

$$I = t^3/12$$

where t is the thickness of the wall of the pipe, and

$$\Delta'_x = \frac{K_e W_e R^3}{Et^3 - 24P_e K_e R^3} \qquad (10.7)$$

If $P_e = 0$, or there is no external pressure

$$\Delta'_x = \frac{K_e W_e R^3}{Et^3} = \Delta_x \qquad (10.8)$$

We determine the maximum bending moment and stress with reference to Figure 10.4.

According to Spangler [10.2] the bending moment for flexible pipe is

$$M_b = K_b W_e R \qquad (10.9)$$

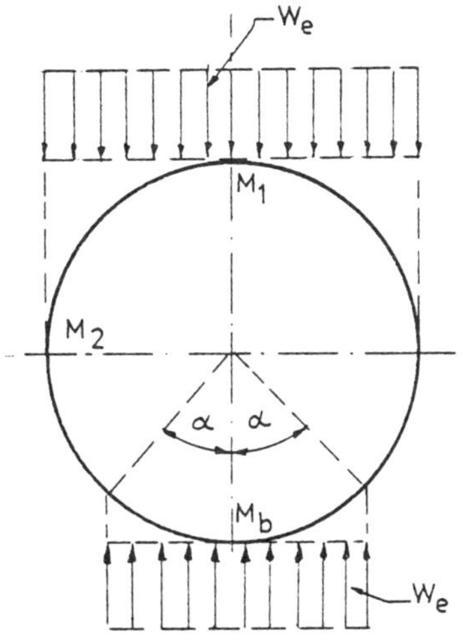

FIGURE 10.4 — Bending moment for flexible pipe.

where

K_b = the moment coefficient

Substituting into Equation (10.9) the value of W_e from Equation (10.3), we obtain the value of bending moment per inch

$$M_b = \frac{K_b}{K_e} \cdot \frac{\Delta_x EI}{R^2}, \quad \text{in.lb/in} \qquad (10.10)$$

By substituting the value of Δ'_x from Equation (10.6), we obtain

$$M_b = M_{max} = \frac{K_b W_e REI}{12(EI-2K_e p_e R^3)} \qquad (10.11)$$

where

K_b = the moment coefficient, Table 10.2.

The maximum bending stress is

$$f_{cb} = \frac{M_{max}t}{2I} = \frac{K_b W_e REt}{24(EI-2K_e p_e R^3)} \qquad (10.12)$$

For a smooth walled pipe laid on the trench bottom, the bedding angle will be about 30° and from Table 10.1, $K_e = 0.108$ and from Table 10.2, $K_b = 0.235$. Then, from Equation (10.12)

$$f_{cb} = \frac{0.117\ W_e RE}{Et^3 - 2.592\ p_c R^3}, \quad \text{psi} \qquad (10.13)$$

The compressive ring stress is

$$f_{cr} = \frac{P_e R}{t}, \quad \text{psi} \qquad (10.14)$$

The maximum total compressive stress is then

$$f_{c\,max} = f_{cb} + f_{cr} \qquad (10.15)$$

which must not exceed $f_y/2$, where f_y is yield stress for steel in a pipe wall.

Then, substituting $t^3/12$ for I in Equation (10.12), and using Equation (10.14), we obtain

$$\frac{K_b W_e REt}{2(Et^3-24K_e p_e R^3)} + \frac{P_e R}{t} = \frac{f_y}{2} \qquad (10.16)$$

whence t may be obtained.

TABLE 10.2 — Moment Coefficient.

Degree 2α	Moment Coefficient Kb
0	0.294
30	0.235
60	0.189
90	0.157
120	0.138
150	0.128
180	0.125

10.4 Thin-Walled Pipe Installed With Controlled Compaction

The method of design described below is a combination of Spangler's method of estimating pipe deflection under external loading with Meyerhof's more recent suggestion [10.4, 10.5] regarding the modulus of soil reaction. The Spangler basic equation for long-term deflection under these conditions is [10.6]

$$\Delta_x = \Delta_y = \frac{D_e K_e W_e R^3}{12(EI\ 0.061\ e\ R^4)} \qquad (10.17)$$

where

D_e = deflection lag factor (1.25 — 1.50)

e = modulus of passive resistance of enveloping earth, psi/in

For the composite action of pipe and soil Meyerhof postulated a ring of soil concentric with the pipe and of an annular width of at least the diameter of the pipe, Figure 10.5.

The average unit vertical pressure at the top of the pipe is

$$P_o = \frac{W_e}{(12)(2)R}, \quad \text{psi} \qquad (10.18)$$

In this method Spangler's value "e" or modulus of passive resistance of enveloping earth is substituted by the identical subgrade modulus "K". Also, substituting $P_o(24)R$ for W_e in Equation (10.17) and ignoring D_e at this stage, for K_e = 0.083, we obtain

$$\Delta_x = \frac{0.167\ p_o R^4}{EI+0.061\ KR^4} \qquad (10.19)$$

and since EI is small for thin-walled pipes compared with 0.061 K R^4 it may be safely ignored.

Then

$$\Delta_x = \frac{2.7p_o}{K}, \quad \text{in} \qquad (10.20)$$

But, as given by Equation (10.4), the effect of external fluid pressure is to increase the deflection of the pipe to Δ'_x by the addition of a load 2 $p_e\Delta'_x$ lb/in.

Then

$$p_o 2R + 2p_e\Delta'_x = \frac{\Delta'_x p_o 2R}{\Delta_x} \qquad (10.21)$$

whence, substituting $\frac{\Delta_x K}{2.7}$ for p_o, we obtain

$$\Delta'_x = \frac{2.7p_o R}{KR - 2.7P_e} \qquad (10.22)$$

Then

$$K_{min} = \frac{2.7(P_o R + P_e\Delta'_x)}{R\Delta'_x}, \quad \text{psi/in} \qquad (10.23)$$

The value of K for a particular soil fill is given as [10.4]

$$K = \frac{E_s}{2(1-\nu^2)R}, \quad \text{psi/in} \qquad (10.24)$$

and since Poisson's ratio for soil, ν = ½ approximately

$$K = \frac{E_s}{1.5R}, \quad \text{psi/in} \qquad (10.25)$$

where E_s is the "modulus of deformation" of the compacted soil as determined by triaxial tests on samples of the fill which have been compacted to a known, or required, density and then saturated under a transverse fluid pressure equivalent to the minimum overburden pressure at the pipe axis over the length of pipeline under consideration.

In weak natural soils it may be necessary to import a stronger soil complying with the above requirements for the fill surrounding pipes in a trench, and to widen the trench sufficiently to ensure that the natural soil is not overstressed by the lateral pressure from the pipe.

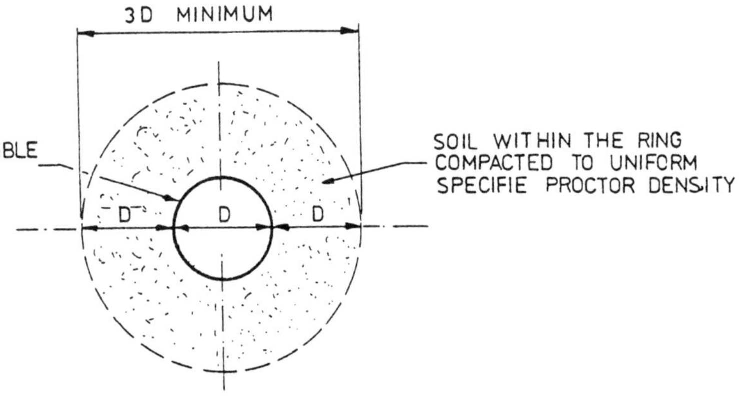

FIGURE 10.5 — Pipe-soil composite structure.

10.5 Ring Stress

For flexible pipelines the "compression ring" theory of White and Layer [10.7] is adopted. This theory postulated that the compressive ring stress in the pipe wall is uniform around the periphery and that the external soil pressure at any point on the periphery is inversely proportional to the radius of curvature of the wall at that point provided that the minimum cover is sufficient, about $B_c/4$, to balance the active pressure imposed on a thin-walled pipe, where B_c is overall diameter of a conduit. Since the radius at the top of the pipe tends to increase with the application of external load, the soil pressure tends to decrease and the compressive ring stress is therefore not likely to exceed

$$f_c = \frac{(P_o+P_e)R}{A}, \quad \text{psi} \qquad (10.26)$$

where

$$P_o = \frac{W_e}{(12)(2)R}, \quad \text{psi}$$

when the pipeline is not submerged; and A = t = the sectional area of the pipe wall, in^2/in.

P_e = 0 = for non-pressure pipes not submerged

P_e = p_{voc} = for pressure pipes not submerged

The allowable ring stress f_r must not exceed the lesser value of:

1. $f_y/2$ to $f_y/4$, psi, depending on the expected efficiency of the compaction of the fill, or the same fraction of the axial seam strength of the pipe if less than f_y.
2. One-half the critical stress computed from Equation (10.28) below.

10.5.1 Critical or Collapse Stress

Meyerhof [10.4] postulates that providing

$$R = 2\sqrt[4]{\frac{EI}{(1-\nu^2)K}}, \quad \text{in} \qquad (10.27)$$

the critical stress may, allowing for practical imperfections in the pipe, be taken as

$$f_c = \frac{f_y}{1 + (f_y A/2)\sqrt{(1-\nu^2)/KEI}} \qquad (10.28)$$

and the permissible stress is

$$f_r = f_c/2$$

10.5.2 Selection of Wall Thickness

For thin-walled pipe which depend for their stability on composite action between the soil and the pipe the selection of the pipe wall thickness, t, is a process of successive approximation as follows:

Given the pipe radius, R, in., the external design load W lb/ft, and the limiting vertical deflection of the pipe, e, in.

The solution should proceed in the following steps:

1. Obtain the vertical pressure at the top of the pipe caused by the external load
$$P_o = \frac{W_e}{(12)(2)R}, \quad \text{psi}$$
2. Obtain the required value of "K" from Equation (10.23), and if less than 20 psi adopt a value of 20 psi or assess the value from known test values of E_s, by Equation (10.24).
3. Assume an allowable ring stress f_n of say 0.25 f_y and from Equation (10.26) obtain a value for A, or
$$A = P_oR/0.25\, f_y, \ in^2$$
4. Adopt the nearest standard plate thickness to give the area A for smooth-walled pipe, and obtain the I-value. Then check that the value of R exceeds the minimum permissible value given by Equation (10.27). Taking E = 30×10^6 psi and ν = 0.3 and K as obtained or assumed in Step (2)
$$R > 2\sqrt[4]{\frac{33x10\ I}{K}}, \quad \text{in}$$
5. If Equation (10.27) is not satisfied, reduce I or increase K as necessary and convenient.
6. Obtain the value "Kt" for a smooth wall and obtain the allowable stress f_r.
7. With this value of f_r if greater than 0.25 f_y, obtain a revised value of
$$A = P_oR/f_r$$
and select a corresponding new value of t.
8. Obtain a new value of "Kt" and repeat Steps (6) and (7) to check that the value of A is satisfactory.
9. Repeat Step (8) if necessary to revise A.
10. Check that the value of Δ_x by Equation (10.22), or
$$\Delta_x = \frac{2.7P_o}{R}$$
does not exceed the specified deflection.

10.6 Stability of the Underground Pipelines

To-date, the problem of the buckling stability in underground pipelines has not been treated widely in any special technical literature. However, it is known that "hot" pipelines under compressive stresses, having relatively low resistance against transverse displacements in soils of low-carrying capacity or built-under embankments, create a problem regarding their longitudinal stability.

It should be recognized that the main purpose in the stability design of underground pipelines under compressive stress is to choose such alternatives as any placements, depths and constructions which will guarantee maximum safety.

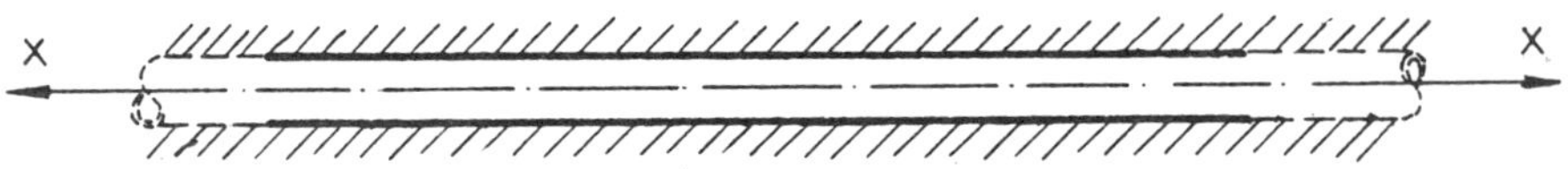

FIGURE 10.6 — Bar in elastic medium.

A loss of pipeline stability results in the development of dangerous stresses in the wall of the pipe due to the bending and damaging of backfill. For this reason, the design of stability includes the determination of not only the critical compressive stresses, but also a corresponding transverse deformation of the pipeline and the bending stresses.

10.7 Stability of Straight Pipeline

The stability analysis of straight pipelines consists of a determination of the critical buckling force, assuming a straight pipeline of infinite length.

In this stage, we consider any small deflections, assuming the elastic resistance of the backfill against the transverse displacements of the pipeline. We therefore consider that the following differential equation or equilibrium of the bar in elastic medium is [10.8], Figure 10.6

$$EI\frac{d^4v}{dx^4} + P\frac{dv^2}{dx^2} + cv = 0 \quad (10.29)$$

where

E = modulus of elasticity of pipeline material

I = moment of inertia of pipe cross-section

P = compressive force acting on the pipeline before loss of stability

v = transverse displacement

x = coordinate along the straight axis of pipeline

c = subgrade reaction

For the pipeline

$$c = K_0 D_0 \quad (10.30)$$

where

K_0 = coefficient of subgrade reaction

D_0 = effective width of the support of the pipeline which is assumed to be equal to the diameter of the pipeline.

Using the designations

$$\frac{P}{EI} = K^2 \ , \quad \frac{c}{EI} = 4r^4 \quad (10.31)$$

and substituting into (10.29) we obtain the following homogeneous linear equation

$$\frac{d^4v}{dx^4} + K\frac{d^2v}{dx^4} + 4r^4v = 0 \quad (10.32)$$

The corresponding characteristic equation has the form

$$p_1^4 + k^2p_1^2 + 4r^4 = 0 \quad (10.33)$$

its roots will be

$$m_1^2 = \frac{K^2}{2} \pm \sqrt{K^4 - 16r^4} \quad (10.34)$$

where

$$p_1 = im, \quad i = \sqrt{-1}$$

At a loss of stability of the infinite pipeline the shape of deflection remains indefinite: any wave-type shape corresponding to the parameter m_1 satisfies the boundary conditions.

From (10.34) we have

$$K^2 = m_1^2 + \frac{4r^4}{m_1^2} \quad (10.35)$$

By equalizing the derivate of the K^2 by m_1^2 to zero, we obtain

$$m_1^2 = 2r^2 \quad (10.36)$$

According to (10.31), the critical loading is

$$P_{cr} = 2\sqrt{EIc} \quad (10.37)$$

10.8 Stability of the Pipeline at Longitudinal and Transverse Flexure

An analysis of different cases of loss of stability of underground pipelines, as well as the experiments, indicates that the critical force given by Equation (10.37) for straight pipelines in elastic medium is an upper limit considering the different positions of non-stability. Also, due to the initial deviation of the pipeline from the ideal straight position there is the possibility of a loss of stability of smaller values of the compressive force.

Let us now consider the longitudinal and transverse bending of the pipeline at the limits of its final deformations and determine the stability criteria at different states of equilibrium.

Let us assume that the backfill at the transverse displacements of the pipeline, is in a state of plastic flow, the ground behaves as a cohesionless body and provides constant resistance during the bending of the pipeline.

The interaction between the pipeline and soil is not limited to the zones in which the bending deformation occurs. Bending of the pipeline produced longitudinal displacements which extend along their adjoining sections, forcing them to participate in the total work in performing the bending of the pipeline.

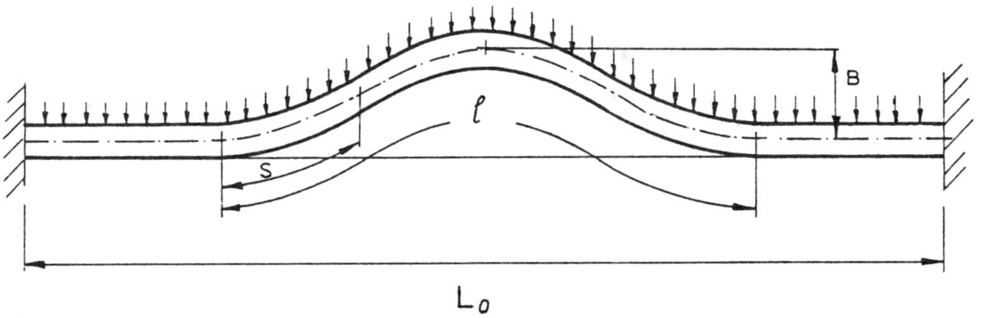

FIGURE 10.7 — Bending of straight pipeline at loss of stability.

During stability analysis of pipeline it is considered that the bulging of the pipe extends only along certain lengths under constant compressive force. However, outside this length there is no change in the stress.

Therefore, the pipeline is considered as an elastic beam on a rigid support, which is under the influence of a longitudinal force, changeable with the deflections, and having a transverse loading equal to the limiting bearing-capacity of the backfill when pulling the pipeline away from the backfill.

A deformed configuration of the pipeline is shown in Figure 10.7, where

ℓ = length of the bent part

L_n = section of the pipeline affected by bulging

q = limiting bearing capacity of the backfill

s = length of the arc

For a solution of the problem we use the Ritz method, applying the energy criteria of stability. The total energy of the system at the deformation stage of the pipeline is

$$U = V - W \tag{10.38}$$

where

V = energy of pipeline deformation consisting of energy of deformation at the bending V_1 and energy of deformation at the compression V_2

W = work of the external forces during the displacement of the pipeline

The initial stresses, acting on the pipeline are reduced in bending. Therefore, the compressive force at the ends of the bent section of each stage of deformation is

$$N = P - \frac{\Delta E A_w}{L_n} \tag{10.39}$$

where A_w is the cross-sectional area of the pipe wall and Δ is the increase in length of the pipeline

$$\Delta = \ell - \int_0^{\ell} \sqrt{1 - (dv/ds)^2}\, ds \tag{10.40}$$

The potential energy at bending is

$$V_1 = \frac{EI}{2} \int_0^{\ell} \chi^2\, ds \tag{10.41}$$

where the curvature χ is

$$\chi = \frac{d^2v/ds^2}{\sqrt{1 - (dv/ds)^2}} \tag{10.42}$$

By expanding the last value into series and substituting into (10.41) we obtain

$$V_1 = \frac{EI}{2} \int_0^{\ell} \left(\frac{d^2v}{ds^2}\right)^2 \left[1 + \left(\frac{dv}{ds}\right)^2\right] ds \tag{10.43}$$

For the energy during compressive deformation of the pipeline, we have

$$V_2 = \frac{1}{2EA_w} \int_0^{L_n} \overline{N}^2\, ds \tag{10.44}$$

where $\overline{N}$ — the force acting in the cross-section of the deformed pipeline. At deflections much smaller than the length of the bent section, the force $\overline{N}$ is constant along the whole length L_n and is equal (10.39).

Using this value, we find for (10.44)

$$V_2 = \frac{P^2 L_n}{2EA} - P\Delta + \frac{\Delta^2 E A_w}{2L_n} \tag{10.45}$$

From this, by extending (10.40) into the series of $(\frac{dv}{ds})$, we obtain

$$V_2 = \frac{P^2 L_n}{2EA_w} - \frac{P}{2} \int_0^{\ell} \left(\frac{dv}{ds}\right)^2 \left[1 + \frac{1}{4}\left(\frac{dv}{ds}\right)^2\right] ds + \frac{EA_w}{8L_n} \left\{ \int_0^{\ell} \left(\frac{dv}{ds}\right)^2 \left[1 + \frac{1}{4}\left(\frac{dv}{ds}\right)^2\right] ds \right\}^2 \tag{10.46}$$

The work of the external forces is produced by transverse loading q along displacements v and is

$$W = - \int_0^{\ell} q\, v\, ds \tag{10.47}$$

The elastic axis of the pipeline at its loss of stability, we approximate by the function

$$v = A \sin^2 \lambda s, \quad (\lambda = \frac{\pi}{\ell}) \tag{10.48}$$

which satisfies the geometrical boundary conditions

$$v = \frac{dv}{ds} = 0 \quad \text{at} \quad s = 0 \quad \text{and} \quad s = \ell \tag{10.49}$$

Therefore, finally

$$U = \frac{EI}{2}\,\pi\{(2A^2\lambda^3 + \frac{1}{2}A^4\lambda^5) - \frac{P}{EI}(\frac{1}{2}A^2\lambda + \frac{3}{32}A^4\lambda^3) + \frac{\pi A_w}{L_n I}\cdot\frac{1}{16}(A^4\lambda^2 + \frac{3}{8}A^6\lambda^4) + \frac{q}{EI}\cdot\frac{A}{\lambda}) + \frac{P^2 L_n}{2EA_w} \tag{10.50}$$

By variation of the total energy of the system by parameters A and λ we introduce equations

$$\frac{\partial U}{\partial A} = 0, \quad \frac{\partial U}{\partial \lambda} = 0 \tag{10.51}$$

By assuming

$$A\lambda = x \tag{10.52}$$

then Equations (10.51) will take the shape

$$2\lambda^3 x(2 + x^2) - \frac{P}{EI}\lambda x(1 + \frac{3}{8}x^2) + \frac{\pi A_w}{L_n I}\cdot\frac{x^3}{16}(4 + \frac{3}{4}x^2) + \frac{q}{EI} = 0 \tag{10.53}$$

$$\lambda^3 x(6 + \frac{5}{2}x^3) - \frac{P}{EI}\lambda x(\frac{1}{2} + \frac{9}{32}x^2) + \frac{\pi A_w}{L_n I}\cdot\frac{x^3}{16}(2 + \frac{3}{2}x^2) - \frac{q}{EI} = 0 \tag{10.54}$$

From the last equations, by neglecting small terms and by designating

$$n = \frac{\pi E A_w}{L_n q} \tag{10.55}$$

we find

$$\lambda = \psi\sqrt[3]{q/EI} \tag{10.56}$$

$$P = \beta\sqrt[3]{q^2 EI} \tag{10.57}$$

where

$$\psi = \sqrt[3]{\frac{1}{[(1+0.375x^2)\dfrac{6.67+3x^2}{1+0.438x^2} - (4+2x^2)]x}} \tag{10.58}$$

$$\beta = \psi^2\frac{6.67+3x^2}{1+0.438x^2} + \frac{0.25+0.156x^2}{1+0.438x^2}\cdot\frac{nx^2}{\psi} \tag{10.59}$$

Considering (10.55) to (10.57), we obtain for the total energy

$$U = \frac{EI}{2}\,\pi\sqrt[3]{q/EI}\;\bar{U} \tag{10.60}$$

where

$$\bar{U} = \psi x^2(2+\frac{1}{2}x^2) - \frac{\beta}{\psi}x^2(\frac{1}{2}+\frac{3}{32}x^2) + \frac{n}{16\psi^2}x^2(1+\frac{3}{8}x^2) + \frac{x}{\psi^2} + \frac{\beta}{n} \tag{10.61}$$

Equations (10.56) to (10.61) describe the multiplicity of the different conditions of equilibrium at the bending of straight pipeline. The corresponding curves "initial compressive stresses — deflection", show in Figure 10.8 their different values of parameter n (10.55). For any value of n, that is different from zero, the curves reach a minimum. Each value of nondimensional coefficient of initial compressive stress β above the minimum, corresponds to two different states of equilibrium of bent pipeline.

The equilibrium conditions of these shapes of deflection may be found by analyzing the limiting values of the total energy (10.60), (10.61).

For values β, x and ψ, corresponding P_ℓ, we obtain

$$\beta_\ell = 3.51\,n^{2/9} \tag{10.62}$$

$$x = \frac{1.415}{\sqrt[3]{n}} \tag{10.63}$$

$$\psi = 0.642\,n^{1/9} \tag{10.64}$$

10.9 The Determination of Critical Stresses

For the determination of initial stresses corresponding to the loss of stability, we establish a relationship between these accepted in this study design scheme and any real

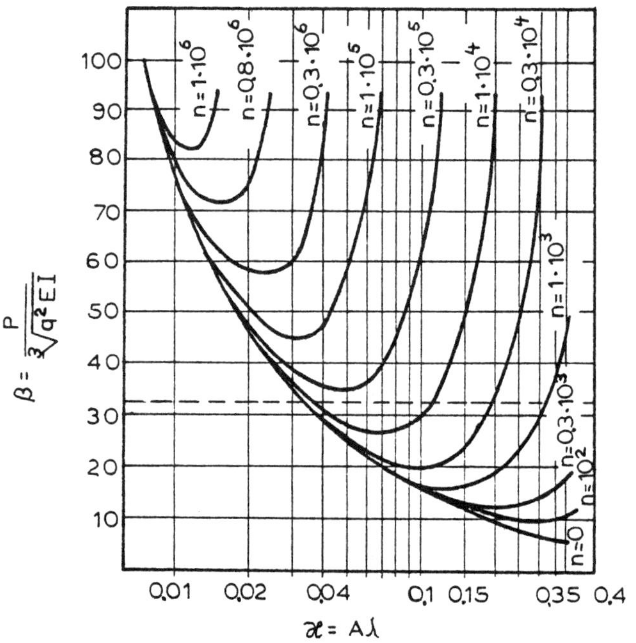

FIGURE 10.8 — Diagram of curves "compressive force — deflection" for straight pipeline.

conditions of tangential interaction between the pipeline and soil.

For this purpose, we consider the conditions of equivalence of the work during compression at a loss of stability

$$W_n = W_p \tag{10.65}$$

and displacements at the ends of bent section

$$U_n = U_p \tag{10.66}$$

where the indices "n" and "p" relate to the accepted schemes of deformations and actual conditions of interaction of the pipeline and soil.

Let us examine these conditions for example, when conditions of limiting equilibrium originate at the boundary pipeline-soil during longitudinal displacements of the pipeline. The resistance against the longitudinal displacement of the pipeline is constant and equal to p.

The compressive energy at the straight section of the pipeline is

$$W_{1p} = \frac{1}{2EA_w} \int_0^L (px)^2 \, dx \tag{10.67}$$

Considering that

$$p = \frac{(P-N_p)}{L} \tag{10.68}$$

we obtain

$$W_{1p} = \frac{1}{2EA_w} \cdot \frac{(P-N_p)^3}{3p} \tag{10.69}$$

The compressive deformation energy at the bent section is

$$W_{2p} = \frac{1}{2EA_w} \cdot \frac{(P-N_p)^2 \ell}{2} \tag{10.70}$$

By analogy, for displacements at the end of the bent section

$$U_p = \frac{(P-N_p)^2}{2pEA_w} + \frac{(P-N_p)\ell}{2EA_w} \tag{10.71}$$

For the scheme shown in Figure 10.7, we obtain

$$W_n = \frac{(P-N_n)^2}{4EA_w} L_n \tag{10.72}$$

$$U_n = \frac{(P-N_n)}{2EA_w} L_n \tag{10.73}$$

We obtain for Equations (10.65), (10.66)

$$\frac{2}{3p}(P-N_p)^3 = L_n (P-N)^2 [1 - \frac{\ell}{L_n} (\frac{P-N_p}{P-N})^2] \tag{10.74}$$

$$\frac{1}{p}(P-N_p)^2 = L_n (P-N) [1 - \frac{\ell}{L_n} \frac{P-N_p}{P-N}] \tag{10.75}$$

Together with Equations (10.39), (10.40), (10.48), (10.55) to (10.64), the Equations (10.74), (10.75) permit determination of the values of the critical stresses.

The reduced length L_n, corresponds to the lower critical force, and is

$$L_n = 0.623 \, f^{9/11} (\frac{\ell}{L_n}) \pi \sqrt[11]{q^4 A_w^2 E^5 I^3 / p^9} \tag{10.76}$$

where the coefficient $f(\frac{\ell}{L_n})$ stems from the ration ℓ L_n.

By substituting the last expression into (10.62) and (10.55), we find for the lower critical force that

$$P_\ell = 3.91 \, f^{-2/11} (\frac{\ell}{L_n}) \sqrt[11]{p^2 q^4 A_w^2 E^5 I^3} \tag{10.77}$$

The value of $f^{-2/11} (\frac{\ell}{L_n})$ for values L_n from 0 to 0.5 varies only 4%.

Therefore, with an accuracy of ±2% the lower critical force is defined by the formula

$$P_\ell = 4.00 \sqrt{p^2 q^4 A_w^2 E^5 I^3} \tag{10.78}$$

By substituting (10.78) into (10.76), we obtain

$$L_n \quad 0.5 \frac{P}{p} f (\frac{\ell}{L_n}) \tag{10.79}$$

where the value $f(\frac{\ell}{L_n})$ changes in limits from 1 to 1.2.

10.10 Stability of the Pipeline Having Initial Flexure

Let us now consider the influence of imperfections, when the pipeline possesses an initial curvature and the compressive force acts under eccentricity. Such imperfections may take place either during the construction or due to bending after placing the backfill into the trench.

Applying the previous assumption, we will consider the scheme shown in Figure 10.9.

The total energy of the system, the potential energy of bending, the energy of deformation of compression along neutral axis and the work of the external forces on the displacements of the pipeline is defined by Equations (10.38), (10.42), (10.46), (10.47) and expression (10.40) for an increase of the pipeline in the length at bending

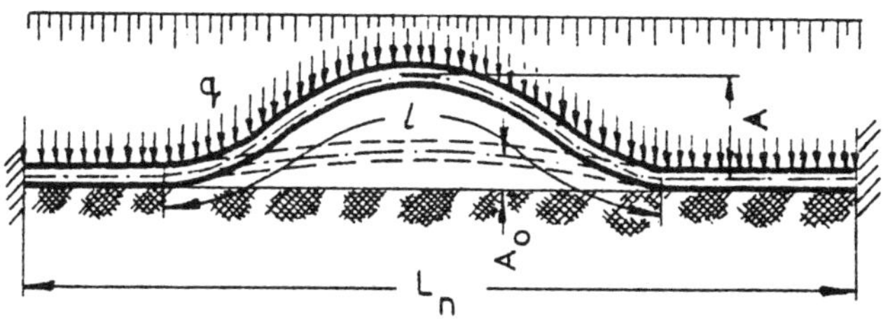

FIGURE 10.9 — Bending of the pipeline having the initial curvature.

$$\Delta = \frac{1}{2} P \int_0^{\ell} \{ (\frac{dv}{ds})^2 [1 + \frac{1}{4}(\frac{dv}{ds})^2] - (\frac{dv_0}{ds})^2 [1 + \frac{1}{4}(\frac{dv_0}{ds})^2]\} ds \qquad (10.80)$$

where v_0 and v are the initial and total deflections of the pipeline. The total deflection v we approximate by the function (10.48).

Let us assume that the initial deflection satisfies a similar pattern

$$v_0 = A_0 \sin^2 \lambda s \qquad (10.81)$$

We may therefore assume that the shape of the initial deflection is "in resonance" with the bending of the pipeline and the only single known parameter is amplitude A_0.

The total energy is defined by the Equation (10.38). At arbitrary point $a > 0$ the bending deformations grow with an increase of the initial stress· the basic state of equilibrium of bent pipeline is stable up to a certain maximum value, after which there are changes in the new equilibrium states. The corresponding value of the initial compressive stress (critical force) we will call critical for the pipeline, and designate as

$$\sigma^0_{cr} (P_{cr})$$

The basic shape of equilibrium of bent pipeline ceases to be the only possible shape at the stress points, corresponding to the maximum curves, and which do not differ from the low critical values determined and shown above for the straight pipeline. Therefore, the low critical force (initial stress) does not depend on the value of initial bending but remains practically non changeable and constant to the characteristics of the buried pipeline.

In the interval $(\sigma^0_{cr}, \sigma_\ell)$ of every value of initial stress corresponds to three states of equilibrium of bent pipeline — two stable and one unstable. The secondary stable state of equilibrium corresponds to the pipeline bent the greatest.

With an increase of amplitude in the initial deflection the upper critical stress σ^0_{cr} is decreased and at a certain value of initial deflection becomes equal to the low critical value; at greater values of the initial deflection only one stable state of equilibrium corresponds to each initial stress.

Therefore, any unstable equilibrium positions during pipeline bending start only at

$$\sigma^0 > \sigma^0_\ell$$

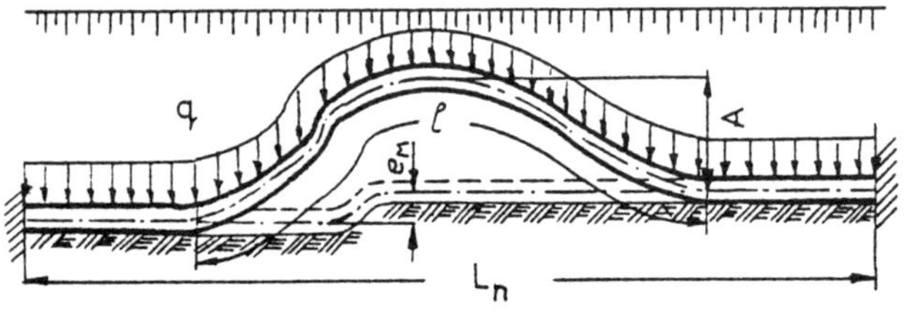

FIGURE 10.10 — Bending of the pipeline under action of the eccentrical compressive force.

And if the initial stress (compressive force) remains smaller than the lower critical stress, and this guarantees the stability of the pipeline.

Diagrams also indicate that with the increase of the parameter n for the transfer of a pipeline from its stability position, it is necessary to have a smaller initial curvature.

Regarding the initial deflections, these correspond to the radii of free bending of buried pipelines.

10.11 Stability of Pipeline at Eccentrical Compression

Due to the possible irregularities in the shape of free bending and also due to the different thicknesses in the walls, the compressive force in the cross-section of pipeline

$$U = \frac{1}{2} EI\pi \{ 2[A-A_0)^2 \lambda^3 + \frac{1}{2}(A-A_0)^4 \lambda^3] - \frac{P}{EI} [\frac{1}{2}(A^2-A_0^2)\lambda + \frac{3}{32}(A^4-A_0^4)\lambda^3] + \frac{\pi A}{16 L_n I} [(A^2-A_0^2)\lambda + \frac{3}{16}(A^4-A_0^4)\lambda^3] + \frac{q}{EI} \cdot \frac{A-A_0}{\lambda}\} + \frac{p^2 L_n}{2EA_w} \qquad (10.82)$$

For equilibrium conditions of the pipeline, a total energy limit is reached.

By assuming

$$A = \frac{a}{\sqrt[3]{q/EI}}, \qquad A_0 = \frac{a_0}{\sqrt[3]{q/EI}} \qquad (10.83)$$

where a, a_0 — are non-dimensional coefficients, according to (10.51), we find the equations for the determination of the parameters of the equilibrium shape of deflection β, ψ, a. At $a \leq 0.2$ in these equations with the error not greater than 2%, we may exclude a number of terms of high order, and obtain

$$\psi^4 = \frac{1}{4} \cdot \frac{3a + a_0}{(2a-a_0)(a-a_0)} \qquad (10.84)$$

$$\beta = 10 \sqrt{\frac{a-a_0}{(3a+a_0)(2a-a_0)}} + 0.1765 n (a+a_0) \sqrt[4]{\frac{(3a+a_0)(a-a_0)^3}{(2a-a_0)}} \qquad (10.85)$$

is acting eccentrically. Apart from this, the pipeline has special structural details which may produce concentrated moments.

In the following, we are considering the scheme of the pipeline as shown in Figure 10.10.

For the total energy and its components Equations (10.38), (10.43), (10.46) and (10.47) are used. In the expression for the work of external forces we add the work of concentrated moments during rotation of the cross-secton W_M.

By an analogy with Eq. (10.47) we then obtain for W_M

$$W_M = Pe_M a_0 - \frac{EA_w}{3L_n} e_M\, a_0\, \Delta \tag{10.86}$$

where

e_M = the eccentricity of the force

α_0 = angle of rotation of the cross-section

To find the total deflection, we then use the former expression (10.48).

The total energy of the pipeline (10.38) is equal

$$U = \frac{1}{2} EI\pi[(2A^2\lambda^3+\frac{1}{2}A^4\lambda^5) - \frac{P}{EI}(\frac{1}{2}A^2\lambda+\frac{3}{32}A^4\lambda^3) + \frac{A_w}{16IL_n}(A^4\lambda^2+\frac{3}{8}A^6\lambda^4) - \frac{P}{EI}e_M\frac{2}{\pi}A\lambda + \frac{Ae_M}{3IL_n}(\frac{1}{2}A^2\lambda+\frac{3}{32}A^4\lambda^3) + \frac{q}{EI}\frac{A}{\lambda}] - \frac{P^2L_n}{2EA_w} \tag{10.87}$$

Let us designate

$$e_M = \frac{e_m}{\sqrt[3]{q/EI}} \tag{10.88}$$

where e_m is a nondimensional parameter.

By applying conditions (10.51), we then determine the parameters of the positions of equilibrium of deflection which after excluding secondary terms are

$$4a\,[\frac{3}{a+\frac{4}{\pi}e_m} - \frac{1}{a+\frac{2}{\pi}e_m}]\psi^4 + \frac{1}{4}na^2\psi^3[-1\,\frac{a+\frac{8}{3\pi}e_m}{a+\frac{4}{\pi}e_m}] - [\frac{2}{a+\frac{4}{\pi}e_m} + \frac{1}{a+\frac{2}{\pi}e_m}] = 0 \tag{10.89}$$

$$\beta = \frac{10a\psi^2+na^2\psi[(\frac{1}{4}a+\frac{1}{2\pi}e_m) + (\frac{1}{8}a+\frac{1}{3\pi}e_m)]}{\frac{3}{2}a + \frac{6}{\pi}e_m} \tag{10.90}$$

10.12 Differential Equation of Stability of Straight Pipeline

In the following we are analyzing the stability of underground pipelines under different conditions to determine the critical compressive forces.

The total energy of the pipeline is

$$U = \frac{1}{2}EI\int_0^{\ell}(\frac{d^2v}{dx^2})^2dx - \frac{1}{2}P\int_0^{\ell}(\frac{dv}{dx})^2dx + \frac{EA_w}{8L_n}[\int_0^{\ell}(\frac{dv}{dx})^2dx]^2 + \frac{P^2L_n}{2EA_w} + \int_0^{\ell}qv\,dx + \frac{1}{2}c\int_0^{\ell}v^2dx \tag{10.91}$$

where the last term indicates the work of elastic deformations of the soil. The first variation of the total energy of the system is

$$\delta U = EI\int_0^{\ell}\frac{d^2v}{dx^2}\,\delta(\frac{d^2v}{dx^2})dx - P\int_0^{\ell}\frac{dv}{dx}\,\delta(\frac{dv}{dx})dx + \frac{EI}{2L_n}\int_0^{\ell}(\frac{dv}{dx})^2dx\int_0^{\ell}\frac{dv}{dx}\,\delta(\frac{dv}{dx})dx + \int_0^{\ell}q\delta v\,dx + c\int_0^{\ell}v\delta v\,dx \tag{10.92}$$

We apply partial integration for

$$\int_0^{\ell}\frac{d^2v}{dx^2}\,\delta(\frac{d^2v}{dx^2}) \quad \text{and} \quad \int_0^{\ell}\frac{dv}{dx}\,\delta(\frac{dv}{dx})$$

$$EI\frac{d^2v}{dx^2}\,\delta(\frac{dv}{dx})\Big|_0^{\ell} - EI\frac{d^3v}{dx^3}\,\delta v\Big|_0^{\ell} + EI\int_0^{\ell}\frac{d^4v}{dx^4}\,\delta v\,dx - [P - \frac{EI}{2L_n}\int_0^{\ell}(\frac{dv}{dx})^2dx]\frac{dv}{dx}\,\delta v\Big|_0^{\ell} + [P - \frac{EI}{2L_n}\int_0^{\ell}(\frac{dv}{dx})^2dx]\int_0^{\ell}\frac{d^2v}{dx^2}\,\delta v\,dx + \int_0^{\ell}q\delta v\,dx + c\int_0^{\ell}v\delta v\,dx = 0 \tag{10.93}$$

Non-integral terms are equal to zero, because at the ends of the bent part of the pipeline the deflections and slope angles are equal to zero, and, therefore, their variations.

Due to the arbitrary variations, we have

$$EI\frac{d^4v}{dx^4} + [P - \frac{EI}{2L_n}\int_0^{\ell}(\frac{dv}{dx})dx]\frac{d^2v}{dx^2} + cv + q\ \text{sign}\ v = 0 \tag{10.94}$$

According to (10.39) the coefficient in the brackets is the force N, acting in every moment of deformation on the bent part of the pipeline, and (10.94) may be written as follows

$$EI\frac{d^4v}{dx^4} + N\frac{d^2v}{dx^2} + cv = -q\ \text{sign}\ v \tag{10.95}$$

Because N depends on the amplitude values of deflection, this equation is linear and not formally connected with the definite interaction of the soil and longitudinal deformation of the pipeline.

10.13 Stability of Underground Pipeline Along Straight Sections

Assume that during a longitudinal displacement of a pipeline, at the boundary pipeline-soil, a state of ultimate equilibrium takes place. According to Equation (10.71) the force acting at the ends of the bent section is

$$P = N - \frac{1}{2}p\ell + \sqrt{\left(\frac{1}{2}p\ell\right)^2 + pEA_W\Delta} \qquad (10.96)$$

Considering the transverse displacement of pipeline in a state of rigid-plastic deformations of backfill and using designations

$$k^2 = \frac{N}{EI} \qquad (10.97)$$

$$m^2 = \frac{q}{EI} \qquad (10.98)$$

we obtain from (10.95)

$$\frac{d^4v}{dx^4} + k^2\frac{d^2v}{dx^2} = -m^2 \text{ sign } v \qquad (10.99)$$

This equation describes any shape of deflection shown in Figure 10.11.

If the ultimate state of backfill is reached in only one direction (for example, the pipeline in a trench), then the shape of deflection corresponds to Figure 10.11 (1). A general solution of Equation (10.99) for this shape of deflection will be

$$v = A\cos\kappa x - B - \frac{m^2}{2\kappa^2}x^2 \qquad (10.100)$$

where A and B are constants of integration.

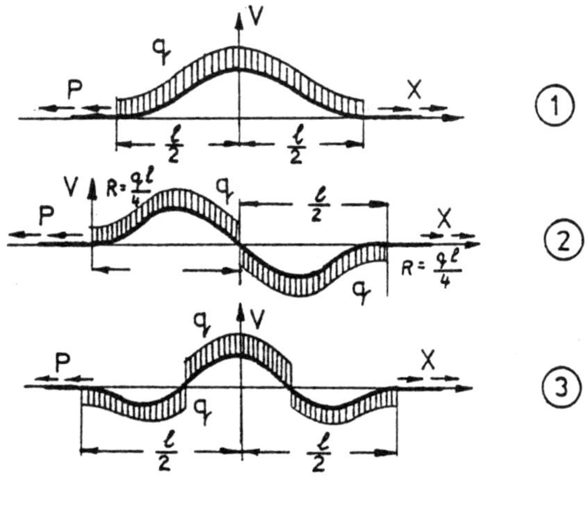

FIGURE 10.11 — Shape of deformed pipelines at a loss of stability.

By applying the following boundary conditions at both ends of bent section

$$v = \frac{dv}{dx} = \frac{d^2v}{dx^2} = 0 \quad \text{at} \quad x = \frac{\ell}{2} \qquad (10.101)$$

we obtain

$$A\cos\frac{k\ell}{2} - B - \frac{m^2\ell^2}{8\kappa^2} = 0 \qquad (10.102)$$

$$-Ak\sin\frac{k\ell}{2} - \frac{m^2\ell}{2\kappa^2} = 0 \qquad (10.103)$$

$$-A\kappa^2\cos\frac{k\ell}{2} - \frac{m^2}{\kappa^2} = 0 \qquad (10.104)$$

From (10.103) and (10.104) we obtain the transcendent equation

$$\tan\frac{k\ell}{2} = \frac{k\ell}{2} \qquad (10.105)$$

The first root to differ from zero is

$$\frac{1}{2}k\ell = 4.493 \qquad (10.106)$$

From this, according to (10.100), (10.102) to (10.104) and (10.102), we find

$$\ell = 6.12\sqrt[4]{A/m^2} \qquad (10.107)$$

$$\kappa = 1.47\sqrt[4]{m^2/A} \qquad (10.108)$$

$$B = -2.42A \qquad (10.109)$$

$$v_{max} = 3.42A \qquad (10.110)$$

$$\Delta = 5.2\, m^{1/2}A^{7/4} \qquad (10.111)$$

After substituting the last equations into (10.96) we obtain the values of the compressive force for the equilibrium position of deformed pipeline.

$$P_{cr} = \frac{2.16m\,EI}{\sqrt{A}} - \frac{3.06p\sqrt[4]{A}}{\sqrt{m}} + \sqrt{\frac{9.35p^2\sqrt{A}}{m} + 5.20p\,EA_W\sqrt[4]{m^2A^7}} \qquad (10.112)$$

The stable positions of equilibrium are realized on the right branches of the curves, corresponding to the Equation (10.112).

We obtain the low critical force, by finding a minimum value of (10.112) with respect to A, or

$$P_\ell = \frac{\partial P_{cr}}{\partial A} = 4.22\left(1 - 0.03\sqrt[11]{p^8/q^6t^2}\right)\sqrt{q^4p^2A_W^2E^5I^3} \qquad (10.113)$$

where t is thickness of the wall of the pipe. The multiplier in the bracket of the last expression is close to 1 for the values of p and q used in practice. Therefore, the following safe value may be used

$$P_e = 4.09\sqrt[11]{q^4 p^2 A_w^2 E^5 I^3} \qquad (10.114)$$

which corresponds to the value (10.78) obtained earlier by the energy method.

When a compressive force changes with an increase of deflections, then according to Equations (10.39), (10.108) and (10.111) we find

$$\frac{P_{cr}}{EI} = 2.16\ mA^{-1/2} + 5.2m^{1/2}\bar{A}^{7/4}\frac{A_w}{IL_n} \qquad (10.115)$$

and the low critical force is then

$$P_\ell = 4.52\sqrt[9]{q^4 E^5 I^3 A_w^2 / L_n^2} \qquad (10.116)$$

By comparing this result with (10.114) we obtain

$$L_n = 1.58\sqrt[11]{q^4 p^{-9} E^5 I^3 A_w^2} \qquad (10.117)$$

$$L_n = 0.39\frac{P}{p} \qquad (10.118)$$

or 20% less than was obtained from Equations (10.65) and (10.66).

Considering that this difference gives, according to (10.116) the error for P_ℓ only 4%, the method for the determination of the length L_n on the basis of Equations (10.65) and (10.66), may be accepted as satisfactory.

10.14 Stability Analysis of Deformed Shapes of Pipeline

The bulging of pipeline in soft backfill in trench bottoms may occur, according to the shapes shown in Figure 10.11. It is necessary to investigate shapes (2) and (3) because the critical forces in such cases may be relatively small.

Let us consider the deformation of pipeline, having an S-shaped curve. For the simplification of this solution we use the 2nd order differential equation, equivalent to Eq. (10.99)

$$EI\frac{d^2v}{dx^2} + Nv = M(x) \qquad (10.119)$$

where M(x) is the bending moment at the cross-section of a pipeline

$$M(x) = \frac{1}{4}q\ell x - \frac{1}{2}qx^2 \qquad (10.120)$$

Considering (10.97) and (10.98), we have

$$\frac{d^2v}{dx^2} + \kappa^2 v = \frac{1}{4}m^2\ell x - \frac{1}{2}m^2x^2 \qquad (10.121)$$

The general solution of this equation is

$$v = A\cos\kappa x + B\sin\kappa x + \frac{m^2}{\kappa^4} + \frac{m^2\ell}{4\kappa^2}x - \frac{m^2}{2\kappa^2}x^2 \qquad (10.122)$$

By satisfying the boundary conditions of the problem

$$v = \frac{dv}{dx} = 0 \quad \text{at} \quad x = 0 \qquad (10.123)$$

$$v = 0 \quad \text{at} \quad x = \frac{1}{2}\ell \qquad (10.124)$$

we obtain

$$\kappa\ell = 4\pi \qquad (10.125)$$

$$\kappa = \sqrt[4]{m^2/A} \qquad (10.126)$$

$$\ell = 4\pi\sqrt[4]{A/m^2} \qquad (10.127)$$

In accordance with Eq. (10.96) we determine the compressive force at the equilibrium states of pipeline as

$$P_{cr} = \frac{mEI}{\sqrt{A}} - 6.27p\sqrt[4]{A/m^2} + \sqrt{39.3p^2\frac{A}{m} + 67.5pEA\sqrt[4]{A^7m^2}} \qquad (10.128)$$

The minimum of this expression represents the lowest critical force

$$P_e = 4.13(1-0.44)\sqrt[11]{p^8/q^6t^2}\sqrt[11]{q^4p^2A_w^2E^5I^3} \qquad (10.129)$$

Therefore, the loss of stability following the S-shape curve occurs at a somewhat smaller value of the critical force, than after shape (1).

It is necessary to note, that at changes of compressive force according to (1.11), we obtain for the S-shape deflection

$$P_e = 4.33\sqrt[9]{q^4E^5I^3A_w^2/L_n^2} \qquad (10.130)$$

By comparing (10.129) and (10.130), we find

$$L_n = 0.382\frac{P}{p} \qquad (10.131)$$

For the deformed shape (3) the total solution of Equation (10.99) is

$$v = A\cos\kappa x + B - \frac{m^2}{2\kappa^2}x^2 + v_1 \qquad (10.132)$$

where

$$v_1 = 0 \quad \text{at} \quad x = x_1$$

(x_1 — coordinate of inflection point)

$$v_1 = -2\frac{m^2}{\kappa^4} + 2\frac{m^2}{\kappa^4}\cos\kappa(x-x_1) + \frac{m^2}{\kappa^2}(x-x_1)^2 \tag{10.133}$$

at

$$x > x_1$$

By satisfying (10.132) and (10.133) at their boundary conditions

$$v = 0 \quad \text{at} \quad x = x_1$$

$$v = \frac{\partial v}{\partial x} = \frac{\partial^2 v}{\partial x^2} = 0 \quad \text{at} \quad x = \ell/2 \tag{10.134}$$

we obtain

$$A = -\frac{m^2}{\kappa^4}\{(\frac{\kappa\ell}{2}\sin\frac{\kappa\ell}{2} + \cos\frac{\kappa\ell}{2}) + $$
$$+ 2[\sin\frac{\kappa\ell}{2}\sin(\frac{\kappa\ell}{2} - \kappa x_1) + \cos\frac{\kappa\ell}{2}\cos(\frac{\kappa\ell}{2} -$$
$$- \kappa x_1)] - 2(\frac{\kappa\ell}{2} - \kappa x_1)\sin\frac{\kappa\ell}{2} + \cos\frac{\kappa\ell}{2}]\} \tag{10.135}$$

$$B = -A\cos\kappa x_1 + \frac{m^2}{2\kappa^2}x_1^2$$

$$(\frac{\kappa\ell}{2}\cos\frac{\kappa\ell}{2} - \sin\frac{\kappa\ell}{2})$$
$$+ 2[\sin(\frac{\kappa\ell}{2} - \kappa x_1)\cos\frac{\kappa\ell}{2} -$$
$$- \cos(\frac{\kappa\ell}{2} - \kappa x_1)\sin\frac{\kappa\ell}{2}] -$$
$$- 2[(\frac{\kappa\ell}{2} - \kappa x_1)\cos\frac{\kappa\ell}{2} - \sin\frac{\kappa\ell}{2}] = 0 \tag{10.136}$$

From the two last equations we find

$$\frac{\kappa\ell}{2} = 7.551, \qquad \kappa x_1 = 2.918 \tag{10.137}$$

and obtain

$$A = 3.888\frac{m^2}{\kappa^4}, \qquad B = 8.042\frac{m^2}{\kappa^4} \tag{10.138}$$

$$v_{max} = 11.93\frac{m^2}{\kappa^4} \tag{10.139}$$

At the maximum deflection of opposite sign we obtain

$$\kappa x_2 = 4.27$$

which is then

$$= 2.476\frac{m^2}{\kappa^4}$$

By an anology with the above we obtain

$$P_{cr} = 1.97\frac{m}{\sqrt{A}}EI - 5.37\frac{\sqrt[4]{A}}{\sqrt{m}}p +$$
$$+ \sqrt{28.85p^2\frac{\sqrt{A}}{m} + 6.00\,EAp\,\sqrt[4]{m^2A^7}} \tag{10.140}$$

From this we have

$$P_\ell = 4.11(1 - 1.135\sqrt[4]{p^8I/q^6A_w^3E^2})\sqrt[11]{p^2q^4I^3E^5A^2} \tag{10.141}$$

Considering (10.39), we now obtain

$$P_\ell = 4.286\sqrt[9]{q^4E^5I^3A_w^2/L_n^2} \tag{10.142}$$

10.15 Assymptotic Forms of Bulging of Underground Pipelines

Apparently the considered forms of deflection do not cover all possible types of bulging. However, the realization of different forms of deflection with a greater number of knots is generally a small possibility. This is because the limiting resistance of the surrounding soil may be reached only after finite transverse displacements.

Also, the difference among the critical forces gradually diminishes with an increase of the possible forms of bulging.

As an example, let us consider as a limiting form of deflection the following expression

$$v = A\,e^{-\lambda_1 x}(\cos\lambda_1 x + \sin\lambda_1 x) \tag{10.143}$$

diminishing to zero.

By applying the method of energy, according to (10.38), (10.43) to (10.47) where the integrals are used on half-infinite intervals, we then obtain

$$U = \frac{1}{2}EI\,[3A^2\lambda_1^3 - \frac{1}{2}\frac{P}{EI}A^2\lambda_1 + \frac{A_w}{IL_n}\cdot$$
$$\cdot\frac{1}{16}A^4\lambda_1^2 + 1.155\frac{q}{EI}\cdot\frac{A}{\lambda_1}] \tag{10.144}$$

From the derivatives

$$\frac{\partial U}{\partial A} = 0, \quad \frac{\partial U}{\partial \lambda_1} = 0 \tag{10.145}$$

we obtain

$$\frac{P_{cr}}{EI} = 5.38\sqrt{\frac{q}{EI}}\cdot\frac{1}{A} + 0.182\frac{A_w}{L_nI}\sqrt[4]{\frac{q}{EI}}\sqrt[4]{A^7} \tag{10.146}$$

The lower critical force will then be

$$P_\ell = 4.306\sqrt[9]{q^4E^5I^3A_w^2/L_n^2} \tag{10.147}$$

which is somewhat higher (10.142) by using the approximate method.

This result indicates that forms 1-3, Figure 10.11, practically speaking, cover all cases of local bulging.

As another limiting case it is necessary to consider the general loss of stability, which is characterized by a uniform sine-shape configuration

$$v = A \sin \frac{\pi}{\ell} x \qquad (10.148)$$

The total energy of the pipeline at a loss of stability is

$$U = \frac{\pi}{2} EI \left[\frac{\pi^2}{2} \cdot \frac{A^2}{\ell^2} - \frac{\pi A_w}{2I} \cdot \frac{A}{\ell} + \frac{n^2 A_w}{16I} \cdot \frac{A^4}{\ell^3} + \frac{q}{A_w I} \cdot \frac{4}{\pi^2} A\ell\right] \qquad (10.149)$$

The first derivation (10.145) yields

$$\frac{P_{cr}}{EI} = \frac{\pi^2}{\ell^2} + \frac{\pi^2 A}{4I} \cdot \frac{A^2}{\ell^2} + \frac{q}{EI} \cdot \frac{4}{\pi^3} \cdot \frac{\ell^2}{A} \qquad (10.150)$$

The lower critical force as the minimum of (10.150) by A and ℓ is

$$P_\ell = 2.25 \sqrt[4]{q^2 E^2 I A_w} \qquad (10.151)$$

The corresponding values of amplitude, the length of the half-wave and the maximum bending stresses will be

$$A = 2\sqrt{I/A_w}; \quad \ell = 4.18 \sqrt[8]{E^2 I^3/q^2 A_w};$$

$$\sigma_{bend} = \frac{1.130}{S_m} \sqrt{E^2 I^2 q^2/A_w} \qquad (10.152)$$

This type of bulging is characterized by small deflections and stresses due to bending. However, for its initiation, the uniformly distributed initial deflections are required.

Therefore, at a loss of stability we may consider only the local forms of deflections of types 1-3, shown in Figure 10.11.

10.16 Stability of Pipeline at the Bends of Greater Radius

A pipeline may be bent into vertical and horizontal planes following the configuration of the terrain, if conditions of structural strength permit. If the plane of the bend coincides with the plane in which the bearing capacity of the backfill is at a minimum, the possibility of stability loss increases.

Let us first consider the equilibrium of a curvilinear element of a pipeline axis, which is under the action of transverse loading q, transverse forces Q, compressive forces $\bar{N}$ and a bending moment, Figure 10.12.

By using normal and tangential projections to the bent axis, we obtain

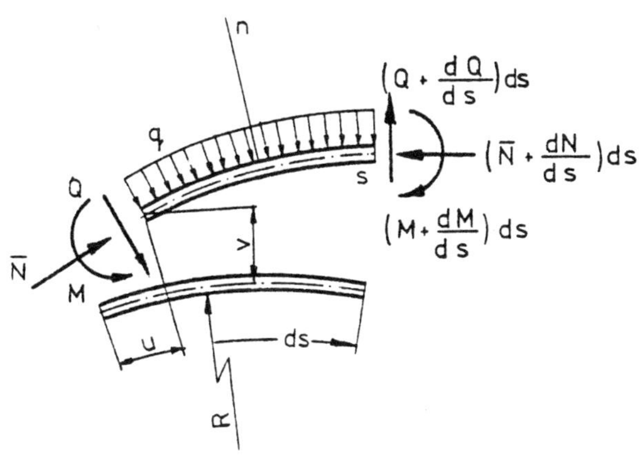

FIGURE 10.12 — Scheme of the equilibrium of curvilinear element of pipeline.

$$-\frac{d\bar{N}}{ds} + Q(\frac{1}{R} + \chi_1) = 0 \qquad (10.153)$$

$$\frac{dQ}{ds} + \bar{N}(\frac{1}{R} + \chi_1) - q = 0 \qquad (10.154)$$

where χ_1 is a change of curvature in a bent pipeline.

The equation of moments for an element under consideration is

$$-Q + \frac{dM}{ds} = 0 \qquad (10.155)$$

Let

$$\bar{N} = N + \delta N \qquad (10.156)$$

where the N-force is at the sections of pipeline holding the δN-changes in force in a bent section, by comparison with N.

For equations of equilibrium (10.153) and (10.154) we obtain

$$-\frac{d(\delta N)}{ds} + Q(\frac{1}{R} + \chi_1) = 0 \qquad (10.157)$$

$$\frac{dQ}{ds} + (\delta N)(\frac{1}{R} + \chi_1) + N(\frac{1}{R} + \chi_1) - q = 0 \qquad (10.158)$$

For bends of pipelines, having a radius of free bending

$$\delta N \leq N \qquad (10.159)$$

and the equations of equilibrium will be

$$\frac{dQ}{ds} + N(\frac{1}{R} + \chi_1) - q = 0 \qquad (10.160)$$

$$-Q + \frac{dM}{ds} = 0 \qquad (10.161)$$

As follows from the scheme of deformation of the element of a pipeline, Fig. (10.12), the change of the curvature of pipeline is

$$\chi_1 = \frac{\partial^2 v}{\partial s^2} - \frac{1}{R}\frac{du}{ds} \tag{10.162}$$

where u is the displacement of the element under consideration.

For these displacements it is possible to introduce the following criteria

$$u \approx \omega v \approx v\frac{dv}{ds} \tag{10.163}$$

where ω-rotation of the cross-section at deformation and, therefore, the order of the 2nd term in equation will be

$$\frac{v}{R}, \frac{d^2v}{ds^2}$$

or the error due to neglect of the second term (10.162) is not greater than $\frac{v}{R}$ by comparison with 1.

Considering that

$$M = EI\chi_1 \tag{10.164}$$

according to Eq. (10.160) and (10.161) we have for the bend of a pipeline the following differential equation

$$EI\frac{d^4v}{ds^4} + \frac{EI}{R^2}\frac{d^2v}{ds^2} + N\frac{d^2v}{ds^2} = -q + \frac{N}{R} \tag{10.165}$$

Putting

$$k_1^2 = \frac{1}{R^2} + \frac{N}{EI} \tag{10.166}$$

$$m_1^2 = \frac{q}{EI} - \frac{N}{EIR} \tag{10.167}$$

We now obtain the differential equation (10.99) already considered for the straight section of pipeline, the solution of which gives the previous relation Eq. (10.100) for the shape of deflection.

Let us introduce several nondimensional parameters after the formulas

$$m = \mu\sqrt{q/EI} \tag{10.168}$$

$$A = \frac{a}{\sqrt[3]{q/EI}} \tag{10.169}$$

$$\frac{1}{R} = \theta\sqrt[3]{q/EI} \tag{10.170}$$

$$z = \frac{\sqrt{P/q \cdot A_W/I}}{\sqrt[3]{q/EI}} \tag{10.171}$$

$$P = \beta\sqrt[3]{q^2EI} \tag{10.172}$$

According to Equations (10.96), (10.107), (10.108), (10.111), (10.166) to (10.172), we find

$$\beta = 2.16\mu p a^{-\frac{1}{2}} - \theta^2 - 3.06\left(\frac{p}{q}\right)\mu^{-\frac{1}{2}}a^{1/4} + 2.28\, z\mu^{1/4}a^{7/8}\sqrt{1+1.8\frac{1}{z^2}\left(\frac{p}{q}\right)^2\mu^{-3/2}a^{-5/4}} \tag{10.173}$$

$$\mu = -1.08a^{-\frac{1}{2}}\theta + \sqrt{1.17\theta^2a^{-1}+(1+\theta^3)} \tag{10.174}$$

Equation (10.173) defines the totality of different states of equilibrium at bending due to rotation of the pipeline. The stable states of equilibrium correspond to values β which are on the right branch of the curve described by Equation (10.173).

The value of amplitude a, corresponding to the low critical force is then determined according to Eq. (10.173) and Eq. (10.174) by Equation

$$z\sqrt{1+1.8\frac{1}{z^2}\left(\frac{p}{q}\right)^2\mu^{-3/2}a^{-5/4}} - 0.54\mu^{\frac{3}{4}}a^{-11/2}\frac{1+0.708\left(\frac{p}{q}\right)\frac{a^{\frac{3}{4}}}{\mu^{3/2}}}{1+\frac{1.23\theta}{\mu\sqrt{a}}} - \frac{1.28}{z}\cdot\frac{a^{-5/4}\mu^{-3/2}\left(\frac{p}{q}\right)^2\left(1+\frac{1.73\theta}{\mu\sqrt{a}}\right)}{\left(1+\frac{1.234\theta}{\mu\sqrt{a}}\right)\sqrt{1+1.8\frac{1}{z^2}\left(\frac{p}{q}\right)^2\mu^{-3/2}a^{-5/4}}} = 0 \tag{10.175}$$

The total maximum curvature of pipeline at a loss of stability is

$$\frac{1}{\rho} = \frac{1}{R} + 2.62\mu\sqrt{a}\ \sqrt[3]{q/EI} \tag{10.176}$$

10.17 Stability of Pipeline at Single Span Crossings

We now determine the upper critical force as the branching point of the equilibrium states of weightless beam having elastically fixed supports.

The deflection of the pipeline in the soil is described by Equation (10.32) having a solution of

$$v = c_1e^{-\sqrt{r^2-k^2/4}\,(x)}\cos\sqrt{r^2+k^2/4}\,(x) + c_2e^{-\sqrt{r^2-k^2/4}\,(x)}\sin\sqrt{r^2+k^2/4}\,(x) \tag{10.177}$$

Deflection of the span is defined by the differential Equation

$$v^{(4)} + k^2 v^{II} = 0 \qquad (10.178)$$

having a solution of

$$v = D_1 \cos kx + D_2 \qquad (10.179)$$

At the boundaries of the sections the deflections of the pipelines and their derivatives up to the 3rd order inclusive, are equal. This permits the writing for constants C and D, of a system of four linear homogeneous equations, the solution of which gives for the parameter kb the following transcendent equation

$$\cot kb = - \frac{1 - \dfrac{k^2}{2r^2}}{\dfrac{k}{r}\sqrt{1 - \dfrac{k^2}{4r^2}}} \qquad (10.180)$$

The upper critical force we find by using one of the formulas

$$P_u = k_1 P_E = k_1 \frac{4\pi^2}{b^2} EI \qquad (10.181)$$

$$P_u = k_2 P_U^* = k_2 2\sqrt{k_0 EID} \qquad (10.182)$$

where P_E and P_U^* are the Euler's force for an embedded pipeline of length b and upper critical force for buried pipeline. k_1 and k_2 are coefficients shown in Figure 10.13 as a function of parameter rb.

The loss of stability at the crossing of values of initial compressive stresses smaller than the upper critical stress, is possible for crossings with small spans and shallow arcs and depends on overcoming the weight of the pipeline and to resistance of the backfill at their banks, Figure 10.14.

The low critical force we find by the Ritz method, using the basic assumptions stated in Section 10.8.

For the deflection of the axis of pipeline at bulging, we use the function (10.48). A total energy of the system will be

$$U = \frac{1}{2} EI\pi\Big[(2A^2\lambda^2 + \tfrac{1}{2}A^4\lambda^5) - \frac{P}{EI}(\tfrac{1}{2}A^2\lambda + \tfrac{3}{32}A^4\lambda^3) + \frac{A_w \pi}{L_n I} \cdot \frac{1}{16}(A^4\lambda^2 + \tfrac{3}{8}A^6\lambda^4) + \frac{q}{EI} \cdot \frac{A}{\lambda} - \frac{q_2}{EI} \cdot \frac{A}{\lambda} \times (b + \tfrac{1}{\lambda}\sin\lambda b)\Big] + \frac{P^2 L_n}{2EA} \qquad (10.183)$$

where

$$q_2 = q - q_1$$

and q_1 is own weight of the span.

By using Equations (10.51), designations (10.55) to (10.57) and $\xi = b\lambda$, we obtain the equation for the determination of the parameters of the equilibrium positions of deflection.

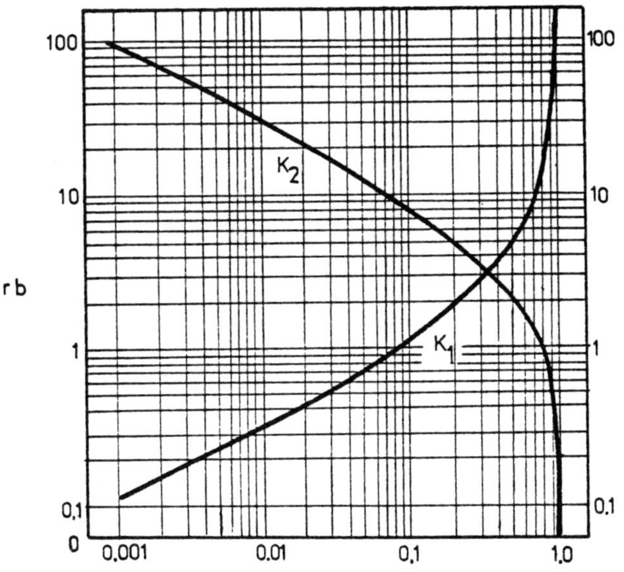

$$K_2 = \frac{P}{2\sqrt{K_0 EID}} \qquad K_1 = \frac{P}{P_3}$$

FIGURE 10.13 — Change of upper critical forces at a loss of stability of single span pipeline in function of coefficient rb.

$$\psi^2 x(10 + \tfrac{9}{2}x^2) - \beta\psi x(\tfrac{3}{2} + \tfrac{21}{32}x^2) + \frac{n}{16}x^3(6 + \tfrac{15}{4}x^2) - \frac{q_2}{q}(\xi + \xi\cos\xi) = 0 \qquad (10.184)$$

$$q^3 x(8 + 3x^2) - 3 - \frac{q_2}{q}(2\xi\cos\xi - 3\frac{\sin\xi}{\pi} - \xi) = 0 \qquad (10.185)$$

From (10.184), (10.185), we obtain the coefficients for the low critical force and the corresponding parameters of the equilibrium shape of deflection

$$\beta = Bn^{2/9} \qquad (10.186)$$

$$x = \frac{X}{\sqrt[3]{n}} \qquad (10.187)$$

$$\psi = \frac{\phi}{\sqrt[3]{x}} \qquad (10.188)$$

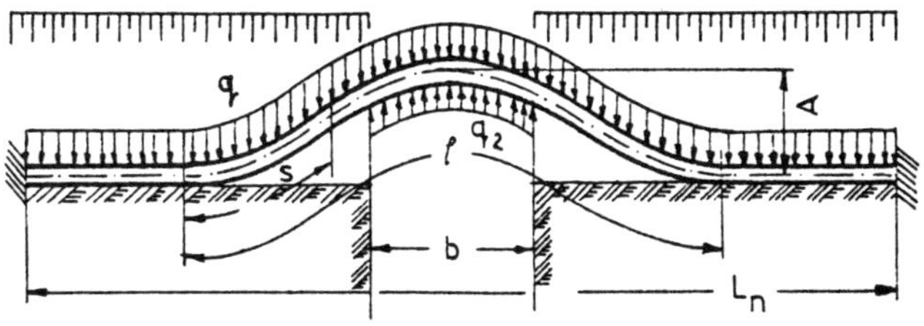

FIGURE 10.14 — Bending of a single span pipeline at a loss of stability.

where the coefficients B, X, ϕ are shown in a function of $\frac{q^2}{q}$ and parameter ξ in Figures 10.15 and 10.16.

The value L_n, entering into Equation (10.55) for n, we determine by using equations (10.65), (10.66). The value obtained in all ranges of coefficients $\frac{q_2}{q}$ differs from the corresponding values for an underground section less than 10%.

This permits, irrespective of the ratio $\frac{q_2}{q}$ with the exactness up to 2% for P_ℓ the use for the L_e Equations (10.76).

With an increase of the span the upper critical force is diminished and simultaneously the lower critical force initially diminishes and further increases. Therefore, in cases where the upper critical force is smaller than the lower value, this value only may serve as the basis for stability design.

10.18 Stability of Underground Pipeline at Bends of Small Radius

In some locations, depending on the terrain configuration, the pipelines are built with small radii or sharp bends.

Let us consider the stability of buried pipeline at a right-angle bend. The bulging of the pipeline on the surface is shown in Figure 10.17.

We assume that the change of the compressive force with an increase of the deflections is defined by formula (10.39).

The potential energy of deformation may be expressed as follows

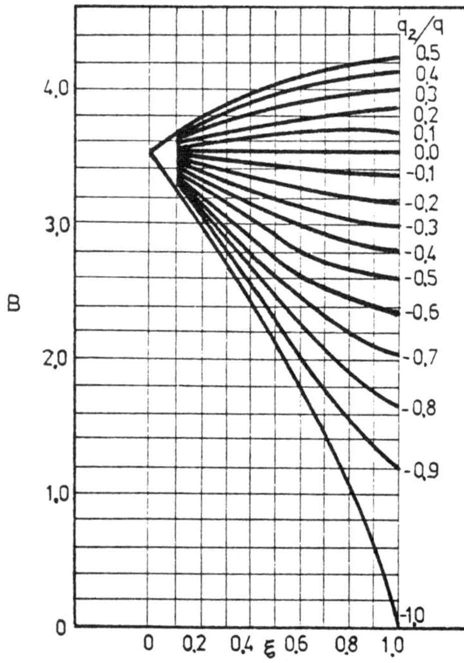

FIGURE 10.15 — Diagram for the determination of low critical force at loss of stability of single span crossing.

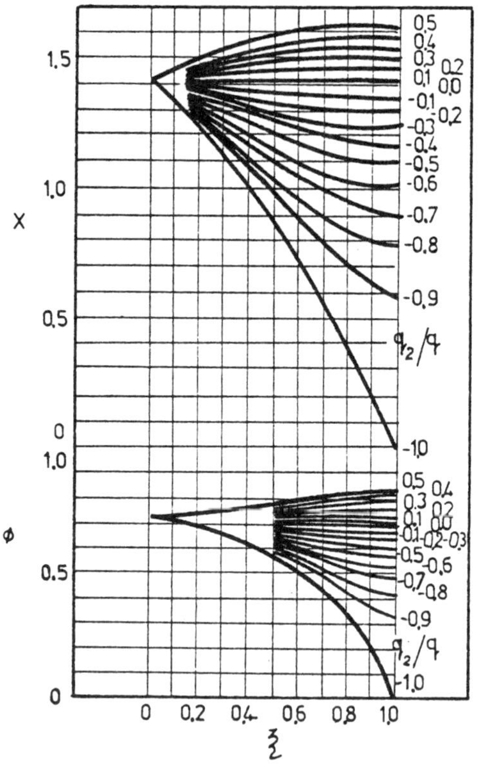

FIGURE 10.16 — Diagrams for the determination of parameters of equilibrium shape of deflection at loss of stability of single span crossing.

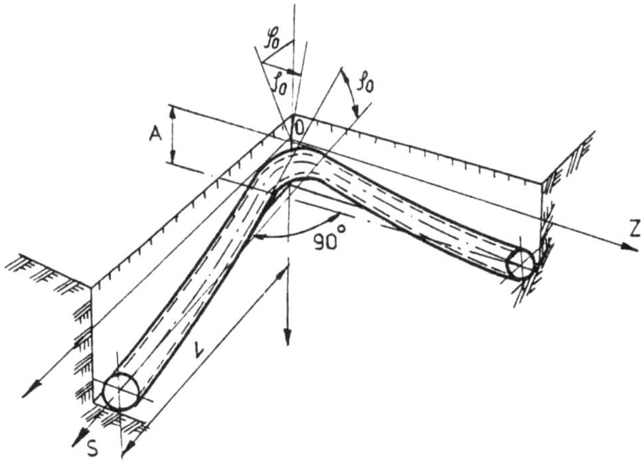

FIGURE 10.17 — Deflection of bend at a loss of stability.

$$U = U_{1b} + U_{2b} + U_t + U_c \tag{10.189}$$

where

U_{1b} = bending deformation energy in the plane of bulging

U_{2b} = energy of bending deformation of branches in the plane of bend

U_t = energy of torsional deformation of branches

U_c = energy of compressive deformation

The bending deformation in the plane of bend and torsion are due to the bending deformation in the vertical plane.

Twisting of the branch of the pipeline at an angle ψ_0 in the vertical plane results in a twist at the same angle ψ_0 and a twist of the bend at an angle $\psi_0^2/2$.

The potential bending energy in the vertical plane and deformation of the compressive energy are defined by relations (10.43) and (10.46).

The potential energy of the deformation at a uniform twist of the bend at the angle ψ_0 is

$$U_t = \frac{1}{2} GJ_0 \frac{\psi_0^2}{\ell_1} \tag{10.190}$$

where

J_0 = polar moment of inertia of cross-section

G = shear modulus

ℓ_1 = reduced length of pipeline, participating in torsional deformation

Considering that

$$G = \frac{E}{2(1+\nu)}, \quad J_0 = 2J$$

we obtain

$$U_t = \frac{EJ}{2(1+\nu)}, \frac{\psi_0^2}{\ell_1} \tag{10.191}$$

where

ν = Poisson's ratio

For the determination of the bending deformation energy the branch of pipelines is assumed at fixed along the length ℓ.

According to the Maxwell-Mohr formula, we find

$$U_{2b} = \frac{\psi_0^4 EI}{2\ell} \tag{10.192}$$

The deflection of the pipeline in the vertical plane we represent by the function

$$v = A \sin \lambda s$$

$$(\text{where} \quad \lambda = \frac{\pi}{2\ell}) \tag{10.193}$$

satisfying the geometric boundary conditions of the problem

$$v = A, \; v' = 0 \quad \text{at} \quad s = \ell \tag{10.194}$$

The total energy of the system at bulging will be

$$U = \frac{EI}{2} \pi\{[\frac{1}{4} A^2\lambda^3 + (\frac{1}{16}+\frac{2}{\pi^2})A^4\lambda^5] - \frac{P}{EI}[\frac{1}{4} A^2\lambda + \frac{3}{64} A^4\lambda^3] + \frac{1}{\pi(1+\nu)\ell_1} A^2\lambda^2 + \frac{A_w\pi}{8L_nI}[\frac{1}{8} A^4\lambda^2 + \frac{3}{64} A^6\lambda^4] + \frac{2}{\pi} \cdot \frac{q}{EI} \cdot \frac{A}{\lambda}\} + \frac{P^2L_n^2}{2EA_w} \tag{10.195}$$

From Equations (10.51) and (10.52), we obtain

$$\lambda^3 x(\frac{3}{4} + \frac{5}{16} x^2 + \frac{10}{\pi^2} x^2) - \frac{P}{EI} \lambda x(\frac{1}{4} + \frac{9}{64} x^2) + \frac{4}{\pi(1+\nu)\ell_1} \lambda^2 x + \frac{A_w\pi}{4L_nI} x^3(\frac{1}{4} + \frac{3}{16} x^2) - \frac{2}{\pi} \cdot \frac{q}{EI} = 0 \tag{10.196}$$

$$\lambda^2 x(\frac{1}{2} + \frac{1}{4}x^2 + \frac{8}{\pi^2} x^2) - \frac{P}{EI}(\frac{1}{2} + \frac{3}{16} x^2) + \frac{4}{\pi(1+\nu)\ell_1}\lambda^2 x + \frac{A_w\pi}{4L_nI} x^3(\frac{1}{2} + \frac{9}{32} x^2) + \frac{2}{\pi} \frac{q}{EI} = 0 \tag{10.197}$$

Using (10.55) to (10.57) and designating

$$\varepsilon = \frac{\ell_1}{\pi} \sqrt[3]{q/EI} \tag{10.198}$$

we obtain finally

$$\psi^3 x + 2.67\psi^3 x^3 + \frac{0.31}{\varepsilon} \psi^2 x - 1.91 = 0 \tag{10.199}$$

$$\beta = \psi^2(\frac{5}{3} + 4.85x^2) + 0.25\frac{nx^2}{\psi} + \frac{0.83}{\varepsilon} \psi \tag{10.200}$$

The given equations define the system of the quilibrium positions at bulging of the pipeline bend.

The value of ϵ has a relatively weak influence on the critical parameter β. From (10.65) and (10.66), for the reduced length L_n and similar relations for ℓ_1, we find

$$\ell_1 = 0.3 L_n \tag{10.201}$$

$$L_n = 0.27 \frac{P}{p} \tag{10.202}$$

The minimum value of $\beta_{cr,\ell}$ exists when $\epsilon = \infty$, and according to (10.199), (10.200) and (10.202) is

$$\beta_{cr,\ell} \simeq 2.47 n^{2/9} \tag{10.203}$$

$$P_{cr,\ell} \simeq 3.25 \sqrt[11]{p^2q^4E^5I^3A_w^2} \tag{10.204}$$

10.19 Stability of Pipelines at Branching

Let us consider the stability of pipeline, having a T-type branch. The bulging scheme of the pipeline is shown in Figure 10.18.

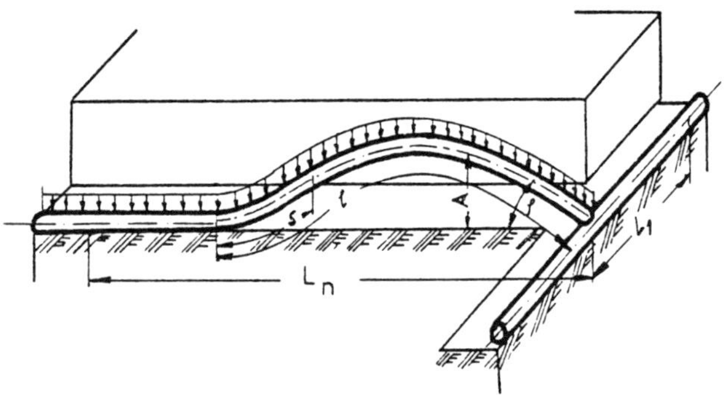

FIGURE 10.18 — Scheme of branch type pipeline at loss of stability.

By an anology of the previous considerations, in the potential energy of deformation we add the twisting work of the adjoining section of pipeline on the angle ψ_0.

The work is defined by the expression

$$U_t = \frac{EI}{2(1+\nu)} \cdot \frac{\psi_0^2}{\ell_1} \qquad (10.205)$$

for the twist of pipeline and

$$U_t = \frac{EI_1}{(1+\nu)} \cdot \frac{\psi_0^2}{\ell_1} \qquad (10.206)$$

for the branch, where

I_1 = moment of inertia of the cross-section of an adjoining cross-section of pipeline under twist

ℓ_1 = reduced length of pipeline, participating in twisting deformation

The deflected axis of the pipeline we represent by the function

$$v = A[(\sin\lambda s - \lambda s) + \frac{1}{2}\pi(1-\cos\lambda s)] \qquad (10.207)$$

where

$$\lambda = \frac{\pi}{\ell}$$

satisfying the boundary conditions at the ends of the bent section

$$v = \frac{dv}{ds} = 0 \quad \text{at} \quad s = 0$$

$$v = 0 \quad \text{at} \quad s = \ell \qquad (10.208)$$

The total energy of pipeline according to (10.189) is

$$U = \frac{1}{2} EI\pi[(1.734A^2\lambda^3 + 1.591A^4\lambda^5) - \frac{P}{EI}(0.734A^2\lambda + 0.333A^4\lambda^3) + \frac{1}{4} \cdot \frac{A_w\pi}{IL_n}(0.538A^4\lambda^2 + 0.488A^6\lambda^4) + \frac{4}{(1+\nu)} \cdot \frac{1}{\pi\ell_1} A^2\lambda^2 + 4\frac{q}{EI\pi} \cdot \frac{A}{\lambda}] + \frac{p^2L_n}{2EA_w} \qquad (10.209)$$

We designate

$$\varepsilon = \frac{\ell_1}{\pi} \sqrt[3]{q/EI} \qquad (10.210)$$

for the turn of a pipeline and

$$\varepsilon_1 = \frac{\ell_1}{2\pi} \cdot \frac{I_1}{I} \sqrt[3]{q/EI} \qquad (10.211)$$

for its branching.

Using the Ritz method and considering (10.55) to (10.57), we obtain

$$\psi^3 x(5.20+7.95x^2) - \beta\psi x(0.734+0.988x^2) + \frac{1}{4}nx^3(1.077 + 1.95x^2) + \frac{8}{(1+\nu)\pi^2\varepsilon}\psi^2 x + \frac{4}{\pi} = 0 \qquad (10.212)$$

$$\psi^3 x(3.47+6.36x^2) - \beta\psi x(1.47+1.33x^2) + \frac{1}{4}nx^3(2.15 + 2.93x^2) + \frac{8}{(1+\nu)\pi^2\varepsilon}\psi^2 x + \frac{4}{\pi} = 0 \qquad (10.213)$$

An energy analysis of the corresponding equilibrium positions leads to the result obtained for the straight sections of the pipeline.

Equations (10.65) and (10.66) yield

$$L_n \simeq 0.3 \frac{P}{p} \qquad (10.214)$$

For an $\epsilon = \infty$ parameter of the lower critical loading, β is

$$\beta_{cr,\ell} = 3.05\, n^{2/9} \qquad (10.215)$$

According to (10.55), (10.57) and (10.214) we obtain

$$P_{cr,\ell} = 3.80 \sqrt[11]{p^2q^4E^5I^3A_w^2} \qquad (10.216)$$

This value of the critical force is greater than (10.204). Therefore, for the twisting of a pipeline bend, it is more probable that the shape of deflection will be that given in the previous section.

10.20 Stability of Parts of a Pipeline on the Banks of a Crossing

The pattern of loss of stability of such sections is similar to the bulging of a T-shape branch pipeline, Figure 10.18.

Instead of the torsional work of an adjoining branch in the equation for total energy, it is necessary to include the work for the span bending at an angle ψ_0.

By an analogy with (10.192), we thus obtain

$$U_{2b} = \frac{2EI}{b} \psi_0^2 \qquad (10.217)$$

where

b = span.

According to (10.210) in this case

$$\varepsilon_2 = \frac{b}{4\pi(1+\nu)} \sqrt[3]{q/EI} \qquad (10.218)$$

Because by the loss of stability, both branches are affected of a pipeline which are on both banks of a crossing, then by an analogy with the straight section, we obtain

$$L_n \simeq 0.5 \frac{P}{p} \qquad (10.219)$$

10.21 Recommendations for Stability Design of Buried Pipelines

The basic purpose of a stability design of pipelines, which are under initial compressive stresses, is the choice of such alternative locations which will guarantee the maximum safety.

The loss of pipeline stability results in the development of dangerous stresses in the walls of the pipe under bending and the destruction of backfill. For this reason the stability analysis includes the determination together with any critical compressive forces, as well as corresponding transverse deformations of the pipeline and flexural stresses.

The results obtained in the previous Section considering the stability analysis permits their application onto different alternate locations of pipeline. An analysis of buried pipeline is carried out, considering their stability, longitudinal bending deformations and destruction of the backfill.

Some design formulas and areas of their applications for the different alternate solutions of pipeline construction are shown in Table Nos. 10.3 and 10.4.

The basic data for the stability analysis of a pipeline under special conditions constitute information regarding the characteristics of the soils along the pipeline. Such data may be obtained from field and laboratory tests during the planning stage of pipeline construction.

10.22 Soil Resistance Against Longitudinal Displacements of Pipelines

According to the results of the previous Section, we introduce as resistance against longitudinal displacements of pipeline, the resistance against shear considering the following cases:

1. For sandy and dry clayey grounds

$$P = fq_a \qquad (10.220)$$

where f is the coefficient of the friction at the wall, depending on the type of isolation and properties of the soil.

For bituminous and polymer types of isolations, the coefficient of the friction is

$$f = 0.8 \tan \psi_0 \qquad (10.221)$$

where ψ_0 is the angle of internal friction.

2. For clayey soils

$$p = \pi Dc + fq_a \qquad (10.222)$$

where

f = $\tan \psi_0$

and

c = cohesion of backfill material in lbs/in²

The loading q_a is the summary of radial components of soil pressure on the surface of a pipeline

$$q_n = \int_0^{2\pi} \sigma_n r d\psi \qquad (10.223)$$

The magnitude of this load is defined by the physical properties of the soil and the following structural parameters of pipeline: the width of the trench at the top edge of the pipeline (B_t), the depth (H), the diameter of pipe (D) and also the ratios

$$\left(\frac{H}{B_t}, \frac{H}{D}, \frac{B_t}{d}\right)$$

In practice, the value of q_n may be substituted by the forces applied along the top and bottom generatrix of the pipe.

A solution by Voelmy [10.9] for the conditions $\frac{H}{D} \leqslant 1$ indicates that for rigid pipelines these loads are practically equal to the weight of the backfill having a width equal to the diameter of pipe.

However, the maximum values of radial stresses at their vertical and horizontal diameters [10.10, 10.11] indicates that for a rigid pipeline, the summary of the vertical components of soil pressure may be substituted by a summary of double vertical soil pressure and the weight of a pipe with liquid.

For rigid pipelines placed in narrow trenches, when

$$\left(\frac{B_t}{D} < 2 \text{ to } 3\right)$$

formulas by Marsten [10.11], Emelianov [10.12], and Klein [10.10], are most often used, or

$$G_{1B} = C_t \gamma_0 B_t^2 \qquad (10.224)$$

$$G_{1B} = K_t \gamma_0 H B_t \qquad (10.225)$$

where the coefficients K_t and C_t are determined by the properties of backfill and base and their relation $(\frac{H}{D})$.

The minimum balue of loading onto pipeline may be estimated after the following formula

$$G_{2B} = \gamma_0 HD \qquad (10.226)$$

This permits the acceptance of the loading, determined by the formula (10.226) as a design load for rigid pipelines under conditions such as "narrow" and "wide" trenches.

According to the classification of Klein [10.10], a pipeline is considered as "flexible" if it satisfies the following condition:

TABLE 10.3 — Formulas for Different Alternatives of Pipe-Laying.

	Alternative	Configuration at a Loss of Stability	Design Formulas: Compressive Force P	Design Formulas: Maximum Deflection v	Design Formulas: Minimum Radius of Curvature
1	Straight sections in trenches. Submarine crossings. Open pipeline in a trench. Straight sections in embankments.	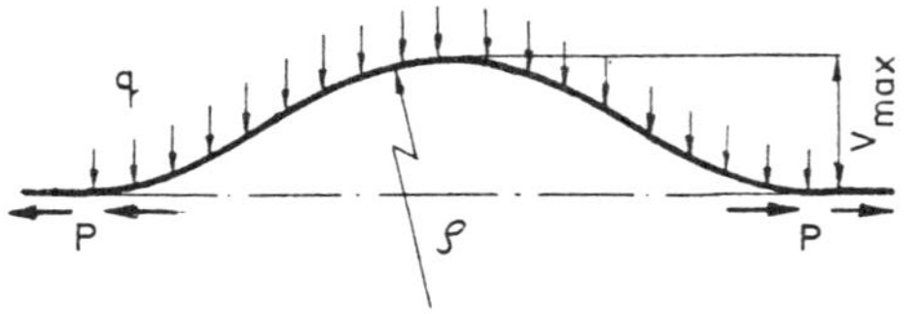 In a vertical plane	$4.09\ \sqrt[11]{p^2q^4A_w^2E^5I^3}$	$2.19\ \sqrt[11]{\dfrac{q^3EI^5}{p^4A_w^4}}$	$0.476\ \sqrt[11]{\dfrac{p^2A_w^2I^3E^5}{q^7}}$
2	Straight sections in embankments, weak soils. Sections of pipelines laid on the ground.	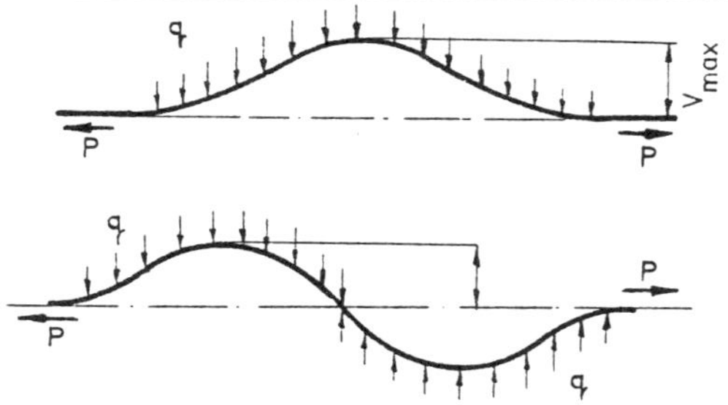 In a horizontal plane	$3.97\ \sqrt[11]{p^2q^4A_w^2E^5I^3}$	$2.19\ \sqrt[11]{\dfrac{q^3EI^5}{p^4A_w^4}}$	$0.476\ \sqrt[11]{\dfrac{p^2A_w^2I^3E^5}{q^7}}$
3	Bends at free bending: in vertical plane in a trench and embankment; in a horizontal plane in weak grounds, embankments; sections laid on the ground.	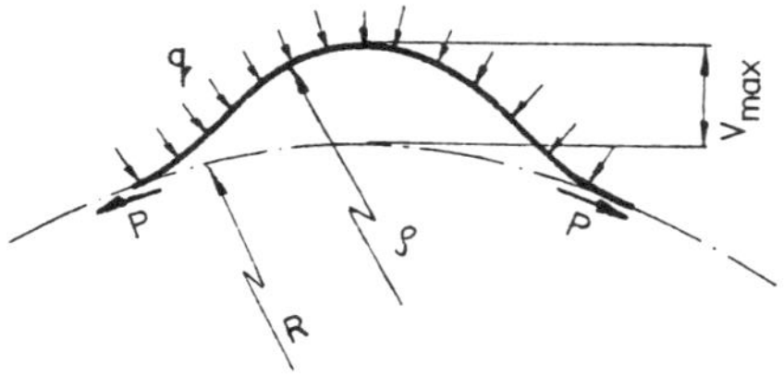 In the plane of turning	$\beta\sqrt[3]{q^2EI}$	$\dfrac{v_{max}\sqrt[3]{\frac{q}{EI}}}{\sqrt[3]{\frac{q}{EI}}}$	$\left(\dfrac{1}{\sqrt[3]{\frac{q}{EI}}}\right) : \left(\dfrac{1}{\rho\sqrt[3]{\frac{q}{EI}}}\right)$
			where coefficients β, $\left(v_{max}\sqrt[3]{\frac{q}{EI}}\right)$ and $\dfrac{1}{\rho\sqrt[3]{\frac{q}{EI}}}$ are found as functions of parameters. $z = \left(\sqrt{\dfrac{pA_w}{q\ell}}\right) : \left(\sqrt[3]{\dfrac{q}{EI}}\right)$; $\theta = \dfrac{1}{R\sqrt[3]{\frac{q}{EI}}}$		

TABLE 10.4 — Formulas for Passing and Turning of Pipelines.

	Alternatives	Configuration at a Loss of Stability	Design Formulas
1	Passing of the pipeline over obstructions.	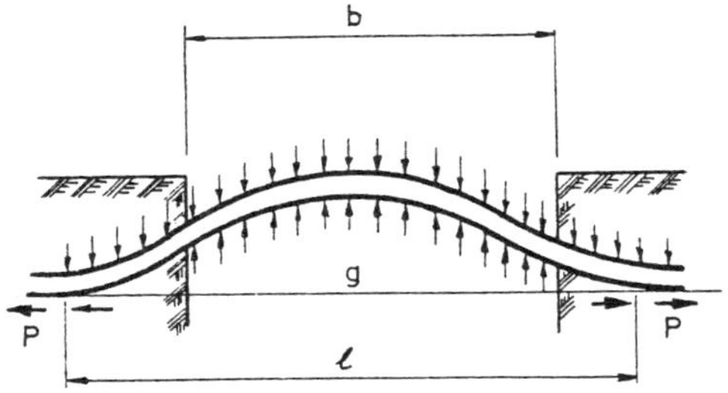	$P_{cr,u} = k_1 \frac{4\pi^2}{b^2} EI$ or $P_{cr,u} = k_2 2 \sqrt{K_0 DEI}$ k_1 and k_2 are obtained from the diagram in Figure 10.13 in a function from the coefficients $rb = b\sqrt[4]{\frac{k_0 D}{4EI}}$ $P_{cr,\ell} = Bn^{2/9} \sqrt[3]{q^2 EI}$; $n = \frac{\pi EA}{qL_n}$; $L_n = \frac{P_{cr,\ell}}{2p}$; Coefficient B is found from the diagram in Figure 10.15 in the functions of $\frac{q_2}{q} = \frac{q-q_1}{q}$ and $\xi = \frac{b}{\ell}$ where $\ell = \frac{\pi\sqrt[3]{x}}{\phi\sqrt[3]{\frac{q}{EI}}}$; $x = \frac{X}{\sqrt[3]{n}}$ and the coefficients ϕ and x are determined by Figure 10.16. Calculation is accomplished by the successive approximation. For the design the smallest value from P_1cru and Pcr is used.
2	Sections of pipelines in a trench, on the bank of a crossing.	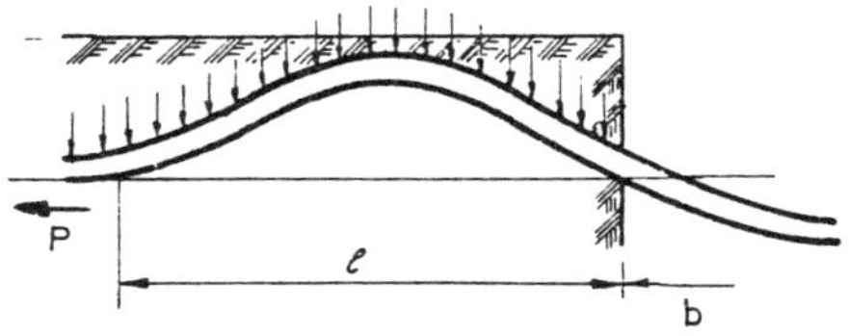	$P_{cr,\ell} = \beta\sqrt[3]{q^2 EI}$ β is found in the function of $n = \frac{\pi EA}{qL_n}$, where $L_n = \frac{P_{cr,\ell}}{2p}$, and $\varepsilon = 0.064b \sqrt[3]{\frac{q}{EI}}$
3	Turning of pipelines into a horizontal plane in a trench at a turning angle of 90°	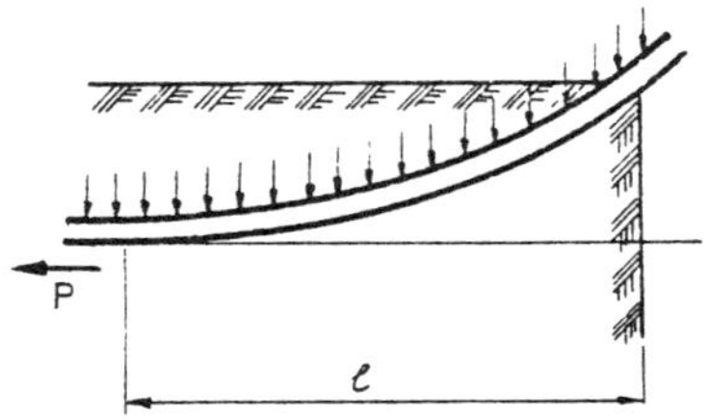	$P_{cr,\ell} = \beta\sqrt[3]{q^2 EI}$ β is found in the function of $n = \frac{\pi EI}{qL_n}$, where $L_n = 0.3\frac{P_{cr,\ell}}{p}$, and $\varepsilon = 0.1 L_n \sqrt[3]{\frac{q}{EI}}$

$$\frac{E}{E_s}\left(\frac{2t}{D}\right)^3 < 1 \qquad (10.227)$$

where E_s is the modulus of deformation of a backfill.

For flexible pipelines the distribution of radial components of pressure are more uniformly distributed with an increase of flexibility, where the pressure of the ground along the upper generatrix is diminishing and side pressures are increasing. The total vertical loading of a pipeline is smaller than that defined in Equation (10.226) and also smaller than the summary of the radial components of pressure.

These considerations permit the acceptance of Equation (10.229) as a design formula for the determination of a total summary of the radial components of soil pressure in practice.

The internal pressure producing the thrust of a backfill leads to an increase of the summary of radial components of the soil pressure.

The recommended value of

$$q_a = 0.8 \times 2\gamma_0 HD + 0.9q_1 \qquad (10.228)$$

where

0.8-0.9 = the coefficient of overloading on the specific weight of the soil and weight of pipeline

q_1 = the weight of a pipeline with liquid

γ_0 = the specific weight of the soil

10.23 The Resistance Against Transverse Displacements of Pipeline

The resistance against transverse displacements of buried pipeline is determined by the limit-bearing capacity of the backfill during "pulling-out" of the pipeline.

The limiting state of the backfill considering its bearing capacity in the function of the depth of a pipeline is reached either by irregular swelling of the soil on the surface, or due to the deformation of the soil around the pipeline.

In the first case, an irregular swelling prism in the backfill is formed which starts at the pipeline and ends at the ground surface.

In the second case, the region of limiting equilibrium does not reach the surface of the ground.

A general expression for the removal of pipeline from sandy soils may be used, according to the following formula

$$q = 0.9q_1 + 0.8\gamma_0 h_0^2 \tan^2\left(45^\circ - \frac{\psi_0}{2}\right)\tan\psi_0 + 0.8\gamma_0 HD \qquad (10.229)$$

where

q_1 = total weight of pipeline including liquid, per unit length

h_0 = depth of pipeline from ground surface to the axis of pipeline

8.0-0.9 = coefficients of overloading considering the weights of soil and pipeline

ψ_0 = angle of internal friction

For wet clayey backfills the formula is

$$q = 0.9q_1 + 0.8\left[\psi_0 h_c^2 \tan^2\left(45^\circ - \frac{\psi^n}{2}\right) \times \tan\psi_0 + \gamma_0 HD + 2h_c c_0\right] \qquad (10.230)$$

where

c_0 = cohesion

At a ratio of $h_c/D \geqslant 5$ it is necessary to check for deep anchorage, according to the formula

$$q = 0.9q_1 + 0.8 \times c_0 D \qquad (10.231)$$

where x should be determined from Table 10.5 [10.13].

From the values determined from Equations (10.230) and (10.231), the smallest is used for the design.

For pipelines in embankments at a calculation of failure in a vertical plane

$$q = 0.9q_1 + 0.8(\gamma_0 V_{pr} + S_{pr} c) \qquad (10.232)$$

where the volume and surface of the prism are determined according to Figure 10.19, and the angle ψp according to Table 10.6.

TABLE 10.5 — Coefficient × in the Function of an Angle of Internal Friction.

Angle of Internal Friction ψ_0 in Degrees	5	10	15	20	25	30	35
Coefficient ×	12	16.5	30	53	97	192	428

TABLE 10.6 — Design Angles of Soil's Prism When a Pipeline is Removed.

Soil Types	Design Angle of a Prism	
	<1.70	>1.70
Sand	0.5 ψ_0	0.8 ψ_0
Clay	0.4 ψ_0	0.6 ψ_0

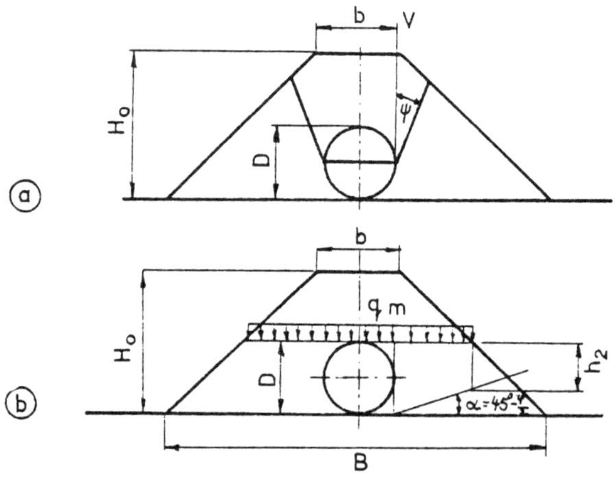

FIGURE 10.19 — The determination of resistance of an embankment against transverse displacements of pipeline at a loss of stability on (a) vertical and (b) horizontal planes.

10.24 Determination of Compressive Stresses in a Pipeline

A method of determining compressive stresses in a pipeline is given in references [10.14, 10.15]. Here, we consider an analysis of the compressive forces due to positive temperature changes. The original temperature change causing compressive stresses in a pipeline is determined by the difference between the temperature of the pipe in service t_s and during the laying of the pipe in a trench t_ℓ. The temperature of the transported liquid or gas is determined by thermal analysis. The temperature of the wall of the pipe, as shown in [10.16] may be used equally to the transported product.

During the laying of the pipeline, the temperature is equal to the temperature of the welded section during backfitting. This temperature depends on the season of the year, however, it is always higher than the air temperature, as shown in Figures 10.20 and 10.21. [10.17] Even in the morning and during the night the temperature of a pipeline is 3 to 5 degrees higher than the temperature of the surrounding air. By analogy, in winter, the temperature of a pipeline may be higher by a few degrees than the air temperature.

The value of this increase changes with the seasons, the intensity of solar radiation and the type of isolation. To increase the factor of safety we consider the temperature of a pipeline during laying, which is equal to the temperature of the air.

In cases where the working temperature of a pipeline is determined by the temperature of the air (during laying of the pipeline above the ground or during repair)

$$t_p = t_0 + \delta t_{max} \tag{10.233}$$

where

t_0 = temperature of the air in F°

δt_{max} = maximum change of the temperature of the pipe to the temperature of the air

According to Figures 10.20 and 10.21 we use δt_{max} = 72° F.

10.25 The Influence of Internal Pressure

The internal pressure caused in the wall of a pipe in the longitudinal direction of the tensile stresses

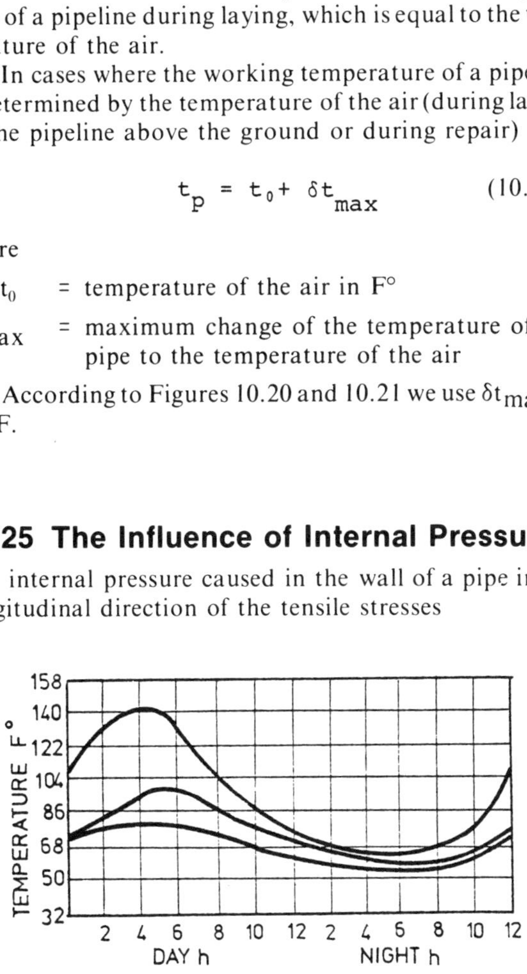

FIGURE 10.20 — Changes of temperature of above the ground pipeline during a 24-hour time limit (1) Top edge of the pipe; (2) bottom edge of the pipe; (3) temperature of the air.

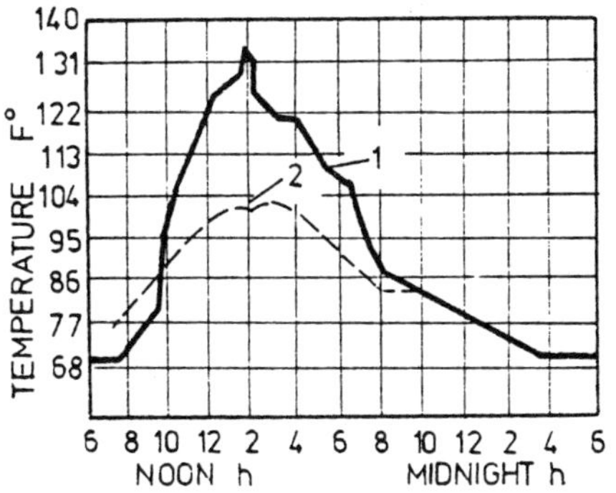

FIGURE 10.21 — Change of temperature of empty pipeline above the ground during a 24-hour time limit (1) Temperature of the pipe; (2) temperature of the air.

$$\sigma_t = \nu \frac{p_i D}{2t} \tag{10.234}$$

where

ν = Poisson's ratio

p_i = internal pressure

D = diameter of pipeline

t = thickness of the wall

In addition, during the bending of the pipeline, which is under internal pressure, there originates a transverse distributed force

$$q_t = \frac{P_i A_i}{\rho} = (p_i A_i)\frac{d^2v}{dx^2} \tag{10.235}$$

where

A_i = cross-sectional area of the pipe opening

ρ = radius of curvature of bent pipeline

Therefore, the influence of internal pressure on the bending, according to Equations (10.234) and (10.235) is equivalent to the axial compression

$$P = (1-2\nu)p_i A_i \tag{10.236}$$

The action of this force diminishes with an increase of deflection due to tensile forces, originating under the resistance of the soil to the longitudinal displacements of a pipeline.

Therefore, the bending of the active pipeline is produced by the force

$$P = \alpha E A_w \delta t + (1-2\nu)p_i A_i \tag{10.237}$$

10.26 Design of Pipelines for Stability

The condition of stability of a pipeline is defined by the formula

$$1.1\,P \leq P_\ell \tag{10.238}$$

where

p = compressive force, defined by Equation (10.237)

1.1 = the safety coefficient

For the pipeline at straight sections and bends, having radii of free bending, we may use

$$1.1\ P > P_{\ell} \qquad (10.239)$$

In the case where the after-critical state of the maximum amplitude of deflection is smaller and the minimum radius of curvature is of greater limit

$$v_{max} < [v_{max}] \qquad (10.240)$$

$$\rho_{min} > [\rho] \qquad (10.241)$$

where

$[v_{max}]$ = limiting permissible transverse displacement of the pipeline

$[\rho]$ = limiting permissible radius of the free bending of pipeline equal to

$$[\rho] = \frac{ED_0}{2\sigma_{all} k_i} \qquad (10.242)$$

where

E = modulus of elasticity

D_0 = outside diameter of the pipe

σ_{all} = allowable stresses of pipe material equal to the minimum value of yield point

k = coefficient equal to 0.4 — 0.5

If

$$P > P_{\ell} \qquad (10.243)$$

and conditions (10.241) and (10.242) are satisfied, then as a critical force we may consider the compressive force, found from the following conditions:

$$\rho = 1.1[\rho] \qquad (10.244)$$

after formula

$$P = \beta \sqrt[3]{q^2 EI} \qquad (10.245)$$

where

$$\beta = 2.16\mu\ a^{-\frac{1}{2}} - 3.06\left(\frac{p}{q}\right)\mu^{-\frac{1}{2}} + $$
$$+ 2.28 z \mu^{\frac{1}{4}} a^{7/8} \sqrt{1 + 1.8\ \frac{1}{z^2}\left(\frac{p}{q}\right)^2 \mu^{-3/2}\ a^{-5/4}} \qquad (10.246)$$

$$\mu = \sqrt{\frac{\bar{x}}{\bar{x} + 5.669}} \qquad (10.247)$$

$$a = \frac{(\bar{x}\ 2.83\theta)^2}{6.82} - 1.17\theta^2 \qquad (10.248)$$

$$\bar{x} = \frac{1}{\rho \sqrt[3]{q/EI}} - \theta \qquad (10.249)$$

$$\theta = \frac{1}{R\sqrt[3]{q/EI}} \qquad (10.250)$$

$$z = \left(\sqrt{\frac{p}{q} \cdot \frac{E}{I}}\right) : \left(\sqrt[3]{q/EI}\right) \qquad (10.251)$$

For the pipeline at the straight sections, in the above formulas

$$\theta = 0 \text{ and } \mu = 1$$

APPENDIX A

Numerical Example No. 1

A pipeline having O.D. = 24 in and a wall thickness t = $\frac{3}{8}$ in is in an open trench. The moment of inertia of a cross-section I = 1,942 in^4, and a cross-sectional area of the wall A_W = 27.82 in^2. Own weight of the pipeline is q_1 = 7.88 lb/in. The soil is a sandy loam having an angle of internal friction of ψ_0 = 23°. Investigate the buckling stability of the pipeline:

1. Resistance of soil against longitudinal displacements of the pipeline using the formula (10.220)

$$p = q_1 \tan \psi_0 = 7.88 \times 0.425 = 3.35 \text{ lb/in}$$

2. Resistance against transverse displacements of the pipeline

$$q = 0.9q_1 = 0.9 \times 7.88 \times 7.88 = 7.09 \text{ lb/in}$$

3. Low critical force (Table 10.3) is

$$P_{cr,\ell} = 4.09 \sqrt[11]{p^2 q^4 A^2 E^5 I^3} =$$

$$= 4.09\sqrt{3.35^2 x 7.09^4 x 27.82^2 x (30x10^6)^5 x1942} =$$

$$= 375{,}462 \text{ lbs}$$

4. The corresponding overfall of temperature is

$$\delta t = \frac{P_{cr,\ell}}{\alpha EA} = \frac{375{,}462}{6.7x10^{-6} x30x10^6 x23.09} = 80.88^{\circ}F$$

5. Maximum deflection of the pipeline (Table 10.3)

$$V_{max} = 2.19\sqrt[11]{\frac{q^3EI^5}{p^4A^4}} =$$

$$= 2.19\sqrt[11]{\frac{7.09^3x30x10^6x1942^5}{3.35^4x27.82^4}} =$$

$$= 87.16 \text{ in}$$

6. Minimum radius of curvature, formula (Table 10.3)

$$\rho_{min} = 0.476\sqrt[11]{\frac{p^2A^2I^3E^5}{q^7}} =$$

$$= 0.476\sqrt[11]{\frac{3.35^2x27.82^2x1942^3x(30x10^6)^5}{7.09^7}} =$$

$$= 6{,}188 \text{ in} = 515.67 \text{ ft.}$$

7. The corresponding stress is

$$\sigma_{max} = \frac{DE}{2\rho_{min}} = \frac{24 \text{ x } 30 \text{ x } 10^6}{2 \text{ x } 6{,}188} =$$

$$= 58{,}177 \text{ psi}$$

At this high compressive stress the pipeline will buckle.

Numerical Example No. 2

A pipeline having O.D. = 12.75 in, a thickness of the wall t = 3/8 in, was laid at a depth of H = 3.5 ft from the ground surface.

The moment of inertia of cross-section is I = 279.33 in^4, the cross-sectional area of the pipe is A = 13.795 in^2, and the total weight of the pipeline including the liquid is q_1 = 0.78 lb/in. The pipeline is under an internal pressure of p_1 = 880 lb/in^2. The temperature overfall is t = 175° F. The soil is dry, clayey sand, having a density of γ = 0.056 lb/in^3. The angle of internal friction ψ_0 = 21°, the cohesion c = 0.

Investigate buckling stability of the pipeline.

1. Embedment of the pipeline in the ground against longitudinal displacements, formulas (10.220) and (10.228)

$$p = 0.8 \tan\psi_0(0.8x2\gamma HD+0.9q_1) = 0.8 \text{ x}$$

$$\text{x } 0.384(0.8x2x0.056x42x12.75+0.9x0.78) =$$

$$= 14.95 \text{ lb/in}$$

2. Resistance against transverse displacements of pipeline, formula (10.229) is

$$q = 0.9q_1 + 0.8\gamma h_0^2\tan^2(45-\frac{\psi}{2})\tan\psi + 0.8\gamma HD =$$

$$= 0.9 \text{ x } 0.78 + 0.8 \text{ x } 0.056 \text{ x } 48^2 \text{ x}$$

$$\text{x } \tan^2(45^o-\frac{21^o}{2}) \text{x } \tan 21^o + 0.8 \text{ x } 0.056 \text{ x}$$

$$\text{x } 42 \text{ x } 12.75 = 33.39 \text{ lb/in}$$

3. The low critical force (Table 10.3) is

$$P_{cr,\ell} = 4.09\sqrt[11]{p^2q^4A^2E^5I^3} =$$

$$= 4.09\sqrt[11]{14.95^2x33.39^4x13.795^2x(30x10^6)^5x279.33^3} =$$

$$= 519{,}430 \text{ lbs}$$

4. A fictitious compressive force, formula (10.237) in the pipeline is

$$P = (1-2\nu)pA_1 + \alpha EA_w\delta t = (1-2x0.30) \text{ x } 880 \text{ x}$$

$$\text{x } 113.04 + 6.7 \text{ x } 10^{-6} \text{ x } 30 \text{ x } 10^6 \text{ x}$$

$$\text{x } 13.795 \text{ x } 175 = 489{,}218 \text{ lbs}$$

We have then

$$1.1P = 1.1 \times 489{,}218 = 538{,}140 \text{ lbs}$$

or

$$538{,}140 > 519{,}430$$

Therefore,

$$1.1P > P_{cr,\ell}$$

This indicates that the longitudinal stability of pipeline is not secured.

For this condition in an after-critical state, we find for the maximum deflection, from Table 10.3.

$$V_{max} = 2.19\sqrt[11]{\frac{q^3EI^5}{p^4A^4}} =$$

$$= 2.19\sqrt[11]{\frac{33.39^3x30x10^6x279.33^5}{14.95^4x13.795^4}} =$$

$$= 83.66 \text{ in} = 6.97 \text{ ft.}$$

Numerical Example No. 3

A gas pipeline having O.D. = 40 in and a thickness of wall t = 15/32 in is laid into sandy soil at a depth of 2 ft from the ground surface, and crosses a creek with a span of 100 ft. Over the creek the pipeline is constructed as a shallow arch. The internal pressure in the pipe is

$$p_i = 880 \text{ lb/in}^2$$

The moment of inertia of the wall of the pipe is

$$I = 11{,}373 \text{ in}^4$$

and the cross-sectional area of the wall is

$$A_w = 58.18 \text{ in}^2$$

The weight of the pipe is

$$q_1 = 16.5 \text{ lb/in}$$

The physical constants of the soil are:

$$\gamma = 0.056 \text{ lb/in}^3 \text{ and } \psi_0 = 30°$$

Investigate buckling stability of the crossing:

1. Resistance of the pipeline against longitudinal displacements of the pipeline according to formula (10.220) and (10.228).

$$p = fq; \quad f = 0.8 \tan\psi$$

$$q = 08.2\gamma HD + 0.9q_1$$

$$p = 0.8 \tan\psi(0.8.2\gamma HD + 0.9q_1) =$$

$$= 0.8 \tan 30^{\circ}(0.8x2x0.056x24x40+0.9x16.5) =$$

$$= 46.56 \text{ lb/in}$$

2. Resistance against transverse displacements along length of the span is

$$q_1 = 0.9 \times 16.5 = 14.85 \text{ lb/in}$$

3. Resistance against transverse displacements at underground sections we determine considering Figure 10-22 and using formula (10.232).

$$q = 0.9q_1 + 0.8\gamma V = 0.9 + 16.5 + 0.8 \text{ x}$$

$$\text{x } 0.056 \text{ x } (24 \text{ x } 40 + \frac{2x6.43x24}{2}) =$$

$$= 14.85 + 49.92 = 64.77 \text{ lb/cm}$$

$$\frac{q_2}{q} = \frac{q-q_1^1}{q} = \frac{64.77-14.85}{64.77} = \frac{49.92}{64.77} = 0.77$$

We accept as the inital values

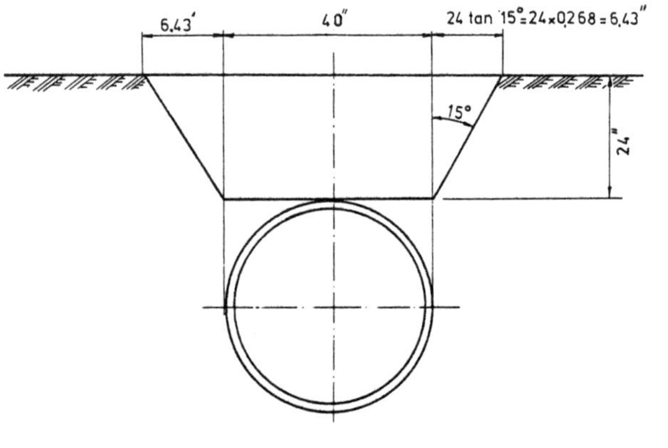

FIGURE 10.22 — Prism of earth pressure on pipeline.

$$\xi_{(0)} = 0.5$$

$$L_{n(0)} = 1{,}000 \text{ ft} = 12{,}000 \text{ in}$$

From diagrams, Figures 10.15 and 10.16, we have

$$B = 2.52; \quad X = 1.06; \quad \phi = 0.62$$

and according to formulas (Table 10.4) we obtain

$$n = \frac{\pi EA_w}{L_n q} = \frac{3.14x30x10^6x58.18}{1.2x10^4x64.77} = 7051$$

$$P_\ell = Bn^{2/9} \sqrt[3]{q^2EI} =$$

$$= 2.52 \text{ x } 7051^{2/9} \sqrt[3]{64.77^6x30x10^6x11373} =$$

$$= 2{,}294{,}680 \text{ lbs}$$

$$x = \frac{X}{\sqrt[3]{n}} = \frac{1.06}{\sqrt[3]{7051}} = 0.055$$

$$\ell = \frac{\pi\sqrt[3]{x}}{\phi\sqrt[3]{q/EI}} = \frac{3.14\sqrt[3]{0.055}}{0.62\sqrt[3]{\frac{64.77}{30x10^6x11373}}} = 3{,}355 \text{ in}$$

Therefore, after the first approximation

$$\xi_{(1)} = \frac{b}{\ell} = \frac{1200}{3355} = 0.3576$$

$$L_{n(1)} = 0.5 \frac{P_\ell}{p} = 0.5 \text{ x } \frac{2{,}294{,}680}{46.56} = 24{,}642 \text{ in}$$

In the second approximation using diagrams, Figures 10.15, 10.16 and Table 10.1, we have

$$B = 3.2; \quad X = 1.35; \quad \phi = 0.750$$

$$n = \frac{3.14x30x10^6x58.18}{2.4642x10^4x64.77} = 3{,}434$$

$$P_\ell = 3.2 \text{ x } 3{,}434^{2/9} \sqrt[3]{64.77^2x30x10^6x11{,}373} =$$

$$= 2{,}483{,}100 \text{ lbs}$$

$$x = \frac{1.35}{\sqrt[3]{3434}} = 0.0894$$

$$\ell = \frac{3.14\sqrt[3]{0.0894}}{0.750\sqrt[3]{\frac{64.77}{30x10^6x11373}}} = 3{,}254 \text{ in}$$

The second approximation yields the following value

$$\xi_{(2)} = \frac{1200}{3254} = 0.3688$$

$$L_{n(2)} = 0.5 \times \frac{2,483,100}{46.56} = 26,666 \text{ in}$$

It is an insignificant difference by comparison with the first approximation. Therefore, it is necessary to repeat the calculation.

The internal pressure fictitious compressive force

$$P = (1-2\nu)A_i P_i = (1-2\times0.3) \times 1197.8 \times 880 =$$

$$= 471,626 \text{ lbs}$$

Therefore, at

$$\delta t = \frac{P_\ell - 1.1P}{1.1\times\alpha EA_w} =$$

$$= \frac{2,483,100 - 1.1 \times 471,626}{1.1\times6.7\times10^{-6}\times30\times10^{6}\times58.18} =$$

$$= 152.7^{\circ}F$$

The stability of pipeline at the crossing is not ensured.

The maximum amplitude of deflection is

$$A = \frac{x}{\lambda} = \frac{x\ell}{\pi} = \frac{0.0894 \times 3254}{3.14} = 92.64 \text{ in}$$

The maximum bending stress is

$$\sigma_{max} = \frac{D}{2} EV''_{max} = DE\lambda^2 A = DEA\frac{\pi^2}{\ell^2} =$$

$$= 40 \times 30 \times 10^6 \times 92.64 \times \frac{3.14^2}{3.254^6\times10^6} =$$

$$= 103,515 \text{ lb/in}^2$$

REFERENCES — CHAPTER 10

[10.1] Marston, A., "The Theory of External Load on Closed Conduits in the Light of the Latest Experiments", Bul. No. 96, Iowa, Eng. Exp. Station, Ames, Iowa, 1930.

[10.2] Spangler, M.G., "Stresses in Pressure Pipelines and Protective Casing Pipes", Proc. Struct. Div. ASCE, Sept. 1956, pp. 1054-1 to 1054-33.

[10.3] Spangler, M.G., Soil Engineering, 2nd Ed., International Textbook Company, 1963, p. 433.

[10.4] Meyerhof, G.G. and Fisher, C.L., "Composite Design of Underground Steel Structures", Eng., Inst. Canada Eng. Journal, Sept. 1963.

[10.5] Sears, E.L., "Ductile Pipe Design", J.A.W.W.A., Jan. 1964.

[10.6] Spangler, M.G., "The Structural Design of Flexible Pipe Culverts", Bulletin No. 153, Iowa Eng. Exp. Station, 1941.

[10.7] White, H.L. and Layer, J.P., "The Corrugated Metal Conduit as a Compression Ring", Proc. U.S. Highway Res. Bd., 1960.

[10.8] Volmir, A.S., "Stability of Elastic Systems", Translation Division Foreign Technology Division WP-AFB, Ohio, 1965.

[10.9] Voelmy, A., Eingebettete Rohre, Zurich, 1937.

[10.10] Klein, G.K., Design of Buried Pipes, Gosstroiizdat, 1957 (in Russian).

[10.11] Pevo, R., Design for Strength of Buried Pipelines, Stroiizdat, 1964 (in Russian).

[10.12] Emelyanov, L.M., "Stressed State of Backfill Between Retaining Walls", Sovetskii Metropoliten, No. 12, 1940 (in Russian).

[10.13] Agamirzyan, L.S., "Design of Anker Foundation for Stability and Displacement", Tbilissi Institut of Railroad Transport, 1957 (in Russian).

[10.14] Kamerstein, A.G. and Skomorovsky, J.Z., "Design of Pipeline Embedded in the Soil", Stroiteljstvo Truboprovodov, No. 4, 1965 (in Russian).

[10.15] Karavaev, J.I., "Construction and Exploitation of Pipelines in Regions of Mountain Mining", Ugletchizdat, 1959 (in Russian).

[10.16] Vinson, D.J. and Burgar, I., "Natural Gas Temperatures in Buried Pipelines", Pipeline News, Vol. 37, No. 2, 1965.

[10.17] Pritula, A.F. and Pritula, V.A., "Transport of Oil", Oil Products and Gas, Part II Pipes, ONTI SSSR, 1937 (in Russian).

CHAPTER 11

underwater pipelines

11.1 Introduction

Submarine pipelines are used extensively as conduits for the transport of oil and gas from offshore fields. Since 1940, a whole industry has developed to provide submarine pipeline installation services.

When submarine pipelines were first laid, offshore, they were laid close to shore in shallow water and were small, flexible pipes. Because of this, there was no particular need for concern with the complexity of the design of these pipelines.

However, progress in offshore oil development has taken pipelines farther offshore and into more hostile marine environments in many parts of the globe. Pipeliners are now faced with deeper water, longer pipelines, larger-diameter pipelines, and the need to use higher yield strength grades of pipe. The result is a tremendous increase in the importance of design integrity, both during construction and during the operating lifetime of the pipeline. This progress has also brought with it new and more complex problems in design and analysis.

The whole topic of design of submarine pipelines has taken on new and greater importance. As the result of a considerable amount of study and research in recent years, submarine pipeline design has made the transition from art to science. Engineers encountered a number of unique and difficult problems, for which solutions had to be developed.

11.2 Pipelaying Methods

The structural analysis and design of underwater pipelines during their construction depend, to a great extent, on the pipelaying methods, which can be classified in the following four general categories.

11.2.1 Bottom Pull Method

When the bottom pull method is used, the pipe is fabricated onshore into one or more sections, launched into the water and pulled along the bottom and into its final position by means of a winch.

The launchway consists of a graded area with dollies, upon which the pipe is placed, mounted on a track. The winch may be located on the far shore, in the case of a river crossing, or upon a barge firmly anchored, in the case of a long underwater line.

Often the pipeline must be divided into several sections because of a limited space on the launchway. When this is the case, the pull must be interrupted for tie-ins. Figure 11.1 illustrates this method.

This method is used principally for narrow water crossings and relatively short pipelines from the coastline to offshore facilities.

11.2.2 Flotation Method

To lay a pipeline by a flotation method, pipe joints are first welded into a number of long strings onshore. Pontoons are attached to provide buoyancy and section-by-section the strings are towed into position. A barge holds one end of a laid section until the next section arrives and is tied-in, Figure 11.2.

Pontoons are released systematically to lower the pipe to the bottom. This procedure is repeated until the line has been completed.

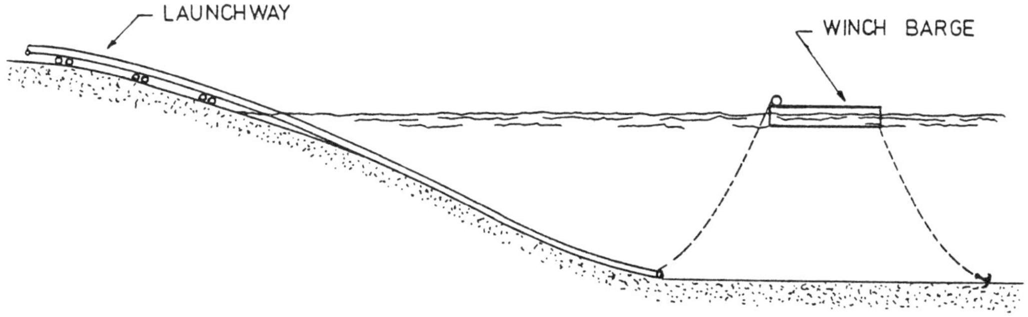

FIGURE 11.1 — Bottom pull method.

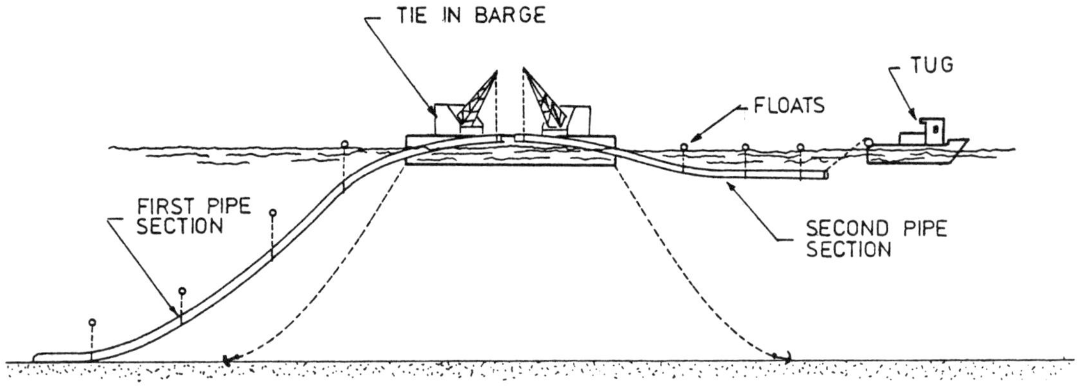

FIGURE 11.2 — Flotation method.

A benefit of the flotation method is that it overcomes the limitations of the length inherent in the bottom pull method. On the other hand, it is highly vulnerable in only moderate seas. The greatest use of the flotation method is for long lines in protected waters.

11.2.3 Reeled Pipe Method

If a pipe is not of large diameter, and the pipeline relatively short, it can be laid by the reel method. The procedure here, is to fabricate the pipe into a continuous length and spool it onto a large diameter reel. The line is laid by simply unspooling it from a moving barge or other vessel, in much the same manner as a cable. Tension is applied to limit the sag in the pipeline while being laid, especially in deep water. A pipeline that is laid by this method can be tested in advance and intalled at rapid speed, Figure 11.3.

11.2.4 Lay Barge Method

The lay barge is used when the pipeline consists of a large diameter pipe, of considerable length and which is to be laid in the open sea.

Pipe joints are welded on the barge, which is fitted with multiple welding stations arranged to handle the pipe in an assembly-line manner. The barge is advanced periodically and each joint progresses through the several stages of welding, radiography and coating, then into the water and over a stinger that allows it to reach the sea floor without undue stress on the pipe.

The stinger is a ramp that extends at a controlled angle from the barge almost to the sea bottom. This limits the sag in the pipe.

A deep water lay barge is usually positioned by means of several anchor lines. As the barge advances, tug boats move the anchors forward, one at a time. Only a small amount of pipe can be stored on the lay barge, so a continuous train of pipe barges feed the operation, Figure 11.4.

11.3 Pipeline Construction

The basic steps involved in the construction of submarine pipeline should be outlined to understand the necessary stages used in the structural analysis and design of such pipelines. The general pulling method is characterized by the extremely short time necessary to actually pull the pipeline into place.

The key to the pulling method of a pipeline installation is the extent of preparatory work which is carried out onshore in order to reduce marine operations. The pulling is done as one of the last activities, and it can proceed rapidly if all of the preparatory work has been accomplished.

The basic steps involved in the construction of submarine pipeline are outlined as follows:

1. *Pipe Stringing*

   Seamless pipe is preferred for marine application, but it is not obtainable in the larger sizes of pipe which

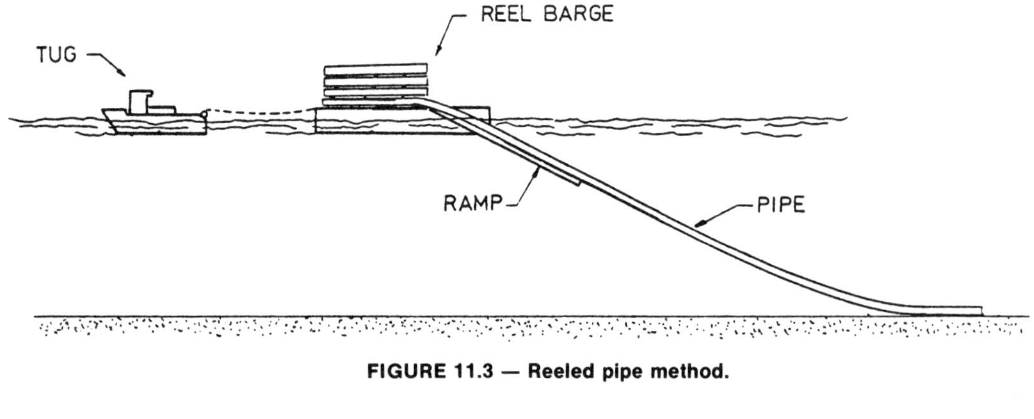

FIGURE 11.3 — Reeled pipe method.

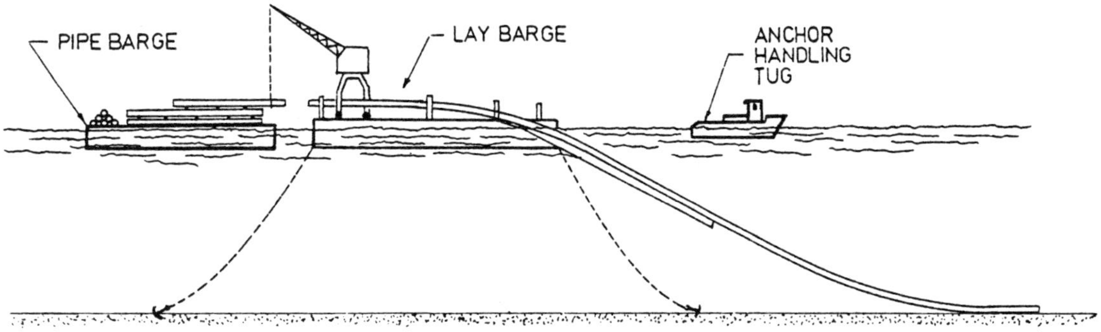

FIGURE 11.4 — Lay barge method.

must be fabricated. The pipe is delivered to the work site and each section is inspected, numbered and weighed. After coating, the pipe sections are welded together into long strings. The area and shape of the available construction site is the factor which may determine the length of the strings, which have ranged from 600 feet to 4,000 feet in length. If is conceivably possible to make up the entire pipeline onshore and then pull it into place. However, with very long pipelines, it is generally more economical to limit the lengths to a mile or less.

2. *Coating and Lining*

   Several methods are available for coating and lining the pipe sections and the selection of the method depends upon the availability of facilities.

3. *Submarine Excavation and Trenching*

   The purpose of submarine trenching is two-fold: first, to provide a smooth grade and eliminate high points which could accumulate air; and secondly, to provide protection for the pipe by burying it below the influence of storms and currents. The first operation is more or less conventional dredging activity, which removes the high points and digs a trench in hard material. When the bottom consists of softer sediments, the pipeline can be buried, using a submarine trencher which rides along the pipe and undercuts it by the use of high pressure water jets fed from a pump and compressor on a floating service barge. By repeated passes of the jet trencher, the pipe is gradually lowered to grade.

4. *Pipe Launching and Installation*

   After the pipe has been welded into long strings, lined, and the field joints made, the strings are stored onto wooden skids along the launchway. The skids are levelled with an instrument, in order to minimize the stresses in the pipe and concrete coating. The pipes are rolled into position onto the launchway by cables attached to the pipe and to synchronized electric winches. The launching units consist of small steel frames on which a rubber tire assembly is attached which permits the pipe string to roll smoothly over the unit. The pipes are then ready for actual installation.

   The next operation is to lay the pulling cables from an offshore anchorage to the launchway. The cables must be aligned precisely under survey control, as the pipe will follow the cable alignment. The cables are laid by a pulling barge which is equipped with a powerful winch. A pulling bridle is connected between the winch and a large diameter sheave block at the end of a pulling sled attached to the first pipe string.

   Once the pulling cables have been laid and the cable attached to the sled, the pulling operation can commence. The pulling operation is done with the barge-mounted winch. The pulling force is transmitted into the anchor-line through a cable arrangement on the barge.

   At the end of each pull, the barge will back up by spooling in on its anchor lines, simply the reverse of the cable-laying operation. A hold-back winch is used to maintain the tension on the landward end of the pipe at all times.

   After each string is pulled, a clamp is attached to the end of the string and connected by a short cable to a standing hold-back anchor. In this manner, the cable from the hold-back winch can be disconnected and carried back to the end of the next string. An epoxy ring internal field joint liner is next installed and the joint welded, connecting up the next string. The pulling can then commence after completion of the coating and wrapping of the field joint.

   The pulling operations should be controlled from a central command point with radio communications maintained between the command point and the pulling barge, the hold-back winch and the welding stations.

11.4 Pipeline Coating

Coatings for underwater pipelines serve three purposes:

1. external protection
2. internal protection
3. weight

External coating is used in conjunction with a cathodic protection system to prevent corrosion and is typically an asphaltic or coal tar material, the same as that used for onshore lines. An internal coating generally is used when the product to be transported is corrosive, or to improve its capacity by reducing pipe roughness.

Weight coating usually consists of a concrete sheath

to give the pipeline the desired specific gravity.

The weight of the pipe must be heavy enough to overcome buoyancy plus any drag or lift forces that might result from currents or storm action, and light enough to avoid sinking into the soil. Wave action on the sea bottom decreases with increased water depth and is barely perceptible at depths to about one-half the wave length. Since hurricane wave lengths frequently exceed 1,000 ft., wave forces should be considered up to a water depth of 500 ft.

Underwater pipelines can be protected against currents and wave action by burial, but sometimes additional weight coatings or anchors are preferred.

A common method of burying a pipeline is with a trenching barge that utilizes high pressure jets. After the pipe is laid on the sea floor, the trenching barge pulls a sled over the pipeline and a trench is excavated by means of the high pressure jets. The pipe then settles into the open ditch and is backfilled by the action of the sea.

11.5 Pipeline Anchoring

Pipeline stabilization, or anchoring, can be defined as an anchoring system designed to maintain the pipeline in a desired position relative to the surrounding environment and subject to the various forces acting on the pipeline.

Anchors currently available for pipelines consist of two basic types:

1. density
2. mechanical.

1. *Density Anchors*

These anchors are usually concrete and take the form of either the bolt-on weights, set-on weights, or on a continuous concrete coating, Figure 11.5.

a. Continuous concrete coating completely encompasses the pipe. The concrete is applied over a protective coating and is usually reinforced with one layer of galvanized wire mesh for every inch of concrete thickness. The design of such anchoring begins by determining the minimum amount of concrete to provide negative buoyancy with the pipeline empty. Added to this is the amount of anchorage required to hold the pipeline in place because of the maximum anticipated current velocities and densities of the water.

The formula for the minimum negative buoyancy required to maintain a pipe resting on the bottom expressed as a function of the current velocity vector perpendicular to the axis of the pipe is

$$W = \frac{V^2 K D \rho}{2} \tag{11.1}$$

where

W = the negative buoyancy of a pipe, lb/ft

ρ = the density of water, slugs/ft³

V = the maximum velocity of water, ft/sec

D = the outside diameter of a pipe including concrete, ft

K = the coefficient of drag, lift and friction between the bottom and the pipe

Other considerations will include the type of service for the pipeline and the depth of burial, if any.

Another consideration is the characteristics of the bottom and backfill. If the backfill material

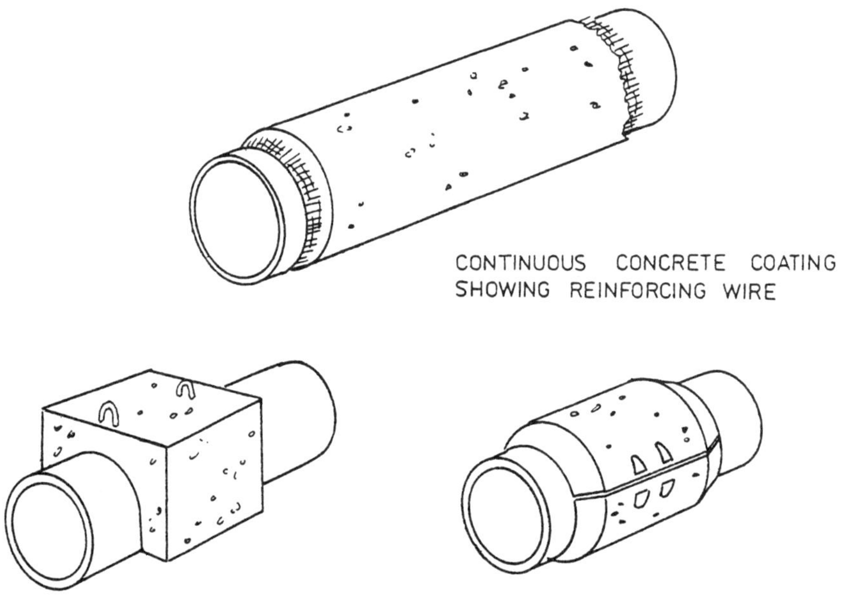

FIGURE 11.5 — Concrete anchors.

becomes a liquid with a minimum of agitation, then the density of the material may become greater than the pipeline, causing it to float out of the ditch. The following formula will provide the minimum concrete thickness required for continuous concrete coating when the minimum specific gravity is known.

$$Sp.Gr = \frac{\text{wt. of Pipe + Concrete}}{\text{wt. of Water Displaced}} \quad (11.2)$$

$$\text{concrete thickness} = \frac{\sqrt{\dfrac{4W_v - \pi (D_i)^2 (D_c)}{\pi (S.G)(D_w) - \pi (D_c)}}}{2} \quad (11.3)$$

where

W_v = weight of the pipe and coating per foot

D_c = density of the concrete

D_w = density of the water

D_i = inside diameter of the concrete

S.G = specific gravity

b. Another type of density anchor is the bolt-on concrete weight, which is built in two halves and designed to be clamped onto the pipeline, and held together with long bolts.

c. The most economical form of density anchor is the set-on weight. These weights are shaped like a "U" and they are set onto the pipeline after the pipeline has been placed in the ditch.

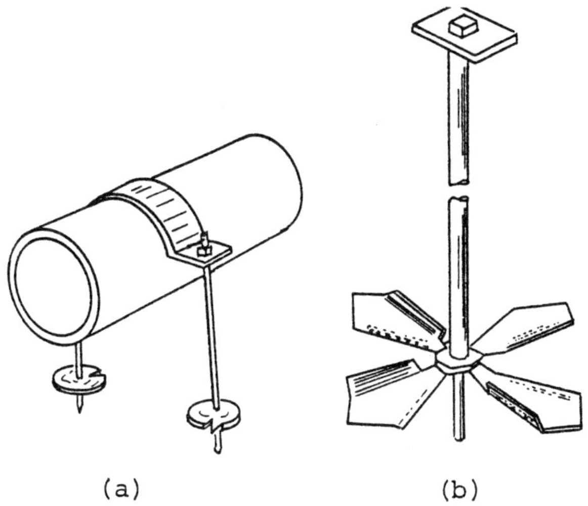

FIGURE 11.6 — Mechanical anchors (a) auger type, (b) expanding type.

The weights are designed with the center of gravity as low as possible. Large diameter thin-wall pipelines can sometimes be overstressed if the pipeline is required to support the set-on weight, as well as maintain the stresses due to pressure and bending.

The following formulas can be of help in determining spacing for concrete bolt-on or set-on weights.

$$Sp.Gr = \frac{\text{wt. of Pipe + Concrete}}{\text{wt. of Water-Displaced}} \quad (11.4)$$

$$S.G = \frac{\text{(Wt.of Pipe/ft)(ft) + (Wt.of conc.wt)}}{\left[\text{(Vol.of Pipe/ft)(ft)} + \left(\dfrac{\text{Wt.of con.wt}}{\text{Den.of conc.}}\right)\right] \text{Density of Water}} \quad (11.5)$$

$$\text{Spacing (ft)} = \frac{(Wt_c) - (S.G.)(D_w)(Wt_c/D_c)}{(S.G.)(D_w)(Vp/ft) - (Wp/ft)} \quad (11.6)$$

where

Wt_c = total weight of concrete weight in air

S.G = the desired specific gravity

D_w = the density of the water

D_c = the density of the concrete

Vp/ft = the volume of the outside diameter of the pipe per foot

Wp/ft = the weight of the pipe per foot

2. *Mechanical Anchors*

Mechanical anchors differentiate from density anchors in that they derive their holding power from the shear strength of the soil. They are inserted into the soil and attached to the pipeline. They are usually made up of steel and are either pile, auger, Figure 11.6 (a) or expanding types, Figure 11.6 (b).

Piles are usually made of steel pipe on wooden posts. The piles are driven alongside of the pipeline and attached to the pipe with some form of strap.

The most commonly used type of mechanical anchor is the auger type. This system consists of two anchors and is strap-shaped to fit the pipeline. Auger anchors come in various sizes from 6-in to 24-in. diameters and can be used in clays, sands, gravels, or any other unconsolidated material.

The anchor spacing will be dependent upon the maximum hold-down strength of the soil and the selected density.

Expanding mechanical anchors are used in the same manner as the auger type. The anchor rod has flukes on one end that are hinged in such a way as to expand outward from the rod. Most are expanded by turning the threaded anchor rod which is run through a nut in the center of the flukes.

The efficiency of the mechanical anchors is far greater than the density type.

The formula for determining the spacing of mechanical anchors is as follows.

$$\text{Specific Gravity} = \frac{\text{(Wt. of Pipe/ft) (Anchor Spacing)} + \text{Anchor Hold Down}}{\text{Wt. of Water Displaced}} \qquad (11.7)$$

$$\text{Anchor Spacing} = \frac{(A_V)}{\frac{\pi}{4}(S.G.)D^2D_W - Wp/ft} \qquad (11.8)$$

where

A_V = minimum anchor hold-down power

S.G = the specific gravity

D = the outside diameter of the pipe plus coating

D_W = density of the water

Wp/ft = the weight of the pipe plus the coating per foot

11.6 Stability of Pipeline During Construction

11.6.1 Lay Barge Method of Construction

The basic elements of the pipe-laying system are shown in Figure 11.7.

These include:

1. The lay barge
2. the mooring system
3. the pipe tension devices
4. the stinger, and
5. the suspended pipe span

The suspended pipe span, characteristically an "S" shaped curve, is divided into a lower portion, the sag-bend region, and on the upper portion, the over-bend region. The transition point divides the sag-bend from the over-bend.

The stinger provides a support for the over-bend while tension devices serve to control the pipe in the sag-bend. The tension devices, located between the various work stations on the barge, develop and maintain a restraining force between the pipe and the lay vessel as the barge moves forward on its anchors.

a. *Control of the Sag-Bend*

Continuous control of the suspended pipe span is the most crucial aspect of the offshore pipeline construc-

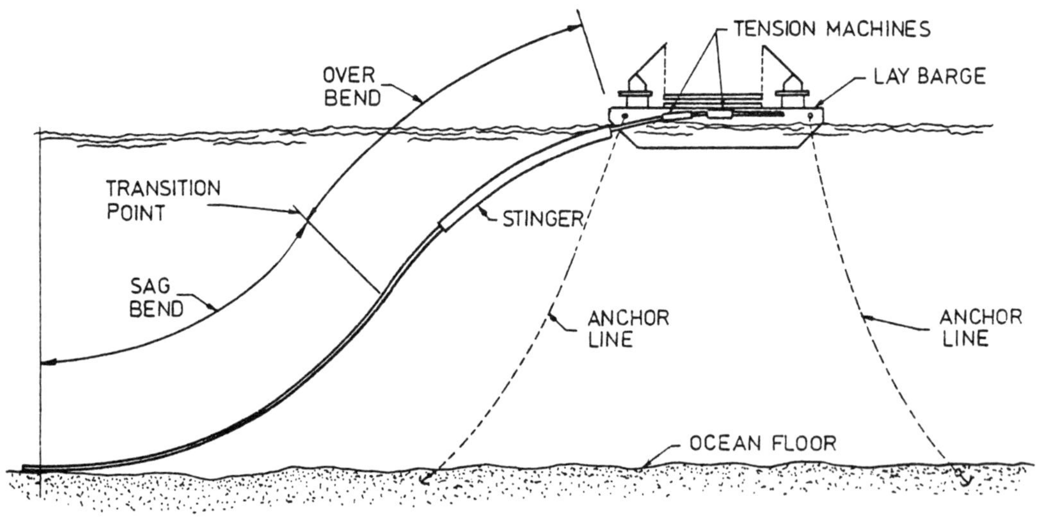

FIGURE 11.7 — Elements of the pipeline laying system.

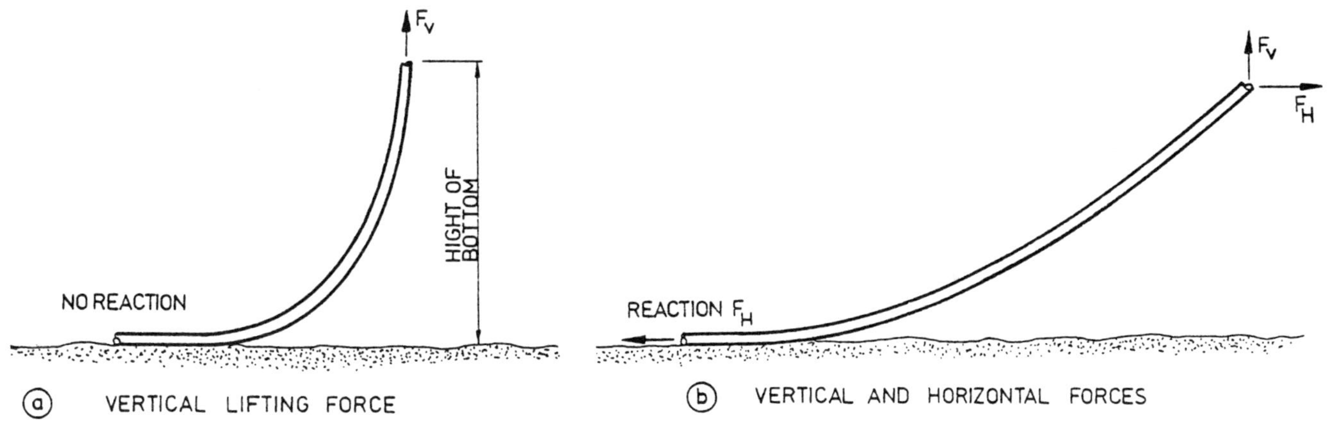

FIGURE 11.8 — Sag-bend configuration under various lifting conditions.

tion. Malfunction of any portion of the pipe-handling system usually leads to catastrophic failure of the pipeline.

The mechanics of the suspended pipe span can best be discussed by starting at the bottom and considering the sag-bend.

First, consider lifting the end of the pipe off the sea bottom by exerting only a vertical lifting force on the end of the pipe, as shown in Figure 11.8.

As the end is lifted in this manner, a sag-bend will develop which for small displacements can be analyzed by ordinary beam theory.

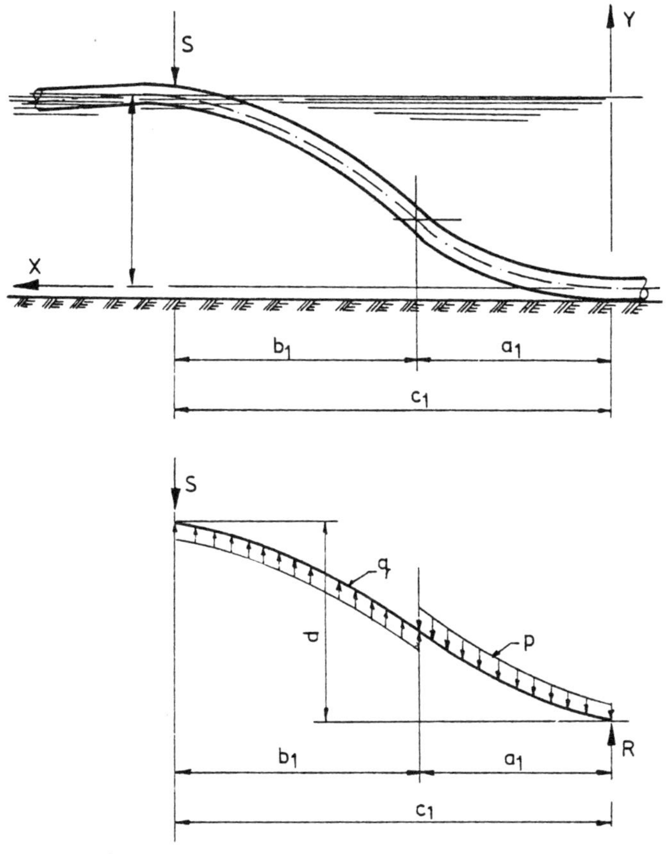

FIGURE 11.9 — Design scheme of the free launching of the pipe by simultaneously filling it with water.

11.6.2 Method of Free Launching of the Pipeline

In the case of free laying of the pipeline, when it is not supported, the pipe is under bending stresses. The magnitude of the bending should be in limits of elastic deformations and will depend upon the external forces and carrying capacity of the pipe, namely, from the dimensions and section modulus.

In the case of launching of the pipeline by filling it with water and in the absence of pontoons, are other supports; the action of external forces on the part of the pipeline which is under stress is shown in Figure 11.9.

To determine the values of bending moments acting upon the pipe, we consider the equilibrium of the launched section of the pipe.

It is assumed that at the contact of the pipe with the ground and at the top level of the bent, the moments and slopes of the pipe are equal to zero.

The equilibrium equations are

$$\Sigma V = R + bq - ap - S = 0 \qquad (11.9)$$

$$\Sigma M = R(a+b) - ap(b+\frac{a}{2}) + \frac{qb^2}{2} = 0 \qquad (11.10)$$

where

R = the reaction of the ground.

At $x = c$; $y = d$ and $\frac{dy}{dx} = 0$, the equation for the "a" section may be written as follows:

$$\frac{d^2y}{dx^2} EI = M(x) = Rx - \frac{pa^2}{2} \qquad (11.11)$$

and after integration

$$\frac{dy}{dx} EI = R\frac{x^2}{2} - \frac{px^2}{6} + C_1 \qquad (11.12)$$

at $x = 0$; $\frac{dy}{dx} = 0$; $C_1 = 0$.

also

$$yEI = R\frac{x^2}{6} - \frac{px^4}{24} + C_2 \qquad (11.13)$$

at $x = 0; \quad y = 0; \quad C_2 = 0.$

For the section "b", the equation of bent pipe is

$$\frac{d^2y}{dx^2} EI = M(x) + Rx - pa(x - \frac{a}{2}) + \frac{q(x-a)^2}{2} =$$

$$= (p+q)\frac{a^2}{2} + [R - a(p\ q)] + \frac{qx^2}{2} \qquad (11.14)$$

After integration

$$\frac{dy}{dx}EI = (p+q)\frac{a^2x}{2} + [R - a(p+q)]\frac{x^2}{2} + \frac{qx^3}{6} + C \qquad (11.15)$$

At $x = a; \quad \frac{dy}{dx} EI = R\frac{a^2}{2} - \frac{pa^3}{6}$

Therefore, $C_3 = -(p+q)\frac{a^3}{6}$

By the integration of Equation (11.15)

$$y\,EI = (p+q)\frac{a^2x^2}{4} + [R - a(p+q)]\frac{x^2}{6} +$$

$$+ \frac{qx^4}{24} - (p+q)\frac{a^2x}{6} + C_4 \qquad (11.16)$$

At $x = a; \quad yEI = \frac{Ra^2}{6} - \frac{px^4}{24}.$

Therefore, $C_4 = (p+q)\frac{a^4}{24}$

From Equation (11.9) it follows

$$R = \frac{ap(b+\frac{a}{2}) - q\frac{b^2}{2}}{a + b} \qquad (11.17)$$

By designating $a + b = c$ and $p + q = \omega$, we obtain from (11.16)

$$R - a(p+q) = -\frac{a^2\omega}{2c} - \frac{qc}{2} \qquad (11.18)$$

After substituting (11.18) into Equations (11.15) and (11.16), using $x = c$, we obtain

$$yEI = \frac{\omega}{24}a^4 - \frac{\omega}{6}a^3c + \frac{\omega}{6}a^2c^2 - \frac{q}{24}c^4 = dEI \qquad (11.19)$$

and

$$\frac{dy}{dx} EI = \frac{\omega}{6}a^3 - \frac{\omega}{4}a^2c + \frac{q}{12}c^3 = 0 \qquad (11.20)$$

By dividing Equation (11.20) with $\frac{qa^2}{12}$ and using $\frac{c}{a} = n$, we have

$$\frac{p}{q} = \frac{2-3n+n^2}{3n-2} \qquad (11.21)$$

This ratio is shown in Figures 11.10 and 11.11. From

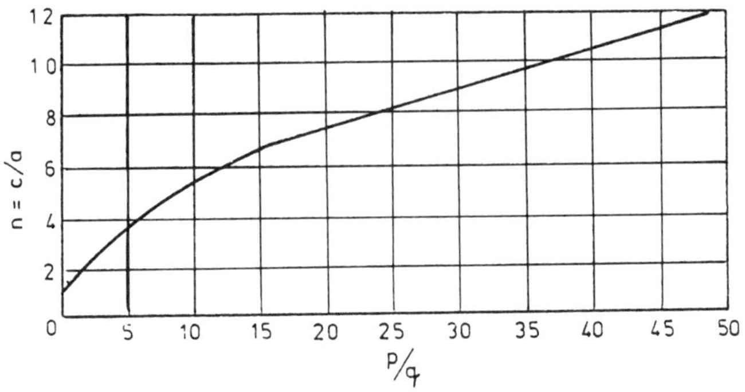

FIGURE 11.10 — Diagram of change of ratio n as a function of ratio p/q at 0 < p/q < 50.

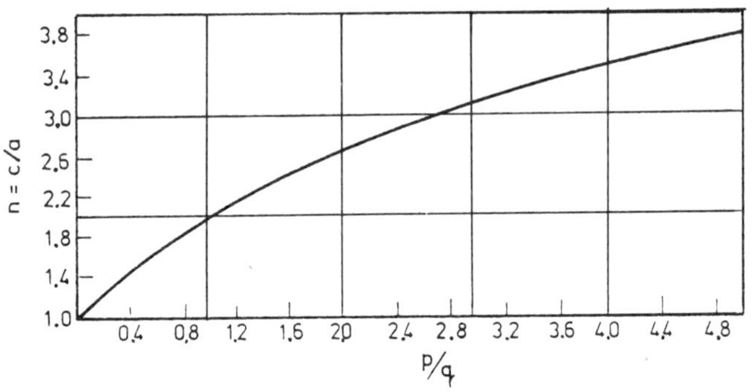

FIGURE 11.11 — Diagram of change of ratio n as a function of ratio p/q at 0 < p/q < 5.

From Equation (11.20), we obtain

$$a = \sqrt[4]{\frac{6DEI}{\omega}}\sqrt[4]{\frac{1}{(n-\frac{1}{2})^2 - \frac{n^4}{4(p/q+1)}}} \qquad (11.22)$$

From Equation (11.21), we have

$$\frac{p}{q} + 1 = \frac{n^2}{3n-2}$$

then Equation (11.22) may be expressed as follows

$$a = \psi_a\sqrt{\frac{6dEI}{\omega}} \qquad (11.23)$$

where

$$\psi_a = \sqrt{\frac{2}{n-1}}$$

Maximum bending moment for section "a" we obtain from Equation (11.11)

$$\frac{dM(x)}{dx} = R - px = 0; \quad x = \frac{R}{p}$$

and

$$M_{max} = Rx - \frac{px^2}{2} = \frac{R^2}{2p} \qquad (11.24)$$

and for the section "b", we have from Equation (11.14)

$$M_{max} = \frac{\omega a^2}{2} - \frac{(R-a\omega)^2}{2q} \qquad (11.25)$$

at

$$x = \frac{a(p+q) - R}{q}$$

Using formulas (11.17) and (11.24), we may find the maximum value of the bending moment for the section "a", when $p > q$

$$M_{max} = \tfrac{1}{2}\sqrt{6dEI\omega} \quad \psi_a(n) \qquad (11.26)$$

where $$\psi_a(n) = \frac{2(n-1)}{n(n+2)} \qquad (11.27)$$

By comparing formula (11.26) and (11.28) it follows that the maximum stresses in the pipe at section "a" are

$$\sigma_a = 1271\sqrt{\frac{r_m}{t}}\ \psi_a\ \sqrt{d} = \theta_a\ \sqrt{d} \qquad (11.28)$$

The value of maximum bending moment at the section "b" may be obtained from the expression (11.25) as

$$\sigma_b = 1271\sqrt{\frac{r_m}{t}}\ \psi_b\ \sqrt{d} = \theta_b\ \sqrt{d} \qquad (11.29)$$

where

$$\psi_b(n) = -\frac{2(n-1)}{n(3n-2)} \qquad (11.30)$$

The maximum depth at which it is possible to lay the pipe by the method of free launching and filling the pipe with water may be obtained from the formulas (11.28) or (11.29).

$$d_{max} = \frac{\sigma^2_{max}}{1615\times10^3\times\psi^2\ \frac{r_m}{t}} \qquad (11.31)$$

The value of ψ may be found as a function of the ratio p/q from formulas (11.27) or (11.30).

At $p > q$ the maximum stresses are at section "a", where the pipe is filled with water, and at $p < q$ at the section "b".

11.6.3 Stresses in the Pipe During Lowering of Its End by Filling with Water

For a stress analysis of the pipe, we are considering a section of pipe between points A and B. The action of the external forces at this section is shown in Figure 11.12.

Under the weight of the water in the pipe, a part of the pipe starts to rise. The weight of the pipe section raising up from the water we shall designate as S, Figure 11.12.

Upon contacting the pipe end with the bottom of the river, the reaction A acts, reducing the stresses in the pipe.

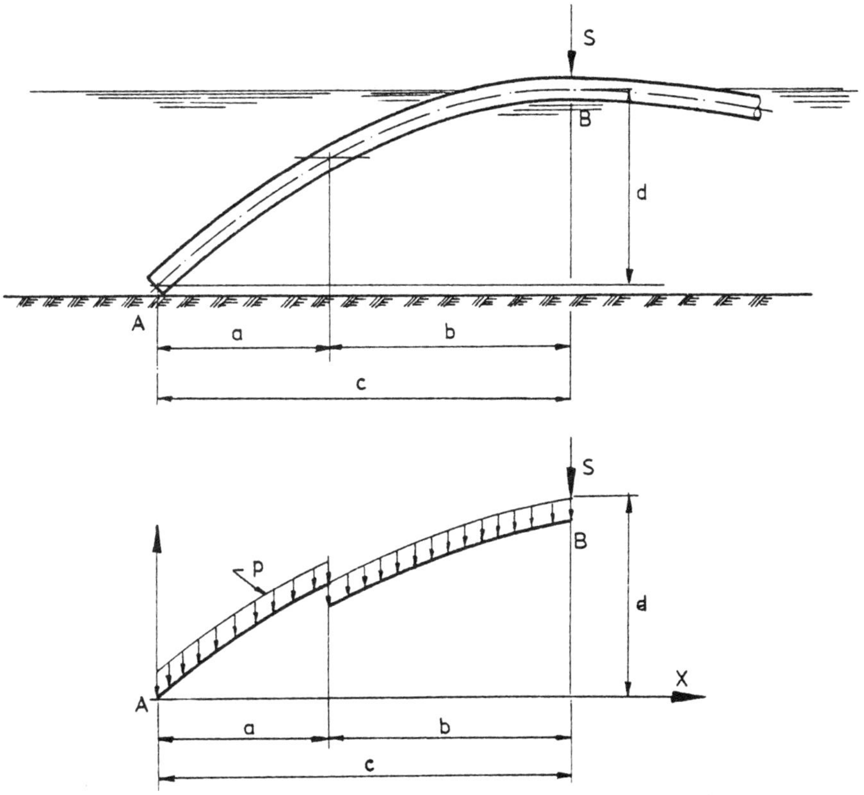

FIGURE 11.12 — Position of the pipeline during lowering of its end.

For a further analysis, the following designations are introduced.

- p = weight of the pipe in the water and filled with water, lb/in
- q = lifting force of the water acting upon an empty pipe, lb/in
- ω = $p + q$ = weight of the water in the pipe, lb/in
- a = length of the pipe filled with water, in
- b = length of the empty pipe, in
- c = total length of the bent section of pipe, in
- d = depth of submerged pipe dt $x = c$, in
- S = weight of the pipe section rising above the water surface, lb
- I_p = moment of inertia of the pipe, in^4
- W_p = section modulus of the pipe, in^3
- D_0 = outside diameter of the pipe, in
- D_i = internal diameter of the pipe, in
- r = internal radius of the pipe, in
- r_m = middle radius of the pipe, in
- γ_w = specific weight of the water, lb/in^3

The lengths of sections of the pipeline filled by the water "a" and of the empty pipe "b", as well as the value of "S" are unknown. For their determination, we apply the conditions of equilibrium of the section under consideration.

$$\Sigma V = bq - pa - S = 0 \qquad (11.32)$$

$$\Sigma M_{x=o} = \frac{pa^2}{2} + qb\left(a+\frac{b}{2}\right) - Sc \qquad (11.33)$$

For section "a"

$$\Sigma M_x = -\frac{px^2}{2} \qquad (11.34)$$

and for section "b"

$$\Sigma M_x = -pa\left(x-\frac{a}{2}\right) + q\frac{(x-a)^2}{2} \qquad (11.35)$$

After substituting the value S from Equation (11.32) into (11.33), and because $b = c - a$, we obtain

$$-\frac{pa^2}{2} - \frac{qc^2}{2} - \frac{qa^2}{2} + acq + acp = 0$$

After substituting into the above equation $\frac{c}{a} = n$, we obtain

$$\frac{p}{q} = \frac{(n-1)^2}{2n-1} \qquad (11.36)$$

from which

$$n = \left(\frac{p}{q}+1\right) \pm \sqrt{\left(\frac{p}{q}+1\right)^2 - \left(\frac{p}{q}+1\right)} \qquad (11.37)$$

The equation of the bent pipe is

$$\frac{d^2y}{dx} EI = M_x$$

From Equation (11.34) for section "a"

$$\frac{dy}{dx} EI = \int M_x dx + C_1 = -\frac{px^3}{6} + C_1 \qquad (11.38)$$

From Equation (11.35) for section "b"

$$\frac{dy}{dx} EI = \int M_x dx + C_2 =$$

$$= -\frac{pax^2}{2} + \frac{pa^2x}{2} + \frac{qx^3}{6} - \frac{qax^2}{2} + \frac{qa^2x}{2} + C_2$$

At $x = c$, the slope of the deflection line $\frac{dy}{dx} = 0$.

Therefore

$$C_2 = \frac{pac^2}{2} - \frac{pa^3c}{2} - \frac{qc^3}{6} + \frac{qac^2}{2} - \frac{qa^2c}{2}$$

and

$$\frac{dy}{dx} EI = \frac{pax^2}{2} + \frac{pa^2x}{2} + \frac{qx^3}{6} - \frac{qax^2}{2} + \frac{qa^2x}{2} +$$

$$+ \frac{pac^2}{2} - \frac{pa^2c}{2} - \frac{qc^3}{6} + \frac{ac^2q}{2} - \frac{qa^2c}{2} \qquad (11.39)$$

We may obtain the value C_1 from Equation (11.38) considering the equality of the slopes at $x = a$:

$$C_1 = \frac{pa^3}{6} + \frac{qa^3}{6} + \frac{pac^2}{2} - \frac{pa^2c}{2} - \frac{qc^3}{6} +$$

$$+ \frac{ac^2q}{2} - \frac{qa^2c}{2}$$

and Equation (11.38) may be written as follows:

$$\frac{dy}{dx} EI = -\frac{px^3}{6} + \frac{pa^3}{6} + \frac{qa^3}{6} + \frac{pac^2}{2} - \frac{pa^2c}{2} -$$

$$- \frac{qc^3}{6} + \frac{ac^2q}{2} - \frac{qa^2c}{2} \qquad (11.40)$$

Then the equation of the bent pipe for the section "a" is

$$y\,EI = -\frac{px^4}{24} + \frac{pa^3x}{6} + \frac{qa^3x}{6} + \frac{pac^2x}{2} -$$

$$- \frac{pac^2x}{2} - \frac{qc^3x}{6} + \frac{ac^2qx}{2} - \frac{qa^2cx}{2} + C_3 \qquad (11.41)$$

At $x = 0$, $y = 0$, and therefore $C_3 = 0$.

By analogy, the equation of the bent pipe for Section "b" from Equation (11.39) is

$$yEI = -\frac{pax^3}{6} + \frac{pa^2x^2}{4} + \frac{qx^4}{24} - \frac{qax^3}{6} + \frac{qa^2x^2}{4} +$$

$$+ \frac{pac^2x}{2} - \frac{pa^2cx}{2} - \frac{qc^3x}{6} + \frac{ac^2qx}{2} -$$

$$- \frac{qa^2cx}{2} + C_4 \tag{11.42}$$

At x = a, from Equation (11.41)

$$yEI = \frac{pa^4}{8} + \frac{qa^4}{6} + \frac{pa^2c^2}{2} - \frac{pa^3c}{2} - \frac{qac^3}{6} +$$

$$+ \frac{qa^2c^2}{2} - \frac{qa^3c}{2}$$

By equalizing this expression to Equation (11.42) at x = a, we obtain

$$C_4 = \frac{pa^4}{24} + \frac{qa^4}{24}$$

Then, the Equation (11.42) at x = c

$$yEI = (p+q)\frac{ac^3}{3} - (p+q)\frac{a^2c^2}{4} + (p+q)\frac{a^4}{24} - \frac{qc^4}{8} \tag{11.43}$$

We may assume that p + q = ω. After dividing both parts of the above equation $a^4/6$ and using c/a = n, we obtain

$$a = \sqrt[4]{\frac{6dEI}{\omega(2n^3 - \frac{3}{2}n^2 + \frac{1}{4}) - \frac{3}{4}qn^4}} \tag{11.44}$$

The maximum bending moment will be on section "b". From Equation (11.35)

$$\frac{dM_x}{dx} = - pa + qx - aq = 0$$

$$x = \frac{a(p+q)}{q} \tag{11.45}$$

After substituting the value x into Equation (11.35), we obtain

$$M_{max} = - \frac{a^2p}{2}\left(\frac{p}{q}+1\right) \tag{11.46}$$

Equation (11.44) may be represented as follows:

$$a = \sqrt[4]{\frac{6dEI}{\omega}}\, f\left(\frac{p}{q}\right) \tag{11.47}$$

where

$$f\left(\frac{p}{q}\right) = \sqrt[4]{\frac{1}{2n^3 - \frac{3}{2}n^2 + \frac{1}{4} - \frac{3}{4}\frac{qn^4}{4\omega}}} \tag{11.48}$$

$$\frac{q}{\omega} = \frac{1}{\frac{p}{q}+1} \tag{11.49}$$

From Equation (11.36)

$$\frac{p}{q} + 1 = \frac{n^2}{2n-1} \tag{11.50}$$

from which

$$f\left(\frac{p}{q}\right) = \sqrt[4]{\frac{4}{2n^3 - 3n^2 + 1}} \tag{11.51}$$

and therefore,

$$a = \sqrt{\frac{6dEI}{\omega}} \cdot \sqrt[4]{\frac{4}{2n^3 - 3n^2 + 1}} \tag{11.52}$$

Then, after certain substitutions, Equation (11.46) may be represented, as follows:

$$M_{max} = \sqrt{6dEI\omega} \cdot \sqrt{\frac{\left(\frac{p}{q}\right)^2}{2n^3 - 3n^2 + 1}} \tag{11.53}$$

Stresses in the cross-section of the pipe under the maximum bending moment is

$$\sigma_{max} = \frac{M_{ax} D_0}{2I} \tag{11.54}$$

For a circular pipe, we may use

$$I = r^3\pi t$$

$$\frac{D_0 r}{2} = r_m^2$$

$$r^2 \approx r_m^2$$

The weight of the water per 1 in. of the pipeline

$$\omega = p+q = r^2\pi\gamma_w$$

Then, after substitutions, Equation (11.54) will be

$$\sigma_{max} = \sqrt{6dE\gamma_w \frac{r_m}{t}} \cdot \psi\left(\frac{p}{q}\right) \tag{11.55}$$

where

$$\psi\left(\frac{p}{q}\right) = \sqrt{\frac{\left(\frac{p}{q}\right)^2}{2n^3 - 3n^2 + 1}} \tag{11.56}$$

From Equation (11.56) the permissible depth of the submerged end of the pipe is

$$d = \frac{\sigma_{max}^2}{6E\gamma_w \frac{r_m}{t}\left[\psi\left(\frac{p}{q}\right)\right]^2} \tag{11.57}$$

Formulas (11.55) and (11.57) are valid for any pipeline covered by coating at the corresponding values of p and q.

In these formulas, the value of n is unknown and may be determined from formula (11.37) for a given ratio p/q.

The dependence between the depth of the launching of an end of pipeline and the stress in the pipe may be presented in a simplified form, using Equation (11.55)

$$\sigma = \sqrt{6E\gamma_w} \cdot \sqrt{\frac{r_m}{t}} \cdot \psi(\frac{p}{q}) \sqrt{d} \quad (11.58)$$

After substituting into this formula the values of $E = 30 \times 10^6$ psi and $\gamma_w = 0.0359$ lb/in³, we obtain

$$\sigma = 2542 \sqrt{\frac{r_m}{t}} \psi(\frac{p}{q}) \sqrt{d} \quad (11.59)$$

Actually, the values of the expressions $\sqrt{\frac{r_m}{t}}$ and $\frac{p}{q}$ for the given section of the pipe are constant. Therefore, formula (11.59) may be expressed as follows

$$\sigma = \theta\sqrt{d} \quad (11.60)$$

where

$$\theta = 2542 \sqrt{\frac{r_m}{t}} \psi(p/q) \quad (11.61)$$

11.6.4 Critical Tension and Inclination of a Deep-Water Pipeline Laid from the Barge

A deep-water pipeline could be made up in an angular position on a lay barge and maintained under a predetermined high tensile force while being lowered to the bottom, as shown by the patented method in Figure 11.13 [11.1]

The axial force is chosen large enough to avoid overstressing and kinking of the pipeline, near its point of tangency with the sea floor. It was necessary to assume that the unsupported pipe would take the form of a natural catenary.

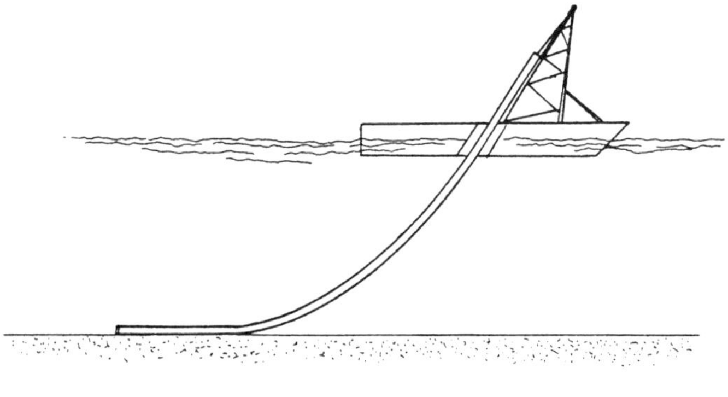

FIGURE 11.13 — Proposed method for laying deep-water pipeline.

Equations for determining the required end conditions are developed herein, under the following assumptions that the pipeline takes the form of a natural catenary.

11.6.4.1 Natural Catenary Method

The shape of the natural catenary is

$$y = \frac{H}{W}[\cosh\frac{xW}{H} - 1] \quad (11.62)$$

where

H = the horizontal component of the axial force,

or

$$H = \frac{cW}{\varepsilon_w} \quad (11.63)$$

W = the buoyant unit weight of the pipeline, lb/ft

and

x and y are the Cartesian coordinates shown in Figure 11.14.

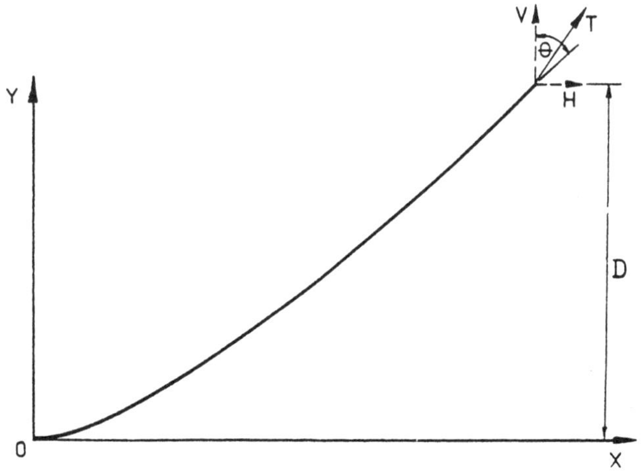

FIGURE 11.14 — Pipeline configuration approximates a catenary between the surface and the bottom in deep water.

The equations for minimum line tension and corresponding ramp angles are

$$T = W[D + (\frac{C}{\varepsilon_b})] \quad (11.64)$$

$$\theta = \cot^{-1}[(\frac{\varepsilon_b D}{C})(2 + \frac{\varepsilon_b D}{C})] \quad (11.65)$$

where

C = the external radius of the pipeline cross-section, ft.

D = the water depth, ft

T = the line tension, lb

ε_b = maximum allowable bending strain, in/in

θ = the ramp angle measured with respect to the vertical angle

It should be noted that the horizontal component of force at the lay barge H is independent of the water depth D, as demonstrated by Equation (11.63).

The maximum allowable bending strain ϵ_b should be chosen so that the corresponding bending stress

$$\sigma_b = \varepsilon_1 E \qquad (11.66)$$

is a fraction of the yielding stress, while the axial stress also exists at the point of maximum curvature. This stress is negligible in most cases, but should be calculated in case the total stress does exceed its permissible range.

As the water depth increases, the tension T increases, H remaining constant until the axial stress at the lay-barge end of the pipeline becomes equal to the allowable bending stress, or

$$\frac{T}{A} = \frac{W}{A} [D + (\frac{C}{\varepsilon_b})] = \varepsilon_b E \qquad (11.67)$$

where

A = the cross-section of the cable

The maximum water depth for which the most exacting stress is at or near the sea floor and not at the lay-barge is therefore,

$$D_{max} = AE (\frac{\varepsilon_b}{W}) - (\frac{C}{\varepsilon_b}) \qquad (11.68)$$

The use of Equation (11.68) as a numerical example has shown that D is of such size that for all practical considerations, the tensile stress at the barge may be disregarded.

11.6.5 Stresses Due to the Lifting of Submerged Pipeline

The scheme of the lifting of a submerged pipeline is shown in Figure 11.15, on the following page.

In this case, it is necessary to determine the stresses in the pipe, depending upon the depth of the water at the location of the pipe lifting. It is considered that there are no horizontal loadings acting on the pipeline.

As a basic design scheme, the pipe is considered as a beam and cross-sections do not turn at the supports or in the middle of the span.

In this case, the conditions of equilibrium are

$$\Sigma V = A + B - 2g\ell + C = 0 \qquad (11.69)$$

$$M_x = Ax - \frac{gx^2}{2} \qquad (11.70)$$

$$\frac{d^2y}{dx^2} EI = M(x) = Ax - \frac{gx^2}{2}$$

$$\frac{dy}{dx} EI = \frac{Ax^2}{2} - \frac{gx^3}{6} + C_1 \qquad (11.71)$$

For $x = 0; \frac{dy}{dx} = 0$ and, therefore

$$C_1 = 0$$

$$yEI = \frac{Ax^3}{6} - \frac{gx^4}{24} + C_2 \qquad (11.72)$$

For $x = 0$, $y = 0$. Therefore,

$$C_2 = 0$$

For $x = \ell$, $y = d$, from (11.70) we obtain

$$d = \frac{A\ell^3}{6EI} - \frac{g\ell^4}{24EI} \qquad (11.73)$$

At $x = \; ; \frac{dy}{dx} = 0$, and from (3) we have

$$A = B = \frac{g\ell}{3} \qquad (11.74)$$

Then from Equation (11.69)

$$C = \frac{4}{3} g\ell \qquad (11.75)$$

and from (11.73)

$$d = \frac{g\ell^4}{72EI} \qquad (11.76)$$

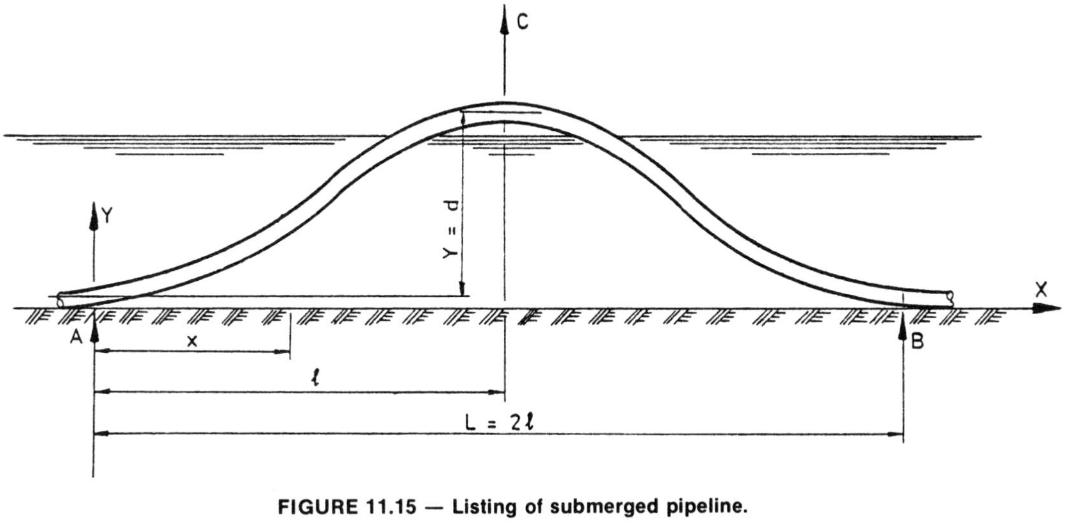

FIGURE 11.15 — Listing of submerged pipeline.

and from this

$$\ell = \sqrt[4]{\frac{72EI}{g}} \tag{11.77}$$

The location of the maximum moment is found from (11.70)

$$\frac{dM_x}{dx} = A - gx = 0$$

$$x = \frac{A}{g} = \frac{\ell}{3} \tag{11.78}$$

Therefore, the maximum span moment is

$$M_{max} = Ax - \frac{gx^2}{2} = \frac{g\ell^2}{18} \tag{11.79}$$

and the maximum support moment at x = ℓ, from Equation (11.70)

$$M_s = -\frac{g\ell^2}{6} \tag{11.80}$$

Therefore the maximum stresses in the pipe at the time of the lifting are

$$\sigma_{max} = \frac{g\ell^2}{6W} \tag{11.81}$$

After substituting the value (11.77) into the value (11.81), we have

$$\sigma_{max} = \sqrt{\frac{EgDd}{W}} \tag{11.82}$$

Therefore, the maximum depth from which the pipe may be lifted is

$$d_{max} = \frac{\sigma^2_{max} W}{EgD} \tag{11.83}$$

11.7 Structural Analysis and Design of Pipelines

The structural analysis of a submarine pipeline is a complex problem due to the nature of its loading, both static and dynamic. The designer of a submarine pipeline must have a knowledge of the anticipated method of installation, as well as the operational conditions. It should be realized that the greatest challenge to the designer of a submarine pipeline is in adequately defining the loading conditions. In knowing this data, the stability of an underwater pipeline must be assured, both during its construction and during its operation.

11.7.1 Stability Requirements

The stability requirements at different depths will depend upon the local conditions, with respect to currents and maximum waves, as well as the probability of simultaneous occurence. A pipeline route direction must be relative to prevailing waves and currents, as well as any frictional resistance between the pipeline. Sea bottom soils are also of importance. The stability requirements to weight will increase with an increasing pipeline diameter. A gas-line may need a higher submerged weight than a liquid line of the same size, due to its permanent light condition.

It is of importance to have accurate stability design methods and criteria, in particular in situations where the submerged weight requirements is close to the lay-barge capability. The friction factor, lift and drag coefficients, may have to be specially investigated, as well as in the actual distributions of the steady-state current and wave-induced current close to the bottom. Further, an accurate design method has to include the inertia effect due to the wave action. Consideration may also have to be given to such problems as the simultaneous action of waves and current perpendicular to the pipeline.

The possibility of flotation of a buried pipeline has to be considered in a submerged weight design. This factor becomes increasingly important in deeper waters, where the submerged weight has to be low. For liquid lines, this problem may be overcome by trenching the liquid-filled line. However, a gas-line may need a weight in excess of its stability requirement, in order to stay buried.

11.7.2 Stability During Construction

The stability of a submarine pipeline during its construction is affected by the following factors:

a. Ocean waves and currents.
b. The method of construction (lay-barge with or without a stinger, reel barge, flotation, bottom-pull, or others.
c. Bottom conditions, and the specific gravity of the pipe.
d. Bottom soil conditions and the bottom currents that will affect the pipe's stability prior to the burying operation, if a pipe is to be buried.

The environmental effect not only influences the behavior of a pipe directly, but also affects the pipe through the motions that are imposed on the construction equipment.

11.7.3 Stability In-Place

In-place or operational stability must be assured for the life of the project. This involves a greater consideration than might be necessitated for short-term construction requirements. Significant factors affecting operational stability are:

a. Bottom soil conditions. It is possible to experience pipeline instability due to unstable soils. This is generally produced by erosion or scouring. Erosion is associated with a general shifting of the bottom soils. Scouring is induced by vortex circulation on the downstream side of the pipe.

   Another type of soil failure is caused by soil liquifaction. If this problem exists, the pipeline must be designed so that it will remain stable in the liquified soil. This requires a careful analysis of the pipeline's specific gravity and the possible use of a form of anchoring.

b. Bottom currents may produce scouring around the pipe. The hydrodynamic forces created by ocean currents passing over and under the pipe can induce sliding and lift-off, especially for exposed pipelines.

c. Bottom topography, especially in the case of uneven bottom conditions, where a pipe may span low areas. Vortex shedding is a common occurrence. The combination of certain span lengths and water velocities can permit a resonance to occur. This condition exists when the frequency of the vortex shedding and the natural frequency of the structural configuration are coincident.

11.8 Hydrodynamic Forces Acting on a Submarine Pipeline

The pipeline must be protected from displacement due to waves, tides and currents. The most positive means of accomplishing this is to bury the pipe below the bottom surface. Burial also keeps the pipes out of the way of dragging anchors which can be a real problem. However, for extremely long pipelines, it may not be economically feasible to bury the pipeline except in the near shore reaches. The designer must then estimate the magnitude of forces which can act on the pipeline while it rests on the bottom. The first step in this process is a study of currents which can appear in the vicinity of the pipeline, particularly during storms. A wave forecast must be made to determine the design wave characteristics at the depth considered. These characteristics are used to estimate the horizontal component of velocity in a non-breaking wave passing over the pipeline. According to Wiegel [11.2]

$$V_p = \frac{\pi H}{T} \cdot \frac{\cosh[2\pi(y+d)/L]}{\sinh \frac{2\pi d}{L}} \cos 2\pi\left(\frac{X}{L} - \frac{1}{T}\right) \tag{11.84}$$

where

V_p = particle velocity

y = vertical distance from the still water surface to the water particle (measured negative downwards)

d = water depth

L = wave length

H = wave height

T = wave period

X = horizontal distance measured in the direction of wave advance from the wave crest

t = time measured as positive from the time the crest of the wave passes the center of the object

Given the design velocity, the force due to this velocity can be estimated by methods proposed by Wiegel [11.2]

$$F = 2\frac{W}{2g} Vf + \frac{1}{2} C_D \frac{W}{2g} A \mu^2 \tag{11.85}$$

where

F = force due to inertia, frictional drag and form drag

$\frac{W}{2g}$ = mass density of fluid

V = volume of fluid displaced by the body

f = acceleration of the undisturbed fluid at the center of the body, in the event that the body did not exist

C_D = coefficient of drag

A = projected area perpendicular to stream velocity

μ = stream velocity

The calculation of forces is generally made for the condition when the pipe is full, and it must resist the largest forces which can be applied. However, the construction condition should also be investigated. While on the one hand, the pulling contractor wants the pipe as light as possible, in order to minimize the pulling lead, it should be so light as to be susceptible to displacement by ordinary currents during installation. The ordinary range of negative buoyancies allowed for the pipe in the construction stage (empty) is from 4 lbs/ft to 10 lbs/ft.

It is necessary to mention at this point, about the use of pontoons to lighten the pipe during its construction. The pontoons provide a considerably additional longitudinal area for the current to work against and their use in certain circumstances could result in considerable extra hazards.

11.9 Preliminary Structural Design

The structural design technique used for a submarine pipeline project must recognize not only the environmental conditions and the operating state, but also the special conditions encountered during the installations. Attention must again be directed to the fact that this chapter deals primarily with long steel submarine pipelines, which are installed under open water conditions by the pulling method. These are the difficult applications which require the closest coordination between the design and construction techniques. The basic objective of the design process is to place the pipeline without damaging it, and hold it in place against the natural forces for its design life.

The initial calculation for the pipe will establish the basic internal and external pressure stresses at maximum points. The hoop stress at a point of maximum internal pressure will be calculated, taking into account any possible surge conditions. The thickness of the pipe needed to resist the internal pressure is given as follows:

$$t = \frac{PD}{f_{max} 2(1-A)} + C_a \tag{11.86}$$

where

t = wall thickness, inches

P = internal pressure, psi

D = outside diameter of pipe, inches

f_{max} = maximum allowable working stress, psi

A = underthickness tolerance of steel pipe expressed as a decimal part of t

C_a = wall thickness allowance for corrosion

The method of construction contemplated requires bulkheading the end of the pipe and pulling the pipe empty through the maximum depth of the water encountered along the profile. The pipe must therefore be designed to resist this full external head during the construction phase, or later, if the pipe is ever emptied again. The formula for the buckling of a thin wall tube is as follows [11.3]

$$P_{cr} = \frac{1}{4}\left(\frac{E}{1-\nu^2}\right)\left(\frac{t}{r}\right)^3 \qquad (11.87)$$

where

P_{cr} = critical pressure, psi

E = modulus of elasticity, psi

ν = Poisson's ratio

t = wall thickness, inches

r = pipe radius, inches

In some instances, involving very deep crossings, this factor may well constitute the basic constraining factor. With special design and construction techniques it is possible, however, to lay a pipe in water as deep as 9,000 feet [11.4].

The next structural condition which must be checked concerns the bending to which the steel pipe will be subjected during installation.

Afterwards, when the pipe is filled with liquid, it may not fit the bottom profile and any possible spanning conditions must be checked.

The sag of a pipe section having a long length, introduces a complex problem, from the standpoint of induced stresses [11.5]. The pipe is in essence, a beam which is so long that when vertical deformation occurs, it is accompanied by a significant elongation. The firm supporting sediment will tend to restrain the movement of the pipe at the ends of the sagging section. This can induce a net axial tension of considerable magnitude, which must be added to the flexural stresses induced by the bending of the pipe.

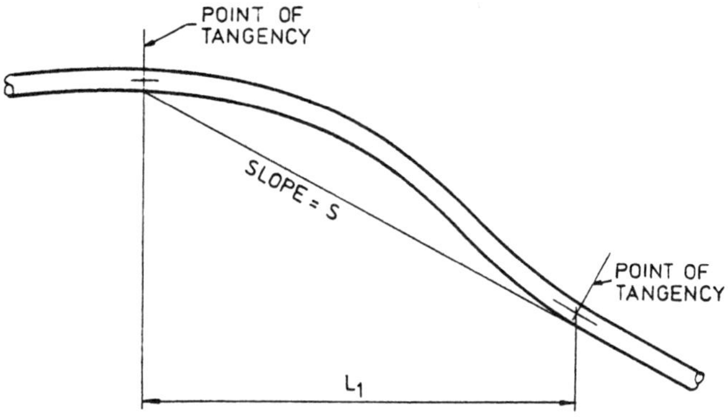

FIGURE 11.16 — Profile "A".

In the case of a very long span, the bending effect can become so small that the sagging pipe can be considered a flexible cable. In this case, the pipe will assume the shape of a catenary. If the span is very short, then the pipe can be designed as a continuous beam with stresses resulting only from bending plus any tension required for pulling the pipe during its construction. In order to assure a safe design where the sag is likely to occur, it is necessary to compute both the flexural and pure tensile stresses induced by the sag.

In the following, the preliminary design equations for the conditions during construction and operational stages are given for the two different types of profiles, A and B, as shown in the following Figures 11.16 and 11.17.

In the construction stage, Figure 11.17, the preliminary design equations are

$$wL_1^3 - 24S\left(\frac{P_1L_1^2}{6} - EI\right) = 0 \qquad (11.88)$$

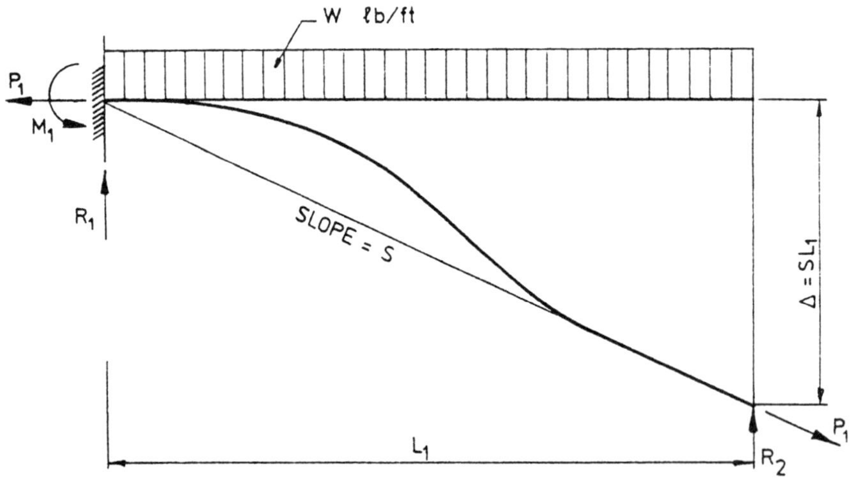

FIGURE 11.17 — Profile "A". Construction stage.

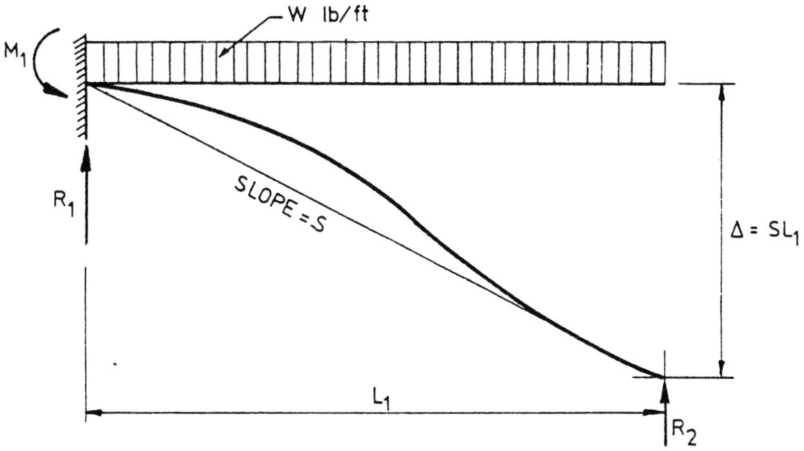

FIGURE 11.18 — Profile "A". Operating stage.

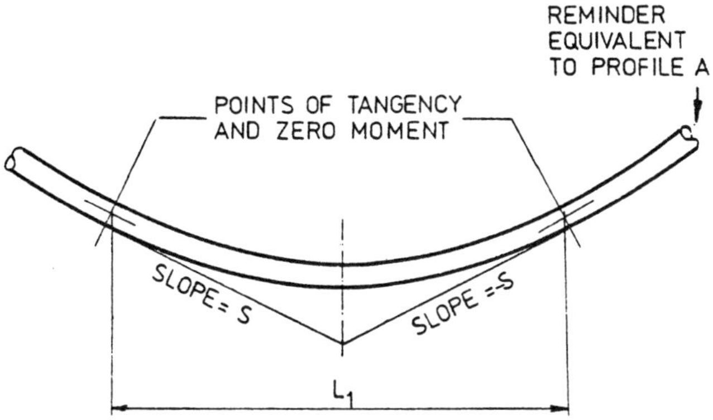

FIGURE 11.19 — Profile "B".

$$R_1 = \frac{3wL_1}{4} + P_1S \quad (11.89)$$

$$R_2 = \frac{wL_1}{4} - P_1S \quad (11.90)$$

$$M_1 = -\frac{wL_1^2}{4} \quad (11.91)$$

In the operating stage, Figure 11.18, the preliminary design equations are

$$L_1^3 = \frac{24S\ EI}{W} \quad (11.92)$$

$$R_1 = \frac{3wL_1}{4} \quad (11.93)$$

$$R_2 = \frac{wL_1}{4} \quad (11.94)$$

$$M_1 = -\frac{wL_1^2}{4} \quad (11.95)$$

In the construction stage, Figure 11.20, the preliminary design equations are

$$wL_1^3 - 3S(P_1L_2^2 + 8EI) = 0 \quad (11.96)$$

$$R_1 = R_2 = \frac{wL_2}{2} \quad (11.97)$$

$$M = \frac{wL_2^2}{8} - \frac{P_1SL_2}{2} \quad (11.98)$$

$$L_2^3 = \frac{24S\ EI}{w} \quad (11.99)$$

$$R_1 = R_2 = \frac{wL_2}{2} \quad (11.100)$$

$$M = \frac{wL_2^2}{8} \quad (11.101)$$

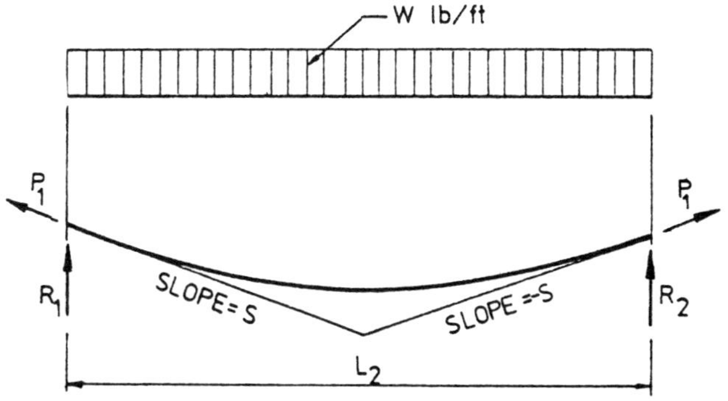

FIGURE 11.20 — Profile "B". Construction stage.

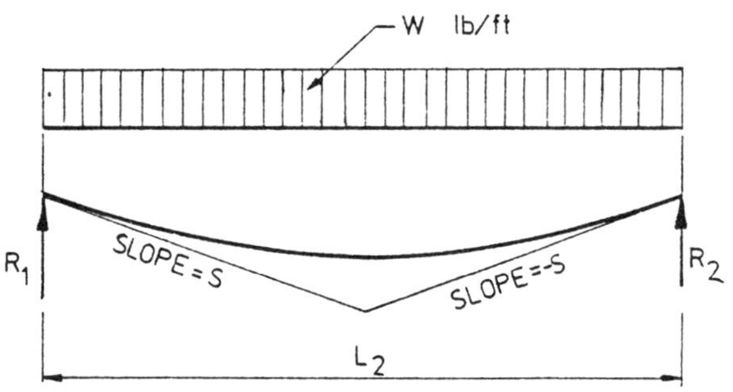

FIGURE 11.21 — Profile "B". Operating stage.

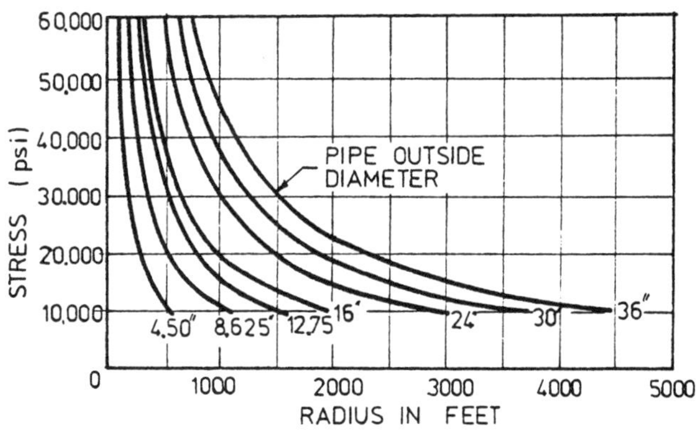

FIGURE 11.22 — Bending radius and stress for a steel pipe.

11.10 Pipe Bending Stress and Radius of Curvature

1. Bending stress is a most important factor in designing an underwater pipeline, frequently being the controlling factor. Since steel assumes the shape of an elastic curve, an analysis may be made by ordinary beam theory methods. A simple but useful way to measure bending stress is in terms of pipe curvature. The maximum fiber stress is given by the following equation

$$f_{max} = \frac{ED_0}{2R} \qquad (11.102)$$

where

f = maximum fiber stress, psi

R = radius of curvature, in

E = modulus of elasticity, psi

D_0 = outside diameter of the pipe

In establishing a safe profile for the bottom pull method or the overbend at a lay barge, examples are given where the radius of the curvature provides a good design criteria. The bending radius and stress for the various diameters of the pipe are shown in Figure 11.22.

For some situations, however, it is not convenient to measure stress in terms of curvature. A pipeline resting on the bottom with one end supported at a certain distance above the sea floor, is shown in a simple illustration, Figure 11.23.

The controlling conditions are the height and the slope of the raised end of the pipe.

An analysis of the unsupported pipe span in Figure 11.23 can be closely approximated from the differential equation for the elastic curve, which is indicated by the known equation

$$EI\frac{d^2y}{dx^2} = M$$

where

x = distance from the left end of the beam to any section, in.

y = vertical deflection, in.

M = bending moment in terms of x, in/lb.

E = modulus of elasticity, psi

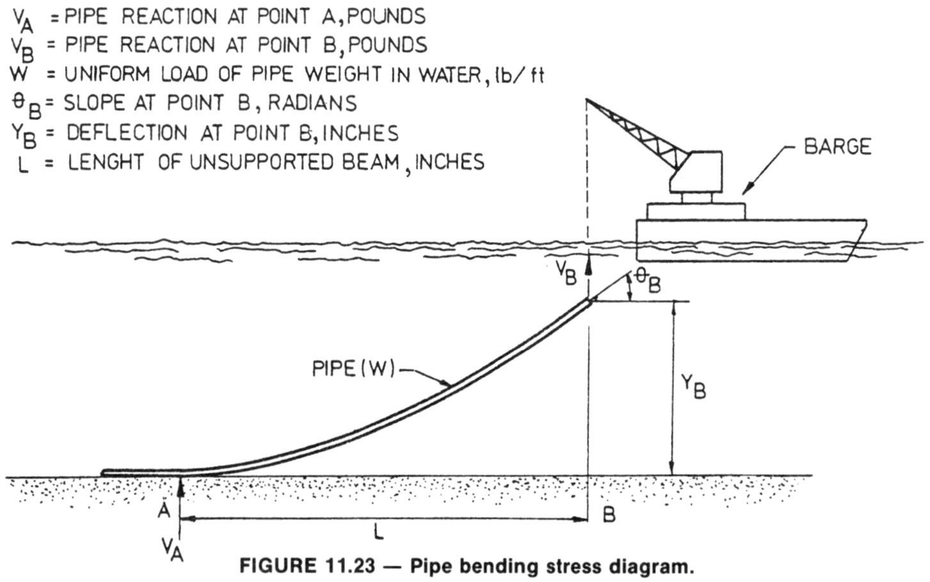

FIGURE 11.23 — Pipe bending stress diagram.

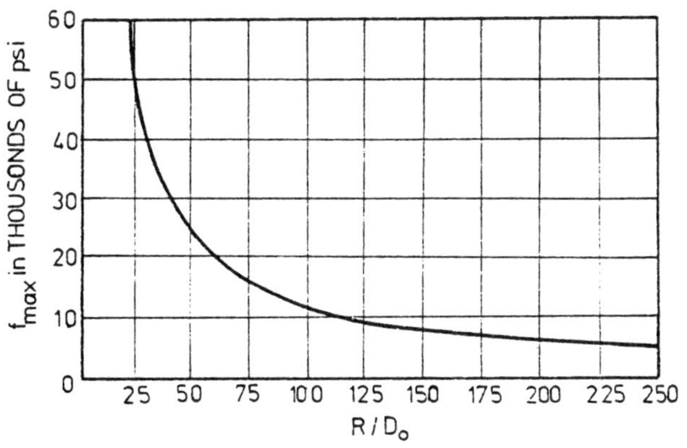

FIGURE 11.24 — Plot of f_{max} versus R/D_0.

I = moment of inertia of the section of the beam in respect to the neutral axis, in^4

2. From Formula (11.102), the radius of the curvature is

$$R = \frac{ED_0}{2f_{max}} \quad \frac{ED_0}{24f_{max}} \text{ (feet)} \qquad (11.103)$$

or using E = 30 × 10^6 psi, and rearranging

$$\frac{R}{D} = \frac{1.25 \times 10^6}{f_{max}} \qquad (11.104)$$

Figure 11.24 is a curve with fmax plotted versus R/D_0.

The radius of curvature is constant over the entire length of the beam only when each section of the beam is acted upon by the same bending moment. The stress will be uniform at each section only when the elastic curve is a circular arch.

11.11 Concept of Structural Stability

Structural stability analysis of submarine pipelines includes the considerations of buckling, collapse, and vibrations.

1. *Local Buckling*

About the worst thing that can happen to a submarine pipeline during its construction is for it to develop a *buckle*. The buckle may be defined as a localized failure in the form of a *wrinkle* or indentation caused by overstress or instability of the pipe wall on the compression side of the pipe subjected to bending. Such *local buckling* is the result of a structural instability of the pipe wall when subjected to compressive stress. Submarine pipelines can experience this type of buckling if the bending stresses exceed critical values.

2. *Collapse*

The only thing worse than local buckling would be the collapse of the pipeline. It is very important to understand the collapse mechanism, since there are indications that under certain conditions the mode of failure may be progressive. A collapse initiated by a critical condition at one point in the pipe may progress or run along the pipe causing collapse at locations where the stress would not otherwise have been critical.

Collapse is defined as a general failure usually in the form of a *flattening* of the pipe cross-section over a considerable length, as the result of the action of external pressure on the pipe.

Both resistance to collapse and resistance to local buckling are functions of the diameter-to-thickness ratio (D/t) of the pipe. The lower this ratio, the greater the resistance to failure in either case. Collapse and local buckling become extremely important during the design of large diameter tanker loading lines, since the D/t ratio is apt to be quite high for economic reasons.

A number of important factors which influence the structural stability of submarine pipelines, are as follows:

a. The effect of initial imperfections such as out-of-rounds, variable wall thickness, etc.
b. The effect of concrete wrap on critical external buckling pressure.
c. The bending moment, axial force, external pressure interaction effects, and
d. The yield strength of the pipe material.

3. *Vibrations*

Although the vibration of submarine spans is caused by hydrodynamic loading due to vortex shedding, this vibration can be considered a structural instability problem in the event that resonance occurs. Resonance can occur if the frequency of vortex shedding is at or near the natural frequency of the pipe span. Under these conditions, the amplitude of vibration may become large enough to create a bending stress condition that could lead to metal fatigue. If an analysis indicates that such a condition may occur, it is necessary to correct this situation either by special bottom preparation prior to its construction, or by supporting the spans after pipe installation.

11.12 Pipe Strength Problem

Pipe strength problems become increasingly important in deeper waters. The pipeline has to be designed to resist the combined action of outside pressure and bending, which for a liquid line will occur during the laying and shortly after. For a gas-line, this condition may also occasionally occur during the lifetime of the line. The resulting heavy wall may give an inside pressure capacity which exceeds the design operating pressure of the pipeline system. Figure 11.25 illustrates the pipe strength design problem.

Being a certain depth the D/t ratio has to be chosen to give protection against the different failure machanisms.

A pipeline with a certain out-of-roundness will have a corresponding collapse pressure. An acceptable design has, therefore, always to be below the collapse region, as

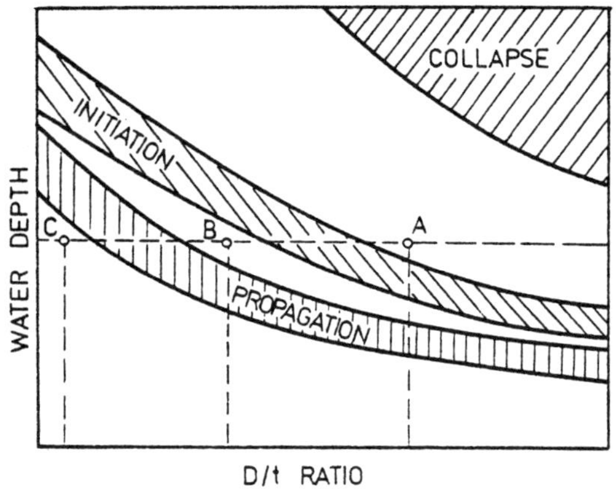

FIGURE 11.25 — Buckling strength design chart.

illustrated in Figure 11.25, which also may include the effect of limited bending. Points A, B and C represent such design, with a decreasing D/t ratio, and therefore increasing the collapse safety.

11.12.1 Buckle Propagation

In the past, the design of submarine pipelines has included consideration of the effect of external pressure, but only to the extent of checking that the critical collapse pressure for the size of the pipe chosen is less than the maximum water depth to be encountered. However, recently, it has been found that if the pipe cross-section is deformed by excessive bending, or on impact, collapse can occur at a much lower pressure than that for round pipe, and furthermore, that the collapse can progress away from the original deformation, resulting in the flattening of a considerable length of the pipeline.

In a practical situation, the outside pressure will be limited to the corresponding water head, and therefore collapse, as a simultaneous action along a longer part of the pipeline, will not take place. However, excessive bending may cause a local buckle in the pipeline, which, depending upon the pressure, may turn into a propagating buckle. The design points, A, B and C, represent three typical situations with respect to buckle propagation.

Point A, is above the initiation pressure, and the water pressure will force a local buckle with a typical transverse orientation to turn into a propagating buckle. This buckle has a typical front, oriented in the longitudinal direction of the pipe, and the front will move with considerable speed from the original buckle to each side. The buckle may be stopped either by a sufficiently reinforced section of the pipe, or if shallower water is reached where the pressure is below the pressure required to sustain the buckle propagation.

Point B is between the initiation pressure and the propagation pressure. The pressure is not sufficient to transform the shape of a typical local buckle into the shape of the propagating buckle front. Propagating buckles should then, theoretically, not occur, even if the pressure is sufficient to cause further propagation of an already formed propagating buckle. However, there is no guarantee that buckle propagation will not take place, as there is always uncertainty with respect to the actual shape of the local buckle.

Point C is below the propagating pressure, and a local buckle will therefore always remain local.

It is seen that certain ranges are given for the initiation and propagating pressure, indicating that the limitations of such design diagrams, are mostly based on experiments. An overlap of the ranges may as well reflect today's knowledge on the subject.

In the case when a buckle may occur, the pipelines designed according to Point A have to be reinforced with stiffening rings at regular intervals, and the rings may be provided also for designs according to Point B. The design aspects of such buckle arrestors then become increasingly complex in deep waters.

The D/t ratio is usually chosen according to a certain buckle strength criterion, taking into consideration the material costs, risk factors, etc., and the possible consequences of a buckle, in particular for the laying operation.

In order to learn more about the exact nature of the problem, and to investigate the means of prevention, a number of tests were carried out in pressure chambers, and on the actual pipelines, both in England and in Japan. A full-scale test was then made at sea, during the pipe-laying trials.

It was found that the problem would only arise in the very deepest water, and an ingenious proprietary method of arresting the propagation of the buckle was developed.

11.13 Buckling of Pipeline Under Hydrostatic Pressure

11.13.1 Long Unstiffened Pipe

According to Timoshenko [11.6], the critical value of the hydrostatic pressure on a long circular pipe is

$$P_c = \frac{E}{4(1-\nu^2)} \left(\frac{t}{R}\right)^3 = \frac{2E(t/D)^3}{(1-\nu^2)} \qquad (11.105)$$

or, for $\nu = 0.3$,

$$P_c = 2.2E\,(t/D)^3 \qquad (11.106)$$

as long as the corresponding compressive stress does not exceed the proportional limit of the material.

The critical stress is

$$\sigma_c = \frac{E}{(1-\nu^2)} \left(\frac{t}{2R}\right)^2 \qquad (11.107)$$

11.13.2 Stiffened Pipe

In the case of very long pipes under hydrostatic pressure, P_c is relatively small and it is necessary substantially to increase the thickness of the pipe wall. It is often economic to provide stiffening rings at intervals. In this case, we consider the design of a short-length pipe, which should be checked against stability, as well as stiffening rings. For the design of a pipe, we apply von Mises' equation for hydrostatic pressure [11.7]

$$P_c = \frac{2E(t/D)}{n^2+(\lambda^2/2)-1}\{\frac{(t/D)^2}{3(1-\nu^2)}[(n^2+\lambda^2)^2 - 2n^2+1] + \frac{\lambda^4}{(n^2+\lambda^2)^2}\} \quad (11.108)$$

where

$$\lambda = \frac{\pi D}{2L} \quad (11.109)$$

For very short spacing of a stiffening ring, the axi-symmetric elastic buckling will occur. As n approaches zero and λ increases, Equation (11.108) at $\nu = 0.3$, reduces to

$$P_c = \frac{2.42E}{(D/t)^2}[\frac{2\theta^2}{\pi^2} + \frac{\pi^2}{2\theta^2}] \quad (11.110)$$

where

$$\theta = \frac{1.818L}{D\sqrt{D/t}} \quad (11.111)$$

An excellent approximation of von Mises formula for

$$2 < \theta < D/t \quad (11.112)$$

was developed by Windenburg and Trilling [11.8]

$$P_c = \frac{2.60E(t/D)^{2.5}}{(L/D - 0.45(t/D)^{0.5}} \quad (11.113)$$

For values

$$10 < \theta < D/t \quad (11.114)$$

the second term in the denominator of Equation (11.113) may be dropped. The resulting equation becomes

$$P_c = \frac{2.60E}{(L/D)(D/t)^{2.5}} \quad (11.115)$$

For

$$\theta > D/t \quad (11.116)$$

the buckling mode changes to n = 2, and Equation (11.108), reduces to

$$P_c = \frac{2E(t/D)}{3+(\lambda^3/2)}\{\frac{(t/D)^2}{3(1-\nu^2)}[(4+\lambda^2)^2 - 7] + \frac{\lambda^4}{(4+\lambda^2)^2}\} \quad (11.117)$$

For

$$\theta > 4(D/t) \quad (11.118)$$

the λ terms become insignificant and Equation (11.117) reduces to the simple formula of Timoshenko for an unstiffened pipe, Equation (11.105)

$$P_c = \frac{E}{4(1-\nu^2)}(\frac{t}{R})^3 = \frac{2E(t/D)^3}{(1-\nu^2)}$$

11.13.3 Critical Length Between Stiffening Rings

1. Equation (11.105) applies to long thin pipes under external pressure without circumferential stiffening rings or with the stiffening rings spaced at or beyond the "critical length". To make allowance for the added restraint offered by stiffeners spaced at less than the critical length, the critical length may first be evaluated.

   The expression for the critical length was first developed by Southwell [11.9]. Southwell's analysis involves a 15-row determinant solution. The relationship resulting from Southwell's analysis is given by the following formula

$$\ell_c = 1.11D\sqrt{D/t} \quad (11.119)$$

where

ℓ_c = critical length, inches
D = diameter of shell, inches
t = shell thickness, inches

11.13.4 Collapsing Pressure of Pipeline with Ring Stiffeners

For pipelines in which stiffening rings are spaced at less than the critical length, the coefficient of Equation (11.105)

$$P_c = \frac{2E}{1-\nu^2}(\frac{t}{D})^3 = 2.2E(\frac{t}{D})^3$$

must be modified according to the proportions of the pipe [11.10], or

$$P_{theor.} = KE(\frac{t}{D})^3 \quad (11.120)$$

Applying a factor of safety 4 gives

$$P_{allow.} = \frac{K}{4}E(\frac{t}{D})^3 \quad (11.121)$$

where

K = coefficient according to the proportions of the pipe

as indicated in Figure 11.26 [11.11]. It should be noted that the minimum value of K is 2.2, as shown in Equation (11.106).

Substituting Equation (11.120) into an expression for circumferential compressive stress

$$t = \frac{pD}{2f} \quad (11.122)$$

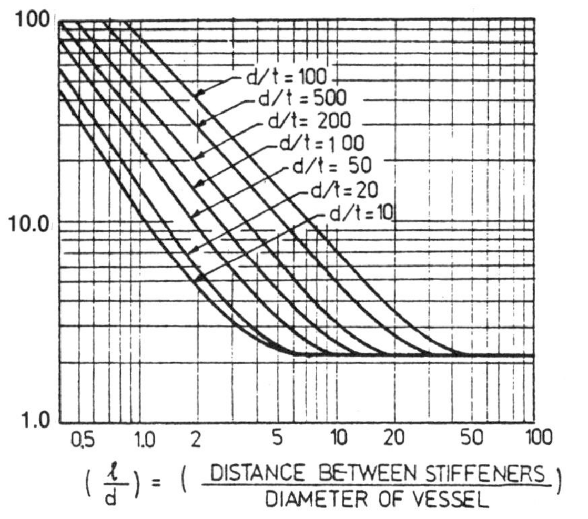

FIGURE 11.26 — Collapse coefficients for cylindrical shells under external pressure.

from external pressure at which collapse occurs is:

$$f = \frac{P_{theor.} D}{2t} = \frac{DKE}{2t} \left(\frac{t}{D}\right)^3 \qquad (11.123)$$

or

$$\frac{f}{E} = \varepsilon = \frac{K}{2} \left(\frac{t}{D}\right)^2 \qquad (11.124)$$

Equations (11.120) and (11.124) may be plotted for convenience of the solution, as shown in Figure 11.27.

The inflections in the parameters occur at the critical lengths which correspond to the critical lengths determined by Equation (11.119). The vertical parameters of D/t above the inflections represent the region where the spacing between the stiffeners exceeds the critical length and the collapsing pressure is independent of the ℓ/D ratio. Equation (11.119) applies in this region. The inclined parameters below the inflection represent the region where the stiffeners have an effect and the collapsing pressure is a function of the ℓ/D ratio, as expressed by the coefficient K in Equation (11.123).

It is significant to note that Figure 11.27 is general and is independent of the material of construction.

11.13.5 Design of Ring Stiffeners

In designing circumferential stiffening rings for pipes under external pressure each stiffener is considered to resist the external load for an ℓ/2 distance on either side of the ring, where ℓ is the spacing between the rings. Thus the load per unit length on the ring at its collapse point is equal to ℓ at P_{theor}.

According to Timoshenko [11.12] the following equation expresses the theoretical or critical load per unit circumferential length of unit width of the circumference

$$P_{theor.} = \frac{3EI}{R^3} = \frac{24EI}{D^3} \qquad (11.125)$$

For a strip of unit width ℓ = 1 the critical load is the pressure at which buckling theoretically occurs. Therefore

$$P = P_{theor}(\ell) = \frac{24EI}{D^3} \qquad (11.126)$$

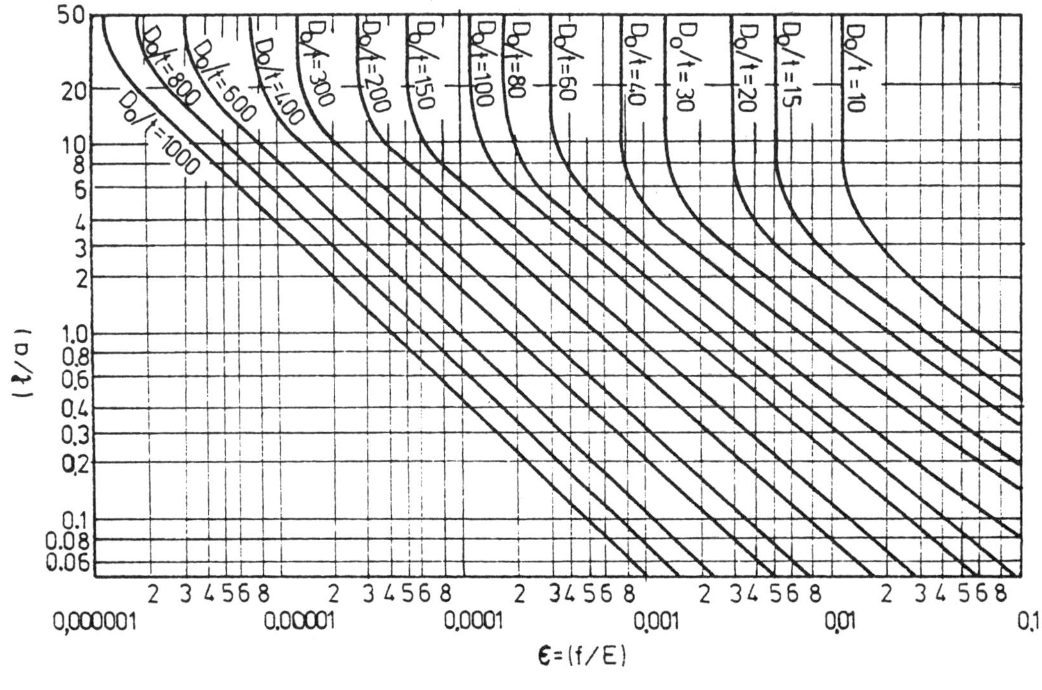

FIGURE 11.27 — General chart for collapse of pipes under external pressure showing relationship between the dimensional ratios D/t and ℓ/D and the physical property f/E [135].

where

P = load on combined shell and stiffeners in lbs per inch of circumferential length

or

$$I = \frac{P_{theor.} D^3 \ell}{24E} \qquad (11.127)$$

multiplying by t/t and rearranging gives

$$I = \left(\frac{D^2 \ell t}{12E}\right)\left(\frac{P_{theor.} D}{2t}\right) \qquad (11.128)$$

Substituting Equation (11.122)

$$f = \frac{PD}{2t}$$

and known formula

$$\epsilon = f/E$$

gives

$$I = \frac{D^2 \ell t}{12E}(f) = \frac{D^2 \ell t \varepsilon}{12} \qquad (11.129)$$

The moments of inertia of the stiffening ring and the shell act together to resist collapse of the pipe under external pressure. Timoshenko [11.12] has shown that the combined moment of inertia of the shell and stiffener may be considered as equivalent to that of a thicker shell, or

$$t_y = t + \frac{A_y}{d_y} = t + \frac{A_y}{\ell} \qquad (11.130)$$

t_y = equivalent thickness of shell, in

A_y = cross-sectional area of one circumferential stiffener, in^2

ℓ = d_y = distance between circumferential stiffeners, in.

Substituting Equation (11.130) into Equation (11.129) gives

$$I = \frac{D^2 \ell}{12}\left(t + \frac{A_y}{\ell}\right)\varepsilon \qquad (11.131)$$

where

I = required moment of inertia of stiffening ring, in^4

In general, the combined moment of inertia of the stiffener and the shell together varies from 30% to 70% greater than the moment of inertia of the stiffener along [11.13]. Using a conservative allowance of a 30% increase in the I of the stiffener when combined with the shell and introducing additional safety in design by increasing the load by 10%, we obtain

$$I = \frac{1.1D}{(1.3)(12)}\left(t + \frac{A_y}{\ell}\right)\varepsilon =$$

$$= \frac{D^2 \ell}{14}\left(t + \frac{A_y}{\ell}\right)\varepsilon \qquad (11.132)$$

where

ϵ = unit strain

11.13.6 Design of Long Pipe Having Imperfections

Timoshenko [11.12] also has developed a formula for determining the elastic critical pressure for pipes of infinite length having imperfections. The hydrostatic pressure P_y at which yielding begins can be determined from the following equation

$$P_y^2 - \left[\frac{2\sigma_y t}{D} + \left(1 + \frac{1.5De_0}{t}\right)P_c\right]P_y + \frac{2\sigma_y t}{D} P_c = 0 \qquad (11.133)$$

where

e_0 = $\frac{D_{max} - D_{min}}{D} = \frac{4e}{D}$ = out-of-roundness, Fig. 11.28

e = the radial eccentricity

P_c = critical pressure determined by Equations (11.105) to (11.118) as appropriate

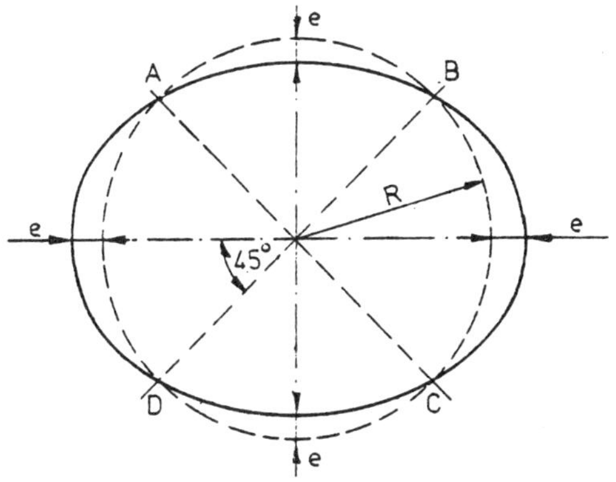

FIGURE 11.28 — Pipe having imperfection.

A value of e_0 = 0.01 has been adopted for fabricated pipes by the American Petroleum Institute [11.14] and ASME [11.15]. The designer should make sure that the value of out-of-roundness used in his design is consistent with tolerances imposed in his situation during construction.

11.14 Vertical Stability of the Sagging Pipeline

The proper design of a submarine pipeline includes a consideration of the vertical stability of the sagging pipeline [11.5].

In the areas where the bearing capacity of the upper sediments is small, a downward sag of a pipeline can occur and entrenchment of the line to considerable depths may be necessary in order that excessive stresses within the pipe be avoided. Because both the flexural and longitudinal tensile stresses, occurring simultaneously, can be important in a sagging pipeline, both must be evaluated.

The sag of a pipe section, having a length of the order of 200 feet or more, introduces a complex problem from the standpoint of computation of induced stresses. In this case, the pipeline is so long that when a vertical deformation occurs, it is accompanied by a significant elongation. The firmer sediment adjacent to the weak zone will tend to restrain the movement of the pipe at the ends of the sagging portion of the pipe so that practically all of the elongation will occur in the sagging section. This can introduce a net axial tension of considerable magnitude. The tensile stress thereby induced in the material is in addition to the tensile and compressive flexural stresses induced by the bending of the pipe.

In the case of a very long pipe, the bending effect can become so small that the sagging pipe can be considered essentially as a flexible cable. In this case, the pipe will assume the shape of a catenary under the action of a uniform load per unit length, with the tensile force carrying the full load. If the pipe section is very short, or if the deflection is very small, then the theory of simple bending may apply. In this case, the net tension would be negligible and the load is carried entirely by shear forces. The situation regarding the pipe sag in the sediments in general, involves both the bending stresses and net tension, and the load is carried partially by shear and partially by tension. In order to insure a safe design where the sag is likely to occur, it is therefore necessary to compute both the flexural and pure tensile stresses induced by the sag.

11.14.1 Criterion for Sinking of the Pipe

An off-shore pipeline which is resting upon the bottom will exert a downward load on the underlying sediment which is simply the submerged weight of the pipe in water. The maximum reaction which the sediment can exert may be referred to as the ultimate load-bearing capacity of the sediment. The criterion for sinking of the pipe is that the net downward gravitational load exerted by the pipe is greater than the ultimate load-bearing capacity of the sediment and the following data is based on the theory proposed by Reid [11.5].

For silty clay sediments, the ultimate load-bearing capacity is evidently independent of the pressure within the sediment, and depends only upon the shear strength and load distribution. The ultimate load-bearing capacity per unit length of pipe, P_b, is

$$P_b = K_b D \tau_u \qquad (11.134)$$

where

K_b = the factor of proportionality

D = the overall diameter of the protected pipe

τ_u = the ultimate shear strength of the sediment

For concrete coated pipe entrenched in silty clay sediment, an approximate value of $K_b = 2$ is indicated from the experience. There is a significant probability that sinking will occur if the net load per unit length, exerted by the pipe, exceeds $2D\tau_u$, and little chance that it will not sink, if the load exceeds $5\ D\tau_u$. If sinking is to be avoided the net load should be less than $2\ D\tau_u$.

11.14.2 Symmetrical Sag of a Pipeline in Plastically Deformed Sediment

If the vertical restraint per unit length, P_m, offered by the plastically deformed sediment is uniformly weak in the zone of pipe sag, and if the conditions of support at the ends of the sagging section are similar, then the vertical deformation of the pipe will be symmetrical with respect to the center of the sag.

The net vertical load on the pipe per unit length, W, is simply the net weight of the pipe in the sediment minus the reaction P_m. If the W_p represents the weight of the pipe per unit length in the air, including the weight of transported fluid, then

$$W = W_p - \frac{\pi D^2}{4} B - P_m \qquad (11.135)$$

where B represents the buoyant force per unit volume of the pipe, or

$$B = \frac{\rho_s g}{1 + \frac{100}{Q}} \qquad (11.136)$$

where

ρ_s = the density of the sediment

Q = the water content of a sediment in terms of the percent of dry mass of sediment

The criterion for sinking is

$$\left(W_p - \frac{\pi D^2}{4} B\right) > P_b \qquad (11.137)$$

Therefore, if the net weight is great enough, then the sediment deformation exceeds the elastic state and its structure is broken down, reducing the possible reaction which it can develop.

11.14.3 The Basic Equations of Combined Flexure and Elongation in a Pipe

An initial tension, N_o, may exist in the pipeline, due to thermal or pressure effects within the pipe, Fig. 11.29 (a) and a schematic diagram of the sagging section of pipe is shown in Figure 11.29 (b).

The equilibrium of forces and moments is represented graphically by Figure 11.30.

The origin of the coordinate system is taken at the point of maximum sag; x represents the horizontal distance measured positively to the right of this point, and y is the vertical distance measured positively upward from this point. The bending moment at the origin is denoted by M_o, and the total axial tension after deformation of the pipe is represented by N. An initial tension may exist in the pipeline, due to thermal or pressure effects within the pipe is denoted by N_o.

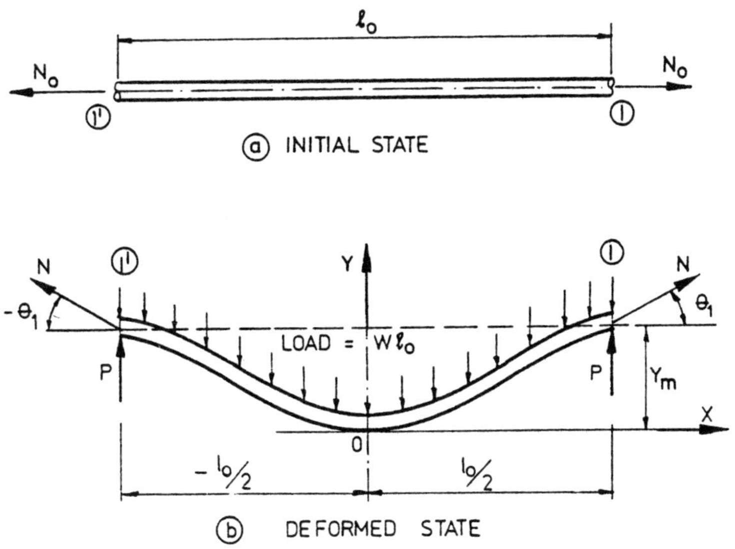

FIGURE 11.29 — Schematic diagrams of symmetrical free sag in a pipe, (a) initial state, (b) deformed state.

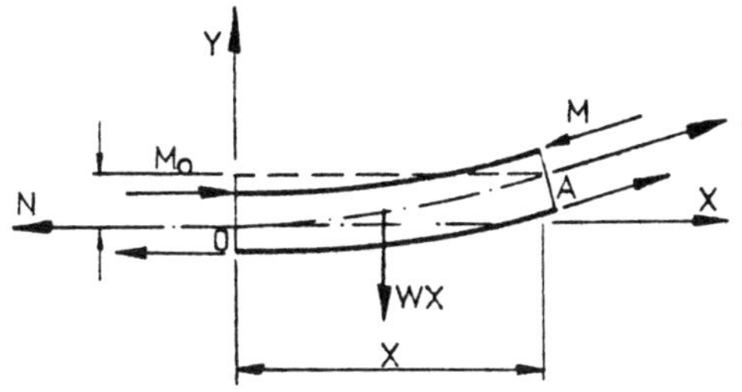

FIGURE 11.30 — Equilibrium of forces and moments in a section of the pipe.

The equation representing the balance of moments about point A, in Figure 11.30 is

$$M = EI\frac{d^2y}{dx^2} = M_o - \frac{wx^2}{2} + N_y \quad (11.138)$$

where

M = the bending moment within the pipe at Section A

I = the moment of inertia of the cross-section of the pipe

E = the modulus of elasticity of the pipe material

$$I = A_s r^2$$

where

A_s = the cross-sectional area of the steel in the pipe

r = the radius of gyration of the cross-section of steel in the pipe

The equilibrium of vertical forces between 0 and A is given by

$$V = \frac{dM}{dx} = -wx + N\frac{dy}{dx} \quad (11.139)$$

where

V = the shear force at section A

Consequently, the shear at the end of the sagging section is

$$V_1 = -\frac{1}{2}w\ell_0 + N\theta_1 \quad (11.140)$$

where

θ_1 = the end slope at $x = \frac{\ell_0}{2}$, given in radians

The solution of (11.138) for the vertical deflection of the pipe is

$$y = \frac{1}{N}\{(M_o - w\lambda^2)[(\cosh\frac{x}{\lambda}) - 1] + \frac{1}{2}wx^2\} \quad (11.141)$$

where

λ = a characteristic length

and is defined by

$$\lambda = \sqrt{\frac{EI}{N}} \quad (11.142)$$

The expressions for the slope, bending moment and shear at any point in the sagging pipe can be derived from Equation (11.141) as follows:

$$\theta = \frac{dy}{dx} = \frac{1}{N}\{\frac{1}{\lambda}(M_o - w\lambda^2)(\sinh\frac{x}{\lambda}) + wx\} \quad (11.143)$$

$$M = (M_o - w\lambda^2)(\cosh\frac{x}{\lambda}) + w\lambda^2 \quad (11.144)$$

$$V = \frac{1}{\lambda}(M_o - w\lambda^2)\sinh\frac{x}{\lambda} \quad (11.145)$$

It can be shown furthermore, that in the limit (11.141) it reduces to

$$y = \frac{1}{EI}\left(M_o\frac{x^2}{2} - w\frac{x^4}{24}\right) \qquad (11.146)$$

when N is extremely small. This equation, representing a special case of more general relation (11.141) is that which the simple theory of flexure yields.

If N is very large then V is small and (11.146) reduces to

$$y = \frac{wx^2}{2N} \qquad (11.147)$$

which is the approximate form of catenary sag associated with tension N. These two limiting cases serve as checks on the more general theory.

The tension can be found by applying Hooke's law to overall extension of the sagging pipe section. If ℓ represents the length of the pipe section between points (1′) and (1) after vertical deformation has occurred, and ℓi is the initial length between the same points in the pipe prior to deformation, then the overall strain is given by

$$\frac{\ell - \ell_i}{\ell_i} = \frac{N-N_o}{A_s E} \qquad (11.148)$$

where

N-N_o = the increase in tension due to the longitudinal strain induced by the sag. The length of the deformed section can be found from the approximate expression

$$\ell = \ell_o + \int_0^{\ell_o/2} \theta^2\, dx \qquad (11.149)$$

which is quite valid as long as $|\theta| \leqslant 1$.

11.14.4 Longitudinal Slippage at the Ends of the Pipe

The value ℓ_i is not necessarily the same as ℓ_o, because if longitudinal slippage of the pipe occurs at the ends, then the original length of the section between points (1′) and (1) will be greater than ℓ_o and the resulting tension in the pipe will be lower than that for the case of no slippage. The amount of slippage will depend upon the longitudinal restraint offered by the stronger sediment adjacent to the zone in which sag occurs. If the sediment exerting this restraining force is perfectly rigid then no slippage will occur and ℓ_i will equal ℓ_o.

The amount of slippage at each end of the sagging section is $(\ell_i - \ell_o)/2$. It can be shown that the following approximation is applicable

$$\ell_i - \ell_o = \frac{(N-N_o)^2}{f_r A_s E} \qquad (11.150)$$

where

f_r = the maximum longitudinal restraining force per unit length of pipe offered by the sediments adjacent to the weak zone

If the pipe is buried in the sediments in the adjacent sections, then f_r can be expressed in terms of the ultimate shear strength, τ'_u, of the relatively strong sediment as follows

$$f_r = \pi D \tau'_u \qquad (11.151)$$

where it is assumed that shear occurs at or near the surface of the pipe.

For a pipe laying on the bottom f_r is equal to the coefficient of friction between the pipe and bottom multiplied by the submerged weight of the pipe per unit length.

11.14.5 The Characteristic Dimensionless Parameters

In order to make the functional relationships existing between the basic variables of the problem as simple as possible, it is convenient to introduce the dimensionless parameters given in Table 11.1, shown on the following page.

The definition (11.150) can be applied to the moments at the middle and at the ends. The quantity

$$m_o = \frac{M_o}{w\ell_o^2} \qquad (11.156)$$

TABLE 11.1 — The Dimensionless Parameters.

Name	Symbol Definition	Equation
Bending moment factor	$m = \frac{M}{w\ell_o^2}$	(11.152)
End shear factor	$p = \frac{V_1}{w\ell_o}$	(11.153)
Tension factor	$n^2 = \frac{N\ell_o^2}{EI}$	(11.154)
Flexibility Parameter	$q = \frac{w\ell_o^4}{EIr}$	(11.155)

is the bending moment factor at the point of maximum sag, and

$$M_1 = \frac{M_1}{w\ell_o^2} \qquad (11.157)$$

is the bending moment factor at the ends of the sagging section.

Similarly, as a special case of (11.156)

$$n_0^2 = \frac{N_o \ell_o^2}{EI} \qquad (11.158)$$

which is the initial tension factor.

By using Equations (11.140) to (11.146) the following relations can be established

$$P = \frac{1}{2} - \frac{n^2}{q} \cdot \frac{\ell_0}{r}\,\theta_1 \qquad (11.159)$$

$$m_o = -\frac{P}{n\tanh\frac{n}{2}} + \frac{1}{n^2} \qquad (11.160)$$

$$m_1 = -\frac{P}{n\tanh\frac{n}{2}} + \frac{1}{n^2} \qquad (11.161)$$

$$\frac{y_m}{r} = \frac{q}{n}\left[\frac{1}{8} + P\,\frac{1 - \cosh\frac{n}{2}}{n\sinh\frac{n}{2}}\right] \qquad (11.162)$$

where y_m is the maximum vertical deflection or simply the sag. Special forms of these expressions are given below for the two common end conditions.

CASE 1 — Rigid ends with zero slope

This condition is illustrated schematically in Figure 11.31, when $\theta_1 = 0$.

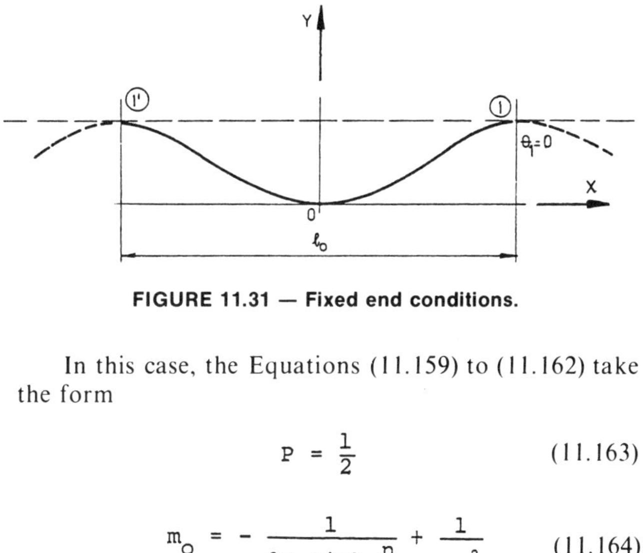

FIGURE 11.31 — Fixed end conditions.

In this case, the Equations (11.159) to (11.162) take the form

$$P = \frac{1}{2} \qquad (11.163)$$

$$m_o = -\frac{1}{2n\sinh\frac{n}{2}} + \frac{1}{n^2} \qquad (11.164)$$

$$m_1 = -\frac{1}{2n\tanh\frac{n}{2}} + \frac{1}{n^2} \qquad (11.165)$$

$$\frac{y_m}{r} = \frac{q}{n}\left[\frac{1}{8} + \frac{1-\cosh\frac{n}{2}}{2n\sinh\frac{n}{2}}\right] \qquad (11.166)$$

The maximum bending moment occurs at the ends for this condition and is given by m_1.

CASE 2 — Ends free to turn

In this case $m_1 = 0$, which indicates a maximum slope at the ends of the sagging section, Figure 11.32.

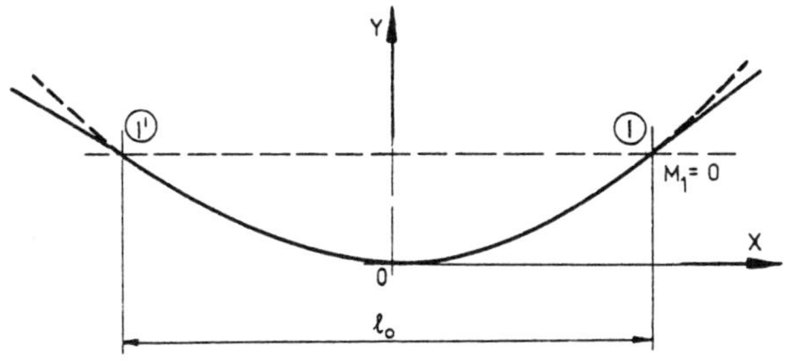

FIGURE 11.32 — Condition of ends free to turn.

For this condition

$$\frac{\ell_0}{r}\,\theta_1 = \frac{q}{n^2}\left(\frac{1}{2} - \frac{\tanh\frac{n}{2}}{n}\right) \qquad (11.167)$$

$$m_o = \frac{1}{n^2}\left(1 - \frac{1}{\cosh\frac{n}{2}}\right) \qquad (11.168)$$

$$P = \frac{1}{n}\tan\frac{n}{2} \qquad (11.169)$$

$$\frac{y_m}{r} = \frac{q}{n}\left[\frac{1}{8} + \frac{1-\cosh\frac{n}{2}}{n^2\cosh\frac{n}{2}}\right] \qquad (11.170)$$

11.4.6 Practical Forms of the Equations for Moments, Slope and Relative Sag

The limiting expressions for M_1 and M_2 in the case of small tensions are

$$M_1 = -\frac{w\ell_0^2}{12}, \quad \text{for} \quad \theta_1 = 0 \qquad (11.171)$$

and

$$M_o = \frac{w\ell_0^2}{8}, \quad \text{for} \quad M_1 = 0 \qquad (11.172)$$

The Equations (11.162), (11.167) and (11.170) can be simplified by the use of the following expressions

$$q = K_1K_2f(n) \qquad (11.173)$$

where

$$K_1 = \sqrt{1 - n_0^2/n^2} \qquad (11.174)$$

$$K_2 = \sqrt{1 + sq^{-3/4}(n^2 - n_0^2)} \qquad (11.175)$$

$$s = \frac{w}{f_r}\left(\frac{EA_s}{wr}\right)^{1/4} = \frac{w}{f_r}\left(\frac{EI}{wr^3}\right)^{1/4} \qquad (11.176)$$

The formula for the slope in Case 2 becomes

$$\frac{\ell_0}{r}\theta_1 = K_1K_2K_3n \qquad (11.177)$$

where

K_3 = a coefficient which depends upon n

It can be shown that for the entire range of n

$$(\text{small } n)\ \sqrt{\frac{70}{n}} < K_3 < \sqrt{6}\ (\text{large } n) \qquad (11.178)$$

The average value of K_3 is about 2.24, and the following approximation will yield values of θ_1 which are never more than 10 percent in error:

$$\theta_1 = 2.24\, K_1K_2 \frac{r}{\ell_0} n, \quad (m_1 = 0) \qquad (11.179)$$

The formula for the relative sag can be simplified in a similar manner

$$\frac{y_m}{r} = K_1K_2K_4n \qquad (11.180)$$

The coefficient K_4 is determined by n, however, for the entire range of n:

$$(\text{large } n)\ \sqrt{3/8} < K_4 < \sqrt{105/256}\ (\text{small } n) \qquad (11.181)$$

The average value of K_4 is about 0.626 for either of the end conditions considered here. That means that the following approximation will yield the value of y_m which is accurate to within about 2 percent.

$$y_m = 0.626\, K_1K_2rn, \quad \begin{pmatrix}\theta_1=0\\ \text{or}\\ m_1=0\end{pmatrix} \qquad (11.182)$$

It will be noted that

$$K_1n = \sqrt{n^2 - n_0^2} \qquad (11.183)$$

so that both θ_1 and ym are proportional to the square root of the increase in tension in the pipe due to sag. Furthermore, if there is no end slippage, the coefficient K_2 = 1. For small values of q the Equations (11.170) and (11.172) reduce to the forms:

$$y_m = \frac{1}{384} \cdot \frac{w\ell_0^4}{EI}, \quad \text{for} \quad \begin{matrix}\theta_1=0\\ m_1=0\end{matrix} \qquad (11.184)$$

provided that there is no initial tension.

11.14.7 Induced Stresses

The bending moment induces a non-uniform normal stress across the section of the pipe. The maximum value of this stress f_b is

$$f_b = \pm \frac{MR_0}{I} = \pm \frac{mw\ell_0^2 R_0}{I} \qquad (11.185)$$

where

R_0 = the outside radius of the pipe

The tension induced by axial elongation of the pipe is

$$f_t = \frac{N}{A_s} = \frac{n\ EI}{A_s\ell_0^2} \qquad (11.186)$$

Thermal stress associated with the restraint of axial elongation or contraction of the pipe is included in this term, since in determining N, the terminal effect must be taken into account in the term N_0.

The vertical shearing force gives rise to a non-uniform shear stress at each section of the pipe. The mean value of this shear stress is

$$f_s = \frac{V}{A_s} = \frac{pw\ell_0}{A_s} \qquad (11.187)$$

The maximum shear stress is approximately $2f_s$. The fluid pressure, within the pipe, will give rise to hoop stress due to the circumferential elongation of the pipe. If ΔP represents the difference in pressure between the inside and outside of the pipe, then

$$f_h = \frac{D_i}{D_o - D_i}\Delta P \qquad (11.188)$$

where

D_o and D_i = the outside and inside diameters of the pipe, respectively

The formula for most severe combined stresses at a given section of the pipe depend upon the value of the criterion parameter $4f_s^2/f_b$. Table 11.2 gives the approximate expression for the governing stress (either $f_{t,m}$ or $f_{s,m}$) for three different conditions imposed upon the criterion parameter.

It will be noted from Table 11.2 on the following page, that condition (C) implies that the maximum shear stress $f_{s,m}$ governs. This presumes that the yield limit of stress in shear is just half that in tension for the pipe steel.

Under conditions (A) the pipe would fall in tension at a point farthest from the neutral surface, provided that $f_{t,m}$ were great enough. Under condition (B), the pipe would fail in tension at the neutral surface, along a plane which forms an angle of less than 90° with the neutral surface. Under condition (C) failure, if it occurred, would manifest itself by shear at the neutral surface, along a plane which forms an angle of less than 90° with the neutral surface.

The position along the pipeline at which the maximum stress occurs depends upon the end conditions. The stress f_b is a maximum at the ends of the sagging section for the case of zero end slope; while in the case of ends free to turn, the maximum value of f_b would occur at the center of the sag. The stress f_s is a maximum at the ends in both cases, but its magnitude depends upon the end condition. The stresses f_t and f_n are presumed to be independent of position along the pipe.

For practical purposes, the condition (A) can be presumed for nearly all cases of pipe sag, and the governing stress therefore is

$$f_{t,m} = f_b + f_t \qquad (11.189)$$

TABLE 11.2 — The Governing Combined Stress.

Condition	Governing Combined Stress
(A) $0 \leq \frac{4f_s^2}{f_b} < f_1$	$f_{t,m} = f_b + f_t$
(B) $f_2 > \frac{4f_s^2}{f_b} > f_1$	$f_{t,m} = \frac{1}{2}[(f_t+f_h) + \sqrt{(f_t-f_h)^2+ 16f_s}$
(C) $f_2 < \frac{4f_s^2}{f_b} > f_1$	$f_{s,m} = \frac{1}{2}\sqrt{(f_t-f_h)^2 + 16f_s^2}$
where	$f_1 = (f_b+f_t) - f_h;\ f_2 = \frac{f_t f_h}{f_b}$

where

f_b is the value occurring at the position of maximum flexure. The validity of this assumption, however, can be checked by computing $4f_s^2/f_b$. This must be less than the value of $(f_b + f_t) - f_h$, otherwise, $(f_b + f_t)$ is not the maximum combined stress.

11.14.8 The Dimensionless Stress Parameters

It is convenient to introduce the following dimensionless stress parameters:

$$f_b = (\frac{\ell_0}{r})^2 \frac{f_b}{E} \qquad (11.190)$$

$$f_t = (\frac{\ell_0}{r})^2 \frac{f_t}{E} \qquad (11.191)$$

From Equations (11.182) and (11.183) it can be shown that the stress parameters are related to the characteristic parameters m, n, and q as follows:

$$f_b = \frac{R_b}{r} mq \qquad (11.192)$$

$$f_t = n^2 \qquad (11.193)$$

$$f_b + f_t = (\frac{r}{\ell_0})^2 E(f_b+f_t) = (\frac{r}{\ell_0})^2 E(\frac{R}{r} mq+n^2) \qquad (11.194)$$

For all practical purposes, the value of R_0/r for most pipes may be taken as $\sqrt{2}$. This is theoretically correct for a circular pipe with thin walls.

REFERENCES — CHAPTER 11

[11.1] Dixon, D.A., and Rutledge, D.R., "Here's How to Calculate the Critical Tension and Inclination of a Deepwater Pipeline Laid From Angled-Ramp Lay Barges", Oil and Gas Journal, Oct. 1967, pp. 114-121.

[11.2] Wiegel, R.L., "Oceanographical Engineering", Prentice-Hall, 1964.

[11.3] Roark, R.J. and Young, W.C., "Formulas for Stress and Strain", 5th Edition, McGraw-Hill Book Company, New York, 1975, p. 556.

[11.4] Anonymous, "New Technique Lays Pipe in 9,000 Feet of Water", Pipe-Line Industry, January, 1962.

[11.5] Reid, R.O., "Some Oceanographic and Engineering Considerations in Marine Pipe-Line Construction", Proceedings of Second Conference on Coastal Engineering, Ed. by J.W. Johnson, Council on Wave Research, University of California, Berkeley, Calif., 1952.

[11.6] Timoshenko, S.P., and Gere, J.M., "Theory of Elastic Stability", 2nd Ed., McGraw-Hill Book Company, Inc., New York, 1961, p. 289.

[11.7] von Mises, R., "The Critical External Pressure of Cylindrical Tubes Under Uniform Radial and Axial Load", (in German), VDI-Z, Vol. 58, No. 19, 1914, pp. 750-755.

[11.8] Windenburg, D.F., and Trilling, C., "Collapse by Instability of Thin Cylindrical Shells Under External Pressure", Trans. ASME, Vol. 56, 1934, p. 819.

[11.9] Southwell, R.V., "On the General Theory of Elastic Stability", Phil. Trans., Vol. 213A, 1913, pp. 187-244.

[11.10] Cook, G., "The Collapse of Short Thin Tubes by External Pressure", Phil. Mag., July, 1914, pp. 51-56.

[11.11] Strum, R.G., and O'Brien, H.I., "Computing Strength of Vessels Subjected to External Pressure", Trans. Am. Soc. Mech. Engrs., Vol. 69, 1947, p. 353.

[11.12] Timoshenko, S.P., and Gere, J.M., "Theory of Elastic Stability", 2nd Ed., McGraw-Hill Book Company, Inc., New York, 1961.

[11.13] API-ASME Code for Unfired Pressure Vessels for Petroleum Institute, 1951.

[11.14] American Petroleum Institute, Division of Production, API Specification for Fabricated Structural Steel Pipe, 2nd Ed., API Spec., 2B, October 1972.

[11.15] American Society of Mechanical Engineers, Boiler and Pressure Vessel Code, A: Sec. III, Nuclear Vessels, 1968; B: Sec. VIII, Pressure Vessels, Div. 1, 1968; C: Sec. VIII, Alternative Rules for Pressure Vessels, Div. 2, 1968.

CHAPTER 12

tubular conveyor galleries

12.1 Introduction

Thin-walled large diameter tubular structures relatively recently found application as conveyor galleries, especially for the mining industry. They are used for the support and housing of conveyors handling bulk materials [12.1, 12.2, 12.3, 12.4].

Tubular conveyor galleries were proposed after studies were made to arrive at an improved conveyor housing and support system. These studies were aimed at finding a system which would incorporate the following improvements:

1. Simpler and more rugged construction, to reduce cost of maintenance.
2. Increased efficiency in the removal of spillage from the gallery floor.
3. Reduction to plant obstruction by supporting bents.
4. Speed of erection.
5. Economy in capital cost.
6. Improved performance and appearance.

It was found that these conditions are best fulfilled by a welded tubular conveyor gallery, shown in Figure 12.1, and illustrated further by detail in Figures 12.2 (a) and 12.2 (b). The tubular gallery is made up of ¼-inch-thick

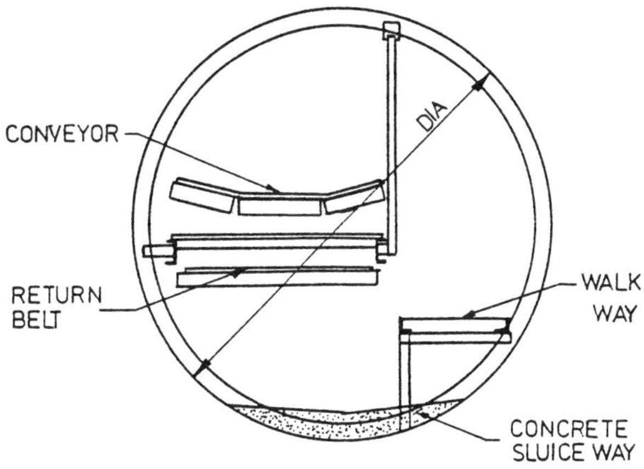

FIGURE 12.1 — Tubular conveyor gallery.

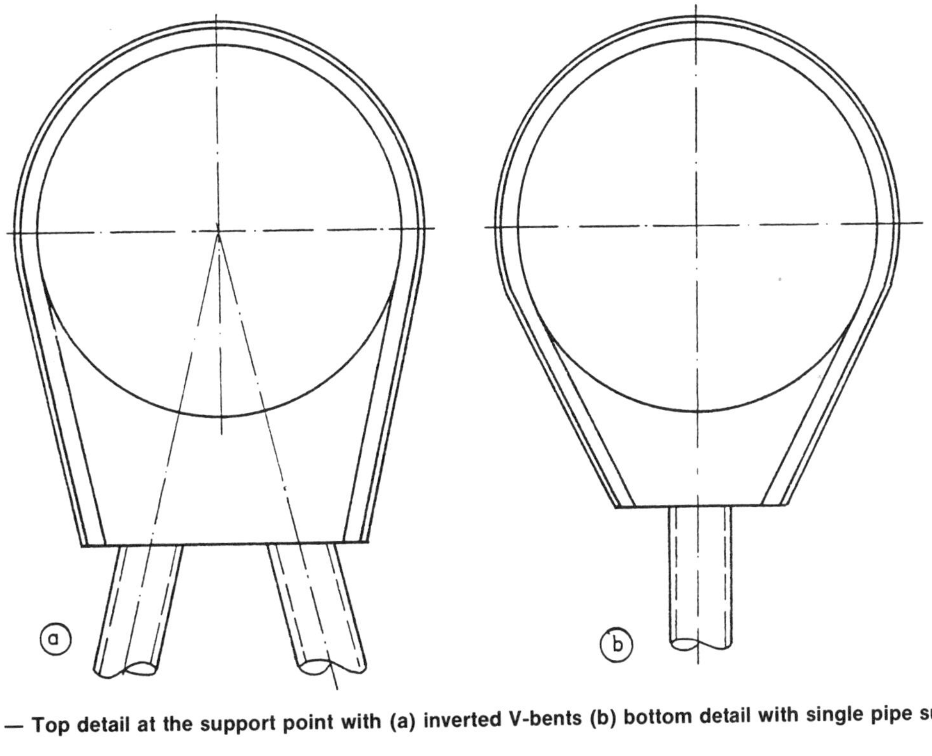

FIGURE 12.2 — Top detail at the support point with (a) inverted V-bents (b) bottom detail with single pipe support posts.

plate welded together to form a tube. The function of the tube is threefold:

1. It houses the conveyor and walkway and protects them from the weather, thus eliminating the need for the siding.
2. It acts as a structural member, carrying the conveyor and walkway above ground.
3. The invert of the tube forms a trough through which spilled material is removed by flushing or hosing it down the inclined gallery. At the lower end of the inclined gallery, the material is removed as slurry by gravity flow, through a pipe.

The favorable wind shape of the circular section, which results in less wind load per linear foot of gallery and the inherent strength of the tube in the horizontal plant make possible longer spans, without the need for intermediate supports to take the wind load.

The tubes are approximately 8 ft — 6 in to 10 ft in diameter, depending upon the size of conveyor they house. A 10 ft tube can house a 54-inch wide conveyor.

12.2 Internal Arrangement and Exterior Protection of the Tube

12.2.1 Sluiceway

Light-weight concrete may be used in the gallery invert to form a sluiceway down which spillage from the conveyor belt could be washed. Sluiceway concrete is usually applied after the gallery is in place, to avoid cracking resulting from handling stresses during erection. Epoxy compound may be applied to the steel surfaces prior to placing the concrete, to insure adhesion of concrete to steel, and after the concrete had cured, to the wearing surfaces of the concrete, to seal the surface and resist abrasion. Drain pipe should be provided at the lower end of the sluiceway to carry spillage slurry to disposal areas.

12.2.2 Thermal Insulation

Thermal insulation for protection against winter temperatures, should be applied to the tubular galleries in which moist concentrate was to be conveyed. Blanket-type Fiberglass insulation, about 3 inches thick should be strapped to the exterior surface of the gallery, and a 24-gauge galvanized metal protective sheathing then applied for weather protection. Heaters may be provided to blow hot air into the gallery at its lower end. Galleries for handling of crude ore and pellets may not be insulated.

12.3 Advantages of Tubular Conveyor Gallery

On the basis of the comparison made between the conventional enclosed gallery and the tubular gallery for the support and housing of conveyor handling bulk material it is possible to conclude that the tubular gallery, supported alternately on inverted V-bents and single pipe posts, offers advantages over the conventional gallery and its supports.

The tubular gallery system permits simplified design and detailing, maximum shop fabrication using modern welding techniques, and rapid field erection, all of which result in lower capital cost for the conveyor system in place.

The tubular gallery system requires little maintenance, and minimum plant space for its supports.

The tubular gallery is designed to:

a. Act as a structural member carrying the conveyor above ground between supporting bents.

b. House the conveyor and walkway to protect them against the weather.

c. Provide a sluiceway, by means of its concrete paved invert, down which spillage can be flushed.

Also, this system possesses certain important structural advantages by comparison with the conventional types of conveyors.

The tubular cross-section has a superior wind shape factor as compared to that of the conventional rectangular section. As a result, the wind load per linear foot on the tubular section will be approximately 40 percent less than the wind load on a conventional enclosed gallery of dimensions to provide equivalent working space. This lesser wind load is a favorable factor in the design of gallery supports and foundations. The tubular section has the same moment of inertia in all lateral directions and provides equal resistance to both horizontal and vertical loads. This uniform moment of inertia or bending strength, makes it possible to carry wind loads over long spans, using inverted V-bent and single post-supports, alternately.

The tubular section is an ideal section for resistance to the torsional stresses imposed by the eccentrically located loads of the walkway and conveyor. The tubular section is an ideal section for taking compression forces imposed by conveyor belt pull. For inclined galleries rising from underground structures and entering buildings, this element may be used to advantage by installing the conveyor head pulley within the tubular section, instead of on building structure. All longitudinal forces are thus transmitted through the tubular gallery section to foundation structure at grade, and the necessity for bracing building structure to take conveyor pull is avoided. In this design, a rolling or sliding bearing is provided between building structure and conveyor gallery to eliminate the transfer of stresses from one structure to the other. This method of transmitting belt pull to foundations was found to be especially advantageous in the design of a mineral processing plant located in an earthquake zone.

In areas where snow loads influence design, snow buildup on the tubular gallery will be minimal, as compared with that on the rectangular gallery of the same width.

12.4 Supporting Structures for Conveyor Gallery

At intervals along the length of a conveyor gallery, supports must be provided to carry the gallery between its

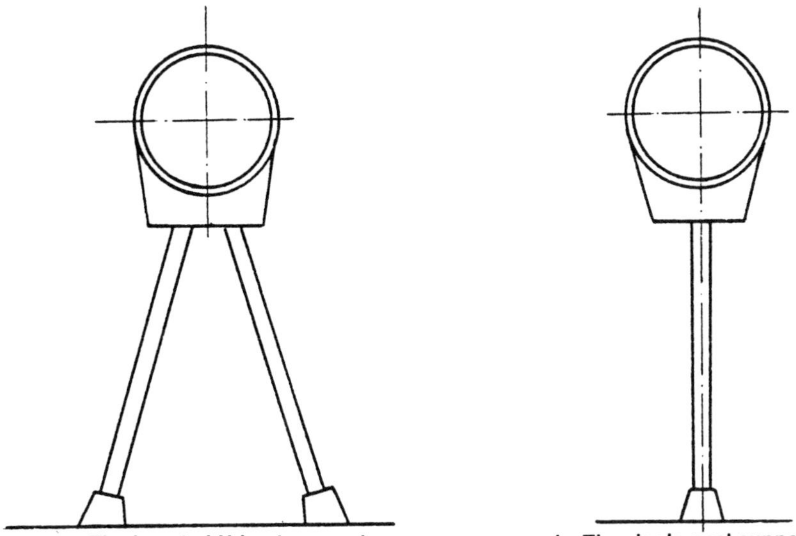

a. The inverted V-bent support. b. The single post support.

FIGURE 12.3 — Tubular conveyor.

junctions with building. The location and height of supports will be dictated by topography, the location of other plant structures, and by other design considerations.

The superior wind shape factor and the uniform bending strength of the tubular gallery section against horizontal and vertical loads permit the use of a design whereby two types of gallery support, both fabricated of steel pipe, may be used. The first type of support is the inverted V-bent Figures 12.3 (a), and is capable of taking both the horizontal and the vertical loads. The second type of support is a single pipe post, Figure 12.3 (b), taking vertical loads only. The latest proposal is to house the conveyor in a clean elliptical-shaped tube of light-weight, high-strength steel, supported by A-frame supports, Figure 12.4.

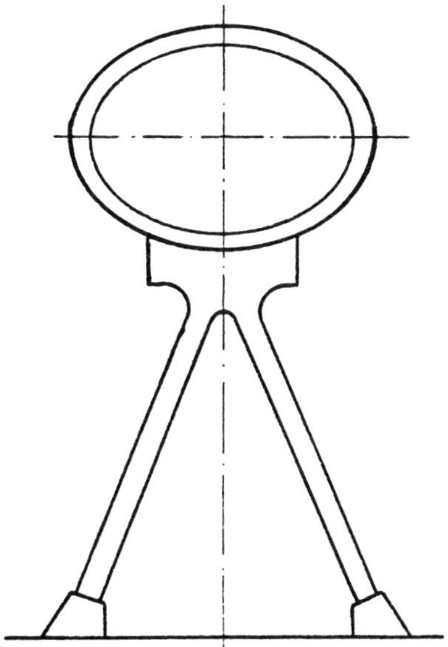

FIGURE 12.4 — Elliptical tube conveyor.

12.4.1 The Inverted V-bent

The inverted V-bent is designed so that the axes of the two pipe legs, extended, will intersect at the point where the vertical dead and live loads of the gallery and the horizontal wind load intersect. All loads are thus transmitted through the pipe legs to concrete footings, as compression and tension forces. The pipe legs are subject to a minimum of bending imposed by the eccentrically located loads of the walkway, walkway traffic, conveyor and conveyor material, and uneven snow accumulations. The V-bents are shop fabricated complete, with a base plate welded to the lower end of each pipe leg, for connection to anchor bolts embedded in concrete footings. At the upper end, a seat weldment is provided, for field-bolted connection to a weldment on the underside of the tubular gallery section.

12.4.2 The Single Post

The single pipe posts, located at intermediate points between the V-bents, are designed to take vertical loads only and are provided with base plates and seat weldments in a manner similar to that described for the inverted V-bents.

Foundations for conveyors are as varied as are the local conditions. They are treated just as foundations for any other project. Normally, the vertical loads are comparatively light and the foundation design is likely to be governed by uplift or lateral loads.

12.4.3 Advantages of Inverted V-bents and Single Posts Over Conventional Supports

The advantages of the V-bent and single post support structure for conveyor galleries over the conventional type support structure include the following:

a. The supports constitute a minimum of obstruction to plant operations. Mobile equipment and plant personnel may move with ease between the legs of V-bents and around single posts.

b. The supports offer a minimum of obstruction to snow removal.

c. The single pipe support requires a very small foundation.

d. The heavy pipe sections used are less vulnerable to damage by mobile equipment than are supports with bracing of the conventional type.

e. Less frequent painting maintenance is required for the supports, for the same reasons given for the maintenance of the tubular gallery section.

12.4.4 Erection of Tubular Galleries

In erecting the structure, V-bents and single pipe posts are lowered on and secured to anchor bolts embedded in the footings. Sections of the tubular gallery, complete with conveyor components mounted in place, of sufficient length to span the distance between a V-bent and adjacent single pipe post, are then lifted into place and secured to supports by bolts.

12.5 Fabrication

Conveyor galleries are designed in accordance with the latest American Institute of Steel Construction Specifications [12.5]. Welding may be according to the American Welding Society Specifications D1.0 for Building [12.6] or D2.0 for Bridges [12.7] depending on the severity of the intended service. Concrete is designed by the current American Concrete Institute Code [12.8].

Nearly all of the steel is shop welded and field bolted using high strength bolts. Normally, the structure is shop-assembled into the largest pieces which can be economically shipped and handled in the field. The tubes are fabricated in the shop by modern welding methods up to 60-foot sections, which can be transported by rail to the plant site for erection. All the conveyor components are assembled inside the tube before erection.

Structural grade steels, ASTM A-36 are used for most work although ASTM A-242 may be used for corrosion resistance and alloy steels for reduction in dead weight.

12.6 Loads and Forces

Loading conditions vary depending on the project requirements, but the following may be considered:

1. **Vertical Loads**

   a. *Dead Load,* which includes the structure itself, the mechanical components, belts, idlers, pulleys, drive, limit switches, wiring, etc.

   b. *Live Load* due to material on the belt. No allowance is made for impact due to the bouncing of lumps as they move along the belt.

   c. *Live Load* on walkways. Normally taken at 100 lb/ft² locally, by 25 lb/ft² for the span.

   d. *Snow Load* according to geographical area.

2. **Horizontal Forces**

   a. *Wind* in any direction. The wind loads are as given in the American National Standard Institute "Minimum Design Loads in Buildings and Other Structures" [12.9].

   b. *Earthquake forces* according to seismic zone.

12.7 Structural Analysis and Design

Longitudinal bending stress rarely governs the design of tubular galleries. What does govern is likely to be transverse bending leading to crushing at the supports. It is necessary to provide transverse stiffening rings at supports. The support may be designed as a saddle or the stiffening ring can be supported at the spring-line.

For the design analysis of circular tube the vertical loads and horizontal forces are considered as follows.

12.7.1 Vertical Loads

Vertical loads generally include, according to Section 12.6 the following components:

1. *Dead Load*

   a. Own weight of tubular structure

$$q_s = \pi D_i t \rho_s + \frac{nW_{st}}{\ell}, \quad \text{lbs/ft}^3 \qquad (12.1)$$

where

D_i = inside diameter of the tube, in

t = thickness of the wall, in

ρ_s = unit weight of steel shell, 0.283 lb/in³

n = number of stiffeners in span

W_{st} = weight of one stiffener, lbs

ℓ = span between bents or supports

   b. Weight of mechanical components and equipment

$$q_c = \text{belts} + \text{idlers} + \text{pulleys} + \text{drive} + \text{switches} + \text{wiring} + \text{other, lbs/ft} \qquad (12.2)$$

   c. Weight of walkway

$$q_w = \text{ch'rd plate} + \text{grating} + \text{expanded metal} + \text{stringers} + \text{posts, lbs/ft} \qquad (12.3)$$

   d. Lightweight concrete on the invert, Figure 12.5.

$$q_i = 60\,R^2\left(\frac{\pi\alpha}{180} - \sin\alpha\right)\text{lbs/ft} \qquad (12.4)$$

2. *Live loads*

   a. Live load due to material on the belt

$$q_{\ell.\ell} \quad \text{lbs/ft} \qquad (12.5)$$

No allowance is made for the impact and due to the bouncing of lumps as they wave along the belt.

   b. Live load on walkway

$$q_{\ell.w} \quad \text{lbs/ft} \qquad (12.6)$$

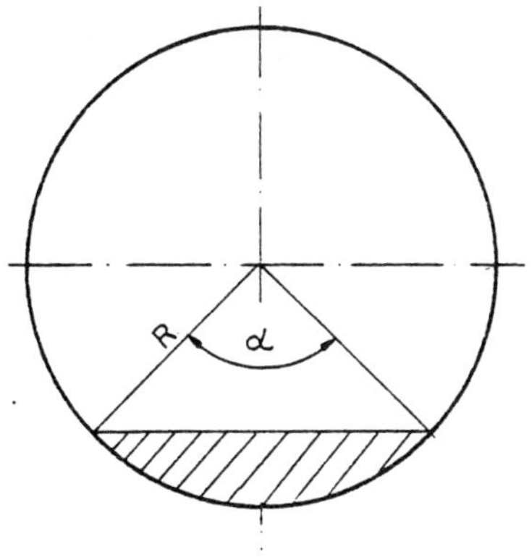

FIGURE 12.5 — Area of the invert.

Normally it is taken at 100 lbs/ft^2 locally, but 25 lbs/ft^2 for the span.

c. Icing and snow (if applicable)

$$q_{i+s} \text{ lbs/ft} \tag{12.7}$$

d. Belt pull during starting, stopping or running.

$$P, \text{ lbs (longitudinal force)} \tag{12.8}$$

12.7.2 Horizontal Forces

1. *Wind Forces*

   The intensity of the wind force acting on the conveyor structure is estimated following the provisions of the American National Standard Institute [12.9].

2. *Earthquake Forces*

   Conveyor structure in earthquake areas shall be designed and constructed to withstand as a minimum, the lateral seismic forces, assuming that these forces can act in any direction. Principles of the design against earthquake forces are treated in Chapter 5, Section 5.2.6.

3. *Longitudinal Forces*

   Sometimes the head pulley of the conveyor is located within the gallery, usually at one end, the highest of an inclined gallery. As a result, a longitudinal compressive force, originated by the head pulley of the conveyor, has to be resisted, by column action, by the tube, and transmitted from one end of the gallery to the other, which is held by braced bent against longitudinal translation. The unsupported length of the tube is 2ℓ, since the single posts are considered as having zero flexural rigidity. This compressive longitudinal force at one end of the conveyor gallery is usually eccentric, with respect to the center of the tube. As a result, we have two moments, one vertical and one horizontal, at this end of the gallery. Usually, the designer tries to minimize this eccentricity and the magnitude of these moments is small. On the other hand, they are applied at the first support of a continuous beam. If these moments are small, they can altogether be neglected, otherwise they may be superimposed to the previous cases.

12.7.3 Temperature Effects

Due to the temperature changes, the inverted V-bents and single posts, being flexible structures allow the tube to expand freely. The lower end of the gallery is usually held against longitudinal movement, and it is designed to take any longitudinal forces such as conveyor pull.

A factor that must be taken into consideration in designing long-span tubular galleries, is the effect of the sun's heat on the tube, one side of which may be in full sun, while the other side is in full shade, with the conse-

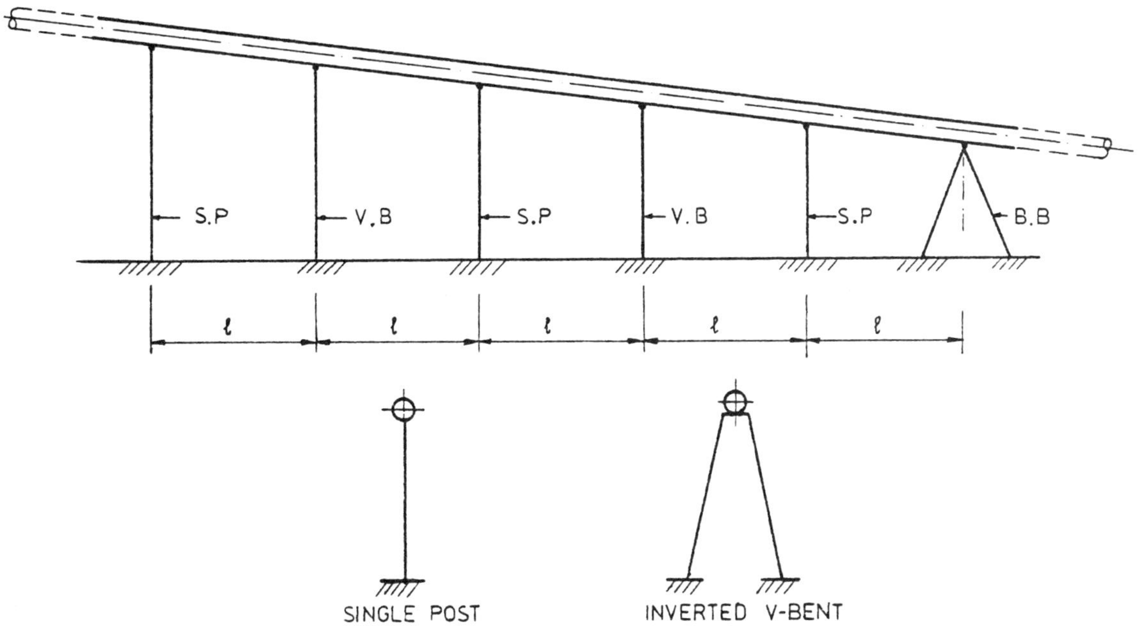

FIGURE 12.6 — General arrangement of conveyor gallery.

quent distortion of the tube to a degree where it may affect the conveyor operation. The temperature differential between the shaded and unshaded surfaces of the tube may be considerably reduced by the application of a paint, selected for minimum reflectance values.

12.7.4 Structural Behavior

The conveyor gallery structure is under the action of three different types of loads, discussed previously. The vertical loads due to gravity, the horizontal forces due to the wind or to earthquakes and the longitudinal loads due to the head pulley. These loads differ, not only in the direction of the application, but also in the degree of importance and frequency of occurrence and, therefore, cause different effects on the structure.

a. *Vertical Loads*

Vertical loads generally include the following components:

q_s, q_c, q_w, q_i, $q_{\ell.\ell}$, $q_{\ell.w}$, q_{i+s}, lbs/ft

Vertical loads are the most important loads, because they prevail in magnitude and frequency of occurrence. They are uniformly distributed along the length of the gallery.

The tubular gallery spannings over the inverted V-bents and single posts have to resist in flexure, as a continuous beam of hollow circular cross-section over multiple supports. In addition to flexure, the vertical loads usually cause uniformly distributed torsional moments.

The resultant of the vertical loads in a typical cross-section of the gallery does not usually coincide with the center of the tube. The eccentricity is often quite small and the tubular section is the ideal section to resist torque.

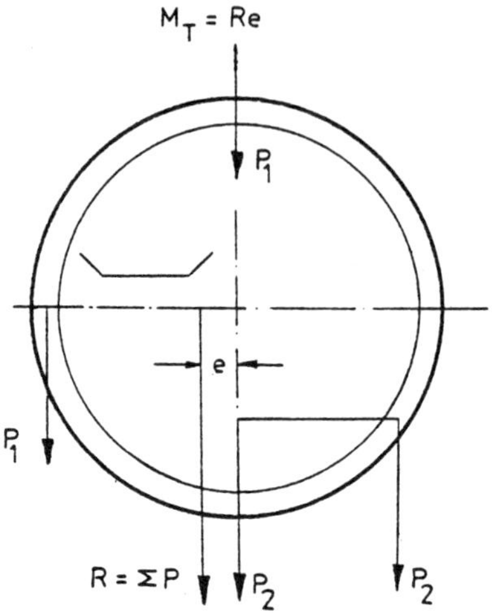

FIGURE 12.7 — Equipment loads producing torsional moment.

In order for the tube to act as a beam and to apply the ordinary theory of flexure, the cross-section of the gallery should remain circular between the supports. This requirement may be achieved by application of the intermediate stiffening rings and ring girders at the supports to prevent the distortion of the tube. Another function of the ring girders is to resist torsion and transmit it through the iverted V-bents to the foundations. The single post supports are designed to take only vertical loads.

In the following analysis we consider the general arrangement of the conveyor gallery as shown in Figure 12.6.

TABLE 12.1 — Maximum Bending Moments, Shears and Reactions Under Uniformly Distributed Vertical Load.

No.	Scheme of Continuous Structure	M_{span}	$M_{support}$	Shear	Reaction
1	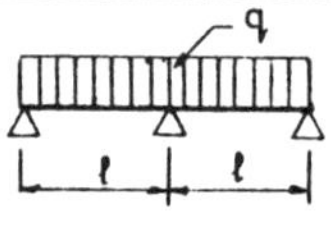	$+0.070q\ell^2$	$-0.125g\ell^2$	0.625qℓ / 0.625qℓ	1.25qℓ
2	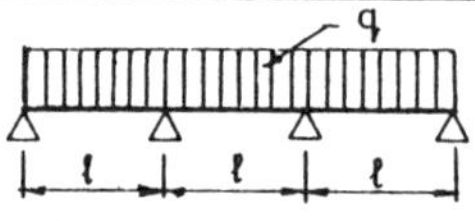	$+0.080g\ell^2$	$-0.100q\ell^2$	0.500qℓ / 0.600qℓ	1.10qℓ
3	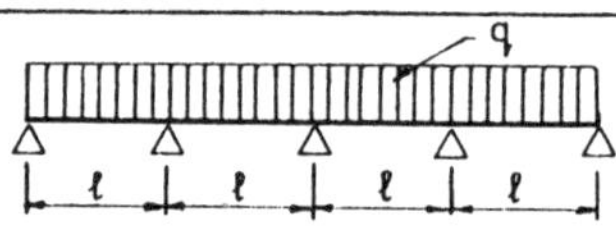	$+0.0773q\ell^2$	$-0.107g\ell^2$	0.536qℓ / 0.607qℓ	1.143qℓ
4	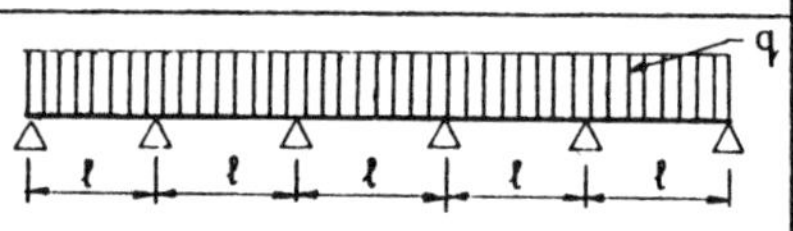	$+0.0789q\ell^2$	$-0.105q\ell^2$	0.526qℓ / 0.605qℓ	1.131qℓ

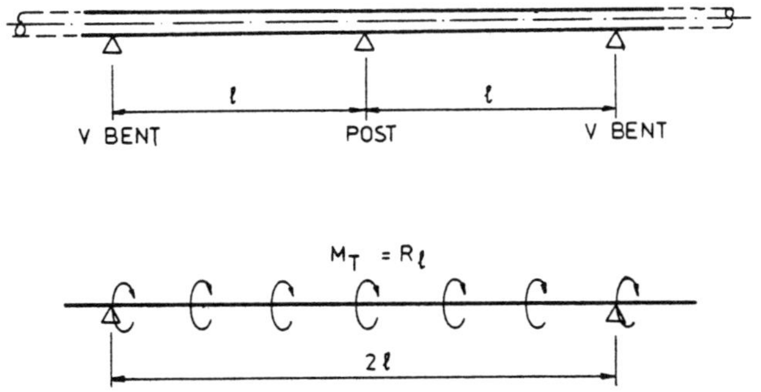

FIGURE 12.8 — Torsional moments.

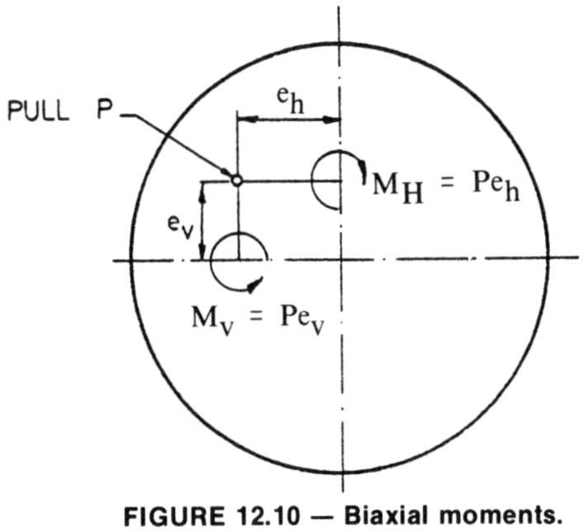

FIGURE 12.10 — Biaxial moments.

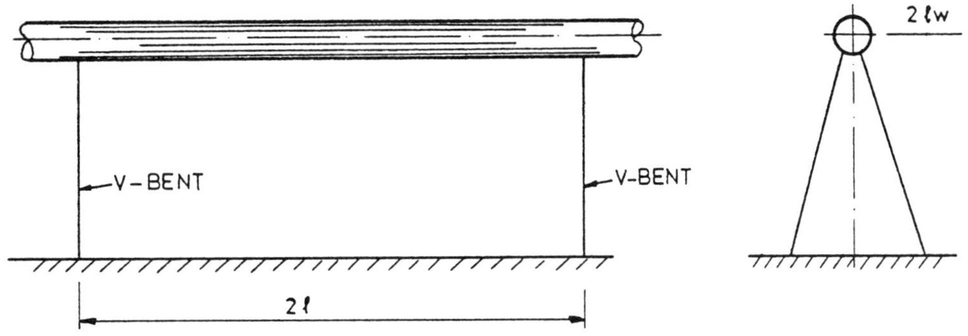

FIGURE 12.9 — Wind action on tubular conveyor.

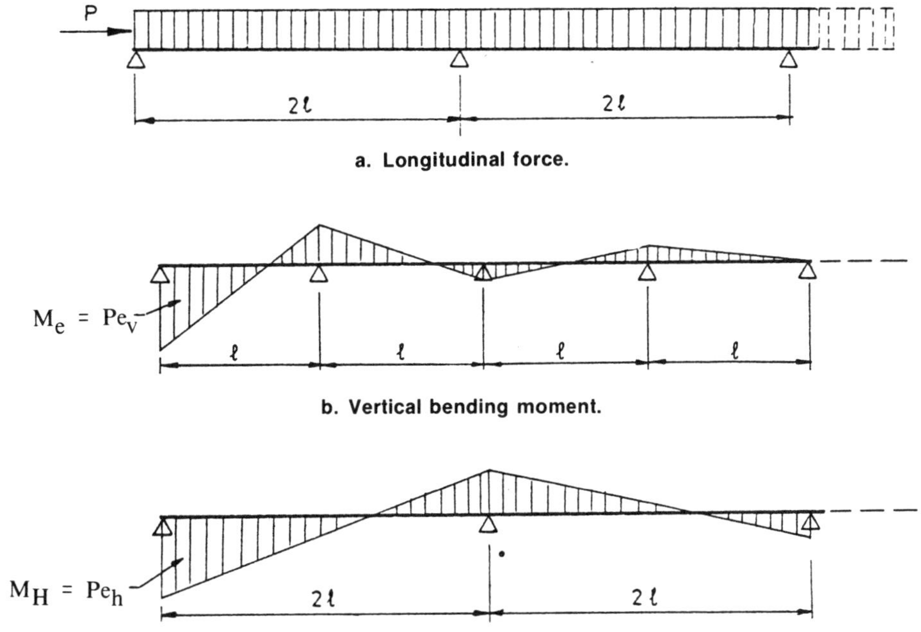

FIGURE 12.11 — Axial force and resulting bending moments.

The statical values, due to the effect of a vertical uniformly distributed load on a continuous structure, are shown in Table 12.1.

b. *Internal torsion due to eccentricity of equipment*

Due to the asymmetric arrangement of the equipment inside the tube, there will originate an internal torsional moment, Figure 12.7.

An internal torsional moment acting on the tube may be transferred only to inverted V-bents spaced at spans of 2ℓ, because single posts due to their flexibility cannot provide the necessary rigidity, Figure 12.8.

c. *Wind forces*

The intermediate posts, due to their flexibility cannot provide rigid support in lateral direction. Therefore, continuous tubular structures have spans of 2ℓ between the inverted V-bents, as shown in Figure 12.9.

d. *Head pulley inside gallery*

Under the action of the head pulley inside the gallery, the tube is under the action of axial compression and biaxial flexure due to internal moments acting in the vertical and horizontal plane, Figure 12.10.

The longitudinal compressive force, due to pull, is acting as an axial force along the continuous gallery. Compressive action is used for 2ℓ spans, between the inverted V-bents, Figure 12.11 (a).

The vertical moment $M_V = Pe_V$ acts on the continuous tube, having multiple spans ℓ, and a horizontal moment $M_H = Pe_h$ acts upon the continuous tube having multiple spans, Figures 12.11 (b) and 12.11 (c).

12.8 Stress Analysis

12.8.1 General Data

The vertical, horizontal and longitudinal loads induce stresses in the tube of conveyor gallery. In the following these stresses will be considered first for each case separately and further for their combined effect. Also, some geometric and material constants of the cross-section of the tube, which will be used in further discussions are given.

The thickness t of the shell is relatively very small and all the computations will be based on the inside diameter of the tube, Figure 12.12.

D_i = inside diameter of the tube, in

t = thickness of the wall, in

c = corrosion allowance, $1/16$ in

D_o = outside diameter of the tube, in

$A = \pi D_i(t - c)$ = cross-section of the tube's wall, in²

$I_x = I_y = \pi \frac{D_i^3}{8}(t - c)$ = moment of inertia of the tube, in⁴

$I_p = I_x + I_y = \pi \frac{D_i^3}{4}(t - c)$ = polar moment of inertia of the tube, in⁴

$S_x = S_y = \pi \frac{D_i^2}{4}(t - c)$ = section moduli of the tube, in³

E = modulus of elasticity of the tube's material, ksi

$G = \frac{E}{2(1 + \nu)}$ = shear modulus of the tube's

ν = 0.3 = Poisson's ratio

F_y = yield stress of the material, ksi

12.8.2 Vertical Loads

The vertical loads induce bending moment, torsional moment and shear in the tube.

a. *Normal stresses due to the bending moment M_v*

Normal stresses due to the bending moment are

$$(\sigma_v)_{max} = \frac{M_v}{S} = \frac{4M_v}{\pi D_i^2 (t-c)} \quad (12.10)$$

or, using more exact expression for section modulus

$$S = \frac{\pi(D_o^4 - D_i^4)}{32D_o} = \frac{\pi D_o^3}{32}\left[1 - \left(\frac{D_i}{D_o}\right)^4\right] \quad (12.11)$$

$$(\sigma_v)_{max} = \frac{32M_v}{\pi D_o^3\left[1 - \left(\frac{D_i}{D_o}\right)^4\right]} \quad (12.12)$$

b. *Shearing stresses due to the torque M_T*

Shearing stresses due to internal torque produced by the eccentric position of the equipment and live load inside the tube, according to formula [12.10]

$$\tau_{max} = \frac{M_T D}{2I_p} = \frac{16M_T}{\pi D_o^3\left[1 - \left(\frac{D_i}{D}\right)^4\right]} \quad (12.13)$$

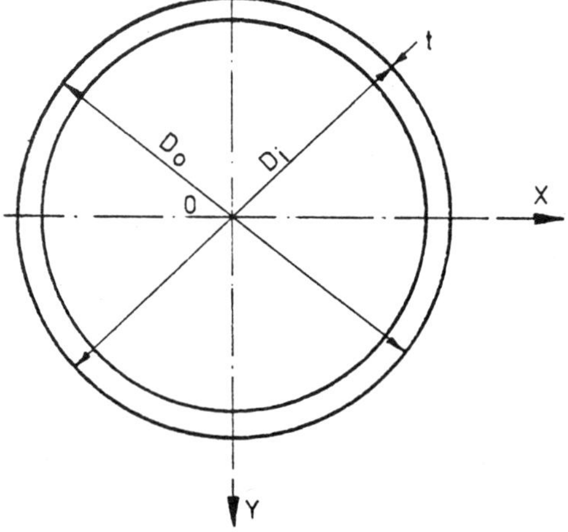

FIGURE 12.12 — Cross-section of the tube.

c. *Shearing stresses due to the shearing force Q_V*

The stress due to the shearing force is usually of only secondary importance. Its maximum value occurs at the neutral axis where the normal stress due to bending is zero. Therefore, the maximum combined stress usually occurs at the point where normal and shearing stresses due to torque are a maximum. In the case under consideration — at the top and bottom surfaces of the tube.

The principal stresses due to the combination of normal and shear stresses are [12.11]

$$\sigma_{max} = \frac{\sigma_V}{2} \pm \frac{1}{2}\sqrt{\sigma_V^2+4\tau^2} \qquad (12.14)$$

Substituting in this equation σ_V and τ, we obtain

$$\sigma_{max} = \frac{16}{\pi D_0^3\left[1-\left(\frac{D_i}{D_0}\right)^4\right]}\left(M_V + \sqrt{M_V^2+M_T^2}\right) =$$

$$= \frac{1}{2S}\left(M_V + \sqrt{M_V^2+M_T^2}\right) \qquad (12.15)$$

12.8.3 Horizontal Loads

a. *Stresses under horizontal loads*

Wind acting on structure. The wind force should be determined according to the National Building Code [12.12].

The horizontal wind force is:

$$q_w = qC_cC_gC_p \quad lb/ft^2 \qquad (12.16)$$

where: q, wind pressure, lb/ft^2; C_c, exposure factor and C_g, gust effect factor. Total force:

$$w = q_wD_o \quad lb/ft \qquad (12.17)$$

where: D_O is the outside diameter of the tube.

The intermediate posts, due to their flexibility cannot provide rigid support in a lateral direction. Therefore, continuous tubular structures have spans of 2ℓ between the inverted V-bents, as shown in Figure 12.9.

After determining the maximum bending moment due to the wind, M_W, the stress in the tube wall is:

$$\sigma_w = \frac{M_w}{S} = \frac{4M_w}{\pi D_i^2(t-c)} \qquad (12.18)$$

b. *Local wind action.*

By indicating the intensity of the horizontal wind pressure as q_W, the distribution of wind pressure around the tube is presented in Figure 12.13.

Due to the local wind action, the bending moment $M_{W\ell}$ and normal force $N_{W\ell}$, acting on the tube, are calculated by the formulae developed by Krupka [12.13].

The bending moment acting on the tube in the interval:

$$\frac{3}{4}\pi \le \varphi \le \pi$$

is:

$$M_{w\ell} = q_wR_m^2(-0.101 + 0.707\sin\varphi - 1.362\cos\varphi -$$

$$- 0.225\,\varphi\sin\varphi)\,lb\,in. \qquad (12.19)$$

The maximum value is at $\varphi = \pi$, or:

$$\max M_{w\ell} = 0.261\,q_wR_m^2 \qquad (12.20)$$

The normal force acting in the wall of the tube in the interval:

$$\frac{3}{4}\pi \le \varphi \le \pi$$

is:

$$N_{w\ell} = q_wR_m(-0.700 - 0.921\cos\varphi + 0.707\sin\varphi -$$

$$- 0.225\,\varphi\sin\varphi)\,lb \qquad (12.21)$$

At $\varphi = \pi$, the value of this force is:

$$N_{w\ell} = 0.212\,q_wR_m\,lb \qquad (12.22)$$

The maximum stress in the wall of the tube under local wind action is:

$$\max\sigma_{w\ell} = \frac{1.566\,q_wR_m^2}{t^2} + \frac{0.212\,q_wR_m}{t}\,psi \qquad (12.23)$$

where: R_m, is the middle radius of tube, in., and t, thickness of tube wall, in.

c. *Earthquake forces.*

Tubular galleries in earthquake areas should be designed to withstand earthquake forces. The seismic action should be considered according to the National Building Code, [12.14] Figure 12.14.

$$F_t = 0.004\left(\frac{h_t}{D}\right)\,kips \qquad (12.24)$$

or:

$$F_t = 0.15\,V\,kips \qquad (12.25)$$

whichever value is greater.

The minimum lateral seismic force, V, is:

$$V = ASKIFW\ kips \qquad (12.26)$$

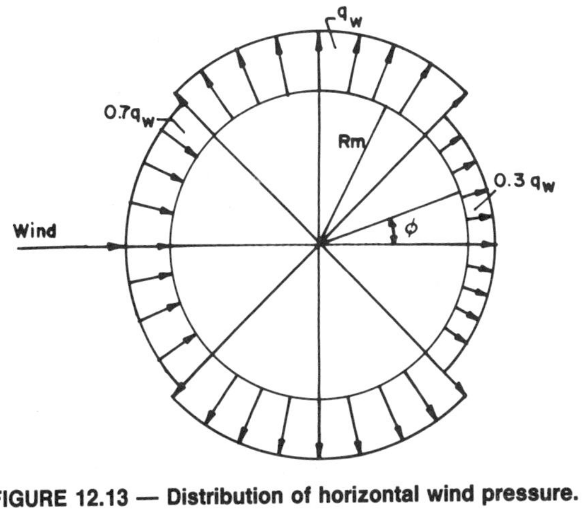

FIGURE 12.13 — Distribution of horizontal wind pressure.

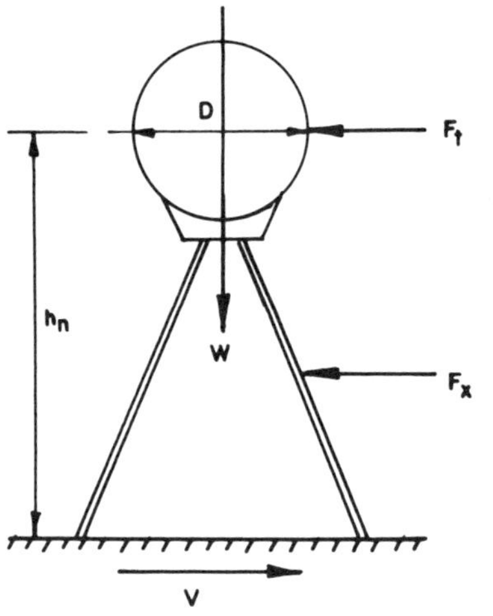

FIGURE 12.14 — Earthquake action.

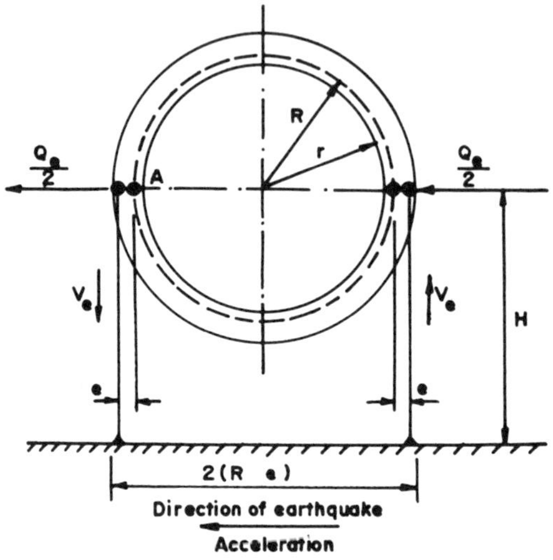

FIGURE 12.15 — Ring-girder under earthquake forces.

where:

h_t = height of force F_t above base, ft
D = diameter of tube, ft
A = acceleration ratio, or ratio of specific ground acceleration to acceleration due to gravity
S = $\frac{0.5}{3\sqrt{T}}$ = seismic response factor
T = $\frac{0.05\,h_t}{\sqrt{D}}$ = fundamental period of vebration of structure in seconds
K = structural coefficient that reflects material and type of structure, damping, ductility and/or energy-absorption capacity of structure
I = importance factor of structure
F = foundation factor depending on type and depth of soil measured from fundament
W = operating weight of structure of combined dead load of tube (including conveyor installations in tube and material on belt) and wieght of supports, kips

The additional lateral forces acting on the structure are:

$$F_1 = \frac{(V - F_t)\,W_t h_1}{[W_t h_1 + (W_t + W_s)\,h_2]} \tag{12.27}$$

$$F_2 = \frac{(V - F_t)(W_t + W_s)\,h_2}{[W_t h_1 + (W_t + W_s)\,H_2]} \tag{12.28}$$

where:

W_t = dead weight of tube, including conveyor installations and material on belt, kips
W_s = dead weight of supports, kips
h_1 = height of force F_1 above base, ft
h_2 = height of force F_2 above base, ft

The tubular gallery is usually stiffened above its supprots by the right-girder. According to Foster [12.15], the ring-girder is subjected to the bending moment and normal force, see Figure 12.15.

The expression of the bending moment is:

$$M_a = \frac{Q_e R}{4\pi}\left(1 + \frac{H\pi}{R + e}\right) \tag{12.29}$$

and the value of the normal force is:

$$N_a = \frac{Q_e}{\pi}\left(\frac{r}{R} - \frac{1}{4} + \frac{H\pi}{4\,(R + e)}\right) \tag{12.30}$$

where:

Q_e = total horizontal reaction of tube support, transmitting earthquake acceleration to tube
R = external radius of tube, in.
r = internal radius of tube, in.
H = height of support, in.
e = eccentricity of vertical reaction, in.

d. *Longitudinal loads*

Stresses due to the longitudinal loads are

$$\sigma_\ell = \frac{P}{A} = \frac{P}{\pi D\,(t - c)} \tag{12.31}$$

The stresses due to bending moments caused by the eccentricity of the longitudinal load will be computed as above.

12.8.4 Combined Cases

a. Bending and shear moment due to vertical loads

$$M_{equiv} = \tfrac{1}{2}\left(M_V + \sqrt{M_V^2 + M_T^2}\right) \tag{12.32}$$

Stress $$\sigma_{max} = \frac{M_{equiv}}{S} \tag{12.33}$$

b. Moments due to vertical and horizontal loads and shear

$$M_{tot} = \sqrt{M_V^2 + M_H^2} \tag{12.34}$$

$$M'_{equiv} = \tfrac{1}{2}(M_{tot} + \sqrt{M^2_{tot}+M^2_T}) \quad (12.35)$$

$$\text{Stress} \quad \sigma_{max} = \frac{M'_{equiv}}{S} \quad (12.36)$$

c. Stresses due to vertical, horizontal and longitudinal loads

$$\sigma_{max} = \frac{P}{A} + \frac{M'_{equiv}}{S} \quad (12.37)$$

In the case of an absence of horizontal loads

$$\sigma_{max} = \frac{P}{A} + \frac{M_{equiv}}{S} \quad (12.38)$$

12.8.5 Allowable Stresses

The tubular steel structures have to be designed to satisfy strength and stability requirements.

According to strength requirements, the maximum stress at a point of a cross-section, thus selected so as the moments and forces due to one loading case of the combination of loading cases are maximum, has to be smaller than the allowable stresses specified by the AISC Manual of Steel Construction.

By an inspection of the diagrams presented in Figure 12.11 and the critically loaded section of the tubular gallery occurs over the inverted V-bents.

The maximum stresses — tensile and compressive shall be computed as presented by formulas (12.18) to (12.24) and checked against the allowable stresses as presented by the AISC Manual.

12.9 Stability Requirements

12.9.1 General Data

A thin-walled cylindrical shell subjected to compression in the direction of its longitudinal axis may fail, either by the instability of the shell as a whole, involving bending of the axis, or by the local instability of the wall of the shell, which may not involve the lateral distortion of the axis, at all. The former type of failure is called the overall or Euler buckling, and the strength depends on the ratio of the length to the radius of gyration of the shell (ℓ/r). The latter type of failure is called the local buckling or wrinkling, and the strength depends upon the ratio of thickness to the radius of the shell wall (t/R).

We can classify a cylinder under axial load into three categories [12.12]

1. Very short cylinders

$$\ell < 1.72\sqrt{Rt}, \qquad N_{cr} = \frac{\pi^2 E(t^3/12)}{(1-\nu^2)\ell^2} \quad (12.39)$$

2. Intermediate length cylinders

$$1.72\sqrt{Rt} \le \ell \le 2.85\sqrt{R^3/t}$$

$$N_{cr} = \frac{Et^2}{R\sqrt{3(1-\nu^2)}}, \quad \sigma_{cr} = \frac{N_{cr}}{t} \quad \frac{Et}{R\sqrt{3(1-\nu^2)}} \quad (12.40)$$

3. Long cylinders

$$\ell \ge 2.85\sqrt{R^3/t}, \qquad N_{cr} = \frac{Et}{2}\left(\frac{\pi R}{\ell}\right)^2 \quad (12.41)$$

or

$$P_{cr} = 2\pi R N_{cr} = \frac{\pi^2 EI}{\ell^2} \quad (12.42)$$

which is the Euler formula.

The mode of buckling of the intermediate length cylinders is characterized by the formation of surface buckles in both the longitudinal and circumferential direction, without involvement of the lateral distortion of the axis. Thus, in this case, local buckling or wrinkling is governing. Long cylinders buckle elastically as ordinary Euler columns by bending of the longitudinal axis in one-half sine wave distortion, or by overall buckling.

In the following Table 12.2, are given the limiting intermediate critical lengths of cylinders having different diameters and wall thicknesses, using formula (12.40)

$$\ell_{cr} = 2.85\ R\sqrt{R/t}$$

TABLE 12.2 — Critical Length for Overall Buckling

	ℓ_{cr} = 2.85 R $\sqrt{R/t}$,ft			
	D, ft	t = ¼"	t = ⅜"	t = ½"
1	8′	158	129	112
2	9	188	154	133
3	10	221	180	156
4	11	255	208	180
5	12	290	237	205

12.9.2 Local Buckling or Wrinkling

Local buckling is the governing consideration in the design of tubular galleries of moderate length. Failure of this type is due to the formation of characteristic wrinkles or bulges, circular or lobed in shape. Wrinkling is local in nature and depends upon the combined compressive stress which occurs at a point of the dangerously loaded cross-section and has to be smaller than the allowable buckling stress.

In designing thin-walled tubular structures, two considerations are of importance. First, the local buckling should be prevented at stresses below yield strength; secondly, a more severe restriction is that the tendency to buckle locally should not reduce the overall buckling load of a whole structure.

There is a serious disagreement between the results of classical and experimental stress for the buckling of isotropic cylindrical shells under axial compression. Similar discrepancies can be observed for other loading conditions, such as bending and torsion. These experiments indicated critical stress levels in the order of ⅓ of those given by classical linear theory. This is due mainly to the nonlinear nature of the buckling process, the initial imperfections of such shapes and the edge effects [12.13]. These facts were taken into consideration by introducing into the recommended design formulae and allowable stresses certain correlation factors, γ, resulting from the experimental tests.

The allowable buckling stresses which will be presented hereafter are for elastic buckling and shells with simply supported edges. The edge of a shell is assumed to be simply supported if at the edge the radial and circumferential displacements are zero and there is no restraint against translation or rotation in the axial direction.

a. *Axial compression*

As given in Chapter 2, the critical buckling stress under axial compression by Plantema

$$\sigma_{cr} = \frac{662}{D/t} + 0.399\ F_y \tag{12.43}$$

where, the ratio D/t is valid for

$$\frac{3,300}{F_y} < \frac{D}{t} < \frac{13,000}{F_y} \tag{12.44}$$

Wilson and Newmark give

$$\sigma_{cr} = \frac{8,000}{D/t}\ ,\ \text{ksi} \tag{12.45}$$

or, by assuming a factor of safety of 1.5

$$\sigma_{cr} = \frac{5,333}{D/t}\ ,\ \text{ksi} \tag{12.46}$$

Finally, according to Baker

$$\sigma_{cr} = 0.6\gamma E(t/R) \tag{12.47}$$

is valid for moderately long cylinders

$$\gamma_1 z > \frac{\pi^2}{2\sqrt{3}} \tag{12.48}$$

where

$$z = \frac{L^2}{Rt}\sqrt{1-\nu^2} \tag{12.49}$$

γ_1 = correlation factor obtained from Figure 12.16

The recommended allowable stresses for local buckling in the function of D/t are shown in Chapter 2, Figure 2.6.

$$N_{cr} = \sigma_c t = 0.6\gamma E(\frac{t^2}{R}) \tag{12.50}$$

$$P_{cr} = \sigma_{cr} 2\pi Rt = 1.2\gamma E\pi t^2 \tag{12.51}$$

b. *Bending*

Buckling tests on cylinders similar to those tested in axial compression indicate that buckling occurred over the compression side of the cylinder in the same wave form, with approximately the same wavelengths, as in the axially-loaded cylinders. A comparison of the axial compression and the pure-bending test results show exactly the same decrease of axial load P with an increase of the ratio R/t, as shown by the axially-loaded cylinders.

The influence of initial imperfections in a shell, considering bending, should not be as great as in the axial compression. However, during the bending of the shells, there occurs a certain flattering of the cross-section, which leads to an increase of maximum stress.

The design-allowable buckling stress for thin-walled circular cylinder subjected to bending as given by Baker is

$$\sigma_{cr} = 0.6\gamma_2 E(t/R) \tag{12.52}$$

which is valid for moderately long cylinders.

γ_2 = the correlation factor obtained from Figure 12.17.

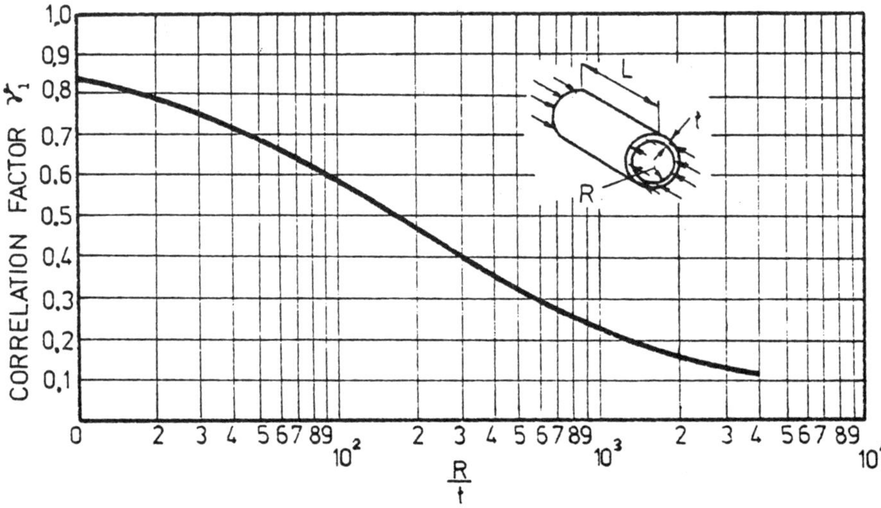

FIGURE 12.16 — Correlation factor for unstiffened circular cylinders subjected to axial compression.

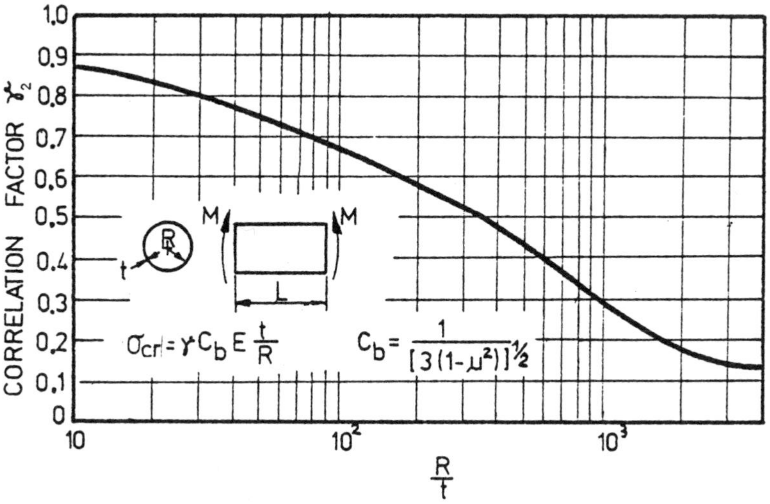

FIGURE 12.17 — Correlation factor for unstiffened circular cylinder subjected to bending.

For elastic stresses, the allowable moment is

$$M_{cr} = \pi R^2 t \sigma_{cr} = 0.6\gamma_2 E \pi R t^2 \qquad (12.53)$$

c. *Ovalling effect under bending*

This theory, known as "Brazier effect" is treated in Chapter 9, Section 9.6.

According to Brazier, the critical bending moment is

$$M_{cr} = \frac{2\sqrt{2}}{9} \, , \; \frac{E\pi R_m t^2}{\sqrt{1-\mu^2}} \qquad (12.54)$$

and the critical buckling stress

$$\sigma_{cr} = 0.33E \; (t/R_m) \qquad (12.55)$$

Therefore, considering the ovalling deformation of the tube, the following condition should be satisfied

$$\frac{M_{cr}}{S} \leq 0.33E \; (t/R_m) \qquad (12.56)$$

d. *Shear or torsion*

As given by Baker [12.12]

$$\tau_{cr} = C_s \frac{Et}{Rz^{1/4}} \qquad (12.57)$$

valid for cylinders of moderate length

$$100 < z < 78(R/t)^2(1-\nu^2) \qquad (12.58)$$

For long cylinders the design-allowable buckling stress is

$$\tau_{cr} = \frac{0.261 \; C_s \; E}{(1-\nu^2)^{3/4}} \left(\frac{t}{R}\right)^{3/2} \qquad (12.59)$$

when

$$z > 78(R/t)^2(1-\nu^2) \qquad (12.60)$$

$$z = \frac{L^2}{Rt}\sqrt{1-\nu^2} \qquad (12.61)$$

The coefficient C_S is given in Figure 12.18.

e. *Combined loading*

The criterion for structural failure of a member under combined loading is frequently expressed in terms of a stress-ratio equation

$$R_1^x + R_2^y + R_3^z = 1 \qquad (12.62)$$

In general, the stress ratio R is the ratio of the allowable value of the stress caused by a particular kind of load in a combined-loading condition to the allowable stress for the same kind of load when it is acting alone. The subscript denotes the stress due to a particular kind of loading such as those of compression, shear, etc., and the exponents, usually empirical, express the general relationship of the quantities for

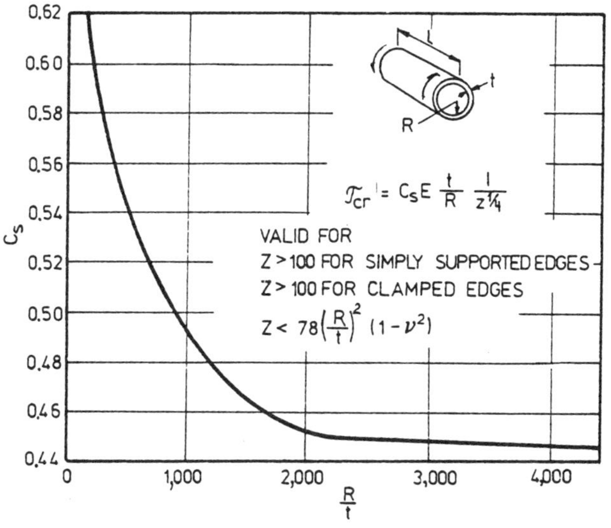

FIGURE 12.18 — Buckling-stress coefficient C_s for unstiffened circular cylinders subjected to torsion.

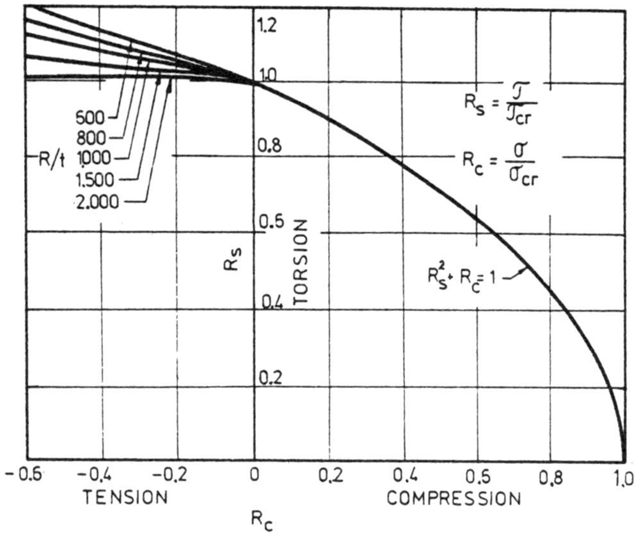

FIGURE 12.19 — Buckling-stress interaction curve for unstiffened circular cylinders under combined torsion and axial loading.

failure of the member. A curve drawn from such a stress-ratio equation is termed a stress-ratio interaction curve. In simple loadings, the term "stress ratio" is used to denote the ratio of applied to allowable stress, (R_a) and if the equation

$$R_{a_1}^x + R_{a_2}^y + R_{a_3}^z < 1 \tag{12.63}$$

is satisfied (or the part falls onto the left side of the stress-ratio interaction curve) the structure is safe.

f. *Combined torsion and axial loading*

A semiempirical interaction curve for circular cylinders is given in Figure 12.19.

σ_{cr}^c is found using Figure 12.16 and τ_{cr} is found using Figure 12.18.

$$\frac{\sigma_{appl}^c}{\sigma_{cr}^c} + \left(\frac{\tau_{appl}}{\tau_{cr}}\right)^2 < 1 \tag{12.64}$$

g. *Bending and torsion*

An empirical interaction curve is given in Figure 12.20.

σ_{cr}^b is found from Figure 12.17 and τ_{cr} from Figure 12.18.

$$\frac{\sigma_{appl}^b}{\sigma_{cr}^b} + \left(\frac{\tau_{appl}}{\tau_{cr}}\right)^2 < 1 \tag{12.65}$$

h. *Axial compression and bending*

The interaction curve is given in Figure 12.21.

σ_{cr}^b is found from Figure 12.17 and σ_{cr}^c from Figure 12.16.

$$\frac{\sigma_{appl}^c}{\sigma_{cr}^c} + \frac{\sigma_{appl}^b}{\sigma_{cr}^b} < 1 \tag{12.66}$$

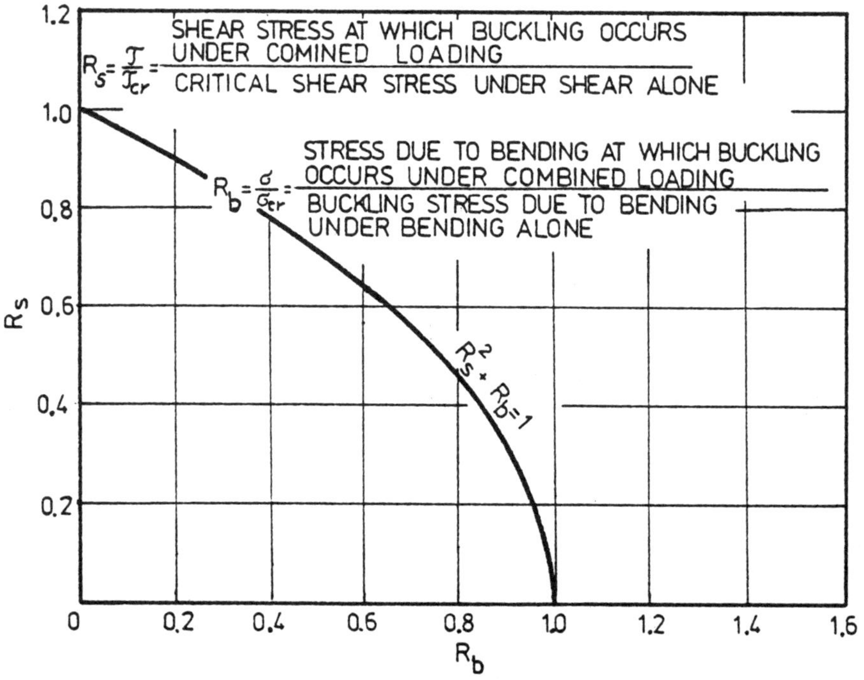

FIGURE 12.20 — Buckling-stress interaction curve for unstiffened circular cylinder under combined bending and torsion.

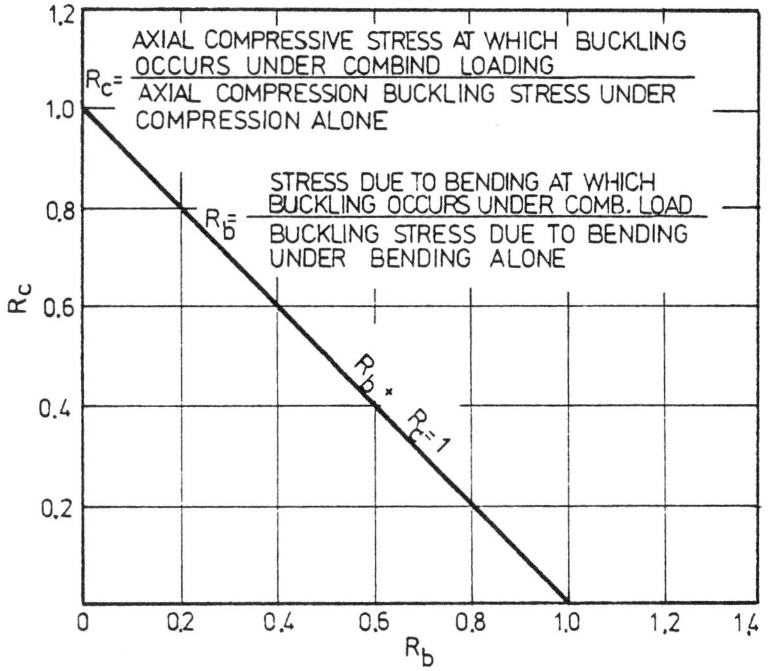

FIGURE 12.21 — Buckling-stress interaction curve for unstiffened circular cylinder under combined axial compression and bending.

i. *Axial compression, buckling, and torsion*

Although a slightly different interaction equation is proposed for each case, Schilling [12.18] has suggested that a single formula can be conservatively applied to all these combinations. He has proposed the relation

$$\frac{\sigma^c_{appl}}{\sigma^c_{cr}} + \frac{\sigma^b_{appl}}{\sigma^b_{cr}} + \left(\frac{\tau_{appl}}{\tau_{cr}}\right)^2 < 1 \qquad (12.67)$$

12.10 Overall Buckling

Although the local buckling is prevailing for axially-loaded tubes of moderate length and diameter (Table 12.2), small variation of radius-to-thickness ratio and span may cause the structure to fail in overall buckling. In addition, in several cases when the D/t ratio of a member is such that the local buckling is probable, the KL/r ratio of the member may be such that the column buckling is also an important consideration. According to AISI Specifications [12.19] purely column-type buckling of tubular members may occur with no local buckling as long as

$$D/t < \frac{3{,}300}{F_y} \qquad (12.68)$$

where

F_y = the yield strength of material in ksi

This formula is very conservative compared with the limits defined in Table 12.2.

In the following, we analyze formulas which are valid for both unstiffened and stiffened shells, having longitudinal stringers and circumferential ring stiffeners. In the case of stiffened shells, the stringers must be included in the calculation of the radius-of-gyration "r" of the tube cross-section, while circumferential stiffeners can be neglected since they have no direct effect on the overall column-buckling mode, Figure 12.22.

Statical values of stiffened cross-sections of the tube are:

The area of the combined section

$$A = \pi D_i (t-c) + nA_{st}, \text{ in}^2 \qquad (12.69)$$

where

A_{st} = cross-sectional area of one stiffener, in^2

n = number of longitudinal stiffeners

D_i = internal diameter of the tube, in

t = thickness of the wall

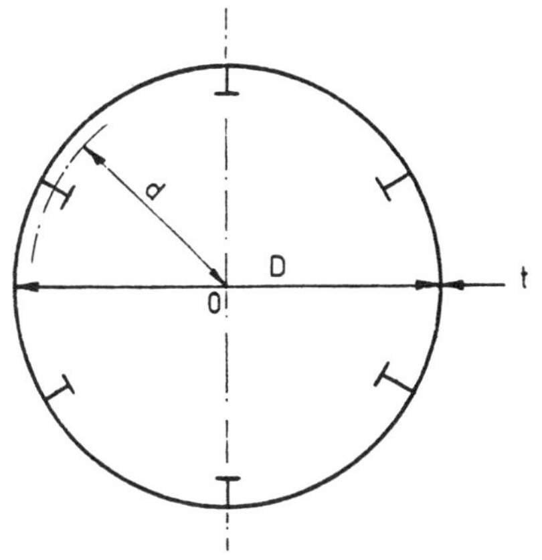

FIGURE 12.22 — Tube reinforced by longitudinal stiffeners.

The polar moment of inertia of the combined section

$$I_p = I_x + I_y = \pi\frac{D^3}{4}(t-c) + A_{st}d^2, in^4 \quad (12.70)$$

where

d = the distance from the center of the tube to the centroid of the stiffener, in

The radius of gyration

$$r = \sqrt{\frac{I_p}{2A}}, in \quad (12.71)$$

The tubes under the axial compression, are classified as in the case of ordinary steel columns into two categories as intermediate tubes, and slender tubes. According to the column Research Council Specifications.

a. *Intermediate columns* $\lambda \leqslant \sqrt{2}$

$$F_a = \frac{\sigma_{cr}}{F.S} \quad (12.72)$$

$$\sigma_{cr} = (1-0.25\lambda^2)F \quad (12.73)$$

$$F.S = 1.67 + 0.265\lambda - 0.44\lambda^3 \quad (12.74)$$

where

λ = $\frac{KL}{r} = \frac{1}{\pi} = \sqrt{\frac{F_Y}{E}}$ = nondimensional slenderness ratio

KL = effective length of circular tube for column-type buckling. In our case the conservative value of K = 1 is used.

F_a = allowable critical stress, ksi

σ_{cr} = theoretical critical axial stress, ksi

F.S = factor of safety

b. *Slender columns,* $\lambda < \sqrt{2}$

$$F_a = \frac{\sigma_{cr}}{FS} \quad (12.75)$$

$$\sigma_{cr} = \frac{F_Y}{\lambda^2} \quad (12.76)$$

where

FS = 1.92

Based on the results of tests with tubes of carbon steels having yield strengths of 45 ksi and 55 ksi, and for which the stub-column tests found proportional limits as low as 50 percent of the yield strength, Wolford and Rebholz [12.16] proposed a more conservative column curve, which is recommended for use in cases of tubular cold-formed structures.

c. *Intermediate columns,* $\lambda \leqslant \sqrt{3}$

$$F_a = \frac{\sigma_{cr}}{FS} \quad (12.77)$$

$$\sigma_{cr} = [1.0 - \frac{2}{3\sqrt{3}}\lambda]F_{Ye} \quad (12.78)$$

$$FS = 1.67 + 0.145\lambda \quad (12.79)$$

d. *Slender columns,* $\lambda > \sqrt{3}$

$$F_a = \frac{\sigma_{cr}}{FS} \quad (12.80)$$

$$\sigma_{cr} = \frac{F_Y}{\lambda^2} \quad (12.81)$$

where

FS = 1.92

A structural cylinder fabricated from plate will typically have high circumferential residual stresses caused by cold-forming processes, as well as from the circumferential and longitudinal stresses due to welding. Because of the exact nature of these stresses and their effect on column strength, they are at this time, still a matter of conjecture, so it is usually conservative to use the equations given by Wolford and Rebholz.

12.10.1 Tube Under Combined Loading

In this case, the usual interaction-type formula for beam-columns in compression and biaxial bending will be used. Allowable-stress interaction equations for design are essentially empirical. The factor of safety depends, in part, on the expressions chosen to define the allowable stresses, F_a (axial) and F_b (bending).

In the following, it is necessary to consider that the local buckling is local in effect — or limits the strength of the structure at a point of a certain cross-section, while the overall buckling has to do with the stability of the structure as a whole.

The interaction equation is [12.21]

$$\frac{P}{P_u} + \frac{C_{mv}\, M_{vT}}{M_{uv}(1-\frac{P}{P_{ev}})} + \frac{C_{mH}\, M_{HT}}{M_{uH}(1-\frac{P}{P_{eH}})} < 1 \quad (12.82)$$

or, in another form [12.21]

$$\frac{f_a}{F_a} + \frac{C_{mv}\, f_{bv}}{F_{bv}[1-\frac{f_a}{F_{ev}}]} + \frac{C_{mH}\, f_{bH}}{F_{bH}[1-\frac{f_a}{F_{eH}}]} < 1 \quad (12.83)$$

It should be noted that the denominator of the second and third term of Equation (12.78) represents the strength in bending at a particular point of a cross-section in vertical and horizontal direction, respectively, magnified by a factor $(1 - \frac{f_a}{F_e})$ to guard against the overall buckling,

where

P = applied longitudinal compressive force, kips

C_{mv}, C_{mH} = reduction factors depending on the edge conditions, transverse loading, and curvature of the bent [12.22]

$M_{vT} = \frac{1}{2}(M_v + \sqrt{M_v^2 + M_T^2})$ = equivalent bending moment due to the vertical loads and torque M_T

M_v = bending moment due to vertical loads

$M_{HT} = \frac{1}{2}(M_H + \sqrt{M_H^2 + M_T^2})$ = equivalent bending moment due to horizontal loads and torque M_T

M_H = bending moment due to horizontal loads

$$f_a = \frac{P}{A}\ ;\quad f_{bv} = \frac{M_{vT}}{S}\ ;\ f_{bH} = \frac{M_{HT}}{S}$$

$$P_e = \sigma_{cr} A = \frac{F_y A}{1.92(\frac{KL}{2})^2 \frac{1}{\pi^2}\cdot\frac{F_Y}{E}} = \frac{\pi^2 E.A}{1.92(\frac{KL^2}{2})}$$

$$= \frac{149{,}000\ A}{(\frac{KL}{2})^2}$$ = the elastic buckling strength of the member

$$F'_e = \frac{Pe}{A} = \frac{149{,}000}{(\frac{KL}{2})^2}$$

$$F'_{ev} = \frac{149{,}000}{(\frac{\ell}{r})^2}$$, ℓ = span between V-bents and single posts

$$F'_{eH} = \frac{149{,}000}{(\frac{2\ell}{r})^2}$$, 2ℓ = span between two V-bents

The loss of stability due to the overall buckling will take place in the horizontal direction when $L = 2\ell$, and in a vertical direction, when $L = \ell$.

a. *Overall buckling*

In the case of

$$L = 2\ell \gg \ell_{cr} = 2.85\ R\sqrt{R/t} \tag{12.84}$$

overall buckling will occur before local buckling. Since local buckling is not present, the full bending strength of the section can be considered

$$F_{bv} = F_{bH} = 0.60\ F_y \tag{12.85}$$

F_a will be taken from the formula (12.77) to (12.81). The interaction Equation (12.83) must be satisfied.

b. *Local Buckling*

In the case of

$$1.72\sqrt{Rt} < L = 2\ell < 2.85\ R\sqrt{R/t} \tag{12.86}$$

local buckling will occur before overall buckling. The interaction Equations (12.64) to (12.67) must be satisfied first. Because of the loss of cross-sectional strength under bending due to local buckling, overall buckling may occur, and interaction Equation (12.83) must be satisfied, too, where

$$F_a = 0.60\,F_y$$

or by Equations (12.77) to (12.81).

$$F_{bv} = F_{bH} = \sigma_{cr} = 0.6\gamma_2\frac{Et}{R} \tag{12.87}$$

or the σ_{cr} given by Brazier

$$\sigma_{cr} = 0.34E(\frac{t}{R})$$

c. *Local and overall buckling*

For many practical applications, when the D/t ratio of a member is such that local buckling is probable, the KL/r ratio of the member may be such that the column buckling is also an important consideration, or

$$L = 2\ell = 2.85\ R\sqrt{R/t} \tag{12.88}$$

In this case, both the local buckling associated with strength locally, and the overall buckling associated with overall stability of the structure must be checked. The requirements are:

a. **Local buckling interaction Equations (12.64) to (12.67) must be satisfied.**
b. **Overall stability interaction Equation (12.83) must be satisfied. In this case, in order to establish the allowable stresses F_a and F_b, the approach suggested by Marshall [12.23] will be followed in a more general form. In this approach, the theoretical local-buckling strength is substituted for Fy in the appropriate beam-column formula. Since the local buckling limits the strength locally, and occurs simultaneously with the overall buckling, we can assume an equivalent material having a yield strength equal to the theoretical local buckling strength**

$$F'_y = \sigma_{cr} = 0.6E(t/R) \tag{12.89}$$

which is identical with the formulae suggested by Baker without the correlation factors γ_1 and γ_2.

$$F_{bv} = F_{bH} = 0.60F'_y = 0.36E(t/R) \tag{12.90}$$

which is almost identical to the formula suggested by Brazier for pure bending

$$F_a = \frac{F'_y}{FS\lambda^2} = \frac{0.6E(t/R)}{1.92\lambda^2} \qquad (12.91)$$

which is derived from Equations (12.82) and (12.83)

From Table 12.2, we see that the ratio R/t fluctuates between 100 and 300. By considering the diagrams giving the correlation factors γ_1 and γ_2 versus the R/t ratio, we notice that $0.70 < \gamma_1 < 0.90$. The correlation factors γ_1 and γ_2 are smaller than the FS of 0.60 and $\frac{1}{1.92\lambda^2}$ employed in this case. In conclusion, the interaction Equation (12.83) yields conservative results.

12.11 Intermediate Stiffening Rings

Along the length of a tubular gallery at equal intervals are located inside the tube intermediate circumferential stiffening rings. Although their spacing is normally such that they should be considered discretely spaced, and have no direct effect on the local and overall buckling mode, they serve a dual structural purpose.

a. To receive the posts and hangers, which support at equal intervals the idlers of the conveyor and stringers of the walkway.

b. To maintain the cross-section of the tube between the supports circular, and thus to allow the tube to act as a beam.

The circumferential stiffness consists of flat plates, angles or T-sections, continuously welded onto the inside surface of the shell of the tube.

If structurally rolled sections are to be used, they are formed to the required radius at lengths of one quarter of the tube's circumference and shop-welded to the shell of the tubular gallery.

12.11.1 Design of the Stiffening Ring

The stiffening rings have to be designed to meet the strength requirements of the AISC Manual of Steel Construction.

The forces which are applied on the ring due to conveyor, walkway, and own weight, are shown in Figure 12.23.

In the design of the ring, the L.L. on the walkway is used at 100 lbs/ft^2. The critical loading condition occurs for D.L. plus L.L. on the conveyor and walkways. These loads can be resolved in a vertical resultant ΣP through the center of the ring and a moment M_0. The analysis of the tube under these loading cases is shown in Table 12.3 based on formulas given by Roark and Young [12.24].

In the design, we are considering the maximum value of the stresses which result from a combination of the loads at point A of the ring.

a. *Due to bending moment*

$$f_b = \frac{0.50\ M_o + 0.24\ \Sigma_p R}{S} \qquad (12.92)$$

b. *Due to axial load*

$$f_a = \frac{0.24\ \Sigma_p}{A} \qquad (12.93)$$

c. *The maximum compressive stress at the section is*

$$f_{max} = f_b + f_a < f_{allow} \qquad (12.94)$$

where

$$f_{allow} = 0.66\ F_y$$

d. *The maximum shear stress*

$$\tau = \frac{0.48\ M_o}{AR}\ ,\ psi \qquad (12.95)$$

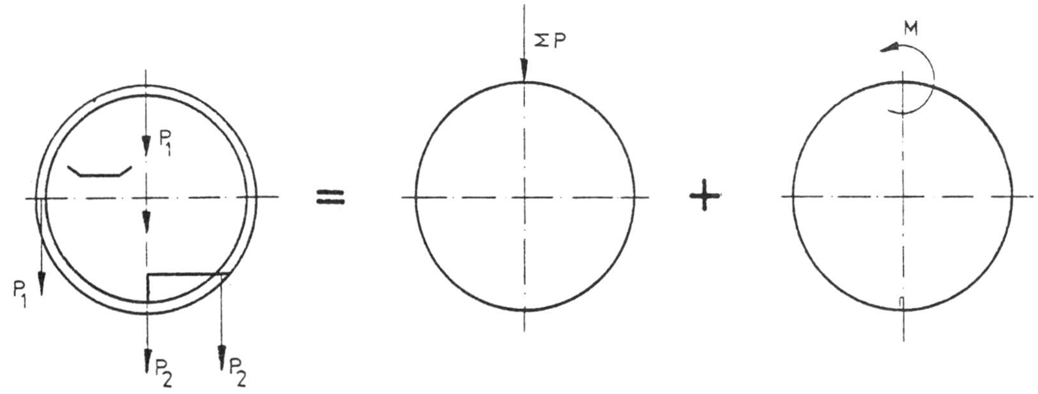

FIGURE 12.23 — Loads and moments acting on stiffening ring.

e. *The maximum shear stress in the shell wall*

$$\tau_w = \frac{\frac{M_o}{2\pi R^2} + \frac{\Sigma p}{\pi R}}{2(t-c)}, \text{ psi} \quad (12.96)$$

12.12 Ring Girders at Supports

Rigid ring girders provide effective support for tubular structures. These girders prevent the distortion of the tube at supports and thus maintain its ability to act as a beam.

Typical ring girders are shown in Figures 12.24 (a) and 12.24 (b).

In Figure 12.24 (a) the ring girder consists of a plate girder having stiffened fixed supports and Figure 12.24 (b) shows the ring girder stiffened by radially located stiffeners.

In Figure 12.25 the girder consists of two stiffening rings continuously welded to the pipe on both sides, and tied together with diaphragm plates welded between the two rings. Two short columns consisting of wide-flange I-beams are bolted between the rings to carry the load to the seat weldments.

12.12.1 Analysis and Design of Ring Girder

The loads are transmitted from the shell of the tube to the ring by tangential shear [12.24].

1. *Vertical Loads*

   The shell transmits by circumferential shear to the ring a concentric vertical load Q — reaction of a continuous beam and a torque M_T — reaction torque transferred to supports at both ends of the fixed beam. This is the typical case for rings over inverted V-bents. The actions of Q and M_T are shown in Figures 12.26 (a) and 12.26 (b).

CASE (a) — Concentric vertical load Q

This case should be considered for rings over the single post supports only.

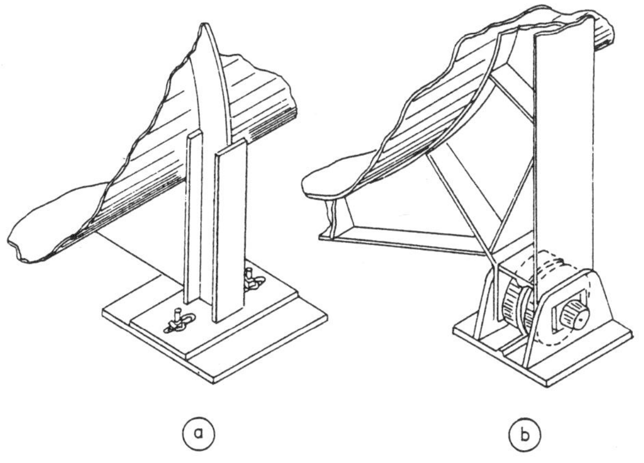

FIGURE 12.24 — Details of ring girder support for tubular conveyor (a) simplest type (b) a roller support.

The bending moment diagram for an eccentricty of the support equal to 0.04R is shown in Figure 12.27.

Figure 12.28 shows the absolute value of M_o, M_i, and M_a plotted in the function of the ratio a/R.

The eccentricity, a, is taken as positive when the reactions are applied outside the neutral axis of the supporting ring. The maximum possible ring moment is obtained for a/R = + 0.04, in which case

$$M_i = M_a = 0.010\ QR \quad (12.97)$$

The maximum normal ring force, or axial force, due to the circumferential shear, occurs at $\alpha = \pi/2$ and is

$$T_a = \frac{Q}{4} \quad (12.98)$$

This force exerts compression just above, and tension just below the horizontal diameter. Since, for a/R $\geqslant$ +0.04, the maximum bending stress is also found at this point, a somewhat smaller combined stress will be obtained with

$$+\ 0.04 > \frac{R}{a} > 0 \quad (12.99)$$

although it is hard to achieve because of fabrication reasons.

From Figure 12.28, for a given ratio of supports, eccentricity to radius, a/R, the values of the moment coefficient, M.C., can be obtained at points "a", "i" and "0"

$$M = MC \times QR \quad (12.100)$$

The maximum combined stresses in the ring for a/R [0.04 occur at the horizontal diameter

$$f_1 = f_{bend} + f_{axial} = \frac{MC \times QR}{S} + \frac{Q}{4A} \quad (12.101)$$

and for a/R = +0.04, the maximum combined stress is

$$f_1 = \frac{0.010\ QR}{S} + \frac{Q}{4A} \quad (12.102)$$

where for the computation of the section modulus S of the ring girder, a portion of the adjacent shell may be considered as acting together with the girder. The effective width of the shell thus acting, is

$$c = 1.56\sqrt{rt} + b$$

where

A = the area of the combined section, Figure 12.29

and

b = the width of the ring girder.

CASE (b) — Torque due to vertical loads

For an analysis of this case, Figure 12.30, use of the tables given in Reference [12.24] was made.

The maximum bending moment due to a combination of these cases occurs at the horizontal diameter

TABLE 12.3 — Analysis of Stiffening Rings.

	CASE I	CASE II	SUPERPOSITION
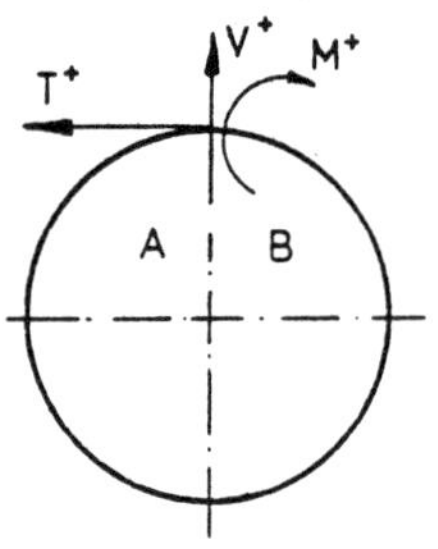	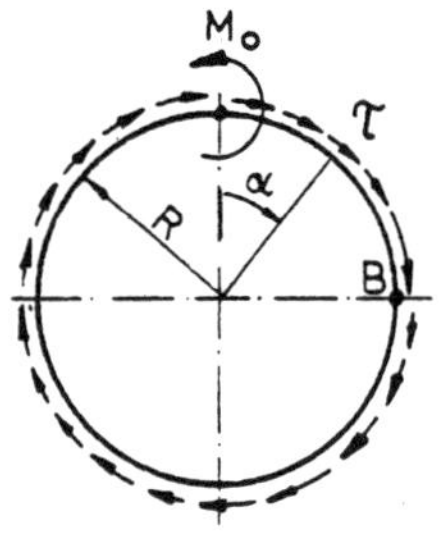 Ring loaded by bending moment M_O, in-kips, and supported by tangential shear.	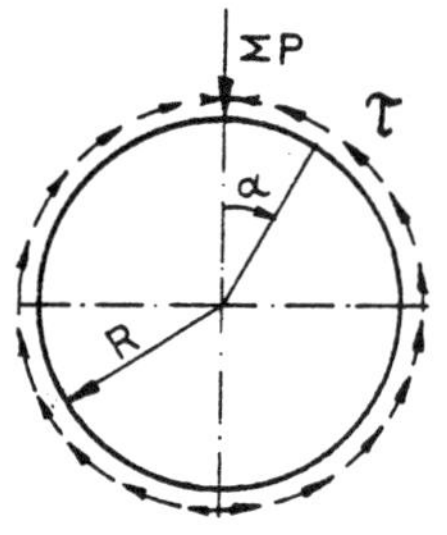 Ring loaded by concentrated load Σp, kips and supported by tangential shear.	Cases I and II at point A $I_A + II_A$
Bending Moment M, in-kips	max $M = 0.500\ M_O$ at point A	max $M = 0.24\ \Sigma p\ R$ at point A	max $M = 0.50\ M_O +$ $+\ 0.24\ \Sigma p\ R$
Axial circumferential force T, kips	$T = 0$, at point A max $T = \dfrac{0.3183\ M_O}{R}$ at point B	max $T = -0.24\ \Sigma p$ at point A	$T = -0.24\ \Sigma p$
Circumferential shear V, kips	max $V = \dfrac{3M_O}{2\pi R} = 0.4775\ \dfrac{M_O}{R}$ at point A	$V = 0$	$V = 0.48\dfrac{M_O}{R}$
Tangential shear τ, K/inch	$\tau = \dfrac{M_O}{2\pi R^2}$	$\tau = \dfrac{\Sigma p \sin\alpha}{\pi R}$ For $\alpha = 90°$ max $\tau = \dfrac{\Sigma p}{\pi R}$	$\tau = \dfrac{M_O}{2\pi R^2} + \dfrac{\Sigma p}{\pi R}$

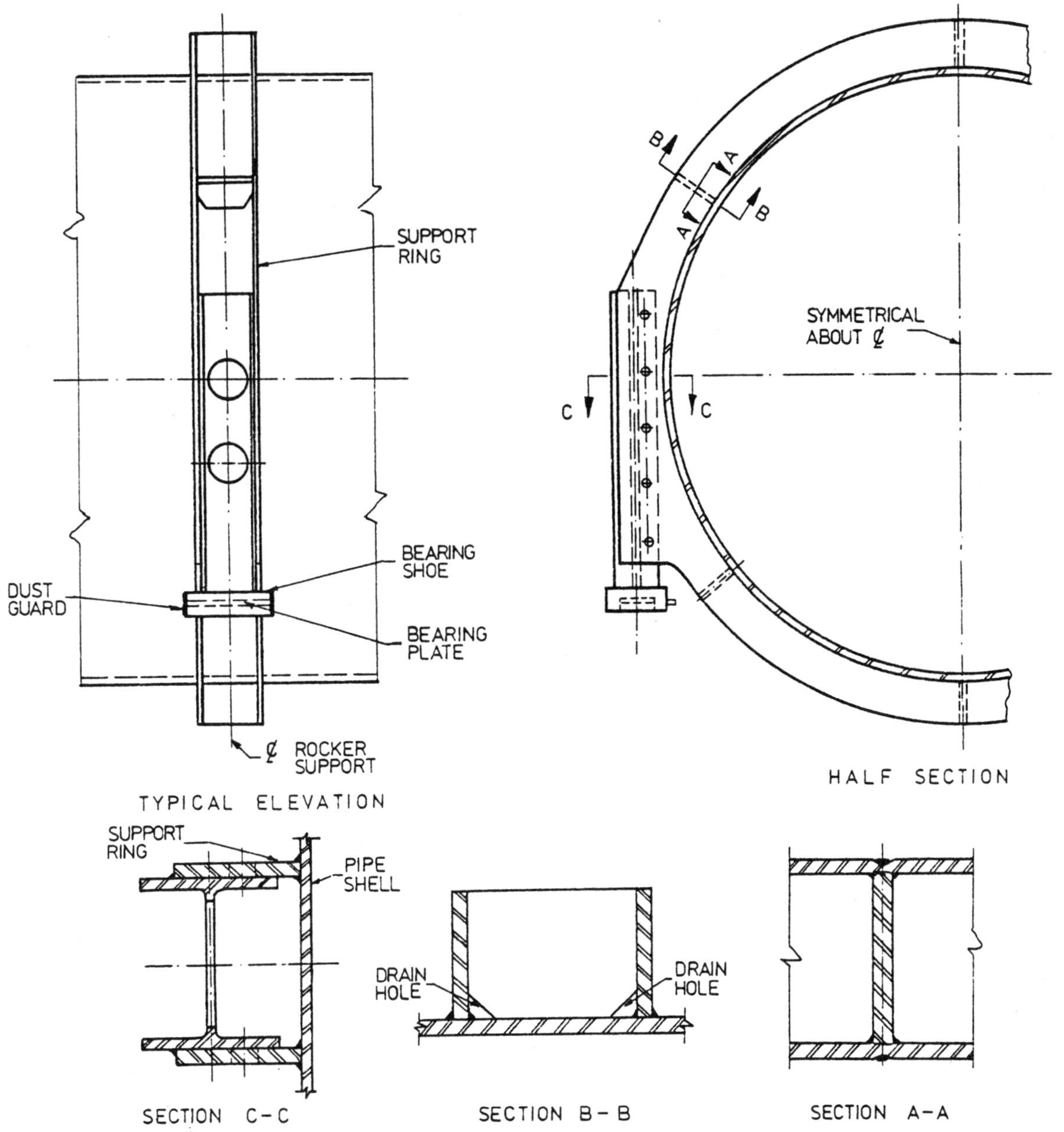

FIGURE 12.25 — Typical ring girder and column support.

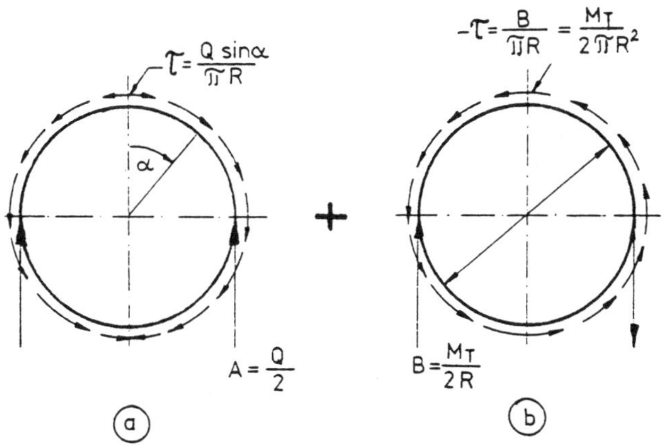

FIGURE 12.26 — Circumferential shear due to reaction and torque.

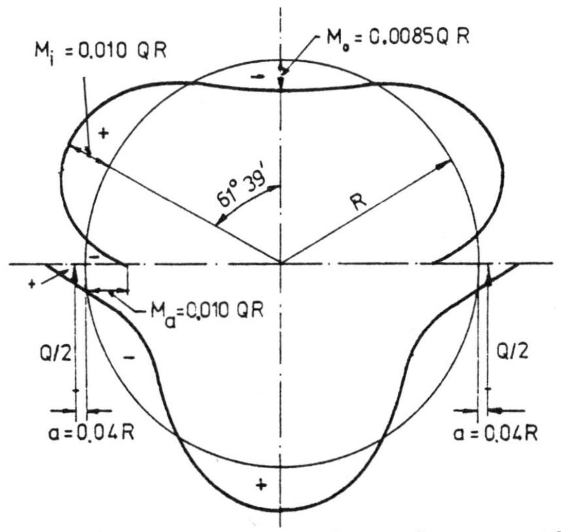

FIGURE 12.27 — Bending moment diagram for an eccentricity of support.

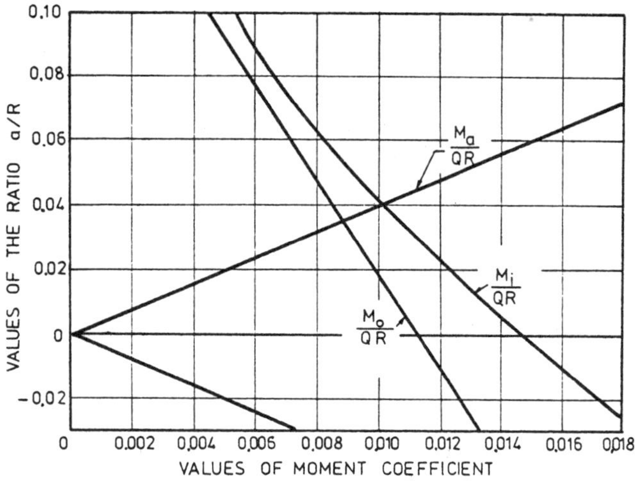

FIGURE 12.28 — Moments M_o, M_i, M_a.

$$\text{Max } M = \frac{M_D}{2} = \frac{M_B}{2} = \frac{aM_T}{4R} \quad (12.103)$$

The maximum axial force occurs at the same point

$$\max N = \frac{B}{2} = \frac{M_T}{4R} \quad (12.104)$$

The maximum combined stress at the horizontal diameter is

$$f_2 = f_{bend} + f_{axial} = \frac{aM_T}{4RS} + \frac{M_T}{4RA} \quad (12.105)$$

12.12.2 Horizontal Loads

The wind reaction W, over the inverted V-bents, is transmitted from the shell of the tube to the ring girders by tangential shear. For an analysis of the ring girder under horizontal loads the tables presented in the Reference [12.24] are used, and the resulting stresses are shown in Figure 12.31.

$$M_C = -M_A = 0.0796\ WR \quad (12.106)$$

$$T_A = -T_C = 0.2387\ W \quad (12.107)$$

$$V_A = 0.0 \quad (12.108)$$

$$V_C = \frac{W}{2} \quad (12.109)$$

The maximum combined stress occurs at the horizontal diameter

$$f_3 = \pm \frac{0.08WR}{S} \pm \frac{0.24W}{A} \quad (12.110)$$

Summary

a. *Vertical Loads*

The ring girder above the single post support

$$f_1 = \frac{0.01QR}{S} + \frac{Q}{4A} < f_{allow.} \quad (12.111)$$

The ring girder above the inverted V-bent

$$f = f_1 + f_2 = \frac{0.01QR}{S} + \frac{aM_T}{4RS} + \frac{Q}{4A} + \frac{M_T}{4RA} < f_{allow.} \quad (12.112)$$

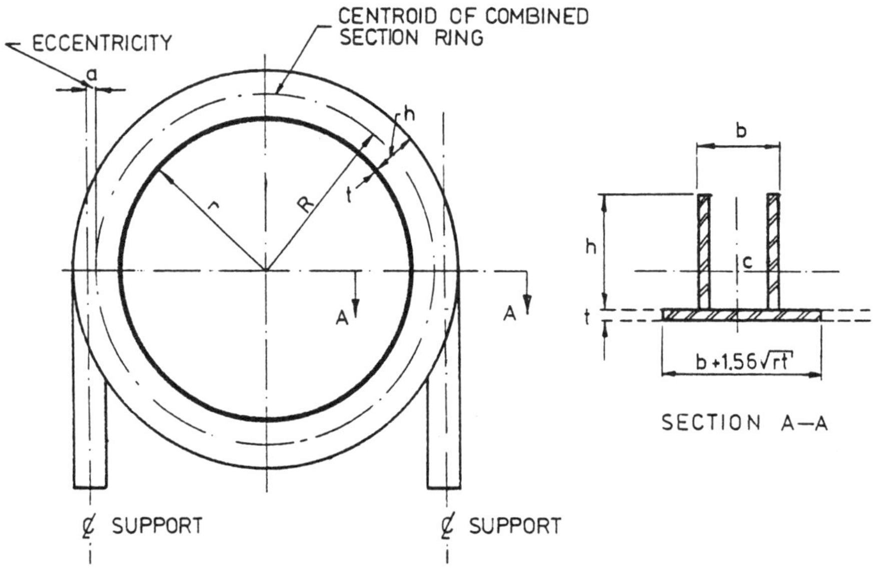

FIGURE 12.29 — Effective width of ring girder.

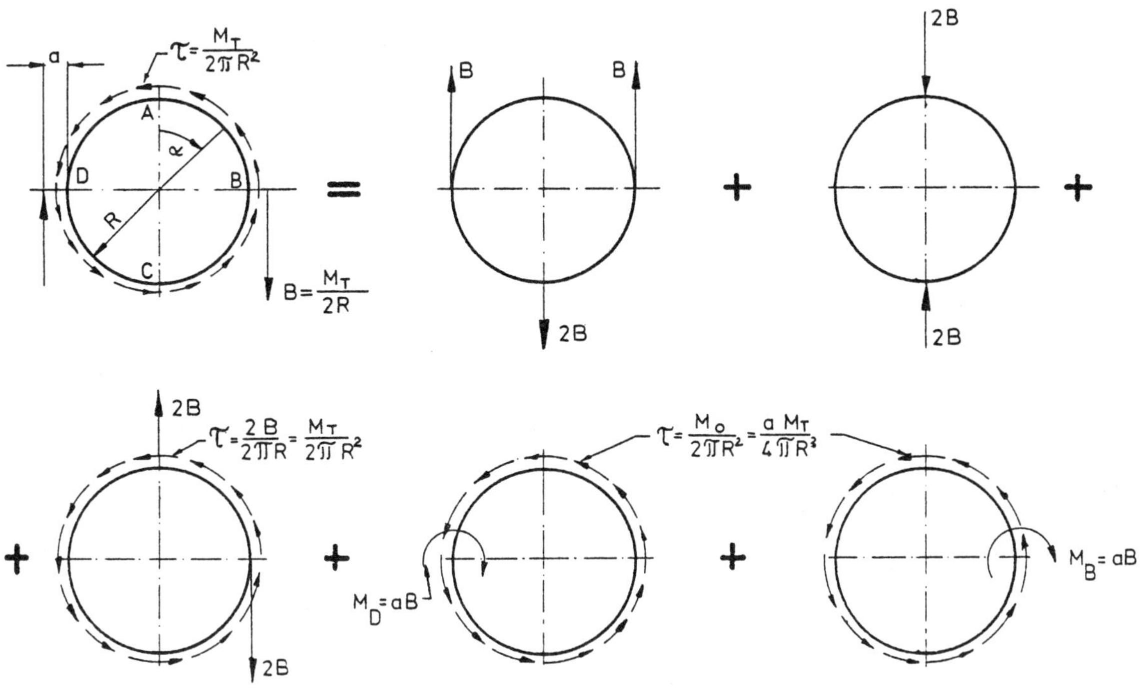

FIGURE 12.30 — Ring girder under torque M_T.

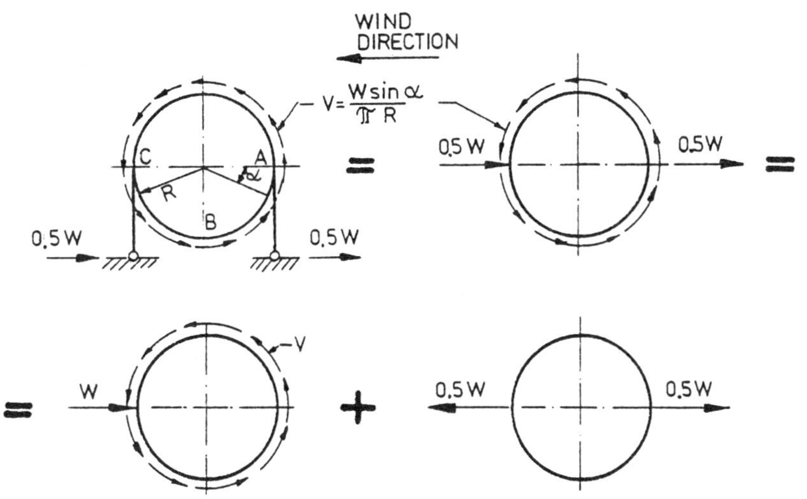

FIGURE 12.31 — Ring girder under wind load.

b. *Vertical and horizontal loads*

The single post support

$$f_1 = \frac{0.01QR}{S} + \frac{Q}{4A} < f_{allow}. \tag{12.113}$$

The inverted V-bent support

$$f = f_1 + f_2 + f_3 = \frac{0.01QR}{S} + \frac{aM_T}{4RS} + \frac{0.08WR}{S} + \frac{Q}{4A} + \frac{M_T}{4RA} + \frac{0.24W}{A} \tag{12.114}$$

12.13 Ring Girders at Supports. Method by Krupka.

Another method of the ring girder analysis, based on elastic theory, was developed by Krupka [12.26].

12.13.1 Vertical Supports

The tube transmits to the ring a concentrated vertical load Q, which is the reaction of a tube acting generally as a continuous beam. In the case of two vertical supports, having reactions Q/2 and acting on the ring girder with the moments and normal (axial) forces along the circumference of the ring girder, see Figure 12.32.

We consider two cases, namely:

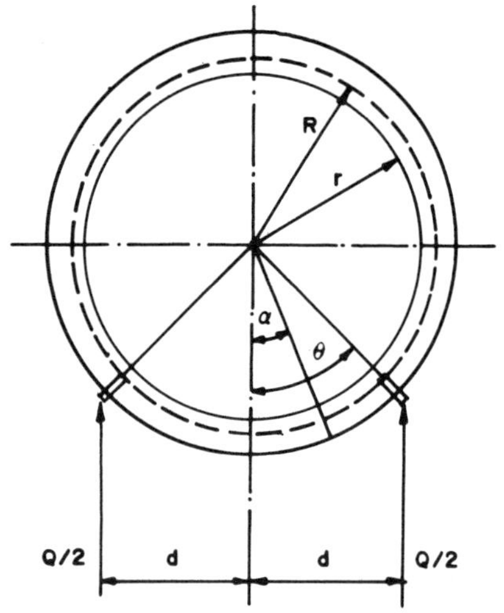

FIGURE 12.32 — Ring girder having two vertical supports.

a. *Bending moment and normal force for any section below support in interval $0 \leq \varphi \leq \theta$:*

$$M_1 = \frac{QR}{2\pi}\left[\cos\theta - \frac{a}{R}(\pi - \theta) + \left(\frac{1}{2} + \frac{a^2}{R^2}\right)\cos\varphi + \varphi\sin\varphi\right] \quad (12.115)$$

$$N_1 = \frac{Q}{2\pi}\left[\varphi\sin\varphi - \left(1.5\frac{a^2}{R^2}\right)\cos\varphi\right] \quad (12.116)$$

b. *Bending moment and normal force for any section above support for $\theta \leq \varphi \leq \pi$:*

$$M_2 = \frac{QR}{2\pi}\left[\theta\frac{a}{R} + \cos\theta + \left(\frac{1}{2} + \frac{a^2}{R^2}\right)\cos\varphi + (\varphi - \pi)\sin\varphi\right] \quad (12.117)$$

$$N_2 = \frac{Q}{2}\left[(\varphi - \pi)\sin\varphi - \left(1.5\frac{a^2}{R^2}\right)\cos\varphi\right] \quad (12.118)$$

where: Q is the reaction at support due to total operating weight of tube, which consists of dead load of tube, conveyor installations in tube and material on belt, kips.

The combination for the simultaneous action of the maximum moment, M_{max}, and the corresponding normal force, N, and also the maximum value of the normal force, N_{max} and the corresponding moment, M, for the typical values of the angle θ, are shown in Table 12.4.

TABLE 12.4 — Values of moments and normal forces.

θ°	φ°	M_{max}	N	φ°	N_{max}	M
45°	90°	−0.049 QR	−0.250 Q	50°	−0.378 Q	0.026 QR
90°	70°	0.014 QR	0.155 Q	90°	±0.250 Q	0

12.13.2 Saddle Support

In a case when the tube is supported by the saddle, see Figure 12.30, the bending moments and the normal forces in the ring girder are calculated, using the following formulae developed by Krupka [12.26]:

a. *Bending moment and normal force for any section below support in interval $0 \leq \varphi \leq \theta$:*

$$M_3 = \frac{QR}{2\pi A_1}\left\{\left(2\sin\theta - \frac{1}{2}\sin 2\theta - \theta\right) + \left[\frac{1}{4}\sin 2\theta - \frac{1}{2}\theta + (\theta - \pi)\sin^2\theta\right]\cos\varphi + \left(\theta - \varphi - \frac{1}{2}\sin 2\theta\right)\pi\sin\varphi + A_1\left[1 + \frac{1}{2}\cos\varphi + (\varphi - \pi)\sin\varphi\right]\right\} \quad (12.119)$$

and

$$N_3 = \frac{Q}{2\pi A_1}\left\{\left[\frac{1}{4}\sin 2\theta - \frac{1}{2}\theta + (\theta - \pi)\sin^2\theta\right]\cos\varphi + (\theta - \varphi + \sin 2\theta)\pi\sin\varphi + A_1\left[(\varphi - \pi)\sin\varphi - \frac{3}{2}\cos\right]\right\} \quad (12.120)$$

b. *Bending moment and normal force for any section above support for $\theta \leq \varphi \leq \pi$:*

$$M_4 = \frac{QR}{2\pi A_1}\left\{2\sin\theta - \theta - \frac{1}{2}\sin 2\theta + \left(\theta\sin^2\theta - \frac{1}{2}\theta + \frac{1}{4}\sin 2\theta\right)\cos\varphi + A_1\left[1 + \frac{1}{2}\cos\varphi + (\varphi - \pi)\sin\varphi\right]\right\} \quad (12.121)$$

also see Table 12.5.

$$N_4 = \frac{QR}{2\pi A_1}\left\{\left(\theta\sin^2\theta + \frac{1}{4}\sin 2\theta - \frac{1}{2}\right)\cos\varphi + A_1\left[(\varphi - \pi)\sin\varphi - \frac{3}{2}\cos\varphi\right]\right\} \quad (12.122)$$

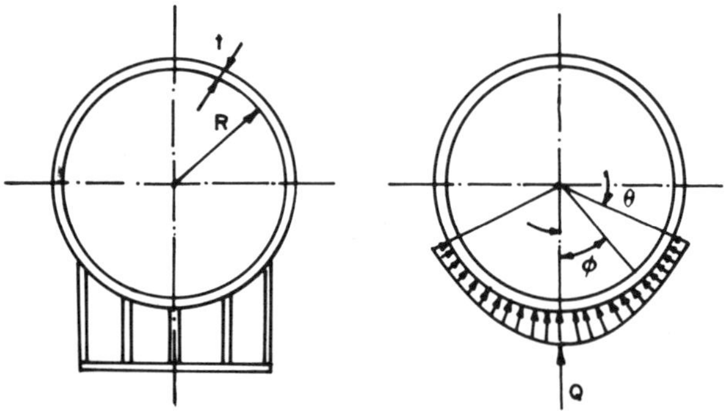

FIGURE 12.33 — Values of moments and normal forces.

where

$$A_1 = \theta + \sin\theta \cos\theta \qquad (12.123)$$

The combinations of the simultaneous actions of the moments and normal forces are shown in Table 12.5.

TABLE 12.5 — Values of moments and normal forces.

θ°	φ°	M_{max}	N	φ°	N_{max}	M
45°	80°	0.079 QR	−0.309 Q	45°	−0.412 Q	0.012 QR
90°	90°	−0.047 QR	−0.250 Q	0°	−0.477 Q	0.043 QR

For both types of supports either vertical or in the shape of the saddle, the maximum stress in the ring girder should be calculated using the formula:

$$f = \frac{M}{W} + \frac{N}{A} \qquad (12.124)$$

where: W is the section modulus of cross-section of ring girder and A is the cross-sectional area of ring girder.

12.14 Elliptical Tube for Conveyor Gallery

In recent years, the attention of some conveyor gallery designers has been directed toward a relatively unexplained area; namely, the analysis and design of conveyor gallery of elliptical shape. It is modern, streamlined, pleasing to the eye and provides lesser obstruction to the wind action. An elliptical tube may provide more convenient space for the placement of a conveyor and walkway than a circular tube. The same types of supports may be incorporated for an elliptical conveyor as those described for a circular conveyor, as well as their procedures for fabrication and erection.

12.14.1 Loads and Forces

1. *Vertical loads*
   a. Own weight of elliptical tube, Figure 12.34.

$$q_s = \pi(3\frac{a+b}{2} - \sqrt{ab})t\rho_s + \frac{nW_{st}}{\ell}, \text{ lbs/ft} \qquad (12.125)$$

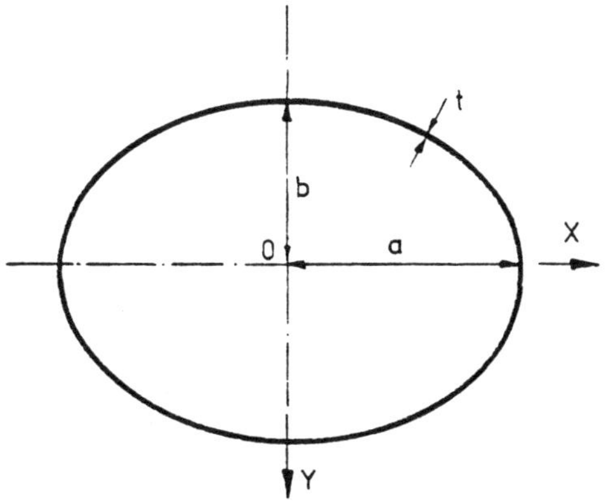

FIGURE 12.34 — Elliptical tube.

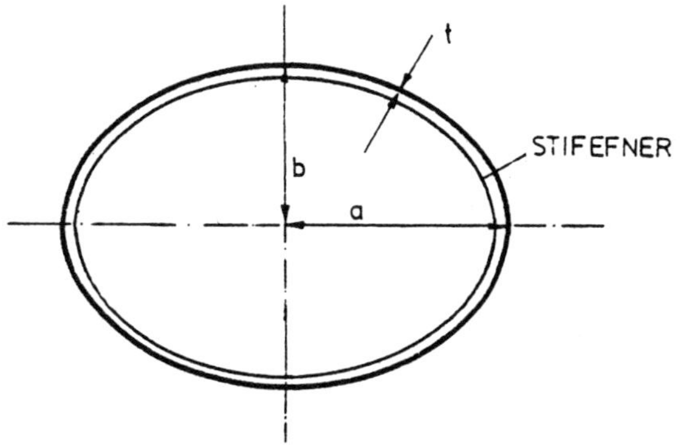

FIGURE 12.35 — Cross-section of the elliptical tube.

where

a,b = semimajor and semiminor axes, respectively

t = thickness of the wall, in

ρ_s = unit weight of steel shell, 0.283 lb/in^3

n = number of stiffeners in a span

W_{st} = weight of one stiffener

= span between bents or supports

All other weights are similar to those given by formulas (12.2) to (12.4).

2. *Live loads*

   The same as those given by formulas (12.5) to (12.8).

3. *Horizontal forces*

   The wind forces are estimated using the same provisions as those given in Section 12.7.2, considering the elliptical shape of the tube. Earthquake forces are then estimated as shown in Chapter 5, Section 5.2.6.

4. *Longitudinal forces*

   The longitudinal forces are as shown under Section 12.7.2 (3).

12.14.2 Stress Analysis

1. *General data*

   In the following, are given some geometric and material constants of the cross-section of the elliptical tube, which will be used in further discussions, Figure 12.35.

a,b = semimajor and semiminor axes, respectively, in.

t = thickness of the wall, in.

c = corrosion allowance, $^1/_{16}$ in.

$$A = \pi(3\frac{a\ b}{2} - \sqrt{ab})(t-c) \qquad (12.126)$$

= cross-section of the tube's wall, in^2.

$$I_x = \frac{\pi}{4}[(a\ t)(b\ t)^3 - ab^3] \qquad (12.127)$$

= moment of inertia of the tube with respect to axis-x, in^4.

$$I_y = \frac{\pi}{4}\left[(a+t)^3(b+t) - a^3b\right] \quad (12.128)$$

= moment of inertia of the tube with respect to axis-y, in^4.

$$S_x = \frac{\pi}{8(b+t)}\left[(a+t)(b+t)^3 - ab^3\right] \quad (12.129)$$

= section modulus of the tube with respect to axis-x, in^3.

$$S_y = \frac{\pi}{8(a+t)}\left[(a+t)^3(b+t) - a^3b\right] \quad (12.130)$$

= section modulus of the tube with respect to axis-y, in^3.

E = modulus of elasticity of the tube's material, ksi

$G = \dfrac{E}{2(1+\nu)}$ = shear modulus of the tube's material, ksi

$\nu = 0.3$ = Poisson's ratio

12.14.3 Vertical and Horizontal Loads

The vertical loads induce bending moment, torsional moment and shear in the tube.

a. *Normal stresses due to the bending moment M_V*

Normal stresses due to the bending moment are

$$(\sigma_V)_{max} = \frac{M_V}{S_x} = \frac{8M_V(b+t)}{\pi\left[(a+t)(b+t)^3 - ab^3\right]} \quad (12.131)$$

b. *Shearing stresses due to the torque M_T*

Shearing stresses due to internal torque produced by the eccentric position of the equipment and live load inside the tube, according to formula given in the Reference [12.24]

$$\tau_{max} = \frac{2M_T}{\pi(a+t)(b+t)^2\left[1 - \left(\frac{a}{a+t}\right)^4\right]} \quad (12.132)$$

at the ends of the minor axis on the outer surface.

At the relatively small thickness of the wall, we may assume a uniform distribution of the stresses along the cross-section, and [12.22]

$$\tau = \frac{M_T}{2At} \quad (12.133)$$

where

$$A = \pi\left(3\frac{a+b}{2} - \sqrt{ab}\right)t \quad (12.134)$$

= cross-sectional area, in^2.

12.14.4 The Principal Stresses

The principal stresses due to a combination of normal and shear stresses are

$$\sigma_{max} = \frac{\sigma_V}{2} \pm \sqrt{\sigma_V^2 + 4\tau^2}$$

Substituting in this equation, σ_V and τ, we obtain

$$\sigma_{max} = \frac{4M_V(b+t)}{\pi[(a+t)(b+t)^3 - ab^3]} + \frac{2}{\pi}\sqrt{\frac{16M_V^2(b+t)^2}{[(a+t)(b+t)^3 - ab^3]^2} + \frac{2M_T^2}{(a+t)^2(b+t)^4\left[1-\left(\frac{a}{a+t}\right)^4\right]^2}} \quad (12.135)$$

12.14.5 Horizontal Loads

Stresses due to the horizontal moments are

$$(\sigma_H)_{max} = \frac{M_H}{S_y} = \frac{8M_H(a+t)}{\pi\left[(a+t)^3(b+t) - a^3b\right]} \quad (12.136)$$

c. *Longitudinal loads*

Stresses due to the longitudinal loads are

$$\sigma_e = \frac{P}{A} = \frac{P}{\pi\left(3\frac{a\ b}{2} - \sqrt{ab}\right)(t-c)} \quad (12.137)$$

12.14.6 Buckling Stability

12.14.6.1 Bending

In case the thin-walled elliptic tube is subjected to pure bending, the flexural rigidity is reduced accompanying the flattening of the cross-section, according to the increase of curvature. Although in the thin-walled cylinder the local buckling occurs before reaching the flattening point generally, the buckling stress due to flattening is shown as follows [12.28].

a. Bending about the major axis

$$(\sigma_{cr})_x = C_x \frac{E}{\sqrt{1-\nu^2}} \cdot \frac{t}{\rho_x} \sqrt{t/t_m} \quad (12.138)$$

b. Bending about the minor axis

$$(\sigma_{cr})_y = C_y \frac{E}{\sqrt{1-\nu^2}} \cdot \frac{t}{\rho_y} \sqrt{t/t_m} \quad (12.139)$$

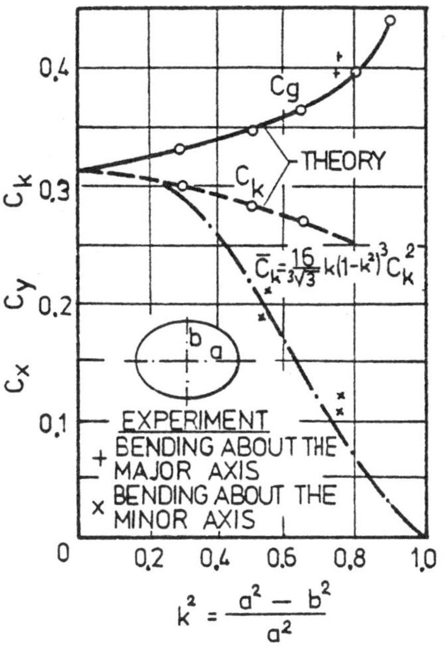

FIGURE 12.36 — Values of C_x and C_y.

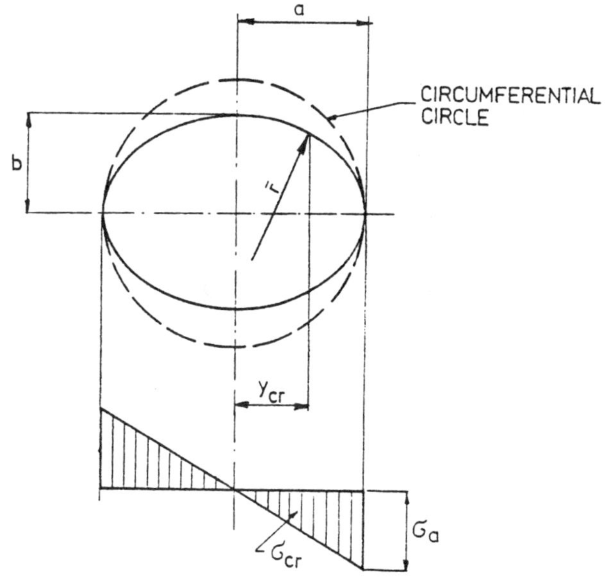

FIGURE 12.37 — Determination of section modulus of the circumscribed circle.

The corresponding critical moments M are, respectively

$$(M_{cr})_x = (\sigma_{cr})_x \frac{\pi}{4} b(b+3a)t_m \qquad (12.140)$$

$$(M_{cr})_y = (\sigma_{cr})_y \frac{\pi}{4} a(a+3b)t_m \qquad (12.141)$$

where

C_x and C_y are the variables varying with

$$K^2 = \frac{(a^2-b^2)}{a^2} \qquad (12.142)$$

as shown in the following Figure 12.36.

$$t_m = t + \frac{A_{st}}{U} \qquad (12.143)$$

where

A_{st} = total of sectional area of the longitudinal stiffeners

U = the circumferential length of the ellips

ρ_x = a^2/b, ρ_y = b^2/a = the radius of curvature at the ends of the minor and major axis, respectively

As is seen in Figure 12.36, in the case of $K^2 \geqslant 0.25$ the experimental results show the local buckling on the compressive side and $K^2 = 0.25$ is a critical point.

Computation of the buckling stress for elliptical tubes under bending [12.29]

1. Compute the section modulus of the circumscribed circle, Figure 12.37.

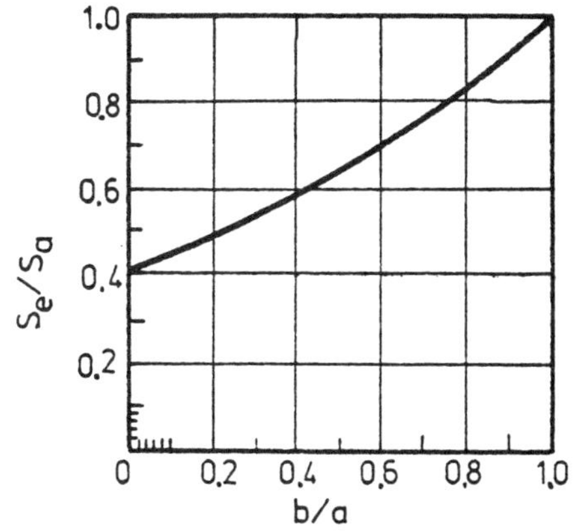

FIGURE 12.38 — Determination of the ratio S_c/S_a in the function of the ratio b/a.

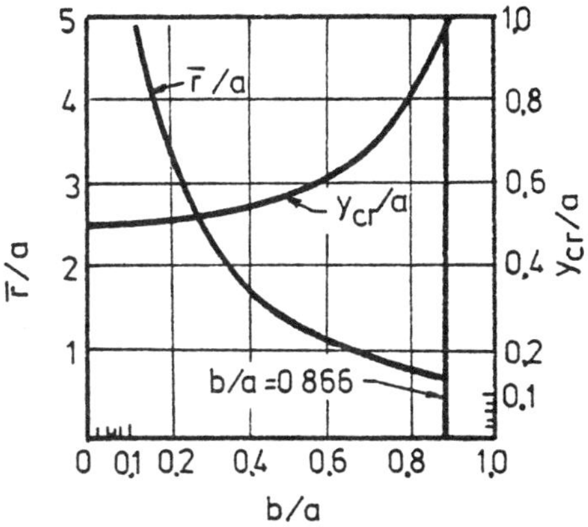

FIGURE 12.39 — Determination of ratios y_{cr}/a and r/a in the function of ratio b/a.

$$S_a = \pi a^2 t \quad (12.144)$$

2. Find the extreme-fiber section modulus of the elliptic cylinder, using Figure 12.38.

$$S_c = \left(\frac{S_c}{S_a}\right) \times S_a \quad (12.145)$$

3. Find y_{cr}/a and $\bar{r}/a$ from Figure 12.39 or Equations (12.146) and (12.147).

$$\frac{y_{cr}}{a} = \frac{0.5}{\sqrt{1 - (b/a)^2}} \quad (12.146)$$

$$\frac{\bar{r}}{a} = \frac{0.649a}{b} \quad (12.147)$$

where

y_{cr} = buckle location on the cross-section

$\bar{r}$ = radius of curvature at y, y_{cr}, i.e., the critical curvature

It should be noted that when $b/a > 0.866$, the buckling must occur at the extreme of the major axis, and $\bar{r} = a$.

4. Compute the applied stress at the location of the critical curvature from

$$\sigma_{y_{cr}} = M \times \frac{1}{S_c} \times \frac{y_{cr}}{a} \quad (12.148)$$

5. Compute the applied stress at this location (for long tube only) using

$$\sigma_{cr} = C_y E \frac{t}{\bar{r}} \quad (12.149)$$

in which C_y is found from the curve of Figure 12.40.

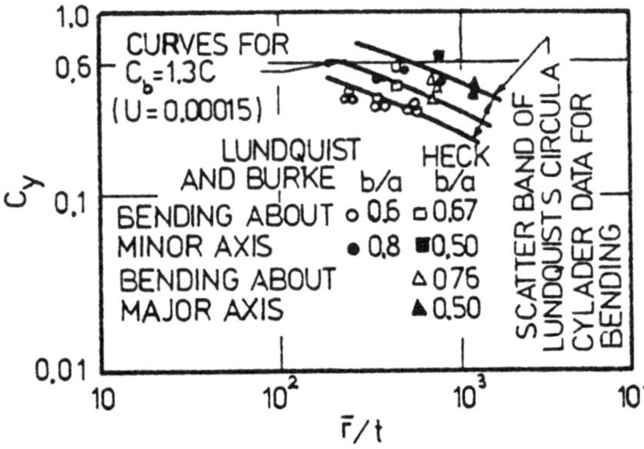

FIGURE 12.40 — Determination of C_y in the funciton of the ratio $\bar{r}/t$.

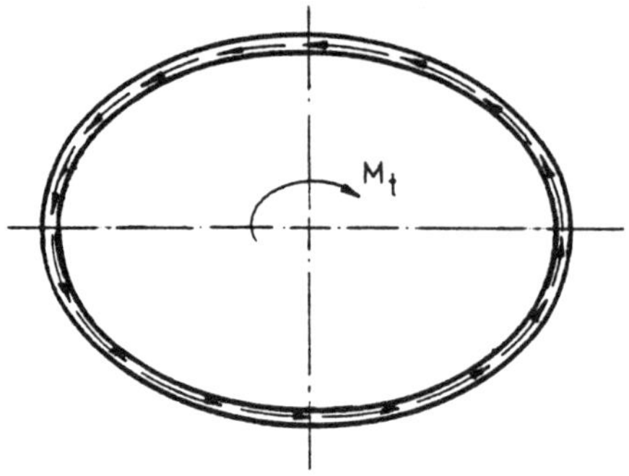

FIGURE 12.41 — Elliptical tube under torsion.

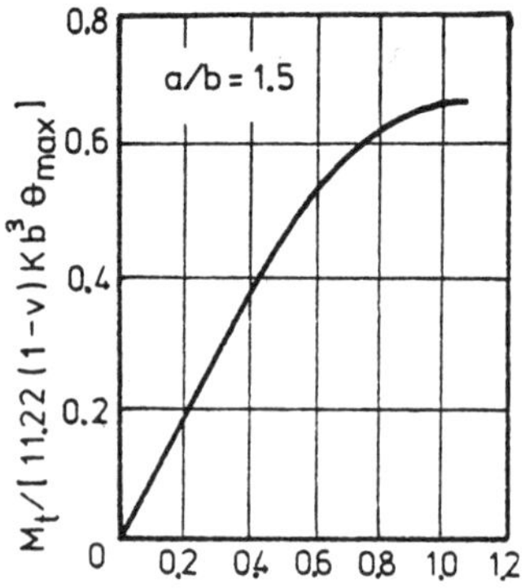

FIGURE 12.42 — Values of M_t in the function of θ/θ_{max}.

12.14.7 Torsion

In the following, we are analyzing the buckling of an elliptic tube under torsion, considering the flattening of the sectional form of the tube, Figure 12.41.

The relation between a twisting angle and a twisting moment before buckling when $a/b = 1.5$

$$M_t = 11.220(1-\nu)K\theta b^3[1-2.585b^2\theta^2(b/t)^2] \quad (12.150)$$

$$\theta_{max} = 0.360\ t/b^2 \quad (12.151)$$

$$M_{t_{max}} = 2.69(1-\nu)Kb^2(t/b) \quad (12.152)$$

Figure 12.42 shows this relation

where

$K = \frac{Et}{1 - \nu^2}$

θ = twisting angle per unit length

M_t = twisting moment

By solving the buckling determinant the relation between

$$2\tau_{cr} tc^2/D\pi^2 \text{ and } \theta/\theta_{max}$$

is obtained, where c is semi-peripherical in length and

$$D = \frac{Et^3}{12(1-\nu^2)}$$

This relation can be described, as indicated by a solid line in Figure 12.43.

The shearing stress τ_{cr} corresponding to any twisting angle may be expressed in the following form

$$2\tau_{cr}tc^2/D\pi^2 = 5.508(1-\nu^2)(b/t)(\theta/\theta_{max}) \times (1-0.167(\theta/\theta_{max})^2 \qquad (12.153)$$

This equation is shown by broken lines, taking (b/t) as a parameter.

The intersections of the solid line and the broken lines give the buckling stress.

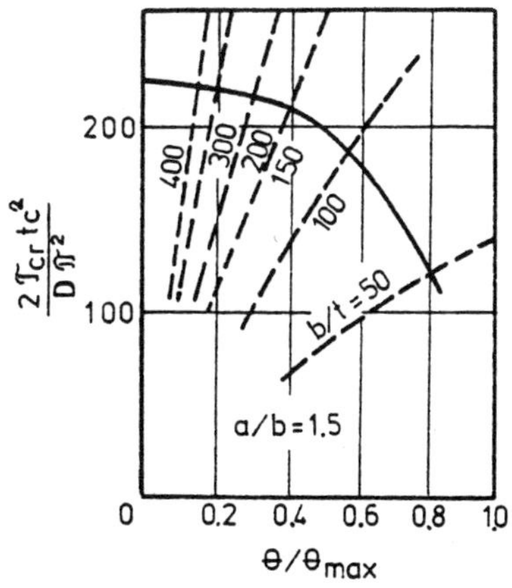

FIGURE 12.43 — Relation between parameter $2\tau_{cr}tc^2/D\pi^2$ and θ/θ_{max}.

REFERENCES — CHAPTER 12

[12.1] Kok, H.G., "Structural Aspects of Bulk Material Handling Systems", paper presented at SME Meeting, Salt Lake City, Utah, Preprint Number 69-B-329, September, 1969.

[12.2] Mylar, D.T., "Belt Conveyor Structures", paper presented at SME Meeting, Seattle, Washington, Preprint Number 71-B-302, September 1971.

[12.3] Fritsch, J.P., "Transiowa Belt Line", National Corn Growers Association, Des Moines, Iowa, April 5th, 1972.

[12.4] Troitsky, M.S., "Design Guidelines for Steel Tubular Thin-Walled Structures", 4th Progress Report, Canadian Steel Industries Construction Council Project No. 724, January 1974.

[12.5] American Institute of Steel Construction, Manual of Steel Construction, 7th Ed., 1970.

[12.6] American Welding Society, D1.1-79, Structural Welding Code.

[12.7] American Welding Society, D1.2-77, Commentary on Structural Welding Code.

[12.8] American Concrete Institute, 318.71 — Building Code Requirements for Reinforced Concrete, ACI Committee 318.

[12.9] American Natural Standards Institute, Building Code Requirements for Minimum Design Loads in Buildings and Other Structures, A58.1-1972, p.20.

[12.10] Timoshenko, S., Strength of Materials, Part I, D. Van Nostrand Company, Inc., New York, 1957, p. 264.

[12.11] McGuire, W., Steel Structures, Prentice-Hall, Inc., Englewood Cliffs, N.J., 1968, pp. 415-418.

[12.12] National Building Code of Canada. The National Research Council of Canada, Ottawa, 1985, p. 163.

[12.13] Krupka, V., "Analysis of cylindrical thin-walled metal pipeline structures", Prague, Ed. SNTL, 1967, pp. 32, 33 (In Czechoslovakian).

[12.14] National Building Code of Canada. The National Research Council of Canada, Ottawa, 1985, p. 165.

[12.15] Foster, H.A., "Formulas indicate earthquake forces in design of ring-girder supported pipes", Civ. Engr. 1949, pp. 19, 45.

[12.16] Baker, E.H. et al., Structural Analysis of Shells, McGraw-Hill Book Company, 1972, p. 232.

[12.17] Troitsky, M.S., "On the Local and Overall Stability of Thin-Walled Large Diameter Tubular Structures", Canadian Structural Engineering Conference, 1976.

[12.18] Schilling, C.G., "Buckling Strength of Circular Tubes", ASCE J. Struct. Div., Vol. 91, No. ST. 5, October 1965, p. 325.

[12.19] American Iron and Steel Institute, Specification for the design of cold-formed steel structural members, 1968.

[12.20] Wolford, D.S. and Rebholz, M.J., "Bean and Column Tests of Welded Steel Tubing With Design Recommendations", ASTM Bull. No. 233, October 1958.

[12.21] Johnston, B.G., Editor, Guide to Stability Design. Criteria for Metal Structures, 3rd Ed., John Wiley and Sons, New York, 1976, p. 215.

[12.22] American Institute of Steel Construction, Manual of Steel Construction, 7th Ed., 1970, pp. 3-7.

[12.23] Marshall, P.W., "Design Criteria for Structural Steel Pipe", Column Res. Council, Proc., 1971.

[12.24] Roark, R.J. and Young, W.C., Formulas for Stress and Strain, 5th Ed., McGraw-Hill Book Company, New York, 1975, pp. 235-237.

[12.25] Schorer, H., "Design of Large Pipe Lines", Trans. ASCE, Vol. 98, 1933, p. 101.

[12.26] Krupka, V., "Analysis of cylindrical thin-walled metal pipeline structures", Prague, Ed. SNTL, 1967, pp. 39, 40, 47-48.

[12.27] Timoshenko, S., Strength of Materials, Part II, 2nd Ed., D. Van Nostrand Company, Inc., New York, 1953, p. 279.

[12.28] Heck, O.S., "Uber die Stabilitat orthotroper elliptischer Zylinderschalen bei reinen Biegung", Jahrbuch der Lilienthal Gesellschaft fur Luftfahrtforschung, 1936, p. 94.

[12.29] Gerard, G. and Becker, H., Handbook of Structural Stability, Part III — Buckling of Curved Plates and Shells, NACA TN 3783, 1957.

new aws code specifies the design of tubular structures

by
Omer W. Blodgett
Design Consultant
The Lincoln Electric Company

Section 10 of the AWS Structural Welding Code D1.1 Rev. 1-81 provides, for the first time, the specifications needed by the designer to work with confidence on structural tubing. It was developed by a special subcommittee and added to the Code to cover all structural use of tubing – including framing for buildings, utility poles, support structures, booms, and tubular sectioning on earth-moving and farm equipment, and especially in response to the need for specifications to govern the construction of offshore drilling rigs.

There is little information on tubular structures in open welding literature. Considerable research has been conducted, however, at the University of Texas, the University of California, and by several large oil companies. Enough data has been developed on the stress distribution in and around connections, on the fatigue testing of connections, and on methods of stress analysis to make possible the AWS specifications.

Welded connections between structural tubes are different from the connections joining pressure piping. With the latter, a hole must be cut into one pipe to permit the flow of a fluid or gas. The strength requirement for the weld is that needed to withstand internal pressure; there is usually little structural load on the piping system. In many cases, some form of backing ring can be used to facilitate making a complete-joint-penetration groove weld.

With structural tubing connections, no holes are cut – a factor that tends to increase the strength and stiffness of the connections. However, the possibility of using a backing seldom exists, and all welding must be made from outside the joint. If the connection is to be complete-joint-penetration groove weld, a new concept of what constitutes such when applied to tubular structures must be developed, since the definition that applies with plate and rolled steel shapes doesn't fit. The AWS has supplied the required definition by specifying that:

> "A complete-joint-penetration groove weld made from one side only, without backing, is permitted where the size and/or configuration prevent access to the root side of the weld."

Special skill is required for single-side welding of complete-joint-penetration welds at tubular connections. Only shielded metal-arc, gas metal-arc, and flux-cored arc welding are prequalified for joints other than butt joints.

For Metric

dimensions	=	mm
stresses	=	MPa
forces	=	N

Figure 1 shows a type of connection likely to be encountered in tubular structures. The dihedral angle is acute, and, at first thought, one might question the strength of this joint. It is apparent that it will be impossible to get sound weld metal all the way to the root of the joint. However, it has been established that for a dihedral angle as small as 30° an effective throat of good weld metal equal to twice the wall thickness of the branch member will give a strong connection – one that will ultimately fail in the tube rather than in the weld under fatigue testing. The weld is made, first, by placing a "back-up" weld into the joint at the root. This is to bridge the gap and provide a base for the remaining fill weld.

Figures 2 (a, b, c, d, e) illustrate loading conditions that may exist with tubular connections. The ideal support for a branch member would be that shown in (2a). Here, the connecting weld is uniformly loaded and the branch member uniformly stressed, two requirements for an ideal connection.

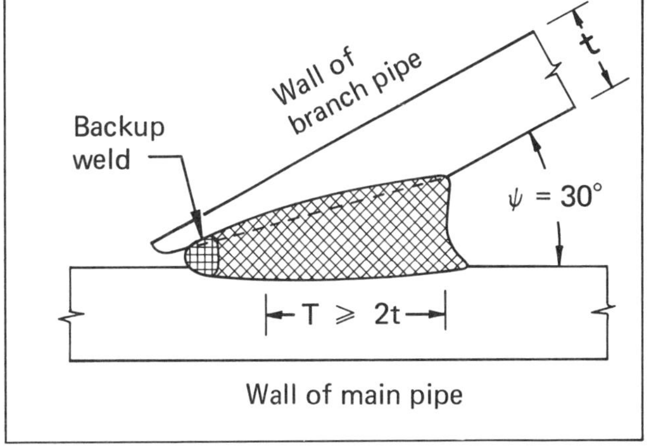

Figure 1

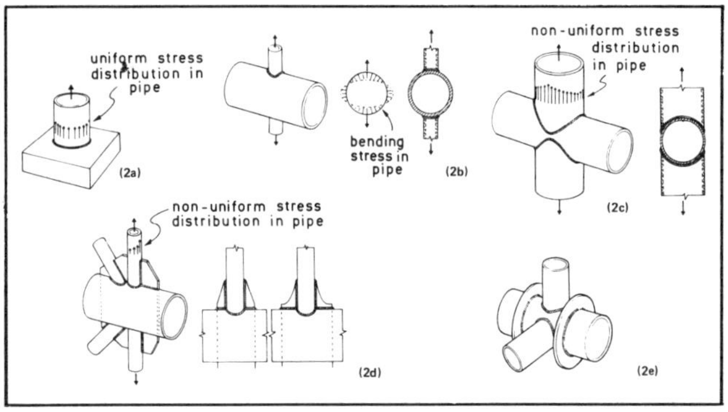

Figure 2

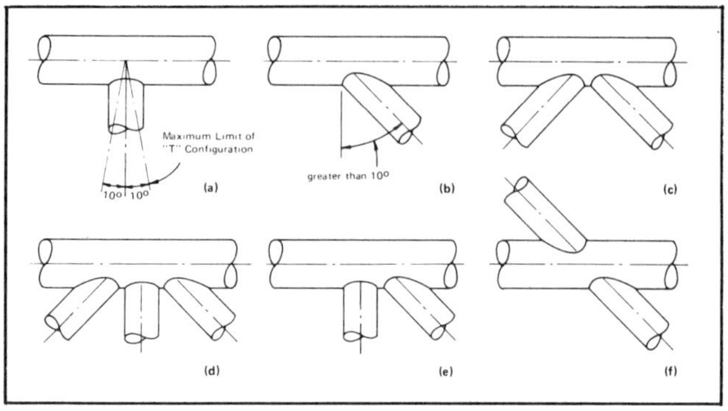

Figure 4

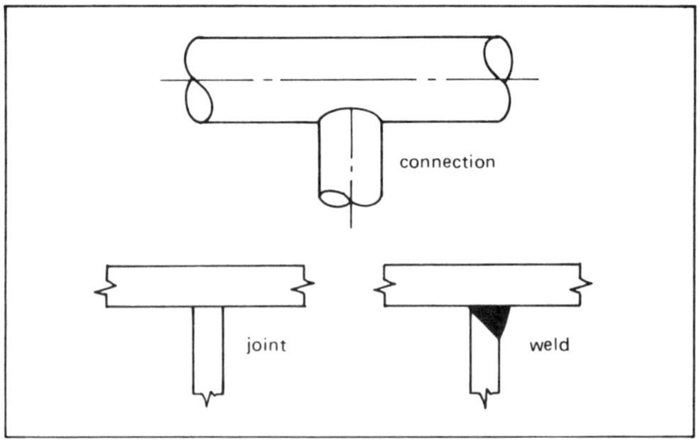

Figure 3

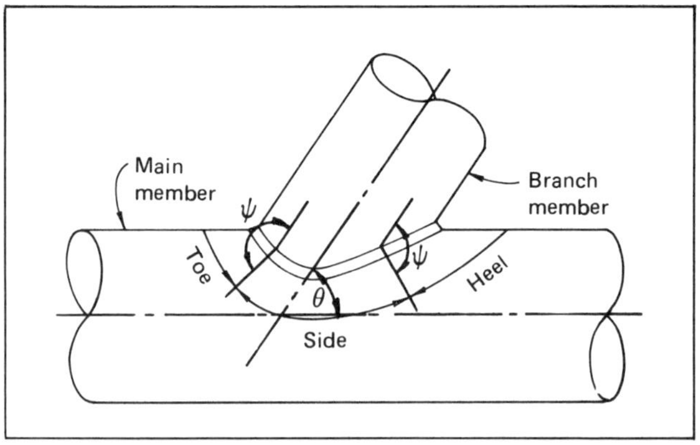

Figure 5

In (2b), the branch members are small in relation to the main member, and the relatively small region of intersection developed also provides fairly uniform support for the branch members. The connecting weld is, thus, uniformly loaded, and the branch members will be uniformly stressed. However, the main member is subjectd to bending as a ring because of the force transferred through it.

When the branch members approach the size of the main member, as in (2c), the axial force in the branch members is transferred more directly through the sides of the connection than through the heel or toe. The central portion of the main member (heel and toe positions of the connection) is more free to bend than the material in the sides – therefore will not transfer as much load between the two branch members. This means there will be an uneven stress distribution in the branch members in the vicinity of the connection.

Placing a plate through a connection (2d) will help in transferring the load through the main member and prevent it from bending. however, this will cause an uneven distribution of stress in the branch members in the vicinity of the plate. Stiffening rings (2e) can be used to give the main member support so it will transfer the load more evenly into branch members.

Figure 3 illustrates the terminology used in reference to tubular structural connections. It is desirable to adhere to these terms closely and not use them interchangeably. A *connection* may be a T, Y, or K. The *joint* may be butt, T, or corner. The *weld* may be a complete-joint-penetration groove weld, partial-joint-penetration groove weld, or fillet weld. Figure 4 shows the various types of connections. The connection is a T if the intersection of the members is 90° ±10° (4a). Any intersection less than 80° on one side of the branch is designated a Y (4b). A K connection involves two branch members, as illustrated in (4c). Combination K and T, and T and Y, connections are possible, as shown in (4d and 4e). A cross joint is shown in (4f).

Figure 5 is a typical connection, with the angles theta (θ) and psi (ψ) indicated. Theta is the angle of intersection between the axes of the two pipes. Psi is the dihedral angle – the angle between the intersecting surfaces to be welded at a given point along the length of the connection, It will determine the type of prequalified groove joint to be used. There are four basic types of prequalified joints, A, B, C, and D, as illustrated at the top of Figure 6.

Joint A, may be used when the dihedral angle is between 180° and 135°. Joint B, is for an angle of 150° to 50°. Joint C, may be used for a dihedral angle of 75° to 30°. Joint D, is for an angle between 45° and 15° – but 30° is the minimum if the joint is considered to be prequalified.

The details for these four types of joints are summarized in this Figure. The table gives the end preparation angle omega (ω), root opening (R), joint included angle phi (ϕ), and resulting throat (T) of the completed weld.

Since the dihedral angle is the governing value, charts – such as the grouping in Figure 7 – have been prepared to indicate when a particular dihedral angle is reached for any given included angle theta and for any rate of beta (β) which if the ratio $\frac{r}{R}$ of the radius of the branch member

FULL STRENGTH TEE-GROOVE WELDS

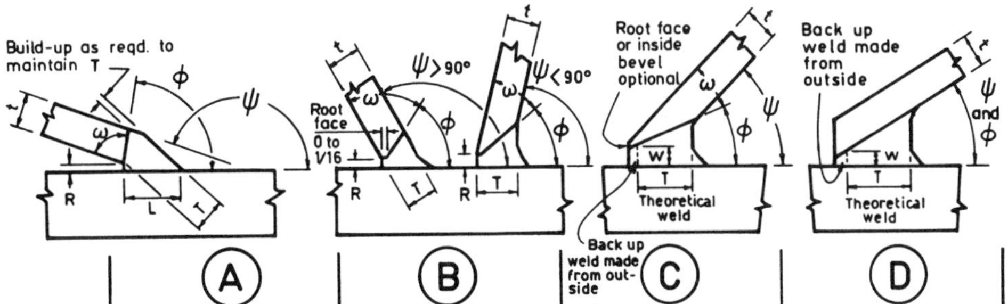

		(A) ψ=180°–135°		(B) ψ=150°–50°		(C) ψ=75°–30°**	(D) ψ=40°–15°**
End preparation (ω)	max.	90°		90° *		*	
	min.	45°		10° or 45° for ψ>105°		10°	
Fitup or root opening (R)		R: SMAW FCAW	GMAW FCAW with gas	R: SMAW FCAW	GMAW FCAW with gas	W *** SMAW FCAW: W max. 1/8" — ϕ 22-1/2°–37-1/2°; 3/16" — 15°–22-1/2°	
	max.	3/16"	3/16"	1/4"	1/4" for ϕ>45°; 5/16" for ϕ≤45°	GMAW FCAW with gas: 1/8" — 30°–37-1/2°; 1/4" — 25°–30°; 3/8" — 20°–25°; 1/2" — 15°–20°	
	min.	1/16"	1/16"	1/16"	1/16"		
Width of weld groove provided by back up weld (W)		No min for ϕ>90°	No min for ϕ>120°				
Joint included angle ϕ	max.			60° for ψ ≤ 105°		37-1/2° if more use (B)	
	min.			37-1/2° if less use (C)		1/2 ψ	
Completed weld	T	≥t		≥t for ψ > 90°; ≥t/sin ψ for ψ ≤ 90°		≥t/sin ψ but need not exceed 1.75t; weld may be built up to meet this	≥2t
	L	≥t/sin ψ but need not exceed 1.75t					

* otherwise as needed to obtain required ϕ

** not prequalified for groove angles (ϕ) under 30°

*** initial passes of back up weld discounted until designated width of groove (w) is obtained

Figure 6

CURVES FOR DETERMINING DIHEDRAL ANGLE

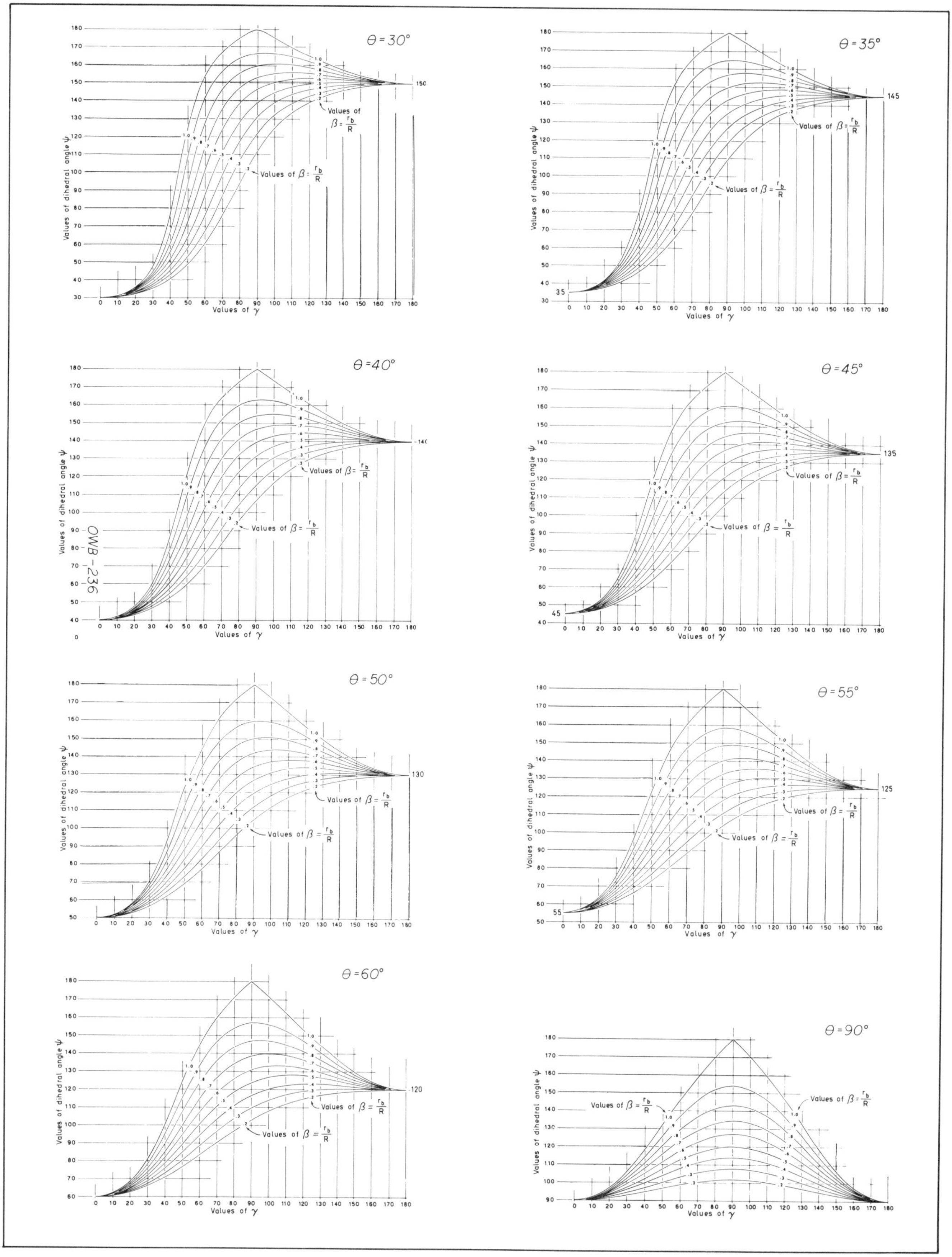

Figure 7

to the radius of the main member. For example, using 60° for th value of theta and a radii ratio of 0.5, a dihedral angle of 75° would be reached at a point on the branch member $\gamma = 36°$ away from the heel of the connection.

A welded tubular connection is limited in strength by four factors:

1. Local or "punching" shear failure (10.5.1).
2. General collapse (10.5.2).
3. Uneven distribution of load on the welded connection (10.5.3).
4. Lamellar tearing (10.5.4).

CIRCULAR SECTIONS 10.5.1 AND 10.8

The allowable punching shear stress for a circular section is found from formula: 1
where σ_y is the specified minimum yield strength of the material (psi or ksi).

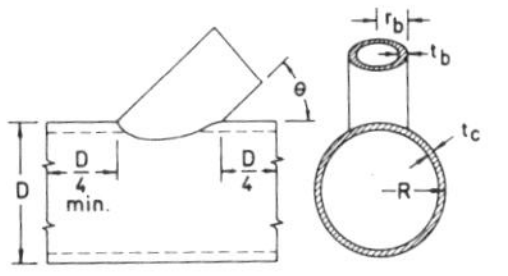

$$\beta = \frac{r_b}{R} \qquad \gamma = \frac{R}{t_c}$$

$$Q_\beta = \underbrace{\frac{.30}{\beta(1 - .833\beta)}}_{\text{if } \beta > 0.6} = \underbrace{1.0}_{\text{if } \beta \leq 0.6}$$

$$Q_f = \underbrace{1.22 - 0.5U}_{\text{if } U > 0.44} = \underbrace{1.0}_{\text{if } U \leq 0.44}$$

allowable punching shear stress

$$(1) \quad \boxed{\tau_p = \frac{\sigma_y Q_\beta Q_f}{0.9\,\gamma^{0.7}}}$$

but not to exceed $0.40\sigma_y$

$$K_a = \frac{1 + \sin\theta}{2 \sin\theta} \qquad K_b = \frac{1 + 3\sin\theta}{4\sin^2\theta}$$

$$\tau = \frac{t_b}{t_c}\left[\frac{\sigma_a \sin\theta}{K_a} + \frac{\sigma_b}{K_b}\right]$$

if out of plane bending use: $\sigma_b = \sqrt{\sigma_{by}^2 + \sigma_{bz}^2}$

actual punching shear stress

$$(2) \quad \boxed{\tau = \frac{t_b}{t_c}\sin^2\theta \left[\frac{2\sigma_a}{1 + \sin\theta} + \frac{4\sigma_b}{1 + 3\sin\theta}\right]}$$

For circular sections in which $\frac{D}{t_c} > \frac{3600}{\sigma_y}$ local* buckling at axial compressive stresses less than yield strength shall be investigated 10.3.1.

If sufficient thickness (t_c) of main member is used a distance of $\frac{D}{4}$ beyond each weld, general collapse need not be checked 10.5.2.2.

The use of fillet welds limited to $\beta \leq \frac{1}{3}$ 10.13.1.3.

* if Metric

$$\boxed{\frac{D}{t_c} > \frac{24800}{\sigma_y}}$$

Figure 8

R is the mean radius of the main member in inches, and

t_c is the wall thickness of the main member in inches.

$\gamma = R/t_c$

Q_β is a shape factor as shown in Figure 8.

Q_f is a utilization factor as defined in Figure 8. It involves a ratio U, which is the ratio of actual to allowable axial stress in the main member.

Had the factor 0.5, rather than 0.9, been used in the denominator, the result would have been the *calculated ultimate* punching shear strength. But a factor of safety of 1.8 has been designated. Thus, $0.5 \times 1.8 = 0.9$, and the use of this factor in the denominator gives the *calculated allowable* punching shear stress. This value, of course, should not exceed the allowable shear stress of the applicable design code.

The term "punching" shear tends to be confusing, since one thinks of punching as "pushing in" rather than "pulling out." However, the forces capable of pulling a connection apart would be equal to the forces capable of pushing a connection apart, should the latter be feasible. The term "punching" relates the phenomenon to pre-existing analytical experience.

The *actual* punching shear stress in the main member caused by the axial force and bending moment in the branch member must be found by formula (2) and compared with the allowable punching shear stress found by formula (1).

The easiest way to do this is to convert the axial and bending stresses in the branch into a unit force around the welded connection, then divide this by the thickness of the main member.

This work is shown in Figure 8.

Since the length of the welded connection as well as its section modulus must be known to treat the axial force and bending member. K factors as defined in Figure 8 are used. K_a is a length factor for axial forces, and K_b is used as a section modulus factor for bending moments.

The terms t_b and t_c are the wall thicknesses in inches of the branch and main members, respectively. The stresses τ_b are the applied axial and bending stresses. θ is the angle between the axes of the intersecting members.

K factors are used for both circular and box sections. K_{by} is sometimes used for in-plane bending and K_{bz} for out-of-plane bending.

If there should be out-of-plane bending (M_z) this can be combined with the in-plane bending (M_y) by using $\sigma_b = \sqrt{\sigma_{by}^2 + \sigma_{bz}^2}$ and then entering this as (σ_b) into the formula.

GENERAL COLLAPSE 10.5.2

As noted previusly, the strength of the connection also depends on what might be termed "general collapse." The strength and stability of the main member in a tubular connection should be investigated by the proper technology in accordance with the applicable design code. If the main member has sufficient thickness as required for punching shear and this thickness extends beyond the branch members at least a distance of D/4, general collapse should be no problem.

Stepped box

$\boxed{\beta < 1.0}$

Punching Shear

10.5.1 10.8

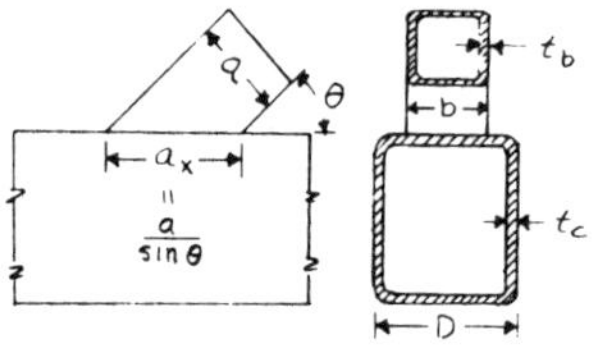

$$Q_\beta = \underbrace{\frac{.25}{\beta(1-\beta)}}_{\text{if } \beta > 0.5} = \underbrace{1.0}_{\text{if } \beta \leq 0.5}$$

$$Q_f = \underbrace{1.22 - 0.5U}_{\text{if } U > 0.44} = \underbrace{1.0}_{\text{if } U \leq 0.44}$$

allowable punching shear stress

③ $$\tau_p = \frac{\sigma_y}{0.6\gamma} Q_\beta Q_f$$

but not to exceed 0.40 σ_y

$$K_a = \frac{a + b\sin\theta}{(a+b)\sin\theta} \qquad K_b = \frac{a + 3b\sin\theta}{(a+3b)\sin^2\theta}$$

$$\tau = \frac{t_b}{t_c}\left[\frac{\sigma_a \sin\theta}{K_a} + \frac{\sigma_b}{K_b}\right]$$

if out of plane bending use: $\sigma_b = \sqrt{\sigma_{by}^2 + \sigma_{bz}^2}$

actual punching shear stress

④ $$\tau = \frac{t_b}{t_c}\sin^2\theta \left[\frac{(a+b)\,\sigma_a}{a + b\sin\theta} + \frac{(a+3b)\,\sigma_b}{a + 3b\sin\theta}\right]$$

fillet welds limited to $\beta \leq .8$ 10.13.1.3

Figure 9

Stepped box — 10.5.1.3

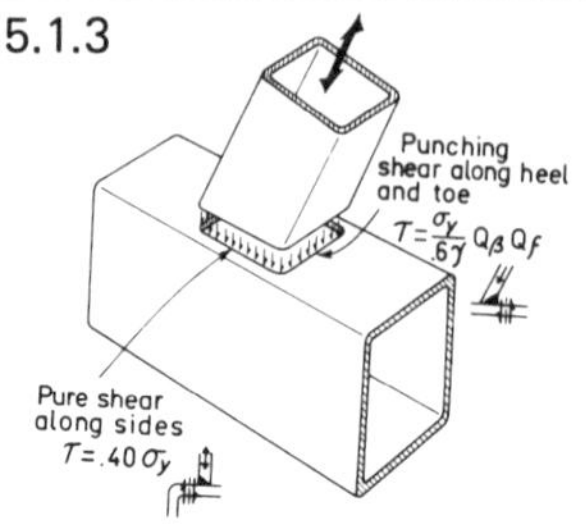

If $\boxed{\beta > .8}$ or $\boxed{\beta > \eta}$ or both

$a < b \sin\theta$

$$Q_\beta = \underbrace{1.25\,(1+\eta)}_{\text{if } \eta < 1.0} = \underbrace{2.5}_{\text{if } \eta \geq 1.0}$$

$$Q_f = \underbrace{1.22 - 0.5U}_{\text{if } U > 0.44} = \underbrace{1.0}_{\text{if } U \leq 0.44}$$

allowable capacity

⑥ $$F = 0.8\, t_c\, \sigma_y \left[\frac{a}{\sin\theta} + \frac{\beta t_c Q_\beta Q_f}{0.12}\right]$$

$$\beta = \frac{b}{D}$$

$$\gamma = \frac{D}{2\,t_c}$$

$$\eta = \frac{a_x}{D} = \frac{a}{D\sin\theta}$$

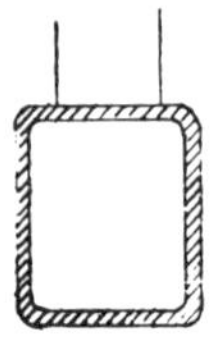

Figure 11

Optional Yield Line Analysis Box Section
10.5.1.4

$\boxed{\beta < .8}$

(from Commentary Table C10.5.1.4b)

	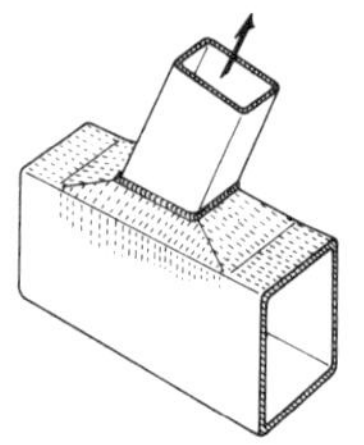 axial load	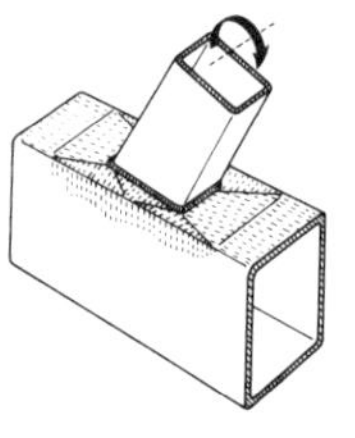 in plane bending
$\beta > 0.625$	$Q_\beta = \dfrac{\frac{\eta}{1-\beta} + \frac{2}{\sqrt{1-\beta}}}{4(\eta+\beta)}$	$Q_\beta = \dfrac{\frac{1}{2\eta} + \frac{\eta}{1-\beta} + \frac{2}{\sqrt{1-\beta}}}{4\left(\frac{\eta}{3}+\beta\right)}$
$\beta < 0.625$	$Q_\beta = \dfrac{\frac{\pi}{2} + \frac{\eta+\beta}{1-\beta}}{4(\eta+\beta)}$	$Q_\beta = \dfrac{\frac{1}{2\eta} + \frac{\pi}{2} + \frac{\eta+\beta}{1-\beta}}{4\left(\frac{\eta}{3}+\beta\right)}$

allowable shear stress

⑤ $$\tau = \frac{K}{SF} Q_\beta \frac{\sigma_y}{0.5\gamma}$$

Q_β = shape factor
Q_f = utilization factor

(for suggested values see commentary Table C10.5.1.4a)

K = reserve strength factor
SF = safety factor
U = ratio actual to allowable stress — main member

Figure 10

BOX SECTIONS 10.5.1 AND 10.8

The definitions and procedures for treating circular sections apply also to box sections. Figure 9 gives the formulas for allowable and actual punching, shear stress.

Applying to box sections only, however, is the optional method of yield line analysis, contained in the commentary of the Structural Welding Code-Steel AWS D1.2-81.

Optional yield analysis, in contrast to punching shear, assumes that under the applied force the main box member will undergo plastic bending as illustrated by the two sketches in Figure 10. By setting external and internal energies equal, and then modifying so as to give conformity to punching shear format, an alternative method of analysis is achieved. The terms used in Figure 10 have been previously described or are described in this and subsequent figures.

In Figures 11 and 12, allowable capacities rather than stresses are calculated, leading to determination of the allowable force (F) that could be placed on the joint. A "stepped" box arrangement is shown in Figure 11, and a "matched" box in Figure 12. The sketches within the figures indicate the difference. With a matched box, the sides would not fail in punching shear but by the web buckling.

Matched Box 10.5.1.1

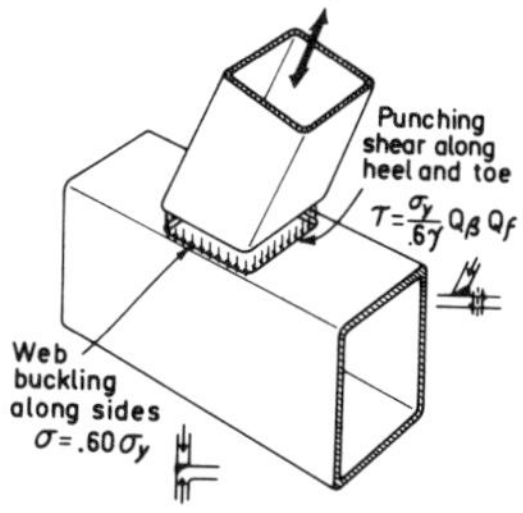

$\boxed{\beta = 1.0}$

Q_β and Q_f as in Fig. 11

allowable capacity

(7) $$F = 1.2\, t_c\, Q_f\, \sigma_y \left[\frac{a}{\sin\theta} + \frac{t_c\, Q_\beta}{0.18}\right]$$

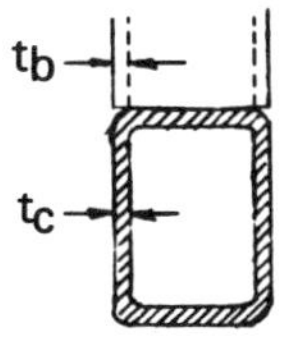

10.5.1.2
Full static load capacity of branch allowed if:

1 Complete joint penetration groove weld and matching weld metal.

2 $\frac{D}{t_c} < 22$ or $\gamma < 11$

3 σ_Y of main not less than that of branch.

4 $\frac{t_b}{t_c} \leqslant 0.72$ or

$\frac{t_b}{t_c} = 1.0$ and $U \leqslant 0.44$ (main member)

5 $\eta \geqslant 1.0$

Figure 12

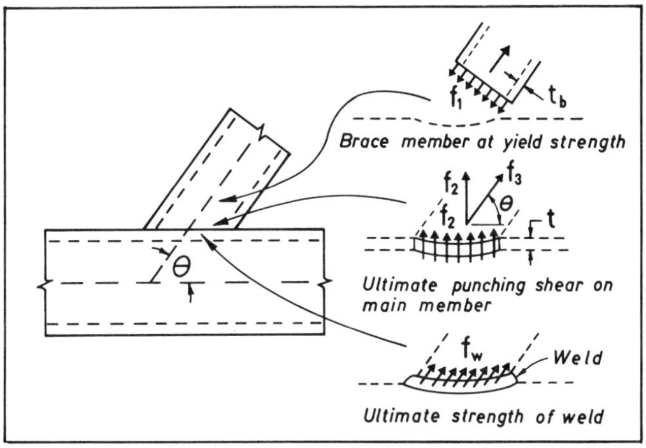

Figure 13

UNEVEN DISTRIBUTION OF LOAD ON THE WELDED CONNECTION 10.5.3

A third condition that will limit strength of the connection is the uneven distribution of load on the welded connection. Under load, some bending of the main member may take place, which will cause uneven distribution of the force applied to the weld. As a result, some yielding and redistribution of force may have to take place for the connection to reach its design load. To provide for this, the welds must be capable – at their ultimate strength – of developing the lesser of (1) the branch member at yield strength or (2) the ultimate punching shear of the shear area of the main member. This is illustrated in Figure 12. This particular part of the design is best handled by working in terms of unit force, pounds per linear inch or kips per linear inch.

The ultimate strength of the weld is found by multiplying its allowable value by the factor 2.67 for E60 and E70 weld metal and 2.21 for higher strengths of weld metal, Figure 14c.

Figure 14a illustrates how to determine the unit yield strength of a branch member – simply by multiplying the yield strength of the material (σ_y) by the wall thickness of the branch member (t_b). The result is in pounds per linear inch or kips per linear inch of weld.

Figure 14b shows how to determine ultimate shear on the main-member shear area at failure. The allowable punching shear stress (τ_a) is multiplied by 1.8. The unit shear force (f_2) is obtained by multiplying this result by the wall thickness (t) of the main member. This unit shear force, however, would be one acting normal to the axis of the weld. The real unit force (f_3) applied to the weld would be obtained by dividing this value by the sine of the angle between the axes of the members. Thus, the unit force (f_3) acting on the weld from the ultimate punching shear stress is:

$$f_3 = \frac{1.8\, \tau_a\, t}{\sin\theta}$$ in pounds per linear inch of weld or kips per linear inch.

LAMELLAR TEARING 10.5.4

At the present time, AWS is suggesting that some thought be given to connections in which forces are to be transferred through the thickness of the material.

OVERLAPPING AND FLARED CONNECTIONS 10.5.1.5 AND 10.5.1.6

In the foregoing text, methods of analysis of the connections of primary interest to designers have been presented. But there are other connections that require special consideration. Two such are overlapping circular joints and flared connections in both circular and box sections.

Figure 15 describes the method of handling overlapping circular joints and Figure 16 the treatment for flared connections. Figure 17 shows how to determine the stress concentration factor at an unreinforced cone-cylinder intersection.

FATIGUE

The fatigue allowables of the AISC have been adapted as a basis for fatigue allowables with structural tubing. Figure 18 shows a specially prepared fatigue table for tubing that is taken off the AISC table for plate material (see James F. Lincoln Arc Welding Foundation Publication D412). This table contains the allowable ranges in stress during fatigue for various categories of joints A, B, C, D, E,

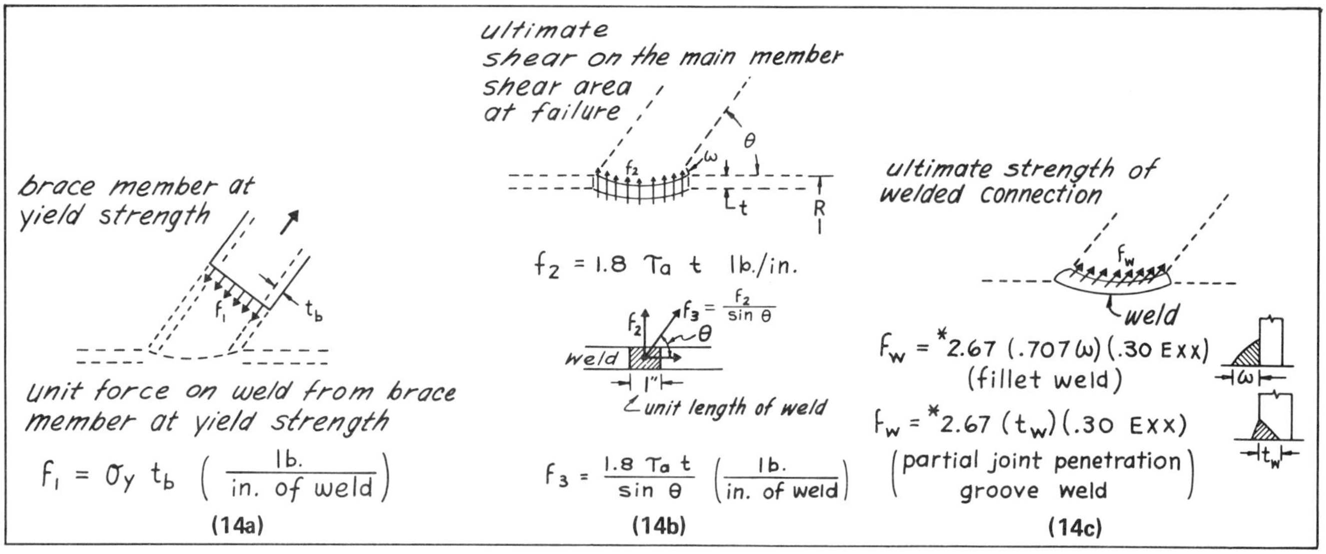

Figure 14

Overlapping Circular Connections 10.5.1.5

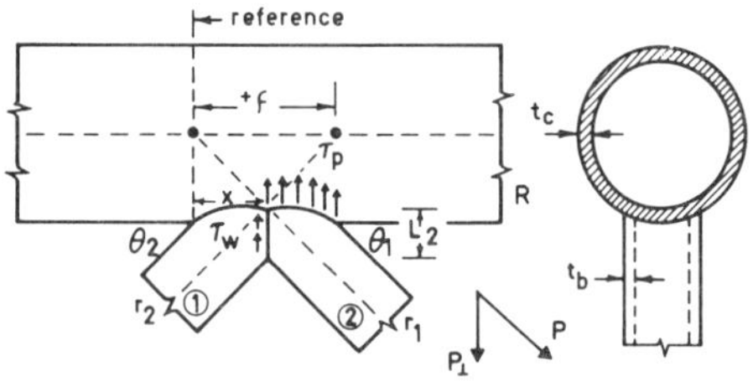

allowable transverse component of branch

$$P_\perp = \tau_p t_c L_1 + 2\tau_w t_w L_2$$

allowable parallel component shall not exceed

$$P = \tau_w t_w \Sigma L_1$$

L_1 = actual weld length for that portion of branch which contacts main member

L_2 = projected chord length (one side) of the overlapping weld measured perpendicular to main member

ΣL_1 = sum of the actual weld lengths for all branches in contact with main member

τ_w = allowable shear stress for the weld between branch members

τ_p = allowable punching shear of main member see 10.5.1

t_w = weld throat, but not to exceed thickness (t_b) of branch member

Overlap (L_2) shall preferably be proportioned for at least 50% of acting $P_\perp$ also $t_b \leqslant t_c$

Figure 15

Flared Connections 10.5.1.6

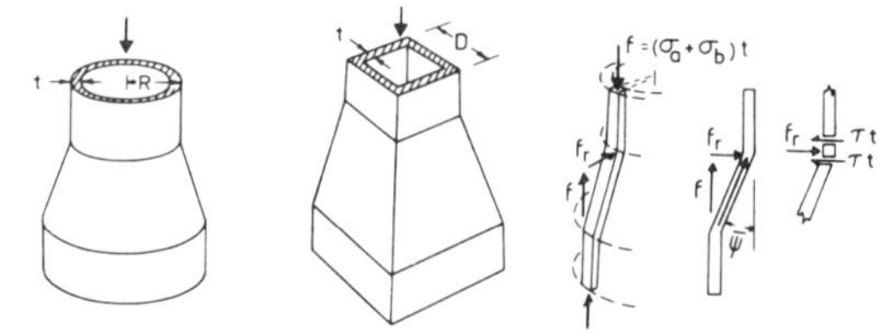

circular section box section

radial unit force

$$f_r = f \tan \bar{\psi} = (\sigma_a + \sigma_b)\ t \tan \bar{\psi}$$

$$2\tau t = f_r = (\sigma_a + \sigma_b)\ t \tan \bar{\psi}$$

or Actual shear stress $\tau = \dfrac{(\sigma_a + \sigma_b) \tan \bar{\psi}}{2}$

Allowable shear stress

$$\tau = \frac{\sigma_y Q_f}{.9\left(\frac{R}{t_c}\right)^{.7}} \qquad \tau = \frac{\sigma_y Q_f}{.6\left(\frac{D}{2t}\right)}$$

circular section box section

where: $Q_f = \underbrace{1.22 - .5U}_{\text{if } U > .44} = \underbrace{1.0}_{\text{if } U \leqslant .44}$

Figure 16

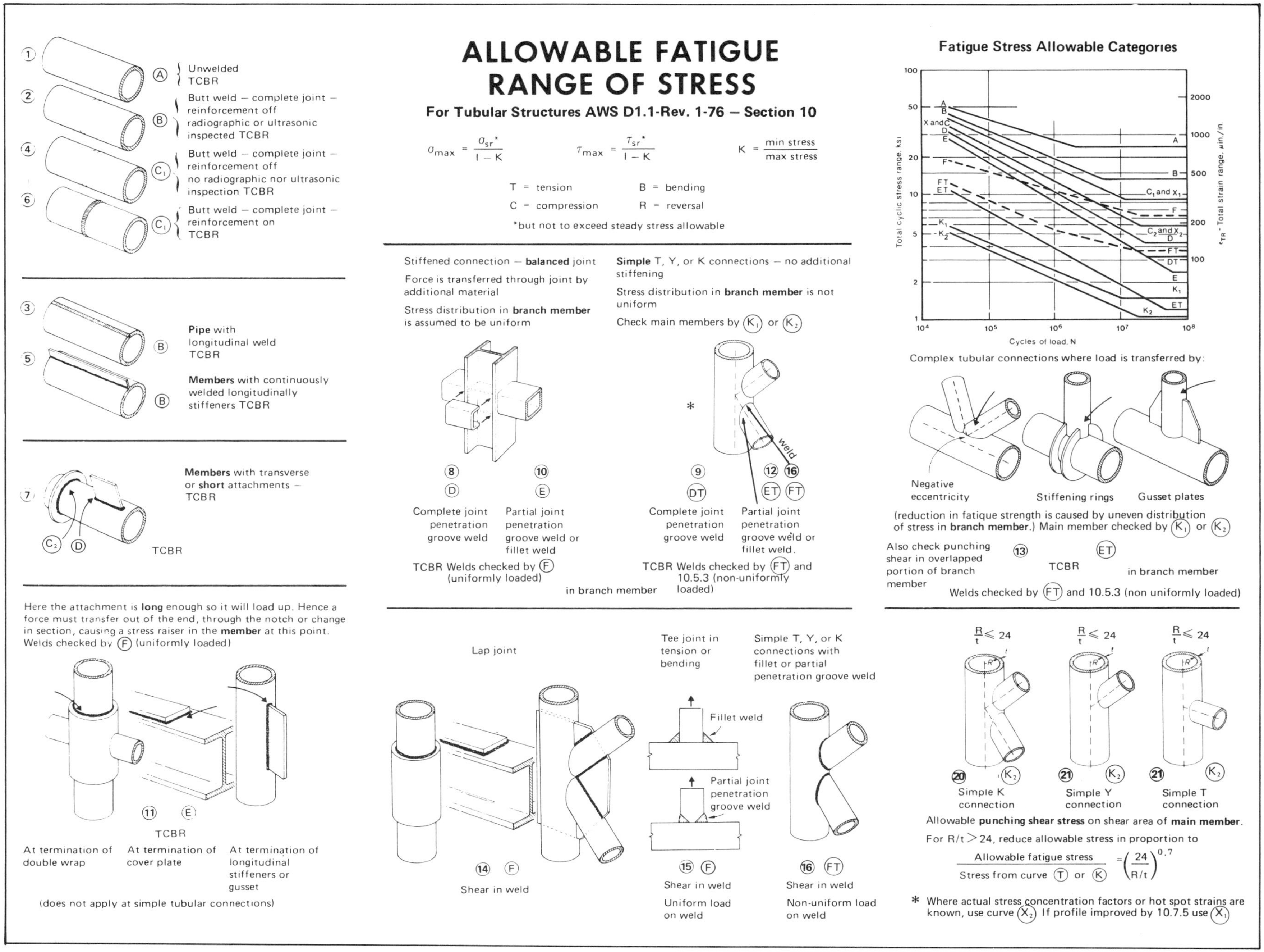
ALLOWABLE FATIGUE RANGE OF STRESS
For Tubular Structures AWS D1.1-Rev. 1-76 – Section 10
$\sigma_{max} = \frac{\sigma_{sr}^*}{1-K}$
$\tau_{max} = \frac{\tau_{sr}^*}{1-K}$
$K = \frac{\text{min stress}}{\text{max stress}}$
T = tension
C = compression
B = bending
R = reversal
*but not to exceed steady stress allowable
1 2 4 6
A Unwelded TCBR
B Butt weld – complete joint – reinforcement off radiographic or ultrasonic inspected TCBR
C1 Butt weld – complete joint – reinforcement off no radiographic nor ultrasonic inspection TCBR
C1 Butt weld – complete joint – reinforcement on TCBR
3 5
B Pipe with longitudinal weld TCBR
B Members with continuously welded longitudinally stiffeners TCBR
7
Members with transverse or short attachments – TCBR
C2 D
TCBR
Here the attachment is long enough so it will load up. Hence a force must transfer out of the end, through the notch or change in section, causing a stress raiser in the member at this point. Welds checked by F (uniformly loaded)
11 E
TCBR
At termination of double wrap
At termination of cover plate
At termination of longitudinal stiffeners or gusset
(does not apply at simple tubular connections)
Stiffened connection – balanced joint
Force is transferred through joint by additional material
Stress distribution in branch member is assumed to be uniform
Simple T, Y, or K connections – no additional stiffening
Stress distribution in branch member is not uniform
Check main members by K1 or K2
*
weld
8 D
10 E
9 DT
12 16 ET FT
Complete joint penetration groove weld
Partial joint penetration groove weld or fillet weld
Complete joint penetration groove weld
Partial joint penetration groove weld or fillet weld.
TCBR Welds checked by F (uniformly loaded)
in branch member
TCBR Welds checked by FT and 10.5.3 (non-uniformly loaded)
Lap joint
Tee joint in tension or bending
Simple T, Y, or K connections with fillet or partial penetration groove weld
Fillet weld
Partial joint penetration groove weld
14 F
Shear in weld
15 F
Shear in weld
Uniform load on weld
16 FT
Shear in weld
Non-uniform load on weld
Fatigue Stress Allowable Categories
Total cyclic stress range, ksi
100 50 20 10 5 2 1
A B X and C D E F FT ET K1 K2
A B C1 and X1 F C2 and X2 D FT DT E K1 ET K2
2000 1000 500 200 100
ϵ_{TR} - Total strain range, µin./in.
10^4 10^5 10^6 10^7 10^8
Cycles of load, N
Complex tubular connections where load is transferred by:
Negative eccentricity
Stiffening rings
Gusset plates
(reduction in fatigue strength is caused by uneven distribution of stress in branch member.) Main member checked by K1 or K2
Also check punching shear in overlapped portion of branch member
13
TCBR
ET
in branch member
Welds checked by FT and 10.5.3 (non uniformly loaded)
R/t ≤ 24
R/t ≤ 24
R/t ≤ 24
20 K2 Simple K connection
21 K2 Simple Y connection
21 K2 Simple T connection
Allowable punching shear stress on shear area of main member.
For R/t > 24, reduce allowable stress in proportion to
$\frac{\text{Allowable fatigue stress}}{\text{Stress from curve T or K}} = \left(\frac{24}{R/t}\right)^{0.7}$
* Where actual stress concentration factors or hot spot strains are known, use curve X2 If profile improved by 10.7.5 use X1

Figure 18

and F. (In general, a descending order in the alphabetic categories means joints of lessening fatigue strength.)

Section 10, for the most part, uses the AISC values for these categories at 100,000 cycles and 2,000,000 cycles. A deviation, however, is the ingnoring of values for 500,000 cycles and the extension of a straight line between 100,000 and 2,000,000 cycles and projection of this line to 20,000 cycles on the low side. The AISC table has values "over 2,000,000 cycles." Section 10 considers this to be 10,000,000 cycles, using such on the chart and extending lines between the 2,000,000 and 10,000,000-cycle values.

Stress Concentration Factor — Unreinforced Cone Cylinder

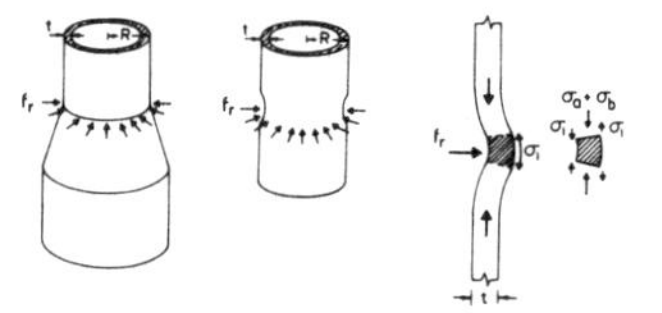

For simplicity, assume the flared connection with its unit radial force (f_r) to be a straight tube — using Roark's "Formula for Stress & Strain" 4th Ed., page 301, case II.

Maximum stress is the meridian stress (σ_1)

$$\sigma_1 = \frac{3f_r}{2\lambda t^2}$$

$$\text{where: } \lambda = \sqrt[4]{\frac{3(1-\nu^2)}{t^2 R^2}} = \frac{1.285}{\sqrt{tR}}$$

$$\text{or} \quad \sigma_1 = \frac{3f_r\sqrt{tR}}{2\,(1.285)\,t^2} = 1.167\, f_r \sqrt{\frac{R}{t^3}}$$

but we found $f_r = (\sigma_a + \sigma_b)\, t \tan\psi$

$$\text{so: } \sigma_1 = 1.167\,(\sigma_a + \sigma_b)\sqrt{\frac{R}{t}}\tan\psi$$

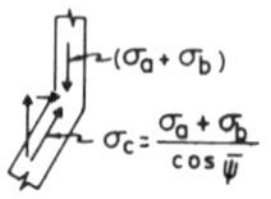

Total stress =

$$(\sigma_a + \sigma_b) + 1.167\,(\sigma_a + \sigma_b)\sqrt{\frac{R}{t}}\tan\psi$$

$$\text{Stress concentration factor} = \frac{\text{total stress}}{\text{applied stress}}$$

S.C.F. =

$$\frac{\dfrac{\sigma_a + \sigma_b}{\cos\psi} + 1.167\,(\sigma_a + \sigma_b)\sqrt{\dfrac{R}{t}}\tan\psi}{\sigma_a + \sigma_b}$$

$$\text{or} \quad \boxed{\text{S.C.F.} = \frac{1}{\cos\psi} + 1.167\sqrt{\frac{R}{t}}\tan\psi}$$

See Fatigue Table 10.7.3 of AWS D1.1-79 — given category (X_1)

Figure 17

Categories FT, ET, K_1, K_2, X_1, and X_2 have also been added in the fatigue chart for structural tubing.

The fatigue chart provides an allowable stress range (and also strain range) as against fatigue life (number of cycles) for several categories of connections and members.

It is intended that this fatigue chart supply the user with a predicated life for each set of given conditions.

Comparing the actual number of cycles to this estimated value will produce a fraction which would indicate the relative amount of life used up for this particular fatigue condition. This would be done for all of the stress ranges applied to the connection or member. The sum of all these fractions must not exceed unity (1.0).

However, it is possible to reverse this procedure and read from this chart the stress range for a given category and life. This stress range then becomes the numerator for the conventional fatigue stress formula to provide a maximum fatigue stress allowable.

$$\sigma_{max} = \frac{\sigma_{sr}}{1-K} \quad \text{or} \quad \tau_{max} = \frac{\tau_{sr}}{1-K}$$

$$\text{where} \quad K = \frac{\text{min stress}}{\text{max stress}}$$

Naturally this maximum fatigue stress allowable should not exceed the steady load stress allowable.

The master chart in Figure 18 can be broken down into smaller charts for discussion purposes. The first, Figure 19, illustrates A, B and C categories:

(1) is an unwelded tube. It has the highest category (A). (The numerical designation refers to the numbers in the master chart.)

(2) is a tube containing a complete-joint-penetration groove weld in a butt joint. The weld reinforcement has been removed and the joint has been radiographically or ultrasonically inspected. It has the same fatigue allowable as (1) and is in category (B).

(4) is the same weld, with reinforcement removed but without nondestructive inspection. Because there has been no radiographic and ultrasonic inspection, the fatigue allowable is reduced to category (C_1).

(6) is the same weld, but with reinforcement intact. The fatigue allowable is reduced to category (C_1).

Figure 20 illustrates the effects of longitudinal welds in a tube on fatigue categories:

(3) is a tube with a longitudinal seam weld. It has category (B).

Capacity of branch member in cross connections 10.5.2.1

For unreinforced circular cross connections, the compressive branch member axial load (P) shall not exceed

$$\boxed{P = \frac{t^2 \sigma_y (1.9 + 7.2\beta)\, Q_\beta\, Q_f}{\sin\theta}}$$

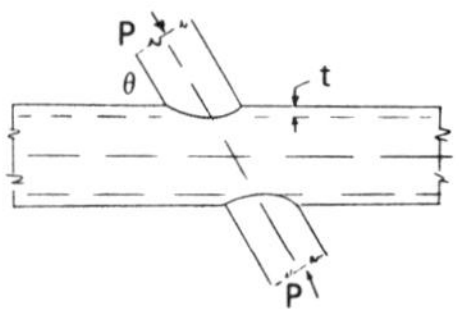

For circular cross sections reinforced by a "joint can" having increased thickness (t_c) and length (L) the allowable load (P) may be

$$\text{for } L < 2.5\,D \quad \boxed{P = \frac{\sigma_y (1.9 + 7.2\beta)\, Q_\beta\, Q_f}{\sin\theta}\left[t^2 + \frac{L}{2.5\,D}(t_c^2 - t^2)\right]}$$

$$\text{for } L \geq 2.5\,D \quad \boxed{P = \frac{t_c^2 \sigma_y (1.9 + 7.2\beta)\, Q_\beta\, Q_f}{\sin\theta}}$$

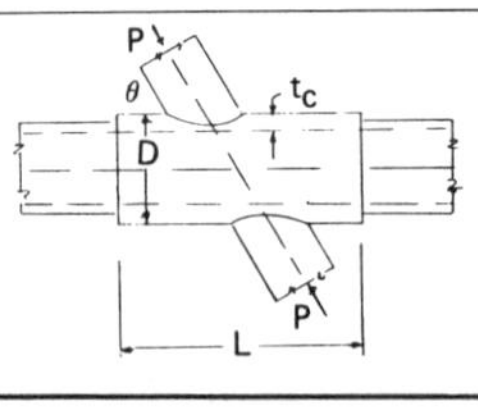

(5) is a tube with longitudinal stiffener welded to it. It has category (B).

In Figure 21, the lowered fatigue allowables for partial-joint-penetration welds are illustrated. In the left-hand sketch there is no additional connecting material to assist in transferring the load from one branch member to the other. With a complete-joint-penetration weld, it has the numerical designation (9) in the master chart and category (DT). If partial-joint-penetration groove welds are used, the designation would be (12) and the category (ET).

The right-hand sketch has additional connecting material to transfer the force through the joint, namely sections of channel welded between and to the flanges. It is a "balanced" joint. With complete-joint-penetration groove welds, the designation is (8) and the category is (D). With partial-joint-penetration groove welds, the designation is (10) and the category is (E).

The stiffened connections (8) and (10) provide a better transfer of force, and the stress distribution in the branch members is uniform; thus, they are given higher fatigue categories than the simpler connections (9) and (12) – in which the prime mark has been added to give them a lower fatigue allowable. Note that categories (D) and (DT) are for complete-joint-penetration groove welds, and (E) and (ET) are for partial-joint-penetration groove welds, which are assigned a lower fatigue allowable.

Welded attachments to a tubular member always create abrupt changes in section that can cause stress risers and, thus, lower the fatigue strength. However, the length of the attachment has much to do with this effect. Research has shown that short attachments do not lower the fatigue strength of members as much as long attachments. Although both cause changes in section or notches to be present, the longer attachment has greater length in which to load up; more stress flows through the notch. In Figure 22, the connection with a short attachment warrants a (D) fatigue category, while the connection with a long attachment is given an (E) category. In Figure 23, the effect of the length of attachments is also illustrated. Note the drop from (D) to (E) in fatigue categories.

In Figure 24, the master-chart designations (13) represent complex tubular connections in which there is a reduction of fatigue strength. These rate category (ET), because there is not uniform distribution of stress in the branch member adjacent to the connection.

In the left assembly, the connection has what is known as "negative eccentricity." The axes of the branch members intersect at a point above the axes of the main member. This means that some of the force component that is normal or transverse to the main member will be transferred through a short path from one branch member to another (or will be transferred directly should the branch members overlap at the connections) – causing regions of the branch members to have higher stress. Non-uniform stress distribution, thus would exist.

Adding ring stiffeners, such as shown by the central sketch, stiffen the main member and also help to transfer forces into a branch member more uniformly. Also, gusset plates, as shown in the right sketch, help in transferring forces. In both of these cases, a greater stress in the branch member will occur in line with the attachments at their terminations.

Referring to Figure 25, note that the K, T, and Y connections have category (K_2).

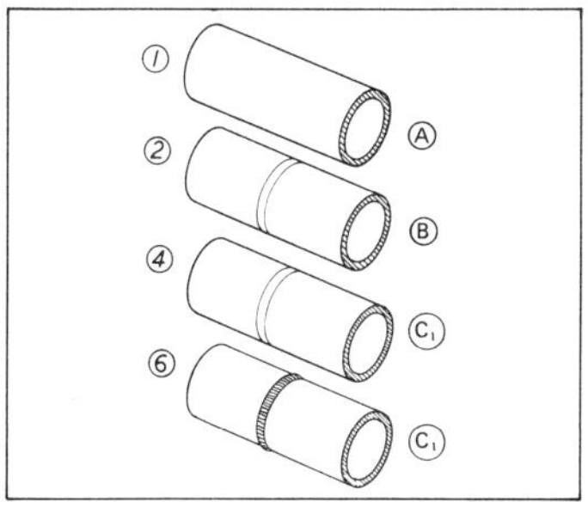

Figure 19

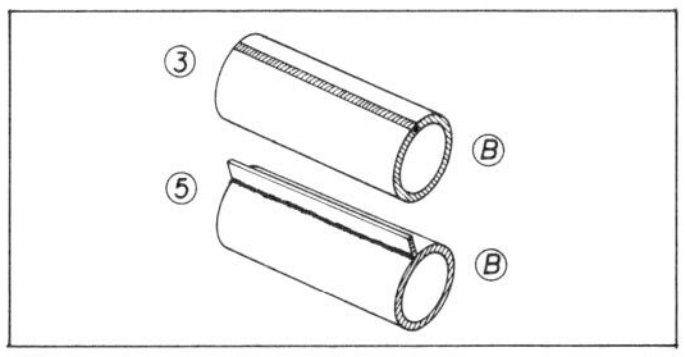

Figure 20

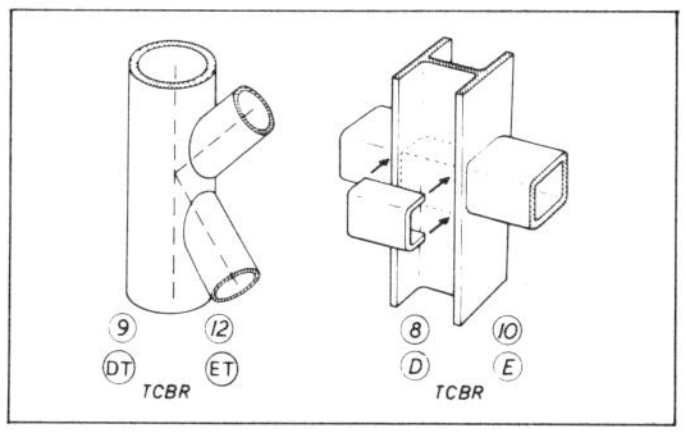

Figure 21

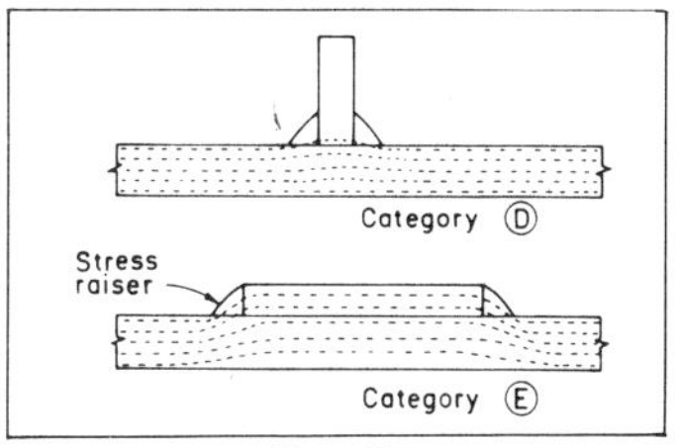

Figure 22

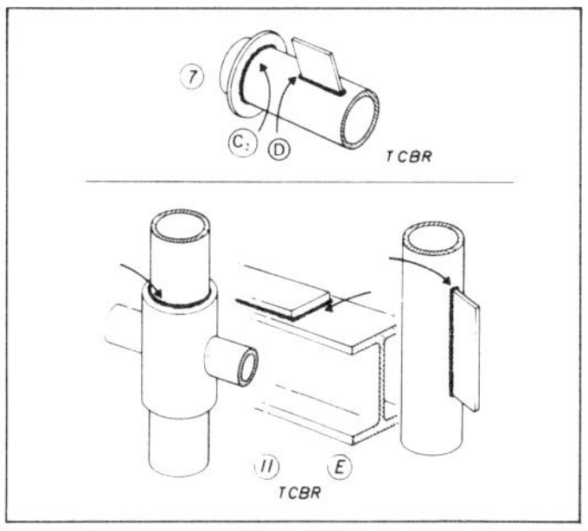

Figure 23

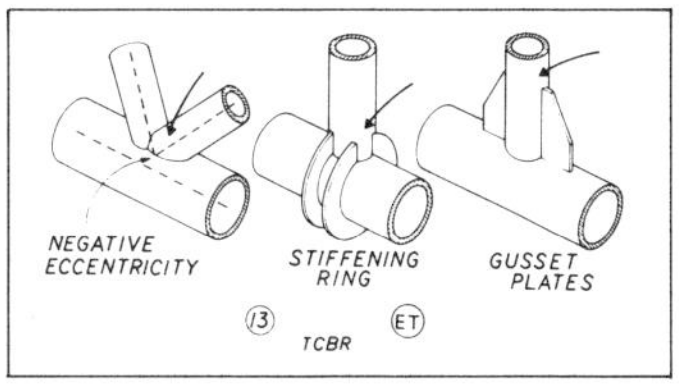

Figure 24

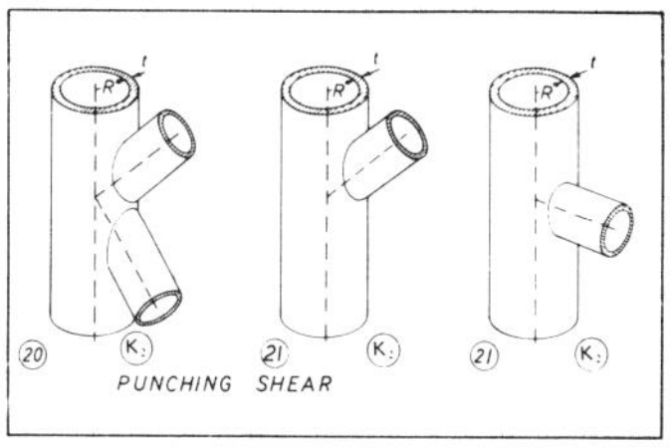

Figure 25

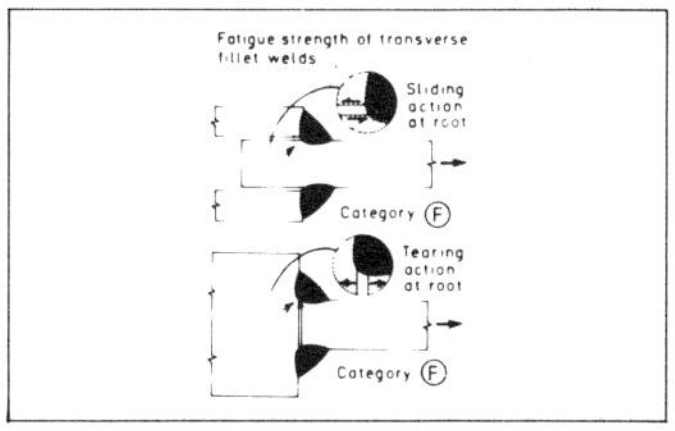

Figure 26

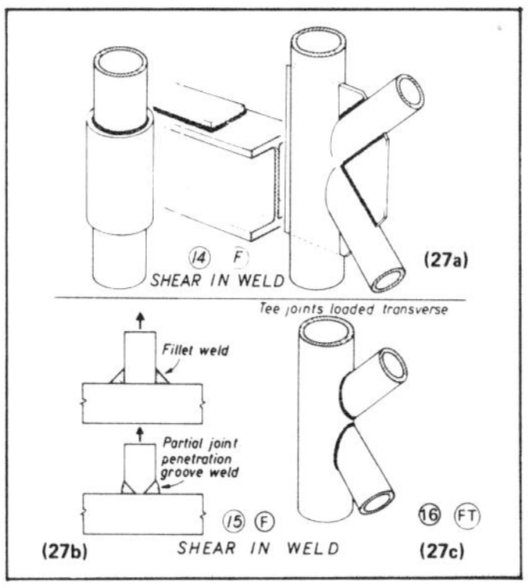

Figure 27

(1)

Consider the following steady load on this tubular connection

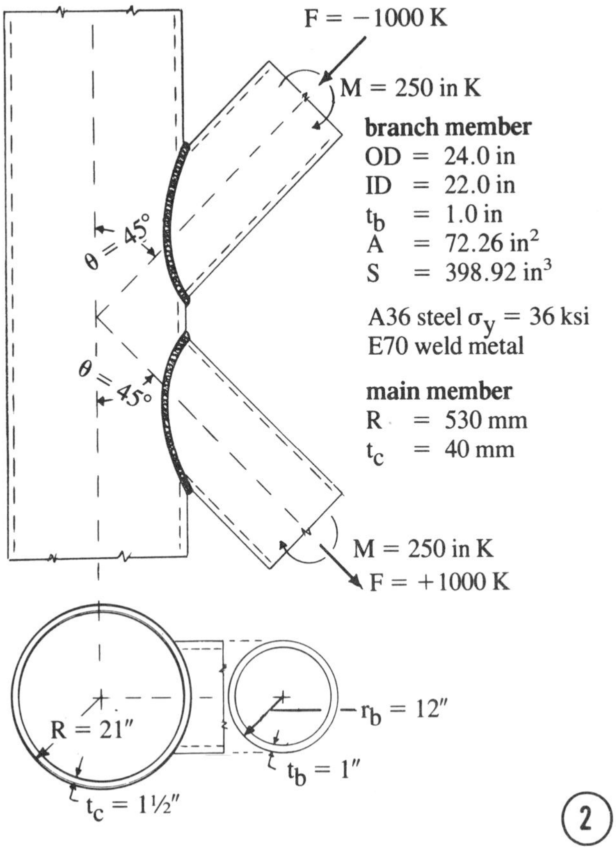

(2)

Allowable punching shear stress in main member 10.5.1

$$\beta = \frac{r_b}{R} = \frac{12 \text{ in}}{21 \text{ in}} = .57 \text{ hence } Q_\beta = 1.0$$

υ = .6 (given) hence

$$Q_f = 1.22 - .5\,\upsilon = 1.22 - .5\,(.6) = .92$$

$$\tau_a = \frac{\sigma_y\, Q_\beta\, Q_f}{0.9\left(\frac{R}{t_c}\right)^{.7}} = \frac{(36 \text{ ksi})\,(1.0)\,(.92)}{0.9\left(\frac{21 \text{ in}}{1\tfrac{1}{2} \text{ in}}\right)^{.7}} = 5.80 \text{ ksi}$$

---

axial stress in branch member

$$\sigma_a = \frac{F}{A} = \frac{1000 \text{ K}}{72.26 \text{ in}^2} = 13.84 \text{ ksi}$$

bending stress in branch

$$\sigma_b = \frac{M}{S} = \frac{250 \text{ in K}}{398.2 \text{ in}^3} = .62 \text{ ksi}$$

treat section as a line

$A = 2\,\pi\, r_m\, t$

$I = \tau\,\pi\, r_m^3$

$S = \frac{I}{C}$

total stress in branch = 13.84 + .62 = 14.46 ksi

Actual punching shear stress in main member

$$\tau = \frac{t_b}{t_c}\sin^2\theta\left[\frac{2\,\sigma_a}{1+\sin\theta} + \frac{4\,\sigma_b}{1+3\,\sin\theta}\right]$$

$$= \frac{1.0}{1\tfrac{1}{2}}\sin^2 45°\left[\frac{2\,(13.84)}{1+\sin 45°} + \frac{4\,(.62)}{1+3\,\sin 45°}\right]$$

$$= 5.67 \text{ ksi}$$

5.67 < 5.80 hence OK

Consider using a fillet weld E70 weld metal. The maximum unit force on the connecting weld (regardless of the type of weld) may be found by multiplying the punching shear stress (τ) by the thickness of the main member (t_c). To change from the normal force component to the actual value, we must divide through by $\sin\theta$.

$$f_w = \frac{\tau\, t_c}{\sin\theta} = \frac{(5.67)\,(1\tfrac{1}{2})}{\sin 45°} = 12.02 \text{ K/in}$$

leg size of fillet weld — E70 weld metal

$$\omega = \frac{12.02 \text{ K/in}}{(.707)\,(.30 \times 70 \text{ ksi})} = .810 \text{ in or use } \tfrac{7}{8}''$$

Do not let weld throat go below .707 ω or .57 inches

---

(3)

Uneven distribution of load on connecting welds 10.5.3

branch member at yield strength

(1) $f_1 = \sigma_y\, t_b = (36 \text{ ksi})\,(1.0 \text{ in}) = 36 \text{ K/in}$

ultimate punching shear on main member

(2) $f_3 = \dfrac{1.8\,\tau_a\, t_c}{\sin\theta} = \dfrac{1.8\,(6.14 \text{ ksi})\,(1\tfrac{1}{2} \text{ in})}{\sin 45°}$

$= 23.44$ K/in

(3) $f_w = 2.67\,(.707\,\omega)\,\tau_w$

$= 2.67\,(.707 \times \tfrac{7}{8} \text{ in})\,(.30 \times 21 \text{ ksi})$

$= 34.69$ K/in

and this is greater than f_3 = 23.44 K/in, hence OK

If a complete penetration groove weld were used, it would not be necessary to check the above uneven distribution of load on the connecting weld.

Also check $\beta = \dfrac{r_b}{R} = \dfrac{12 \text{ in}}{21 \text{ in}} = .57 > \tfrac{1}{3}$ 10.13.1.3

since β exceeds ⅓, the branch member along the side portion and toe of the connection shall have a groove weld.

Do not let the throat of the weld fall below .707 ω or .57 in.

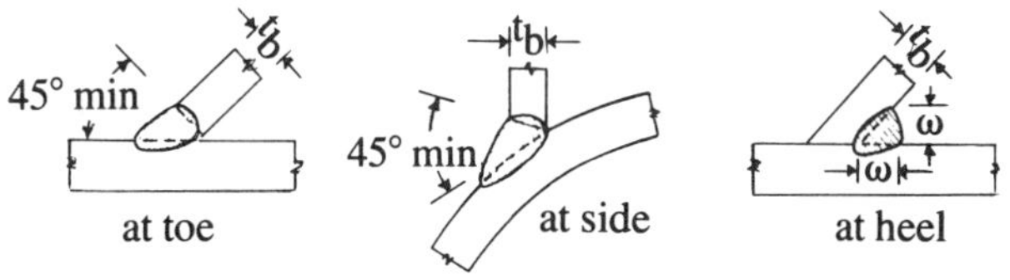

Now consider the same connection under fatigue (4)

N = number of cycles = 10^{+6} $K = \frac{\text{min}}{\text{max}} = +\ 1/4$

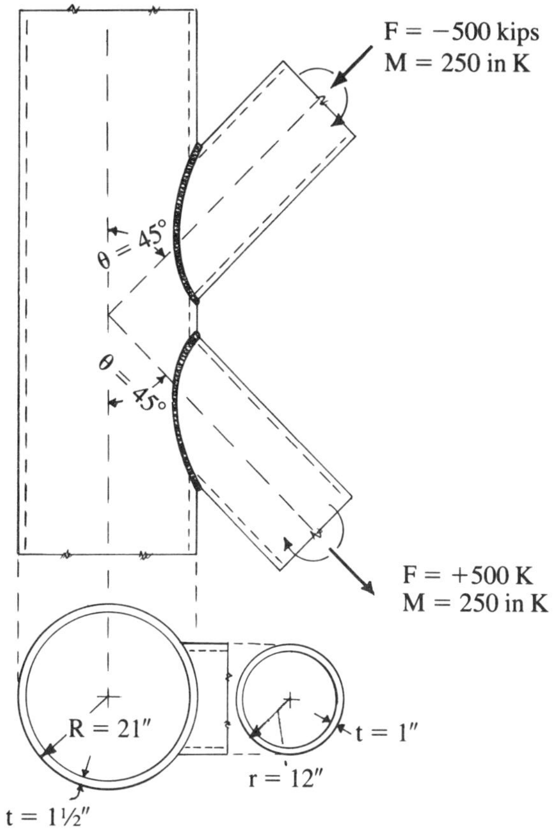

(5)

Allowable punching shear stress in main member – Fatigue loading 10.7

illustration (20) category (K_2)

$$N = 10^{+6} \text{ cycles} \quad K = \frac{\text{min}}{\text{max}} + 1/4$$

from figure 10.7.4 read stress range $\tau_{sr} = 2.5$ ksi

$$\text{so } \tau_{max} = \frac{\tau_{sr}}{1 - K} = \frac{2.5 \text{ ksi}}{1 - 1/4} = 3.33 \text{ ksi}$$

but not to exceed the steady allowable of

$$\tau_a = \frac{\sigma_y\, Q_\beta\, Q_f}{0.9 \left(\frac{R}{t_c}\right)^{0.7}} = \frac{(36 \text{ ksi})\ (.92)\ (1.0)}{0.9 \left(\frac{21 \text{ in}}{1\frac{1}{2} \text{ in}}\right)^{0.7}} = 5.80 \text{ ksi}$$

hence use 3.33 ksi

axial stress in branch member

$$\sigma_a = \frac{F}{A} = \frac{500 \text{ K}}{72.2 \text{ in}^2} = 6.92 \text{ ksi}$$

bending stress in branch

$$\sigma_b = \frac{M}{S} = \frac{250 \text{ in K}}{398.2 \text{ in}^3} = .62 \text{ ksi}$$

Actual punching shear stress in main member

$$\tau = \frac{t_b}{t_c} \sin^2\theta \left[\frac{2\ \sigma_a}{1 + \sin\theta} + \frac{4\ \sigma_b}{1 + 3\ \sin\theta}\right]$$

$$= \frac{1.0}{1\frac{1}{2}} \sin^2 45° \left[\frac{2\ (6.92)}{1 + \sin 45°} + \frac{4\ (.62)}{1 + 3\ \sin 45°}\right]$$

$$= 2.97 \text{ ksi}$$

(6)

Fatigue allowable in branch member

(assume fillet welds or partial penetration groove welds are to be used)

illustration (12) category (ET)
from figure 10,7,4 read stress range $\sigma_{sr} = 2.8$ ksi

$$\text{so } \sigma_{max} = \frac{\sigma_{sr}}{1 - K} = \frac{2.8 \text{ ksi}}{1 - 1/4} = 3.73 \text{ ksi}$$

however the actual stress is 6.92 + .62 = 7.54 ksi hence fillet welds or partial penetration groove welds cannot be used, unless the branch member is increased in thickness.

therefore use complete penetration groove welds and move to:

illustration (9) category (DT) (complete penetration groove welds)

from figure 10.7.4 read stress range $\sigma_{sr} = 7.0$ ksi

$$\text{so } \sigma_{max} = \frac{\sigma_{sr}}{1 - K} = \frac{7.0 \text{ ksi}}{1 - 1/4} = 9.33 \text{ ksi}$$

9.33 > 7.54 hence OK

Since complete penetration groove welds are to be used, it will not be necessary to check the uneven distribution of load on the connecting welds as in 10.5.3

$$\beta = \frac{b}{D} = 0.6$$

(14)

$$\gamma = \frac{D}{2t_c} = 10 \qquad \eta = \frac{a}{D \sin\theta} = .846$$

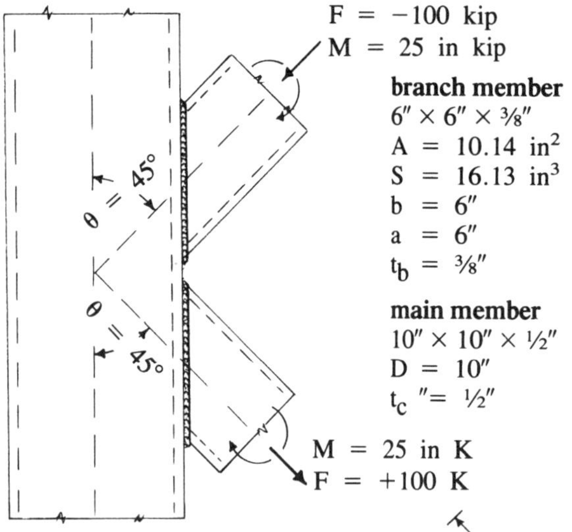

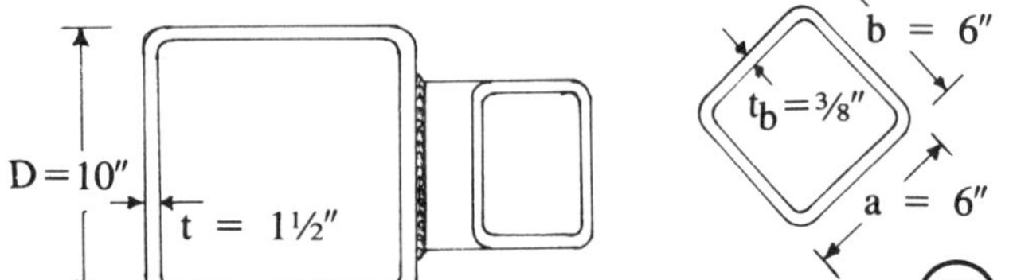

(15)

Allowable punching shear stress in main member 10.5.1

$$\beta = \frac{b}{D} = \frac{6 \text{ in}}{10 \text{ in}} = .6$$

$$Q_\beta = \frac{.25}{\beta\,(1 - \beta)} = \frac{.25}{.6\,(1 - .6)} = 1.04$$

$\upsilon = 0.6$ (given)

hence $Q_F = 1.22 - 0.5\upsilon = 1.22 - 0.5\,(.6) = .92$

$$\tau_a = \frac{\sigma_y\, Q_\beta\, Q_f}{.6\left(\frac{D}{2\,t_c}\right)} = \frac{(36 \text{ ksi})\,(1.04)\,(.92)}{.6\left(\frac{10}{2 \times 1/2}\right)} = 5.74 \text{ ksi}$$

---

axial stress in branch member

$$\sigma_a = \frac{F}{A} = \frac{100 \text{ K}}{10.14 \text{ in}^3} = .986 \text{ ksi}$$

bending stress in branch member

$$\sigma_b = \frac{M}{S} = \frac{25 \text{ in K}}{16.13 \text{ in}^3} = 1.55 \text{ ksi}$$

total stress in branch member = 9.86 + 1.55 = 11.41 ksi

Actual punching shear stress in main member

$$\tau = t_b \sin^2\theta \left[\frac{(a + b)\,\sigma_a}{a + b \sin\theta} + \frac{(a + 3b)\,\sigma_b}{a + 3 \sin\theta}\right]$$

$$= \frac{3/8}{1/2} \sin^2 45° \left[\frac{(6 + 6)\,(9.86)}{6 + 6 \sin 45°} + \frac{(6 + 3 \times 6)\,1.55}{6 + 3 \times 6 \sin 45°}\right]$$

$= 5.08$ ksi

$5.08 < 5.74$ hence OK

Consider using a fillet weld

The maximum unit force on the connecting weld (regardless of the type of weld) may be found by multiplying the punching shear stress (τ) by the thickness of the main member (t_c). To change from the normal force component to the actual value, we must divide through by $\sin\theta$ (by doing this we will also include the parallel component).

$$f_w = \frac{\tau\ t_c}{\sin\theta} = \frac{(5.08 \text{ ksi})\,(1/2 \text{ in})}{\sin 45°} = 3.59 \text{ K/in}$$

leg size of fillet weld E70 weld metal

$$\omega = \frac{3.59 \text{ K/in}}{(.707)\,(.30 \times 70 \text{ ksi})} = .242 \text{ in} \quad \text{or} \quad (1/4\,\Delta)$$

weld throat must be at least (.707) (.242 in) = .17″ or 3/16″

Use optional Yield Line Analysis 10.5.1.4

$$\beta = \frac{b}{D} = \frac{6 \text{ in}}{10 \text{ in}} = 0.6 \quad \text{is less than } 0.8$$

– can use yield Line analysis

$$\eta = \frac{a}{D \sin\theta} = \frac{6 \text{ in}}{10 \text{ in} \sin 45°} = 0.85$$

For axial load

$$Q_\beta = \frac{\frac{\pi}{2} + \frac{\eta + \beta}{1 - \beta}}{4\,(\eta + \beta)}$$

$$Q_\beta = \frac{\frac{\pi}{2} + \frac{.85 + .6}{1 - .6}}{4\,(.85 + .6)} = .896$$

allowable punching shear stress due to axial stress in branch member

$$\tau = \frac{K\ \sigma_y\ Q_\beta}{SF\ .5\left(\frac{D}{2\,t_c}\right)} = \frac{1.5\,(36 \text{ ksi})\,(.896)}{1.8\,(.5)\left(\frac{10 \text{ in}}{2 \times 1/2 \text{ in}}\right)} = 5.38 \text{ ksi}$$

For in-plane bending

$$Q_\beta = \frac{\frac{1}{2\eta} + \frac{\pi}{2} + \frac{\eta + \beta}{1 - \beta}}{4\,(\eta/3 + \beta)}$$

$$Q_\beta = \frac{\frac{1}{2\,(.85)} + \frac{\pi}{2} + \frac{.85 + .6}{1 - .6}}{4\,(.85/3 + .6)} = 1.64$$

allowable punching shear stress due to bending stress in branch member

$$\tau = \frac{K\ \sigma_y\ Q_\beta}{SF\ .5\left(\frac{D}{2\ t_c}\right)} = \frac{1.5\ (36\ \text{ksi})\ (1.64)}{1.8\ (.5)\left(\frac{10\ \text{in}}{2 \times 1/2\ \text{in}}\right)} = 9.84\ \text{ksi}$$

Actual punching shear stress

$$\tau = t_b \sin^2\theta \left[\frac{(a + b)\ \sigma_a}{a + b \sin\theta} + \frac{(a + 3\ b)\ \sigma_a}{a + 3\ b \sin\theta}\right]$$

$$= \frac{3/8}{1/2} \sin^2 45° \left[\frac{(6 + 6)\ (9.86)}{6 + 6 \sin 45°} + \frac{(6 + 3 \times 6)\ 1.55}{6 + 3 \times 6 \sin 45°}\right]$$

$$= 5.08\ \text{ksi}$$

$$\tau = .375\ (11.552 + 1.986) = \underset{\text{axial}}{4.33\ \text{ksi}} + \underbrace{.75\ \text{ksi}}_{\text{bending}}$$

$$= 5.08\ \text{ksi}$$

using:

$$\frac{\text{actual } \tau \text{ from } \sigma_a}{\text{allowable } \tau \text{ from axial}} + \frac{\text{actual } \tau \text{ from } \sigma_b}{\text{allowable } \tau \text{ from bending}}$$

$$\leqslant 1.0$$

$$\frac{4.33}{5.33} + \frac{.75}{9.84} = .886 < 1.0 \quad \text{hence OK}$$

Since $\beta < 0.8$ and $\beta < \eta$ cannot satisfy 10.5.1.3

so *cannot* calculate allowable capacity.

21

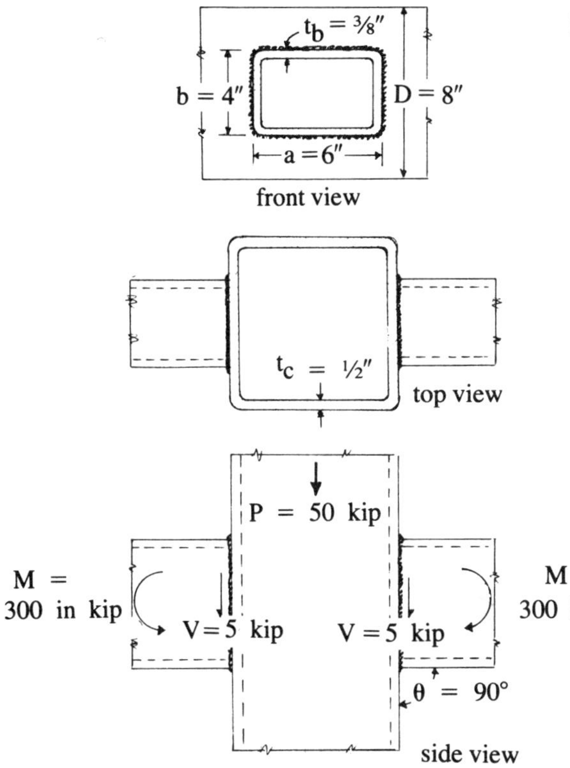

axial stress in column $\sigma = 4.32$ ksi

allowable stress $\sigma = 18.0$ ksi $\quad \upsilon = \frac{4.32}{18.00} = .24$

$$\sigma_b = \frac{M}{S} = \frac{300\ \text{in kip}}{14.5\ \text{in}^3} = 20.70\ \text{ksi}$$

$$\gamma = \frac{D}{2\ t_c} = \frac{8\ \text{in}}{2\ (1/2\ \text{in})} = 8$$

$$\beta = \frac{b}{D} = \frac{4\ \text{in}}{8\ \text{in}} = .50$$

$$Q_\beta = 1.0 \qquad Q_f = 1.0$$

allowable punching shear stress

$$\tau = \frac{\sigma_y\ Q_\beta\ Q_f}{.6\ \gamma} = \frac{(36\ \text{ksi})\ (1.0)\ (1.0)}{.6\ (8)} = 7.50\ \text{ksi}$$

actual punching shear stress

$$\tau = \frac{t_b}{t_c} \sin^2\theta \left[\frac{(a + b)\ \sigma_a}{a + b \sin\theta} + \frac{(a + 3b)\ \sigma_b}{a + 3b \sin\theta}\right]$$

$$= \frac{3/8}{1/2} \sin^2 90° \left[0 + \frac{(6 + 3 \times 4)\ 20.70}{6 + 3 \times 4 \sin 90°}\right] = 15.53\ \text{ksi}$$

is excessive

22

To increase the allowable punching shear stress and reduce the actual punching shear stress on the box column, an 8″ × 6″ × 5/8″ end plate will be used.

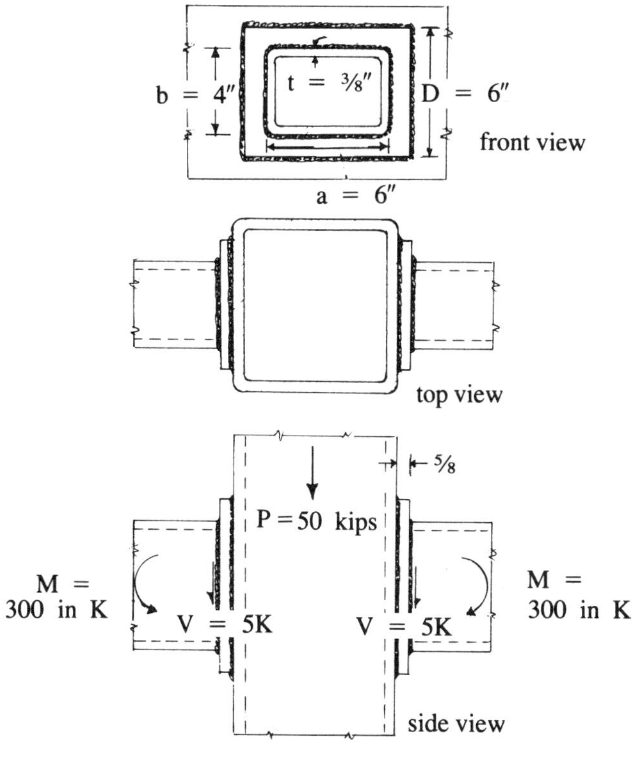

6″ × 4″ × 3/8″ box to 8″ × 6″ × 5/8″ Plate

$$\gamma = \frac{D}{2t} = \frac{6 \text{ in}}{2 \ (5/8 \text{ in})} = 4.8$$

$$\beta = \frac{b}{D} = \frac{4 \text{ in}}{6 \text{ in}} = .67$$

$$Q_\beta = \frac{.25}{\beta\ (1 - \beta)} = \frac{.25}{.67\ (1 - .67)} = 1.13$$

$$Q_f = 1.0$$

allowable punching shear stress

$$\tau = \frac{\sigma_y\ Q_\beta\ Q_f}{.6\ \gamma} = \frac{(36 \text{ ksi})\ (1.13)\ (1.0)}{.6\ (4.8)} = 14.13 \text{ ksi}$$

actual punching shear stress

$$\tau = \frac{t_b}{t_c} \sin^2\theta \left[\frac{(a + b)\ \sigma_a}{a + b \sin\theta} + \frac{(a + 3b)\ \sigma_b}{a + 3b \sin\theta}\right]$$

$$= \frac{3/8}{5/8} \sin^2 90° \left[0 + \frac{(6 + 3 \times 4)\ 20.70}{6 + 3 \times 4 \sin 90°}\right] = 12.42 \text{ ksi}$$

this portion of 6″ × 4″ × 3/8″ box to 8″ × 6″ × 5/8″ plate is OK

---

(23)

8″ × 6″ × 5/8″ Plate to 8″ × 8″ × 1/2″ box column

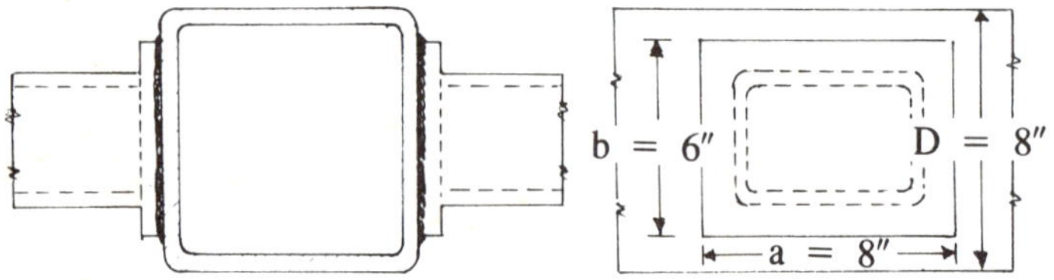

$$\gamma = \frac{D}{2t} = \frac{8 \text{ in}}{2\ (1/2 \text{ in})} = 8$$

$$\beta = \frac{b}{D} = \frac{6 \text{ in}}{8 \text{ in}} = .75$$

$$Q_\beta = \frac{.25}{\beta\ (1 - \beta)} = \frac{.25}{.75\ (1 - .75)} = 1.33$$

$$Q_f = 1.0$$

allowable punching shear stress

$$\tau = \frac{\sigma_y\ Q_\beta\ Q_f}{.6\ \gamma} = \frac{(36 \text{ ksi})\ (1.33)\ (1.0)}{.6\ (8)} = 9.98 \text{ ksi}$$

actual punching shear stress

property of punching shear zone

$$I = 2\ (6 \text{ in})\ (4 \text{ in})^2 + \frac{2\ (8 \text{ in})^3}{12} = 277.33 \text{ in}^3$$

$$S = \frac{I}{C} = \frac{277.33 \text{ in}^3}{4.0 \text{ in}} = 69.33 \text{ in}^3$$

unit bending force

$$f_b = \frac{M}{S} = \frac{300 \text{ in K}}{69.33 \text{ in}^2} = 4.33 \text{ K/in}$$

unit punching shear stress

$$\tau = \frac{t_b}{t_c} = \frac{4.33 \text{ K/in}}{1/2 \text{ in}} = 8.75 \text{ ksi}$$

this portion of 8″ × 6″ × 5/8″ plate to 8″ × 8″ × 1/2″ box column is OK

---

(24)

check fillet weld size between 6″ × 4″ × 3/8″ box and 8″ × 6″ × 5/8″ plate

① $f_1 = \sigma_y\ t_b = (36 \text{ ksi})\ (3/8 \text{ in}) = 13.50 \text{ K/in}$

branch at yield

② $f_3 = \dfrac{1.8\ \tau\ t}{\sin\theta} = \dfrac{1.8\ (14.13 \text{ ksi})\ (5/8 \text{ in})}{\sin 90°}$

$= 15.90 \text{ K/in}$

ultimate punching shear of plate

③ $f_w = 2.67\ (.707\ \omega)\ .30 \text{ EXX}$ must equal the smaller of the above two values — hence set equal to $f_1 = 13.50 \text{ K/in}$

$$2.67\ (.707\ \omega)\ (.30 \times 70 \text{ ksi}) = 13.50 \text{ K/in}$$

$$\omega = \frac{13.50 \text{ K/in}}{2.67\ (.707)\ (21 \text{ ksi})} = .34$$

or use $\omega = 3/8''\ \Delta$

---

check fillet weld size between 8″ × 6″ × 5/8″ plate and 8″ × 8″ × 1/2″ box column

① $f_1 = \dfrac{M}{S}$ where $S = 69.33 \text{ in}^2$

$$* f_1 = \frac{300 \text{ in lb}}{69.33 \text{ in}^2} \times \left(\frac{36.0 \text{ ksi}}{36.0 \text{ ksi}}\right) = 7.53 \text{ K/in}$$

② $f_3 = \dfrac{1.8\ \tau\ t}{\sin\theta} = \dfrac{1.8\ (9.98 \text{ ksi})\ (1/2 \text{ in})}{\sin 90°} = 8.98 \text{ K/in}$

③ $f_w = 2.67\ (.707\ \omega)\ .30 \text{ EXX}$ must equal or exceed the smaller of the above two values — hence set equal to $f_1 = 7.53 \text{ K/in}$

$$2.67\ (.707\ \omega)\ (.30 \times 70 \text{ ksi}) = 7.53 \text{ K/in}$$

$$\omega = \frac{7.53 \text{ K/in}}{2.67\ (.707)\ (21 \text{ ksi})} = .17$$

or use $\omega = 1/4''\ \Delta$